A NATURAL HISTORY *of* CALIFORNIA

A NATURAL HISTORY *of* CALIFORNIA

Second Edition

Allan A. Schoenherr

UNIVERSITY OF CALIFORNIA PRESS

University of California Press, one of the most distinguished university presses in the United States, enriches lives around the world by advancing scholarship in the humanities, social sciences, and natural sciences. Its activities are supported by the UC Press Foundation and by philanthropic contributions from individuals and institutions. For more information, visit www.ucpress.edu.

University of California Press
Oakland, California

Library of Congress Cataloging-in-Publication Data

Names: Schoenherr, Allan A., author.
Title: A natural history of California / Allan A. Schoenherr.
Description: Second edition. | Oakland, California : University of California Press, [2017] | Includes index.
Identifiers: LCCN 2017005894 (print) | LCCN 2017008534 (ebook) | ISBN 9780520295117 (cloth : alk. paper) | ISBN 9780520290372 (pbk. : alk. paper) | ISBN 9780520964556 (epub and ePDF)
Subjects: LCSH: Natural history—California.
Classification: LCC QH105.C2 S36 2017 (print) | LCC QH105.C2 (ebook) | DDC 508.794—dc23
LC record available at https://lccn.loc.gov/2017005894

Manufactured in United States of America

25 24 23 22 21 20 19 18 17
10 9 8 7 6 5 4 3 2 1

CONTENTS

ACKNOWLEDGMENTS

For fear of forgetting someone, I am tempted not to acknowledge anyone's help on this book. However, with a project of this size, it is obvious that quite a few people had something to do with the outcome. With this in mind, I hope that the friends, colleagues, and acquaintances that were inadvertently overlooked will understand how much I appreciate their help.

For starters, I am indebted to the administration of Fullerton College for providing me with a sabbatical leave, during which time I wrote the first draft. Second, Debbie Horrocks and Amy Gigliotti converted the typewritten first draft to word-processed versions. I am also deeply indebted to Fullerton College computer wizards such as Kent Gordon, Co Ho, Vinh Ho, Geoff Clifton, and Bill Dalphy, who tirelessly gave their time to jury-rig equipment and write programs enabling me to incorporate numerous revisions with minimal effort and convert all my computer files to a single format.

Unless otherwise noted, all photographs are mine. For contributions of art, I appreciate the work of Karlin Grunau Marsh, Phil Lingle, Geoff Smith of Fullerton College, Pat Brame of Eaton Canyon Nature Center, and Philip Brown of the Southwestern Herpetologists Association. Further thanks go to those many persons cited in text who gave me permission to use work from previously published materials.

For their sage advice and comments on writing style, I am deeply appreciative to the following persons: Diana Cosand from Chaffee College and Chuck Leavell of Fullerton College, for their help with the sections on basic ecology; Peter Tresselt, formerly of Fullerton College, Rick Lozinsky of Fullerton College, and N. King Huber of the US Geological Survey, for their help with the sections on geology; Alan Romspert of the California State University Desert Studies Consortium and Lenny Vincent of Fullerton College, for help with Chapter 9, on deserts; and Phil Pister of the California Department of Fish and Wildlife for his extremely helpful review of Chapter 11, on inland waters. In preparation of the second edition, I am indebted to colleagues such as Mick Bondello, Chuck Leavell, Doug Allan, Rick Lozinsky, and Lenny Vincent who kindly read and made beneficial suggestions.

Finally, I am indebted to Art Smith and Ernest Callenbach, editors from the University of California Press, who provided invaluable assistance and encouragement at all stages in the development of the first edition. I am thankful for the encouragement of Blake Edgar from the University of California Press, without which I would not have undertaken the effort to do a second edition.

In 1992, when the first edition of this book was published, the University of California Press, in celebration of entering the second century of publishing, honored *A Natural History of California* among 100 "Centennial Books" published between 1990 and 1995. A special imprint opposite the title page declared that special honor. It has been 25 years since the first edition of this book was written. During that time, several things have changed that should be addressed. For starters, associated with a revolution in DNA technology, there has been a serious realignment of biological taxa. Not only have many species names been changed, but there has also been significant reassignment within families. Several strange relationships have appeared. For example, Maples and Buckeyes are now in the Soapberry family, Sapindaceae. Some familiar families have been broken up. For example, the Figwort or Snapdragon family, Scrophulariaceae, is now broken up into three families. Bee Plant, *Scrophularia*, and Mullein, *Verbascum*, are still "Scrophs," but the Bush Monkeyflowers, *Mimulus* spp., are in the Lopseed family, Phrymaceae. The Broomrape family, Orobanchaceae, now includes Broomrape (*Orobanche*), Bird's Beak (*Cordylanthus*), Lousewort (*Pedicularis*), and Paintbrush and Owl's-clover (*Castilleja* spp.). The Plantain family, Plantaginaceae,

includes Snapdragons (*Antirrhinum*), Chinesehouses (*Collinsia*), Ghost Flower (*Mohavea*), Bush Penstemon (*Keckiella*), and all the herbaceous Penstemons (*Penstemon* spp.). The Lily family, Liliaceae, has been split into at least six families. Remember the Lily family? About all that remains in the Lily family are Mariposa Lilies (*Calochortus* spp.). Now, Agaves, Yuccas, and Desert Lilies (*Hesperocallis*) are in the Agavaceae. Onions and garlics are in the Alliaceae. Beargrasses (*Nolina* spp.) are in the Butcher's-broom family, Ruscaceae. The Brodiaea family, Themidaceae, includes Brodiaeas (*Brodiaea* spp.), Goldenstars (*Bloomeria* spp.), and Blue Dicks (*Dichelostemma* spp.). Most of the garden varieties such as Daffodils, Paper Whites, Narcissus, and Naked Ladies are now in the Amaryllis family, Amaryllidaceae.

There have been many changes in animal taxonomy as well. Not only have many of the familiar genera and species names been changed, but there have been family realignments as well. For example, in the past, many of our familiar lizards have been in the Iguana family, Iguanidae. Desert Iguanas (*Dipsosaurus*) and Chuckwallas (*Sauromalus*) are still Iguanas, but the Collared Lizards (*Crotaphytus*) and the Leopard Lizards (*Gambelia*) are now in the Crotaphytidae. All the rest, at least seven genera of common lizards, including Fence Lizards (*Sceloporus*) and Side-blotched Lizards (*Uta*), are now in the Horned Lizard family, Phrynosomatidae.

Essentially all of the traditional field guides are now out of date. In order to compensate for that problem, I often will include the new scientific name followed, in parentheses, by the former scientific name of the various plants and animals that I discuss. The concept, for naming purposes, of making scientific names permanent by using the "dead" languages Greek and Latin, in recent years, has been upended with so many name changes. It is a bit ironic that the unofficial "common names" of many species have become more permanent than the scientific names.

Another big change for California is that many of the regions and parks that I discuss have been reclassified by federal and state agencies. For example, Death Valley, Joshua Tree, and Pinnacles have been enlarged and upgraded from National Monuments to National Parks, and a large region of the east Mojave Desert has been established as the Mojave National Preserve. Furthermore, a number of lands administered by the US Forest Service and the Bureau of Land Management (BLM) have been reclassified. For example, the San Gabriel Mountains, Berryessa Snow Mountain, and Giant Sequoia in the Sierra National Forest have been upgraded to National Monuments and will be administered by the National Park Service.

In 1994, with passage of the Desert Protection Act, the total area of California classified as wilderness was brought to 14 million acres (56.656 km²). In addition to the national parks mentioned above, 69 Wilderness Study Areas on BLM land, many of them scenic isolated mountain ranges, officially became classified as wilderness. In February 2016, three new national monuments in the Mojave Desert added 1.8 million acres (7284 km²) of federal protection.

The specter of climate change also has changed much of the information about California's natural areas. Drought and warmth have stressed our ecosystems to the point that the future of California and its landscapes must be discussed. Fire regimes and intensity have changed. Species composition of ecological communities has changed, and will continue to change. Uphill and northward distribution of plant communities is already underway. Invasion of nonnative species has altered our natural plant communities, which also changes species composition and fire frequency. The epilogue addresses these issues.

Although not necessarily new, there have also been changes in ecological thinking. As the distribution and composition of natural communities becomes changed, there is a group of ecological concepts that need to be addressed. Subjects such as ecological islands, patch size, habitat fragmentation, keystone species, mesopredator release, and trophic cascades will be incorporated where appropriate.

Finally, research on animals, plants, and their interactions in ecosystems has continued during the interval since publication of the first edition of *A Natural History of California*. There are many new stories to tell about natural history in California and those stories will be incorporated as much as possible in this new edition. Hopefully readers will find this material as useful and interesting as they have in the past.

INTRODUCTION

A natural history is an account of natural phenomena. Over 300 years before Christ was born, Aristotle wrote his *Historia Animalium*, a series of nine books on the anatomy and habits of many animals native to Greece. In A.D. 77, Pliny the Elder, a Roman, wrote the 37-volume *Natural History* in an attempt to compile an encyclopedia of all known natural phenomena. For over 1500 years, this work served as the basic reference for information about "nature." Ever since then, the expression "natural history" has been used to refer to a description of living organisms, their habits, and how they relate to the environment.

A more modern term used to describe organisms and their relationships with the environment is "ecology." The literal meaning of this word is "study of the house." In this context, "house"' means the environment. This book easily could have been entitled *The Ecology of California*, because it is about the creatures and the environment in the state of California. In recent years, however, ecology has become a highly theoretical and technical science, and it does not encompass a study of rocks and geologic history. This book includes a description of organisms, rocks, and the environment, and an attempt to explain what factors through time created what we see today.

The facts behind the concepts that are stressed in this book have been drawn from ecology, zoology, botany, biogeography, climatology, paleontology, and geology. Therefore, it seems best to call it a natural history.

California is a highly complex geographic unit. There is more climatic and topographic variation in California than in any other region of comparable size in the United States. The highest and lowest points in the lower 48 states are less than 100 mi (160 km) from each other in eastern California. Mount Whitney, figure 4.11 located on the crest of the Sierra Nevada, at an elevation of 14,505 ft (4421 m), is only 84.6 mi (136.2 km) west of Badwater in Death Valley, which is 282 ft (86 m) below sea level. Interestingly, the US Geological Survey brass benchmark on the summit of Mount Whitney indicates an elevation of 14,494 ft (4418 m). The new designation is the product of new techniques for estimating elevation.

California also has a great range of climates. The *Sunset Western Garden Book* (Lane Publishing Co.) describes 24 different climatic zones within the state. Outside of California, few states have more than four zones, and most have only one.

Total precipitation per year of more than 120 in (300 cm) may commonly occur in the northwestern forests. Honeydew in northern California has recorded the highest average annual precipitation of 104.18 in (264.6 cm), although the official record for annual precipitation, 161 in (403 cm), occurred in the Santa Lucia Mountains of the southern Coast Ranges. In the southwestern deserts, it is not uncommon for some locations to go several years without measurable precipitation. Bagdad, a now-deserted community on old Route 66 in the Mojave Desert, was reported to be the driest place in the United States. No measurable precipitation was recorded for 25 months. The lowest official record for annual precipitation of 1.6 in (4.06 cm) is held by Death Valley.

The range of temperatures in California is also extreme. Subzero temperatures may continue for many days at high elevations in the mountains, and Furnace Creek in Death Valley, at 134°F (57°C), holds the record for the highest official air temperature ever recorded. A slightly higher temperature for a location in Libya is sometimes touted as the highest, but in recent years it has been determined that the record was erroneous.

The geologic picture of California is also very complex. Most of the state is composed of a mélange of rock units from different sources, and of different ages, that became attached to its western border as North America slid westward over the floor of the Pacific Ocean. Tremendous forces have stretched and warped the land, so that California today is a mixture of mountains and valleys of diverse origin cut through by major fault systems. Lands west of the San Andreas fault, for example, have slid to their present position from a point adjacent to mainland Mexico, perhaps as much as 300 mi (480 km) to the south. The dominant topographic feature of California, the Sierra Nevada, which in eastern California runs on a predominantly north to south direction for approximately 400 mi (640 km) is composed primarily of granitic rocks. The great variety of rock materials degrade to form a corresponding variety of soils. Soils of different mineral content and texture have a profound influence on

plants and animals. A particular soil type may have its own specially adapted community of organisms.

The combination of diverse climate and diverse soil is responsible for the development of a diversity of habitats. The total number of habitat types varies, depending on whether the classifier is a splitter or a lumper. For example, the California Natural Heritage Section recognizes about 300 natural communities, and the California Department of Fish and Wildlife recognizes 178 major habitat types. This book shall consider about 35 terrestrial communities modified from a system developed by Munz and Keck in 1959. Furthermore, about 15 different aquatic communities will be discussed.

Many specialized plants and animals are found in California. There are more endemic species in California than in any area of equivalent size in North America. There are more than 6000 native plants in California, and about a third of them occur naturally nowhere else on earth. Among vertebrate animals, there are nearly 1000 native species. Depending on how a person classifies the species, that number includes about 540 birds, 214 mammals, 77 reptiles, 47 amphibians, and 83 freshwater fishes. Of these, 65 species are considered to be endemic. Among animals as a whole, including insects and other invertebrates, at least 50% of the species and subspecies are confined within the borders of the state.

The reason for writing a book of this type is to familiarize readers with this special place called California. After reading it, a person should be able to describe the climate, rocks, soil, plants, animals, and biogeography of any area in California. The reader should be able to explain how those things got there and the ways in which they relate to each other.

Another, no less important reason for writing a book of this type is to foster appreciation for California's natural diversity. Much of California's nature is threatened. The Nature Conservancy has reported that about 25% of the state can no longer support its original communities of plants and animals. Since 1900, 65% of the state's coastal wetlands have been dredged, filled, or drained. Riparian forests of the Great Central Valley, which once covered hundreds of square miles, are nearly gone, and some specialized communities, such as Southern Coastal Dune Scrub, have been reduced to only a few hundred acres. Not only does California have the greatest number of unique organisms in the continental United States, but it also has the highest number of threatened species, many of which are among the unique or endemic ones. Over 600 kinds of plants are listed as threatened with extinction. The greatest threat to California's natural ecosystems is growth of the human population. California already is the most populous state. About 70% of these people live within a few miles of the coast, with the greatest concentration of humans in the San Francisco Bay area and South Coast from Los Angeles to San Diego. The Great Central Valley, while mostly associated with agriculture, also has population centers such as Fresno and Sacramento. So, in spite of the threats mentioned above, over half of California's landscape is comparatively undisturbed. According to the California Protected Areas Database, 52% is public land, and 46.7% of California is classified as "protected." California has more officially designated wilderness, 14.36% of its area, than any state outside of Alaska.

This book is organized around geographic regions using, as a starting point, 12 geomorphic provinces described by the California Geological Survey (previously known as California Division of Mines and Geology). Superimposed on these regions are the natural biotic provinces based on climate and living organisms. The result is a sequence of chapters in which various regions of the state are characterized with respect to climate, geology, and biotic communities. Animals and plants are discussed using recent standardized common and scientific names wherever possible. Standardized vocabulary is also used for geologic formations.

In no part of this book is the treatment of plants, animals, and geologic units meant to be exhaustive. In each region, the emphasis is on those things that are conspicuous, distinctive, and/or interesting. In cases where an organism, a community, or a rock unit occurs in more than one geographic area, it is discussed in the context in which it seems to be most significant.

Every attempt has been made to ensure that facts and concepts presented here are based on the most up-to-date references. However, there is a good deal of disagreement among experts, particularly about evolutionary relationships, biogeography, and geologic history. If errors or discrepancies seem to occur, or if more information is desired, the reader is encouraged to consult the references at the back of the book.

California's Natural Regions

THE CALIFORNIA GEOLOGICAL SURVEY (previously known as the California Division of Mines and Geology) has divided the state into 12 geomorphic provinces based on rock type and topography (figure 1.1). The words *topography* and *geomorphic* refer to the shape of the land: *topography* means "place picture," and *geomorphic* means "earth form." These geomorphic provinces represent natural units within which the boundaries of landforms are remarkably consistent with those of biological communities. That is, the shape of the earth influences climate, and climate influences the distribution of plants and animals. I have adapted geomorphic provinces, with some modifications, into "natural regions" which shall provide the framework of organization for this book (figure 1.2).

Reference to landforms is way to refer to geographic features. They can be illustrated by means of aerial photographs or they may be depicted in the form of a relief map that shows topography, highlighting mountains and valleys (figure 1.2). Outlines of the landforms on a relief map show how California can be divided into "natural regions" which will be the subject of most of the chapters in this book. Other authors have divided the state into natural regions using different criteria. Some of these divide the state into fewer regions and others include more categories. Peter Berg and Raymond Dasmann in the 1970s divided the state into "bioregions," which roughly correspond to the natural regions to which I refer (Table 1.1). M. D. F. Udvardy in 1975 described biogeographical provinces of the world, and Robert Bailey in 1976 introduced the idea of "ecoregions." In 1983, he defined 19 ecoregions in California. The concept behind bioregions or ecoregions is that ecological boundaries need not correspond to political boundaries. Bioregions are supposed to be based on a series of criteria including humans living in harmony with the natural environment. W. J. Barry in 1991 described 24 ecological regions and Hartwell Welsh, who was associated with the US Forest Service, in 1994, divided the state into 16 bioregions, while other authors refer to 10 or 6. Adoption of the bioregion system, therefore, has been inconsistent. Some authors would

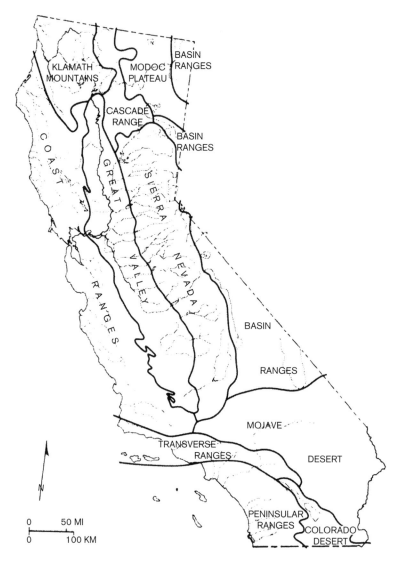

FIGURE 1.1 Geomorphic provinces (from Hill 1984).

divide the coastal mountains into three parts, north, central, and south, using the term South Coast to refer to the region I call Cismontane Southern California, which includes the Transverse and Peninsular Ranges. Other authors would split the Coast Ranges in the San Francisco area with a region called Bay/Delta. I recognize three desert regions, Great Basin, Mojave, and Colorado, from north to south. Others do not recognize the Great Basin Desert in California, instead referring to the areas as Eastern Sierra and/or Modoc in reference to the Modoc Plateau in the northeastern corner of the state. Most authorities refer to the southernmost desert as the Colorado Desert, which is the California part of a larger desert unit, the Sonoran Desert, which also occurs in southern Arizona, Baja California, and the state of Sonora in northwestern Mexico. Some would not recognize the Colorado Desert at all, but instead refer simply to the Sonoran Desert in California. Bailey divided the Colorado Desert of California into Sonoran and Colorado subdivisions.

LANDFORMS AND NATURAL REGIONS OF CALIFORNIA
1. SIERRA NEVADA
2. KLAMATH MOUNTAINS   } PACIFIC
3. CASCADE RANGE   } NORTHWEST
4. COST RANGES
5. TRANVERSE RANGES   } CISMONTANE
6. PENINSULAR RANGES   } SOUTHERN CALIFORNIA
7. GREAT BASIN DESERT   }
8. MOJAVE DESERT   } DESERTS
9. COLORADO DESERT   }
10. GREAT CENTRAL VALLEY

SCALE
0 25 50 75 100
MI

FIGURE 1.2 Landforms and natural regions of California (courtesy of California Insect Survey, Department of Entomology and Parasitology, University of California).

CLASSIFICATIONS OF BIOTIC COMMUNITIES

Dividing the state into regions based on climate and geography leads to classifying its biotic communities. It has been recognized for many years that different vegetation is associated with different types of climate. Various classifications of vegetation, based on climatic data, have pervaded scientific and lay literature. Table 1.2 compares three such systems that apply to California's flora.

One of the earliest attempts was the life zone system of C. Hart Merriam. This system, published in 1892, was based primarily on temperature. The zones were named for geographic areas that had temperature regimes similar to that which Merriam observed in Arizona as he traveled from the floor of the Grand Canyon to the summit of nearby San Francisco Peaks. Merriam worked for the US government, and his system became entrenched in government communications. It is still used by many naturalists who are employed by various federal and state agencies. The correlation between elevation and latitude has merit, but the fact that the system is based primarily on temperature and infers precipitation makes it truly applicable only to the southwestern United States.

TABLE 1.1
A Comparison of Natural Regions with Bioregions

Biome	California Community	Life Zone
Grassland	Valley Grassland	Lower Sonoran
Desert	Cactus Scrub	
	Sagebrush Scrub	
	Creosote Bush Scrub	
	Shadscale Scrub	
	Alkali Sink	
	Blackbrush Scrub	
	Joshua Tree Woodland	
Scrub	Coastal Sage Scrub	
	Lower Chaparral	
	Upper Chaparral	Upper Sonoran
	Desert Chaparral	
	Oak Woodland	
Coniferous Forest	Pinyon-Juniper Woodland	
	Mixed Coniferous Forest	Transition
	Montane Forest	Canadian
	Subalpine Forest	Hudsonian
Temperate Rain Forest	Coast Redwood Forest	–
	Mixed Evergreen Forest	
Tundra	Alpine	Arctic-Alpine
Temperate Deciduous Forest	Riparian	–

Two ecologists, Clements and Shelford, popularized the biome concept. Biomes are large ecosystems in which variations in temperature and precipitation have created a characteristic assemblage of plants and animals. On a broad scale, this system is appropriate, and its vocabulary is in widespread use.

In 1947, L. R. Holdridge proposed a scheme for classifying world plant formations based on evapotranspiration ratios. It is an eloquent system that applies raw climatic data more precisely than the biome system. The Holdridge scheme, however, lacks simplicity and has not gained widespread use.

California has been named one of the world's top 25 biodiverse regions of the world. In California, the range of climate is so extreme and the vegetation so diverse that broad classification schemes such as world plant formations or biomes have proved inadequate. Instead, the system of plant communities, published by Phillip Munz and David Keck in 1959, became a system of choice used by many ecologists in California. This system is based on one or more dominant plant species that occupy each area. It is broad enough that there are not too many categories to remember, and it is precise enough that its categories encompass meaning-

TABLE 1.2
A Comparison of Three Systems of Community Classification

Biome	California Community	Life Zone
Grassland	Valley Grassland	Lower Sonoran
Desert	Cactus Scrub	
	Sagebrush Scrub	
	Creosote Bush Scrub	
	Shadscale Scrub	
	Alkali Sink	
	Blackbrush Scrub	Upper Sonoran
	Joshua Tree Woodland	
Scrub	Coastal Sage Scrub	
	Lower Chaparral	
	Upper Chaparral	
	Desert Chaparral	
	Oak Woodland	
Coniferous Forest	Pinyon-Juniper Woodland	
	Mixed Coniferous Forest	Transition
	Montane Forest	Canadian
	Subalpine Forest	Hudsonian
Temperate Rain Forest	Coast Redwood Forest	–
	Mixed Evergreen Forest	
Tundra	Alpine	Arctic-Alpine
Temperate Deciduous Forest	Riparian	–

NOTE: The California communities listed here are modified from those described in 1959 by Munz and Keck in *A California Flora*.

ful units. Originally, the system included only 29 plant communities. Throughout this book, the Munz and Keck system of plant communities, with some modification to include about 35 categories, will provide a basic frame of reference.

Robert Bailey in 1983 developed a system of "ecoregions" that is used today by the US Forest Service. He divided California into 19 ecoregions. Using categories that are similar to ecoregions, the *Jepson Manual of the Vascular Plants of California* divides the state into 50 "geographic subdivisions." These categories are arranged hierarchically into Provinces, Regions, and Subregions. The categories are defined on the basis of geography, climate, and vegetation. The manual uses these 50 subdivisions to describe the distribution of California plants. These categories are a bit too refined for use in this book, but they will not be ignored entirely.

According to the *Jepson Manual*, there are three floristic provinces: California, Great Basin, and Desert. The California Floristic Province is the primary group of plants in the state, and is distinctly Californian. It includes the most diverse group of plants in the state.

The province essentially refers to everything west of the deserts, and it encompasses that part of California which is profoundly influenced by our Mediterranean climate, characterized by winter precipitation and dry summers.

A chain of mountain ranges extending the length of California forms a significant block to winter storms off the Pacific Ocean. From north to south, these ranges are the Cascade Mountains, Sierra Nevada, Transverse Ranges, and Peninsular Ranges. Deserts lie to the east of these ranges. The Great Basin Province is the western edge of the largest desert in North America, which extends eastward across Utah and Nevada all the way to the Rocky Mountains. This is a cold desert in which winter precipitation often arrives as snow. In general, the dominant vegetation is composed of cold-tolerant shrubs such as various forms of sagebrush (*Artemisia* sp.). The Desert Province is subdivided into the Mojave and Sonoran (Colorado) Deserts. These are warm deserts characterized predominantly by drought-tolerant shrubs such as Creosote Bush (*Larrea tridentata*). The Mojave gets primarily winter precipitation with snow at higher elevations. California's Colorado Desert has two rainy seasons with approximately half its precipitation arriving as summer thunderstorms.

People who classify things could be splitters or lumpers. Lumpers are folks who tend to categorize things on the basis of similarities. Splitters like to see differences between things, and tend to assign a name to each of the categories. Phillip Munz and David Keck were lumpers. In recent years, splitters have developed various systems of plant community classification which may be referred to as "vegetation types." In association with the University of California's Natural Reserve System and the California Department of Fish and Game, Daniel Cheatham and J. R. Haller in 1975 developed a list of "habitat types" that included about 145 communities. In 1986, Robert Holland working on the California Natural Diversity Database for the California Department of Fish and Game developed a list of 260 "terrestrial natural communities." In 1995, the California Native Plant Society published *A Manual of California Vegetation* by John Sawyer and Todd Keeler-Wolf. This manual was designed as an "evolving system" that would recognize rare or unrecognized vegetation types as well as refining categories into more specific plant associations and series. Originally, the manual included about 250 vegetation types. In the second edition in 2009, in collaboration with Julie Evans, the number of units called alliances was upped to nearly 500. Many hours of scientific surveys, mapping, and analysis went into preparation of these publications, and they are meant to be the basic reference for descriptions of California vegetation. Unfortunately, identification of many of these alliances in the field requires a certain amount of botanical knowledge, and alliances often are composed of a single species, which makes using the alliance system unwieldy for a book of this type. For example, the common forest community known as Mixed Coniferous Forest (formerly Yellow Pine Forest) is composed of about six common species of trees, each of which may grow in clumps. In the *Manual of California Vegetation*, this community is split into 10 alliances, basically associated with clumping of individual species. Therefore, without ignoring the needs of botanists, but describing vegetation that is understandable to a lay person, in this edition of the book, I will continue to use a system of recognizable plant communities (with some reference to alliances) that has been used by ecologists and nature writers for many years. To enhance understanding of California's diverse vegetation, a color-coded map of generalized vegetation cover is depicted in plate 1.

CLIMATE AND WEATHER

California's Mediterranean climate primarily is responsible for its great diversity of biotic communities, although other factors such as soil type and slope exposure are also involved. Unique among climates of the world, in a Mediterranean climate, precipitation occurs primarily during winter months, the coldest time of year and the time with short daylight hours. Since photosynthesis in green plants provides the nutrient basis for food chains and ecosystems, it should be obvious that winter is generally not the optimum time to promote plant growth. Thus, most ecosystems associated with a Mediterranean climate must be considered relatively food-poor, and that dictates specialized adaptations among the plants and animals that inhabit the area.

The study of climate and weather is known as meteorology. *Climate* refers to the overall combination of temperature, precipitation, winds, and so forth that any region may experience. *Weather* refers to daily variations in these phenomena. Changes in temperature and precipitation are functions of many variables. As the earth rotates, light strikes its surface at different angles in different latitudes, and because the earth is tilted on its axis by 24° these angles change with the seasons. During summer in the northern hemisphere, the North Pole is tilted toward the sun so the sun's rays hit the surface more directly, and the photoperiod (day length) is longer. These differences cause uneven heating, which is expressed as wind. Winds carry water vapor, the amount of which is influenced by differing rates of evaporation and precipitation, depending on the temperature. Furthermore, wind circulation patterns are influenced by the earth's rotation on its axis and differences in local topography. All this translates into major generalizations about what causes climate and weather for a specific geographic region.

Major Movement of Air Masses

Near the equator, sunlight hits the earth directly. On the first day of spring in the northern hemisphere, or the first day of autumn in the southern hemisphere, the sun is directly over the equator, at which time daylight and darkness are 12 hr long all over the world. Wherever the sun shines straight down, there will be a greater degree of warming than anywhere else. Air will rise at that point. Meteorologists indicate that for every 1°F (1.7°C) increase in temperature the atmosphere can hold 4% more moisture. Air on the surface of

the earth will move as winds toward the point of rising air from the north and south. As the air rises, it becomes less dense and thus cooler. This is called adiabatic cooling. The precise amount of cooling varies with the amount of water vapor in the air, but on the average it is about 3–5°F/1000 ft (1°C/100 m). A drop in temperature of 20°F (11°C) will decrease the amount of water held by the air by one-half. When the air is cooled, water vapor condenses and falls in the form of rain. As a result, it rains nearly every day, usually in the afternoon, in a belt around the earth at the equator. As summer progresses in the northern hemisphere, the photoperiod gradually increases until the sun is shining directly on the Tropic of Cancer, 24° north of the equator. On the same day the sun is shining on the Tropic of Capricorn in the southern hemisphere, it shines down at its greatest angle. On this day, June 22, it is the summer solstice in the northern hemisphere and the winter solstice in the southern hemisphere: the longest day of the year in the north and the shortest day of the year in the south. The climate in this region between the tropics is therefore said to be tropical.

Where air descends to the earth, the process reverses. Descending air becomes compressed and heats at about the same rate as it cools when rising. This process causes the air to absorb water vapor. Descending air therefore tends to be dry and causes liquid water to evaporate from the environment. This dry air descends in two belts around the earth, approximately 30° north and south of the equator, just outside the tropics (figure 1.3). A quick glance at a world map will show that deserts lie along this belt, and along the southwestern coasts of each continent lie regions with a Mediterranean climate. When it is summer in the northern hemisphere, this belt of dry, descending air is pushed farther north. It therefore seldom rains in southern California in the summer. Los Angeles lies at about 35° north latitude. Meanwhile, the belt of precipitation has also moved farther north of the equator, creating regions of summer rainfall in the vicinity of the Tropic of Cancer.

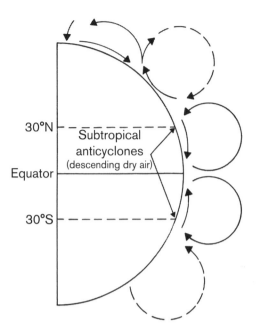

30°N

Subtropical
anticyclones
(descending dry air)

Equator

30°S

FIGURE 1.3 Simplified model of major movement of
air masses (after Barry, R. G., and R. J. Charles. 1982.
Atmosphere, Weather, and Climate. London: Methuen.
Illustration from Hensson and Usner 1993, p.34).

The Influence of Ocean Currents

The influence of the ocean on California's climate is profound. California's coastline is about 1100 mi (1760 km) long. This great range of latitude causes the climate to vary considerably along the coast from north to south. From Crescent City to San Diego, the range of average annual precipitation is from about 80 in (200 cm) to 10 in (25 cm). On the other hand, the proximity of the ocean means that air temperatures are not extreme. At the coast, freezing is rare, and the temperature rarely exceeds 100°F (38°C). At San Francisco, September temperature averages 62°F (17°C), and January temperature averages 51°F (11°C).

Ocean water along the California coast is generally cold. Off the northern California coast, winter water temperature is usually around 50°F (10°C). In its journey southward, it warms about 7–8°F (3–4°C). During summer, it is only 8–10°F (5–6°C) warmer. Cold ocean water causes frequent fog, and relative humidity remains high most of the time. Plants water themselves with fog drip, and evaporation rates are low. Along the coast, plant communities range from moist Mixed Evergreen Forest in the north to dry Coastal Sage Scrub in the south.

Ocean water swirls clockwise in the northern hemisphere and counterclockwise in the southern hemisphere (figure 1.4). This swirling, a product of the earth's rotation, is known as the Coriolis effect. The earth rotates in an easterly direction. That is why the sun appears to move westward, rising in the east and setting in the west. The effect of this eastward motion is that the water at the equator appears to flow westward. As the water moves westward relative to the earth's surface, it begins to turn toward the right in the northern hemisphere and toward the left in the southern hemisphere. This phenomenon occurs because the earth's motion is fastest at the equator, where the circumference of the earth is greatest. To make one complete rotation, the land at the equator has to travel farther than at any other place on earth. Because the surface of the earth moves

A climate marked by periods of summer precipitation is called a monsoonal climate. That includes southern Florida in the United States.

It is not really accurate to use the terms *warm* and *cool* with respect to air masses. What actually causes air to rise or descend is a change in density. Warm air is less dense, so it will rise. Cool air is comparatively dense, so it will sink. As it sinks and becomes warmed by compression, it gets even denser. Even though it has become warmed, it will continue to descend. It is best, therefore, to compare air masses in terms of density. Dense air pushes downward with greater pressure; thus, it is said to represent a region of high pressure. Conversely, rising air represents a low-pressure region. Meteorologists explain changes in weather as changes in pressure, which are measured by a barometer. They translate predicted changes of pressure into a weather forecast. A reduction in air pressure might imply that a storm, or a mass of rising air, is approaching. Satellite photos show low-pressure areas covered by clouds and high-pressure areas are clear.

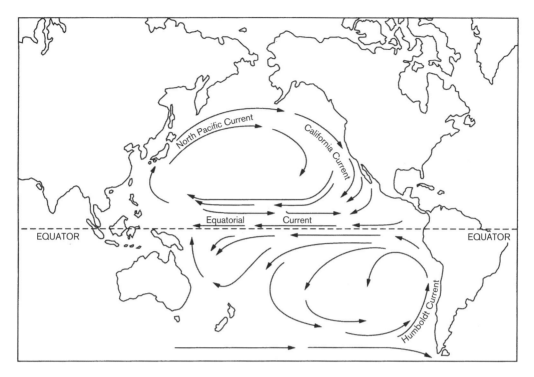

FIGURE 1.4 Pacific Ocean currents (from Caughman, M., and J. S. Ginsberg. 1987. *California Coastal Resource Guide.* Berkeley: University of California Press).

more slowly north or south of the equator, the increased drag makes flowing water turn to the right, clockwise, in the northern hemisphere and to the left, counterclockwise, in the southern hemisphere.

As the water moves westward at the equator, it picks up heat in the region of direct sunlight. Currents of water that come from the equator are comparatively warm, and they flow northward along the eastern edge of a continent in the northern hemisphere. As this water flows through the arctic region, it cools considerably. When the water flows southward again, toward the equator, it flows along the western edge of a continent. A cold current of this type flows southward along the coast of California.

In the southern hemisphere, ocean water swirls in an opposite direction. Warm currents from the equator flow southward on the eastern edge of a continent. Cold currents from the Antarctic flow northward along the western edge of continents. One bit of consistency to emerge from all this is that water off the east coast of

any continent is comparatively warm regardless of the hemisphere and water off the west coast of any continent is comparatively cold.

A climate dominated by the influence of the ocean is known as a maritime climate. On a daily basis, water temperature fluctuates very little. Therefore, air temperature fluctuations on the nearby land are also moderated. When the water offshore is comparatively cold, as it is on the west coast of the United States, fog forms over the water and adjacent land nearly every day, a condition that is accentuated as you go farther north. In southern California, rain is typically confined to the ocean and a short distance inland. For rain to occur farther inland, the land itself must be significantly cooler than the air. When the offshore current is comparatively warm, as it is on the east coast, it rains frequently on the land, and fog seldom forms.

Air masses tend to follow the ocean currents because they are also under the influence of the Coriolis force. The circles of flow, however, are interrupted by the descending air at latitudes about 30° north and south. Where air is descending, very little wind flows transversely across the surface of the earth. Years ago, seamen in wooden sailing ships would have to row when they got becalmed in this part of the ocean, so they would throw over some of the cargo to lighten the load. So many horses from the old world were floating "belly up" that this part of the ocean became known as the Horse Latitudes. A comparable area of little wind, known as the Doldrums, occurs at the equator. Here, the air is rising. In the northern hemisphere, air flowing toward the earth at the 30th parallel turns to the right, forming a clockwise flow. The effect is that air south of the 30th parallel tends to flow southwestward, forming what is known as the Trade Winds, so named because sailing ships could use these winds to carry goods across the Atlantic to the new world. North of the 30th parallel, winds tend to flow from west to east. These winds are known as the Prevailing Westerlies.

Storms develop as low-pressure areas over the ocean, and are highly dependent on the water temperature. Warm air in a low-pressure cell rises, carrying with it evaporated water. Wind on the surface of the earth therefore blows toward a low-pressure cell from all directions. In the northern hemisphere, air moving across the surface to replace the rising air of a low-pressure cell must turn to the right, which gives the mass a counterclockwise spin. This spinning mass is similar to a spinning top. It moves randomly unless other forces give it a direction. Prevailing winds provide this force. Storms generated in the tropics tend to be blown westward by the Trade Winds, and storms generated over the northern part of the ocean tend to be blown eastward by the Prevailing Westerlies. In North America, major winter storm tracks come from the north Pacific and flow southeastward (figure 1.5A). They spin counterclockwise as they travel down the coast, so they continue to pull moisture off the ocean and dump it on the land. As one moves from north to south, the land warms faster than the water because air changes temperature more

rapidly. During summer, when the air and land in southern California is considerably drier and warmer than the water, the storm fizzles because water vapor won't condense (figure 1.5B). During winter, storm after storm potentially makes its way down the coast before moving eastward across continental North America (figure 1.5A). This is basic reason that the northern half of the state gets two-thirds of California's precipitation. Fewer storms cause precipitation in southern California and they rarely continue as far south as the center of Baja California. A high-pressure cell is like a dome of dense air over the southern part of the state and that deflects the storms northward and eastward. Baja California still has cold water offshore, but the land seldom experiences measurable precipitation. Because condensation in moist air occurs over cold water, the coast of Baja California is frequently foggy. Plants water themselves with fog drip but cannot depend on rainfall to provide adequate water. In Baja California, therefore, the desert occurs on the coastline.

At high elevation, there are two jet streams, a polar jet stream and a subtropical jet stream, which move eastward around 100 mph (160 kph). These jet streams tend to push storms eastward, carrying weather systems across the continental United States. The polar jet stream usually has more influence on California weather, but these masses of flowing air meander back and forth similarly to flowing water in a river. When the polar jet stream meanders southward, it pushes Pacific storms toward southern California and precipitation increases. During summer, this is an important force that brings precipitation to the forests of the Pacific Northwest and the northern Sierra Nevada. The subtropical jet stream also can meander northward and bring tropical (monsoonal) storms to southern California, a condition that is accentuated when the eastern Pacific Ocean is exceptionally warm. This is known as an El Niño condition, a Spanish word that is derived from a change in weather that also influences the coast of South America.

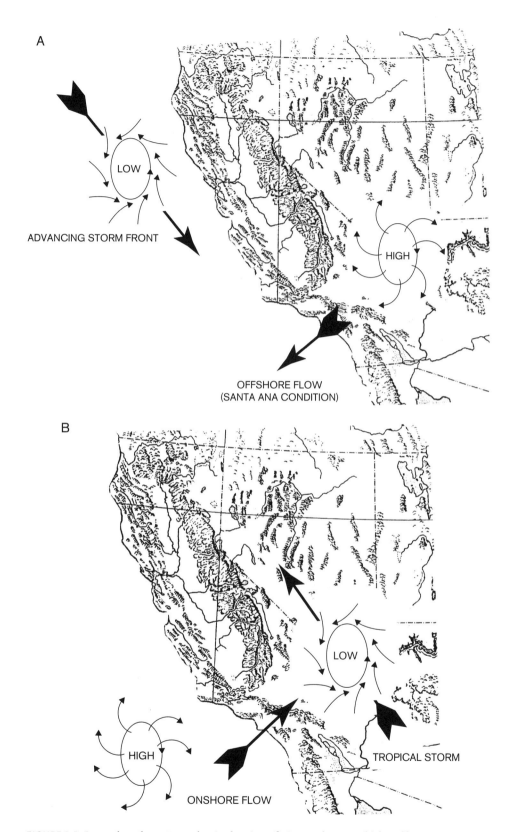

FIGURE 1.5 Seasonal weather patterns showing locations of winter and summer high- and low-pressure zones.

(A) Winter weather pattern.

(B) Summer weather pattern.

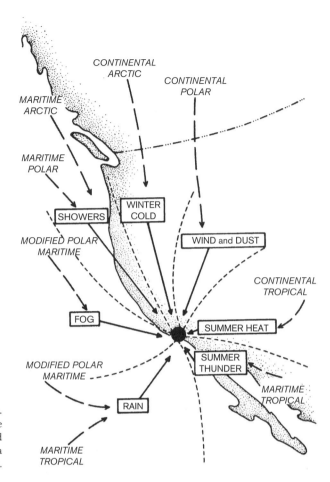

FIGURE 1.6 Different weather conditions. Arrows indicate the direction and source of different weather conditions and seasons for the southern part of California (from Bailey 1966).

Heavy rain would arrive around Christmas time along the coast of Peru and Chile where climate is desertlike. The name is in reference to the birth of the Christ child at that time of year. El Niño conditions brought heavy rains, essentially double of normal amounts to parts of California in 1978, 1983, 1993, 1998, 2016 and 2017.

In the summer, it seldom rains in southern California (figure 1.6). The sun has moved northward so that it shines straight down on the Tropic of Cancer. That is where the belt of rising air generates tropical storms. Because the rising air is farther north at this time, so is the descending air, and that keeps southern California dry in the summer. Even in the absence of an El Niño, during late August or September occasional tropical storms move into southern California from the south. These

storms are usually carried westward from the east Pacific toward Hawaii, but some of them move into the belt of descending air where there is no surface wind to give them a westward push. These storms can spin erratically (like a top) and may move over the land, dumping a considerable amount of water in a short time. These sporadic storms, known in Baja California as *chubascos*, are able to move northward. Such thunderstorms over the Colorado Desert and Peninsular Ranges may cause intense local flooding. Some 5–10 in (12–25 cm) of rain may fall in a few hours, representing a large portion of the annual share of precipitation. Some of these tropical storms also come all the way from the Gulf of Mexico.

The moist forests of the Pacific Northwest contain a variety of relict species that depend on the relatively predictable supply of summer

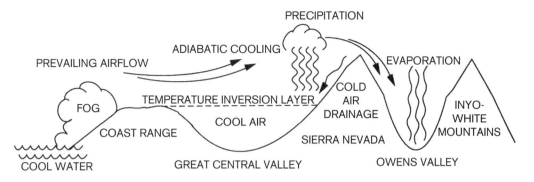

FIGURE 1.7 The influence of local topography on climate and weather, as indicated by an exaggerated cross section of California at about the latitude of Fresno.

rain from the tail end of Alaskan storms. Summer rain in southern California, however, may not occur every year. Nevertheless, certain plant species, such as different populations of relict cypress (Cupressaceae), to some degree owe their existence to summer rain.

The center of the North American continent tends to be dry because it is a long way from the ocean, the main source of water. Pacific storms, during summer, are deflected over the Midwest by the southern belt of high pressure. In winter, strong storm systems carry precipitation all the way to the interior, where the moisture may be added to the storms traveling eastward from the Pacific. If the difference in pressures between the two storm fronts is extreme, at the contact point between the warm southern storms and the cold Pacific storms, there may be a belt of tornados. Furthermore, during summer, thunderstorms from the southern storm belt carry northward across the Gulf of Mexico all the way to the plains states. They may carry as far as California's Colorado Desert. In late August, these storms often reach as far as Arizona, where locally heavy precipitation is common. The Arizona portion of the Sonoran Desert experiences considerable summer precipitation. This rain falls during the hottest time of the year, however, so that evaporation rates are very high and the water does not sink very far into the ground. The area around Tucson is a desert, whereas the area around Los Angeles is not. Both areas receive approximately the same amount of precipitation each year.

The Influence of Local Topography

Rain-Shadow Effect

Mountain ranges have a profound influence on climate. Figure 1.7 is an exaggerated cross section through the state of California at about the latitude of Fresno. It shows how the Sierra Nevada intercepts precipitation on its western slope. This phenomenon is known as the rain-shadow effect. As moisture-laden air flows inland from the ocean, it passes eastward, rising over the Sierra Nevada. Air is chilled adiabatically as it reaches higher elevations, like rising air at the equator. This chilling causes precipitation on the side of the mountain toward the coast. As the air descends from the mountain into Owens Valley, on the eastern side of the Sierra, it becomes heated by compression and causes evaporation. Thus, the Great Basin Desert occurs in the rain shadow of the Sierra Nevada.

On the western side of the Sierra Nevada, the influence of elevation is similar to that of changes in latitude. As air moves up the western slope of the Sierra, its temperature decreases adiabatically by about 3–5°F/1000 ft (1°C/100 m), which is roughly equivalent to moving about 300 mi (480 km) northward. Precipitation increases accordingly, and the form of precipitation switches to snow. Moving northward, biotic zones occur at lower elevation. In northern California, coniferous forests occur at sea level. Climate above 11,000 ft (370 m) is roughly equivalent to that of the Arctic. A mountain is thus a microcosm of many types of climate.

Temperature Inversion Layers

Valleys also affect climate. Dense air flows downhill like water, a phenomenon known as cold-air drainage. Also like water, cold air collects in low spots that have limited or no drainage, such as the Great Central Valley of California. Cold air flows from the mountains at night and collects in the valley, filling it to the brim. In this case, the brim is the top of the Coast Ranges. Sometimes this cold air is so dense that winds are unable to disturb it. These circumstances produce dense fog that may remain in the valley for many days. In the Great Central Valley, this is known as a tule fog. Episodes of tule fog are famous for causing massive traffic jams and huge chain-reaction collisions. What is unique about these puddles of cold air is that normal temperature relationships seem to be inverted. It is supposed to become gradually cooler as one goes higher in an air mass, but when cold-air masses are trapped in a valley, it becomes suddenly warmer at the top. This paradoxical phenomenon is known as a temperature inversion layer.

The same kind of temperature inversion occurs in the Los Angeles Basin, where air is trapped by the Transverse and Peninsular Ranges (figure 1.7). Cool marine air, which flows inland from the ocean and joins cool air flowing off the mountains, is held in the basin by the onshore flow of marine air.

A serious consequence of an inversion layer is that exhaust gases or other pollutants rise until they reach a layer of equal density. The difference in density is so great at the inversion layer that pollutants are unable to rise above it. Below the inversion layer, sunlight effects a chemical change on these materials, causing photochemical smog, a mixture of haze and oxidized chemical vapors. Smog is not restricted to the Los Angeles Basin or the Central Valley, but it is especially bad there because of the amount of industry and vehicular traffic, which produce the vapors. On nearly any summer day, from a position in the mountains above the inversion layer of the Great Central Valley or the Los Angeles Basin, one can see the flat top of a yellowish brown mass of air trapped in the valley below.

In the Los Angeles Basin, the only relief from the inversion layer occurs when the prevailing airflow is offshore rather than onshore. The flow of air is commonly onshore, however, because inland air heats up more quickly than air over the ocean. This creates a low-pressure cell, and air flows toward a low. The inland air rises and draws the marine air onshore, where it becomes trapped by the mountains. Occasionally, though, when air becomes cooler or denser in the Great Basin and the Mojave, a high-pressure cell is formed. Air flows away from a high, so a flow toward the coast can occur (figure 1.5A). To escape from the desert, air must travel through Cajon Pass, one of the few gaps in the surrounding mountains. Situated between the San Gabriel and San Bernardino Mountains, Cajon Pass allows air to flow from the high desert at approximately 3500 ft (1200 m) elevation downhill toward the coast, where the Los Angeles Basin lies just above sea level. This flow occurs at irregular intervals, during autumn and winter when the air in the desert is cooler than it is in the basin. At the extreme, a strong wind from the desert blows through Cajon Pass and becomes heated by compression as it descends to the Los Angeles Basin. This is known as a Santa Ana wind because air follows the course of the Santa Ana River as it pushes pollutants out to sea.

Unlike the Los Angeles Basin, the Great Central Valley has little chance to be cleared out by a wind. It is too thoroughly surrounded by mountains. The only gap is where San Francisco Bay has formed an opening in the Coast Ranges. Air usually flows onshore through that gap, and when it flows toward the coast, it tends to flow over the dense air of the Great Central Valley.

Seasons in California

People acquainted with traditional eastern seasons find seasons in California to be confusing.

The pattern of the eastern United States has been popularized for years by romantic poets, novelists, and songwriters. Winter snow is followed by a spring awakening. Green summer forests are followed by autumn color. This pattern, associated with eastern deciduous forests, is caused by a climate in which some form of precipitation can be expected at any time of year. Water is seldom scarce, and the vegetation responds to marked seasonal change in temperature. A person from the east who yearns to see that sort of change in California must visit a community where vegetation borders a permanent stream. This is called a Riparian (riverside) community. Canyons all over the mountains of California experience seasonal change similar to that of the eastern United States. Water is always present, and temperature fluctuates markedly. Autumn color in aspen groves on the eastern side of the Sierra Nevada provides a scene of great beauty that may be reached by driving Highway 395 northward from Owens Valley (plate 8B). In southern California, autumn color is provided by sycamores and cottonwoods in canyons bordering the Los Angeles Basin.

The claim that southern California has no seasons is simply untrue. In southern California, the most significant variable component of the climate is precipitation rather than temperature. In lowlands, the "awakening" season is winter, when hillsides and valleys turn green as vegetation responds to winter precipitation. If precipitation has been heavy enough, spring wildflowers may cover many acres, even in the desert. Summer is the season of warm, sunny days. It is the time to visit the beach or go hiking in the mountains. Autumn brings the Santa Ana winds, and trees along streams take on red and yellow colors. Clear skies and warm days extend into October. In winter, rainstorms alternate with clear, warm days associated with mild Santa Ana conditions. Football games, viewed by fans in summer clothing, are played in stadiums surrounded by snow-capped peaks—all on national television! Spring may seem to be the least spectacular of southern California's seasons because of many foggy days along the coast. A short trip inland or to the mountains, however, may reveal carpets of wildflowers, a reward for any Californian willing to drive 1 hr from the coast (plate 14B). Critics of southern California's climate emphasize the worst: spring fog followed by summer smog, autumn fires followed by winter floods. Nevertheless, seasons are distinct. If you are willing to travel, you can find examples of any sort of climate somewhere in the state.

Northern California may experience precipitation any time of year, although it is heaviest in winter. Along the coast, fog is frequent as a result of persistently cold water offshore. The influence of a maritime climate is felt over the entire coastline, but it seems to be stronger in northern California because of the persistent fog. Coast Redwoods (plate 11A), which require a great deal of moisture, do not do well beyond the influence of maritime climate. Heavy winter precipitation is followed by persistent summer fog, which condenses on the foliage and continues to wet the soil by dripping.

A chain of mountains formed by the Cascades, Sierra Nevada, Transverse Ranges, and Peninsular Ranges forms a rain shadow, east of which lies a vast desert area. In California, the Great Basin Desert (plate 15A) covers the Modoc Plateau and extends southward to Owens Valley. The Great Basin is a region of winter precipitation, mostly snow. South of the Great Basin, the Mojave Desert (plates 4A and 15B) gets snow at higher elevations. Still farther south, the Colorado Desert (plates 4B and 16A) rarely gets snow, but it does get about half of its precipitation as summer rain in the form of thunderstorms that move northward from the tropics.

California has a surplus of water in the north and in the mountains. Toward the south, evaporation and transpiration rates exceed precipitation, creating a net deficit of water. Two-thirds of the water is in the north, but two-thirds of the people live in the south. California's massive water transport systems are an attempt to solve this discrepancy.

Basic Ecology

TO FULLY APPRECIATE THE SPECIAL place called California, one must first understand the basic principles and processes of ecology and geology as they relate to natural systems. It is also necessary to become familiar with the vocabulary that scientists use to describe natural phenomena. This chapter and the next, on basic geology, provide the background information needed for a better understanding of California.

ECOSYSTEMS

An ecosystem is an interacting unit in nature made up of living and nonliving components. While the word "ecosystem" is relatively new in science, the concept is not. Alexander von Humboldt, the person for whom several places in California are named, wrote about nature in the late 1700s and early 1800s. He regarded nature as a web of life in which the earth was one great living organism and everything was connected. That sentiment was repeated by John Muir in his now famous quotation, "When we try to pick out anything by itself,

we find it hitched to everything else in the universe."

In a healthy ecosystem, matter is recycled over and over, and the system is powered by energy from the sun. Living organisms are dependent on a one-way flow of energy, but when they are in a balance with nature, they do not use up their resources. It is traditional to divide organisms into three categories: producers, consumers, and decomposers (figure 2.1). Producers are usually green plants, which convert light energy to food energy by means of photosynthesis. Because green plants produce the food for the ecosystem, they are known as producers. The amount of photosynthesis per unit of time is known as primary production. Animals are the consumers in the system. Herbivores are primary consumers because they are the first to consume food. Carnivores are secondary consumers because they eat the animals that eat the plants. Omnivores are able to function as either primary or secondary consumers, an important adaptation if food is scarce. Decomposers include bacteria, fungi, or other microorganisms that break down

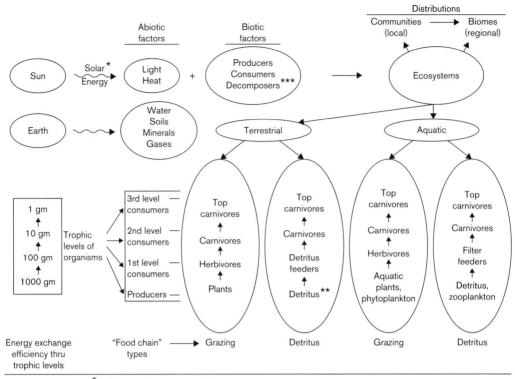

FIGURE 2.1 Summary of ecosystems.

detritus, the remains of other organisms. These products of decay are, in turn, used by green plants to carry on more photosynthesis. In this way, matter in nature is recycled.

Although it is convenient, and generally correct, to employ these three categories, there are many exceptions. For example, some special ecosystems contain microorganisms that produce food without relying on photosynthesis. Instead, they use a process that is sometimes called chemosynthesis. These organisms may live in mud, in caves, in hot-water vents of the deep ocean, or in salt flats. Conspicuous examples are red bacteria that are so numerous on salt flats that they make the salt appear red. This phenomenon occurs in various salt pans in southwestern deserts, but the most visible example is Owens Lake on the eastern side of the Sierra Nevada. The red color of these bacteria is created by a pigment similar to the light-sensitive pigment in the human eye. This pigment enables the bacteria to convert light to food in the absence of chlorophyll. Although chemosynthesis is not as efficient as photosynthesis, the bacteria that use it are, indeed, producers rather than decomposers.

Other exceptions to the three simple categories of organisms are scavengers or detritus eaters. For example, it seems logical to consider vultures as carnivores even though, like bacteria, they consume carrion. Another way to think of scavengers is to consider them as sharing the food left over by the carnivore, without harming the predator. In terms of life together or symbiosis, this relationship would be called commensalism. Similarly, it seems logical to consider mushrooms as decomposers even though they are not microorganisms.

Actually, it is simplest to view the organisms of an ecosystem in terms of two rather than three categories. The first category, called *autotrophic* (self-nourishing), contains organisms that produce their own food. The second, called *heterotrophic*, contains organisms that do not produce their own food. As far as an ecosystem is concerned, there is little difference between consumers and decomposers. They all eat one another. In fact, they compete with one another for a fair share of the same food. When humans put food into a freezer, they are putting their share where microorganisms can't get to it. Microorganisms, on the other hand, get their share by producing foul odors or toxins that discourage humans from eating the food.

A simple ecosystem might have only two kinds of organisms. A sewer outflow, for example, might have only algae (autotrophic) and bacteria (heterotrophic). The important thing is that one organism makes the food, and the other recycles matter by consuming it.

The nonliving components of an ecosystem include matter that is cycled and the energy that powers it. Energy in an ecosystem is represented by heat and light, the ultimate source of which is the sun. Fluctuations of these two energy forms have a profound influence on the system. The matter in a system may be thought of in terms of pure elements such as carbon, hydrogen, oxygen, nitrogen, and phosphorus. These elements may combine to form various materials such as water (H_2O), carbon dioxide (CO_2), carbohydrate (CH_2O), nitrate (NO_3), and phosphate (PO_4). A simple way to describe the nonliving components of an ecosystem is to lump all these factors into five categories: light, heat, air, water, and soil (minerals). If one factor is deficient, it becomes ecologically limiting and exerts a powerful influence on the ecosystem. This phenomenon is described by the *law of the minimum*: that is, whichever factor occurs at a minimum becomes limiting, and the entire ecosystem must adapt to it. Light is limiting on the ocean floor, heat is limiting on a mountaintop, water is limiting in the desert, and so forth. In fact, because water is a requirement for pho-

tosynthesis, in California's Mediterranean climate, characterized by summer drought, water becomes the primary limiting factor, dictating that most California ecosystems are food poor. Matter can also be limiting. We therefore refer to nitrate, phosphate, carbon dioxide, water, and carbohydrates as nutrients: plants and animals cannot live without them.

The Flow of Energy and Nutrients

A characteristic of ecosystems is that energy and matter (nutrients) seem to flow through them (figure 2.1). It is important to realize, however, that materials cycle, whereas energy flows through the system in one direction. Matter is used over and over, but energy dissipates.

Energy powers all life; all ecosystems therefore depend on a constant transfer of energy. At its simplest, we can think of energy transfer as a three-step process:

1. The sun is the source. Nuclear activity causes the sun to emit solar radiation, upon which life on earth depends.

2. Green plants convert light energy to chemical energy (food). In other words, the process of photosynthesis converts solar radiation to the energy locked up in food molecules. This step may be symbolized in this way:

$$CO_2 + H_2O \xrightarrow[\text{chlorophyll}]{\text{light}} CH_2O + O_2$$

More specifically, the chlorophyll in green plants absorbs light energy and uses it to rearrange molecules of carbon dioxide gas from the air and water from the soil to form energy-rich carbohydrate molecules. The chemical notation for carbohydrate is CH_2O, which indicates that it is composed of carbon and water. The specific carbohydrate produced is glucose, a sugar that is symbolized as $C_6H_{12}O_6$. In order to produce glucose, the above equation is balanced by multiplying it through by a factor of six.

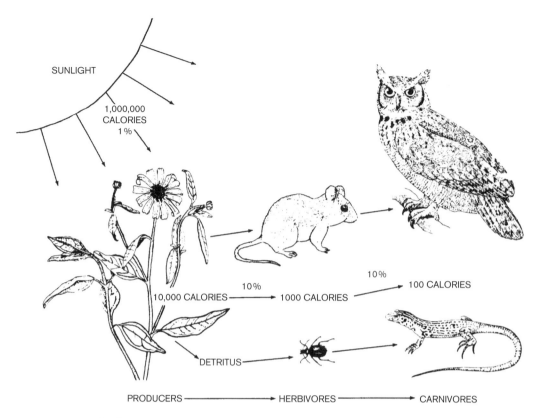

SUNLIGHT

1,000,000
CALORIES
1%

10,000 CALORIES —— 10% → 1000 CALORIES

10%

100 CALORIES

10%

DETRITUS

PRODUCERS ——————→ HERBIVORES ——————→ CARNIVORES

FIGURE 2.2 Transfer of energy in ecosystems (animal illustrations by Ellen Blonder, from Verner and Boss 1980. Plant illustration: Coastal Brittlebush by Karlin Grunau Marsh from The Mountains and the Wetlands, Orange County Land Environment Series, no. 2, 1975).

3. Consumers, through cellular respiration (metabolism), convert the energy in carbohydrate to the energy that makes them move, grow, or think. This step may be symbolized this way:

$$CH_2O + O_2 \longrightarrow CO_2 + H_2O + energy$$

The cyclical nature of this sequence of steps should be obvious. The end products of photosynthesis are the raw materials for cellular respiration, and the end products of cellular respiration are the raw materials for photosynthesis.

The raw materials for cellular respiration are easy to remember. You eat to obtain carbohydrate (glucose), the source of energy. You breathe to obtain oxygen. It is also easy to remember the end products of cellular respiration. You exhale carbon dioxide, and you exhale water, which appears in the form of a small cloud when you breathe out in cold air. All living organisms,

including bacteria and plants, depend on cellular respiration as a source of energy. Some microorganisms are able to carry on metabolism without oxygen, but they still depend on glucose as their energy source. Likewise, in order to promote growth of roots stems and leaves, green plants carry on cellular respiration, but they produce more glucose and oxygen than they use. The oxygen is excreted, and the carbohydrate is stored.

In summary, then, energy is transferred from solar radiation to green plants, and then to animals and microorganisms. Without this transfer, there would be virtually no life on earth.

Laws of Thermodynamics

The transfer of energy through organisms in an ecosystem is not very efficient. An ecosystem will therefore support only a limited number or mass of organisms (figure 2.2). Not all organisms are equally efficient, but they are similar enough that we may generalize. Because water is a requirement for photosynthesis, and green plants use a portion of their end products for their own metabolism, autotrophic (photosynthetic) organisms are only about 1% efficient. Heterotrophic organisms are about 10% efficient. This means that on average, only 1% of the solar radiation that falls on a green plant is converted to food and only 10% of that energy is incorporated into the animal that eats it. Simply stated, this means that 100 lb of vegetation will support only 10 lb of herbivore, which will in turn support a mere 1 lb of carnivore.

This limitation exists because two "laws" of nature dictate energy relationships: the first and second laws of thermodynamics. *Thermodynamics* means "heat power," and heat, of course, is a form of energy. Because energy may be converted from one form to another, it has become convenient to discuss energy relationships in terms of heat. For example, a calorie is a unit to measure heat energy. To express the energy available in a given food item, we assign it a value in Calories. A food Calorie is spelled with a capital *C*. The Calories used to express food values in human nutrition are the same as the kilocalories used by scientists. A Calorie (kilocalorie) will raise the temperature of 1 L (1.06 qt) of water 1°C (1.8°F). This means that 1 oz (28.35 g) of milk chocolate has enough energy in it to raise the temperature of 147 L (more than 25 gal) of water 1°C.

The first law, loosely translated, states that the total amount of energy in the universe is constant, and we cannot add to it or take away from it. Another way to think about it is to realize that the total amount of energy that falls on the earth in a given year is fairly constant, which means that the total amount of photosynthesis on the earth is also fairly constant. If food production on planet earth is fairly constant, it means that the total amount of animal life supported on the earth is also constant. The message is, "There are limits!"

The second law or the law of entropy, perhaps even more loosely translated, states that energy can be converted from one form to another, but transformations are never completely efficient. Disorder (entropy) tends to occur during energy transformations; therefore, to create order or put things in a precise arrangement, it requires more energy than can be reclaimed later. This is the reason that energy transfers in ecosystems (i.e., from producers to consumers) are inefficient. Unused energy tends to escape as heat. Nature tends to move toward randomness or disorder. If humans try to fight that tendency by putting things in order, we expend a great deal of energy that we cannot reclaim. For example, planting crops in rows requires effort, and weeding them, which is fighting randomness, also requires effort. By the time the crop is harvested, processed, and transported to our homes, as a society we have expended more total energy than we can hope to reclaim by eating the food. *You can't break the laws of nature: there are limits, and our use of energy is not very efficient.*

Food Pyramids

A food chain is the sequence by which energy is transferred from sunlight to green plant, to herbivore, to carnivore. The total amount of energy that flows along a food chain in an ecosystem can graphically be depicted as a food pyramid, sometimes called an Eltonian pyramid after Charles Elton, an English ecologist. Food pyramids are simple applications of the laws of thermodynamics: in any given ecosystem, plants are more common than herbivores, and herbivores are more common than carnivores, by a factor of 10%.

Pyramids can be based on three types of data: numbers of individuals, amount of energy, or total weight (biomass; figure 2.3). A pyramid of numbers is a simple expression of what a person sees on a walk in nature. There are more plants than squirrels, and more squirrels than coyotes. A more quantitative pyramid is based on energy. Given the efficiencies previously described, 100,000 Calories of light will support 1000 Calories of vegetation because

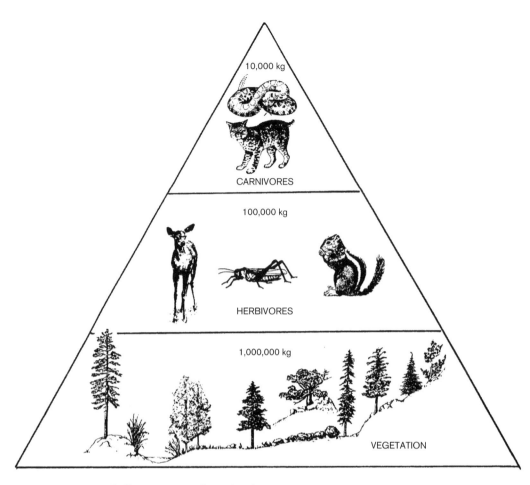

FIGURE 2.3 A pyramid of biomass. Energy flow is from bottom to top.

photosynthesis is only 1% efficient. That vegetation in turn will support 100 Calories of herbivores, which will support about 10 Calories of carnivores. A biomass pyramid would show the same pattern, with only one-tenth the mass at each successive level. It should be obvious that it takes a tremendous biomass of vegetation to support a large population of carnivores.

Because these pyramids are based on who eats whom, the steps are called *trophic* (nourishment) levels. Sometimes, food pyramids consist of four trophic levels, including two levels of carnivores. However, it would take an extremely large ecosystem, such as the ocean, to produce enough food to support such a scheme. For example, a tuna eats sardines, which eat krill, and krill eat diatoms. Most ter-restrial ecosystems are unable to support a tertiary (third-level) consumer because only a small amount of energy or mass would be available to feed an animal that routinely feeds four steps removed from sunlight. For the same reason, in food-poor ecosystems, such as those in our Mediterranean climate, very few animals are exclusively carnivorous. Animals such as bears, foxes, and Coyotes, which have reputations for being fierce predators, are actually omnivores that survive primarily on plant material. It is also noteworthy that in nations where human populations have become too large, people are obliged to be vegetarians. Ten times as many humans can be fed by the same food if the calories are not pushed through animals first. This explains why the largest animal that has ever lived, the Great Blue Whale (*Balaenoptera musculus*), eats plankton. It also explains why "big, fierce, animals" are scarce in nature.

Food Webs

A diagram illustrating many animals and plants in a given ecosystem is known as a food web because it illustrates the interweaving of many different food chains. For example, a simple food web may be superimposed upon a food pyramid to illustrate that, no matter how many species occur at a given trophic level, the biomass of those species cannot exceed the limits imposed by the laws of thermodynamics.

Stability as a Function of Complexity

Although there are exceptions, the population sizes of complex ecosystems, with many species on each trophic level, tend to be more stable than those of simple systems. Scientists refer to the value of biodiversity as an important factor for maintaining stability in ecosystems. Biodiversity refers to the numbers of species that occupy ecosystems, the numbers of trophic pathways through which energy passes, and also to the genetic diversity of populations that maintain lots of variability in their gene pools. Research on fossil ecosystems supports the concept that the large periods of extinction, for example, a major extinction event at the end of the Permian Period about 250 million years ago, at which time some 70% of terrestrial species and 96% of marine species disappeared, are correlated with a prior major loss of diversity in ecosystems. Similarly a great loss of dinosaur diversity at the end of the Cretaceous Period about 66 million years ago, also associated with an asteroid impact, was coupled with a major extinction event. Ancient Egyptian paintings of animals, over a period of 12,000 years in caves and tombs, also document increasing instability of ecosystems as midsized mammal species were reduced from 37 to 8, a decline associated with climate change and increased human population density.

Suppose that a given area has 100,000 lb of vegetation. It will support 10,000 lb of herbivorous species and 1000 lb of carnivores. Suppose there are five kinds of carnivores at the top of the pyramid. These would be called *apex predators* or *top-order carnivores*. If one species were eliminated, the remaining four probably would increase in number until collectively they reached a total of 1000 lb. If four species were eliminated, the remaining population of a single apex predator would increase in size dramatically until it reached the limit of 1000 lb. The temporary surplus of food created by the elimination of four species would, however, stimulate reproduction to the degree that more offspring would be produced than could be sustained over a long period. As a result, the population might become too large for long-term existence, in which case a number of the animals would have to either die or emigrate. If this occurs, we say that the population has overshot *carrying capacity*, and this would be an example of *density-dependent* regulation of population growth. The higher the density of a population, the more likely natural processes such as reduced birth rate and increased emigration would occur. These sorts of population fluctuations have a minor effect on the ecosystem when there are many kinds of animals on each trophic level, but they tend to have extreme effects in simple ecosystems.

Humans have a tendency to cause simplification of ecosystems and thus produce unstable ecosystems. Agriculture, for example, creates a simple ecosystem that is not only energetically inefficient because of the laws of thermodynamics, but unstable as well. Suppose a logging company clears and replants a forest. This is an example of agriculture because frequently only the economically important species is replanted, and they are the same size and age. The crop is then weeded and sprayed with pesticides to keep out other species. In the end, it would take an outbreak of only one resistant pest species to wipe out the whole crop. A managed forest is therefore not very stable and takes considerable effort to maintain.

Ecological Islands

In the late 1960s, R. H. MacArthur and E. O. Wilson in their now classic book *The Theory of Island Biogeography* established a set of principles about populations on islands. Essentially they pointed out that the diversity and abundance of species on islands was proportionate to the size of the island and its degree of isolation. Once an island becomes saturated, the number of species remains fairly constant, and the number of colonizations is roughly equivalent to the number of extinctions. It should be clear that this is a phenomenon associated with *carrying capacity*. In other words, the density and abundance of species is a function of available food, which is dictated by the amount of photosynthesis. Another phenomenon associated with islands is that many organisms are unique; they may be relictual, i.e., leftovers from a time when they were distributed more widely, or they may have evolved on site in association with a unique ecological niche.

The California Floristic Province associated with its unique Mediterranean climate may be considered an ecological island. Isolated from the desert to the east and the ocean to the west, the diversity of California's biota is legendary, greater than any other part of the United States. There are over 1500 endemic plant species in California representing over 30% of the total number of species. Reference to California as an island is not new. Sometime around 1510, Garci Rodríguez Ordóñez de Montalvo of Spain wrote about a population of Amazon black women who kept few males for stud service and fed the extras to trained griffins. His lyrical description of the area is as follows: "Know ye that on the right hand of the Indies there is an island called California, very near the Terrestrial Paradise." Apparently Montalvo had heard about the Baja California Peninsula, separated from the rest of Mexico by the Sea of Cortez, but he couldn't have known about the great chain of mountains that separates most of California from the great deserts of North America. Nevertheless, short a population of female

Amazons, his description of California as an island is appropriate. In 1971, Elna Bakker promoted the island theme even farther in her now classic work *An Island Called California*.

Important in the concept of an ecological island is that the number and density of species is proportionate to its size: a small island would have low numbers and diversity. Encroachment of civilization, with its envelope of noxious gases, heat, light, and pavement, breaks our natural ecosystems into smaller and smaller patches. Consequently, as these patches get smaller, diversity of species is reduced, primary production is diminished, the larger animals are eliminated, and the ecosystems become simplified and less stable. California is fortunate in that much of its landscape, 52%, is public land managed by the US Forest Service, Bureau of Land Management, National Park Service, and the California State Parks system; 56.7% is classified as protected and 14.36% is wilderness. These regions constitute large undisturbed areas where natural systems are allowed to prevail. Nevertheless, associated with increased levels of tourism and development, it has been demonstrated that our public parks are losing their diversity as civilization closes in their borders. In a sense, California's famous national parks are being "loved to death."

Game Management

The total number of animals in an ecosystem is controlled by the amount of food produced by photosynthesis or it can be controlled by the abundance of predators, either of which is primarily dictated by the overall size of undisturbed habitat. In reference to a food pyramid, the former system is said to be controlled from the bottom up, and the latter by top-down control. Competition for food is an important regulating device at every trophic level. Competition tends to be reduced at any level because animals have slightly different food requirements, a phenomenon known as *niche partitioning* or *resource partitioning*. Coyotes (*Canis latrans*, figure 8.50) and Bobcats (*Lynx rufus*, figure 8.52) may share a high trophic level, but they coexist by having different food habits. Bobcats are obligate carnivores, and coyotes are omnivores, feeding largely on plant food. They both control the herbivore population from the top down, but Coyotes usually are more abundant by being able to feed lower on the food chain, where food is more available. Coyotes have demonstrated their adaptability and varied food habits by moving into urban areas where they feed on pet food, garbage, and may prey upon domestic pets. A study of fecal remains (scats) on the Palos Verdes Peninsula in Los Angeles County demonstrated that coyotes and Gray Foxes (figure 8.51) were eating primarily domestic vegetation and some domestic animals, but no rodents or rabbits. While there is little doubt about the existence of urban coyotes, most of the discussion is based on anecdotal observations. There seem to be little data as to their numbers, habits, and significance in urban areas.

A controversial war on the Coyote continues, in spite of evidence showing that coyotes eat mostly vegetable material, not livestock. The US Fish and Wildlife Services acts as a pest management service and as recently as 2012 was responsible for killing 76,070 Coyotes in the United States. Apparently the federal government spends more money killing coyotes than it would if it paid ranchers directly for their losses of livestock. Wild Coyotes eat an abundance of meat only when small mammals such as mice and rabbits become too numerous. When rabbits become too common, they eat crops. Coyotes thus serve as a natural control mechanism of the whole ecosystem from the top down. One scientist, using published information on the diets of coyotes, rabbits, and cattle, determined that killing five coyotes increased the jackrabbit population to the degree that it ate an amount of forage equivalent to that which would be consumed by a steer.

Sometimes, in an attempt to make a natural ecosystem "better" for a certain species, humans wind up doing more damage than if they had left it alone.

For example, humans have declared war on predators (varmints) that might attack livestock. The California Grizzly Bear (*Ursus arctos*) was exterminated by a state-supported program of hunting, poisoning, and trapping. The bounty for a grizzly was $10. By 1922, they were gone. The California Grizzly Bear is still the state mammal, and its image adorns the state flag. California is the only state in the union that has as its symbol an animal that humans forced to extinction. The Gray Wolf, *C. lupus*, was extirpated in 1924, but amazingly, a family of seven Gray Wolves was photographed in southern Siskiyou County in August 2015, raising hopes among environmentalists that after nearly 100 years the wolf has returned to California.

Interestingly, after the demise of the California Grizzly Bear in the absence of competition, a subsequent increase in the number of Mountain Lions (*Puma concolor*; figure 8.49) occurred to fill the ecological niche of the apex carnivore. A controversy rages today over state protection of the Mountain Lion, which was voted into law with the Mountain Lion Protection Act of 1990. Critics, particularly hunters, claim that Mountain Lions have become too common and they are eating too many deer.

The history of the Mule Deer, *Odocoileus hemionus* (figure 4.40), on the west side of the Sierra tells a different story. Changes in the size of the Jawbone deer herd have been recorded since the days of the forty-niners. Prior to disturbance by humans, a mosaic of formerly burned and recently burned chaparral vegetation provided adequate food, so that deer did not occur abundantly in forest communities. They were also controlled from the top down by predators, such as Mountain Lions and Gray Wolves. During the first few years following a fire, annuals and herbaceous perennials are common. From 4 to 10 years after a fire, California-lilacs, *Ceanothus* spp., and Mountain-mahoganies, *Cercocarpus* spp., are choice browse. New growth on these common shrubs contains about 14% protein and an abundant supply of calcium, which is necessary for bones and antlers.

Gold was discovered in 1848, and the Gold Rush was on. Thus began a 50-year decline in the deer herd. Deer were hunted for food and hides; in 1880, one firm in San Francisco marketed 35,000 deer hides. The amount of winter forage, one of the most significant limiting factors on the size of deer herds, was severely reduced by the introduction of sheep and cattle. By 1876, 6 million sheep were reportedly grazing in the Sierra foothills. If hunting and competition were not enough, between 1879 and 1907 a series of hard winters further reduced the amount of winter forage in the foothills, and the remaining deer were pushed lower into the valleys. By the first decade of the 1900s, deer were considered scarce in the Sierra foothills.

The decline of the Jawbone deer herd was not unnoticed. In the late 1800s, a new environmental awareness became significant. National parks were established in Yosemite and Sequoia. John Muir and the newly formed Sierra Club fought for regulations to control the killing of big game. Parks were to be refuges for plants and animals alike. In 1883, the "buck law" was born: does were not to be hunted. In 1893, the buck season was reduced to six weeks. In 1901, the concept of limits was begun: no more than three bucks could be taken during a single hunting season. In 1905, the legal kill was reduced to two. In 1907, the state of California began two programs that were to have a profound effect for many years to come. A hunting license became required, and this not only generated revenue, but also enabled the state to enforce hunting laws. Also, during 1907 a bounty on Mountain Lions was initiated and was to remain in effect, at least locally, until 1964.

The influence of government control in the early 1900s had a profound effect, and the deer herd began to recover. By 1940, the deer range was fully stocked, but a new pattern of deer behavior began to emerge. Fire suppression in the foothills had allowed a thick vegetative cover to develop, and without fire, the primary food for deer became more and more scarce. In order to find adequate food, and because wading through snow is energetically expensive, greater numbers of deer began migrating into the forest communities in spring and moving back to the foothills and valleys with the onset of winter.

By the 1950s, the Jawbone deer herd had reached 6000 animals. In the summer, they were feeding in Yosemite National Park, where there were over 250 mi² (625 km²) of range available to them. In contrast, there were only about 40 mi² (100 km²) of winter-feeding grounds in the foothills. A density-dependent pattern began to emerge in which there was a sizable die-off of fawns, nursing does, and aged animals each winter. A 20% winter die-off was not atypical, and in a harsh winter 40% of the animals would starve to death. The overabundance of deer also impacted the summer range. In 1951 and 1952, studies in the national parks showed that wildflowers and seedling oaks were being heavily browsed. Juvenile California Black Oaks (*Quercus kelloggii*) had disappeared, leaving only mature trees. Meadows that formerly abounded with spring wildflowers were becoming dominated by noxious species, such as milkweed and thistles, not fed upon by deer.

Overpopulation by deer in the Jawbone herd had created a problem similar to the famous episode documented for the Kaibab deer herd in Arizona. Human tampering with population dynamics once again proved to cause more harm than good. Professional game management took over and in the 1960s a series of corrective measures was begun in an attempt to restore deer herds to a more realistic size, one regulated by more natural environmental phenomena. In 1963, national parks began translocating thousands of deer to other localities. In 1964, special doe hunts and late season hunts were initiated. Also, in 1964 the last bounty for a Mountain Lion was paid, allowing Mountain Lion populations to recover. In 1968, grazing fees for cattle in national forests were raised. Reduced grazing could increase the available food for deer and restore winter range to a more normal condition. In 1970 Mountain Lions were declared game animals, so that a season and a limit could be established if they became too abundant, and presumably Mountain Lions could be removed from an area of overabundance to one where the deer were too abundant; however, as of 1990, with the Mountain Lion Protection Act no hunts may be authorized, nor is it legal to translocate Mountain Lions. Could Mountain Lions overpopulate under total protection?

According to state records, during the 56 years that a bounty on Mountain Lions was in effect, 12,461 lion bounties were paid. According to some authorities, there were fewer than 600 Mountain Lions left in California at the time the bounty was dropped. Year after year since 1964, the moratorium on lion hunting was renewed over the objection of ranchers and sheepherders, even though lions that harm livestock may legally be killed. The moratorium remained until 1987, when, under pressure from hunters and ranchers, a limited lion hunt was approved. Environmentalists, however, initiated a series of legal blocks to the lion hunt. As of this writing, that lion hunt has never taken place. To this day, very little clear-cut evidence has been presented that the government expense in controlling predators has been rewarded with equal value in livestock saved. On the contrary, a healthy herd of deer kept in check by a healthy number of predators has repeatedly proved to be a more stable, cost-efficient program. The balance of approximately 200 deer per 200 acres (80 ha) in association with one Mountain Lion is a figure that seems to prevail throughout the west. The lessons learned from the Jawbone and Kaibab deer herds are part of the basis for game management principles applied on a nationwide scale. Game management involves not only protection for a single species; it must involve entire ecosystems. Species do not occur in isolation.

Bobcat (*L. rufus*) trapping is another issue involving game management. In response to public outcry that trappers were taking Bobcats in and around Joshua Tree National Park in 2013, the Bobcat Protection Act was passed which provided full protection for the animals in that area. Bobcat pelts are valuable items in the export trade. Apparently Russian and Chinese dealers are willing to pay over $500.00 for a high-quality pelt, and the demand for spotted cat fur has not waned since the early 2000s. In 2006, nearly 50,000 pelts were exported from the United States. In 2014, 1639 Bobcats were taken in California with most of those taken from five counties, San Bernardino, Riverside, Kern, Inyo, and Mono. In response to increased value, that number increased from about 1000 a year since the early 2000s. In 2015, reacting to continued pressure from environmentalists, the California Fish and Game Commission voted to ban all commercial Bobcat trapping. At that time, there were only 100 active trappers in the state, so the financial impact was considered minimal. The influence on the Bobcat population and its favored prey remains to be seen.

Keystone Species

It appears that changes in abundance of certain species, rather than others, have a greater effect on population dynamics. These are known as *keystone species*, and they are often predators. Clearly, whatever food is available at the top of a food pyramid must be shared by the predators there, and if one disappears, the others will increase in abundance. The stories above demonstrate that selective removal of Grizzly Bears and Wolves increased the statewide number of Mountain Lions. Similarly, available food will also control the number of lions. It can be concluded, and has been demonstrated, that patches of habitat that are too small will not provide enough food for a population of Mountain Lions and the lions will disappear. Large animals require large patches of undisturbed habitat. Further research has shown that the home range of a male Mountain Lion is about 200 mi² (518 km²). Large predators, through competition and aggression, limit the number of smaller predators in an ecosystem. If the size of the system is too small for large predators such as Mountain Lions, the smaller ones will move in and increase in abundance in proportion to the size of the habitat. This phenomenon is called *mesopredator release* and is a consequence of *habitat fragmentation*. In patches of habitat too small for lions, Coyotes and Bobcats become more common, which has been demonstrated at urban interface areas throughout California. In 1988, Michael Soulé and others conducting population studies in fragmented habitats in San Diego County also found that many patches of habitat between housing tracts were too small for Coyotes, in which case the smaller predators, Gray Foxes (*Urocyon cinereoargenteus*) and Domestic Cats (*Felis catus*), moved in to share the position of apex carnivores. In this case, a significant change or *trophic cascade* in the ecological relationships of the whole ecosystem was manifested. Both foxes and cats prey on birds, which resulted in an overall decline of birds in the area. The predator that has a stabilizing effect on the ecosystem is called a *keystone species*, in this case in San Diego County it was the Coyote. In the example of the Jawbone Deer Herd, in a larger patch of habitat, it was the Mountain Lion. Removal of the apex carnivore, the Mountain Lion, causes an increase in abundance of the next lower trophic level, the herbivores, which in turn causes a decrease in the vegetation. In this case, when removal of the top carnivore is involved, it is called a top-down trophic cascade.

A recent study in southern California demonstrated again the significance of habitat fragmentation. It has been estimated that from 17 to 27 Mountain Lions inhabit the Cleveland National Forest, most of which includes the Santa Ana Mountains. The study area extends from the Santa Ana Mountains southward through the Peninsular Ranges to the Mexican border and east to the Salton Sea. However, Interstate 15 (I-15) lies along the eastern flank of the Santa Ana Mountains and bisects the range of Mountain Lions, preventing their migration back and forth to the southern Peninsular Ranges. During the 13 years of the study, it was determined that the lion population had only a 56% survival rate. In general, the roads simply represent barriers to dispersal, but most of the deaths were caused by vehicles, and most of those were on the 241 Toll Road in Orange County. Other causes of population decline included one lion that crossed into Mexico, public safety removals, deaths due to depredation permits, illegal shootings, and wildfires. Sometimes kittens are killed by adult male lions. In 2015, in the Santa Monica Mountains a female named P-23 had five kittens. Two were killed by a male lion, two others were killed by unknown predators, and one survived at least six months. In the Santa Ana Mountains, only one lion was documented to cross I-15 from east to west. That male lion, named M86, successfully bred and produced four offspring before he died. Of the four, one was killed by a car, one was poisoned, and one was taken into captivity because it had become too accustomed to humans. The fourth went on to produce two kittens that were known to survive. It can be seen that the genetic diversity of the Santa Ana Mountain population of Mountain Lions is threatened by a lack of outbreeding.

Similarly, in December 2016, P-39 a mother of three kittens, was killed while attempting to cross the 118 Freeway. Subsequently, in separate incidents, two of her kittens were killed crossing the same freeway. Furthermore, one of the few males in the Santa Monica population killed several domesticated Alpacas in a fenced corral. In order to protect biodiversity in the population, the owner chose not to use a depredation permit but instead build more secure fencing. Perpetuation of large corridors of contiguous habitat is important for dispersal of wildlife. Conservation organizations attempt to purchase properties that will complete these corridors. A corridor connecting the San Joaquin Hills in Orange County with the Cleveland National Forest is currently being established. Part of it involves the "Great Park," the former El Toro Marine Base that was protected from development by a vote of the people. A recent land acquisition in San Diego County preserves a corridor between Anza-Borrego Desert State Park and Cuyamaca Rancho State Park in the Laguna Mountains.

A similar situation of genetic isolation has been documented in the Santa Monica Mountains where 41 lions have been studied, although the present population is estimated to contain two or three adult males, four to six females, and several kittens. These animals are isolated on the north and east by the 101 and 405 freeways, where most deaths occur. It also has been shown that inbreeding is becoming a problem. One "famous" lion, known as P-22, managed to cross both freeways and took up residence in Griffith Park, after which it was captured with a serious case of mange. Its poor health was blamed on anticoagulants found in the blood from consuming animals that had been killed with rat poison. This lion is apparently quite healthy now, and in 2015 it apparently was responsible for killing a koala in the Los Angeles Zoo. Another, P-32, the only known male to leave the Santa Monica Mountains, was killed on August 10, 2015, in an attempt to cross Interstate 5 near Castaic. This lion previously had successfully crossed the 101 Freeway, as well as State Highways 23, 118, and 126. P-34, a female sibling of P-32, in December 2014, was photographed lounging under a trailer in a Newbury Park mobile home community. Then, in November 2015, its body was discovered next to a trail in Point Mugu State Park. In this case, it was a victim of anticoagulant rodent poison. Similarly, in 2012, also at Point Mugu, P-25 was also found dead, and rat poison was suspected as the culprit. Other predators, such as raptors and Bobcats have been affected. In 2014, three Bobcats were found dead at UC Santa Cruz. During 20 years of research in the Santa Monica Mountains, 88% of 140 Bobcats, Coyotes, and Mountain Lions, alive and dead, tested positive for anticoagulants. Such toxins are used in urban areas to kill rats, but unfortunately it works its way up the food chain and in this case it killed the apex carnivore. As of 2014, these poisons were no longer available in bulk form for over-the-counter purchase, but they are available to professional exterminators. They are still available in containment traps; presumably the dead animal would be trapped in a container and not available for easy pickings by a predator.

Constructing highway overcrossings or undercrossings have been proposed to facilitate the movement of animals to and from these threatened populations. The 241 Toll Road and State Highway 91 in the Santa Ana Canyon each have one undercrossing, and fences are being built to direct animals to them. In an attempt to facilitate highway crossings and promote genetic diversity, a multimillion dollar landscaped overcrossing of Highway 101 has been proposed to connect the Santa Monica Mountains population with the Santa Susana Mountains and Los Padres National Forest to the north. Another option could be translocating animals into the threatened populations from the outside. At the present time, however, this option is prohibited by the Mountain Lion Protection Act.

Wildlife managers have discovered that the genetic diversity of individual species is important because of its role in the protection and management of endangered species. Size of a reserve is of basic importance, but even if a species returns from the brink of extinction, the new population may be fostered by too few parents, and therefore all its members would be very much alike, a phenomenon known as a *genetic knothole*. It is feared that this phenomenon could occur in the Mountain Lions of the Santa Ana and Santa Monica Mountains. This also could be a problem with the present herds of Tule Elk (*Cervus elaphus nannodes*), numbering about 2500, all of which are descended from as few as 2 individuals. This lack of diversity in the gene pool could mean that a single environmental change or disease would wipe them all out. The greater its diversity, the more likely a species is to survive a crisis. The reason that all California Condors (*Gymnogyps californianus*) were removed from the wild was to increase the number of individuals, and hence the genetic diversity in the captive breeding program. When a population of proper size and diversity was achieved, captive California Condors were released into the wild.

TABLE 2.1
A Comparison of Primary Production in Selected Communities

Community	C, g/m^2/yr	C, lb/acre/yr
Oligotrophic lake	25	222
Eutrophic lake	250	2225
Intertidal kelp	14,600	129,940
Upwelling	6000	53,400
Open ocean	200	1780
Estuary	3200	28,480
Salt marsh	3000	26,700
Tropical forest	5000	44,500
Coniferous forest	1400	12,460
Chaparral	1056	9400
Coastal sage scrub	335	2980
Coastal grassland	527	4690
Alpine	140	1250
Desert	100	890

NOTE: Primary production refers to the amount of carbon (C) that is added to a plant each year through the process of photosynthesis. There is a high degree of variability for these communities, so the numbers should be considered as rough estimates only.

Productivity

In ecosystems, productivity refers to the production of food. The first step is photosynthesis, so the amount of food produced by green plants is known as primary production. The total amount, or gross primary production, however, is not all available for herbivores to use because cellular respiration and other processes in plants use some of the energy. Net primary production is the amount of energy available for herbivores. That amount, measured over a certain period of time, may be measured by weight (mass) or by energy (calories) and is often expressed in proportion to a square meter of surface per unit of time. For example, primary production in a typical desert region, the least productive of all ecosystems, can be expressed as approximately 100 g/m^2/yr or 400 Calories/m^2/yr. The reason that the

caloric value is usually about four times the mass is that the product, carbohydrate, yields 4 Calories/g.

Because values of productivity vary considerably from one study to another, they must be considered estimates. But, as in many forms of ecological research, tendencies are more important than precise numbers. To put the value for desert productivity in perspective, compare it to that of other ecosystems (Table 2.1). The most productive terrestrial ecosystem is a tropical forest, at 5000 g/m^2/yr. Ecosystems with comparable productivity include marshes and estuaries. The most productive ecosystem on earth seems to be the intertidal kelp community. Causes for this remarkable productivity are discussed in Chapter 11. Ecosystems with low productivity include deserts, mountains, lakes, and open ocean.

Cycles of Matter

Hydrogen (H), Oxygen (O), and Carbon (C) recycle in ecosystems primarily within the gases carbon dioxide (CO_2) and water (H_2O), which recycle through photosynthesis and cellular respiration. In addition, nitrogen (N) and phosphorus (P) recycle in living systems as well. Nitrogen is a component of amino acids, the building blocks of proteins. Proteins are used in living cells of all living organisms as enzymes that control chemical reactions. In animal bodies, skin, hair, and muscle are composed of protein, and protein can be a source of energy for metabolism, yielding 4 Calories/g, the same as carbohydrates. Phosphorus is an important component of nucleic acids, which are the building blocks of DNA and RNA, the genetic makeup of cells. Phosphorus is also an important component of adenosine triphosphate (ATP), the universal source of energy for most metabolic processes inside cells. Nitrogen and phosphorus also recycle as abiotic (nonliving) components of the soil. Nitrates (NO_3) and phosphates (PO_4) enter the living world dissolved in water and are absorbed as water soluble minerals taken up by the roots of plants in the soil. They enter the living component of ecosystems as they are absorbed by the roots of plants. Invariably fertilizers are sources of nitrate and phosphate. In addition, nitrogen enters ecosystems by means of symbiotic bacteria that live in the roots of certain plants. This type of symbiosis in which two organisms depend on each other is called *mutualism*. These nitrogen-fixing bacteria are common in the roots of plants in the Legume family, and also in the roots of California-lilacs (*Ceanothus* spp.) and Mountain-mahoganies (*Cercocarpus* spp.), which explains why these plants are important sources of protein and nucleic acids in animal diets, and why they are pioneer plants after fires. Beans and peas in the Legume family are obviously important in the diets of humans as well. The process of soil formation and the significance of different types of soil will be discussed in Chapter 3.

3

Basic Geology

GEOLOGY IS THE STUDY of the earth—more specifically, it is the study of rocks, the nonliving components of the earth. Scientists have divided matter into 102 pure inorganic substances known as elements. Living organisms consist primarily of four elements: hydrogen (H), oxygen (O), nitrogen (N), and carbon (C). Ninety-nine percent of all rocks are composed of some combination of eight elements: silicon (Si), oxygen (o), aluminum (Al), iron (Fe), calcium (Ca), sodium (Na), potassium (K), and magnesium (Mg). All the other elements are present in the earth in small amounts. The earth is composed of these elements, either alone or in a myriad of combinations called compounds.

A mineral is composed of a single element or compound. By definition, a mineral is a naturally occurring inorganic substance with a definite chemical composition and ordered atomic structure. Table salt, for example, is a mineral called sodium chloride (NaCl). Its ordered structure is apparent because it occurs in crystals shaped like small cubes. Another common mineral is quartz, or silicon dioxide

(SiO_2). Its crystals have a specific hexagonal shape. Gold (Au) and silver (Ag) are minerals composed of a single element. Coal is a mineral composed entirely of carbon, originally trapped by living organisms through the process of photosynthesis. The carbon in coal is therefore of organic origin, which leads some authorities to object to the definition of a mineral as an inorganic substance. The controversy over the true definition of a mineral, however, is beyond the scope of this book.

Rocks are usually composed of several minerals. Granite is a rock with a speckled appearance caused by different minerals in crystal form, such as quartz, mica, and feldspar. Limestone is a rock composed of a single mineral, calcite ($CaCO_3$). On the basis of their origin on earth, rocks may be divided into three primary rock groups: igneous, sedimentary, and metamorphic (plate 2).

IGNEOUS ROCKS

Igneous rocks are formed from cooling and solidification of molten rock. The term *igneous*

TABLE 3.1

Mineral	Chemical Composition	Appearance
Quartz	Silicon dioxide	Glassy, clear, cloudy, white, gray, pink
Feldspar		
Plagioclase	Calcium or sodium aluminum silicate	Blocky, dark gray to white
Orthoclase	Potassium aluminum silicate	Blocky, pink
Mica		
Biotite	Complex iron silicates	Thin, shiny black sheets
Muscovite	Complex potassium silicates	Thin, shiny clear sheets
Ferromagnesian minerals		
Pyroxenes	Complex iron, magnesium silicates	Short, stubby crystals, green to black
Amphiboles	Complex iron, magnesium silicates	Grains or long crystals, light green to black
Olivine	Complex iron, magnesium silicates	Glassy to grainy, light green

TABLE 3.2

Types of Igneous Rocks

Texture	Color		
Fine-grained (volcanic)	Rhyolite (light)	Andesite (medium)	Basalt (dark)
Coarse-grained (plutonic)	Granite (light)	Diorite (medium)	Gabbro (dark)

NOTE: Darker rocks indicate increased amounts of ferromagnesian materials and decreased amounts of quartz.

refers to fire; it comes from the same root as *ignite*. The high internal core temperature of the earth causes convection of heat energy, which melts rock to produce magma. Upon cooling, magma becomes igneous rock. The length of time magma takes to cool determines the size and arrangement of the crystals of each mineral. If it remains deep in the earth, it will cool slowly, and the crystals will have a long time to form. These rocks will be coarse-grained; the individual mineral crystals will be visible to the naked eye. Granite is the most common of these coarse-grained rocks. If magma comes to the surface and cools rapidly, as in a volcano, it will be fine-grained because there has been too little time for large crystals to form. The most common fine-grained igneous rock is basalt, a heavy, black volcanic rock. Rocks that have been cooled slowly, deep beneath the earth, are called intrusive or plutonic rocks. Rocks that have been formed by molten material that flowed out upon the surface are called extrusive or volcanic rocks.

Igneous rocks also differ in color and composition. Some minerals, such as quartz and feldspar, are light colored. Minerals that include iron and magnesium are dark colored and are called ferromagnesian minerals. The relative proportions of quartz and feldspar and of ferromagnesian minerals are responsible for the principal color of the rock. Table 3.1 is a list of minerals found in igneous rocks. On the basis of relative proportions of orthoclase feldspar, plagioclase feldspar, and quartz, geologists have named 15 different kinds of plutonic rocks alone, which is beyond the scope of this book. For purposes of simplicity, six main types of igneous rocks based on chemical composition (color) and texture (Table 3.2) will be considered here.

In California, granite and diorite are the common rocks that make up much of the Sierra Nevada. A large intrusive block such as that of the Sierra Nevada is known as a *batholith* (deep rock) in reference to its origin deep beneath the surface. Granite, associated with the Southern California batholith, is found in the Transverse Ranges, the Peninsular Ranges, and some of the ranges in the Mojave Desert. Gabbro is a dark-colored plutonic (intrusive) rock formed either in recent continental plutons or in mid-ocean spreading centers. It is found in the Klamath Mountains, the Peninsular Ranges in San Diego County, the Bodfish area south of Lake Isabella, and the foothills of the Sierra Nevada in El Dorado County, where it has degraded into a dark, iron-rich soil upon which many specialized (endemic) plants live.

Igneous rocks that are low in silica (quartz), high in plagioclase feldspar, and high in ferromagnesian minerals are classified as "mafic." Gabbro and basalt, which are rich in iron and magnesium, are thus called mafic rocks. Rocks that are similar in composition, but are also low in feldspar are said to be "ultramafic." Sometimes these rocks are also called ultrabasic, a name that refers to rocks that are less than 45% silica. Ultramafic rocks derived from deep mantle sources become deposited on the earth's surface during subduction events. Layers of these rocks, known as extant terranes, have become deposited along the coast from San Francisco northward as the ocean floor slipped beneath the North American continent. Suites of such rocks are sometimes called "ophiolites." Peridotite is an ultramafic igneous rock, rich in olivine and pyroxene, that becomes metamorphic serpentinite when subjected to heat and pressure in the presence of water. Serpentinite is composed of serpentine minerals including asbestos. The terms ophiolite and serpentinite both refer to a snakelike appearance of patterns in these rocks. The Greek work *ophidion* means "serpent," and ophiolites and serpentine are named for a mottled pattern that seems to resemble a snake's skin. In 1965, in order to honor two entities that typify California's geologic wonders, gold was declared the state mineral and serpentine was named the state rock. It is a beautiful waxy green material that is found at various localities in the western foothills of the Sierra Nevada and the Coast Ranges. It degrades into a specialized soil that harbors many unique plants. The term serpentine in reference to those soils is entrenched in literature, so these will be called serpentine soils in this book. Because a number of parental rock materials are rich in magnesium and iron (gabbro, for example), many authorities refer instead to ultramafic soils.

Of the volcanic (extrusive) rocks, basalt is highest in ferromagnesian minerals. It is the dark black rock that forms large flows on the eastern side of the Sierra Nevada, for example, in Devil's Postpile, near Mammoth Mountain. Basalt is most common, however, as the main component of oceanic islands such as the Hawaiian Islands. Andesite, lighter in color than basalt, is the primary volcanic rock found on the borders of continents. The name comes from the Andes, the large mountain range on the western border of South America. In North America, the Cascades are composed primarily of andesite. Rhyolite, the extrusive rock with the lightest color, has approximately the same chemical composition as granite, but differs because it cooled quickly as it flowed out upon the surface. In California, rhyolite is found in the Mojave Desert, where it forms layered mesas, flat-topped buttes that project hundreds of feet above the desert floor. Rhyolite is viscous or sticky when it flows. Therefore, it often has many stones imbedded in it, and when it cools it can trap large bubbles of gas. One popular campground in the Mojave, called Hole-in-the-Wall, gets its name from large holes formed by trapped gas in the rhyolite. The Mono Craters, on the eastern side of the Sierra, often considered the youngest mountain range in the lower 48 states, is composed of rhyolite.

Of the extrusive rocks, obsidian has the smoothest surface. It is noncrystalline, similar to glass. It lacks ordered atomic structure; therefore, by definition, there are no minerals in obsidian. It is an amorphous mixture of the same elements found in granite or rhyolite, but, due to rapid cooling, atoms did not have time to become arranged in an ordered structure. Black obsidian cooled in the absence of oxygen. Its color is due to nonoxidized (reduced) iron.

Large deposits of obsidian are found east of the Sierra Nevada north of the town of Bishop. A large mountain, prominent on the horizon there, is known as Glass Mountain, and a large butte, known as Obsidian Dome, is found north of Mammoth Lakes. Obsidian is also found on two buttes south of the Salton Sea. Obsidian was of great importance to early California Indians, who carried it great distances in order to make their tools. It was used to make knives, arrowheads, and spear points. It is not uncommon to find large numbers of obsidian flakes high in the Sierra Nevada many miles from the nearest source.

Mahogany obsidian is brown because its iron is oxidized. Large amounts of high-quality mahogany obsidian are found in the Warner Mountains, on the eastern edge of the Modoc Plateau. Although mahogany obsidian is found in limited amounts at other localities, Modoc Indians must have traded obsidian with peoples farther south because flakes of high-quality mahogany obsidian are found in chipping sites as far south as the Kern Plateau of the southern Sierra Nevada.

Obsidian commonly is found with pumice, a lightweight volcanic "froth." The gray-colored soil that extends for miles in the Mammoth Lakes area is composed of pumice. The rock is so light that the wind is able to pick up pea-sized pieces of gravel. If the gravel lands upon a lake, it floats because of all the air trapped in it.

SEDIMENTARY ROCKS

Sedimentary rocks make up only 5% of the earth's crust, but they make up 75% of the rocks on the surface. Usually they are formed by the consolidation of particles transported from another source. The formation of sedimentary rocks is a significant part of what is known as the rock cycle. Mechanical and chemical weathering causes rock materials to break up into smaller particles. These particles are transported from their source by forces of erosion such as wind, water, and glaciers. (These processes are discussed in more detail in Chapter 4.) When particles come to rest, they represent sediment that in time may become cemented or consolidated into rock in a process called lithification.

Most sedimentary rock is clastic, that is, composed of clasts or particles. These rocks are named according to the size of the particle that makes up more than 50% of its framework. If the sediment is composed of fine particles (i.e., clay), the rock is called shale. Silt makes siltstone, and sand makes sandstone. If the particles are gravel sized (2 mm) or larger, the consolidated stone is known as a conglomerate. If the rocks imbedded in the conglomerate are sharp or angular, the rock is known as a breccia.

For many millions of years, the area we now call California was covered by seawater, and during that time sediments accumulated at the bottom of the ocean. Today those sediments are above sea level, many having been uplifted thousands of feet by mountain-forming processes. Mountains composed of sedimentary rock have a distinctive banded appearance. Often the bands are of different colors. Mountains of this type are found throughout the Basin-Range Province, extending from the eastern side of Owens Valley and the Modoc Plateau all the way to the Rocky Mountains. Formerly, the Sierra Nevada also was covered with sedimentary rock, but erosional forces have carried almost all of it away. Small remnants of the overlying altered sediment, called roof pendants, are still found along the crest of the Sierra Nevada in the area near Bishop (figure 4.5). Most of the sediments have been redeposited as sand and gravel thousands of feet deep on the floor of the Great Central Valley. The rain shadow provided by the Sierra Nevada has been a major factor in the preservation of all the old ocean sediments we now see exposed throughout the deserts: there has not been enough precipitation to carry those materials away.

The size of the particles in the sediment depends on the amount of energy in the water at the time of deposition. Fast-moving water is capable of carrying most small particles in suspension. The size of the particles left behind, those that settle out, is directly proportional to the motion of the water. As water slows down, finer particles settle out. Conglomerates indicate a high-energy environment of deposition, such as a river bottom or beach. A sand bottom indicates slow-moving water, and a mud bottom is deposited in still water. Marine sediments are deposited under seawater. The nature of the sediment is a good indicator of the depth of the ocean at the time of deposition. Sandstone was deposited close to the shoreline. Shale was deposited in deeper water, indicative of a mud bottom. Trilobite fossils may be found in shales as old as 600 million years. These creeping crablike creatures, apparently among the most common organisms in these ancient seas, have been extinct for nearly 300 million years.

Sedimentary rocks also may have a chemical or biological origin. Those rocks with a chemical origin are produced by the precipitation of minerals from solution. Precipitation occurs because the solution is saturated; it holds all it possibly can, and the extra material precipitates. Evaporation of water also causes minerals to precipitate from solution. Four common sedimentary rocks have a chemical origin: limestone (calcium carbonate), gypsum (calcium sulfate), halite (sodium chloride), and chert (silicon dioxide).

Salt flats in California deserts are produced by evaporation, and the minerals are known as evaporites. Certain locally common minerals are valuable. Some of the lakes have very rare minerals or minerals of economic importance. Trona is a town in the Mojave that gets its name from a rare mineral mined from Searles Dry Lake. Potash, a potassium mineral also mined from the salt of Searles Lake, is the primary source of potassium for commercial fertilizers.

Borates are also important minerals found in Searles Lake. Boron is used as an additive in jet fuel, and some borates are used in detergents. Boron became infused into lake water by volcanic gas emissions that occurred millions of years ago, when huge lakes filled many basins in the desert. Due to evaporation, crystalline borate minerals are also found among the salts of some lake beds. Other lake deposits with borate minerals are found in the Mojave Desert. In Death Valley are borate minerals known as colemanite and ulexite, and near the town of Boron is a large mine where borate minerals such as ulexite, colemanite, and kernite are profitably mined.

Plaster of Paris is gypsum, an evaporite composed of calcium sulfate ($CaSO_4$). It is an important component of the various kinds of plasterboard used in construction. Plaster City, in the Colorado Desert, is a location where gypsum is mined. During the Renaissance, a solid form of gypsum known as alabaster was a common material used in the carving of statuary.

Halite is the same substance as common table salt, or sodium chloride (NaCl). It has been mined from Bristol Dry Lake in the Mojave Desert, and it is also extracted by evaporation from seawater. Up until the 1950s, table salt evaporation ponds could often be found at the upper ends of estuaries such as Newport Bay and San Francisco Bay.

Chert is a form of silicon dioxide (SiO_2), or quartz. It occurs in many colors. Depending on its color and the mechanism by which it was formed, chert might be called flint, jasper, agate, or petrified wood. Geodes are nearly spherical deposits of this material in which quartz crystals have grown inward to a hollow cavity. Well-known geode localities occur in the Colorado Desert south of Blythe.

Sedimentary rocks of biological origin are also common. Most limestone is produced this way. Many organisms extract calcium carbonate from water and use it to form shells or other skeletal parts. Among the organisms with this ability are mollusks, corals, sponges, and microorganisms such as foraminifera and blue-green algae (Cyanobacteria). Many of these creatures live in clear, shallow, warm seawater. Most limestone in the desert area was formed in this environment. These desert limestones commonly contain fossils including crinoids, corals, brachiopods, and mollusks.

Foraminifera are ameba-like microorganisms that are extremely important predators at mid-water depths in the ocean. They produce calcium carbonate "shells" that accumulate in great abundance in fairly deep water. In some parts of the world, there are limestones 250–350 million years old made up of great numbers of foraminifera. For example, the pyramids near Cairo, Egypt, are composed of solidified foraminifera skeletons. In California, these deep-water limestones are now uplifted in the Coast Ranges.

Some organisms are able to extract silicon dioxide from seawater to make skeletons of glass. Among these organisms are certain sponges, microscopic amoebas known as radiolaria, and photosynthetic planktonic algae known as diatoms. Diatoms are among the most abundant organisms in the world. Their contribution to photosynthesis is unrivaled, and they probably contribute more oxygen to the atmosphere than any other living organism. Their microscopic glass skeletons may accumulate in great numbers in deep water forming a sedimentary rock known as diatomite. Below 18,000 ft (6000 m), extreme pressure causes all calcium carbonate to dissolve, and thus diatoms form pure deposits. These deep-water deposits of diatomite are now displaced above sea level in various regions along the coast of California. One deposit near Lompoc is reported to be 700 ft (230 m) thick.

Calcium phosphate is incorporated into the skeletons of some organisms, including modern crabs, lobsters, and shrimps, as well as many extinct organisms. Where calcium phosphate accumulates, it forms a mineral known as apatite. Apatite deposits are not widespread, but where they occur, they are mined as a source of phosphate for fertilizers. In the United States, such deposits occur in Wyoming, Florida, and Georgia. The largest apatite deposit occurs in Morocco. This is one of the most important sources of the world's fertilizer.

Coal is also a sedimentary rock of biological origin. It is composed of carbon material produced by vegetation that accumulated in an environment with very slow decay rates. Such an environment is warm, acidic water, as in bogs or swamps. These environments were common in North America about 350 million years ago, at which time the continent was located in an equatorial position on the earth. As a result, abundant coal resources are now found in the United States, particularly in Appalachia. The eastern Rocky Mountain deposits are the result of a similar climate of deposition that occurred on the fringes of an inland sea about 100 million years ago.

The western part of North America was submerged by seawater for about 400 million years, during which time the ocean level fluctuated and successive layers of sediment were deposited. The result was an accumulation of about 23,000 ft of limestones, shales, and sandstones, which are now exposed in mountain ranges of the Great Basin. A similar series of sediments is exposed in the Grand Canyon, Arizona, where the Colorado River has cut through to basement rocks more than a billion years of age.

Lake sediments are also found throughout California's deserts. Many of these silts and clays were deposited during periods of heavy precipitation. Remains of these large freshwater lakes are testimony to the changes in climate and topography that took place over millions of years in western North America. One

TABLE 3.3
A Partial List of Metamorphic Rocks

Metamorphic Rock	Appearance	Original Rock
Foliated		
Slate	Fine-grain, laminar, splits easily	Shale
Schist	Splits easily, many fine layers, often rich in mica	Slate
Gneiss	Thick banding, resembles granite	Granite
Nonfoliated		
Quartzite	Tough, visible grains, many colors	Quartz sandstone
Marble	Many colors, variously banded, fizzes with acid	Limestone
Serpentinite	Green to black, waxy surface	Peridotite

fascinating area is in northeastern Arizona, where preserved in lake sediments are the remains of a forest that existed some 100 million years ago. This area is now known as Petrified Forest National Park. In these lake sediments, entire trees have been turned to chert by mineral replacement of wood. These logs are so well preserved that even the cell structure is visible under a microscope. Trees similar to these still live today in the Andes, but they are no longer native in the Northern Hemisphere. California has no freshwater lake beds that are this old, but in the beds uplifted in the Red Rock Canyon Area of the northern Mojave are preserved important vertebrate fossils over 60 million years of age. This fossil assemblage is important because that epoch, known as the Paleocene, has a very poor record in the west.

METAMORPHIC ROCKS

Rocks that have been altered from their original form by heat and pressure are called metamorphic. The term means "after-form," in reference to its altered form. In foliated metamorphic rocks, high pressure causes minerals to line up in visible layers. Examples include slate, schist, and gneiss. Nonfoliated metamorphic rocks such as marble, quartzite, and serpentinite, which are formed under high heat, show no foliations. Table 3.3 lists some common metamorphic rocks, indicating also the rock from which each was derived.

Heat and pressure are generated as large blocks of land, known as plates, move around on the surface of the earth. These phenomena associated with plate movements are known as plate tectonics. Movement of plates also produces the force that causes rock to become molten. In the case of metamorphic rock, enough force is supplied to cause the rock to become "plastic" but not molten. As a result, some characteristics of the parent rock remain after the deformation.

Many of the oldest rocks on the continent are metamorphic, and some of these are found in California. In the San Gabriel Mountains are metamorphic rocks that have been dated in excess of a billion years (figure 8.5). Included among these rocks are gneisses at 1.05 billion and 1.7 billion years of age. There is also a schist known as the Pelona Schist that formerly was assigned an age of 1.1 billion years, but new evidence indicates that it may not be older than 130 million years. Another place in California where corresponding rocks occur is 100 mi (160 km) south, on the other side of the San Andreas fault. This site is in the Orocopia Mountains, east of Indio, in the Colorado Desert. This displacement is important evidence that land west of the San Andreas fault has been displaced considerably northward since its time of origin. Other ancient metamorphic rocks well represented in the Mojave Desert are exposed in the Newberry-Ord Mountains, Old Woman Mountains, and Marble Mountains, all of which may be visited by traveling on Interstate 40 east of Barstow.

Another ancient schist that is of interest lies at the bottom of the Grand Canyon. It has been dated at 1.1 billion years. Overlying it are thousands of feet of sedimentary rock-limestones, shales, and sandstones, with the oldest at the bottom, on top of the schist. This layering illustrates a geological principle known as the law of superposition, which simply states that unless the land is deformed by mountain-building processes, the youngest rocks are on top. If the schist is 1.1 billion years of age and the oldest sedimentary layer is about 606 million years of age, this means that there is nearly a half-billion-year period of time for which the record is missing. Such an absence of rock layers is called an unconformity. It is assumed to represent a period of erosion, during which materials were carried away. It is interesting that rocks representing this time period are missing all over the West. It has been proposed that much of the material is represented by sediments thousands of feet thick that lie in the Nopah Mountains of the eastern Mojave Desert.

SOILS

Soil is a mixture of organic material and sediments from the weathering of rocks. It is the substratum upon which plants grow. It also provides homes for animals and is a water reservoir. The nature of the soil is dictated by the parent rock material from which it is derived and the climate in which it is formed. The texture of soil is based on particle sizes and is classified according to the characteristics of sediment described above. Thus, an area may be characterized by sandy soils or clay soils, etc., characteristics that have a profound effect on water-holding capacity and percolation rates.

Soil scientists classify soil into layers called horizons, each of which has certain characteris-

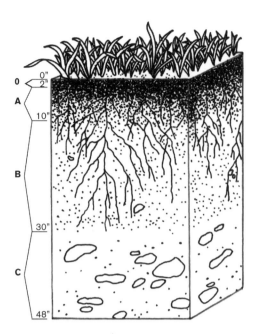

FIGURE 3.1 Characteristics of soil horizons (adapted by Math/Science Nucleus [www.msnucleus.org] from a US Department of Agriculture Bulletin).

ABBREVIATIONS: O, surface composed of detritus: organic material from dead plants and animals; A, topsoil containing roots of plants, water, minerals, and decaying organic material; B, mineralized subsoil containing plant roots, fewer decay organisms, source of water, and minerals for plant growth; C, weathered mineral layers with reduced organic component.

tics based on age and degree of composition of organic material. Figure 3.1 is an idealized soil profile illustrating the nature of four soil horizons.

The parent rock determines the chemical composition of a soil. Soils derived from limestone are rich in calcium carbonate. Soils derived from gypsum are rich in calcium sulfate, and so on. Calcium is an important plant nutrient which is taken up by roots. Soils derived from serpentinite have little calcium, but lots of iron and magnesium. A total of 215 species and varieties of plants are restricted to (endemic on) serpentine soils in California. Because of its common usage, the term serpentine will be used here. Serpentine soils are high in magnesium and low in calcium, and they contain heavy metals such as cobalt, nickel, and iron. They are often claylike and low in nutrients. Chamise (*Adenostoma fascicula-*

tum) is not endemic, but it is able to tolerate serpentine soils and thus is an indication that these soils represent a dry substrate for plants.

Plants absorb both calcium and magnesium indiscriminately, but when magnesium is high it poisons the plant. Serpentine soils also contain heavy metals such as cobalt, nickel, and iron. They are often claylike and low in vital nutrients. As a consequence, most plant species do not grow well on serpentine soils; thus, such soils are characterized by a number of specialized serpentine endemics, plants that often grow nowhere else. Leather Oak (*Quercus durata*) is such an endemic that seldom grows anywhere else. Various Cypresses and Knobcone Pines are also frequently associated with serpentine soils, but most of the serpentine endemics are herbaceous species in the Sunflower family (Asteraceae), Lily family (Liliaceae), and the Mustard family (Brassicaceae). California has serpentine soils in the Klamath Mountains, the Coast Ranges, and the foothills of the Sierra Nevada. They also contain a carcinogenic form of asbestos. A more thorough discussion of serpentine endemics will be addressed where those regions are discussed in later chapters.

If soils are very high in calcium carbonate, it poses a different problem for plants. Such soils could be too alkaline or salty and interfere with the uptake of water from the soil. Plants adapted to such soils are called halophytes and they are often concentrated in desert basins where evaporation rates are high. Calcium carbonate layers called caliche are common in desert soils. Soils derived from limestone or marble are also high in calcium carbonate and have a basic pH. Plants associated with these soils are often limited in distribution. There are several areas of limestone soils in the California deserts, and each of them is characterized by endemic plants. Limestone is not common in the Sierra Nevada, but there are a few regions, such as Convict Creek south of Mammoth, where the specialized soils derived from marble have a significant number of endemic plants.

Another interesting soil is that derived from gabbro. This red-colored soil, high in iron and magnesium, has its own list of endemic plants. It is found in the Klamath Mountains, Laguna Mountains in the Peninsular Ranges in San Diego County, near Bodfish south of Lake Isabella, and in the foothills of the Sierra Nevada in El Dorado County. Interestingly, Leather Oak is also found there. Cuyamaca Cypress, *Hesperocyparis stephensonii* (= *Cupressus stephensonii*) is endemic to the King Creek drainage on the west side of Cuyamaca Peak in the Laguna Mountains.

The climate in which a soil matures has a profound effect on its characteristics. This phenomenon is very important to agricultural scientists, so it is not surprising that the Department of Agriculture has developed a classification in which soils have been divided into at least 12 categories known as soil orders. Some soil vocabulary, still in use, originally was developed by Russian scientists primarily based on color, but those terms are largely being replaced by the USDA system. A complete discussion of those soil orders is beyond the scope of this book, but the most common of them are significant to California ecosystems.

Mollisols

Mollisols form where there is abundant sunlight and moderate precipitation during the spring growing season. These are typically soils associated with grasslands and are highly favored for agriculture. They are rich in organic material and they crumble easily. Russians called them chernozems, which means nut brown. The process by which these soils are formed usually involves grasses which die off on an annual basis and that supplies the organic material that contributes to a rich humus layer. In California, these soils are associated with coastal and valley grasslands.

Spodosols

Also known as podzols, a Russian word that means ash gray. Podzolization is a process associated with a cold-wet climate and coniferous forests. Formed in mountain areas and northern latitudes, this soil is found where snow is the dominant form of precipitation. Acid humus increases the solvent power of percolating water, which carries away minerals such as iron, aluminum, and carbonates from surface layers, depositing them in deeper soil horizons. This is a coarse, light-colored soil with low fertility and an acid pH.

Histosols

Bogs contain soils that are water logged. They contain high levels of organic material, including peat. Bacterial decomposition rates are high, leading to extreme podzolization. They are quite acidic and sterile. Soils lack horizons with organic material that often leads down to an impermeable hardpan or thick subsurface of organic material. Bogs are found in certain areas of the Pacific Northwest where heavy precipitation allows accumulation of standing water in low lying areas of the forest floor. There is an area in Mendocino County near Fort Bragg where coastal terraces have been elevated by tectonic activity. On the oldest terrace, there is a region of white-colored, highly acidic soil which remains from a former bog. Here the nutrient-poor ancient histosol supports a stand of dwarfed cypress and pine trees known as a pygmy forest. This interesting region is discussed in detail in the chapter on the Coast Ranges.

Aridisols

Formed in a warm, dry climate, these are soils often associated with deserts. Also known as pedocals, these are soils developed over long periods of calcification where a high rate of evaporation, makes the soil alkaline or "salty." Stringers of caliche may be present. In the Imperial and San Joaquin Valleys of California, where farmers grow crops in a desert environment, over-irrigation is often practiced in order to carry minerals away from surface layers. Evaporation subsequently may deposit an impenetrable "hardpan" in lower soil horizons

which further inhibits percolation. In the long run, this will cause accumulation of salts in upper layers and fields become abandoned as unsuitable for agriculture. Abandoned fields, invaded by nonnative, salt-tolerant weeds, and saltbushes are characteristic of many acres in these formerly productive agricultural lands. This alkalinization process is avoided in some areas where perforated draining pipes are buried deep in the soil. These pipes pick up the leach water and drain it away. In the Imperial Valley, this water, laden with salts, pesticides, and fertilizer, ends up in the Salton Sea. In the San Joaquin Valley, the water ends up in the San Joaquin River or various low spots such as Lake Webb, a recreation area near Bakersfield. One of these low spots was known as Kesterson National Wildlife Refuge. A drain to carry wastewater to San Francisco Bay was never completed, and its water drained into the Kesterson marsh. Unfortunately, selenium, a toxic mineral that usually is rare in nature accumulated in this area and it caused congenital deformities in migratory water birds. Repeated efforts to bury the area reached only marginal success. The Kesterson story is explained in more detail in the chapter on the Great Central Valley.

Andisols

Soils resulting from volcanic activity are found in California along the eastern side of the Sierra Nevada and north through the Cascade Mountains. These soils, often basaltic in origin, are high in aluminum, iron, and phosphate minerals, and may be colored red to black. They may have high organic content and be fairly fertile. Volcanic ash deposits are often light colored. North of Bishop, there is a large deposit of light-colored Bishop Tuff, which resulted from a massive volcanic explosion that occurred about 700,000 years ago in the present location of the Long Valley Caldera. In a blast that was a couple of thousand times larger than the 1980 Mount Saint Helens eruption, layers of tuff and volcanic ash were deposited on the landscape. Because the material was evacuated from the side, the top collapsed forming the caldera. After Lake Crowley Reservoir was completed in 1941, as part of the Los Angeles Aqueduct system, waves washing against the eastern bank in a region called Crowley Cliff Columns began eroding the soft ash sediments and exposed a bizarre set of columns that resembled an ancient Greek temple. Lowered lake levels in recent years have provided overland access to the columns. Scientists have determined that the icy water from snowmelt percolated down through the hot ash layers. When the water boiled, it created evenly spaced convection cells resembling heat pipes which got cemented together as erosion-resistant columns. One estimate claims there may be thousands of these columns buried in the area and the exposed columns are just the edge of the formation.

In the region extending from Mammoth Mountain and Devil's Postpile north to Mono Lake, there are vast deposits of light gray–colored pumice, a unique soil that supports a significant stand of Jeffrey Pine (*Pinus jeffreyi*). Mammoth Mountain was once known as Pumice Mountain. Nearby, a coarse silica-rich soil, derived from rhyolite associated with the eruption of Mono Craters, harbors a unique wildflower, a Mono Craters Blazing Star, *Mentzelia monoensis*, that is found nowhere else in the world.

Oxisols and Ultisols

These are brick-red, clay soils formed in warm, wet climates, an environment not normally found in California today. However, there was a time when such a climate did exist here, and there are some residual soils from the time period about 10–15 million years ago. Also known as laterites, these soils were formed by a process known as laterization, a process still occurring in tropical climates. Characterized by abundant rainfall, soil becomes oxygenated and encouraged by warm temperatures, rates of decomposition increase. Percolation leaches silica and alkaline minerals from the soil leaving behind oxidized aluminum and iron, the latter causing the red color. These are sterile, fine-grained soils not suitable for agriculture; however, the clay from these soils was an important ingredient in the production of adobe bricks used in the construction of buildings throughout the Southwest. Because they are high in ferromagnesian minerals, some authorities refer to serpentine and gabbro derived soils as laterites.

GEOLOGIC TIMESCALE

Figure 3.2 shows a geologic timescale. The dates and terminology used here are those agreed upon by The Geological Society of America in 2012. They may not jibe exactly with other published versions. A geologic timescale combines absolute age determinations, in years, with relative age based on sequences of rock strata. Estimations of absolute age are derived by combining known rates of sedimentation and erosion with decay rates of radioactive minerals, the latter being much more reliable. A complete sequence of rock strata is not found at any single locality, although the Grand Canyon is one of the best places in the world to observe repeated layers of rocks representing hundreds of millions of years of absolute time.

Certain fossil organisms are associated with certain time periods. If these index fossils are found in an area of unknown age, fairly reliable age estimations can be made even though a complete fossil assemblage is not present. The law of superposition dictates that fossils found in deeper rock layers are older than those found in layers above. This layering of fossils is the basis for relative age determinations. Many different techniques have been discovered for making estimations of absolute age and relative age. These techniques can be combined to produce a picture of events that have occurred during the earth's history. Unfortunately, there is considerable disagreement among geologists about the absolute ages of certain events, particularly since the beginning of the Miocene, about 24 million years ago. In this book, every attempt has been made to use the most recent interpretation, or at least the most commonly held belief, for these ages.

RELATIVE TIME			ATOMIC TIME (in millions of years)
Era	Period	Epoch	
Cenozoic	Quaternary	Holocene	0.01
		Pleistocene	
			2.6
	Neogene	Pliocene	5.3
		Miocene	
			23
	Paleogene	Oligocene	
			34
		Eocene	
			56
		Paleocene	
			66
Mesozoic	Cretaceous		145
	Jurassic		
			201
	Triassic		
			252
Paleozoic	Permian		
			299
	Carboniferous Systems	Pennsylvanian	323
		Mississipian	
			359
	Devonian		
			419
	Silurian		
			444
	Ordovician		
			485
	Cambrian		
			541
Precambrian	Proterozoic Eon		2500
	Archean Eon		
			4000

FIGURE 3.2 Geologic timescale (vocabulary and data from Geological Society of America 2012).

The different time subdivisions are named for life-forms and localities where certain rocks are found. For example, the terms *Proterozoic, Paleozoic, Mesozoic,* and *Cenozoic,* respectively, mean "first animals," "ancient animals," "middle animals," and "recent or new animals." Many of the periods are named for localities; for example, *Cambria* is the Roman name for Wales, *Mississippian* refers to the Mississippi Valley, and so on. Within the Cenozoic Era, epochs are named for their relative age. Paleocene is the oldest epoch; the word means "ancient-recent," or "oldest of the new." *Pleistocene* means "most recent."

In the mid-eighteenth century, based on biblical history, time was divided into four periods. The third and fourth periods were known as the Tertiary and the Quaternary. The Tertiary, supposedly the time of the "great flood," was used until recently to refer to the Paleogene and Neogene, and the Quaternary is still used to refer to the Pleistocene (the "Ice Ages") and the Holocene (recent time). Noting a series of human-caused (anthropogenic) documentable changes in sediments, tree rings, sea level, and species composition of ecosystems, geologists today are proposing a new Epoch called the Anthropocene, which will mark the point at which humans began driving the future history of the earth. Tentatively, the line between Holocene and Anthropocene will fall in the mid-1900s, although many scientists point to significant changes that go back to the mid-1800s.

The geologic history of California may be explained in part by reference to the geologic timescale. Precambrian rocks in California include the billion-year-old metamorphic rocks of the San Gabriel Mountains and desert ranges. Paleozoic rocks are represented by the sedimentary rocks of the Mojave and Basin-Range complex. Mesozoic rocks include plutonic rocks (granitics) in the Sierra Nevada, the Transverse Ranges, and the Peninsular Ranges. Cenozoic rocks include mostly sediments, but many volcanic rocks are also recent, including those of the Cascades and Modoc Plateau. In order to understand why certain rocks are associated with these different eras, it is necessary to understand what was happening in California during each of these major periods of time. To this end, it is necessary to recreate the processes of rock formation in relative time and to study the forces responsible for these phenomena.

PLATE TECTONICS

No other concept in geology is as important to understanding how California came to be so diverse than that of plate tectonics, plates sliding about on the surface of the earth. Two or more plates interact on continental margins, but the type of interaction is not always the same. Two plates may converge on each other, but the speed of convergence may differ. On the other hand, two plates may diverge from each other. Each of these types of relative motion produces a different topography.

Figure 3.3 illustrates four types of plate boundaries along with a summary of geologic events in California. California is on the western margin of the North American plate. At first, California was underwater as part of a diverging margin similar to that which is present today on the Atlantic side of North America. About 400 million years ago (Devonian), that motion reversed, and California became a converging margin. The floor of the Pacific Ocean began to slide under the advancing edge of the North American plate in a manner known as subduction. Heat and pressure caused by the subducting Farallon plate caused an offshore series of volcanic islands. At this time, the shoreline ran from about southern Idaho southward, cutting across what is now eastern California. One or more island arcs occurred offshore about where the Klamath Mountains lie today. The topography of the continental margin at that time resembled the present-day east coast of Asia in the vicinity of Japan. As the Farallon plate continued to slide beneath North America, the rocks of the island arcs became attached (accreted) to the western margin of the continent. By means of these

A Californian type

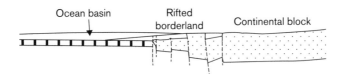

Ocean basin   Rifted borderland   Continental block

B Andean type

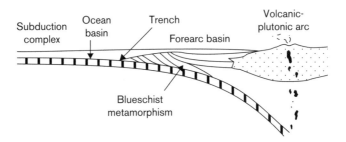

Subduction complex   Ocean basin   Trench   Forearc basin   Volcanic-plutonic arc

Blueschist metamorphism

C Japanese type

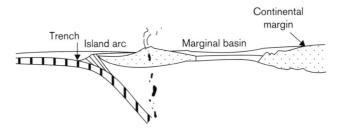

Trench   Island arc   Marginal basin   Continental margin

D Atlantic type

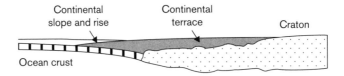

Continental slope and rise   Continental terrace   Craton

Ocean crust

FIGURE 3.3 Geologic events in the history of California. Historical plate boundaries. See table (opposite) for summary of geologic events.

A. Californian type (Miocene to Holocene).

B. Andean type (Cretaceous).

C. Japanese type (Triassic).

D. Atlantic type (Paleozoic).

SOURCE: Four types of plate boundaries after Ernst, W.G. 1979. California and Plate Tectonics. *California Geology* 32(9): 187–196.

Time Frame (oldest at bottom, most recent at top)	Geological Characteristics
Holocene	This epoch was characterized by continued motion along fault lines associated with earthquakes and volcanism. Rain-shadow effects continued with increased drying of deserts.
Pleistocene	This, the "Ice Age," was characterized by repeated glacial episodes in the mountains and heavy rains elsewhere. Sedimentation occurred in coastal and desert basins, and in the Great Central Valley. Formation of the Cascade volcanoes occurred. Uplift of southern Coast Ranges closed direct access from the sea to the San Joaquin Valley. Major uplift and rotation of fault-block mountains occurred in association with northward translocation along the San Andreas fault system. Coastal terraces formed by uplift and changing sea level (see figure 3.3A).
Pliocene	Sedimentation occurred in the Great Central Valley, and marine sandstones continued to be uplifted along the coast. Early Cascade and Modoc Plateau volcanism occurred, and Sutter Buttes erupted.
Miocene	This epoch was characterized by a Californian type of continental margin as the San Andreas fault system began to move. In the early parts of the epoch, coastal southern California was located many miles to the south. Large islands occurred offshore composed of uplifted Franciscan rocks. A large inland sea covered most of the present coastal plain. Northern Coast Ranges began to be uplifted. Volcanism was widespread, including the first basalt flows of the Modoc Plateau and andesite flows that formed mesas in the southwest deserts. A period of heavy erosion transformed the ancestral Sierra Nevada to a gentle rising plain. Red Rock Canyon exhibits 6100 ft of sediments deposited at this time. The climate was warm and moist.
Oligocene Eocene Paleocene	These epochs were characterized by long periods of erosion. The only terrestrial Paleocene fossil record in California is in the El Paso Mountains along the Garlock fault. Freshwater sediments of the Peninsular Ranges were deposited at this time. Seawater covered most of the present coastal plain.
Cretaceous Jurassic	These periods were characterized by an Andean type of continental margins associated with shallow, rapid subduction. A large coastal mountain range composed of volcanic rocks occurred, remnants of which form the Ritter Range northwest of Mammoth. Subduction of the ocean floor into the offshore trench formed metamorphic rocks of the Franciscan complex. Orocopia-Pelona-Rand schists of southern California were formed. Melting and intrusion of plutonic batholiths occurred. The Klamath Mountains separated from the ancestral Sierra Nevada and probably formed a large island separated from the mainland by a sea that extended eastward over the present Modoc Plateau (see figure 3.3B).
Triassic	This period was characterized by a Japanese type of continental margin associated with deep subduction. Subduction caused scraping of marine volcanic rocks and ocean floor sediments on to the western borders of the Klamath Mountains and Sierra Nevada, causing the continental margin to grow westward (see figure 3.3C).
Paleozoic	This era was characterized primarily by an Atlantic type of continental margin. California was underwater as part of a shallow continental slope. The era began with a long period of erosion represented by sediments in the Nopah Mountains of the northeastern Mojave Desert. Limestones, shales, and sandstones found in desert ranges, as well as in the Transverse and Peninsular Ranges, were deposited under seawater at this rime. Roof pendants of the Sierra Nevada and Klamath Mountains are also composed of these materials. In the latter part of this era, rapid subduction began and a Japanese type of continental margin appeared. Volcanic islands occurred offshore in the vicinity of the present Klamath Mountains and northwestern Sierra Nevada (see figure 3.3D).
Precambrian	Most of the very old rocks representative of this era are found where they have escaped erosion. Examples include a 1.7-billion-year-old gneiss in the San Gabriel Mountains and billion-year-old igneous and metamorphic rocks in desert ranges such as the New York, Providence, Kingston, Clark, and Marble Mountains.

accreted terranes, the continent began to grow westward. More island arcs would form, and portions of the western margin would break and thrust eastward along thrust faults over the top of the land. Mountains were uplifted in this way.

About 210 million years ago (Jurassic), the speed of convergence on the west coast increased rapidly, presumably as North America collided with Europe, obliterating what may have been a former Atlantic Ocean. In the west, episodes of continental accretion accelerated and subduction rates increased as the Farallon plate was shoved rapidly, often obliquely, beneath the North American plate. The heat generated as the ocean floor squeezed beneath the land caused the rock to melt, forming a deep pool of magma that, upon cooling, would become the granites, diorites, and gabbros of the batholiths that now make up much of the Sierra Nevada and southern California mountains. A deep trench formed offshore west of the zone of subduction. Sedimentary and metamorphic rocks from this trench, including shale, diatomite, and serpentinite, are now uplifted and exposed in the Coast Ranges. Pieces of the Farallon plate were being accreted along western California as it was being subducted. These pieces ultimately would become known as the Franciscan assemblage of the Coast Ranges. During this period of maximal subduction, the topography resembled what is found today on the west coast of South America. There was a large mountain range similar to the Andes, the proto-Sierra Nevada (nongranitic). A narrow continental shelf extended steeply to a deep submarine trench.

Beginning about 65 million years ago, with the onset of the Cenozoic Era, roughly 40 million years of erosion from the highlands and deposition westward occurred. Deposition covered most of the western half of the state. The Klamath Province, which broke away from the northern Sierra, moved to its present position about 60 mi (100 km) northwestward, where it sat as a large island. The Great Central Valley was underwater, and the ocean extended northward to cover what is now the Cascades and

Modoc Plateau. Volcanism continued in the areas now occupied by deserts and the Coast Ranges.

California as we know it today originated about 27 million years ago, during late Oligocene. The relative motion of the two plates changed from head-on convergence to a sideways motion known as a transform system. The Pacific and North American plates made direct contact for the first time when the closest part of the Farallon plate was subducted beneath the North American plate. Deep cracks or faults formed in the earth's crust. Motion began to take place along these faults as some blocks sank and others rose. Some blocks were carried sideways. The western part of the United States began to be stretched in a manner known as extension.

Earthquakes occur when these blocks move (figure 3.4). If most of the motion is up or down along a fault, it is called dip-slip. If most of the motion is sideways, it is called strike-slip. Most of the time, motion along a fault is oblique, but the major direction will be either dip-slip or strike-slip. The modern Sierra Nevada, Transverse Ranges, and Peninsular Ranges were uplifted along dip-slip faults and are known as fault-block mountains. Other fault-block mountains are those of the Basin-Range Province. In California, conspicuous examples include the Warner Mountains, Inyo-White Mountains, and Panamint Mountains, but the topography of alternating basins and ranges continues all the way across Nevada to the Rocky Mountains.

Prior to 27 million years ago, everything west of the San Andreas fault was a good deal farther south, some authorities contend—as much as 300 mi (480 km). Most authorities agree on at least 200 mi (320 km) of displacement. This translocated region includes most of the Coast Ranges south of San Francisco, the Transverse Ranges (except the San Bernardino Mountains), and the Peninsular Ranges. The Salinian Block, which formed near the southern end of the Coast Ranges, was carried northwestward along the San Andreas fault to its

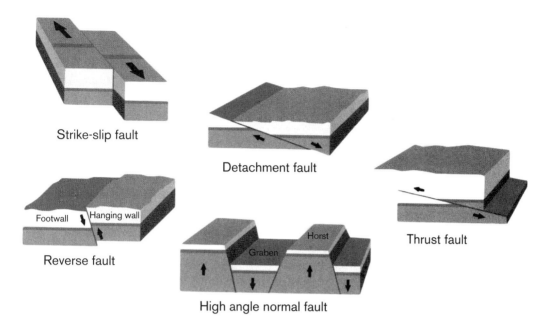

Strike-slip fault

Detachment fault

Footwall   Hanging wall

Reverse fault

Graben   Horst

Thrust fault

High angle normal fault

FIGURE 3.4 Types of faults (from Schoenherr 2011, after Teacher Feature in California Geology, July/August 2000).

present position along the central California coast. The plutonic rocks (e.g., granite) of these ranges were formed by subduction along the continental margin, but the entire unit was south of the Sierra Nevada, forming one long sequence of batholithic igneous rocks. The Salton Trough and the Gulf of California did not exist prior to 25 million years ago because the entire Peninsular Ranges sequence including Baja California was actually the west coast of what is now mainland Mexico. However, some authorities place the formation of the Gulf of California as recently as 5–10 million years ago.

EARTHQUAKES AND FAULTS

The San Andreas is but one of many faults in California (figure 3.5). Motion on the San Andreas fault is right-lateral strike-slip. That means that the land is moving primarily sideways, and if a person stood on either side of the fault during an earthquake and watched the other side, the land would move to the right. The Garlock fault along the north side of the Mojave Desert is a left-lateral strike-slip fault.

During an earthquake on the Garlock fault, the relative motion of the land would be to the left. Total displacement on the Garlock fault has been calculated at 40 mi (64 km).

The type of continental margin found in California today has no modern counterpart in the world. It is known as a rifted borderland. Right-lateral motion on the San Andreas is responsible for much of the rifting, but not all of it. Considered in total, it is known as the San Andreas Transform System.

Parts of the San Andreas fault move at different rates, expressed by different degrees of displacement at various points along the fault (figure 3.6). Some of this discrepancy may be explained by major motion along one part of the system shifting to a parallel but different fault. For example, the San Gabriel fault, on the south side of the San Gabriel Mountains, used to be the major fault. It has been inactive for some time, and now the major fault is on the north side of the mountains.

Another discrepancy is that some parts of the system experience many small earthquakes each year, whereas others go many years between quakes. Areas where there are many

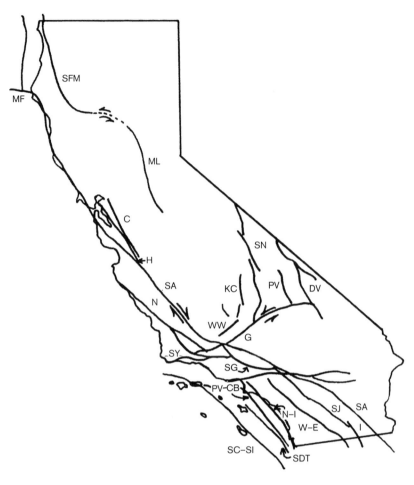

FIGURE 3.5 Some principal California faults.

ABBREVIATIONS: C, Calaveras; DV, Death Valley; G, Garlock; H, Hayward; I, Imperial; KC, Kern
Canyon; MF, Mendocino Fracture; ML, Mother Lode; N, Nacimiento; N-I, Newport-Inglewood;
PV-CB, Palos Verdes-Coronado Bank; PV, Panamint Valley; SA, San Andreas; SC-SI, San Clemente-
San Isidro; SDT, San Diego Trough; SG, San Gabriel; SJ, San Jacinto; SY, Santa Ynez; SN, Sierra
Nevada; SFM, South Fork Mountain; WW, White Wolf; W-E, Whittier-Elsinore.

quakes may move 2 in (5 cm) a year, but there is seldom a serious earthquake. Regions such as these are known as active areas or creep zones. For example, the town of Hollister, about 100 mi (160 km) south of San Francisco, is bisected by the Calaveras fault, just east of the San Andreas. In Hollister, fences bend, sidewalks buckle, foundations on buildings crack, and streets become displaced as the western part of the town is carried slightly farther northward each year.

In contrast to creep zones are locked zones. The southern locked zone is in the area of the Transverse Ranges where there is a decided northwestward bend in the fault zone. The San Andreas passes between the San Gabriel and San Bernardino Mountains in the vicinity of Cajon Pass. These are known as Transverse Ranges because they lie in a mostly east-west line, whereas most ranges in California lie in a north-south line. Apparently the Transverse Ranges have been rotated about 90° clockwise by motion along a locked zone of the San Andreas fault. These areas are bounded by hard rock materials such as granite, so they do not move easily or often. When they do move, a

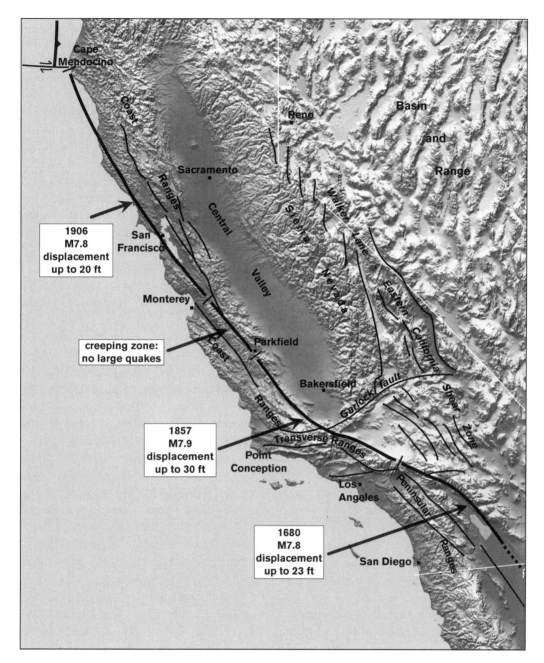

FIGURE 3.6 Historical earthquakes on the San Andreas fault showing the importance of locked zones and creep zones (Meldahl 2015).

large earthquake is the result. Apparently, tension or strain in a creep zone is released a little at a time, but in a locked zone it builds up for many years. When the earth finally does move, a tremendous amount of energy is released all at once, causing severe damage to buildings and reshaping the surrounding terrain. In California, the most famous and devastating such earthquake took place in 1906 offshore from San Francisco. It flattened buildings, started fires, and caused about 3000 deaths. It resulted in 21 ft of right-lateral displacement, most of which took place in a sparsely inhabited area near Point Reyes. The effect of that earthquake may be seen today at Point Reyes National Seashore. In this area, the National Park Service maintains a trail along the fault trace with signs and markers at appropriate points of interest. Damage to buildings in San Francisco was widespread, largely because many of them were constructed of brick (nonreinforced brick buildings suffer the most damage in an earthquake), and damage to water delivery systems was so severe that citizens were helpless to put out fires. What the earthquake didn't topple, fire burned to the ground.

In 1857, a break took place in another locked zone at Fort Tejon, near the present town of Gorman, about 50 mi (80 km) north of Los Angeles. The amount of right-lateral slip from that earthquake has been measured at 35 ft (12 m), and the line of disturbance was visible along the San Andreas fault for 225 mi (360 km). Two people were killed.

An elevated face occurring along a fault is known as a scarp. Some fault scarps are visible for many years; others become buried by sediment. By digging a trench across the San Andreas fault near Pallett Creek on the north side of the San Gabriel Mountains, scientists were able to locate the 1857 fault scarp as well as 11 others dating back to AD 260. Estimation of the intervals between earthquakes at Pallett Creek indicated that in that part of the locked zone, breaks occurred on the average every 140 years: some had occurred as close together as 75 years, and some were as far apart as 275 years.

Research such as that on the fault scarps of Pallett Creek is necessary if we are to attempt to develop a system for predicting earthquakes. Another technique is to measure tension or strain in rocks along a fault. Regions of high strain are likely candidates for an earthquake. Recent evidence indicates that a region of high strain occurs along the San Andreas fault north of the Salton Sea. By combining several techniques of analysis of seismic gaps, scientists have concluded that the locked zones on the San Andreas might be expected to move about every 100–140 years, and that movement in these zones has averaged 2 in (5 cm) per year over the last 25 million years. Consensus among scientists is that there is a 50% probability of an earthquake of major magnitude in the near future, and it probably will occur near San Bernardino.

In the northern locked zone, there was an earthquake in 1865, another in 1906 (only 41 years later), and another in 1989, 83 years later. Another earthquake could occur at any time, although using the 100- to 140-year average, one might expect the next major quake to occur sometime after the year 2100. Strain in the rocks near San Juan Bautista, near Monterey Bay, has accumulated to such a degree that scientists are predicting an earthquake to occur there at any time.

The first earthquake for which there is a recorded witness took place in southern California in 1769 (Table 3.5). Gaspar de Portola, a Spanish explorer, while camped along the Santa Ana River witnessed what might have been a magnitude 7.3 earthquake. His account describes the Santa Ana River changing direction, and he named the river "Nombre Dulce Jesus de la Temblors," which means "Sweet Jesus of the Earthquake." Recent evidence indicates the quake actually took place under the San Joaquin Hills and the Orange County coastline may have risen 11 ft (3.1 m) in places.

Earthquakes do not always occur on faults with a recent history of activity. In 1952, one of the largest earthquakes in California history took place on the White Wolf fault near the town of Tehachapi, in southern Kern County. This fault was presumed to be inactive. Its total length was mapped at only 34 mi (55 km), and it had never been known to move. Yet the earthquake ranks as the largest in southern California since 1857. Twelve people died. Crustal shortening near Bealville was measured at 10 ft (3.3 m). Similarly, in 1987 on a blind thrust fault near the town of Whittier, in Los Angeles County, a damaging earthquake took place on a virtually unknown extension of the Whittier-Elsinore fault, and eight people died.

TABLE 3.4
A Comparison of Magnitude and Intensity

Magnitude	Intensity	Effect
1	I	Only observed instrumentally
2	II	Is barely felt near epicenter
4.5	VII	Slight damage at epicenter, felt 20 mi (36 km) away
6+	VIII	Moderately destructive
7+	IX	Considerable damage—a major earthquake
8+	X+	All buildings suffer some damage, panic is general—a great earthquake

NOTE: There may be some discrepancies in the magnitude of older earthquakes because they are based on the Richter scale as opposed to the more modern moment magnitude scale. The Modified Intensity Mercalli Scale measures the intensity of an earthquake where a person feels it. Intensity varies with the type of ground (bedrock vs. sand) and the degree of saturation by water. The two scales therefore are difficult to equate.

The largest earthquake in California during recorded history took place in 1872 on the Sierra Nevada fault near the town of Lone Pine in Owens Valley. Earth movements were enormous: 23 ft of vertical displacement and 20 ft of right-lateral shifting were recorded. A mass grave just north of town contains the remains of 27 people that died in that event. There are reports by native peoples of a major quake in the same area in the 1790s. The validity of these reports is subject to question, but if a 100-year interval is assumed, that fault is also due to move at any time.

If a major earthquake were to occur today near a major population center, the probable damage and loss of life would be immense. Collapsing structures is one problem, but all major utilities cross faults. Imagine gas lines, power lines, and water lines bursting. All the major aqueducts that deliver water to southern California cross the San Andreas fault, and the California Aqueduct crosses it several times. Ever since 1933, when an earthquake on the Newport-Inglewood fault caused severe damage and killed over 100 people in Long Beach, building codes have required that new structures be built to resist earthquakes of certain magnitude. A great deal of progress has been made also on developing plans to deal with large-scale emergencies. Nevertheless, the specter of the next big quake is a fear with which all Californians must live.

There are two ways to measure the size of an earthquake. The traditional Richter scale (M_L) is a mathematic measure of magnitude based on a logarithmic scale. Each number on the scale indicates an increase in magnitude by a factor of 10. A magnitude 8 earthquake is 100 times more severe than one with a magnitude of 6. An instrument known as a seismograph measures vibrations in the earth. Data from many seismographs are coordinated to locate the epicenter of a quake. Based on the amounts of vibration at various distances from the epicenter, a magnitude value is assigned to the quake.

In the 1970s, the Richter scale was modified to include energy release, rock strength, surface length of rupture, and amount of rock displacement. The new measure of magnitude is called *Moment Magnitude Scale* (M_w or MMS), but the units of 1–10 have been retained to reduce confusion. Often the initial report of magnitude is on the Richter scale, but the Moment magnitude, because it takes longer to calculate, may not be released for a day or two after the quake.

Because logarithmic scales are difficult to understand by laymen, intensity values are assigned to earthquakes as well. Intensity

TABLE 3.5

A Partial List of Major Earthquakes in California

Date	Location	Magnitude	Intensity
1769	Santa Ana Canyon	8.0	XI
1812	Wrightwood	7.0	X
1836	Hayward	7.0	X
1838	San Francisco	7.0	X
1857	Fort Tejon	7.7	X
1861	Livermore	7.0	VII
1872	Lone Pine	8.3	XI
1906	San Francisco	8.3	X
1918	San Jacinto	6.8	IX
1925	Lompoc	7.5	X
1929	Whittier	6.8	XI
1933	Long Beach	6.3	IX
1940	Imperial Valley	7.1	X
1952	Tehachapi	7.7	XI
1971	San Fernando	6.6	XI
1979	Imperial Valley	6.4	IX
1980	Eureka	7.0	VII
1983	Coalinga	6.7	VIII
1986	Palm Springs	6.0	VII
1987	Whittier	5.9	VIII
1987	Imperial Valley	6.3	IX
1989	Loma Prieta (Santa Cruz Mountains)	7. 1	IX
1992	Joshua Tree	6.2	VII
1992	Cape Mendocino (3)	6.5–7.2	VIII
1992	Landers	7.3	IX
1992	Big Bear	6.5	VIII
1994	Northridge	6.7	IX
1999	Hector Mine	7.1	VII
2003	San Simeon	6.6	VIII
2010	Cape Mendocino	6.5	VII
2014	Napa	6.0	VIII

scales are based on the amount of damage caused by the earthquake or its effect on people. Intensity scales are subjective, but they are intended to be understood more easily by people with limited mathematic skills. The most popular intensity scale is known as the Modified Mercalli Intensity Scale. Table 3.4 is a comparison of the two scales. The levels of magnitude and intensity of early earthquakes has to be inferred, but a list of major California earthquakes with magnitude and intensity values is presented in Table 3.5. The largest conceivable earthquake is one of magnitude 9.0. If such a quake were to occur, damage would be total, and it would be assigned the highest intensity value, XII.

GEOLOGIC MAP OF CALIFORNIA

The geologic map depicted in plate 2 is color-coded to illustrate rocks of different age and origin, making California appear to be a patchwork quilt. Such illustrations are invaluable in demonstrating that California is one of the most diverse areas in the world. Superimpose on this picture a diversity of soils and climate, and it becomes possible to understand why California has more endemic species of plants and animals than any region of equivalent size in the United States.

SUMMARY

There are three kinds of rocks: igneous, sedimentary, and metamorphic. Igneous rocks are formed by the cooling of magma. If the cooling takes place deep in the earth, the resulting rocks are said to be plutonic or intrusive. Plutonic rocks are coarse-grained because minerals have had a long time to accumulate into visible crystals. The most common plutonic rock is granite, a light-colored form that is rich in silica but low in iron and magnesium. Volcanic rocks that have cooled on the surface are said to be extrusive. The most common extrusive rock is a dark, heavy form known as basalt. The glassy variety of extrusive rock is known as obsidian. Volcanic rocks are fine-grained due to rapid cooling.

The great diversity of rocks in California is due to the area's long history as a continental borderland. During most of the Paleozoic Era, California was the trailing edge of a landmass moving eastward. During this time, thousands of feet of marine sediments were laid down over what is now California. These sediments are now visible in mountain ranges uplifted in the southwest deserts. During the Mesozoic Era, subduction of the Pacific Ocean floor caused a pool of magma to form; this is now represented by the Sierra Nevada and Southern California batholiths. Volcanics that were built up as offshore island arcs became accreted to the western border of North America as the sea floor slid beneath the continental margin. Volcanic rocks and sediments from an offshore deep marine trench are now uplifted in the Coast Ranges.

During the Cenozoic Era, a long period of erosion and sedimentation was followed by evolution of the San Andreas Transform System. California, due to a plate boundary change, became a rifted borderland crossed by many faults, most of which are dip-slip and some of which are strike-slip. Right-lateral displacement has carried everything west of the San Andreas fault northward from its position prior to the Miocene (24 million years ago), which was at the western edge of what is now the Mexican mainland. Fault-block uplift of the Sierra Nevada, the Transverse Ranges, the Peninsular Ranges, and the Basin-Range Province took place during the last 10–12 million years, but the greatest amount of uplift probably took place during the last 3 million years.

4

Sierra Nevada

FIGURE 4.1 Pioneer Basin, a subalpine basin in the John Muir Wilderness, Sierra National Forest (drawing by Geoff Smith).

FIGURE 4.2 The Sierra Nevada.

THE SIERRA NEVADA is the most conspicuous geographic feature of the state of California. It is essentially a single mountain range approximately 400 mi (640 km) long and 50 mi (80 km) wide (figure 4.1). It is one of the largest of its kind in the world. In Spanish, *sierra* means "saw" or "jagged range," and *nevada* means "frozen" or "snowed upon." The plural form, *sierras*, is a common but incorrect reference to this single large range of mountains. Its spectacular glacier-carved scenery is frequently the subject of photographs or artwork that depicts California's natural beauty (figure 4.2). Its distinctive scenery has captured the minds and hearts of naturalists and photographers such as John Muir and Ansel Adams, whose writing and photography have made the landscape of the Sierra Nevada familiar all over the world.

By virtue of its geologic history, the Sierra Nevada is unique among mountain ranges of the world. It is a large, jagged, glacier-carved mass of granite. Because uplift has taken place mostly on the eastern edge of the Sierra, the western slope is, in fact, the top of a large rock mass. This mass angles gradually westward (about 2°) to pass beneath sediments that form the floor of the Great Central Valley. The eastern slope, a high scarp that runs on a north-south line along the western border of Owens Valley, may be seen as the edge of the rock mass. From the tops of high peaks along the crest of the Sierra, 13 of which are higher than 14,000 ft (4268 m), the eastern slope drops nearly 2 vertical miles into Owens Valley (plate 15A). At 14,505 ft (4421 m), Mount Whitney, west of Lone Pine (figures 4.11 and 5.1), is the highest mountain in the contiguous 48 states.

The high country at the southern end of the range is bisected by the Kern River, which flows southward along a fault. At the southern end, therefore, there are two high divides: the main divide, just mentioned, and the western divide, which extends from Sequoia National Park southward. Many peaks of the western divide exceed 13,000 ft (3900 m) in elevation.

As air masses move off the Pacific Ocean, they must rise if they are to cross the Sierra Nevada. As the air rises, it becomes cooled and water vapor condenses. The western slope of the Sierra therefore receives abundant precipitation. The steep eastern slope is, however, considerably drier. To the east of the Sierra, Owens Valley is a desert land that lies in the Sierran "rain shadow." Owens Valley is the westernmost part of a vast desert region known as the Great Basin, which extends eastward all the way to the Rocky Mountains.

During summer, the polar jet stream often swings southward bringing precipitation to the northern Sierra. From about Yosemite northward, annual precipitation is increasingly greater.

Large rivers flow down the western slope of the Sierra Nevada. For the most part, these rivers have been impounded to provide a water supply for massive agricultural enterprises of the Great Central Valley and for the San Fran-

cisco Bay area. By means of the Sacramento River and the California Aqueduct, some of this water is carried from its source in the northern Sierra over 500 mi (800 km) to southern California. Most of the water in rivers that cascade down the steep eastern slope of the Sierra is diverted into the Los Angeles Aqueduct and transported to the city of Los Angeles. The Sierra Nevada and its water are responsible not only for great scenic beauty, but also for the wealth of California as a whole and the crowding of southern California in particular. Agriculture and industry are California's economic base, and they depend on the water that comes from the Sierra Nevada.

The ruggedness of the Sierra has made large portions of the range inaccessible to motor vehicles. To allow people to experience the extraordinary beauty of the Sierra Nevada, the federal government has set aside large tracts of land as wilderness areas that may be visited only by foot or on horseback. These areas today are divided among three national parks—Yosemite, Sequoia, and Kings Canyon—as well as national forest wilderness areas such as those named for John Muir and Ansel Adams. Altogether, these wilderness areas represent the largest roadless area in the contiguous 48 states.

Along the crest of the Sierra is the 170 mi (270 km) stretch of alpine wilderness known to hikers and packers as the "high Sierra," or simply "the high country." From Sonora Pass on State Highway 108 southward to the road up Nine Mile Canyon to Kennedy Meadows, the Sierra is traversed by only one road, the road through Yosemite National Park. It crosses Tioga Pass at 9941 ft (3031 m) elevation. This spectacular region of alpine meadows, sculpted rock, twisted trees, and thousands of alpine lakes is the home of Bighorn Sheep, ground squirrels, marmots, Pikas, and a unique species of salamander. The high lakes of this country are legendary for their beauty and their trout fishing. It is interesting to note, however, that all species of trout but two were introduced by humans. The unique Golden Trout (*Oncorhynchus mykiss aguabonita*), native to the Kern Plateau of the southern Sierra, is the state freshwater fish. This highly colored descendant of the Rainbow Trout is the only true high-country fish, and its distribution was limited originally to those southern Sierra streams that escaped glaciation.

CLIMATE

Because of a profound rain-shadow effect, the western slope receives much more precipitation than the eastern slope. During an average year in the Montane Forest at middle elevations, over 80 in (200 cm) of precipitation may fall. In the Yosemite area at 5500 ft (1800 m) elevation, the western slope receives approximately 75 in (188 cm) of precipitation annually, whereas the eastern slope receives only 20 in (50 cm). Average annual precipitation for Bishop in Owens Valley is only slightly over 5 in (13 cm). There is also a decline in precipitation from north to south. At 5000 ft (1600 m) in the north, average annual precipitation is 90 in (225 cm). Farther south, in Kern County, the average annual precipitation at that elevation is only 30 in (75 cm).

A person ascending a mountain will observe certain trends in climate. On the west side of the central Sierra, precipitation increases from about 10 in (25 cm) in the Great Central Valley to a high of 90 in (225 cm) at about 8000 ft (2400 m). There are no official weather stations above that elevation, but precipitation generally declines above 9000 ft (3000 m).

The lower snow line is at about 3000 ft (1000 m). Above this point, the proportion of precipitation that falls as snow increases with elevation to as much as 90% of the total. The snowfall and the resulting snowpack in the Sierra Nevada is the highest in the state. Although there is considerable variation from year to year, during the winter of 1982–1983 there was a snowpack of 86 ft (29 m) on Donner Pass. Ten inches (25 cm) of snow is equivalent to about 1 in (2.5 cm) of rain; thus, 80 ft of snow roughly equates to 8 ft or 96 in (240 cm) of rain. Snow surveys are taken on April 1 each year. During the record drought of 2015, the sierra snowpack on that date was only 5% of normal. At Phillips Station, during the annual snow survey, at 6800 ft (2073 m), east of Sacramento, there was bare ground on that date. The average for that location is 66.5 in (169 cm). The snow depth at Tahoe City on that date was 19.5 in (49.5 cm), approximately 12 ft (366 cm) below the average.

Over half of Sierra precipitation falls during January, February, and March. Less than 3% typically is received in summer, although during the drought of 2015, May was the wettest on record, and an unseasonal snowstorm dropped snow on high peaks in July. Even though summer thunderstorms seem to dump a lot of water, precipitation is patchy, and the contribution to the yearly total is not great. Precipitation decreases abruptly below about 3000 ft (1000 m), which is the average elevation of the crest of the Coast Ranges. The rain shadow formed by the Coast Ranges influences precipitation across the Great Central Valley and into the foothills of the Sierra Nevada.

Temperature also is influenced by elevation. Due to adiabatic cooling, the air temperature tends to drop 3–5°F per 1000 ft (ca. 1°C/100 m). The combination of lower temperature and greater amounts of snowfall reduces the length of the growing season at higher elevations in the alpine zone to about 6–8 weeks during the summer. Conversely, at lower elevations, particularly on the desert side, the growing season is reduced by summer drought. The shift from a winter growing season to a summer period of growth occurs at about the lower snow line.

Air becomes drier at higher elevations. Adiabatic cooling causes condensation and a loss of water vapor, but also, air at higher elevations has a lowered capacity to hold water because it has fewer molecules per unit volume. There are fewer oxygen molecules, too. This is the reason that humans experience difficulty breathing at high elevation. This dry air at high elevation changes temperature rapidly, which creates strong winds. Furthermore, the wind travels at a greater rate as it moves over peaks and passes, a phenomenon known as the Venturi effect. Air passing through a narrow place, or a constriction, must increase in speed for all of its molecules to go through in the same length of time. This windy, dry air tends to increase evaporation rates, which makes it even drier for plants in exposed places and increases the effect of evaporative cooling. In total, at high elevation wind causes drying, chilling, and abrasion.

The prevailing wind in the Sierra Nevada blows from west to east (figure 1.7). Air tends to heat more rapidly in the desert, and the rising desert air draws air from the coast over the top of the Sierra Nevada.

During summer months, the intensity of this wind is increased, particularly in the afternoon. As air in the Great Central Valley is heated, it begins to rise. This rising air is caught by the eastward flow over the mountains and is drawn up the canyons of the western slope. Because canyons become narrower near the top, this wind moves faster as it ascends. This phenomenon is known as the chimney effect, but it is merely an expression of the Venturi effect. This air increases in speed as it moves over passes and ridges, and it descends into the Owens Valley, becoming heated by compression. The result of this daily airflow is a predictable afternoon period of wind, uphill on the west and downhill on the east.

When temperatures become equalized at night, a gentle downhill flow of cool air prevails on both sides of the Sierra. It is this cold-air drainage that contributes to the inversion layer, which characterizes the Great Central Valley year-round. It is also the reason that nighttime temperatures are considerably colder in canyons and valleys. Campers who enjoy sleeping next to streams in the high country soon learn to place their sleeping bags with the foot toward the cold airflow. If the air is too cold, moving away from the stream to slightly higher ground can increase the temperature by several degrees.

Large lake basins in the Sierra are also subject to the inversion layer phenomenon. Lake Tahoe (Plate 3B), for example, has an inversion layer complicated with smog, which is caused by the enormous number of automobiles carry-ing tourists to and from casinos and other resorts. The photochemical production of ozone has caused a loss of chlorophyll (chlorotic decline) in pine trees around the shoreline of Lake Tahoe. Another place where chlorotic decline is evident, for the same reason, is in Ponderosa Pines (*Pinus ponderosa*) in Yosemite Valley. Chlorotic decline can also be observed in any number of slopes and canyons on the western side, where chimney effect during summer draws smoggy air up the canyons on a daily basis from the Great Central Valley.

Another effect of thin air at high elevations is an increase in ultraviolet radiation, high-energy light that is invisible to the naked eye. Its short wavelength gives it penetrating power beyond that of other forms of light, which makes it especially harmful to the tissues of plants and animals. Ultraviolet light causes chromosome breakage and mutations in irradiated tissue. Alpine organisms therefore have special adaptations to protect them from this potentially harmful radiation.

Changes in soil are also apparent during the ascent of a mountain. On any slope, the soil tends to become coarser toward the top, so water-holding capacity is greater at the bottom. Soil also is influenced by a variety of parent rock types, such as serpentine and limestone. The effect is a mosaic of soil types, some of which are notable for specialized vegetation. This mosaic is particularly evident in the western foothills.

At higher elevations, almost all soil is decomposed granite. Mountain climate results in podzolization of soils. These spodosols become increasingly acidic at higher elevations. Alpine soil with a pH reading of 4 is not uncommon. In boggy areas, the soil is even more acidic.

These spodosols (podzols) also tend to be nutrient deficient, which poses several problems for plants in mountain environments. Symbiotic relationships between nitrogen-fixing bacteria and members of the legume family (Fabaceae) have developed in response to the problem. Species of lupine and vetch are common legumes in all communities in the mountain environment. In forests, members of the genus *Ceanothus*, often known as California lilacs, have nitrogen-fixing bacteria in their roots. Along streams, these bacteria are found in the roots of alders.

Conifers have a symbiotic relationship with fungal filaments known as mycorrhizae. These filaments originate as part of the absorptive structure of a number of mushroom species, but they become intimately associated with the root system of the conifers. The fungus improves the absorption of nutrients by the tree, and the tree produces excess carbohydrates for the fungus. Ultimately, the mycorrhizal filaments connect the root systems of the trees with each other, allowing them to share various chemicals, including those that may discourage herbivores.

At high elevations, soil is coarse and thin. In low places, the soil tends to become waterlogged because percolation is prevented by bedrock or permafrost. In the Sierra, bedrock is usually the complicating element. Soggy soil slips on slopes, and it freezes solid in cold temperatures. This type of soil is a poor substratum for most plant species and is not a good place for burrowing animals to build a home.

GEOLOGY

Rocks of the Sierra Nevada (figure 4.3) can be divided into three groups: batholithic, prebatholithic, and postbatholithic. The Sierra is made up primarily of granitic rocks of the Sierra batholith. These rocks were formed during the Mesozoic Era, which ended about 65 million years ago. The batholith resulted as subduction of the ocean crust beneath the North American plate caused formation of a pool of magma about 10 mi (16 km) beneath the surface (figure 4.4). This process of magma formation must have lasted more than 100 million years, during which time numerous major subunits (plutons) developed. These plutons are north-south-trending units of homogeneous material that grow gradually younger toward the east. This age relationship

is interpreted as a result of the North American plate moving gradually westward over the Pacific plate during the time of magma formation.

Uplift of the batholith probably began about 80 million years ago. The overlying prebatholithic rocks were old ocean-floor sediments. As these sediments were raised, a proto-Sierra began to form, and erosion carried rock materials westward, gradually thinning the 10 mi (16 km) layer of materials overlying the granitics. The actual height to which the sediments were raised has been the subject of considerable debate, but by 40 million years ago, near the close of Eocene, the Sierra Nevada consisted of a relatively flat terrain of low, rolling granitic hills and broad valleys.

From about 20–5 million years ago, there was extensive volcanism along the northern crest of the Sierra. These volcanoes were most likely the southern extension of the Cascade Range. Examination of these lava flows shows that they covered an essentially level terrain. Fossil remains to the east show no evidence of a rain shadow until some 10–12 million years ago. Fossil plants show a drying trend at about that time to the east of the Sierra, implying that uplift and rain-shadow activity had begun. Drainage was still primarily to the west, adding more sediments to what is now the Great Central Valley. Major uplift and tilting westward of the Sierra batholith began about 3 million years ago and is still continuing. Uplift was recorded as recently as 1872, when an earthquake in the vicinity of Lone Pine raised the eastern front of the Sierra 23 ft (8 m).

Erosion is believed to have removed about 1.5 ft (46 cm) of earth every 1000 years. Thus, nearly all of the overlying sediments were removed, exposing to modern-day weathering the granitics of the batholith. Also, based on comparison of erosion rates to rates of uplift, it appears that the Sierra will one day be considerably higher than it is now.

Westward tilting of the Sierra is evident because uplift along the eastern flank was considerably greater than it was on the west. Like-

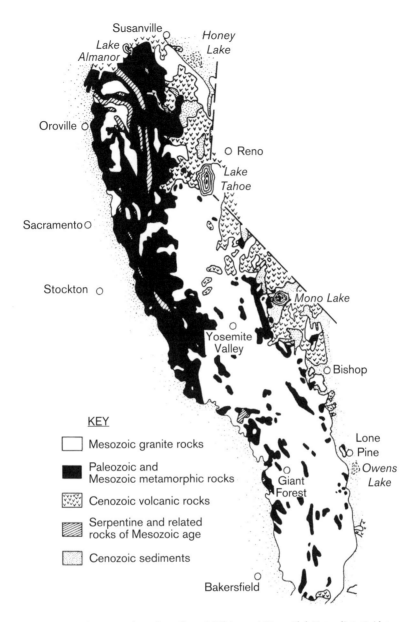

KEY

☐ Mesozoic granite rocks

■ Paleozoic and
Mesozoic metamorphic rocks

▨ Cenozoic volcanic rocks

▨ Serpentine and related
rocks of Mesozoic age

▨ Cenozoic sediments

FIGURE 4.3 Sierra Nevada geology (from S. Whitney. *A Sierra Club Naturalist's Guide to the Sierra Nevada.* San Francisco: Sierra Club Books, 1979; reprinted by permission).

wise, uplift in the south was greater than it was in the north. A consequence of higher elevation was the stripping of nearly all old rock sediments from the southern part of the range. As the Sierra rose, it pushed aside a portion of the prebatholithic rocks. On the east side of the Sierra, south of Bishop, all of that metamorphic rock has been carried away. In that area today, the east face of the Sierra is composed of a series of staggered parallel fault scarps exposed to further weathering. Granitics rise sharply from the floor of Owens Valley to form the highest peaks of the Sierra, 13 of which are higher than 14,000 ft (4230 m). This great granitic wall represents one of the most abrupt elevational gradients in the world, rising about 2 mi (3.5 km) above the valley floor (plate 15A).

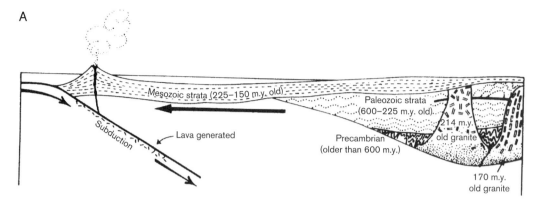

A

Mesozoic strata (225–150 m.y. old)

Subduction

Lava generated

Paleozoic strata
(600–225 m.y. old).

214 m.y.
old granite

Precambrian
(older than 600 m.y.)

170 m.y.
old granite

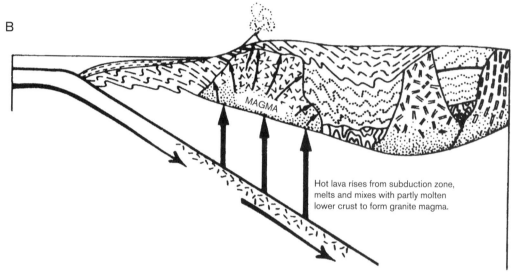

B

MAGMA

Hot lava rises from subduction zone,
melts and mixes with partly molten
lower crust to form granite magma.

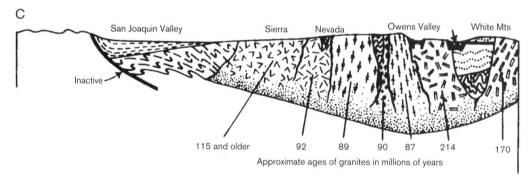

C

San Joaquin Valley   Sierra   Nevada   Owens Valley   White Mts

Inactive

115 and older   92   89   90   87   214   170

Approximate ages of granites in millions of years

FIGURE 4.4 (Opposite) A summary of geologic events leading to the formation of the Sierra Nevada (illustrations from Bateman, R.C. 1974. Model for the Origin of Sierran Granites. *California Geology* 27(1): 1–5).

A. *Late Paleozoic.* Toward the end of the Paleozoic Era, about 250 million years ago, subduction of the Pacific plate beneath the North American plate had begun. At this time, there was a Japanese type of continental margin with island arc volcanoes that spewed a vast amount of volcanic debris into an inland sea. This debris mixed with the material eroding from the continent to form sediments that are found today as roof pendants in some parts of the Sierra Nevada.

B. *Early Mesozoic.* During the early parts of the Mesozoic Era, over 200 million years ago, subduction deepened, and an Andean type of continental margin formed. At this time, magma was generated beneath the surface, and this would ultimately become the granite of the Sierra Batholith. Also at this time, in an event known as the Nevadan orogeny, a "proto-Sierra Nevada" of faulted, folded mountains was formed. Large amounts of volcanic rock were also deposited along the continental margin. Metamorphosed remains of the old mountains are still found as roof pendants at various points along the Sierran crest.

C. *Present.* By the end of the Mesozoic Era, over 60 million years ago, a long period of erosion had removed most of the volcanic rocks, and the granite batholith was becoming exposed. By middle Cenozoic, some 30 million years ago, the ancestral Sierra Nevada was eroded to a range probably no more than a few thousand feet high. By early Miocene, about 25 million years ago, the Californian type of margin with its rifted borderland evolved. The San Andreas fault, with its lateral motion, created a set of stresses causing major uplift of the Sierra batholith that is still occurring today. During late Miocene, particularly north of Yosemite, there was a new period of volcanism associated with the formation of the Cascade Range. This volcanism, which terminated about 5 million years ago, covered the range north of Yosemite with a variety of lava flows and formations, most of which have since been eroded. Now, after more than 150 million years of erosion, the Sierra Nevada stands as a range over 14,000 ft (4230 m) high, composed mostly of granite that was formerly buried by 5–10 mi (8–16 km) of rock. That eroded material now makes up the 5000–10,000 ft (1800–3000 m) of sediment in the Great Central Valley.

Prebatholithic rocks in the Sierra today are mostly metamorphic rocks that lie on the eastern and western flanks of the range north of the San Joaquin drainage. Folding and metamorphism of the prebatholithic sediments is closely tied to the forceful emplacement of the plutons, at depths where high pressures and temperatures would allow plastic deformation of the sediments. Throughout the high country are isolated outcrops of prebatholithic rock perched on top of the granite. These outcrops, known as roof pendants (figure 4.5), represent portions of metamorphic rock that have not yet been eroded. Roof pendants are highly visible because the uppermost portion of the peak is a darkly colored mass of rock, distinctly different from the underlying granitics. A number of high peaks north of Mount Whitney have roof pendants.

Many metamorphic outcrops are composed of metasediments (altered sedimentary rock). Sandstones become quartzite, limestones become marble, and shales become slate and schist. These metasediments are of Paleozoic age. They are visible as deep red or chocolate brown rocks on top of grayish granitics. On the east side of the Sierra from Bishop northward, nearly to Bridgeport, these red and brown rocks are conspicuous (figure 4.6). Entering the high country by any of the canyons in this portion of the Sierra, a person cannot help but be struck by the marvel of colors and folding in the rocks. Mount Banner, Mount Ritter, and the Minarets, near Mammoth, are composed of these old metamorphic rocks, too.

Metavolcanics are also included in this prebatholithic assemblage, although these are considerably younger than the metasediments. The metavolcanics are of Mesozoic age and may have been established in place by magma coming to the surface explosively, or they may have been island-arc volcanics scraped off the ocean floor as subduction proceeded to consume the Pacific plate. The reddish rocks of Mount Dana, Mount Gibbs, and the Ritter Range in eastern Yosemite are chiefly metavolcanic rocks of explosive origin. Roof pendants composed of volcanics are often black. The Kaweah Peaks Ridge (figure 4.7) is a series of metavolcanics located on the western divide in Sequoia

FIGURE 4.5 View of the east side of the Sierra Nevada from Sherwin Grade north of Bishop. Note the metamorphic roof pendant on Mount Tom at left center, and the staggered fault scarps on the eastern face of the range to right of center.

FIGURE 4.6 Folded metamorphic rock along Convict Creek, Inyo National Forest.

National Park. It is visible from all over the southern Sierra as a ridge of dark-colored peaks.

The largest sequence of metasediments and metavolcanics occurs along the western foothills on the northern flank of the range. These long belts of metamorphic rock are oldest to the east and become progressively younger westward. Interbedded with these Paleozoic meta-

sediments are slices of Mesozoic rocks similar to those known as the Franciscan Formation in the Coast Ranges. Conspicuous among the Franciscan rocks are the serpentines and sandstones that form specialized soils with specialized plants. It is assumed that these metasediments and Franciscan rocks were deposited by accretion against the western edge of the North American continent as the sea floor slipped beneath. Many of these ancient sediments became folded and compressed as they were layered onto the western margin of the continent. In the foothills east of Merced is a "tombstone pastureland," where thin layers of metavolcanic rock stand vertically amid grasses and livestock.

Important mineral-bearing rocks are found at the point of contact between the metamorphic rocks and the granites. Where sedimentary rocks are intruded by large masses of granite, deposits of metals such as gold and platinum may be present. At the time of intrusion, hot water driven to the surface by the heat of magma carried gold and quartz in solution. Where the silica solutions cooled in cracks of the overlying rocks, veins formed, and erosion carried minerals from these deposits out to the

FIGURE 4.7 Metavolcanic rocks of Kaweah Peaks Ridge from Little Five Lakes, Sequoia National Park.

valley floor. In 1848, it was in the American River near Sutter's Mill that gold was discovered in the gravel, initiating the Gold Rush. Miners moved into the foothills and dug out the quartz veins hoping to find the source of the gold that had washed out into the rivers. This dream discovery, known as the "mother lode," was never realized, nor was it likely. The action of rivers carrying away light materials concentrates heavy minerals such as gold in the gravel left behind. Searching through gravel is known as placer mining, and this technique has proved to be the most fruitful. A drive on Highway 49 along the foothills through the mother lode country is a drive back into history. Along the way, town names such as Placerville bear witness to the days of the forty-niners.

Although gold is the most valuable metal that has been mined in the Sierra Nevada, other metals have accumulated along these contact zones, and their occurrence is not limited to the western foothills. In 1916 on the east side of the Sierra, north of Bishop, scheelite, a tungsten mineral, was discovered in a similar contact zone between the intrusive granite and the old metamorphic rocks. This led to construction of one of the largest tungsten-processing mills in the United States. This mill, in Pine Creek, a few miles northwest of Bishop, got its ore from a series of mines located between 9500 and 12,000 ft (2800 and 3650 m) near the headwaters of Pine Creek. This mill produced tungsten for 70 years, and it was large enough that a layoff of personnel in the late 1980s influenced the economy of Bishop. There are also many old tungsten mines in the low range known as the Tungsten Hills just north of Bishop Creek.

Postbatholithic rocks of the Sierra are sedimentary and volcanic. Sediments washed westward as the proto-Sierra was eroded, and this sedimentation has continued to the present day. About 20 million years ago, there began a major period of volcanism that spilled lava over the top of the northern Sierra Nevada and filled valleys of the undulating lowland. Volcanic rocks of the Modoc Plateau were formed at that time, too. Most of that volcanic material has since eroded away from the Sierra Nevada, but a few peaks north of Yosemite, such as Mount Rose, Squaw Peak, Sonora Peak, and Dardanelles Cone, are remnants of that period. The uplift of the Sierra and accumulation of sediment in the Great Central Valley during that period pushed back the sea on the west.

Recent volcanism of the Sierra occurred along the eastern flank, where lava flows and cinder cones dot the landscape. A major volcanic field lies in the area from Mammoth Mountain to Mono Lake (plate 6A). Major eruptions in this area have occurred in the last million years, and the activity along the Sierra Nevada fault system during this time caused a major portion of the uplift of the Sierra Nevada range.

A long, steep grade on Highway 395 north of Bishop is known as Sherwin Grade. It progresses up a large deposit of rhyolite and volcanic ash known as the Bishop tuff. The tuff is represented by 30–40 mi³ (126–168 km³) of material that resulted from a massive volcanic eruption about 700,000 years ago. Ash from this eruption is found in sediments as far east as Kansas and Nebraska. This eruption formed a huge crater that covered about 75 mi² (195 km²). That crater today, known as the Long Valley caldera, extends from about the summit of Sherwin Grade, near Tom's Place, to Glass Mountain Ridge, east of Mammoth Lakes, a town that lies along the southwestern edge of the caldera. This valley is a collapsed block or caldera that was formed by subsidence when the material that became the Bishop tuff was forcibly evacuated from beneath. During glacial times, this caldera was filled by a huge lake about 300 ft (90 m) deep. Today in the bottom of the caldera is Lake Crowley, a reservoir on the Los Angeles Aqueduct. The Long Valley caldera represents the headwaters of the Owens River. The river formerly cut a gorge in the rim of the caldera to drain south, ultimately all the way to China Lake, near the present town of Ridgecrest in the Mojave Desert.

Mammoth Mountain is a quiescent volcano that formed about 370,000 years ago. It is 11,053 ft (3370 m) high. Recent research has determined that it is a pile of 25 overlapping lava domes. From the top of Mammoth Mountain, a person looking toward the northeast can see a chain of volcanic peaks that lies in a straight line for about 40 mi (64 km). Beneath these features is a major fault or fissure that is part of the Sierra Nevada system. Wilson Butte

FIGURE 4.8 Columnar basalt smoothed off by a glacier, Devil's Postpile National Monument, Mono County.

(Plate 6A) and Obsidian Dome lie along this line, as do the Inyo and Mono Craters. Ages of the Craters range from 6500 to 600 years, making this chain the youngest range in the continental United States. The islands in Mono Lake are the northernmost points of this volcanic system. Steam eruptions were recorded in Mono Lake as recently as the early 1900s.

Less than 100,000 years ago, a basalt flow occurred southwest of Mammoth Mountain. As the basalt cooled, it cracked into long hexagonal columns. A glacier smoothed off the top, and the result is what is known as Devil's Postpile (figure 4.8). It is now managed by the National Park Service.

Recent earthquakes in the Mammoth area indicate that this area is still active. In May 1980, four earthquakes within 48 hr, registering about 6 on the Richter scale, jolted Mammoth Lakes. Tremors of lesser magnitude continued through the summer of 1983. Geologists studying the area since 1975 have found that in the 2 year

FIGURE 4.9 Hot Creek, Mono County. An area of hot springs east of Mammoth Mountain.

period following the 1980 earthquakes, the magma pool beneath the Mammoth area rose from a depth of 6 mi (10 km) to about 2 mi (3.2 km). Steam vents and hot springs became realigned. It also was discovered that the terrain within the Long Valley caldera northeast of Mammoth Lakes had risen about 12 in (30 cm). In 1982, the US Geologic Survey issued a "notice of potential hazard" to warn people of the imminent threat of eruption. The Mammoth Lakes-Long Valley area became the greatest threat of volcanic activity in the lower 48 states.

The volcano notice sent the local chamber of commerce into a frenzy. They predicted that skiers and other tourists would stay away in droves, and they feared that all the newly built condominiums would lie vacant and unsold. The chamber of commerce was critical of the US Geologic Survey, who pointed out that they had accurately predicted the Mount Saint Helens eruptions, and that it would be criminal to keep such a threat a secret for the sake of business. Instruments in the area will continue to monitor potential threat, in the hope that if an eruption appears certain, there will be adequate time for evacuation.

The presence of active volcanism on the east side of the Sierra Nevada has spawned a controversy of a different kind. In association with the

effort to establish alternate forms of generating electricity, a number of geothermal facilities have been proposed. The largest, associated with Coso Hot Springs in Inyo County, was begun in 1987. As the reservoir of heated water began to decline, the Coso facility now pipes freshwater from a well in Rose Valley west of Coso Junction. Because that area is not frequented by large numbers of tourists, and because the project generates tax dollars for Inyo County, it has been the source of little controversy. On the other hand, similar facilities at Casa Diablo Hot Springs in the area east of Mammoth Lakes have been the subjects of heated debate. First built in 1985 and expanded in 1990, there are more plans for expansion. In particular, transporting hot water from Hot Creek (figure 4.9) and drilling several new wells have been proposed. Also proposed are plans for pumping hot water to the town for the purpose of heating buildings and construction of a Geothermal Interpretive Center and Living Laboratory. Locals fear contamination of their freshwater aquifer by the brine from the geothermal system, and potential release of hydrogen sulfide gas into the air. The issue is one of tourism versus industry, with possible implications for the economy of Mono County and the city of Mammoth Lakes.

FIGURE 4.10 Granitic landforms. View eastward from Glacier Point, Yosemite National Park. Note half dome and glacier carved valleys.

The appearance of the Sierra today is largely a function of geologic processes. It is unique among major mountain ranges because it is composed almost entirely of granitics. It is also geologically young and therefore characterized by steep, jagged landforms. Domes are also a typical feature of granitic landforms.

Certain characteristics of granite influence weathering processes. Granite is very hard, and it weathers slowly. But fissures develop in granite along points of structural weakness. These fissures, known as joints, tend to run parallel to the north-south axis of the range and also perpendicular to that axis. Granite thus has a tendency to break up into large cubes. Furthermore, granite appears to expand from the inside, which causes domes to form, within which there are more horizontal fractures. The cause of this expansion is not fully understood, but the present theory is that as the overlying sedimentary rock was peeled away by erosion, it relieved pressure from the underlying granite. Loss of this overburden allowed the granite to expand from within, causing it to crack in a spherical or onion-peel pattern. Intersections of joints and horizontal cracks have caused the granite to break into large cubes. Freezing of water or growth of roots, forces of mechanical

weathering, expands these cracks. Gravity and other erosive forces cause portions of each of these layers to fall away in a process known as exfoliation. These processes—expansion, cracking, weathering, and erosion—have caused the dome shape so characteristic of granitic landforms (figure 4.10).

Tree roots and frost wedging are not the only sources of mechanical weathering. Fires, by causing expansion and contraction, also break up rocks. In arid areas, growth of salt crystals in cracks can hasten the breakup of rock materials. In general, these processes work more rapidly on sedimentary and metamorphic rocks.

Chemical weathering is more subtle than mechanical weathering. The dark color on the surface of granitics in desert areas is a consequence of the oxidation of metals such as iron and manganese. Water dissolves some minerals and leaves others behind. In granite, the feldspars tend to be more water soluble; the quartz minerals have low solubility in water. Water dissolves, forming grooves and creases on the surface. If subterranean water saturates a layer of limestone, it can form caverns. Large caverns such as those near Carlsbad, New Mexico, were formed in this way. The western slope of the Sierra Nevada has a few limestone outcrops that contain caverns, including Boyden Cave, in Kings Canyon, and Crystal Cave, in Sequoia National Park.

Another form of chemical weathering is the action of weak acids, such as carbonic acid (H_2CO_3). This can produce small pits on flat granite surfaces, in which organic debris and water may accumulate. As bacteria decompose the organic material, they excrete carbon dioxide, as do all heterotrophic organisms. The carbon dioxide combines with the water to produce carbonic acid. Over millions of years, the pits become larger as this weak acid wears away at the rock. The tops of many Sierra peaks are flattened, and on these flat surfaces are many such pits, most of which are fist-sized or slightly larger. Such pits occurring along streams were used for thousands of years as ready-made *morteros*. Using a rounded stone, Native Americans would grind acorns and buckeyes to a mush in these pits.

Another way in which carbonic acid is produced is by the fungus component of lichens. Lichens growing on the surface of rocks are called crustose lichens because they seem to form a colorful crust on the rocks. These lichens adhere to the surface by microscopic fungal filaments, which excrete carbon dioxide (the fungus is a heterotrophic organism). The weak carbonic acid formed helps to break up the rock, and it also provides small cracks and fissures to aid attachment of the lichen to the rock surface.

Mechanical and chemical weathering break up the rocks into smaller and smaller particles. Erosion carries away the rock fragments and exposes new surfaces to weathering. The forces of erosion include gravity, wind, water, and glaciers. They transport rock materials from their source and deposit them somewhere else.

Gravity is an important force on all steep slopes. Rock falls, landslides, and avalanches are gravity driven as are streams, waterfalls, and glaciers. The rubble at the base of a cliff, known as talus, is the result of weathering processes, which loosen the materials on the cliff face, and gravity, which does the rest.

Wind is an important factor in the alpine zone. Strong winds loaded with sediment produce the sand-blasted look on the sides of high-elevation trees. In the winter, the wind picks up ice particles, which have the same effect. This abrasion not only works on vegetation, but on rock surfaces as well. Hollows on the sides of otherwise smooth rock can be the product of this abrasion, although it is less likely to be seen on high-country granite than it is in the desert, and it is less likely to be seen in granite than in softer materials such as sandstone or limestone.

Water is the primary erosive force in the Sierra today. It carries away rock materials before they have a chance to show the strong color change associated with surface oxidation. The granitics of the high country are blue gray, but they are of the same chemical composition as those in the Alabama Hills near Lone Pine (figure 4.11). There, in the rain shadow of the Sierra, granitics are coarse and brown on the surface, showing the effects of thousands of years of chemical weathering by oxidation, with limited influence by water or snow.

The granitics of the Alabama Hills are noteworthy for other reasons as well. The rounded boulders are a product of chemical weathering of the large cubes produced by block jointing. Because there is proportionately more surface at the corners of a cube than there is in the center of each side, oxidation works on the corners more rapidly. Over the course of thousands of years, the cubes become rounded.

It is also interesting that the Alabama Hills lie on a small down-dropped block located between two parts of the Sierra Nevada fault system. Most of the Sierra uplift has taken place to the west of the Alabama Hills, but the 1872 earthquake took place on the eastern side. Nevertheless, the crest of the Sierra and the top of the Alabama Hills were formerly at the same elevation. The Alabama Hills and the rest of the Owens Valley have been dropping, while the mountains on both sides have been rising. This type of valley is called a graben, which means "grave" in German.

FIGURE 4.11 Granitic landforms in the Alabama Hills near Lone Pine, Inyo County. Mount Whitney is in the background. Note that Sierran granite in the background has a lighter color due to weathering by ice and water rather than oxidation by sunlight as in the foreground.

Another area of scenic oxidized boulders occurs on the eastern side of the Sierra west of Bishop. Here, along the lower reaches of Bishop Creek are large outcrops of block jointed granitic rocks that have the brownish coating produced by chemical weathering. This is a popular area for rock climbers.

The major force responsible for the appearance of the Sierra today is glaciation. It has carved and shaped the terrain and removed most of the soil. During the Pleistocene, the Ice Age, there were many periods when glaciers covered large portions of the high country. Four of these major intervals correspond to periods of continental glaciation in the Midwest. The oldest episode was about 3 million years ago. Other major episodes can be dated to 1 million, 75,000, 20,000, and 4,000 years ago. Also, a short period of glacial advance, a little ice age, was recorded about 600 years ago. The present 60–70 glaciers of the Sierra Nevada date back to that time. The Palisade Glacier near Big Pine (figure 4.12) is the largest of these glaciers today, only about 1 mi² (2.6 km²) in surface area. In association with global warming, research today indicates that Sierra glaciers

have retreated 30–70% in the last century. The only glacier in California that seems to be expanding is on Mount Shasta, which will be discussed in Chapter 6 on the Pacific Northwest.

A glacier (figure 4.13) occurs when snow accumulates faster than it melts. As the snowpack becomes thicker, it becomes compressed into ice by its own weight. The ice then flows slowly downhill, often in a canyon, to a point at a lower elevation where the ice melts. If snowfall at the upper end is greater than the melt at the lower end, the glacier advances. If snowfall is less than the amount of melt, the front retreats. Rocks that fall on the surface of the ice gradually sink to the bottom of the glacier. As the glacier slides downhill, friction melts the ice on the lower surface. Water on the underside helps to lubricate the flow, and the rocks trapped at its lower surface grind against the bedrock. This grinding shapes the canyon into a U-shaped trough, unlike canyons cut by water, which are V-shaped. The grinding of rock on rock also produces a fine powder that remains suspended in the water flowing from the toe of the glacier. Rivers and lakes fed by

FIGURE 4.12 Palisade Glacier, the largest in the Sierra Nevada, and Fourth Lake, Big Pine Creek, Inyo National Forest.

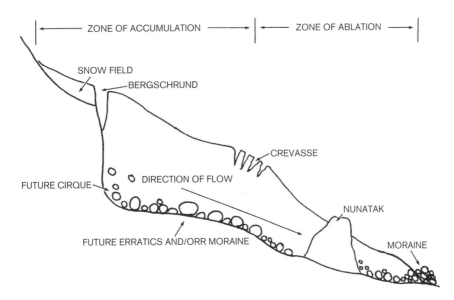

FIGURE 4.13 Diagrammatic representation of a valley glacier.

FIGURE 4.14 Glacial erratics at Olmstead Point, Yosemite National Park. Note half dome in the background.

glacier melt have an azure color due to this suspended rock material.

When a glacier retreats, it leaves behind a ridge of rubble known as a moraine. The moraine left at the toe of the glacier is a terminal moraine, and those left along the sides of the glacier are lateral moraines. Sometimes, isolated boulders are left stranded upon a surface that has been smoothed by a glacier. Boulders of different composition from the rock upon which they lie are known as erratics (figure 4.14). The enormous size of some erratics bears witness to the erosive power of glaciers.

The upper end of a glacier freezes into cracks and fissures, a process that causes more rocks to break off. Sliding of the glacier away from its upper wall of rock carries off the quarried material, making room for more ice. This process causes the glacier to "eat" its way into the rock at its upper end, and the result is a basin with a circular cliff wall. This type of basin, resembling an amphitheater, is known as a cirque (figure 4.15). Many lakes and meadows in the Sierra today lie in such basins. Often these cirque lakes occur in a series, one above the other in a canyon. These series are known as paternosters (plate 17B). Paternoster means "Lord's prayer"; the term refers to a string of lakes as if they were a string of beads in a rosary.

FIGURE 4.15 A cirque lake in Fourth Recess, upper Mono Creek, Sierra National Forest.

The mechanism by which paternosters form is subject to some debate. One theory is based on the fact that all rock is not of equal hardness. As the glacier slides downhill in the canyon, it cuts more deeply in some areas than in others,

FIGURE 4.16 Glacial landforms: horn silhouetted on the skyline, and an arête left of center near Lost Lake, Kings Canyon National Park.

and the low spots become the cirques. Another theory implicates waves in the ice. A third theory assumes that a glacier forms only one cirque at a time and that a series of cirques is formed by different glaciers, each occurring during a different glacial episode. Uplift of the mountain range between glacial episodes carries the older cirques upward, so the lowest cirque would be the youngest. Two bits of evidence support this theory. First, there are often four or five lakes in a Sierra paternoster, each presumably corresponding to a major glacial episode. Second, glaciers form best at a single elevation, at the zone of heaviest precipitation. The tops of the high peaks in the Sierra were never glaciated. They emerged above the glacial ice, like islands.

Large glacial systems flow down canyons like rivers of ice and are joined by ice from side canyons or tributaries. Since these side canyons have less ice in them, the main glacier cuts into the rock to a depth that is below the side canyons. Ice from the side therefore tends to flow out on top of the larger mass. After the ice melts, a deep canyon remains where the major ice flow was, but the side canyons, which were not cut as deeply, appear as valleys high up on the wall of the main canyon. These side tributaries are known as hanging valleys. When hanging valleys contain flowing water, they dump into the larger canyon as waterfalls. The high waterfalls of Yosemite Valley are examples of this phenomenon. The Yosemite Falls plunge in two steps from a hanging valley. The total drop is 2565 ft (782 m), making it one of the highest waterfalls in the world.

The process of a glacier eating backward into the rock can produce other landforms besides cirques. If two glaciers work toward each other back to back, the result could be a high, thin ridge known as an arête. An arête can lie between two cirques, or it may be a ridge formed by two glaciers that flowed parallel to each other. If two back-to-back glaciers eroding headward merge across a mountain divide, they will carve a low saddle known as a col. The major east-west passes across the Sierra, such as Donner and Tioga, were formed by this process. When three glaciers move toward each other, they can leave a pyramid-shaped peak known as a horn. The Matterhorn of the European Alps is the most famous such horn. In the Sierra, the Palisades, the Minarets, and the Kaweahs are large arêtes. In the Yosemite area, Clark Peak, Cathedral Peak, and Unicorn Peak are horns. Figure 4.16 shows a single location featuring both horns and an arête.

The surface over which a glacier slid also shows distinctive features. Grooves and striations show where rocks were dragged. Some surfaces have been polished to a high shine, and this glacial polish can be dangerous to walk on when wet. Walking over polished surfaces should never be attempted next to a cliff, a rapids, or a waterfall.

Maximum glaciation occurred in the Sierra about 60,000 years ago. An ice cap about 275 mi (445 km) long and 40 mi (64 km) wide stretched across the high country, with tall peaks protruding above the ice. The longest glaciers flowed through what are now Yosemite and Hetch Hetchy Valleys. The glacier of Hetch Hetchy followed the course of the Tuolumne River for a distance of 60 mi (100 km). At its thickest, it filled a 4000 ft (1300 m) gorge with ice. On the east side of the Sierra, the glaciers were much shorter due to steeper terrain and less precipitation. The large moraines from these glaciers are found at the mouths of numerous canyons from Big Pine northward. Some of these moraines are hundreds of feet high.

John Muir first entered the Sierra Nevada in the 1860s, and he shocked the geologists of the time by suggesting that the landforms of the Sierra were the result of repeated glaciations. Josiah Whitney, state geologist at the time, for whom Mount Whitney was named, postulated that Yosemite Valley was a graben, a down-dropped terrain between two faults. It was many years before Muir's ideas were verified. Yosemite Valley, John Muir's home, is the ultimate glacial landscape. The floor of Yosemite Valley is not U-shaped today because it is filled with sediment, but other glacial features, such as hanging valleys, cirques, moraines, and erratics, are all evident. Yosemite Valley is estimated to have about 2000 ft (650 m) of sediment in its bottom. That means that the glacier there about 60,000 years ago was at least 4000 ft (1300 m) deep.

At the valley's eastern end, Half Dome and Cloud's Rest stood above the glacial ice. Half Dome (see figures 4.10 and 4.14) was once a typical dome, but its northern half was probably heavily jointed, and the ice flowing down Tenaya Creek carried away the rocks of the dome's northern half. Tenaya Creek joins the Merced River at the eastern end of the valley. At the juncture is a medial moraine that still separates the flow of the two rivers for over a half mile (1 km).

Largely because of its glacier-scoured beauty, Yosemite was established as a state park in 1866. It was not until 1890 that it and Sequoia were established as national parks. With the addition of Kings Canyon National Park in 1940, the largest region of glaciated mountain terrain in the lower 48 states was set aside to remain unspoiled forever.

The southernmost glacier in the Sierra flowed southward along the Kern Canyon. The Kern River today flows nearly due south for about 50 mi (80 km), following the course of a right-lateral strike-slip fault that is the longest fault in the Sierra. Offset on this fault up to 8 mi (13 km) has been measured. The fragmented nature of the rock in the fault zone represents a "soft" spot along which the ice could gouge and water could cause erosion. At its upper end, the Kern Gorge is a long, straight, U-shaped valley (figure 4.17). At its lower end, south of the influence of glaciation, it is V-shaped. The gorge and its adjacent plateaus bisect the Sierra high country into two long ridges known as the western divide, at 13,000 ft (3900 m), and the main divide, at 14,000 ft (4230 m).

The plateaus on each side of the Kern Gorge are believed to be erosion surfaces that predate the glaciers. The oldest surfaces are represented by slanting, flat peaks such as Mount Whitney and Mount Langley, both of which are higher than 14,000 ft (4230 m). These surfaces were eroded by events over 12 million years ago that predate most of the Sierra uplift. Alternating phases of uplift and erosion formed other table-lands at later intervals. These plateaus were presumably formed at successively lower elevations as cycles of erosive forces stripped away the rock in between periods of uplift. South of Mount Langley is Cirque Peak at 12,863 ft (3828 m), which could represent the next oldest

FIGURE 4.17 View southward along Kern Canyon, a glacial carved U-shaped valley, Sequoia National Park.

period of erosion, the next period of quiescence. Flat-topped Diamond Mesa, near Forester Pass, probably formed at the same time. Following this, uplift occurred again, followed by another period of quiescence during which the Boreal Plateau, near Rocky Basin, was formed. This plateau, at 11,000 ft (3300 m), roughly matches a plateau on the west side of the Kern Gorge near the Kaweahs. Other plateaus occur at the same elevation all around the Kern Basin. To the north, these include Siberian Flat, Guyot Flat, Sandy Meadow, and Bighorn Plateau, landmarks along the Pacific Crest Trail.

Uplift then took place again, perhaps amounting to a gain in elevation of 2500 ft (760 m). The following erosional period formed the 9000 ft (2750 m) Chagoopa Plateau, on the western side of the Kern Gorge. The Kern Basin lava flow partially covers this plateau. These volcanics have been dated at 3.5 million years of age, making the Chagoopa surface Pliocene in age. Following this, the glacial episodes of the Pleistocene, coincident with the bulk of the Sierran uplift, raised the mountains to their present elevation. Glaciers sliding along the Kern fault cut the U-shaped Kern Gorge a full 2000 ft (600 m) deeper into the surface of the Chagoopa Plateau. From the

west, a small glacier cut a 1000 ft (300 m) gorge known as Big Arroyo. At the confluence of Big Arroyo and the Kern Gorge, there is a spectacular hanging valley and waterfall.

BIOTIC ZONATION

The distribution of communities in the Sierra Nevada (figure 4.18) is influenced by elevation, latitude, rain-shadow effect, and slope effect. Changes in environment that are associated with elevation are superimposed upon the differences in latitude between the northern and southern parts of the Sierra. As one moves to the north, precipitation increases and temperature decreases. The trend, therefore, is for biotic zones to be displaced downward as a person goes northward. Upper timberline is about 1000 ft (300 m) lower at the northern end of the Sierra than at the southern end.

Because the eastern slopes are drier than the west, the result is that zones tend to occur at higher elevations on the east side of the range. The cooler temperatures at higher elevations reduce evaporation; therefore, the water goes farther. Near the crest of the Sierra, east-west relationships are reduced because precipitation and temperature are similar. The effect on east-

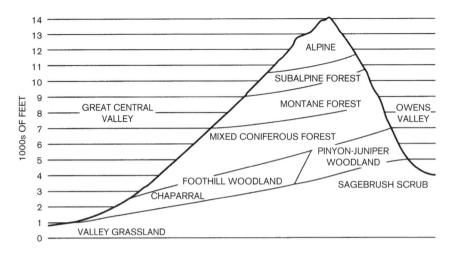

FIGURE 4.18 Biotic zonation of the central Sierra Nevada. Corresponding zones are elevated toward the south and on the east side of the Sierra Nevada.

side biotic zonation is that each zone is displaced to a higher elevation and also confined to a narrower belt of distribution.

Slope effect is more pronounced at lower elevations, and also toward the south because that is where the growing season is limited by summer drought. On the west side in the vicinity of Three Rivers, at about 3000 ft (1000 m) elevation, slope effect is pronounced. Drought-tolerant scrub vegetation appears brown on south-facing slopes during the summer, and oaks appear green on north-facing slopes. The result is a conspicuous color mosaic illustrating slope effect. On the southeastern side of the Sierra, in the vicinity of Olancha, a similar color mosaic is obvious at about 9000 ft (3000 m) in elevation. North-facing slopes are covered by Singleleaf Pinyon Pines (*Pinus monophylla*), drought-tolerant evergreen trees. South-facing slopes are covered by desert vegetation, blue-gray Great Basin Sagebrush (*Artemisia tridentata*). Principal trees and shrubs of the Sierra Nevada biotic zones are depicted in figure 4.19.

Valley Grassland

The Valley Grassland community, which will be discussed in more detail in the chapter on the Great Central Valley, is a highly disturbed ecosystem characterized by introduced Mediterranean grasses and weeds. Of primary significance to the Sierra Nevada is that these introduced species have invaded the Foothill Woodland as well.

Grassland is a community associated with deep alluvial soils. Sands and gravels thousands of feet thick have washed out of the mountains and lie in the valley. Water percolates deeply, soils dry out, and these plants tend to die out every summer. They resprout from seed or dormant root systems with each new surge of winter precipitation. Approaching the perimeter of the Great Central Valley, soils become thinner, and water is closer to the surface. At this point, deep-rooted oak trees become scattered among the grasses. This mixture of trees and grasses often is referred to as a savannah. This is where the Valley Grassland meshes with the Foothill Woodland.

Foothill Communities

Foothill Woodlands

The Foothill Woodland surrounds the Great Central Valley, occurring in the Sierra foothills and the Coast Ranges. Also known as Oak

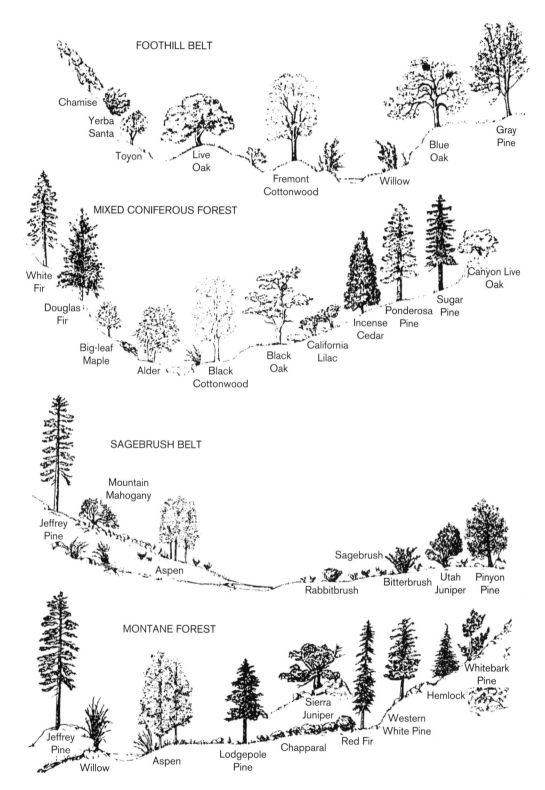

FIGURE 4.19 Principal trees and shrubs of the Sierra Nevada biotic zones (Storer, T. I., and R. L. Usinger. 1963. *Sierra Nevada Natural History*. Berkeley: University of California Press).

FIGURE 4.20 Blue Oak, *Quercus douglasii*, the dominant species of oak in the Foothill Woodland of the Eastern Sierra Nevada and Coast Ranges.

Woodland, this is one of California's most widely distributed communities. Excluding brushy habitat such as Chaparral, Oak Woodland covers about 10 million acres (4,000,000 ha) of California's foothill or low mountain terrain. Foothill Woodland is composed of a mixture of drought-tolerant species that are truly Californian in distribution. The two predominant trees found throughout this belt are Gray Pine, *Pinus sabiniana*, and Blue Oak, *Quercus douglasii* (figure 4.20). The discussion of pines and oaks in California may begin here, for no two groups of conspicuous woody plants more clearly illustrate the diversity associated with the California Floristic Province. Although the total number of species and/or subspecies varies with the person doing the classifying, there are at least 28 kinds of pines and 36 kinds of oaks in California. The greater portion of these taxa is truly Californian in distribution, and many occur nowhere else. Oaks and pines will be mentioned as indicator species as each biogeographic province and its communities are discussed. Pines are more widely distributed in mountain communities, but oaks are more common in communities associated with summer drought. Because many oaks and pines are relatively easy to identify, and because they are conspicuous, attention will be directed to their appearance and characteristics of their distribution.

The first pine encountered by a person driving into the Sierra Nevada from the Great Central Valley is the Gray Pine. Formerly known as Digger Pine, it also is known as Ghost Pine or Foothill Pine. Authorities objected to the name "Digger" because that was a derogatory term used for the Miwok Indians, who ate the seeds of these pines. These Indians also ate various bulbs and tubers they dug from the soils; hence, they became known as "Digger Indians." The name Gray Pine refers to its blue-gray foliage.

The Gray Pine is characterized by long needles that occur in clumps (fascicles) of three. They are 7–13 in (18–33 cm) in length. The grayish foliage is sparse enough that the tree is nearly transparent. One early biologist exclaimed, "Behold the Digger Pine, its foliage is so thin, one can see right through it." Because the trees seldom grow in a crowded forest situation, the crown is seldom composed of a single trunk from which other branches grow outward. Rather, the crown tends to be rounded, and commonly there are several main branches growing upward. Cones are quite large, ranging from softball to volleyball size. Scales making up the base of the cone have large hooks on the outer edge. At the base of each of these scales are found large seeds, about the size of kidney beans. These are the seeds that were gathered by the Indians. In fact, these seeds, along with acorns from a variety of oaks, are an important food for many foothill animals.

An unexplained gap occurs in Gray Pine distribution. It should be an associate of Blue Oak all around the Great Central Valley. However, it is missing from the Kaweah drainage near Three Rivers, at the southern entrance to Sequoia National Park. Otherwise, components of Foothill Woodland in this region appear typical. On the other hand, on the southeastern part of the Sierra, in the vicinity of Owens Peak and Chimney Peak, Gray Pine is associated with Pinyon Pines in a more desertlike setting.

Blue Oak is a member of a subgroup known as white oaks. It is winter-deciduous. Its leaves are about 1–3 in long (2.5–7.5 cm), and the margin of the leaf is often composed of seven large lobes, but this is highly variable. Some leaves seem to have only undulations. These slow-growing trees mature to rugged, stately specimens. Data for primary production for blue oaks indicate carbon assimilation at about 700–1100 gC/m²/yr. This places their productivity at about the same as Chaparral. Their ability to tolerate summer drought is related to their ability to close stomates during periods of high evaporation and their large root systems, which continue to absorb water during drought. Apparently, evergreen oaks such as Coast Live Oaks and Interior Live Oaks are slightly less productive.

Acorns of Blue Oaks are among the most palatable of all the oaks. They are eaten by a variety of birds and mammals and formerly were a staple for Native Americans. Because Foothill Woodland contains lots of grass as an understory, the habitat has been used extensively for grazing of cattle and sheep. To protect livestock, a predator extermination program imposed by humans has caused an unnatural abundance of acorn-eating animals such as gophers, mice, ground squirrels, rabbits, and deer. The consequence of this imbalance is that throughout much of the Foothill Woodland, in the Sierra Nevada and the Coast Ranges, Blue Oaks are not regenerating. Acorns and seedlings are being eaten out of existence, which ultimately could cause Blue Oaks to disappear.

Plants that are fed upon heavily by animals seem to have developed chemicals (phytotoxins) that discourage herbivores. Tannins in acorns discourage herbivory by binding to protein molecules such as digestive enzymes. This action interferes significantly with chemical digestion in animals, such as cattle, that are unaccustomed to acorns as natural food. On the other hand, native foothill animals, such as squirrels and Mule Deer, have enzymes in their saliva that bind to the tannins before they are swallowed, enabling digestive enzymes farther along in the digestive tract to function correctly. Native Americans solved the problem by making a mush of the acorns and leaching out the water-soluble tannins with running water.

Other large oaks found in Foothill Woodland include Valley Oak, *Q. lobata* and Interior Live Oak, *Q. wislizeni*. Valley Oak is another white oak. It differs from Blue Oak by having larger, greener leaves, with 9–11 lobes, and long, narrow acorns. The largest of our western oaks, it is often the first oak to invade the grassland, but it requires deep soil. Its distribution therefore is localized. This stately oak was a major component of the savannahs and gallery forests that were formerly widespread on deep soils throughout the Great Central Valley. They fell victim to the damming of water courses and drying of lowland habitats, activities associated with agriculture. Large numbers of these trees also have been cut down for firewood and, to promote cattle and sheep grazing, with the mistaken notion that the amount of grass would be increased if oaks were removed. Recent research shows that certain species of grass grow taller, are more nutritious, and last longer into the dry season under the large canopy of the aged oaks.

Valley Oaks are also winter-deciduous. Like Blue Oaks, they drop their leaves in response to cold. Their huge root systems reach deeply to tap underground water supplies to enable them to remain active during the long, hot summer. Blue Oak is more drought tolerant than Valley Oak. Blue Oak forms savannahs, but it also occurs in association with Gray Pine on slopes, close together, in a community known as a woodland rather than a savannah.

Interior Live Oak retains leaves in the winter. It is an evergreen oak and its leaves are dark green on both sides and smooth and hairless on the underside. Leaf margins may be smooth or spine tipped. Interior Live Oak commonly occurs with Blue Oak and Gray Pine (plate 11B), but it tends to occupy rocky outcrops where its roots tap deeply into cracks and fissures that hold water. Interior Live Oak is more common at higher elevation, where it is responsible for the green canopy on north-facing slopes. Interior Live Oak is sometimes confused with Canyon Live Oak (Golden-cup Oak or Maul Oak), Q. chrysolepis, which grows primarily in canyons where the soil remains wet near the surface. Canyon Live Oak, the most widely distributed oak in the state, is more of a riparian or streamside species and occurs at middle elevations. The underside of its leaves is gray with golden fuzz.

At slightly higher elevations than Blue Oak, California Buckeyes, Aesculus californica and Western Redbuds, Cercis occidentalis, become common on north-facing slopes. Buckeyes are small, shrub-like trees, and although they do not resemble them, along with maples, they are in the in the Soapberry family (Sapindaceae, formerly Hippocastanaceae). Buckeyes are widely distributed in the Northern Hemisphere, but California Buckeyes are endemic, with the exception of one unexplained population in Oregon. Another species, the Ohio Buckeye, A. glabra, is the state tree of Ohio. California Buckeye are distinctive in the spring because of long, 4–6 in (10–15 cm) clusters of white flowers. Unique among foothill trees, they are drought-deciduous, dropping leaves in the summer. These trees are easily spotted in early summer, because as their leaves turn brown they appear to be dying. In late summer and fall, they are further distinctive because they bear large, pear-like seed pods on naked stems. From these pods come the buckeyes or "horse chestnuts," an important food for some animals. The name comes from the appearance of the large fruit when a thick husk cracks open revealing the glossy brown "bucks' eye" within. In addition, the fruits were formerly harvested by Native Americans, who treated them by leaching, as they did for acorns. Without leaching, they would be toxic to humans. In fact all parts of Buckeyes are toxic. In early spring, the foliage is edible, and again after leaves fall off the dead leaves are once again consumed by deer and ground squirrels. It also is interesting that the nectar is poisonous to non-native Honey Bees (Apis mellifera), whereas native bees are not affected.

At scattered localities in the foothills, there are Western Redbuds. They are particularly conspicuous along the Merced River west of Yosemite National Park. During late winter and early spring, these shrub-like trees are covered with small, reddish pink flowers. They are winter-deciduous, so there are no leaves on them when the flowers appear. The flowers are quite small and give the appearance that the tree is covered with small, reddish buds. Close inspection, however, reveals that the flower has the distinctive appearance of a member of the legume or pea family (Fabaceae). A pair of small, fused petals, the banner, projects upward as an attractant for pollinators. The other petals are variously modified to enclose the stamens and pistil, the reproductive parts. An insect pollinator must pry open the flower to expose these parts, at which time they spring upward to dust its lower abdomen with pollen; the insect at the same time transfers pollen from the previously visited flower to the pistil of this flower. Large, heart-shaped leaves remain on the tree through the summer, and long, brown beanpods persist in to the next winter. These beanpods are also an important source of food, particularly protein. Nitrogen-fixing bacteria inhabit the roots of these plants.

Oak woodlands also include a large component of ephemeral or annual plants that grow vigorously and bloom after winter rains (plates 11C and 14B). Some of California's most attractive wildflowers, such as Purple Owl's-clover, Castilleja exserta (plate 20H), and Harlequin Lupine, Lupinus stiversii (plate 20C), are found here also. Unfortunately, many of these are being eliminated by livestock grazing and the invasion of Mediterranean weeds.

FIGURE 4.21 Flannelbush, *Fremontodendron californicum.*

Foothill Chaparral

There are several kinds of scrub or shrub communities that go by the name of chaparral. These variations will be discussed in detail in the chapter on Cismontane Southern California, where scrub communities dominate. In the northern part of the Sierra Nevada, chaparral is an important component of the foothill community only on impoverished soils. South of Yosemite, however, it becomes increasingly abundant on south-facing slopes, which, due to heat and evaporation, have a shortened growing season. Most of the shrubs keep their leaves all year, but the leaves are variously adapted to resist water loss. Leaves are often small and coated with resinous or waxy secretions. They typically bear fine hairs, and they roll when they dry out. Many species bear berries. All species are fire adapted. Chamise, *Adenostoma fasciculatum* (figure 8.19), is the dominant species. Several species of California lilac, *Ceanothus* spp., occur here. Two varieties of Scrub Oak also occur here: California Scrub Oak, *Quercus berberidifolia* (figure 8.21) and a scrub form of Interior Live Oak, *Q. wislizeni.*

One of the most spectacular plants in Sierran Chaparral is Flannelbush, *Fremontodendron californicum* (figure 4.21), so named because its leaves are covered with a myriad of fine hairs. They are known also by the name of the former genus, Fremontia. In spring, these large shrubs produce showy yellow flowers in great abundance. Along the Kern River, in the southern Sierra, entire hillsides appear yellow because of the abundance of these shrubs. The shrub was named in honor of John C. Fremont, the explorer-army officer who became one of California's first politicians. The journal of the California Native Plant Society is called Fremontia in honor of the plant. A peculiarity of its distribution is that, in addition to its occurrence on the western side of the Sierra, it occurs as a component of Desert Chaparral on the north side of the Transverse Ranges.

Foothill Edaphic Communities

Edaphic communities occur where soil is the principal factor to which the community must adapt. The western foothills from about Yosemite northward include a belt of metamorphic and sedimentary rocks more typical of the Coast Ranges. Patches of soil, such as serpentines, with peculiar mineral composition and texture form ecologic islands in the Sierra Nevada and Coast Ranges upon which specialized plants have evolved (figures 7.10 and 7.11).

Included among the serpentine endemics are Interior Silk Tassel, *Garrya congdonii*, and a special scrub oak known as Leather Oak, *Quercus durata*, which differs from other scrub oaks in that the leaves are distinctly curled or convex. Often they are quite hairy on both surfaces as well. Knobcone Pine, *Pinus attenuata* (figures A and B, p. 348), is also typical of serpentine soils, although it is more common in the Coast Ranges. It is one of the closed-cone pines that require fire to open its cones. Its distribution in the northern Sierra is patchy, as it is along the coast, where it waters itself with fog drip.

Serpentine endemism is not restricted to plants. No one knows how many insect species are restricted to these areas, but at least nine species and subspecies of butterflies are associated with serpentine endemic plants.

Pine Hill in El Dorado County has a soil derived from gabbro, a dark plutonic rock. This soil is also high in magnesium. It is interesting to note that Leather Oak occurs here also. Several endemic species occur here. Notable among these are Pine Hill Flannelbush, *Fremontodendron decumbens*, and Pine Hill Ceanothus, *Ceanothus roderickii*. South of Lake Isabella, near the town of Bodfish is a region of gabbro soil on which grows Piute Cypress, *Hesperocyparis nevadensis*. Near the town of Ione, in Amador County, there is a localized patch of oxisol, a red clay laterite formed in the Eocene during a period of tropical climate. Among the peculiarities of the soil here is its acidic pH, less than 4.0. Ione Buckwheat, *Eriogonum apricum* var. *apricum*, and Ione Manzanita, *Arctostaphylos myrtifolia*, are found nowhere else in the world. The few other chaparral shrubs that occur in this region are dwarfed, indicative of the nutrient-poor nature of this soil.

Patches of limestone and marble also occur in the western Sierra. Limestone endemics are not as common in the Sierra as they are in desert ranges; one notable exception is on the east side of the Sierra along Convict Creek. Here, in a mixed situation of Desert Chaparral, Sagebrush Scrub, and Riparian vegetation, is a cluster of Rocky Mountain and Great Basin species that are found nowhere else in California. Included are a Bear-berry *A. uva-ursi*, a willow, *Salix brachycarpa*, and a Dwarf Bulrush, *Scirpus pumilus*. A lousewort, *Pedicularis crenulata*, is an herbaceous species that grows along the stream here and nowhere else in California.

Animals of the Foothills

The foothill community is drought adapted. Deep-rooted species such as oaks are able to remain active during seasons lacking in precipitation. Consequently, a greater mass of animals can be supported in a foothill community than might be expected based on raw climatic data. The bulk of these animals, as in the desert or scrub communities, are ectothermic and of small size. There are many kinds of insects and reptiles, whose low metabolic rates means that they require less food, and whose long periods of dormancy enable them to endure the long, hot summer or cold winter. Dormancy during a cold period is called hibernation, and dormancy during a warm or dry period is called estivation; however, there is little difference physiologically. The strategy is for the animal to lower its metabolic rate, which significantly reduces its requirement for food. It then lives on organic molecules, such as fat stored within its body.

The great variety of foothill insects and reptiles will be discussed thoroughly in the chapter on southern California communities. Likewise, a host of nocturnal, seed-eating rodents are associated with Foothill Woodland, and they will be discussed more thoroughly in the section on southern California chaparral communities. In this chapter, attention will be on conspicuous mammals and birds that cope with their harsh habitat through one of two basic strategies. Either they harvest resources when they are abundant, storing some for future periods of scarcity, or they change diet with the seasons, feeding on what is abundant.

Another concept to be reviewed here is that of niche partitioning. A niche refers to the role played by a particular species as it interacts with all the other species of an ecosystem. Ecologic and evolutionary theories predict that an ecologic niche can be occupied by only one species at a time. A broad ecologic niche occupied by two similar species may become divided or partitioned so the species avoid head-to-head competition for the same resources. An alternative to niche partitioning is extirpation of the species that is more poorly adapted. A combination of niche partitioning and extirpation leads to an assemblage of plants and animals best suited to the vagaries of climate and/or resources at a particular locality.

A consequence of niche partitioning, well illustrated by communities in the Sierra Nevada, is that as one moves up in elevation, an orderly replacement of similar species seems to take place (figure 4.22). This replacement is particularly evident in conspicuous families,

Community	Ground Squirrels	Tree Squirrels	Jay-Crow Family	Finch Family
Valley Grassland	California Ground Squirrel		Crow	House Finch
Foothill Woodland	California Ground Squirrel	Gray Squirrel	Yellow-billed Magpie (north)	House Finch
Yellow Pine Forest		Gray Squirrel	Steller's Jay	Purple Finch
Lodgepole-Red Fir Forest	Golden-mantled Ground Squirrel	Chickaree	Steller's Jay	Cassin's Finch
Subalpine	Golden-mantled Ground Squirrel	Chickaree	Clark's Nutcracker	Cassin's Finch
Alpine	Belding's Ground Squirrel			Gray-Crowned Rosy Finch
Pinyon Pine Woodland	White-tailed Antelope Squirrel		Pinyon Jay	House Finch
Great-Basin Sagebrush	White-tailed Antelope Squirrel		Raven, Black-billed Magpie	House Finch

FIGURE 4.22 Distribution of related species according to communities.

such as squirrels (Sciuridae). Among birds, members of the crow-jay family (Corvidae) and members of the finch-sparrow family (Fringillidae) also show this replacement. The effect is that similar ecologic niches at different elevations seem to be occupied by different but related forms. Thus, for a biotic zone, animals as well as conspicuous plants can be listed among indicator species.

A more subtle form of niche partitioning is illustrated by groups of similar species that inhabit the same community but utilize it in slightly different ways. Examples of this type of partitioning are found within a number of bird families, such as the wood warblers (Parulidae) and woodpeckers (Picidae).

BIRDS Foothill Woodland is the home for large variety of bird species for which oaks, in particular, are important for food and shelter. Daytime retreats for Great Horned Owls, *Bubo vir-ginianus* (figure 8.45), and California Quail, *Callipepla californica* (plate 19A), are provided by the canopy. Owls hunt for small rodents at night, and they rest in the trees in the daytime. Quail escape the heat of the day by remaining in the trees, emerging to search for seeds and insects when it cools off.

Cavities in oak trees are homes for a variety of animals. Western Gray Squirrels, *Sciurus griseus* (plate 18B), and owls use large cavities. Woodpeckers, Western Bluebirds, *Sialia mexicana* (plate 19D), and the White-breasted Nuthatch, *Sitta carolinensis*, use smaller holes for homes. The Western Bluebird is a characteristic bird of the foothills on a year-round basis, although they may move up or downslope with the seasons. They make their homes in the oaks, but they forage daily into adjacent communities for food. These small birds are blue on the back and have buff-colored sides. The iridescent blue of bluebirds depends on the angle at which the light strikes, a phenomenon that is shared with other blue-colored birds such as jays. Western Bluebirds are conspicuous on fence posts or high perches, where they wait for flying insects. When insects become scarce in autumn, they switch to eating berries from a variety of shrubs, but particularly those of Western Poison Oak (*Toxicodendron diversilobum*) and California Mistletoe (*Phoradendron serotinum tomentosum*) often found on oaks. An interesting behavior pattern is that wintering birds often take shelter in clumps of mistletoe, and the number of birds overwintering in some areas is proportionate to the number of mistletoe clumps. Western Bluebirds nest in cavities in oak trees, or they also will take up residence in nest boxes placed by humans. Females typically lay five eggs in the spring and incubation lasts about two weeks during which time her mate gathers insects and feeds her. Chicks usually fledge about a week later. Parents may have two or three broods each year. A male is kept busy feeding the fledglings while the female incubates a new clutch of eggs. An extended family of fledglings from multiple broods stays with the adults for the rest of the summer. Bluebird pairs remain together for many years. Nevertheless, it has been determined that about a fourth of the offspring in any nest are fathered by a different male, usually one from a nearby nest. In this way, outcrossing helps perpetuate genetic diversity. Also, in late summer young females leave the group to join other families, presumably locating future mates among the males of a new family. During winter, two parents with a dozen or so sons and foster daughters may stick together in flocks. In 1914, a huge flock of 125 Western Bluebirds, apparently eating mistletoe berries, was reported in Yosemite National Park.

The white-breasted nuthatch is a small, acrobatic bird that walks headfirst up and down tree trunks, even clinging upside down on lateral branches. It uses its long bill to snap up insects found on the surfaces of bark, a foraging technique known as gleaning. Nesting usually takes place in abandoned woodpecker holes in which males and females spend time enlarging and lining with moss, fur, or feathers. Clutches of five to eight eggs are incubated by the female, and the male continues to gather insects to feed her. Both parents share the duty of carrying out fecal envelopes during the time nestlings are developing.

Most birds of the foothills are omnivores. It is a disadvantage to be too specialized, because favored food is not always abundant. For example, acorns are not available year-round, nor are they common every year. When the acorn store has been exhausted, as in the summer, animals turn to other forms of food—particularly insects, which are the most abundant animals in the foothills.

Acorns are probably the most important food in the foothills. The problem with predation by a host of small mammals on acorns and seedlings of Blue Oak was mentioned earlier. Several bird species conspicuous in the foothills also depend heavily on acorns for food. Acorns are fed upon by Band-tailed Pigeons, *Patagioenas fasciata*; California Scrub-jays (Western Scrub-jays), *Aphelocoma californica* (plate 19F); and Acorn Woodpeckers, *Melanerpes formicivorus* (figure 7.29). Band-tailed pigeons and scrub jays feed heavily upon acorns when they are abundant, but they switch to other foods, such as berries and insects, when the acorns are gone. Scrub-jays and acorn woodpeckers store the acorns and eat them later in the season, when the source on trees is gone. Scrub-jays bury their acorns, and in so doing help to plant oak trees.

Acorn woodpeckers place acorns in holes they have drilled into dead or dying tree trunks. It is noteworthy that they also store the large seeds of gray pines in these holes. They return sometime later to eat the stored food. Spaniards referred to them as "carpintero" in reference to their woodworking activity, which can appear destructive. They put holes in cabin walls, and they can make a telephone pole look like Swiss cheese. The soft wood of a telephone pole is a favored place to store acorns, and utility companies have taken to painting them with a green-

FIGURE 4.23 Northern Flicker, *Colaptes auratus* (photo by Richard L. Bellomy).

colored woodpecker repellant, which works only temporarily.

The Acorn Woodpecker has been the subject of considerable behavioral research, which will be discussed in Chapter 7, on the Coast Ranges. They are conspicuous, gregarious birds that are easy to watch. They are dark with a white chin and a small red cap. When they fly, they display a conspicuous white rump and white patches in the wings. Typical of woodpeckers, they flap their wings a few times and then glide a short distance. This pattern makes them seem to move up and down as they fly. They also attract attention because they make considerable noise. Their call is an easily recognized "wack-a, wack-a."

Niche partitioning among similar species is particularly true among birds that feed upon the abundant insects in foothill areas. In particular, woodpeckers are specialized to avoid competition by partitioning food resources in space and time. Acorn Woodpeckers apparently do not eat insects that invade their stored acorns, but when the acorns are gone they behave as though they were flycatchers: they perch on a high branch and catch flying insects in midair. The Northern Flicker, *Colaptes auratus* (figure 4.23), is also a woodpecker. Western birds are also called Red-shafted Flickers. It prefers ants and termites, but it also eats berries. In agricultural areas, much to the dismay of farmers, Northern Flickers are known to hollow out fruit while it is still on the tree. Nuttall's Woodpecker, *Picoides nuttallii*, eats insects that bore in bark, avoiding surface insects, which are eaten by White-breasted Nuthatches. Downy Woodpeckers, *P. pubescens*, eat the same way as Nuttall's Woodpecker, but they prefer to feed on the bark of riparian trees such as alders and sycamores. The Lewis's Woodpecker, *M. lewis*, is a vagrant; it doesn't remain in an area long unless food is abundant. It eats acorns when they are available, but it seems to prefer feeding on beetle larvae in areas that have been burned. Sapsuckers (*Sphyrapicus* spp.) drill parallel holes in the bark of trees, usually in the riparian community. They return to feed on sap, soft wood, and small insects in the holes.

Genetic studies on woodpeckers have produced an interesting question. Why do several species pairs resemble each other? Nuttall's Woodpecker and the closely related Ladder-backed Woodpecker, *P. scalaris*, have horizontal bars across their backs, and generally they are similar in appearance, a condition that can be attributed to having a recent common ancestor.

On the other hand, the distantly related Downy Woodpecker and Hairy Woodpecker, *P. villosus*, also look very much alike. These look-alikes, although differing in size, share forested habitats. Other than a large white patch in the center of their backs, they are mostly solid colored. Lacking bars and spots has been considered an adaptation for living in a habitat in which the birds are often concealed. Since they are not closely related, the similarity is considered to be an example of convergent evolution.

Flying insects are an abundant food source, but mostly in the summer. The Black Phoebe, *Sayornis nigricans*, is a year-round resident of foothill communities. In the winter, it is one of the few species that continues to rely on flying insects. Acorn woodpeckers, which become flycatchers in the summer and fall, seem to prefer eating winged ants. In the summer, however, a number of migratory insect eaters appear on the scene in order to share the wealth. Several species of flycatchers, including the Western Kingbird, *Tyrannus verticalis*, hunt from a perch. Swallows and swifts catch insects high in the air by continuously flying. Nighthawks (*Chordeiles* spp.) are large-mouthed birds that appear at dusk, catching insects on the wing when other insect eaters are at rest. These wide-mouthed birds are in the Goatsucker family, Caprimulgidae. The name goatsucker comes from an erroneous notion that these birds might creep up on sleeping goats at night and suck on the nannies teats. A mythical creature in Latin American folk lore is also known as "chupacabra." For a similar reason, the Spanish translation is goat sucker, although the creature is supposed to suck blood from its victim. The peculiar, slow wingbeat and jerky flight of the nighthawks make them appear bat-like as they move among the insects in midair. Bats avoid the competition of nighthawks by feeding after dark and locating insects with sonar, inaudible squeaks that bounce off objects in the air. The acute hearing of bats enables them to locate insects and accurately judge distance by the length of time it takes for the sound to return to their ears.

In addition to the White-breasted Nuthatch, there are numerous other gleaners. Gleaners among the foliage include the Oak Titmouse, *Baeolophus inornatus*, a small, gray bird with a crest, which is inconspicuous in outer foliage. Large territories are defended by pairs of Oak Titmice. Because they are inconspicuous, they have a large repertoire of vocalizations with which they can communicate while remaining unseen. Other gleaners include bushtits, vireos, and warblers, each of which avoids competition by feeding in a different part of the foliage.

A number of migratory birds are conspicuous at different times of year. In winter, American Robins, *Turdus migratorius*, feed on worms and grubs in the soil. With their acute hearing, they apparently are able to hear the digging noises of the subterranean burrowers. In autumn, beautiful Cedar Waxwings, *Bombycilla cedrorum*, move through in large flocks, feeding on berries.

It may seem that for a "food-poor" ecosystem there is an inordinate number of animals sharing oak trees. The number, however, is minimal compared to an eastern deciduous forest, where there may be hundreds of species of birds sharing the canopy. In an eastern forest, there may be 10–20 species of warblers sharing the food resources of a single tree.

MAMMALS A conspicuous member of the Grassland and Foothill Woodland communities is the California Ground Squirrel, *Otospermophilus beecheyi* (plate 18A). Their natural range extends from coastal bluffs to the foothills. These grizzled-looking squirrels typically dwell in hillsides or grassy areas, where they build complicated burrows. Burrows may be 6 ft (2 m) deep and extend horizontally over 30 ft (10 m). The burrows include many dead ends and blind alleys that presumably are to confuse potential predators. The complex may include sleeping chambers, food storage chambers, and nurseries. Communal burrow systems may be the homes of several dozen squirrels from multiple generations. Other squirrels may live by themselves and many burrows may also form homes for other species, including salamanders, frogs, and Burrowing Owls. Where food is too scarce in summer, they estivate. In areas where it is cold, ground squirrels hibernate in winter.

In agricultural areas, humans have warred against ground squirrels, shooting and poisoning them. In cultivated areas, their numbers have become considerably reduced. Where squirrels are not preyed upon by humans, or where their natural predators are controlled, their numbers have increased, encouraged by an abundant seed crop associated with introduced Mediterranean weeds. In situations where poison baits are put out for ground squirrels, there is an unfortunate food chain effect where predators that eat the dead or dying squirrels also become poisoned.

Ground squirrels have moved to considerably higher elevations by invading road cuts, a vacant niche created by humans. In areas such as Yosemite Valley, where visitors insist on sharing their junk food, California Ground Squirrels have become tame and conspicuous. They make burrows along roads and trails and subsist on an unnatural diet of potato chips, corn chips, and cheese puffs. In their natural habitat, their diet consists of seeds, herbaceous vegetation, acorns, and buckeyes gathered from the ground. The California Ground Squirrel is one of the few animals known to be immune to the Buckeye toxin. The appetite for acorns, coupled with their increase in numbers, is considered to be one of the contributing factors leading to the decline in the number of oak seedlings.

In areas where oaks are close together, typically at upper elevations in the Foothill Woodland, Western Gray Squirrels, *Sciurus griseus* (plate 18B), harvest acorns by climbing in the trees. The Gray Squirrel is typically a tree squirrel. In the absence of Gray Squirrels, California Ground Squirrels are also known to climb trees and harvest acorns.

An entire community of animals is associated with California Ground Squirrels. These squirrels are preyed upon heavily by Red-tailed

FIGURE 4.24 Burrowing Owl, *Athene cunicularia*. These small, long-legged owls often inhabit abandoned ground squirrel burrows.

Hawks, *Buteo jamaicensis*, that swoop down upon them as they venture from their burrows. American Badgers, *Taxidea taxus*, dig them out from their burrows and also make homes out of enlarged, abandoned ground squirrel burrows. Another inhabitant of abandoned burrows is the Burrowing Owl, *Athene cunicularia* (figure 4.24), which raises its young in underground nests. These quail-sized owls decorate the burrow entrance with the dung of livestock, apparently to disguise their odor. In this way, the predators, including American Badgers, seem not to recognize the burrows as being occupied. An interesting instinct in juvenile Burrowing Owls helps to further protect them from potential predators. If an intruder is heard entering the burrow, young owls clack their bills together, which collectively makes a sound that resembles the rattling of a rattlesnake. Rattlesnakes do use these burrows for refuge, and they also enter burrows in search of food. A mammal usually makes a hasty retreat upon hearing the warning of the rattlesnake. Unfortunately Burrowing Owls have become victims of rodent poisoning and cultivation practices in agricultural areas and are now classified as a Species of Special Concern. Their primary habitat has become reserves such as the Carrizo Plain National Monument and ironically abandoned ground squirrel burrows near agricultural fields in the Central and Imperial Valleys.

California Ground Squirrels have a special relationship with Southern Pacific (Western) Rattlesnakes, *Crotalus viridis* (= *C. oreganus helleri*). After millions of years of coevolution, the squirrels are immune to the rattlesnake venom and the venom of the snakes has become stronger and stronger. Squirrels also have developed behavioral patterns that help confuse a threatening snake. One tactic is called "tail flagging." The squirrel wags its tail back and forth when faced with an approaching snake, which also could include a Gopher Snake, *Pituophis catenifer*. Because California Ground Squirrels are known to attack snakes, tail wagging acts as a warning. Sometimes several squirrels at a time will attack a snake by biting and scratching. Sometimes a squirrel will kick sand and gravel at the snake's face. Another interesting trick is called "tail flooding." While its tail is wagging back and forth, it becomes engorged with blood, which elevates the temperature of the tail. Somehow this is detected by the heat-sensitive pits on the upper jaw of the snake, either causing the snake to strike at the tail instead of the body or perhaps fooling the snake into thinking the squirrel is larger than it really is. Another tactic is for the squirrel to chew on discarded snakeskin and then lick its fur, and in so doing conceal its squirrel odor from a snake. This is particularly important for ground squirrel pups, which are not resistant to rattlesnake venom. Rattlesnakes are known to invade burrows in spite of aggressive adults and feed heavily on the pups. About 70% of the diet of Southern Pacific (Western) Rattlesnakes in the spring consists of young California Ground Squirrels.

Ground squirrels have a warning system that not only alerts each other, but every other animal in the vicinity as well. A member of the squirrel community, upon noticing danger, emits a shrill squeak, and all potential prey species duck for cover. It is interesting that the alarmist, by attracting attention to itself, is often the only squirrel that is captured. It seems, however, that the alarmist is usually one of the oldest squirrels in the colony and, as such, already has reproduced. By sounding an alarm, it protects its offspring. In so doing, it accomplishes what it set out to do by reproduction; it contributes a genetic endowment to the squirrel population.

Mixed Coniferous Forest

Different authorities refer to the Mixed Coniferous Forest by different names. Munz and Keck referred to it as the Yellow Pine Forest, which originally might have been a logger's term. In the old life zone system, this community was known as the Transition Zone. The terms mixed coniferous and transition refer to a mixture of coniferous and winter-deciduous trees. The name Western Yellow Pine refers specifically to Ponderosa Pine, *Pinus ponderosa*, but yellow pine is used also in a general sense to refer to a closely related species, the Jeffrey Pine, *P. jeffreyi*. Throughout this book, the term Mixed Coniferous Forest shall be used for a forest in which either Ponderosa Pine or Jeffrey Pine is a major component.

Mixed Coniferous Forest is one of the most widely distributed communities in California, occurring in every mountainous region with sufficient elevation and climates (plates 3A, 6A, 9A, and 14A). Authorities differ in their estimates of total coverage, partly because of the controversy in naming the community, but it is probably safe to say that there are around 14 million acres (5,700,000 ha) of Mixed Coniferous Forest in California.

Mixed Coniferous Forest occurs on both east and west slopes of the Sierra Nevada. It is composed principally of coniferous trees requiring at least 25 in (63 cm) of precipitation, much of which falls as snow. Adapted to a summer growing season, the community requires sufficient groundwater remaining from snowmelt and winter rains. From about 2000 ft (600 m) in the north and 5000 ft (1500 m) in the south, the Foothill Woodland grades into Mixed Coniferous Forest. Average precipitation in the Mixed Coniferous Forest is 40 in (100 cm) per year, twice that of the foothills. On the east side of the Sierra, the lower elevation reached by this forest is about 4000 ft (13,300 m; plate 6A) to the north, and about 10,000 ft (3300 m) to the south.

Depending upon which species predominate, Mixed Coniferous Forest has been divided further by various authorities into subgroups, which overlap across broad regions. In the *Manual of California Vegetation*, there are at least 10 alliances. Separation of the subgroups occurs in nature primarily in association with changes in available soil moisture, although temperature plays a role as well. In this book, we shall distinguish two subgroups, dry Mixed Coniferous Forest and moist Mixed Coniferous Forest. At most elevations, the distinction between the two groups is as simple as the difference between north- and south-facing slopes. On the east side of the Sierra, dry Mixed Coniferous Forest prevails.

Members of the dry Mixed Coniferous Forest include Ponderosa and Jeffrey pines, as well as California Black Oak, *Quercus kelloggii*, and Incense Cedar, *Calocedrus decurrens*. Black Oak replaces Blue Oak above the foothills, and like Blue Oak or Valley Oak, it is winter-deciduous. Leaves of Black Oaks turn yellow in the fall, adding a nice dose of autumn color to the forest. California Black Oak differs from Valley Oaks and Blue Oaks by having large leaves with seven pointed lobes. It is characteristic on open slopes and dry ridges, where it often grows in clumps; hence, a grove of oaks may be referred to as the California Black Oak alliance. This clumping may result from unclaimed acorns left behind by Gray Squirrels, who bury them in sizable caches for winter sustenance. Black

Oaks also are able to resprout after a fire, which can give a single tree the appearance of multiple trunks.

Black Oak is not the only deciduous oak in Mixed Coniferous Forest. A variety of Oregon Oak, also known as Shin Oak, *Q. garryana* var. *semota*, occurs in the southern Sierra Nevada in Kern County, where it is a component of dry Mixed Coniferous Forest and Upper Chaparral communities. It is conspicuous on dry exposed sites in the Kern River drainage. It also occurs on Pelona Ridge in the Transverse Ranges north of Santa Clarita. Its distribution is disjunct from populations in the Pacific Northwest where Oregon Oak is an important forest species. Such a disjunction is not totally unique for forest species. For example, there are also similar gaps in the distribution of Gray Pines and Foxtail Pines.

Instead of needles, Incense Cedar has flattened, scalelike leaves arranged in branching chains (figure 4.25). Its color is bright yellow green, which is particularly visible in winter. The bark is reddish and fibrous, resembling the bark of giant Sierra Redwoods, and many people confuse large old cedars with redwoods. One important difference, however, is that redwood bark is fireproof. Bark of Incense Cedar, when shredded, makes excellent tinder for starting fires. It was favored by Native Americans for that purpose.

It is not easy for most people to tell Ponderosa Pine from Jeffrey Pine (Figure 4.26). The needles of both pines occur in clusters of three and are yellow green. They differ from those of Gray Pine by being darker and shorter, 5–10 in (12.5–25 cm) in length. They also differ from Gray Pines by having a single vertical trunk in the crown. The bark of Ponderosa and Jeffrey pines is similar. In older trees, Ponderosa bark tends to be a brownish yellow, whereas Jeffrey bark tends to be a reddish brown. Bark of young trees tends to consist of a series of narrow ridges, whereas in larger, older trees the bark is often characterized by large, flat plates. The odors of the bark are markedly different. Jeffrey Pine is noted for its pleasant vanilla or

FIGURE 4.25 Incense Cedar, *Calocedrus decurrens.*

butterscotch odor; Ponderosa bark has no distinctive odor. Cones of the two trees are similar in shape, but Jeffrey cones are usually much larger. Ponderosa cones are seldom longer than 5 in (12.5 cm), whereas Jeffrey cones can be as long as 10 in (25 cm). An important distinguishing feature is that a Ponderosa cone has small spines that stick out of the ends of each scale. On Jeffrey cones, the spines are turned under. A person who rolls a Ponderosa cone between his or her hands will be pricked by the spines (prickly Ponderosa!); spines of the Jeffrey cone do not prick (gentle Jeffrey!).

Ponderosa and Jeffrey Pines can grow to a very large size. In Plumas County, there is a Ponderosa Pine that stands 223 ft (69.7 m) high. It is almost 8 ft (2.4 m) in diameter, and the crown has a 68 ft (21 m) spread. The largest Jeffrey Pine is on Sonora Pass. Standing 175 ft (54.6 m) high, it is almost 8 ft (2.4 m) in diameter, but the spread of its canopy is a mighty 87 ft (27 m).

Gray Pine, Ponderosa Pine, and Jeffrey Pine are similar in appearance, and they may grow together. What range of tolerances and requirements can be outlined to explain their differences in distribution? One important factor is the ratio between photosynthesis and cellular respiration, which is not the same for all plants. Temperatures above or below the optimum tend to decrease photosynthetic rate, and the rate of cellular respiration tends to increase as temperature increases. This increase continues beyond the temperature for optimal photosynthesis, and the result is that cellular respiration requires more carbohydrate than can be produced by photosynthesis. In this way, a lower elevational limit or a limit in southward distribution of any species could be controlled by temperature. Ponderosa Pine might be subject to such limitation. It occurs below 3000 ft (1000 m) only in northern latitudes, but it is widespread throughout the west from the Pacific coast to the Rockies, to the exclusion of Gray Pine or Jeffrey Pine. In California, Ponderosa Pine has only limited distribution south of Riverside County, but it reaches its southern limit in the Peninsular Ranges of San Diego County. Specimens previously reported from Baja California apparently were misidentified. The upper limit of Ponderosa Pine seems to be controlled by cold temperatures. It seems not to occur where average January temperature is below 30°F (−1°C).

Photosynthesis also requires water; therefore, precipitation could limit photosynthetic

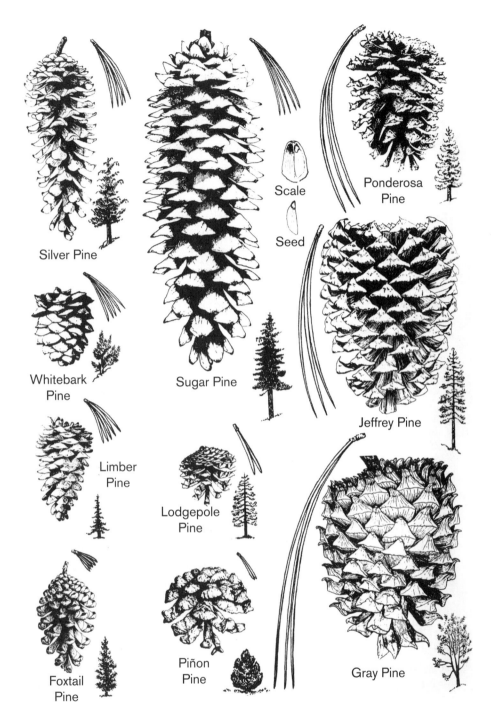

Silver Pine

Whitebark
Pine

Limber
Pine

Foxtail
Pine

Sugar Pine

Scale

Seed

Lodgepole
Pine

Piñon
Pine

Ponderosa
Pine

Jeffrey Pine

Gray Pine

FIGURE 4.26 Pines of Mixed Coniferous Forest showing cones and needles (from Storer, T.I., and R.L. Usinger. 1963. *Sierra Nevada Natural History.* Berkeley: University of California Press).

FIGURE 4.27 Jeffrey Pine, *Pinus jeffreyi*, on rock outcrop along Mono Creek, Sierra National Forest.

rate. The distribution of Ponderosa pine seems to coincide roughly with those areas that get more than 25 in (62.5 cm) of precipitation, most of which is snow.

Jeffrey Pine (figure 4.27) of the western Sierra tends to be restricted to rocky outcrops at higher elevations than Ponderosa Pine. It also grows on serpentine soils at lower elevations. The implication is that it can tolerate more cold and less water. Distribution of Jeffrey Pine has been correlated with areas where the average temperature in January is no warmer than 30°F (−1°C). Its tolerance for dry conditions is substantiated by its widespread distribution along the eastern side of the Sierra to the exclusion of Ponderosa Pine. A nearly pure stand of Jeffrey Pine occurs over many acres on pumice soil just north of Mammoth Lakes (plate 6A). Jeffrey Pine also extends along the Peninsular Ranges into Baja California far south of the southernmost Ponderosa Pines. North of the Sierra, however, in areas east of the Cascades that seem to resemble the eastern Sierra, Ponderosa Pine occurs to the exclusion of Jeffrey Pine. In the Warner Mountains, in the northeastern corner of the state, Ponderosa Pine occurs at higher elevations than Jeffrey Pine. This is exactly reversed from the elevational relationship on the western side of the Sierra. This distribution seems to imply that water is more limiting to the distribution of Ponderosa Pine than it is to Jeffrey Pine. Jeffrey Pine has a more extensive root system than Ponderosa Pine. Perhaps this feature, related to water absorption capability, is what separates the species in nature.

Another theory is that Jeffrey Pine is more able to survive lightning strikes, which are more likely at higher elevation. Jeffrey Pine in southern California commonly survives lightning, maybe due to its open crown. Perhaps the structure of its thick bark enables the tree to tolerate lightning without shattering, or the unique chemical nature of Jeffrey Pine bark may provide the answer. It has been reported that about 80% of Ponderosa Pines are killed by bark beetles after being struck by lightning, whereas a much smaller fraction of Jeffrey Pines are so afflicted. Jeffrey is the only pine whose bark contains paraffin-like compounds and aldehydes rather than terpenes. This different chemical nature may provide a resistance to bark beetles, or it may influence the way lightning is conducted by the tree. A logical question, of course, is how do other species of high-elevation conifers survive lightning?

What is it that prevents Gray Pines from moving upslope into the realm of Mixed Coniferous Forest? Could it be the photosynthesis-to-respiration ratio as a function of temperature? Perhaps it is simply a function of how deeply roots are able to penetrate. Gray Pines and Jeffrey Pines are known to have elaborate root systems. If the answer simply is drought tolerance, why does not Gray Pine occur farther south or more abundantly along the eastern Sierra in conjunction with Jeffrey Pine? Tolerance for drought or heat certainly is demonstrated by

FIGURE 4.28 Pandora Pinemoth, *Coloradia pandora*, on bark of Jeffrey Pine.

the intermingling of Gray Pines and Pinyon Pines, a desert species. In the southern Sierra, from the Kern River drainage eastward to the Owens Valley, there is an intermingling of Gray Pine, Pinyon Pine, and Jeffrey Pine. In this area, Jeffrey Pine clearly occupies the moister sites.

The question of whether temperature or water, or a combination of the two, limits the respective distributions of these pines has not been answered, but microclimate differences do occur, as illustrated in localities where several species grow in close proximity. Such an area occurs in the southern Sierra on the western side of the Kern River, in the vicinity of Peppermint Creek. This area at 6400 ft (2100 m) elevation has nice groves of Ponderosa Pine on relatively flat, sandy alluvium along the stream. On slopes and rocky outcrops above the stream are Jeffrey Pines. On exposed south-facing slopes and the tops of granite domes are Gray Pines. The total distance of this gradient is about 0.25 mi (400 m). Approximately 0.5 mi to the east on a dry, exposed, south-facing slope, there are also a few Pinyon Pines. Separation of species along a cool-moist to warm-dry gradient is clearly evident in this area.

Yellow pines are known to be associated with a wide variety of insects. Over 200 species have been catalogued that may, on occasion, become pests. Bark beetles (Scolytidae) are important, but they seem to attack primarily weak or old trees. Drought and/or ozone are known to weaken trees, after which they become susceptible to bark beetle damage. One of the most common and best-known pests is the larva of a butterfly known as the Pine White, *Neophasia menapia*. The adults of this species are about 2 in (5 cm) across. They are white, with heavy black markings around the tips of the front wings. They are difficult to see close-up because they spend most of their time fluttering high in the trees. During rare outbreaks, the caterpillars of this species can completely defoliate a variety of conifers.

An interesting pest that attacks in alternate years is the Pandora pinemoth, *Coloradia pandora* (figure 4.28). The adult is one of the silk moths (Saturniidae). It is a heavy-bodied moth with a wingspan of about 2 in (10 cm). It has a drab, brownish color and a single black spot on each wing. When they are abundant, they can be seen by the hundreds clinging to the bark of Jeffrey Pines. It is the larvae that do the damage. These caterpillars are a greenish color with six broad, interrupted longitudinal stripes. They eat pine needles, and although their cyclical infestations seem to create a lot of damage, they are an important part of the forest ecosystem. They recycle nutrients by converting pine needles, which normally decay slowly, to fertilizer that is easily available from the forest floor. They also provide a short-term abundant food supply for birds and other wildlife. Native Americans also ate the caterpillars, particularly in the Jeffrey Pine forests of the eastern Sierra Nevada. As late as the early 1900s, Native Americans in the Mono and Modoc areas gathered the larvae by digging trenches, which trapped them beneath the trees as they crawled down. They were baked in hot soil and dried out for later consumption. A form of biological control is that tiny parasitic wasps lay their eggs in the eggs and larvae of the moth.

In the Warner Mountains, in extreme northeastern California, there is a third member of the yellow pine group. At high elevations, usually above 8000 ft (2600 m), are found Washoe Pines, *P. ponderosa* var. *washoensis*. These unique pines are found only in a few localities, at high elevations on the western edge of the Great Basin in California and Nevada. In the Warner Mountains, Jeffrey Pines are found on the lower slopes, typical of the tree's distribution bordering desert areas. Above that, and more commonly toward the north, are the Ponderosa Pines. Washoe Pines grow above the Ponderosas.

Washoe Pines are probably of hybrid origin. It appears that hybridization between Ponderosa and Jeffrey pines has led to the Washoe Pine. This large pine has needles in clusters of three, similar to the other two species. The cone is intermediate in appearance. Washoe Pine cones have the spines turned inward, similar to those of Jeffrey Pine, but the cones are smaller than either of the parental species. They look like small versions of Jeffrey Pine cones and are about 3 in (7.5 cm) in length.

Washoe Pines in the Warner Mountains grow in a woodland that includes Quaking Aspen, *Populus tremuloides*. This community more nearly resembles what a person would expect to find in the Rocky Mountains. In California, aspens tend to occur as a riparian species, highly dependent upon abundant soil moisture. Their presence in a woodland setting in the Warner Mountains seems to indicate a reasonably moist habitat, and it is probably dependent on a good supply of summer rain. This implies that the Washoe Pine also requires a good supply of moisture. This beautiful forested area is also the location for abundant deposits of high-quality mahogany obsidian, which was so useful to Native Americans. The entire area is included in the Warner Mountains Wilderness, probably the least-visited wilderness area in the state.

The moist Mixed Coniferous Forest occurs on north-facing slopes and at higher elevations than the dry Mixed Coniferous Forest. The moist forest is dominated by White Fir, *Abies concolor* (figure 4.29), and Sugar Pine, *P. lambertiana* (figure 4.30, plate 14A), although Ponderosa Pine is not uncommon. White Fir has stiff, blunt needles distributed evenly along the stem. On the parts of the tree where needles are exposed to abundant sunlight, particularly near the top, they grow upward on the stem. Small firs are popular, expensive Christmas trees. Any of a number of firs that grow in western states can be sold as Christmas trees, but they often are sold as "Silver Tips," the implication being that the Christmas tree comes from the top, or tip, of a Pacific Silver Fir, *A. amabilis*, a species of the Cascades. California Red Fir, *A. magnifica*, because it actually may have a silver tip, is perhaps the most common fir sold as a "Silver Tip."

In contrast, inexpensive Christmas trees are Douglas-firs, *Pseudotsuga menziesii*. Although they are not common in most of the Mixed Coniferous Forest, they do occur in moist locations, particularly on the west slope of the Sierra north of Yosemite Valley. Douglas-fir is darker green and has long, hanging side branches. If cones are visible, the firs are easy to tell apart. White Fir cones grow upward on the tips of the branches, and Douglas-fir cones grow downward. Furthermore, White Fir cones are composed of tightly layered scales that are deciduous; the cones come apart when they dry

FIGURE 4.29 White Fir, *Abies concolor.*

FIGURE 4.30 Sugar Pine, *Pinus lambertiana.*

out. Douglas-fir is an important component of coastal forests from northern California to Washington, but overall, its presence is minor in the Sierra. They occur no farther south than the San Joaquin River drainage, and only on the western slope of the Sierra Nevada.

Sugar Pine is the tallest and largest of all pine species in the world. The largest known living specimen is on the western slope of the Sierra in Tuolumne County. It is 216 ft (64.8 m) high and 10 ft (3 m) in diameter. Cones of Sugar Pines dangle from the tips of long, horizontal branches. The cones are usually over a foot (30 cm) in length and often reach lengths of 18 in (45 cm). Large Sugar Pine cones are often

prized as table decorations. Needles are 3–4 in (33–45 cm) long, much shorter than those of other pines of the Mixed Coniferous Forest, and occur in clumps of five. When viewed against the horizon, Sugar Pines are a majestic sight. They seem to be rather flat-topped, with the long branches extending horizontally away from the trunk. The short needles give the branches a "furry" appearance, like huge spider arms.

In general, five-needle pines are common at high elevations, but Sugar Pine occurs at lower elevations and requires more water than the other five-needle species. It is moderately shade tolerant, has thick, fireproof bark, and may

FIGURE 4.31 Giant Sequoia (Sierra Redwood), *Sequoiadendron giganteum*. The Grizzly Giant, Mariposa Grove, Yosemite National Park. Oldest Sierra Redwood.

basins and on north-facing slopes. There are 75 such groves scattered along the western slope of the Sierra Nevada. Unlike the groves toward the south, however, northern groves tend to occur on south-facing slopes. Toward the north, precipitation is heavier so water is less limiting, and trees grow on south-facing slopes, where light is more intense. Most of these groves are fully protected in Yosemite National Park, Sequoia-Kings Canyon National Park, and Giant Sequoia National Monument administered by the US forest Service as part of Sequoia National Forest. Established by President Clinton in 2000, the Giant Sequoia National Monument contains over half of the protected groves.

In spite of their high demand for water, recent research indicates Sierra Redwoods are remarkably drought tolerant. During the drought years of the early 2000s, while large numbers of pines were dying in the Mixed Coniferous Forest of the western slopes of the Sierra, Sierra Redwoods were remarkably tolerant, losing many leaves, but large numbers of trees were not dying. Sierra Redwood leaves are scalelike, similar to Incense Cedar or various junipers. At the tops of the trees, where sunlight is more abundant, leaves are smaller and succulent, representing significant water storage capacity. Interestingly, while most of the leaves of Coast Redwoods resemble those of White Firs, the leaves at the tops of the trees often are scalelike, similar to those of Sierra Redwoods.

The combination of light and moisture required by these trees exemplifies the climate that was formerly widespread on the west coast of North America; hence, these groves are only relics of a former widespread distribution in the Northern Hemisphere. Historical factors such as glaciation may also have influenced present-day distribution. In the western United States, fossil redwoods are known from as far east as Colorado, Wyoming, and Nevada, and as far south as the Santa Monica Mountains. Studies of pollen 30 ft down in meadows indicate that Sierra Redwoods did not grow at their present localities until 4500 years ago. On the other hand, mummified remains associated

exceed 500 years of age. If the bark is cut, it exudes a thick, sweet sap. John Muir is reported to have preferred Sugar Pine sap over that from the Sugar Maple. Native Americans ate the sugary pitch and used it as a laxative. They also harvested and ate the seeds.

Giant Sierra Redwoods or Giant Sequoias, *Sequoiadendron giganteum* (figure 4.31, plate 5), are the most moisture demanding of all the species in the Mixed Coniferous Forest; hence, they are absent from the east side of the Sierra as well as most of the west side. They require year-round soil moisture of at least 15–20%, and mature trees can suck up 800 gal a day, whereas average Ponderosa or Jeffrey Pines use about 130 gal a day. In the southern Sierra Nevada, soil moisture is maintained at this level by runoff from higher elevations. The contribution of summer thundershowers is particularly important. Runoff accumulates in granite basins or where bedrock is near the surface, and the patchy distribution of these basins means that favorable sites for redwoods are limited. Redwoods therefore tend to occur in groves in with woodrat nests in caves indicate that 10,000–35,000 years ago, during the height of glaciation, Sierra Redwoods occurred as low as 3000 ft (900 m) in elevation. Apparently, they moved upslope to their present localities as warming and drying occurred following the end of Pleistocene. Because the distribution of climatic features required by Sierra Redwoods is so localized, it may be stated that they grow in ecological islands, like the foothill plants mentioned previously that are restricted to certain soils. Plants and animals of ecologic islands are often unique, or they are relicts, which is the case for redwoods. In either situation, distribution is limited, and endemism is the result. Ecological islands are an important feature that makes California a special place.

Redwood bark is resistant to fire. The wood will burn, but it is impregnated with tannins that are toxic to termites and other wood-consuming organisms. This resistance to both fire and insects allows the trees to reach great age, with many surpassing 2000 years. At 2400 years, the Grizzly Giant in the Mariposa Grove (figure 4.31) is the oldest in Yosemite National Park. It has a striking appearance in that it leans to the side. It is said that its lean is greater that the Leaning Tower of Pisa. Given abundant moisture, the trees grow rapidly. It has been said that Sierra Redwood is the fastest-growing tree in the United States. Interestingly, recent studies of redwood tree rings indicate that both Sierra Redwoods and Coast Redwoods may be thriving on climate change. Apparently they have grown more rapidly during the last century than ever before. Scientist speculate that drier weather may mean more sunlight which has lengthened the growing season.

Rapid growth and old age lead to great size. Sierra Redwoods are the largest living things in the world. General Sherman, the largest of the redwoods, is 275 ft (84 m) tall and 36.5 ft (11.1 m) thick at its base (plate 5). It is estimated to be between 2300 and 2700 years of age. If it were a useful lumber tree, it has been estimated that it would contain the equivalent of 175 mi (280 km) of 2 × 4 (in) planks, enough wood to build 40

five-room houses. Its lower branches are individually larger than any single tree east of the Mississippi River. General Sherman (plate 5) grows in a grove at Giant Forest in Sequoia National Park. A few miles to the north in Kings Canyon National Park is General Grant, the second largest living thing. General Grant is 268 ft (81.5 m) high and 40.3 ft (12.3 m) thick at its base.

The practice of naming these giants after generals began at the turn of the century with the establishment of the first national parks. Military heroes were honored by naming trees after them. Up until that time, General Sherman was known as the Karl Marx tree, named after a hero of another sort by a group of Communists that had established a cooperative colony in the area now known as Sequoia National Park. The Friedrich Engels tree stood nearby.

The names of these two species are also of historical interest. Sequoyah was an Indian who established an alphabet, consisting of 86 characters, for the Cherokee language, making it possible to write in that language. Redwood trees were named for this Native American, and Sequoia National Park was named for the trees. It was later determined that the Sierra Redwoods were different enough from Coast Redwoods to warrant their own genus. The name *Sequoia* was retained for Coast Redwoods, because they were named first, and the name for the Sierra Redwoods was changed to *Sequoiadendron*, which means "sequoia tree." The common name for the Sierra Redwood, "Giant Sequoia," became inappropriate with the new genus name, but neither it nor the name of the park was changed. Would it make it any less spectacular if the park were called Sequoiadendron National Park? In 1937, the California legislature named the California Redwood the state tree, but it neglected to distinguish between the two species of redwood. However, in 1951 the California Attorney General in an attempt to clarify the issue declared that both species were the state tree. Following this decision, a report was submitted to the California legislature by the State Park Commission and the State Forester in which they recognized both species as redwoods by officially adopting the common names of Coast Redwood and Sierra Redwood.

A great deal is understood about Sierra Redwood ecology. They are conspicuous, inspirational trees that have long been studied and eulogized. One of the most significant discoveries about Sierra Redwoods did not emerge until the early 1960s, although it had been known for some time by scientists. Redwoods require fire if they are to perpetuate. Ground fires do not burn through redwood bark, but the rising heat opens the cones, which remain on the trees up to 20 years. When the cones open, the seeds fall on the soft, recently burned earth (figure 4.32). When it rains, these seeds germinate in sunlight that shines to the ground, uninhibited by an overstory of vegetation.

In the past, fires were started by lightning or by natives who knew the value of such fires to the economy of the forest ecosystem. These frequent fires would burn off litter and duff, open the canopy to allow light to enter, and open the cones. Fires would loosen and soften the soil and release nutrients. In contrast, fire suppression allowed a long accumulation of flammable materials that could produce a very hot conflagration if a fire did start. Fire suppression changed the forest fire regime from frequent low-intensity ground fires to infrequent high-intensity fires. In contrast, chaparral fire regimes have also changed in association with human activity. Fire suppression increased not only fuel load, but also human activity, through either accidental ignitions or deliberate arson; and associated with climate change, it also increased the number and intensity of fires, and extended the length of the fire season.

There is also evidence that long periods of fire suppression enabled a deadly fungus to invade the root systems of the conifers, particularly Ponderosa and Jeffrey pines. This disease, known as root rot or Annosus root disease, is killing a significant number of trees in Yosemite Valley.

A fungus that seems to be a fire follower is the Morel, *Morchella angusticeps*, a conical mushroom with deeply pitted cross veins between gray to black vertical ridges. This edible fungus is most common on recently burned soils, where it usually appears as soon as the snow melts. It is particularly common in areas of Mixed Coniferous Forest where California lilacs, *Ceanothus* spp., have become established. Morels are highly prized for their flavor and may cost $250 a pound in gourmet food stores.

By suppressing fire, the National Park Service and the US Forest Service were "loving their trees to death." White Firs began to encroach upon the redwood groves, posing a problem because they shade out redwood seedlings and they are flammable from top to bottom. A fire in tall White Firs can carry flames to the foliage of the redwoods, high above the forest floor, and large, hot crown fires can kill the redwoods, too. In 1955, a 13,000 acre (5200 ha) fire threatened Grant Grove. In 1963, the A. Starker Leopold report indicated that downed wood and other fuels could cause a holocaust if they became ignited. In the late 1960s and early 1970s, the National Park Service began to view fire as a natural component of the ecosystem. Fires at elevations above 8000 ft (2650 m) were allowed to burn themselves out. In redwood groves, small prescribed burns were initiated to reduce fuel buildup, burn small White Firs, kill root rot, and otherwise simulate more natural conditions. For a variety of reasons, the US Forest Service has been slow to recognize the value of fire. Mountain cabins and resorts occur in national forests, and timber is a marketable resource. Suppressing forest fires helps to preserve these outside interests. Because national parks and national forests are adjacent to each other, at the present time it is possible literally to move in one step from a philosophy of prescribed burning to a philosophy of fire suppression. In recent years, however, in certain circumstances the Forest Service has also adopted a program of prescribe burning.

During the summer of 1987, there were an inordinate number of fires in California. In less than a month, more than 12,000 lightning strikes ignited fires that burned for 50 d. By the time the fires were put out, a total of 775,000 acres (315,000 ha) had burned. About 90% of the affected land was managed by the US Forest Service. During the same period of time, lightning-caused fires also occurred in the national parks. Where there was no danger to people or structures, these fires, particularly in the subalpine zone, were allowed to burn themselves out. The US Forest Service viewed the fires as a disaster, even in areas a great distance from the nearest roads or habitations. The National Park Service, while justifiably concerned, considered the fires as part of the natural scheme of things. Some authorities contend that one of the reasons the damage was more severe in the national forests is that too much fuel had accumulated over too long a period of time, and the consequence was predictable. It has been determined that humus alone accumulates at the rate of about 1000 lb/acre (185 kg/ha) per year.

Nevertheless, the US Forest Service has calculated that 1.4 billion board feet of marketable timber was damaged. This is the equivalent of 140,000 three-bedroom homes, with a market value of $140 million. On the other hand, 84% of the burned timber was salvageable. Furthermore, in 1987, 155 mi (248 km) of Riparian Woodland was burned, 274 mi (438 km) of trails were destroyed, and 100,000 acres (40,000 ha) of designated wilderness was burned. The US Forest Service would have to rebuild the trails, relocate campgrounds, and plant about 100 million seedlings to revegetate the most seriously burned land. Salvage logging, in which burned trees are removed, is in itself an ecological problem, particularly if the area is treated as a future tree farm in which only valuable tree species are replanted, and the area is treated with herbicides to discourage growth of competing shrub vegetation. Some environmentalists feared that the amount of true damage was overestimated, and that the US Forest Service would use the salvageable lumber as an excuse to put roads into formerly roadless, wilderness-quality areas. Ecologists also point out that the remaining burned snags provide homes for animals, in particular cavity-nesting birds.

Since the summer of 1987 in association with climate change and the encroachment of civilization, the number of wildfires has been increasing. There were 10 major fires in the 1990s and 11 in the 2000s. The largest fire in California history, the Cedar Fire, occurred in 2003. It burned 280,278 acres (113,424 ha) of San Diego county, mostly chaparral, much of it in the Cleveland National Forest. In October of that year, 800,000 acres (3200 km²) were burned in the state. In 2007, fires in October alone burned 970,977 acres (392,940 ha). Summer wildfires in 2008 burned 1,161,197 acres (469,920 ha). During the summer of 2014, the Rim Fire, the largest ever recorded in the Sierra Nevada, burned over 250,000 acres (101,171 ha) of Stanislaus National Forest and Yosemite National Park; 77,000 acres (31,161 ha) were in Yosemite. The fire also burned habitat for a distinct clan of California Gray Owls, as well as threatening habitat for rare Red Foxes and Pacific Fishers. During the summer of 2015, one of the worst fire seasons on record, three large fires were ignited. A lightning strike triggered the Rough Fire which burned 150,000 acres (60,703 ha) of two National Forests and Kings Canyon National Park. The Valley Fire in Lake, Napa, and Sonoma Counties burned 76,000 acres (30,756 ha), and destroyed 1958 homes. The Butte Fire burned 70,000 acres (28,327 ha) in Amador and Calaveras Counties and destroyed 1293 structures. The smoke from these fires caused respiratory difficulties for people in the Owens and Central Valleys for several months.

Foresters throughout the state claim that prescribe burning is the best deterrent for future severe wildfires. The increased number of dying trees, including Sierra Redwoods, at the lower reaches of Mixed Coniferous Forest, seems to indicate that as a response to increased heat and drought, forest zones may gradually move up mountains to higher elevations. One researcher in Sequoia National Park estimates that about a third of the Black Oaks in the park are dying. In August 1969, four redwood trees fell over in the Hazelwood picnic area of Sequoia National Park, one killing a 60-year-old lady who was eating her lunch. A hierarchy of park supervisors began to worry about liability, and the resultant flurry of research uncovered another interesting fact about redwood groves. Giant Carpenter Ants, *Camponotus laevigatus*, make nests in redwood trees by hollowing them out, a phenomenon that weakens the structure of the trees. One investigator discovered a nighttime trail of these ants 2 in (5 cm) wide. These ants passed by, 200 of them per minute, on their way to consume honeydew, an exudate of aphids that live in the White Firs. The reason that Giant Carpenter Ants were so common is that there were too many White Firs, and the reason there were too many White Firs is that fire had been suppressed for nearly 50 years.

In December 1973, during the night, a large lower branch fell off General Sherman. Had it fallen at midday in July, untold numbers of people might have been killed. It is hoped that the program of prescribed burning will reduce the number of Giant Carpenter Ants so that such catastrophes will not occur again. Large oaks also drop branches from time to time. In August 2015, two youngsters sleeping in a tent in Yosemite Valley were killed by a branch that fell from an oak tree. As it turns out, similar incidents have occurred before. In 1985, two people were killed and nine were injured when a 25 ft oak branch fell on an open-air tram, and again in 1992 a falling oak branch injured seven tourists in a tour bus. In 2012, a concession worker was killed when a limb fell on his tent cabin. Outside the park, in 2013 a camp counselor was killed by a tree that fell into a campfire. There is no explanation for these events other than oak limbs often fall from trees. Old oak trees with the scars from fallen limbs are found throughout forested areas.

Redwood trees have few enemies. In addition to Giant Carpenter Ants, there are Spruce Budworms, Bark Beetles, fungi, and Dwarf Mistletoe. Periodic fires reduce the fuel; at the same time, they reduce the pests. Pests are typically associated with weak or sickly trees. It has been estimated that over 140 species of insects live in the crown of a redwood grove without harming the trees. Of particular importance

are Green Lacewings (*Chrysoperia* spp.) which feed upon aphids. Of course, there is also a variety of organisms that feed on the lacewings.

A Long-horned Beetle, *Phymatodes nitidus* (Cerambycidae), helps cones release their seeds (figure 4.32). An average Sierra Redwood produces about 1500 cones per year, with about 200 seeds per cone. The cones remain unopened on the tree for about 20 years, or until some mechanism such as heat from a fire or sunlight causes them to open. The larvae of the long-horned beetles chew into food-conducting channels in the stems of the living cones, which causes the cones to die. The cones dry out when they die, and the seeds are released. Heat, however, remains the major mechanism for opening cones. Among mammals that are considered beneficial to redwoods are Chickarees, *Tamiasciurus douglasii* (plate 18C), also known as Chickaree or Douglas's Squirrels. They feed on 2- to 5-year-old redwood cones, but as they feed, many tiny seeds are released. This mechanism also contributes to the seed bank that could become future Sierra Redwoods.

Under appropriate conditions, seedlings grow from 6 in (15 cm) to 2 ft (60 cm) per year. If litter or duff on the ground is thicker than 0.5 in (12.5 mm), seeds will not be able to reach bare mineral soil. Stored carbohydrate in the tiny seeds is not adequate to sustain root growth beyond that distance, and the seedling dies. After germination and for 600–700 years, the trees retain a spire shape that is typical of many species adapted to shed snow. After that, the uppermost branches reach the canopy of the forest about 250 ft (80 m) from the ground. Old trees have rounded tops, and all their branches are near the top. These tall trees are often the victims of lightning strikes, but they often survive the strike with various degrees of damage to the crown. Flaming branches from a lightning strike land on the ground and roll downhill. If a flaming log rests against the trunk of another Sierra Redwood, by continuously burning it may "eat away" the bark. Without its protection of bark, the wood is able to ignite, and the fire may actually hollow out the tree. The resulting loss of

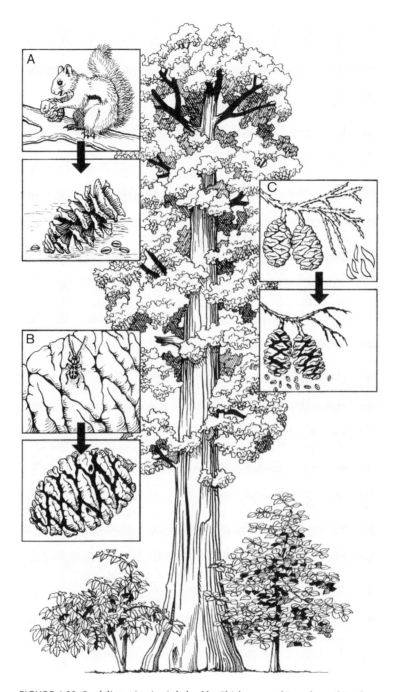

FIGURE 4.32 Seed dissemination is helped by Chickarees and Long-horned Beetle larvae. A. Chickaree eats green cone releasing seeds. B. Beetle lays eggs on cone. Beetle larvae cause cone to open. C. Fire updrafts dry and open cones.

SOURCE: Johnston 1998.

wood tissue reduces the ability of the tree to transport water to its crown, and the uppermost branches begin to die, also rounding the top. Lightning thus causes the rounding directly and indirectly. Drying winds also play a role in shaping the crown of an old tree.

Another effect of rolling, burning logs is that the upper sides of many old trees become burned. The root system of redwoods is remarkably shallow: approximately 95% of the trees have root systems no deeper than 3 ft (1 m). On the other hand, the base of an old redwood is a good deal larger than the upper portion of the trunk, and this "buttresses" the tree to help it remain standing in spite of its shallow root system. The buttresses at the bases of many old trees are burned off on the uphill side, which is why these trees often fall uphill when they do go over. Another pattern is for trees to fall into meadows because they lean toward spongy ground. Where Sierra Redwoods grow, the easiest way across a meadow is often by walking along one of these fallen trees. Because these trees are so resistant to decay, some of them have been used as bridges for hundreds of years.

The first national park was Yellowstone National Park in Wyoming, established in 1872. The second and third oldest parks are Yosemite and Sequoia, which were established in 1890. Kings Canyon was added in 1940. The effort to preserve Sierra Redwoods in these national parks of the Sierra Nevada is largely attributed to the legendary John Muir, who popularized these trees and the Yosemite Valley in numerous writings that appeared in the late 1870s. Of the huge redwoods he exclaimed, "God has cared for these trees ... but He cannot save them from fools. Only Uncle Sam can do that." John Muir and the Sierra Club can be thanked for preserving many of these ancient treasures for posterity. The 150th anniversary of John Muir's birthday was celebrated throughout the National Park Service on April 21, 1988.

Sierra Redwood does not make good lumber. Huge Sierra Redwoods tend to break up when they are felled, as if they are brittle. Nevertheless, the ancient giants were threatened by logging in the 1870s; the wood was used principally for fenceposts and shake shingles. The new logging threat is that, in some national forests, all trees but the redwoods are logged, leaving them to stand as silent sentinels without the benefit of the forest ecosystem.

Photochemical smog is having an influence on the composition of Mixed Coniferous Forest. Ozone causes loss of chlorophyll and early death for needles on some species. This degradation, known as chlorotic decline, particularly affects Ponderosa Pines and White Firs. When trees become weakened by ozone, particularly during periods of drought such as occurred during the mid-1970s, late 1980s, and mid-2000s, bark beetles (Scolytidae) invade, forming channels beneath the bark. This invasion may kill the trees. During the droughts, many dead trees of the Sierra Nevada bore witness to the insidious effect of the photochemical smog produced in the San Joaquin Valley. Nearly every afternoon, prevailing wind drew the smog up the slopes. Selective destruction of one or two species can alter significantly the makeup of an ecosystem. During the extensive drought of the 2000s, an unprecedented number of trees were dying. Using sophisticated instruments to document water in foliage of the canopy it was shown that from 2013 and 2015, 58 million trees, covering 2.4 million acres (9600 km²) had lost more than 30% of their canopy water. Furthermore, 9.9 million acres (39,600 km²) of lower elevation forests in the Sierra Foothills and Coast Ranges showed significant canopy water loss. By 2015, 565 million large trees over 16.3 million acres (65,200 km²) were showing signs of stress from the drought.

One estimate shows more than 35 dead trees per acre, particularly large Ponderosa and Sugar Pines in the forests on the south-western slope of the Sierra. There is also a shrub component in Mixed Coniferous Forest. Several species of manzanita (*Arctostaphylos*) grow on sunny sites. In Spanish, manzanita means "little apple" in reference to the plant's berries, which are an important food for forest animals. Manzanitas are in the heath family (Ericaceae) and are char-

FIGURE 4.33 Sierran Gooseberry, *Ribes roezlii*. A. Flowers. B. Fruit and foliage.

acterized by shiny red bark, urn-shaped flowers, and oval leaves.

Another common shrub on dry sites, particularly south-facing slopes, is Kit-kit-dizze, *Chamaebatia foliolosa*, a member of the rose family (Rosaceae). In spring, the shrubs are covered with small white flowers resembling those that appear on fruit trees of the rose family, such as apricots, peaches, plums, and apples. Kit-kit-dizze is the name given to this fernlike resinous shrub by the Native Americans, who made a tea from the leaves and used it for a variety of maladies. Old-timers, particularly those that rode horses, referred to it as Mountain Misery; apparently, the resin irritated the horses when they walked through it. It is also known as Bear Clover. It is believed that bears roll in it to help prevent ticks from attaching to them, and people who work with bears in Yosemite National Park say that the odor of Kit-kit-dizze reminds them of the odor of bears. Perhaps the horses become irritable because the odor makes them think a bear is in the area!

Various kinds of wild currants (*Ribes* spp.) also occur in shady areas. In autumn, these shrubs produce sweet berries that are an

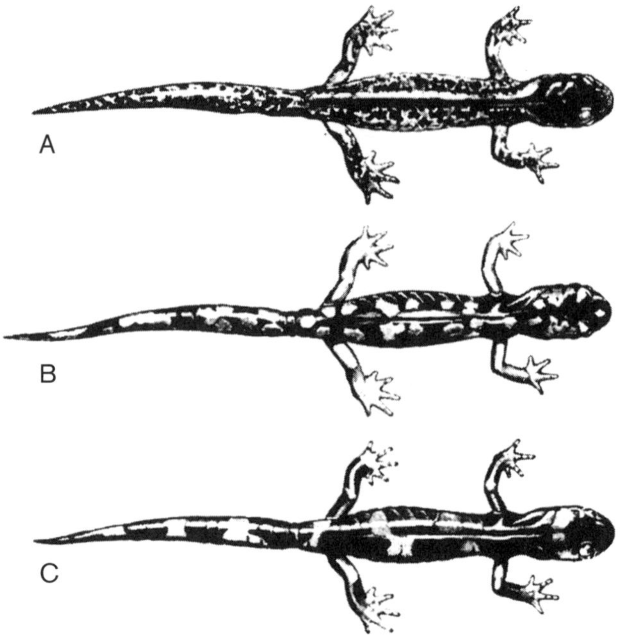

FIGURE 4.34 Some salamanders of the Ensatina complex.

A. Painted Salamander, *Ensatina eschscholtzii picta*, from the Klamath Mountains.

B. Sierra Nevada Ensatina, *Ensatina eschscholtzii platensis*.

C. Yellow-blotched Ensatina, *Ensatina eschscholtzii croceator*, from the Sierra Nevada. For range map and complete discussion of this group, see Chapter 8 (figure 8.32).

SOURCE: Stebbins, R.C. 1972. *California Amphibians and Reptiles*. Berkeley: University of California Press.

important food for animals and are particularly relished by bears. The spiny-fruited members of the group are called gooseberries, the flowers of which resemble small fuchsias (figure 4.33). They are among the most attractive of all forest shrubs.

In moist places, either along streams or in redwood groves, California Azalea, *Rhododendron occidentale*, is conspicuous. Showy white blossoms cover the shrubs in late spring and early summer. The foliage is poisonous, another example of a plant equipped with phytotoxins to discourage herbivory.

In areas that have experienced a recent fire, various members of the California lilac group (*Ceanothus* spp.) occur in thick stands. This shrub has nitrogen-fixing bacteria in its roots and is an important browse for deer. Hence, the names of certain members of the group, such as Buckbrush, *C. cuneatus*, and Deer Brush, *C. integerrimus*, reflect this dietary association. In spring, they produce small, fragrant white flowers in clumps that resemble miniature lilacs.

Animals of the Mixed Coniferous Forest

At the elevational level where Foothill Woodland grades into Mixed Coniferous Forest, there is a major environmental change. Snow becomes a significant part of the precipitation. Along with the change in vegetation, many animals shift from a period of winter activity to a period of summer activity. Some animals store food to make it through the winter; others become dormant. Many animals migrate south or to the foothills in the winter. Ectothermic animals, such as amphibians, reptiles, and insects, are far more common in Foothill Woodland than they are in areas where low winter temperatures restrict activity.

AMPHIBIANS Amphibians are neither conspicuous nor common above the foothill belt, or away from water. Amphibians include salamanders, toads, and frogs. Among salamanders, that which occurs farthest from water is the Sierra Nevada Ensatina, *Ensatina eschscholtzii platensis* (figure 4.34), which may be found under rocks and logs in moist places. Depending on moisture conditions, these salamanders may remain active all summer. If they estivate during summer months, it is in holes or under woodrat nests. They are brownish with large orange spots on their backs.

At the southern end of the range, the Sierra Nevada Ensatina is replaced by the Yellow-blotched Salamander, *E. e. croceator*. It occurs from the Kern River Canyon south to Mount Pinos. On this form, the spots may be 0.125–0.25 in (0.5 cm) across. At the northern end of the range for this species, as in the Klamath Mountains, the spots are smaller—in some cases mere speckles. In the Coast Ranges, these salamanders have no spots at all, and they are known as Monterey Salamanders, *E. e. eschscholtzii* (figure 8.33). They are common in the Foothill Woodland of the Coast Ranges, and in southern California they range up into Mixed Coniferous Forest. The solid-colored form has limited distribution in the Sierra foothills east of Sacramento; otherwise, it is missing from the Sierra Nevada. Distribution of this highly variable species is interesting from a geographic perspective. The fact that both the spotted and the solid-colored forms may be collected at the same locality in southern California makes it a fascinating subject for biogeographic and evolutionary studies (figure 8.34). The story of this salamander and its peculiar distribution will be discussed in the section on the Peninsular Ranges (figure 8.33). Most of the other amphibians will be discussed in the sections on Oak Woodland and Riparian communities.

REPTILES The most common snakes and lizards of the Mixed Coniferous Forest are also found in the foothills. As typical chaparral dwellers, they will be discussed in a later chapter. One lizard of the Mixed Coniferous Forest is particularly interesting—Gilbert's Skink, *Plestiodon gilberti*. As a juvenile, its tail is an iridescent blue. If a predator goes for the lizard, the tail breaks off very easily. The bright blue tail then quivers and bounces, distracting the predator and allowing the lizard to escape. The lizard is subsequently able to grow a new tail. As an adult, the lizard loses the blue color, and its head takes on an orange-red color. Gilbert's Skink may be 8 in (20 cm) in total length, and when it moves it appears to be a snake. It uses its small legs for locomotion primarily when it

moves slowly, stalking insects on the forest floor. Most often, it is nocturnal, spending daylight hours beneath logs or rocks.

The Western Skink, *P. skiltonianus*, has a relict population on the Kern Plateau. Juveniles have the bright blue tail of Gilbert's Skink, with which it easily can be confused. In fact, in old collections, specimens of the Western Skink were frequently misidentified because they were not expected to occur in the Sierra Nevada. In general, where the ranges might overlap, as in the southern Sierra Nevada, Gilbert's Skink occurs in Foothill Woodland, and the Western Skink occurs in the pine forests. Both species have been recorded in Yosemite Valley. The southern Sierran form of the Western Skink is markedly smaller than forms from the Coast Ranges or southern California. Where ranges of the two species overlap in southern California, such as the Oak Woodland near Gorman, juvenile Gilbert's Skinks have red tails.

There are also two kinds of alligator lizards in the Mixed Coniferous Forest. The Northern Alligator Lizard, *Elgaria coerulea*, is a northern species that occurs throughout the forest belt of the Sierra Nevada. It is replaced to the southeast, and to some degree in the foothills, by the Southern Alligator Lizard, *E. multicarinata* (figure 7.28). On the southeastern part of the Kern Plateau and in canyons on the southeastern side of the Sierra, the Southern Alligator Lizard occurs to the exclusion of the northern form. Along the west slope of the Sierra, ranges of the two species overlap. In canyons in the lower parts of the Mixed Coniferous Forest, both species can be found within a few feet of each other.

Alligator lizards are among the most sedentary of lizards. They prefer not to bask in the sun, seeming to prefer the shade of the forest. Unlike many lizards, they avoid temperatures above 86°F (30°C). They seem to prefer temperatures around 72°F (22°C) and are able to maintain digestive efficiency and other physiologic activities at this low body temperature. Apparently, they are ambush foragers, lying in wait for insects and other small prey items. These are live bearers; rather than laying eggs, they give birth to living young. During breeding, males grab females by the head and will remain that way until the female is receptive, which may be hours.

Two interesting snakes may be visible in the daytime, although neither is common. The California Mountain Kingsnake, *Lampropeltis zonata*, and the Northern Rubber Boa, *Charina bottae*, are more common in the Mixed Coniferous Forest than in the foothills. Both snakes occur particularly where there are large rock outcrops that serve as retreats and/or sites in which to hibernate.

The kingsnake gets its name from the fact that it will eat other snakes, including rattlesnakes. Most often, it eats lizards, skinks and fence lizards in particular. The southern populations of the California Mountain Kingsnake are threatened, and there are restrictions regarding the collection and sale of these beautiful snakes. They are colorful, with red, white, and black bands. A threat to these docile reptiles comes from people who mistake them for venomous coral snakes, which do not naturally occur in California. However, even venomous snakes, such as rattlesnakes, are protected in national parks, and it is illegal to collect or kill them there.

The Rubber Boa has a habitat similar to that of the California Mountain Kingsnake. It is yellowish below and greenish above. Its head and tail are similarly rounded, so it appears to be a large worm. It feeds on insects, salamanders, lizards, and small mammals. When molested, it rolls into a ball and wiggles its tail, seemingly to perplex a predator who would attempt to bite it and swallow it headfirst. Its tail is often scarred, and some authorities contend that the scars are from the bites of predators. Other authorities point out that the Rubber Boa crawls into the burrows of mice and feeds on the babies, and they believe the scars are mouse bites. These snakes are very docile. Their greatest threat also appears to be snake collectors. Rubber Boas are live-bearers, producing litters of two to eight young. Home ranges are rather small, and individuals are sighted time and again in the same area, which in part has led to estimates of

remarkable age. One account records an individual that was frequently sighted in the field and finally kept in captivity for 30 years. In this instance, the snake was estimated to be up to 69 years of age, and it remarkably gave birth to a litter of young at that ripe old age.

The Rubber Boa of the Sierra Nevada is the Northern Rubber Boa, *C. b. bottae*. In the San Bernardino and San Jacinto Mountains, there occurs a smaller subspecies, the Southern Rubber Boa, *C. umbratica*. Because of habitat reduction in southern California, the state of California lists this subspecies as threatened, and it is illegal to collect them. The two subspecies seem to intergrade in the Mount Pinos area, but neither is present in the region between Mount Pinos and the San Bernardino Mountains, even though appropriate habitat seems to be present. This unexplained disjunct pattern of distribution seems to apply to other animals as well, such as the spotted salamanders (Ensatina) discussed earlier.

BIRDS Among Sierran birds (figure 4.35), there are no known hibernators at higher elevations. Most birds migrate in the winter, but some, such as Steller's Jay, *Cyanocitta stelleri*, remain active in the winter, living on stored food. Steller's Jay gathers and stores acorns and pine seeds. Similar to the Western Scrub-jay, *Aphelocoma californica* (plate 19F), of the foothills, Steller's Jay is the alarmist. It is a brightly colored bird with a loud, raucous call, which it uses to alert the entire forest community when there is danger. Steller's Jay differs in appearance from the Scrub-jay by having a large, dark crest on its head. It is the only crested jay in the west. Scrub-jays and Steller's Jays eat acorns, but both are omnivorous, feeding also on available insects.

Why do their ranges not overlap to a greater degree? Scrub-jays can tolerate higher temperatures, and they can survive without liquid water as long as they eat insects or berries. Steller's Jays must drink water. Comparative tests of the two jays, in chambers where the temperature is gradually raised, show that Scrub-jays are

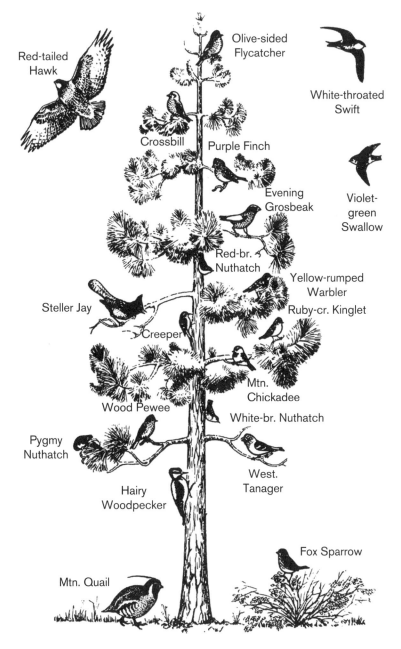

FIGURE 4.35 Niche partitioning among some forest birds (from Storer, T. I., and R. L. Usinger. 1963. *Sierra Nevada Natural History*. Berkeley: University of California Press).

able to keep a constant body temperature of 107°F (42°C) if they remain in the shade, even if the air temperature is raised from 50°F to 90°F (10–32°C). Apparently, Scrub-jays are better able than the Steller's Jays to lose heat by radiation from the unfeathered portions of their feet.

Perhaps the most common bird of the Mixed Coniferous Forest is the Dark-eyed Junco, *Junco hyemalis*. This is a small gray bird with a dark head. It has been mentioned that the name junco refers to the medieval executioner who wore a black hood. Actually the origin of the name is from a Spanish word that means reed

FIGURE 4.36 Mountain Chickadee, *Poecile gambeli.*

or rush. A seedeater, the junco in winter moves downslope below the snowline, where it continues to forage for seeds. In summer, it nests on the ground among Dwarf Bulrushes or grasses. When threatened, it does a broken-wing act, leading a potential predator away from its nest.

Another common bird of the Mixed Coniferous Forest is the Mountain Chickadee, *Poecile gambeli* (figure 4.36). In its niche, it replaces the White-breasted Nuthatch above the foothills. Interestingly, while we think of the Mountain Chickadee as a gleaner, eating small insects on bark and needles, people in mountain communities find the chickadee visiting feeders to eat seed, and also drinking the liquid from hummingbird feeders. The Mountain Chickadee may not be often seen, but it is often heard. Its call is the mournful sound of the Mixed Coniferous Forest. It utters three slow notes—two high tones and one somewhat lower. They also utter a chattering sound that is described as "chick-a-de-de-de." They are small birds with gray bodies and black-and-white stripes on their heads. Their chins are black, and the sides of their heads are white. They nest in old woodpecker holes.

The jolly or rollicking sound of the Mixed Coniferous Forest is that of the Black-headed Grosbeak, *Pheucticus melanocephalus.* This large member of the Cardinal family (Cardinali-dae) has a very thick bill, implying that it is a seedeater, but it also eats buds and insects in the upper canopy of trees. It also enjoys eating fruit if it is available. This bird migrates to Mexico in the winter, but in spring and summer its melodic song of rising and falling notes is one of the common sounds of the forest. This bird has a rusty-colored body with a black head and black-and-white wings. It is replaced at higher elevations by Evening Grosbeaks, *Coccothraustes vespertinus,* and Pine Grosbeaks, *Pinicola enucleator.* In spite of their similar appearance to the Black-headed Grosbeak, these two species are in the Finch family (Fringillidae).

A list of summer visitants is long, but three conspicuous birds are worth mentioning: the Western Tanager, *Piranga ludoviciana* (plate 19C), the Bullock's Oriole, *Icterus bullockii,* and the American Robin. The Western Tanager with its bright yellow body and red head are sure to attract attention. But, even more spectacular is the male Bullock's Oriole. It has a bright yellow body, orange head, and a black chin. The Western Tanager is a gleaner in foliage, whereas the oriole drinks nectar from the same sorts of tubular flowers visited by hummingbirds. The tanager and the robin have songs that resemble that of the Black-Headed Grosbeak. The American Robin is a conspicuous ground feeder in the foothills and the forest. It has a brownish back and an orange breast, and it feeds on worms and grubs that it locates by sound.

An assortment of woodpeckers is also characteristic of the Mixed Coniferous Forest. As in the foothills, they are partitioned according to food habits. The Northern Flicker (figure 4.23) ranges from the foothills to the forest. Other woodpeckers fill niches similar to related forms of the foothills. The Downy Woodpecker of the foothills is replaced by a larger cousin, the Hairy Woodpecker, *Picoides villosus,* in the forest. Nuttall's Woodpecker is replaced by the White-headed Woodpecker, *P. albolarvatus,* the most commonly seen woodpecker in the Mixed Coniferous Forest. It often works the bark of the lower portions of the trees in search of insects.

FIGURE 4.37 Pileated Woodpecker, *Dryocopus pileatus* (photo by Chuck Leavell).

FIGURE 4.38 White-crowned Sparrow, *Zonotrichia leucophrys*.

The most spectacular woodpecker is the Pileated Woodpecker, *Dryocopus pileatus* (figure 4.37). This crow-sized bird is black with a bright red, pointed crest on its head. It seems to prefer feeding on White Firs, although it is found on pines and oaks as well. The Pileated Woodpecker can be heard seemingly for miles, hammering on dead or diseased trees. They feed mainly on Giant Carpenter Ants and large beetle larvae (grubs). For a time it was thought that these spectacular woodpeckers were on the decline in the Sierra, but numerous sightings in recent years, particularly in the Yosemite area, indicate that they are still with us.

Niche partitioning among warblers and sparrows is also evident, if one takes the time to attempt positive identification. These busy little creatures spend most of their time out of sight in the foliage. Warblers are all gleaners. As many as 10 different species can be identified at certain times in the Yosemite area. Because most warblers are yellow to some degree, they are often called wild canaries. Different seasonal population and foraging habits prevent these birds from competing with each other.

Perhaps 15 different kinds of sparrows may be sighted at various elevations in the Sierra Nevada, most of them lowland or foothill species. In the Mixed Coniferous Forest or above, one is likely to see only four kinds. The sparrow that typically hangs around campgrounds is the Chipping Sparrow, *Spizella passerina*. In the summer, it has a distinctive reddish crown. In willows along water, the incessant chipping noise that one hears is usually the call of the White-crowned Sparrow, *Zonotrichia leucophrys* (figure 4.38), which can be identified by three white stripes outlined in black on its head. Often joining the White-crowned Sparrow in the willows is a more typical-looking brown sparrow, the Lincoln's Sparrow, *Melospiza lincolnii*. Another common-looking brown sparrow is the Fox Sparrow, *Passerella iliaca*, which forages in shrubs located away from water. Sparrows feed primarily on seeds, but many of them rely heavily on insects when seeds become scarce.

MAMMALS Among mammals, the most conspicuous is the Western Gray Squirrel, *Sciurus griseus* (plate 18B). Subsisting largely on acorns, gray squirrels inhabit oaks in the upper foothill belt and in the Mixed Coniferous Forest, where they occur in association with California Black Oaks. Black oaks are a component of the dry Mixed Coniferous Forest; therefore, gray squirrels are more common there. In moist Mixed Coniferous Forest, they eat seeds from Sugar Pine and White Fir, but where snow becomes deeper, gray squirrels are replaced by

FIGURE 4.39 California Black Oaks growing in a clump.

Chickarees (Douglas Squirrels), *Tamiasciurus douglasii* (plate 18C).

Western Gray Squirrels remain active through the winter and find shelter in cavities, particularly in oaks. They bury caches of acorns or pine seeds in many places on the forest floor, and during the winter they dig through the snow to retrieve their stored food. Apparently, they locate the caches through a combination of odor and memory. If a squirrel dies, or for some other reason fails to dig up its cache, the seeds may germinate. It is in this way that oak trees become dispersed uphill, and it also explains why they may grow in clumps (figure 4.39). Another reason for clumped oaks may be regeneration of sprouts from a single trunk after a fire.

In areas where snow lies deeply on the ground for a long time, gray squirrels are absent. Presumably, the energy required to dig through deep snow would not be rewarded by the calories supplied by a small cache of food. On the other hand, the Chickaree stores its food in a few large caches, mostly above ground near its home, or at the base of a tree where it is able to defend the stores. In digging through deep snow, the energy expended is replaced by the much larger amount of food in the cache. Chickarees replace gray squirrels in redwood groves and at higher elevations in the Montane Forest. Chickarees are also notable because they raid bird's nests and eat the eggs.

In Yosemite Valley, at least three squirrel species seem to coexist. These include the California Ground Squirrel, the Western Gray Squirrel, and the Chickaree. It is also possible that, under certain circumstances, the Golden-mantled Ground Squirrel, *Callospermophilus lateralis* (plate 18D), may join the group.

Most of the floor of Yosemite Valley would be characterized as dry Mixed Coniferous Forest. Dominant trees include Ponderosa Pine, California Black Oak, and Incense Cedar. The Merced River meanders broadly through Yosemite Valley. On its banks, and in shady localities, are trees of riparian affinity that also are sources of food and shelter. Among these are Canyon Live Oak, *Quercus chrysolepis*; California Bay Laurel, *Umbellularia californica* (figure 6.8); Mountain Dogwood, *Cornus nuttallii*; and Big-leaf Maple, *Acer macrophyllum* (figure 6.9). Even so, in other areas where this assemblage of trees occurs, there does not seem to be the overlap of squirrels that occurs in Yosemite.

Perhaps the situation is a consequence of humans. A great deal of food is available, either directly from handouts, from untended picnic tables, or from trash. In campgrounds, predator species are normally wary and may be reduced in abundance. Coyotes in Yosemite Valley also depend heavily on handouts and may be well fed without having to rely on prey species. Furthermore, the numerous roads and trails have provided abundant open, cliff-like habitats for ground squirrel burrows. It seems that the parklike environment of Yosemite Valley, with all its human trappings, is responsible for this atypical assemblage of squirrel species.

The Yosemite area also provides a setting for the discussion of Mule Deer (figure 4.40)

FIGURE 4.40 Mule Deer, *Odocoileus hemionus*, with antlers in velvet.

and American Black Bears. The history and present distribution of these two large mammals is deeply rooted in the story of human impact on natural habitats. The Mule Deer population in the Jawbone Deer Herd and its interaction with the Mountain Lion is described in Chapter 2 on Basic Ecology. It is a classic study of fluctuations of an animal population in association with human interference.

The story of the American Black Bear, *Ursus americanus*, is also closely related to game management practice in Yosemite National Park. Grizzly Bears are no longer found in California; all Sierra bears are American Black Bears. They are smaller and less ferocious than grizzlies, although people tend to overestimate their size. Large males may exceed 400 lb (190 kg), but the average weight of males is 190 lb (86 kg). Average weight of females is 128 lb (58 kg). Most American Black Bears spotted in the Sierra Nevada are about the size of a Saint Bernard dog. They are omnivorous, but they rely heavily on vegetable matter as a staple in their diet and eat a large amount of grass. For protein, they rely mostly on ants and termites. In season, they feed heavily on berries.

Normally, American Black Bears are shy, retiring creatures that seldom interact with humans. In national parks, however, campers who are careless with their food have caused bears to become dependent on human sources of food. It is almost as if the bears have become junk-food addicts. High-calorie food has become a motive for the animals to dump ice chests and break into cars. Humans are not harmed by American Black Bears unless they try to take their food back. To protect foolish campers from possible harm, the National Park Service has developed a practice of capturing errant bears, tagging them, and translocating them to a portion of the park where natural food is abundant. If the bear returns to pester campers, it is moved again. It will be moved several times before drastic action is taken. Unfortunately for the bear, if it returns several times to steal food or break into cars, or if it develops a pattern of bluff-charging humans, it will be killed. Although this killing is justified in order to protect campers, ironically, the campers who feed bears are actually leading the bear to its death.

Bears are able to travel long distances in search of food. If a bear raids a camp in the backcountry, it is for the same reason it raids a camp in a popular car-camping area. Aggression or competition for food can cause subdominant bears to be forced out of major campgrounds to fend for themselves in the backcountry. In 2015, about 40 juvenile bears, an inordinate number, were found roaming about in the town of Three Rivers, where the headquarters for Sequoia National Park is located. It is believed that the bears were feasting primarily on acorns, although they raided trash cans where they were not bear proof.

If a tagged bear appears outside of national park boundaries, it does not mean that the National Park Service carried the bear by helicopter and dropped it off in the hopes of being rid of it. Helicopter flying time is expensive and is never used to transport American Black Bears. Bears are released in remote parts of the park deliberately to avoid human interactions. They are transported in a trailer, towed by a pickup truck.

During the winter, American Black Bears go into a state of hibernation. Some authorities contend that American Black Bears do not really hibernate, because their body temperatures do not drop as low as that of the environment, and they do not produce high-calorie brown fat. It is unlikely, however, that a mammal of such large body size would lower its body temperature as much as a small mammal. American Black Bears lower their body temperature about 10–12°F (6–7°C) during winter torpor. Energy demand is reduced by half for every 10°C (14°F) of lowered body temperature. They remain at about 88°F (31°C) for most of the winter, which lowers their heart rate and metabolic rate to about one-half. Most small mammals that hibernate waken every few days, burning brown fat to raise their body temperatures, so that they can move around and urinate. Some eat and defecate. American Black Bears seldom waken, they do not urinate, and they do not defecate during hibernation. They recycle nitrogenous wastes to form amino acids, and they live on stored fat during their dormancy. When an animal lowers its body temperature, its metabolic rate is reduced; therefore, hibernation is important for energy conservation.

American Black Bear cubs do not hibernate. They are born in the den during January while the mother is hibernating. Cubs weigh about a pound at birth. They sleep against the mother's warm, sparsely haired underside and suckle while she remains dormant. From time to time, the mother arouses and eats the cubs' feces. Gestation lasts about 335 d, and cubs are dependent on mother's milk for about 30 weeks.

In spring, the mother and cubs emerge from the den. The cub at this time weighs about 4 lb, having increased its weight about fourfold since birth. The mother has lost from 15% to 30% of her body weight during winter, and she is ready to eat. Bears mate in the spring after they emerge from hibernation, but the fertilized egg does not implant in the uterus until fall. Whether or not it does implant depends on how much food the female has eaten during the summer; in lean years, the egg is reabsorbed. When food is abundant, twins or triplets may be born. In national parks, where campground food has added calories to the diet, bears often have multiple births, which act to further increase the number of bears in the campgrounds. The American Black Bear population in the Sierra Nevada continues to increase. They are now common on the dry east side of the Sierra, where they formerly were rare. According to recent estimates, more than 25,000 Black Bears roam the state today.

The problems of human and American Black Bear interaction are exacerbated by ignorant humans. American Black Bears' "addiction" to human food is not a natural situation. The National Park Service is charged with the responsibility of protecting wildlife, and rigid enforcement of no-feeding laws is a means to that end. Fortunately for Californians, these are not Grizzly Bears that crave high-calorie food.

Montane Forest

The Montane Forest is the snow forest of the Sierra Nevada. Trees here are spire-shaped in order to shed snow. Precipitation in this forest, most of which is snow, is the heaviest in the range, averaging from 35 to 65 in (90–165 cm) per year. It is not uncommon for snowfall of 50 ft (13 m) to fall in a given year. Within the Red Fir zone of the western Sierra, the greatest amount of precipitation occurs in the central portion of the range, east of Sacramento. More precipitation falls here because at this latitude there is a "window" in the rain shadow of the Coast Range. The opening in the Coast Ranges is where San Francisco Bay is located. Air masses that flow through this gap are allowed to carry their moisture all the way to the Sierra before adiabatic cooling causes precipitation.

In the Merriam system of classification, the Montane Forest community was known as the Canadian Life Zone. It was formerly known as Lodgepole-Red fir Forest. Some authorities insist on separating the community into two realms, the Lodgepole Pine Forest and Red Fir Forest. This distinction might also be a function of moisture, with Red Firs occupying the wet zones. Unlike the Mixed Coniferous Forest where moist and dry regions are primarily a function of slope aspect, it appears that the distinction in the Montane Forest is more a function of elevation and rain-shadow effect.

On the west side of the Sierra, from Yosemite National Park northward to the latitude of Lake Tahoe, there is a magnificent forest composed of nearly pure stands of California Red Fir, *Abies magnifica*. As mentioned earlier, Red Firs are among the species sold as Christmas trees under the name "Silver Tip." These are moisture-loving trees. Depth of the winter snow averages 10–30 ft (3–10 m), as evidenced by the lower level of lichen growth on the trunks. These lichens (*Letharia* spp.) are known as Staghorn Lichens in reference to their branched appearance. Also known as Wolf Lichens, they are of a variety known as fruticose lichens, in reference to their cup-shaped fruiting bodies. These lichens are active all yearlong, deriving their moisture entirely from the air. Because these lichens require

light, the lowest point on a trunk where they can grow is dictated by snow depth.

Red Fir requires deep, moist but well-drained mature soil. It can tolerate cold temperature as well as any conifer, but availability of soil moisture is an important limiting factor. For this reason, Red Firs are scarce on the eastern slope of the Sierra, and south of Mammoth, with the exception of upper Sawmill Canyon, they are absent on the east slope. Red Firs in the southern Sierra are now considered to be a separate variety known as Critchfield Red fir, *A. magnifica* var. *critchfieldii*. On the other hand, waterlogged soil near a meadow also cannot be tolerated. Red Firs are not fire tolerant. Lightning commonly kills old Red Firs, but fires are seldom large due to the moist nature of litter beneath the forest canopy. Nevertheless, these small fires are important for opening the forest canopy and allowing germination of new trees. Decay at maturity also provides openings. The resulting mosaic of trees of different ages provides a variety of habitats for animals as well. This is a major reason for preservation of old-growth forests.

Red Firs grow at higher elevations than White Firs, and there is a zone of overlap, although the two species are not known to hybridize. They are difficult to tell apart without practice. Red Firs usually have redder bark, but unless the trees are large and old, the color difference may be difficult to discern. Cones are visible at the tops of trees in late summer and autumn. If cones are apparent, the trees are easy to identify. Red Fir and White Fir (figure 4.29) cones both grow upward on the stem, but Red Fir cones are twice as large. They are fatter, and they are 4–8 in (10–20 cm) long. If no cones are available, it is difficult to distinguish a Red Fir from a White Fir, and it requires close inspection of the needles. Needles of Red Fir are shorter than those of White Fir, but on both species they grow upward on the stem when there is abundant light. White Fir needles are flattened and have a quarter turn, or slight twist at the base. Red Fir needles are round in cross section and have no such twist. Red Firs often reach great size. The tallest tree, known as Big Red, is 203 ft (61.9 m) high and 7.4 ft (226 cm) in diameter at breast height. It is found at the headwaters of the South Fork of the Tuolumne River. Another giant of the species at 172 ft (52.4 m) is not as tall, but at breast height it is 9.7 ft (295 cm) across. This tree known as the Leaning Tower is distinctive not only because it leans, but also because it has a double crown. It is found along the Tioga Pass Road in a grove of large Red Firs.

The most common species of tree in this forest is the Lodgepole Pine, *Pinus contorta murrayana* (plate 6B). Where conditions are appropriate, other associated tree species may include Mountain Hemlock, *Tsuga mertensiana* (plate 6B), Western White Pine, *P. monticola*, and Sierra Juniper, *Juniperus grandis*. Lodgepole Pine is easy to identify. It is the only native two-needle pine of the high mountains. Its bark is whitish and scaly. It appears as if hundreds of corn flakes are stuck to the bark.

Lodgepole Pine is a pioneer tree species. At higher elevations, it is the first tree to move into an area denuded by fire, flooding, or avalanche. These are also the trees that ring meadows and through time will replace a meadow. This is a rapidly growing species, but it is not long-lived. Large specimens over 3 ft (1 m) in diameter are often less than 300 years of age. The largest Lodgepole Pine in the Sierra Nevada is found in the Emigrant Basin Wilderness. It is 7 ft (2.2 m) in diameter, it is 91 ft (28.4 m) tall, and its canopy covers 36 ft (12 m). A Lodgepole Pine of similar size also occurs near Big Bear Lake, in the San Bernardino Mountains of southern California. This tree is 110 ft (34.3 m) high, but it is slightly smaller in diameter.

In the Sierra Nevada, Lodgepoles occur in small, even-aged clumps. These clumps are seldom larger than 900 ft² (80 m²). The primary sites for regeneration are areas of treefall, usually associated with winds or avalanches. In contrast, in the Rocky Mountains, where summer lightning is more common, crown fires may occur that wipe out extensive stands. The

Rockies therefore contain large stands of Lodgepole Pine trees of similar size and age.

Over a period of thousands of years, Lodgepole Pines gradually invade and replace high-country meadows. Their roots can tolerate waterlogged soil, so they probably aid in the drying of this wet soil. As soil and humus accumulate and the soil becomes better drained, Red Firs gradually replace Lodgepoles. Because the Sierra Nevada is geologically young, Lodgepole Pine is abundant and widespread throughout higher elevations. In many areas, it occurs all the way up to tree line. Backcountry hikers and backpackers are familiar with Lodgepole Pines. The needles are everywhere on the ground, and the small, sharp cones always lie in great abundance exactly where you want to lay out your tent or sleeping bag.

The shape and density of Lodgepole Pines are highly variable. The name refers to its use by Native Americans for the structural support of their homes. In locations where the trees grow close together, the trunks are straight and thin—exactly what an Indian would want. In exposed sites, old trees are short—less than 30 ft (10 m) tall—and the trunk may be 2 ft (60 cm) thick (figure 5.4). Commonly, old trees are twisted and may show the effect of lightning. On the east side of the Sierra, many lakes are rimmed by pure stands of Lodgepoles (plates 6B and 11.3). These rather widely spaced trees form an open forest approximately 30 ft (10 m) high. In contrast, on the west side of the Sierra there are many acres of Lodgepoles where the trees grow close together and reach heights of 50–80 ft (15–25 m; figure 11.3). The difference between these two types of forest is simply the amount of moisture. Of all pine species studied, Lodgepole Pine has the most remarkable ability to regulate its internal water balance over a wide range of soil moisture.

During the last 30 years or so, many acres of Lodgepoles have been defoliated by a small moth larva known as the Lodgepole Needle Miner, *Coleotechnites milleri*. These defoliated trees are particularly conspicuous in Yosemite

National Park along the Tioga Pass Road through Tuolumne Meadows. Every other year, during August, millions of tiny white moths hatch from their pupae. They mate and lay their tiny yellow eggs on the needles of mature Lodgepoles. During this time, millions of the moths may be in flight at any time, giving the air the appearance of being filled with blowing ashes. When the eggs hatch, a tiny caterpillar bores into the needles and hollows them out. They go dormant in the winter, but otherwise they continue to feed until the spring of their second year of life, after which they metamorphose into moths, and the cycle is repeated. Normally, they are kept under control by heavy predation from gleaning birds such as the Mountain Chickadee. There are also many species of parasitic wasps and other insects that feed on the Needle Miner larvae. However, for some unknown reason, there are occasional population explosions, and the moths become extremely abundant. At first, the National Park Service, in an attempt to save thousands of acres of Lodgepoles, used DDT (dichlorodiphenyltrichloroethane) and malathion to kill the moths. The experiment was a disaster, because the insecticides killed many of the moth's natural enemies, including birds. Furthermore, the DDT began to appear in tissues of trout living in high-country lakes and streams. Today, the National Park Service regards Lodgepole Needle Miner infestations as a natural phenomenon. Educational exhibits along the Tioga Pass Road explain how defoliation of Lodgepole Pines by Needle Miner larvae is a part of the grand scheme of things.

In regions where there is slightly more moisture and the soil is cold, Lodgepole Pines are joined by Mountain Hemlocks. This is particularly true for the west-slope forest from Yosemite northward (plate 6B). Southernmost Mountain Hemlocks are located near Silliman Lake in Sequoia National Park. Toward the south and on the east side of the Sierra, Mountain Hemlocks become a part of the upper forest community, the Subalpine Forest, where the trees grow in clumps in moist, cold pockets (figure 5.3). In exposed sites, they appear shrub-like. Most often, Mountain Hemlocks occur in even-aged stands, and they seem to die all at the same time, usually in association with a root fungus.

Hemlocks resemble firs by having needles all along the branches. They differ from firs in that the needles project outward from all sides of the branches. Cones grow downward from the tips of the branches and superficially resemble those of Douglas-fir. In fact, the scientific name for Douglas-fir, *Pseudotsuga*, means "false hemlock." Mountain Hemlock is not as large a tree as Douglas-fir, but the largest specimen, near Tyron Meadows in Alpine County, is a healthy 113 ft (35.3 m) tall. It is over 7 ft (2.2 m) in diameter, and its crown spreads to 44 ft (14 m).

Foliage on Mountain Hemlocks is flexible and droops downward, giving the tree a beautiful, graceful appearance. Often the top nods over, making the tree easy to recognize from a viewpoint overlooking the forest canopy. In his first book, *The Mountains of California* (1894), John Muir raved about Mountain Hemlocks, calling them "the most singularly beautiful of all the California coniferae."

Western White Pine, also known as Silver Pine (figure 4.26), replaces Sugar Pine at higher elevations. It is a five-needle pine with needles about 2–4 in (5–10 cm) long, and it has the same sort of appearance as Sugar Pine, with long horizontal branches that resemble spider arms. However, White Pine branches tend to grow upward at the outer end, sort of like a candelabra. Cones of Western White Pine resemble those of Sugar Pine, but they are only one-quarter to one-half the size; they are seldom over 8 in (20 cm) in length. The bark of Western White Pine is easy to recognize because it breaks up into squarish plates 2–4 in (5–10 cm) across. On dry sites such as slopes with coarse soil, where water percolates rapidly, there may be pure stands of Western White Pine. In this respect, they resemble a high-elevation version

FIGURE 4.41 Sierra Juniper, *Juniperus grandis*.

of Jeffrey Pine. Large, old trees may stand in solitary splendor on rocky outcrops. In one study in the Emerald Lake Watershed of Sequoia National Park, Western White Pine represented 71% of the individual trees, and 83% of the basal area.

Sierra Juniper, *J. grandis* (figure 4.41), grows in the most exposed sites of the Montane Forest and is the tree most commonly associated with rocky outcrops here. Most species of juniper in California are shrubs, but this is a true tree. Old trees on exposed sites have a distinctive appearance. The trunk is very thick in proportion to the height of the tree, and it is often twisted and contorted by wind and lightning. Many a photographer has used the twisted shape of a Sierra Juniper to accent a picture of the Sierra Nevada. The largest known juniper in the world is a Sierra Juniper that grows near Sonora Pass. This tree, known as the Bennett Juniper, is estimated to be at least 3000 years old. It is nearly 14 ft (3.5 m) in diameter and 87 ft (28 m) in height. As in the other junipers, the cone is fleshy and berrylike. These cones are an important autumn food for many birds and mammals in spite of their turpentine flavor.

On the east side of the Sierra Nevada, Sierra Junipers are common on south-facing slopes in canyons at elevations lower than the forest belts. This distribution applies particularly from Lee Vining southward to Lone Pine. They also grow in isolated localities as far south as the Transverse Ranges.

To the north of the Sierra Nevada, Western Junipers, *J. occidentalis* (figure 6.13), are common on upper slopes of the Great Basin, where they grow amongst Great Basin Sagebrush. These close relatives of Sierra Juniper are particularly common on the east side of the Cascades in northern California, Oregon, and Washington.

There is not much of a shrub component in Lodgepole-Red Fir Forest. Deep shade precludes low-growing vegetation in many areas, and deep snow makes it difficult in others. When there is a shrub component, it is likely to be dominated by Bush Chinquapin, *Chrysolepis sempervirens* (figure 4.42), a member of the beech and oak family (Fagaceae). The fruit has a spiny husk enclosing three bitter nuts. Humans seldom get a chance to taste these nuts because they are harvested by squirrels and bears as soon as they are ripe. On dry slopes in the forest, there may be thickets of one or more species of manzanita. Of particular note is a prostrate (ground-covering) species known as Pine-mat Manzanita, *Arctostaphylos nevadensis* subsp. *nevadensis*.

The deep shade of the Lodgepole-Red Fir Forest makes it difficult for understory plants

FIGURE 4.42 Bush Chinquapin, *Chrysolepis sempervirens*.

in general. Some plants, however, have developed a strategy of deriving most if not all of their nutrition not from photosynthesis, but from decay of humus. These plants are known as saprophytes. Many fungi, such as mushrooms and puffballs, live this way.

A number of members of the Heath family (Ericaceae) possess little or no chlorophyll. These plants have solved the problem of limited light by tapping into the mycorrhizal fungi that thrive in the forest soil and deriving nutrition and water in this way. The beneficial association of the roots of conifers with mycorrhizal fungi was mentioned previously, but these saprophytic plants survive with little or no photosynthetic activity. The degree to which the saprophytes are associated with or depend upon the fungi is variable. Shinleaf or White-veined Wintergreen, *Pyrola picta*, contains some chlorophyll, but its preference for deep shade dictates that it relies to some degree on nutrition from the mycorrhizae.

Snow Plant, *Sarcodes sanguinea* (figure 4.43), is perhaps the most spectacular of these plants. Apparently its relationship is parasitic with a host plant, usually White Firs with which they share mycorrhizae. It is a bright red, thick-stemmed plant that appears early in the spring. Its name refers to the fact that it erupts mushroomlike from the forest floor when there is still a good deal of snow on the ground. It often occurs in clusters about a foot (30 cm) high. Later in the spring, a cluster of red flowers appears at the tip of each stem. It is not uncommon to see hummingbirds or butterflies drawing nectar from these flowers.

Pinedrops, *Pterospora andromedea*, looks like a tall, skinny Snow Plant. Some authorities prefer to call this plant a parasite because it has no root system or fungi of its own. Instead, it draws its nutrition from the mycorrhizae that are associated with adjacent plants.

The Corralroots (*Corallorhiza* spp.) are members of the orchid family (Orchidaceae) and are also dependent for nutrition on symbiotic fungi. The name refers to the coral-like appearance of the reduced root system. These are thin-stemmed plants that usually grow individually from humus or a rotting log. The color of the stems ranges from pale red or yellow to brown or purple. At blooming time, close inspection will reveal a long cluster (raceme) of up to 40 orchid-shaped flowers, each of which is about 0.25 in (6–8 mm) across.

Phantom Orchid, *Cephalanthera austiniae*, so called because it may appear to be an all-white plant, is a Eurasian saprophytic orchid that arises from a branched, creeping rootstock. Its flowers, which may be tinged with yellow, are about twice the size of those of the Corral-roots; therefore, it is more easily recognized as a flowering plant. It is closely related to Stream

FIGURE 4.43 Snow Plant,
Sarcodes sanguinea.

Orchid and Broad-leaved Helleborine (*Epipactis* spp.) (plate 20 C), beautiful flowering orchids that occur in wetlands.

The situation with paintbrushes (*Castilleja* spp.) is somewhat different (plate 20A). These spectacularplantswithtubular,red,hummingbird-pollinated flowers are hemiparasites, meaning they are half-parasitic. They have green leaves and produce their own sugars, but they tap into nearby shrubs from which they get an additional water supply. In the case of Giant Red Paintbrush, *C. miniata*, it has been demonstrated that it obtains alkaloid defense compounds from Silvery Lupine, *Lupinus argenteus*, its host plant, which helps protect it from insect herbivores.

Animals of the Montane Forest

BIRDS Many bird species of the Montane Forest are shared with Mixed Coniferous Forest. However, some good examples of replacement within similar species are found within the finches (Fringillidae; figure 4.22). In particular, there are three species of finches in which the male has a red head and brown streaking on the breast; the females all resemble sparrows but can be distinguished because they have thicker bills. At lower elevations, in the western foothills and the Great Basin on the east, the House Finch or Linnet, *Carpodacus*

mexicanus, is a common bird seen in foliage on the outer edge of trees. Males defend their territories and court females with a loud warbling song. They eat buds, seeds, and small insects. In the Mixed Coniferous Forest, the same niche is occupied by the Purple Finch, *C. purpureus*. It forages similarly and sounds the same. It also looks like a House Finch, but streaking on the breast is less distinct. In the Montane Forest Cassin's Finch, *C. cassinii*, occurs in this niche. It also looks and sounds similar, but the red color is concentrated on its crown. It takes considerable skill to tell these birds apart by sight. Most amateurs identify them according to elevation and assume they are correct. Designation by elevation is helped by the fact that these finches tend to remain resident in each zone regardless of season, although a bit of downslope movement in winter tends to confuse the issue somewhat.

Another source of confusion is that the male Pine Grosbeak also looks and sounds like the other three finches. However, the Pine Grosbeak is much larger, and the male is redder. The small red-headed finches are seldom over 6 in (15 cm) long, whereas the Pine Grosbeak is about 8 in (20 cm) long. The female Pine Grosbeak is yellow and is so different in appearance from the male that it is difficult to believe they are related. More confusion can occur if

Evening Grosbeaks are in the area. Both male and female Evening Grosbeaks are yellow, and, in contrast with the female Pine Grosbeak, they have larger, light-colored beaks and distinctive white wing patches. Evening Grosbeaks are not good singers. The Black-headed Grosbeak is the large, brightly colored singer of the Mixed Coniferous Forest. Large finches, the grosbeaks feed and act similarly to the smaller red-headed finches. Grosbeaks are also somewhat separated by elevation, but the boundaries are less distinct. The Black-headed Grosbeak occurs in the foothills and the Mixed Coniferous Forest, Evening Grosbeaks tend to be most common in White Fir forests, and the Pine Grosbeak is a component of the Montane Forest. These six species are separated by elevation and, to some degree, the size of their food. Grosbeaks are much more dependent on pine seeds as a source of food, but they also eat buds, insects, and fruit. Habitat partitioning keeps these species from competing severely with each other.

Large finch-like birds observed in nervous flocks are probably Red Crossbills, *Loxia curvirostra*. Males are red, and females are yellow, like the Pine Grosbeaks. Furthermore, Red Cross bills are grosbeak sized. Careful inspection, however, will reveal the large crossed bill tip that is used to pry seeds from unopened Lodgepole Pine cones. These birds are year-round residents of the Lodgepole forests, but they are not always common. They migrate in large flocks, congregating where food is abundant.

If the Mountain Chickadee is the mournful voice of the Mixed Coniferous Forest, the Hermit Thrush, *Catharus guttatus* (figure 4.44), may be its equivalent in the Montane Forest. Its clear, liquid whistle of alternating long low and short high notes is common from May through July. Similar to other thrushes, such as the American Robin, it feeds on the ground. It is a solitary bird, and its brown plumage and reddish tail make it difficult to spot. Seldom seen but often heard, the Hermit Thrush is one of the most common summer visitors of this forest zone. During winter, it migrates to Mexico.

FIGURE 4.44 Hermit Thrush, *Catharus guttatus.*

A large bird that remains resident year-round is the Sooty Grouse or Blue Grouse, *Dendragapus fuliginosus* (figure 4.45). Even though these birds are 20 in (50 cm) long, they are difficult to see. Their bluish, mottled appearance is a perfect disguise as they walk slowly or poise motionless in filtered light on the forest floor. They feed primarily on needles of pines and firs, although they take some berries and insects as well. Apparently, the diet of conifer needles has provided a distinct flavor to the meat. Those that have tasted the flesh of a Blue Grouse indicate that the meat tastes like turpentine. Perhaps this combination of a food that few others eat and its residue in the flesh is what makes the grouse so common. They are well concealed, but they move slowly. It seems that they would be a favored prey species.

The sound of the courting male grouse, a deep-throated booming noise that carries for a mile or so, is familiar to hikers. A female moving along slowly, herding and clucking quietly to her brood of six or seven small grouse, may be a surprise sight along a trail. In the summer, the birds are dispersed widely in the upper forests. In winter, they become concentrated at the lower fringe of the forest among Red Firs on the west side of the Sierra. There are also

FIGURE 4.45 Sooty Grouse, *Dendragapus fuliginosus.*

reports of large winter flocks of grouse in the Mixed Coniferous belt of canyons on the eastern side of the Sierra.

Among predator birds, in general, owls are nocturnal and hawks are diurnal. There are at least nine different kinds of owls and seven kinds of hawks that range through Sierran forests. Owls hunt at night, using their excellent vision and hearing to locate their prey. An owl's eyes are about 100 times more sensitive than a human's; therefore, owls are able to spot their prey in very dim light. Furthermore, the large ring of feathers around each eye gathers sound in the same way as the external ear of a mammal. By having a sound-gathering device around each eye, an owl is able to locate its prey very accurately even if it is pitch dark. The most common owl is the Great Horned Owl, *Bubo virginianus* (figure 8.45). It ranges through a variety of habitats from the desert to the mountains, wherever there are trees for nests and daytime refuge. Hooting of this large owl from dusk to dawn is a familiar sound to campers.

Perhaps the most spectacular Sierran owl is the Great Gray Owl, *Strix nebulosa*. The largest of all the Sierran species, it has a wingspan of 4.5 ft (1.5 m). Unlike most owls, it tends to be diurnal, although the sounds produced by prey are still important for hunting. In fact, they can locate pocket gophers (*Thomomys* spp.), which are among their favorite prey, by sound alone while they move underground. A Great Gray

Owl can swoop down and take a hapless gopher from its burrow in soil or snow by grasping it with its talons while it is still beneath the surface. Great Gray Owls are not common in California, although they range throughout Arctic forests of the Northern Hemisphere. In California, there are probably no more than 30 pairs, and these are concentrated in the Central Sierra Nevada, particularly in the Yosemite area.

Forest-dwelling hawks are not common either, but the Northern Goshawk, *Accipiter gentilis* (figure 4.46), is a year-round inhabitant of the forests. It resembles a Red-tailed Hawk in size, but it does not soar like a Red-tail, and it has a narrow, black-and-white-banded tail. A Northern Goshawk may circle a few times, but it tends to pursue prey through the trees with uncanny speed and agility, sometimes crashing through branches to make a capture. It feeds on squirrels, chipmunks, and birds, including grouse. Birds are plucked before they are eaten, leaving a telltale pile of feathers as evidence of the capture.

MAMMALS In this forest, deep snow is the most important feature to which mammals must adapt. Squirrels are the most conspicuous mammals here. They cope with deep snow either by hibernating or by relying on stored food. The most common squirrel is the Golden-mantled Ground Squirrel, *Callospermophilus lateralis* (plate 18D). Why it seldom ventures

FIGURE 4.46 Northern Goshawk, *Accipiter gentilis.*

into the Mixed Coniferous Forest is an unanswered question. Perhaps it is a matter of competition from nocturnal rodents, which are more common in the lower forest belt. Unlike those of the chipmunk, this squirrel's stripes do not extend up to the head and face. The head is a solid bronze color. Although most predators see only black and white, a Golden-mantled Ground Squirrel is difficult to see among patches of light and dark on the forest floor. Many hikers unknowingly call these squirrels chipmunks because of the stripes, but chipmunks are much smaller and spend much of their time in trees. Golden-mantled Ground squirrels seldom climb trees.

These are familiar squirrels to backpackers because they boldly search for food in campgrounds. In areas where there are many visitors, the squirrels have become very tame as they beg for handouts. These squirrels have learned that plastic bags contain food, and when given the chance, they will gnaw a hole in anything wrapped in plastic expecting to find it edible. Many a backpacker has neglected to tie down the top flap on the pack or to close a zipper, only to find that a Golden-mantled Ground Squirrel has crawled inside and gnawed holes in everything plastic, including the bag that contains dirty clothes.

Normal food for these ground squirrels includes grasses, Dwarf Bulrushes, bulbs, and berries. They seem to eat the greens early in the season, and later, during the time they are nursing young, they vigorously search for fungi. They are particularly fond of a variety of underground mushrooms known as truffles. The spores of these fungi pass undigested through the digestive tracts of the squirrels. The squirrels therefore are very important in dispersing the fungi, which add considerably to the mycorrhizae that help the forest trees absorb nutrients. The squirrels also eat some insects, an important source of protein.

In autumn, the Golden-mantled Ground Squirrel retreats to its burrow under a rock or log and curls up to hibernate for the winter. When autumn days become too cold or snow becomes too deep, it would be energetically too expensive to keep foraging for food. The squirrel's body temperature drops considerably, during which time its lower rate of metabolism enables it to continue on the energy provided by stored high-calorie brown fat. Every few days the squirrel arouses, and urinates and perhaps nibbles on a bit of stored food. When snow melts in the spring and daytime temperatures warm up, the squirrel comes out of hibernation and begins to forage, eagerly awaiting the arrival of the first backpackers.

The tree squirrel of the Montane Forest is the Chickaree or Douglas Squirrel, *Tamiasciurus douglasii* (plate 18C). Its principal food is seeds from pines, firs, and redwoods. It gathers cones and seeds in great numbers and stores them at the base of trees or in hollows near its home. Chickarees are noisy squirrels, and make a considerable racket in the early morning, particularly as they chase each other in circles, up and down the trunks of trees. Even when they are not seen, they are frequently heard. They make a rapid series of high-pitched, birdlike chirps, each series lasting 2 or 3 s.

Except in places such as Yosemite Valley, where there is an abundance of food, Chickarees seldom venture into the realm of the Gray Squirrel. Apparently, Gray Squirrels must have their acorn supply. Other limitations to distribution of Gray Squirrels into the realm of the Chickaree include deep snow and the predatory Marten. As mentioned earlier, Gray Squirrels bury small caches in many locations. The Chickaree, however, makes a few large caches near its home, which it defends vigorously. In deep snow, there is more reward for digging out a larger cache of food.

The American Marten, *Martes americana*, is a large weasel, up to 20 in (50 cm) from nose to rump. It is an agile predator that hunts on the ground and in the trees. Its prey is largely squirrels and chipmunks. Small Chickarees frequently escape from the Marten, but presumably the large Gray Squirrel is no match for it. Why the Marten does not go down the mountain in search of plump, succulent Gray Squirrels is a question that has not been answered, but two hypotheses are worth considering. First, perhaps competition from other predators such as Bobcats, *Lynx rufus* (figure 8.52), and Coyotes, *Canis latrans* (figure 8.50) excludes them. Second, it is suspected that the thick fur of the Marten would cause it to overheat during times of maximum energy expenditure, such as when it is running down its prey.

Martens are seldom seen, although they tend to maintain a rather small home range, 0.04 mi² (0.01 km²). They hunt at night and early morning. When they are seen, they are usually running, probably across a snowfield. Their short legs and long bodies give them a very distinctive appearance when they run. The only animal with which it might be confused is the Fisher, *M. pennanti*, which is rare in the Sierra.

Martens remain active during winter. They grow additional hair on their feet to aid locomotion on soft snow, and they feed on winter-active mammals such as pocket gophers, Pikas, and mice. They dive and/or dig rapidly in the snow to chase their prey. It is said that they will dive under the snow and attack Porcupines from beneath, thus avoiding the quills.

The largest predator in the weasel family is the Wolverine, *Gulo gulo*. Over the years, anecdotal observations have placed Wolverines at various localities in the Sierra Nevada. The only known Wolverine in the state was spotted in 2008 in the Lake Tahoe area. In November 2014, this animal was captured by a motion-sensitive camera near Truckee, and scat analysis has verified that this is the only Wolverine in the area. It is assumed that this animal is the last remaining example of the species in the state and will probably die of old age without reproducing.

Interestingly, the Sierra Nevada Red Fox, *Vulpes vulpes necator*, had not been observed in Yosemite since 1915, and it was assumed to be extirpated. However, in December 2014, another motion-sensitive camera in Yosemite National Park captured two images of a Red Fox on different dates. Previously, in 2010, at Sonora Pass, a Red Fox was photographed chewing on a bait that triggered the camera. Saliva analysis revealed it was indeed a Sierra Nevada Red Fox. While it is not as serious as the status of the Wolverine, the Sierra Nevada Red Fox is still considered one of the rarest species in California. Nonnative Red foxes were introduced to California many years ago for fur farming. They subsequently escaped and became established at various points around the state, particularly in coastal southern California. A well-established population at the Seal Beach National Wildlife Refuge was finally extirpated, as they were a threat to the endangered Least Tern. Partly in association with a growing population of Coyotes, the Red Foxes appear to be gone. Similarly, at Point Mugu, Red Foxes disappeared after Coyotes were introduced. Now, the Coyotes are moving into residential areas in Seal Beach.

Sierra Nevada Chipmunks, *Neotamias* (= *Tamias*) spp., are small squirrels, seldom exceeding 6 in (15 cm) from nose to rump. There are eight different species in the Sierra, and they look very much alike. As in the case of woodpeckers and warblers, each has its own ecologic niche. They are separated by behavioral interactions, elevation, latitude, and slope exposure. Some remain in trees, some seldom climb trees, and others prefer rocks and fallen timber.

The complex of factors responsible for separation of chipmunks into different ecologic niches can be illustrated with three species that occur in Lee Vining Canyon, on the eastern slope of the Sierra Nevada, in the Yosemite area. The Lodgepole Chipmunk, *N. speciosus*, primarily inhabits Lodgepole Pine forests where it spends most of its time on the ground foraging for seeds, insects, and fungi. It is restricted to forested sites because it is vulnerable to heat stress. It excludes other chipmunks by aggression. The Yellow-pine Chipmunk, *N. amoenus* (plate 18F), occurs in the Jeffrey Pine and Pinyon Pine forests, but can range up into brushy areas in the subalpine zone. It digs burrows for refuge and storage of food. It is more tolerant of heat stress than the Lodgepole Chipmunk. The Least Chipmunk, *N. minimus*, occupies open Sagebrush Scrub habitats. Throughout the Great Basin, the Least Chipmunk occurs in all habitats that are available in the eastern Sierra Nevada. In the Sierra, it is kept from entering adjacent forested habitat by aggression from the Yellow-pine Chipmunk, but it is able to inhabit hot, dry habitat because it possesses thermoregulatory adaptations absent in other species. For example, the Least Chipmunk can tolerate a body temperature of 109°F (43°C) and a range of 13°F (7°C). So, in general, dispersal up the mountain is restricted by aggression, and movements down the mountain are restricted by physiological adaptations to heat stress. Recent evidence indicates that an increase of temperature associated with climate change is causing chipmunks to move to higher elevations. This has been documented particularly for Least Chipmunks in Yosemite National Park. Another chipmunk, the Inyo Chipmunk or Uinta Chipmunk (*N. umbrinus*), a high-elevation, tree dwelling chipmunk, seems to have disappeared from its former range in the eastern Sierra.

Chipmunks are conspicuous and abundant. They feed mostly on small seeds, which they store in small holes in the ground. They also eat a surprising number of insects, including grasshoppers. One of the important controls on the abundance of termites seems to be predation by chipmunks, particularly when termites in winged form are dispersing. Chipmunks hibernate during winter, but unlike Golden-mantled Ground Squirrels, they are unable to rely on stored brown fat. Chipmunks become aroused frequently, during which time they urinate and eat. They dig through the snow to their food caches, which they locate by memory and odor.

Northern Flying Squirrels, *Glaucomys sabrinus*, are not common. They occur mostly in moist forests from Yosemite northward on the west side of the Sierra, but they are locally abundant on the east side of the Sierra as well. One interesting population occurs at Sagehen Basin north of Truckee, and residents of Mammoth, south of Yosemite, report that Northern Flying Squirrels beg for food in residential areas. These are small squirrels, about the size of chipmunks. They have a large web of loose skin between their forelimbs and hind limbs, which they use to glide from tree to tree. Apparently they triangulate to judge distances between trees as they often lean out and pivot from side to side before leaping. They are nocturnal, and perhaps this is why they seem more scarce than they really are. Apparently, they are omnivorous, as many of them have been caught in meat-baited traps set for carnivores.

Riparian Woodland

Riparian communities occur along watercourses. Abundant water and cold-air drainage provide a cold, moist climate that is unique in California, where hot, dry summers dictate the nature of most communities. Characterized primarily by small trees and large shrubs, this community is often called a woodland. At higher elevations, however, the vegetation is typically all shrubs. At lower elevations, the

FIGURE 4.47 Leaves of Black Cottonwood, *Populus trichocarpa*, and Quaking Aspen, *Populus tremuloides*. Most species of riparian trees are winter-deciduous.

trees are so large and dense that the community may be referred to as Riparian Forest. Discussion here will emphasize elevational differences in the community. Where it is relevant, a comparison will be made with riparian communities in other parts of the state.

Streamside vegetation also consists of a variety of conspicuous wildflowers which put on a show of color in the spring. Conspicuous among these wildflowers are purple Larkspurs (*Delphinium* spp.), Monkshoods (*Aconitum* spp.), which resemble Larkspurs, and orange Leopard Lilies (*Lilium pardalinum;* plate 20D). In wetter areas, there are a variety of Monkeyflowers (*Mimulus* spp.) and a beautiful orchid, the Broad-leaved Helleborine (*Epipactus Helleborine;* plate 20G). The density and diversity of species in a riparian community are greater than in any other community in California. This occurs for two reasons. First, a riparian community is very productive. Lots of food means lots of animals. Second, the riparian community is a transitional community between water and land. The zone where two communities overlap, called an ecotone, shares characteristics of both communities and therefore is diverse. That is, the edge of a community is more diversified than its center, a phenomenon also known as "edge effect."

Trees and shrubs that grow along a watercourse are highly water dependent. Abundant water on a year-round basis dictates that many

species of plants will be broad-leaved. Because there are seasonal fluctuations in temperature, and winter is particularly cool due to cold-air drainage, most of these species are winter-deciduous. Unlike conifers, broad-leaved plants are particularly sensitive to low temperature. Photosynthesis in these plants during winter is unable to keep up with cellular respiration. The net effect is that, rather than use up carbohydrate stores during the winter, the plants respond by dropping their leaves and becoming dormant. This is the strategy followed by forests of hardwoods in the eastern United States.

Another characteristic of riparian species is that they are often wind-pollinated and wind-dispersed. Wind dispersal is an effective way for a riparian plant to get its seed from one body of water to another. To enhance dispersal by wind, fruits or seeds may bear wings or plumes. Also, some riparian species have floatation devices or fleshy fruits to encourage animal dispersal. Flowers are usually small, borne in dense clusters that hang down, and many are wind-pollinated. Flowers often occur in catkins, a name that refers to the resemblance of the clusters to a cat's tail. Common woody plants are willows, poplars, alders, maples, ashes, and dogwoods.

Notable among riparian species are members of the willow family (Salicaceae), which includes willows (*Salix*) and poplars (*Populus*). Willows are distributed throughout the state from the alpine zone to the coast, invariably along water. California has 29 species, most of which are shrubs, and many of which have very localized distribution. They can be recognized by their simple, elongate leaves. Flowers are in unisexual catkins borne on different plants. The fruit is a furry capsule with many hairy seeds.

In the high country, growing season is short, and soil is thin. Riparian plants at high elevation therefore are deciduous shrubs, not trees. Willows appear where water first appears. At high elevation, along streams leading away from snowbanks, Rocky Mountain Willow, *S. petrophila*, may be found. The scientific name means "rock-loving." This common species occurs as a carpet-like shrub from Mount Lassen to Mount Whitney. Of limited distribution in similar habitat is Snow Willow, *S. nivalis*, and Snow Willow is found only on metamorphic rock on the east side of the Sierra Nevada between Koip Peak and Mount Dana.

Going down the mountains along watercourses, one observes a regular replacement of species. At lower elevations, the shrubs are taller, and in the valleys, willows are trees. Throughout the state, from about 10,000 to 5000 ft (3000–1500 m) a common species is Lemmon's Willow, *S. lemmonii*. Common between 10,000 and 8000 ft (3000 and 2400 m) is Sierra Willow, *S. eastwoodiae*. Lemmon's Willow has narrow, lance-shaped leaves about 3 in (7.5 cm) long, whereas Sierra Willow has shorter, broader leaves. Of limited distribution at the same elevation in the Sierra Nevada are at least four other species. Common below 5000 ft (1500 m) is Arroyo Willow, *S. lasiolepis*, and common below 3000 ft (900 m) is Sandbar Willow (Hinds' Willow), *S. exigua* var. *hindsiana*, both of which reach tree size, 20–30 ft (7–10 m) in height.

Tree species of willow do not occur higher than about 8000 ft (2400 m). They are not always easy to tell apart, but they differ in characteristics such as leaf or twig color. Pacific Willow, *S. lasiandra* occurs from the foothills to about that elevation. A shrub version of Pacific

Willow occurs at higher elevations. In the foothills and Great Central Valley, common willows are Godding's Black Willow, *S. gooddingii*, and Red Willow, *S. laevigata*. Red Willow may occur as high as 5000 ft (1500 m). It has reddish brown twigs.

A very interesting phenomenon associated with various species of willows is that they produce a plant toxin known as salicin, a name derived from the name of the genus *Salix*. This chemical is similar in some ways to aspirin, or acetylsalicylic acid, also named for the willow genus. Salicin is apparently produced by willows to discourage herbivores from eating them. In harsher climates, such as high elevation, the plants appear to produce greater concentrations of the chemical. An interesting side effect of this production is that some insects have developed immunity to the toxin and have taken deliberately to eating willow leaves in order to use the toxin for their own protection. Similarly, larvae of the Monarch Butterfly consume milkweed to make themselves distasteful. In the case of willows, the insect involved is the larva of a Leaf Beetle, *Chrysomela aeneicollis*. Studies of Gray-leafed Sierra Willow, *S. orestera*, between 8000 ft (2400 m) and 9500 ft (2950 m) along Big Pine Creek of the eastern Sierra showed that the plants with more salicin were more heavily preyed upon by beetle larvae. Furthermore, beetles on those willows suffered less predation. In this case, it appears that the production of the toxin has backfired, as it has encouraged predation by the beetle larvae.

Trees called poplars include cottonwoods and aspens, and they occur all across North America, Europe, and Asia. They are moisture-loving trees occurring in forests in some areas. In California, they are primarily riparian species, nicely segregated by elevation. At lower elevation, from the coast to about 2500 ft (750 m), there is Fremont Cottonwood, *P. fremontii*. On the desert side of mountain ranges, it occurs in canyons up to 6500 ft (2000 m) in elevation. This is the common cottonwood with serrated, heart-shaped leaves that are bright green on both surfaces, turning yellow in autumn. When female catkins come into fruit, long white hairs produce cottony masses that give the tree its name. Black Cottonwood, *P. trichocarpa* (figure 4.47), occurs at higher elevation, up to about 9000 ft (2700 m), and along streams in forests along the coast. This cottonwood differs from Fremont Cottonwood by having finely serrated leaves that are dark green above and rusty or silvery beneath. At higher elevations, leaves become more lance shaped than heart shaped. This is the largest of all native poplars, up to 160 ft (50 m) in height. A third species of cottonwood, Narrow-leaved Cottonwood, *P. angustifolia*, occurs in a few small groves at about 6000 ft (1800 m) on the eastern slope of the Sierra near Lone Pine and Independence. This species is more common to the east, into the Rockies. It is a small, slender tree with lance-shaped leaves.

Quaking Aspen, *P. tremuloides*, has the widest distribution of any tree species in North America. This tree gets its name from the fact that its leaves tremble because they have a flattened, flexible stem (petiole). Even the slightest breeze sets the pendulous leaves quivering. This characteristic is shared by many cottonwoods as well, but it seems more pronounced in Quaking Aspens. Leaves differ from cottonwoods by being smaller, about the size of a half-dollar (figure 4.47). They are light green on top and silvery underneath. Quaking Aspen shoots, rich in calcium, are a very important browse for Mule Deer and Bighorn Sheep.

Throughout most of its range, Quaking Aspen is considered a successional species. It is an aggressive pioneer, one of the first species to return after a disturbance. Sprouts return from an extensive root system.

In California, Quaking Aspens are usually a high-elevation, riparian species. They occur in groves bordering streams at elevations higher than Black Cottonwoods. At lower reaches of their distribution they occur as good-sized trees, up to 80 ft (25 m) tall. These large trees are easily recognized by their white bark, which becomes broken up into black, warty bands. At the upper reaches of their distribution, at about 10,000 ft (3000 m), trees are much shorter, sometimes shrub-like. Young trees often have greenish bark.

In the northern Sierra Nevada, at upper elevations in the Great Basin, and in the White Mountains, Quaking Aspen groves are not restricted to riparian sites. They may occur in sunny places where soil remains wet, such as at the foot of talus slopes. In the Warner Mountains, above 8000 ft (2400 m), they occur in groves along with Washoe Pines and are not riparian trees in this setting. The situation in the Warner Mountains resembles that of the Rocky Mountains, where summer precipitation keeps the soil moist throughout the year.

The most vivid displays of autumn color that occur in California are associated with Quaking Aspens (plate 8B). Leaves range from yellow to gold, orange, and red. Many people have autumn pilgrimages to the canyons on the eastern side of the Sierra from Bishop to Bridgeport in order to gawk and photograph the spectacle of color exhibited by Quaking Aspens as they enter dormancy.

Because of the sprouting behavior of this species, groves are often clones of the same color. Because all clones do not enter dormancy at the same time, some groves remain green while others have turned. This obvious difference in behavior for different clones has led scientists to study the mechanism that initiates dormancy. The stimulus to turn color is not the first hard frost of the winter, as is often thought. Rather, the change is a hereditary response to day length, to shortening of the photoperiod as winter approaches.

There is some evidence to support the contention that Quaking Aspen never goes fully dormant during winter. Bark tissues contain a high chlorophyll content, which may allow photosynthesis to continue on sunny winter days. Furthermore, it has been determined that Quaking Aspen's killing temperature of −100°F (−80°C) represents the lowest such temperature among winter-deciduous trees.

Quaking Aspen wood is light and brittle. Dry snags of Quaking Aspen are easy to break, and

FIGURE 4.48 White Alder, *Alnus rhombifolia*, with female cones.

they form an ideal wood for a campfire. The wood burns hot and without smoke. An unfortunate characteristic of Quaking Aspen (as well as some other poplars) is that it also burns when it is green. People hard up for firewood will cut down green Quaking Aspens. These are the same people who complain that Quaking Aspen wood burns with a foul odor. It serves them right.

Other winter-deciduous species grow along streams with willows and cottonwoods. At lower elevations from the foothills to the coast, Western Sycamore, *Platanus racemosa*, is a common species. This is California's only native member of the plane tree family (Platanaceae). It is essentially a California species, but it extends into Baja California on the coastal side of the Peninsular Ranges. In California, its distribution occurs along the coast from the southern Coast Ranges southward through the Transverse and Peninsular Ranges. To the interior it occurs along the southwestern side of the Sierra Nevada and along the Sacramento and San Joaquin Rivers of the Great Central Valley, where it ascends the main tributaries to low elevations in the Sierran foothills. In the Sierra Nevada foothills it is more common in the south, but it occurs northward as far as Tehama County.

Sycamores are characterized by large, palmate leaves and whitish, scaly bark. Often they do not grow straight but twist back and forth, giving the trunk a gnarled appearance. These are long-lived trees that may become quite large, up to 100 ft (30 m) in height.

Alders are in the birch family (Betulaceae). There are four species in California. They play an important role wherever they occur, because they are equipped with nitrogen-fixing bacteria in their roots. The White Alder, *Alnus rhombifolia* (figure 4.48) has the most extensive distribution, but they are all similar in appearance. Their elm-like leaves are oval and serrated, up to 5 in (12 cm) in length. Flowers are borne in catkins, but male and female catkins occur on the same tree. After the seed is set, the remains of a dry, hard fruit that resembles a small redwood cone stays on the tree.

White Alders occur throughout California in Riparian Woodland. They are easy to distinguish by their straight, grayish trunks. A dark, downturned mark appears on the trunk just above each branch. Where old branches have fallen off, the dark marks remain, giving the tree trunk the appearance of having many mustaches.

On wet slopes and along streams between 5000 and 8000 ft (1500 and 2500 m) in elevation, from Tulare County north, there is a shrub-like alder known as Mountain Alder, *A. incana*. It replaces White Alder at higher elevations. These small trees have leaves similar to other alders, but the serrations are of two sizes. Fine teeth along the margin are superimposed on larger sawlike teeth.

On the eastern slope of the Sierra Nevada from about 5000 to 9000 ft (1500–2700 m), there is another member of the birch family: Water Birch, *Betula occidentalis*. It looks very much like Mountain Alder, but its leaves lack the large sawtooth serrations. Furthermore, its distribution is patchy. It is conspicuous along Lone Pine Creek, Big Pine Creek, and Bishop Creek, in Inyo County, and it also occurs on Mount Shasta and westward through the Klamath Province to the interior of the northern Coast Ranges. A disjunct distribution of this sort is similar to distributions of some of the subalpine plants, such as the Foxtail Pine, which will be discussed in the chapter on mountaintops (Chapter 5). Although Water Birch is essentially a Great Basin species, it also occurs along watercourses in the mountains of the western Mojave Desert.

FIGURE 4.49 Mountain Dogwood, *Cornus nuttallii*.

Big-leaf Maple, *Acer macrophyllum* (figure 6.9), is California's largest member of the Soapberry family (Sapindaceae, formerly Aceraceae). Like the sycamores, maples have large, palmate leaves, but the leaves are arranged opposite to each other on the stem. A winter-deciduous species, leaves turn yellow in autumn, adding to the display of autumn color in locations such as Yosemite Valley. In spring, they produce small, greenish-yellow flowers in hanging clusters. The flowers are large enough that five small petals can be seen. The fruit is the type known as a samara. Fruits are winged and, when dry, fall from the tree like little helicopters.

In the northwest, Big-leaf Maple is a forest species, growing with Douglas-fir and Coast Redwood. Its range extends southward along the western Sierra Nevada and down the coast on the western side of the Transverse and Peninsular Ranges. In the southern part of its range, it is a riparian species growing at middle elevations along with White Alders. Big-leaf Maple is one of California's commercially valuable hardwoods.

At lower elevations in the Sierra Nevada, Big-leaf Maples are replaced by another member of the Soapberry family that occurs in California: the Box Elder, *A. negundo*. It occurs frequently in association with Western Sycamores and Fremont Cottonwoods. It is not present in all streams, but where it occurs it may be common. Box Elders resemble maples, but the leaves are compound, composed of three palmately veined leaflets, giving the appearance of a maple with small leaves.

Often co-occurring with White Alders and Big-leaf Maples is one of California's most beautiful native trees, Pacific or Mountain Dogwood, *Cornus nuttallii* (figure 4.49). This is the only member of the dogwood family (Cornaceae) that is a tree in California. Because its distribution is not limited to mountains, closely paralleling that of Big-leaf Maple, it seems best to refer to this species as Mountain Dogwood. In moist forests, such as in association with Sierra Redwoods, it becomes an understory tree growing in deep shade. It rarely grows in full sun, and maximum photosynthesis in this species occurs at only one-third the intensity of full sun.

During spring, Mountain Dogwoods are most conspicuous when leafless branches become laden with white flowers up to 6 in (15 cm) across. Close inspection of the blossoms will reveal that what are thought to be petals are actually large white bracts, and what appears to be the center of a single large flower is a cluster of small flowers. There are usually five to six of these white bracts, whereas the eastern Flowering Dogwood, *C. florida*, has only four. The eastern variety often has reddish notches at the end of each bract, leading to the story that dogwood was used to build Christ's cross. Lest we forget, cross-shaped flowers with blood spots remind us of that event. Although the dogwood family does occur in Europe, the genus *Cornus* does not, and the eastern Flowering Dogwood is the only species with four showy bracts.

Leaves of dogwoods resemble alder leaves in shape, but they are not serrated. They are opposite on the stem, and the veins of the leaf turn and run parallel to the midrib. If a dogwood leaf is torn, elastic threads within the veins remain intact, often holding the torn parts of the leaf together.

If the beautiful flowers are not enough, the autumn display provided by Mountain Dogwood leaves is spectacular. The color ranges from orange to scarlet. Bright red berries may also be seen on the tree at this time. They are quite bitter, but Band-tailed Pigeons are reported to feast upon them every autumn.

A shrubby dogwood of the Sierra Nevada and northern forests is known as American Dogwood, Creek Dogwood, or Red-osier Dogwood, *C. sericea*. (The term osier refers to pliable twigs used in basket making.) This shrub often grows in association with alders and willows. The leaves have the same type of venation as the other dogwoods, but the flowers do not have large, showy bracts. Instead, small white flowers grow in clusters at the tips of the stems. The fruits are whitish and are relished by many species of birds. The most distinctive feature of this shrub is its bright red bark. Presumably, Native Americans scraped, dried, and smoked the inner lining of this bark, which is reported to have hallucinatory properties. Perhaps this helped them weave the pliable branches into the intricate patterns found in their baskets.

Perhaps the most beautiful of the riparian shrubs is California Azalea, *Rhododendron occidentale*. These shrubs occur in similar locations as Mountain Dogwoods, although they are absent from the southern Coast Ranges and Transverse Ranges. Among members of the heath family (Ericaceae), these are unique by being winter-deciduous and having large flowers. Manzanitas and Pacific Madrone are also in this family, but they are characterized by their small, urn-shaped flowers. California Azaleas, in contrast, have large, white, funnel-shaped flowers, fully 2 in (5 cm) long. During spring, when Mountain Dogwoods and California Azaleas are in full bloom, a person cannot help but admire the beauty of these plants. In moist localities within the forest, these shrubs also become a part of the understory. When they are not flowering, they are more easily overlooked among willows and other riparian species of shrubs. Azaleas may be recognized by their thin, light green, lance-shaped leaves, up to 3 in (8 cm) in length, which grow in whorls on the stem. Of special interest is that they are poisonous, particularly to domestic livestock.

One more group of deciduous trees occurs in the Riparian Woodland. These are the ashes, deciduous members of the olive family (Oleaceae). These small, shrub-like trees are usually characterized by pinnately compound leaves, with about five to seven lance-shaped leaflets. The most common is California Ash, *Fraxinus dipetala*. The name dipetala refers to the branched clusters of white flowers that each bear two petals. Fruits of these ashes are single-winged samaras. This species occurs throughout the foothill areas along streams or on north-facing slopes. Its range extends outside of California into appropriate habitat in Baja California.

California's only large treelike ash is Oregon Ash, *F. latifolia*. This is a species found on rich alluvial soils. It occurs along the big rivers in the Great Central Valley and into the Sierra Nevada foothills along with willows and Fremont Cottonwood.

The Mixed Evergreen Forest of northwestern California includes broadleaf evergreen species, relicts of southern affinities that also occur to various degrees in riparian situations. The presence of permanent water compensates for the long, dry summer. Included among these species are broad-leaved trees such as California Bay or California Bay Laurel, *Umbellularia californica* (figure 6.8), and Pacific Madrone, *Arbutus menziesii* (figure 6.7).

The odoriferous laurel, also known locally as Pepperwood, is common in canyons as far south as the Laguna Mountains in San Diego County. In the Sierra Nevada, it occurs primarily below 5000 ft (1500 m), particularly where streams cut through chaparral. Shrubby forms of this species may be associated with moist serpentine soils in the northern Sierra Nevada foothills.

Pacific Madrone occurs in the northern Sierra Nevada primarily between Butte and Calaveras Counties, where it also becomes a component of Mixed Coniferous Forest, particularly in association with White Firs. As a broad-leaved tree with a twisted trunk, it stands in stark contrast to the conifers.

Among the conifers of Mixed Evergreen Forest that occur in riparian communities are Douglas-fir, *Pseudotsuga menziesii*, and California nutmeg, *Torreya californica*, both of which occur in canyons of the western Sierra Nevada. Particularly striking are the large Douglas-firs that occur along the famous Mist Trail to Vernal and Nevada falls in Yosemite National Park. This is also one of the localities where California Bay Laurel is viewed by millions of people each year. Progressing northward from about Lake Tahoe, Canyon Live Oak, Douglas-fir, California Bay Laurel, Pacific Madrone, and Tanoak, *Notholithocarpus densiflorus*, become important components of the Mixed Coniferous Forest.

In summary, so many dominant species are found over such a wide elevational range that it is difficult to get a picture of species assemblages within the riparian communities. On an elevational basis, we may think of four assemblages. Along larger streams in the valleys and foothills, Western Sycamore, Box Elder, Fremont Cottonwood, and one or more of the tree-sized willows are dominant species. Up through the Mixed Coniferous Forest are Riparian Woodlands that include species also found in the northern forests and Coast Ranges. For example, Big-leaf Maple, Mountain Dogwood, California Bay Laurel, and Black Cottonwood occur along the coast and reappear along streams in the mountains where they are joined by White Alder. In the Montane Forest and on the eastern side of the Sierra, Quaking Aspen dominates. At high elevations, into the alpine zone, riparian vegetation becomes shrubby. Various species of willow dominate this community.

Animals of Riparian Woodland

The diversity of animals is very great in Riparian Woodland. Most of the animals come to the streams to find water or food or to escape the heat. These animals are discussed in other parts of the book. The aquatic animals will be discussed in Chapter 11, on inland waters. In this section, emphasis will be on conspicuous butterflies and birds.

BUTTERFLIES Butterflies often occur along watercourses because they commonly feed on riparian plants. For many butterflies, the larvae are restricted to a single plant species. Adults seem to be less specialized but may be associated with certain species as well. Adult butterflies rely entirely on liquid food, usually nectar obtained from flowers by means of a long, hollow proboscis.

In foothill canyons, as in the valleys, the most common butterflies are admirals and swallowtails. Some of them, such as Lorquin's Admiral, *Limenitis lorquini*, and the Western Tiger Swallowtail, *Papilio rutulus*, are discussed in the chapter on the Great Central Valley. Also common in the foothills are two powerful fliers: the Two-tailed Swallowtail, *P. multicaudata*, and the Pale Swallowtail, *P. eurymedon*. The Two-tailed Swallowtail has two tails on each wing and is the largest western butterfly. The Pale Swallowtail looks like the Western Tiger Swallowtail, but its background color is whitish rather than bright yellow.

The larvae of these butterflies are bright green with a yellow line on each side. They have two black-bordered yellow spots that resemble eyes on each side of the body. When at rest, they tuck their heads under so that the segment with the large spots looks like a large head with big eyes. Larvae of the Western Tiger Swallowtail feed on a variety of riparian plants such as willows, cottonwoods, alders, and sycamores. Larvae of the Pale Swallowtail feed on chaparral plants that border the Riparian Woodland, such as Coffee Berry (*Frangula californica*), California lilacs (*Ceanothus* spp.), and Holly-leafed Cherry (*Prunus ilicifolia*). Larvae of the Two-tailed Swallowtail feed on ash, Service-berry (*Amelanchier* spp.), and Hop Tree (*Ptelea crenulata*).

One of the most common and distinctive spring butterflies is the Pacific Orangetip, *Anthocharis sara*. This is a small butterfly with a wingspan of about 2 in (5 cm). Its wings are white with orange tips. The adults fly up and down canyons, usually close to the ground. The larvae are green with wide yellow stripes on each side. They feed on members of the mustard family (Brassicaceae).

The Sonoran Blue Butterfly, *Philotes sonorensis*, is one of California's most beautiful butterflies. It occurs in local colonies in rocky canyons of the Sierra Nevada foothills from Placer County southward to the desert margins, and in the Coast Ranges from Santa Clara County southward to Baja California. The wingspan on this butterfly is barely over an inch (26 mm). Males are metallic blue with a bright orange spot on the forewing. Females have the orange spot on both wings. Adults fly close to rocky cliffs and steep canyon walls. Larvae of these butterflies are variable in color, from pale green to mottled rose, and they feed on various Liveforevers (*Dudleya* spp.).

During late spring, a common butterfly in foothill canyons is the Variable Checkerspot or Chalcedon Checkerspot, *Euphydryas chalcedona*. The adult of this butterfly is frequently seen on wildflowers. It has a wingspan slightly over 2 in (6 cm), and its wings are marked by numerous yellowish and red spots on a background of black. The larvae are black with branched spines and have a row of red-orange spots along the back. They feed on various members of the figwort or monkeyflower family (Scrophulariaceae).

Associated with willows in canyon bottoms are small butterflies known as hairstreaks. The Sylvan Hairstreak, *Satyrium sylvinus*, is a small gray butterfly with a pale blue cast. It is a reddish brown near its tail. The caterpillars are green with cream-colored side lines. Larvae appear to be restricted to willows, but adults will visit other flowers, such as milkweeds, for nectar.

Higher in the mountains, where streams wind through coniferous forests, you may see the Zephyr Comma, *Polygonia zephyrus*. The adults have a 2 in (5 cm) wingspan. The ragged-looking wings are orangish with a brown border interrupted by large tawny crescents. They flit about as if they were propelled by little spurts of wind, or zephyrs. Adults feed on streamside wildflowers; the larvae feed on various wild currants and gooseberries (*Ribes* spp.). The larvae are black with reddish spines in front and white spines in back.

BIRDS Birds are the most conspicuous animals of riparian communities. Many species of small birds coexist in Riparian Woodland through mechanisms of niche partitioning. However, because they spend so much time among the leaves, they are seldom seen. Included here are swallows, flycatchers, vireos, goldfinches, warblers, and sparrows. They partition the resources by feeding on different plants and nesting in different places. The greatest threat to them is habitat destruction associated with drying and diversion of water.

Another threat that is becoming increasingly more serious is the invasion of introduced species. Such species are common in areas of habitation, but they also invade woodlands, particularly Oak Woodland and Riparian Woodland. For example, House Sparrows or English Sparrows, *Passer domesticus*, compete with native birds for food, nesting materials, and nesting sites. The more House Sparrows that occur in an area, the fewer other birds that can coexist. European Starlings, *Sturnus vulgaris*, first appeared in California in 1942. Since that time, they have become common near habitation and have invaded many woodland habitats. Both House Sparrows and

European Starlings have moved into woodlands in the Sierra Nevada, particularly in the vicinity of the national parks. Both species are examples of birds introduced from Europe that have been able to expand their ranges because they have no natural enemies here.

European Starlings resemble blackbirds, but they are short-necked and short-tailed and have yellow bills. They are very aggressive, displacing all sorts of birds that nest in tree cavities. Cavity-nesting birds that are being replaced include woodpeckers, bluebirds, and small owls. The aggressive European Starlings move into the nests and physically remove the juveniles.

A problem that is a bit more subtle than that of introduced European birds is what has happened with Brown-headed Cowbirds, *Molothrus ater*. These birds are in the blackbird family and look like Brewer's Blackbirds, *Euphagus cyanocephalus*, with stubby bills. The male cowbird is the same color as a male blackbird, but its head is brown and it does not have yellow eyes. Female cowbirds differ from female blackbirds by being browner and by often having a streaked breast. It is sometimes difficult to tell these birds apart, particularly because the two species occur in mixed flocks.

Cowbirds get their name from their frequent association with livestock. Prior to the introduction of domesticated livestock, they were "Buffalo-birds," accompanying herds of bison on the prairies. They then followed humans and their domesticated animals to California. They eat seeds and other edible items found in droppings, and they may be seen feeding in pastures, corrals, and feed lots.

The problem with Brown-headed Cowbirds is that they are nest parasites. Female cowbirds remove the eggs of native riparian birds from their nests and replace them with their own. The native birds then incubate the eggs and raise young cowbirds instead of their own offspring.

Recent studies of Brown-headed Cowbirds in the Sierra Nevada have revealed an interesting pattern of daily migration. Cowbirds spend the night in the mountains roosting in riparian vegetation. In the morning, the birds are solitary, and females do their nest robbing. Usually each female lays a single egg per day in another bird's nest. Then the cowbirds make their way down the mountain to pastures or pack stations, where they spend the day feeding in flocks. In the evening, they fly back up mountain to repeat the procedure, a round trip that may cover 15 mi (20 km). This is a very large home range for a songbird, which typically has a range measured in meters, not miles. Cowbirds must expend an incredible amount of energy on their long daily migrations, and such energy expenditure is possible only if the food is rich enough to make it worthwhile. It does not seem logical that the long trips would be worth it, but one cannot argue with success. Cowbirds continue to expand their ranges to the detriment of native songbirds. They have even invaded Santa Catalina Island, where they feed in association with the bison herd there.

A question arises: Why do Brown-headed Cowbirds parasitize the nests of some birds but not others? The answer seems to lie in the color of the eggs. Species that lay blue eggs, such as the American Robin or the Hermit Thrush, easily recognize cowbird eggs and remove them from the nest. It is thus apparent that the true value of specific colors and/or patterns on eggs is parental recognition. Nest parasites such as Brown-headed Cowbirds must depend on their eggs bearing some resemblance to those of the host species. Those birds that invest a bit of energy in producing uniquely colored eggs enjoy the benefit of not being parasitized.

Another blackbird associated with wetland habitats is the Red-winged Blackbird, *Agelaius phoeniceus* (plate 19E). This bird and its close relative, the Tricolored Blackbird, *A. tricolor*, are conspicuous in marshy habitats from the coast to forest-bordered meadows. Males have a conspicuous red patch on their shoulders and defend their territories by calling from a perch as high as possible in shrubs or tules. Their distinctive call is not pretty, resembling the sound made by a squeaking, rusty gate.

Experiments with Red-winged Blackbirds have revealed an interesting thing about the

FIGURE 4.50 House Wren, *Troglodytes aedon*.

roles of color and sound in the defense of territories. If the red patch is covered, birds in forest habitats are able to defend territories with their calls, but birds in open habitats, such as marshes, are not. On the other hand, if the birds are made mute, those in open habitats are still able to defend territories, but forest birds are not. The conclusion is that in habitats where birds may be concealed by heavy vegetation, sounds are more important signals. In habitats where visibility is not a problem, color is more important. These experiments tend to verify the oft-quoted hypothesis that shrubland and forest birds need not be colorful, but they commonly are equipped with distinctive voices.

Typical of cavity-nesting birds that inhabit streamside woodlands, and particularly susceptible to invasions of nonnative birds is the House Wren, *Troglodytes aedon* (figure 4.50). The generic name of this bird refers to a hermit or cave man, in reference to its nesting behavior in the hollows of trees.

In the Sierra Nevada, this bird is particularly associated with Quaking Aspens and cottonwoods, where nests are often found in abandoned woodpecker holes. It is especially common on the east side of the mountains and the western foothills. As its name implies, a House Wren also may nest in or around houses and barns. In the spring or summer, the bubbling song of a male from a prominent perch is one of the familiar sounds of east-side aspen groves. Males and females forage for insects in adjacent brushy habitat. Sierran birds migrate south to a

warmer climate for the winter, but southern California birds may remain resident.

One of the most striking of the cavity nesters is the Violet-green Swallow, *Tachycineta thalassina*. The rich green and deep violet colors on the backs of these swallows flash in and out of view as they dodge acrobatically to gather insects on the wing. The term *thalassina* comes from a Greek word *thalassinos* meaning sea green from the sea. These are also migrators, flying south to Mexico and Central America around August. While they are in the Sierra, unlike lower elevation swallows that build mud nests, Violet-green Swallows nest primarily in abandoned woodpecker holes in riparian trees. They vigorously defend these nests against other cavity nesters such as Mountain Chickadees and bluebirds, and fortunately they are also successful in defending their nests against European Starlings. They also will nest on cliffs or in artificial nest boxes. At Mono Lake, they have taken to making nests on the tufa towers. They are conspicuous flying about open meadows, above waterfalls on both sides of the Sierra up to the subalpine zone in the vicinity of Riparian Woodlands.

Perhaps the most strictly riparian bird is the American Dipper or Water Ouzel, *Cinclus mexicanus*. These small, gray, stubby-tailed birds inhabit mountain streams throughout the state and never leave the company of water. They fly up and down the stream course, following every bend. Once they mate, each pair will maintain a territory of about 150 ft (50 m) along the stream. They build nests, often of moss, right next to the water, and sometimes behind waterfalls. Imagine this little bird flying kamikaze-like into a waterfall.

Dippers get their name from their habit of standing on a boulder near a stream and bobbing up and down. It is believed that the bobbing helps them see beneath the glare of surface water. They feed on aquatic insects and larvae, which they gather by entering the water. They swim well, and they can walk on the bottom of a stream. They do this by holding their wings partway open; turbulence pressing on the wings helps to hold them down. They have been observed to go as deep as 20 ft (6 m) and to remain under for as long as a minute.

Dippers are year-round residents of forest streams. As long as the water remains unfrozen, they continue to feed there. During winter, they sometimes become more common in lower reaches of the streams on the western slopes of the Sierra Nevada. This tendency implies a downslope migration. The highest elevation at which they have been recorded during the winter is 7500 ft (2250 m), an observation made by John Muir at the turn of the century.

The Belted Kingfisher, *Ceryle alcyon*, is also an obligate waterside bird. Unlike the Dipper, Belted Kingfishers also frequent estuaries and lagoons, especially in winter. They are more common along lower reaches of streams, where they hunt for small fish from a streamside perch, usually an overhanging tree branch. When the water becomes murky, they rely mostly on crayfish as a food source. Kingfishers are rarely seen above 7000 ft (2300 m) elevation. They do move into desert areas where permanent water occurs.

The call of a Belted Kingfisher is a well-known sound along streams. It is a loud, screeching rattle. During breeding season, males pursue females up and down the stream, calling loudly, and females battle for good nesting sites. These are large birds, up to 13 in (33 cm) in length. Their heads appear to be extra large; its size is accentuated by a large, ragged, often double-pointed crest and by the bird's very small feet, which are a clue that these large fish eaters are actually somewhat related to hummingbirds.

Belted Kingfishers are blue gray above and white beneath. Both males and females have a blue-gray band across the upper breast; the female also has a rust-colored band across the lower breast. The additional color present on females is rare among birds: the female is more colorful! When this is the case, as in phalaropes, there is a reversal of sex roles: the female defends the territory. This seems to be the case with Belted Kingfishers, although it is not as extreme as in the case of a phalarope.

The birds are solitary except during breeding season. They construct a nest in a stream bank, consisting of a tunnel that may go back 6 ft (2 m). At this time, the sex-role reversal becomes evident: males do most of the nest building. The advantage of the male doing most of the digging is that it allows the female to conserve her energy for laying eggs. Five to eight young are raised in the tunnel, and feeding the young keeps both parents busy. In the Pacific Northwest, Kingfishers have a mythical significance for some Indian nations. Perhaps it is the habit of a bird disappearing into an underground burrow that makes them mysterious. In open habitats, they hunt from a hovering position, similar to a kite or kestrel. Once they grasp a fish, they return to a perch, where they thump the fish repeatedly on a branch until it stops squirming. Only then will they swallow it, head first.

Another advantage to conservation of energy for the female is that these nests may be wiped out by flooding. If this occurs, the female must have enough stored energy to lay another clutch of eggs. The point is that to maximize the number of offspring he produces, the male should do more than his share of nest building and caring for the young in order to keep his mate healthy.

5

Mountaintops

FIGURE 5.1 Alpenglow on Mount Whitney at dawn. At 14,505 ft (4421 m) in elevation, this is the highest peak in the contiguous 48 states.

MOUNTAINTOPS ARE ISOLATED from each other by regions of low elevation and different climate (figure 5.1). Just like regions of specialized soil, mountaintops are ecologic islands. On mountaintops, therefore, relicts and endemics are found. Species may be remnants of former widespread distribution, or they may be products of evolution during a long period of isolation. Similar, closely related species are found on adjacent mountaintops, just as they are on adjacent oceanic islands.

Mountaintop climate is similar in every mountain range throughout the state of California. Plants and animals have evolved similar strategies for these communities, regardless of their location. The climatic trends that one observes upon ascending a mountain reach their extremes on the mountaintop. Here is the coldest climate, the most intense sunlight, and the strongest wind. Precipitation is reduced from that of the Montane Forest, and, with the exception of an occasional thunderstorm, nearly all precipitation falls as snow. Except for the northern Sierra Nevada, only about 3% of total precipitation in the Sierra Nevada comes from summer rainfall. Snow is late to melt, and the summer growing season may be less than seven weeks. Frost can occur at any time of year. Soil is shallow, nutrient-poor, and coarse, and it has an acidic pH.

BIOTIC ZONATION

The Sierra Nevada has the greatest abundance of alpine and subalpine communities, but other areas in the state also are of sufficient elevation to support these types of communities. Subalpine and alpine communities occur in the Klamath Province, the Cascades, the Great Basin, the Transverse Ranges, and the Peninsular Ranges.

Pinyon-Juniper Woodlands are also mountaintop ecologic islands. This is a community that is both drought tolerant and snow tolerant. Although these woodlands represent desert communities to some authorities, their localized distribution and evolutionary history qualify them as mountaintop communities.

Subalpine Forest

The alpine zone has no trees. The subalpine zone is the region referred to by C. Hart Merriam as the Hudsonian life zone. It is the upper limit for trees. The term alpine is in reference to the Alps, a European mountain range; it does not mean "all pines." Precipitation, falling almost entirely as snow, measures from 30 in (75 cm) to 50 in (125 cm) per year. On the average, precipitation is 10 in (25 cm) per year less than that which falls in the Montane Forest, at lower elevation. It has been determined that the upper limit of tree growth on a worldwide scale is controlled by temperature. Tree line seems to be determined by an isotherm where average daily temperature is at least 43.7°F (6.5°C) for 100 d during the growing season. This is the elevation where trees are able to grow upright and not where they may take a twisted or ground-hugging shape.

At upper timberline, trees are bent and twisted by the wind and weather (figure 5.5). This twisted form known as a *krummholz*, means "bent wood" in German (figures 5.2 and 5.5). The height to which these trees grow is closely correlated with snow depth and wind intensity. Three winter phenomena shape these trees: snow loading, blowing ice particles, and dehydration. Snow loading refers to heavy snow that can break and bend vegetation. Many trees have "snow knees," bent-over portions of the trunk that heal, followed by new erect growth.

During winter, branches that protrude above the snow are subjected to ice blast and drying by the wind. The abrasion from windblown ice particles damages the outer, waterproof cuticle of conifer needles and the resultant water loss causes death of the tissue. Furthermore, ice from chilled water vapor may form on needles exposed to fog or clouds. The intensity of the wind may be so great that it causes ice to change to water vapor without ever entering a liquid phase. This extremely rapid rate of evaporation causes an extreme drop in temperature, which kills tissue by supercooling.

Most conifers protect themselves from cold weather by withdrawing water from cells in

FIGURE 5.2 A krummholzed and dying Jeffrey Pine, *Pinus jeffreyi*, on top of Sentinel Dome, Yosemite National Park.

twigs and stems. In most cases, it is the formation of ice crystals within plant tissues that does the damage. There are two theories of what causes cell death when ice crystals form. The oldest theory implies mechanical damage to membranes from freezing of water inside the cells. The alternative explanation, known as the vital water hypothesis, implies that all water is pulled from the cells except that which is mechanically bound to structural molecules, such as those in membranes. With a continued decrease in temperature, particularly in association with supercooling, the membrane proteins become altered, allowing irreversible dehydration of cells and/or the accumulation of toxic dissolved materials inside the cells. In cases of supercooling, even withdrawal of water cannot protect tissues. Compounded by desiccation and abrasion, supercooling damages new foliage projecting above the snow level. If several years of deep snow occur, new foliage might be able to reach a height where the action of wind is less severe. Therefore such trees may have one or two tall branches with all the needles near the top, with the remainder of the needles forming a flat "skirt" of foliage near the ground.

The upper elevation achievable by any tree species seems to be dictated by temperature. It is true that lack of soil may be limiting, as most of the subalpine zone experienced extensive Pleistocene glaciation, but the upper tree limit seems to be correlated with a line where the temperature may drop to −40°F (−40°C). This is the temperature at which pure water freezes instantly. This particular line, known as the −40° isotherm, seems especially important in limiting northward distribution of trees in the Arctic where soil or moisture conditions seem to be adequate. Dryness no doubt also has an influence. Whatever is the cause of upper tree line, it follows that photosynthesis is limited by cold temperatures. Where the annual rate of photosynthesis is so slow that cellular respiration would use all the plant's carbohydrates, growth is unable to occur. A tree cannot long endure under these circumstances.

Intuitively, a person would expect that photosynthesis cannot occur as long as snow is present, but this may not be the case. Experiments show that as long as the ground is not frozen, water transport and photosynthesis may begin in spring long before the snow is melted. Furthermore, it has been demonstrated that blue wavelengths of light required for photosynthesis may penetrate snow to a depth of over 30 in (80 cm). This observation means that some plants may be able to carry on photosynthesis under the snowpack!

Under the best of conditions at tree line, however, photosynthesis barely exceeds cellular respiration, and growth rate is slow. John Muir counted the tree rings of a Whitebark Pine that was only 6 in (15 cm) in diameter, and he found the tree to be 426 years old. One of its twigs,

FIGURE 5.3 Mountain Hemlock, *Tsuga mertensiana*, in Subalpine Forest near Anne Lake, Sierra National Forest.

about 0.125 in (3 mm) in diameter, was 75 years old.

In California, most of the trees of the subalpine zone are five-needle pines that occur in single-species stands. In some areas of the northern Sierra, however, Mountain Hemlock, *Tsuga mertensiana*, is the most common species (figure 5.3; and in many locations it is mixed with Lodgepole Pines in the Montane Forest [plate 6B]). A member of the pine family (Pinaceae), Mountain Hemlock looks like a fir, but its needles are blunter and stick out all around the branchlets. It is distributed in cold climates throughout northwestern North America from Alaska to California. In California, it is the most common subalpine tree in the Klamath Province (plate 9C) and is common in the northern Sierra Nevada as far south as Yosemite National Park. In the Cascade Province, it is most common on Mount Shasta. Where it is common in the subalpine zone, soil is thick enough and remains moist enough to support these cold-tolerant but moisture-requiring trees. South of Yosemite, it is restricted to pockets of moist soil at high elevation. At its southernmost locality, near Silliman Lake, it is localized on a spring-fed, rocky slope.

Two species of juniper also occur in the subalpine zone. Junipers are in the cypress family (Cupressaceae). Their leaves are not needlelike, as are those of the pines, but are small, flat scales that adhere tightly to the branches. The cones are berrylike and provide an important food for subalpine animals. In the central portion of the Sierra Nevada, high-elevation pines may be joined by Sierra Juniper, *Juniperus grandis* (figure 4.41), although this species is more common on dry, exposed sites. This tree is discussed in Chapter 4 as a component of Montane Forest.

The true subalpine juniper is known as Pygmy or Common Juniper, *J. communis*. It occurs at scattered localities from Mono Pass, in Yosemite, to Mount Shasta. In California, this juniper is rather rare, although its distribution is circumpolar, including Alaska, Canada, Greenland, and Eurasia. In more favorable localities, outside the harsh subalpine zone, this species may occur as a small tree; in these locations, it is known as Common Juniper. Similar to Western Hemlock, this common northern species reaches its southernmost distribution at high elevation in the Sierra Nevada.

The two subalpine junipers, Sierra Juniper and Common Juniper, clearly are relicts. Both species enjoy widespread distribution to the north. A related species, Western Juniper, *Juniperus occidentalis,* may grow in extensive stands (figure 6.13). Their patchy distribution in California may be due partly to the apparent requirement that in order to germinate, their seeds first must be passed through the digestive tract of an animal. This provides an acid treatment known as scarification. It is a requirement in other plants as well, and will be mentioned later. It is noteworthy that in the Rocky Mountains, a bird known as the Bohemian Waxwing, *Bombycilla garrulus*, was observed to pass 900 juniper berries through its digestive tract in a period of 5 hr!

The most characteristic pines of the subalpine zone are the five-needle pines. These trees dominate the Subalpine Forest throughout most of the state, including Mount Shasta, the Warner Mountains, the Sierra Nevada from Yosemite south, the Transverse and Peninsular Ranges, and mountaintops in the Great Basin. These trees have short needles, less than 3 in (8 cm) in length, that occur in clumps of five. There are four species of these pines in the subalpine zone of California. True to the ecologic-island concept, their ranges barely overlap. Each species usually occurs in a pure stand or alliance.

The four species of subalpine pines may be divided into two subgroups composed of closely related species. One subgroup includes White-bark Pine, *Pinus albicaulis*, and Limber Pine, *P. flexilis*. The other includes Foxtail Pine, *P. balfouriana*, and Bristlecone Pine, *P. longaeva*. The two groups may be distinguished by the appearance of the needles. Trees in the latter group are characterized by short needles, seldom longer than 1.5 in (4 cm), that grow for a considerable distance along each stem. The former group has needles up to 2.5 in (8 cm) in length, and they occur mostly on the terminal portions of the stems, giving each stem the appearance of a bottlebrush.

Other Pines that may occur in the subalpine zone are Lodgepole Pines and Western White Pines, both typical components of Upper Montane Forest. Lodgepole Pine (plate 6B; figure 5.4) may look very much like Whitebark or Limber Pine. All three species have the same "corn flake" bark, and needles are concentrated on the tips of the stems. Inspection of the needles will distinguish the Lodgepole Pine, whose needles occur in clumps of two. Western White Pine has needles in clumps of five, but the tree is not likely to appear dwarfed or krummholzed. Western White Pine cones are large, up to 8 in (20 cm) in length, and the bark is usually distinctly checkered.

Limber Pines and Whitebark Pines are similar in appearance. In their zone of overlap, the only sure way to tell them apart is by examining the cones. Cones of Limber Pines are small versions of the cones on Sugar Pines and Western White Pines, except that they have thick scales, and they seldom exceed 6 in (15 cm) in length.

FIGURE 5.4 Lodgepole Pine on shoreline of middle Crabtree Lake, Sequoia National Park. Old specimens of Lodgepole Pine in the Subalpine Forest exhibit a growth pattern of thick trunk and short branches, which is typical of other subalpine tree species, such as Sierra Juniper, or Foxtail and Bristlecone Pines.

Cones of Whitebark Pines are seldom seen because they are deciduous when dry; they fall apart. Furthermore, most White bark Pine cones are consumed while they are green by squirrels and birds, as will be discussed later. A five-needle pine occurring near timberline with no evidence of cones is usually a Whitebark Pine. If it looks like a Whitebark Pine, but cones are visible on the tree, it usually is a Limber Pine. Limber Pines seem to have a different growth form as well. They often have many branches in the crown, which gives them a candelabra appearance.

Whitebark Pine (figure 5.5) is probably the most common subalpine pine. It is widely distributed from Lake Tahoe to Mount Whitney, as well as the high Klamath Mountains, Cascades, and Warner Mountains. These trees frequently grow in clumps at or near tree line (plate 6B). Dis-

FIGURE 5.5 Krummholzed Whitebark Pine, *Pinus albicaulis*, on an exposed site near the summit of Mount Lassen.

FIGURE 5.6 Whitebark Pine cone partly consumed while still on the tree.

tances between clumps vary with the terrain, but 20–30 ft (6–10 m) is common. Whitebarks grow in open woodland with very little understory. They often are found growing on coarse, bare soil at the edge of a meadow. At lower elevation, or in protected sites, they are capable of growing straight and tall, up to 50 ft (15 m) in height. In the Whitebark's most common location, in areas where glacial scour has removed most of the soil, the trees are seldom over 10–15 ft (3–5 m) tall. At timberline, they are often krummholzed (figure 5.5). At high elevation, they may reach great age, although they do not reach the age of the other subalpine Pines. The oldest Whitebark Pine recorded in the Sierra Nevada was a full 1000 years of age. One study at the Eastern Brook Lakes on the east side of the Sierra, in a region occupied only by Whitebark and Lodgepole Pines, found that 39% of the trees were krummholzed, and that 90% were Whitebark Pines.

FIGURE 5.7 Limber Pine, *Pinus flexilis*, on Mount Baden-Powell, San Gabriel Mountains, Angeles National Forest.

Few Whitebark cones escape being consumed by squirrels or birds (figure 5.6). The ones that do, and that do not fall apart, will be observed to be small and circular with very thick scales. Golden-mantled Ground Squirrels, *Callospermophilus* (= *Spermophilus*) *lateralis* (plate 18D), climb the trees and gnaw on the green cones as if they were apples. Often the central core is all that is not eaten. Birds such as Clark's Nutcrackers, *Nucifraga columbiana*, tend to peck large holes in the cones, nearly hollowing them out as they gather the seeds. These high-elevation members of the jay family bury the seeds and retrieve them the following spring. Seeds that are not retrieved may germinate, which appears to be one of the reasons that Whitebark Pines often grow in clumps.

Limber Pines grow mostly in small groves on dry sites in association with Lodgepole Pines. These sites are most common in steep eastern-slope canyons between Mammoth and Cottonwood Creek near Lone Pine. Limber Pines seem to grow by default at high, dry locations where nothing else will grow—analogous, perhaps, to Whitebark Pines growing on cold, dry localities with very little soil.

Limber Pine is widely distributed throughout desert ranges in the Great Basin, where they may be the only tree species in the subalpine zone. In southern California, groves of Limber Pines occur at patchy localities on high peaks in the Transverse and Peninsular Ranges. The southernmost trees apparently occur on Toro Peak in the Santa Rosa Mountains. Two localities where hikers are apt to see them are on Mount San Gorgonio, in the San Bernardino Mountains, and on Mount Baden-Powell, in the San Gabriel Mountains. One of the most spectacular Limber Pines is on the ridge near the summit of Mount Baden-Powell (figure 5.7). Erosion by the wind had carried away most of the coarse soil, leaving its root system elevated. Approaching the tree from below, one observes blue sky between the ground and the root system. This tree has been dated in excess of 1200 years. The oldest recorded Limber Pine was a 2500-year-old specimen near Robinson Lake, in the Onion Valley area of the eastern Sierra Nevada, and a 2300-year-old specimen has been found on Mount San Jacinto. Most specimens, however, are no older than 250 years.

In the White Mountains, Limber Pines occur in the Subalpine Forest in association with Bristlecone Pines. In the southern Sierra Nevada southwest of Lone Pine, they occur in association with Foxtail and Lodgepole Pines near Cottonwood Creek, and on Sirretta Peak. Farther north, in Onion Valley, for example, Limber Pines may occur with lower populations of Whitebark Pines. On the west slope of the Sierra near Mount Whitney, there is a region near Crabtree Meadow where Limber Pines, Whitebark Pines, and Foxtail Pines grow in close proximity. Here, Foxtails are on deep, coarse soils, and Limber Pines are on south-facing slopes with less soil. Limber Pines are not common.

The other subgroup of subalpine five-needle pines includes Foxtail and Bristle-cone pines. If these trees grew together, they would be difficult to tell apart. Their distribution, however, never overlaps. In California, Bristlecone Pines are common only in the White Mountains on the east side of Owens Valley (plate 7B). Foxtail Pines are five-needle pines of the southern Sierra Nevada and Klamath Province. Both species have needles that are very stiff and short, seldom longer than 1.5 in (4 cm). The needles remain on the tree for many years, which gives it a fir-like appearance because the needles are not restricted to tufts on the ends of stems. The advantage of retaining needles for many years is that photosynthetic tissue is retained in adequate supply to provide food even though the amount of new growth for any year might be quite small. On Foxtail Pines, needles have been reported to remain on trees up to 30 years. On Bristlecone Pines, needles are reported to remain on the trees up to 45 years.

Foxtail Pines differ in appearance from other trees they may grow with. Needles grow all along the branchlets, and the weight of these needles may cause the branchlets to droop. It is the droopy branchlet, resembling fox's tail that is the source of the tree's name. The bark of these trees differs from the "corn flake" bark of other subalpine trees. Resembling most closely the bark of Western White Pines, Foxtail bark breaks up into squares that are usually about 1–1.5 in (2.5–4 cm) across, smaller than those of Western White Pine.

Unlike other subalpine species, Foxtail Pines seldom become krummholzed (plate 7A). In particularly exposed sites on ridgetops, however, they may assume prostrate forms. Krummholzed Foxtail Pines may be viewed on the ridge above Rocky Basin Lakes in the Golden Trout Wilderness. In spite of the harsh climate and fierce wind, these trees usually remain erect, with a single stout trunk. The effect of the wind is that branches may all grow out of one side of the tree, a phenomenon known as flagging or banner effect (figure 5.8). Another effect of abrasive wind is that all the bark containing the living tissue needed to nourish a few living branches remains on the leeward side of the

FIGURE 5.8 Foxtail Pine flagged by the wind. Nine Lakes Basin, Sequoia National Park.

trunk. This sometimes causes what are known as "piggyback" trunks, a phenomenon that occurs when a branch on the leeward side of the trunk begins to grow erect. After many years, perhaps hundreds, this branch may form a new trunk. These trees have few enemies other than the climate and may be the oldest trees in the Sierra. The oldest specimen, aged at 3300 years, occurs on the south side of Split Mountain, near the northern end of its distribution in the Sierra. In the Sierra Nevada, only giant Sierra Redwoods are known to rival Foxtail Pines in age.

In spite of their great age, Foxtail Pines never reach the size of Sierra Redwoods. However, they are quite large for subalpine species. The largest Foxtail Pine is located at about 10,000 ft (3000 m) on a ridge above Timber Gap, in Sequoia National Park. The diameter of the trunk is 8.5 ft (2.5 m), and it is over 70 ft (23 m) tall.

Most of the very old Foxtail Pines are only partly living (figure 5.9). They inhabit the most

FIGURE 5.9 Old Foxtail Pine, Cottonwood Lakes Basin, Inyo National Forest.

exposed sites, where the wind has denuded most of the bark. In more protected localities, they grow straight and tall and are conspicuous only because of their drooping, needle-covered side branches.

Their distribution is most common on coarse well-drained soils. On south-facing slopes, they form pure stands of thick-trunked trees up to 30 ft (10 m) in height. Foxtails have a well-developed root system. Roots are widespread and penetrate deeply in order to absorb water from the well-drained soil. It may be this requirement for relatively deep soil that keeps these trees from expanding their distribution northward into the region occupied mainly by Whitebark Pines, where repeated glacial scouring has removed most of the soil. Foxtail Pines are most common at higher elevations south of the Middle Fork of the Kings River, and they occur in nearly pure stands in the Cottonwood Basin on the east side of the Sierra. Seedlings are fairly common at lower elevations and they decline markedly at upper tree line.

The distribution of Foxtail Pines is the enigma of subalpine vegetation. Foxtails do not occur outside the state of California, and within the state there is a 300 mi (480 km) disjunction in their distribution between the Klamath Mountains and the southern Sierra Nevada. In the Klamath Mountains, they occur at scattered localities, but the bulk of the population is in the upper Trinity Alps on south-facing, serpentine soils. Interestingly, while there is a significant morphological difference between the trees from the widely disjunct populations, those from the Klamath Mountains show a greater genetic variability within the group than between the two populations. Such variability indicates long periods of isolation between various stands in the Klamath Mountains.

The disjunct distribution of Foxtail Pines is typical of a number of plant species. It is similar, for example, to the disjunct distribution of the Shasta subspecies of Red Fir between the southern Sierra and the Cascade and Klamath Ranges. The ecologic features responsible are not understood. One hypothesis associates these disjunctions with the distribution of summer rain. This may be true for Foxtails on southern exposures in the Kern and Cottonwood basins, but summer rain seems unlikely to be an important feature for populations on the west slope of the Sierra in the upper Kings River drainage. Summer rain may be a factor in the Trinities, but Foxtails grow there in the rain shadow, in drier localities.

A second hypothesis relates to light and/or temperature. Foxtails grow best on south-facing slopes with deep soil. These slopes are desertlike in terms of sunlight and dryness. The presence of deep, penetrating roots is a desertlike adaptation not shared by other pine trees. Long retention of needles probably enables the Foxtail to optimize intense light in order to shift the photosynthesis-metabolism balance in its favor while its deep roots tap the meager water supply.

Perhaps the critical factor in Foxtail distribution involves seed germination or seedling survival. Seedlings tend to be shade intolerant. Seed germination may require more heat, which would be optimized on these southern exposures. Once seedlings begin to appear, the intense light could favor root growth with a minimum of foliage, another desertlike adaptation. The similarity between Foxtail and Bristlecone Pines attests to drought tolerance, and the Foxtail's frequent association with other Great Basin species supports the concept that adaptations to a cold, desertlike environment are important.

In California, Bristlecone Pines, *P. longaeva*, occur only in a few localities (plate 7B). They are common in the subalpine zone of the White Mountains and at high elevations in the Inyo Mountains, the Last Chance Mountains, and the Panamint Mountains. In Nevada and Utah, they occur at scattered high-elevation localities. In Colorado and New Mexico, in suitable subalpine sites, the species is replaced by the Rocky Mountain Bristlecone Pine, *P. aristata*.

Bristlecone Pines are distinctive because their cones bear sharp spines on the end of each scale. Otherwise, they resemble Foxtail Pines. The nearest the two species come to each other is 20 airline miles (32 km), across Owens Valley. In the White Mountains, Bristlecones grow in groves. A few Limber Pines grow with them, at lower elevations. Limber Pines are easily distinguished by their longer needles, clustered in tufts at the ends of the stems.

In California, Bristlecone Pines are notable because of their great age. Several specimens have been dated at more than 4000 years. The oldest standing tree, dated at 4600 years, has been named Methuselah and is billed as the oldest living thing. The oldest Bristlecone ever dated was 4900 years of age. Because its size was not appropriate for its great age, it inadvertently was cut down in order to date it. The usual dating technique is to remove a narrow core from the trunk that represents a cross section of the tree rings. On Bristlecone or Foxtail Pines, however, the core may not represent a true cross section of tree rings because new tissue is laid down only under healthy bark, and only a fraction of the outer trunk may be covered with bark.

Bristlecone Pines are notable for other reasons. They grow in one of the harshest climates possible. The White Mountains lie in the rain shadow of the Sierra Nevada, and where Bristlecone Pines grow, precipitation averages only about 12 in (30 cm) per year. Snow makes up 99% of the precipitation; therefore, growth is delayed until spring, and the growing season is rarely more than a few weeks. Bristlecones grow on the summit of the range, where wind is extreme. Furthermore, they grow on nutrient-poor alkaline soil derived from dolomite, a type of limestone. At the same elevation, on different soil, there is a scrub forest of Curl-leaf Mountain Mahogany, *Cercocarpus ledifolius*, and Great Basin Sagebrush, *Artemisia tridentata*.

In the White Mountains, Bristlecone Pines usually are not krummholzed (plate 7B). In exposed sites, they may be short in stature, but they do not exhibit prostrate form. In eastern Nevada, on Mount Washington in the Snake Range, they are krummholzed at timberline. It is speculated that better drought resistance occurs in the White Mountains populations because growth of new needles occurs during summer drought. Because the eastern populations grow where there is summer rain, new needles are not produced during a time of drought stress. When winter wind desiccates the needles, they are poorly prepared and become damaged, forming a typical krummholz growth form.

Growth of these trees is so gradual that tree rings must be counted with a microscope. This slow production of new tissue also makes the wood very dense. At this elevation, decay rates of this wood are very slow, and fallen wood lies around for thousands of years. By overlapping known sequences of narrow and wide tree rings between standing and fallen wood, researchers have identified an absolute sequence of tree rings dating back about 8000 years. Because ring width is an indicator of climate, an estimation of each year's precipitation has been obtained.

An interesting side effect of this 8000 year record is that it has provided a means of checking the accuracy of the ^{14}C dating system. ^{14}C is formed in the atmosphere by cosmic radiation and is then distributed randomly. The proportion of ^{14}C present in the atmosphere should also appear in plants, because photosynthesis absorbs CO_2 without discriminating against the radioactive form. The same proportion should also be present in the animals that eat the plants. When the plant or animal dies, the amount of ^{14}C within the organism gradually disappears. By measuring the amount of ^{14}C left in a specimen, one can estimate how many years ago the plant or animal died. It had been claimed that organic material can be so dated with reasonable accuracy to ages of 40,000 years. This claim is based on two assumptions. First, the rate of ^{14}C decay must remain consistent through time. Second, the amount of ^{14}C in the atmosphere must remain constant.

When tree rings from Bristlecone Pines were dated using the ^{14}C method, a surprising result occurred: ^{14}C dating consistently underestimated the age of material older than 2500 years. Tree rings actually 7000 years old were dated radiometrically at only 6000 years of age. ^{14}C ages were 1000 years too young! Of the two assumptions regarding ^{14}C dating, one of them must be wrong. Either decay rate is not consistent or the amount of ^{14}C in the atmosphere has not been constant. The rate of decay does not alter: the half-life of ^{14}C is a consistent 5800 years. This means that, no matter how much ^{14}C is present at any given time, it will lose half of its radioactivity in 5800 years. Therefore, the second assumption, that the amount of radioactive carbon in the atmosphere has always been the same, must be in error. Evidence from other sources has confirmed this conclusion. The amount of cosmic radiation that forms ^{14}C varies with changes in the composition of the earth's magnetic field and with alterations in sunspot activity.

An important facet of this discovery was that it allowed European artifacts to be dated more accurately. Evidence of ancient man in Europe is older than was formerly thought. Western European structures such as Stonehenge were re-dated, and their construction was determined to have begun around 2600 B.C., a full 800 years earlier than was sup-

posed. This means that Stonehenge-type structures were constructed far too early to have been influenced or copied from structures in the Middle East. Likewise, clay tablets from the Balkans were found to be nearly 7000 years old—a full 2000 years older than the oldest known form of writing in ancient Mesopotamia. What all this means is that because Bristlecone Pines grow in California, we now have a better idea of how ancient civilization developed in Europe! Older radiocarbon dates are reported as a certain number of years BP (before present). The newer calibrated dates are reported as years calBP.

History of Subalpine Vegetation

The distribution of Subalpine Forest vegetation in the Sierra Nevada can be summarized as follows. Whitebark Pine is the most widely distributed, occupying the central Sierra on glacier-scoured terrain with little soil. North of Yosemite, Whitebark Pine is replaced by Mountain Hemlock. The northern Sierra is not as high as the rest of the range, so subalpine vegetation here occurs at lower elevation, where soil is more abundant. Mountain Hemlock is cold tolerant but thrives in moist soil, so it is common in the north. The subalpine tree of the southern Sierra is the Foxtail Pine. This deep-rooted species occupies coarse, well-drained slopes with southern exposure and seems to require more soil than Whitebark Pines. In the Sierra, Limber Pines grow in dry canyons on the eastern slope, where they occur in groves among Lodgepole Pines. Their distribution seems to be limited primarily by moisture. Apparently, these dry sites resemble conditions on mountaintops in the Great Basin and in southern California, where Limber Pine is the distinctive subalpine tree. Its present discontinuous distribution is probably relict. They formerly occurred more commonly with lower-elevation species that were widespread about 10,000 years ago, during the moister conditions of the late Pleistocene.

Both Whitebark and Foxtail Pines have disjunct distributions between the Sierra Nevada and areas to the northwest. To the north of the Sierra, Whitebark Pine appears in the Warner Mountains and in the Cascade and Klamath Mountains. Because of the northerly distribution, the elevation of the subalpine zone is remarkably low. On ridge crests and summits above 6700 ft (2200 m), particularly on non-glaciated southwestern exposures, there are dwarfed forests of Whitebark Pines. In the Cascades, only Mount Lassen and Mount Shasta have a subalpine zone. Krummholzed Whitebark Pines occupy southwestern exposures on these volcanic peaks (figure 5.5). In these northern provinces, subalpine sites are not numerous and seem to be characterized by soil and exposure rather than high elevation.

Subalpine vegetation has not always been distributed in an island-like mosaic. During the Miocene, 10–12 million years ago, the Sierra began to rise. Regions to the east began to get drier as the rain shadow increased, and, for other reasons, summer precipitation began to decrease. From pollen records and fossils of certain moisture-loving plants, it can be inferred that abundant summer precipitation continued throughout the Sierra and into southern California until about 1.5 million years ago. Up to that time, subalpine vegetation must have been distributed widely across the western United States.

During periods of extensive glaciation, from about 1 million to 10,000 years ago, ice sheets scoured terrains now occupied by subalpine vegetation. Fossils from the Pleistocene, the Ice Age, indicate that forest belts occurred at considerably lower elevation during that time. Extensive glaciers covered much of the Sierra, but they also were present in the Trinity Alps of the Klamath Province, on Mount Shasta and Mount Lassen in the Cascade Province, in the White Mountains, and as far south as Mount San Gorgonio in the San Bernardino Mountains. Following the end of the Pleistocene, about 10,000 years ago, subalpine vegetation presumably reinvaded the high country from low-elevation refugia. For long-lived plants such as subalpine pines, 10,000 years is not

enough time for much speciation to occur. Only four to five generations of Foxtail and Limber Pines may have occurred since the end of the Pleistocene. It is possible that there have been only three generations of Bristlecone Pines in the White Mountains since the retreat of the glaciers.

Foxtail and Bristlecone Pines have been experimentally hybridized, indicating they are very closely related. In fact, evolution of the two species probably resulted from the forest separating into two populations as the Owens Valley sank and the Sierra Nevada and the White Mountains rose on each side. This geologic event and the coincident speciation probably began about 10–12 million years ago.

Studies of Whitebark and Limber Pines indicate that they do not hybridize even though they may grow together, as they do in several localities at the lower fringe of Whitebark Pine distribution in the central Sierra. This lack of hybridization implies that the speciation event occurred a long time ago. Whitebark Pine probably is a northern species, and Limber Pine is a Rocky Mountain, or southern, form accustomed to more summer precipitation.

These pines must be considered relicts. Perhaps their continued existence is threatened by increasing dryness since the end of the Pleistocene. Studies of fallen wood from Bristlecone Pines show that in the last 1000 years, upper tree line has been lowered by nearly 3000 ft (1000 m), perhaps in association with a change in summer temperature or a decrease in annual precipitation. In eastern Nevada, it has been speculated that a similar lowering of upper tree line may be due to an increase in summer precipitation over the last 4000 years. This trend certainly seems not to apply in the White Mountains.

Past climate change associated with increasing temperatures and drought has been responsible for a change in tree line elevation through time. Tree-ring records for the last 3500 years in the southern Sierra show that tree line remained higher than today for most of that time. Tree line decline occurred from 950 to 550 years ago during a warm period, but again from 450 to 50 years ago during a cool period, confounding understanding of the causes for change. Droughts and warming during the 2000s seem to show most tree mortality occurring at the lower elevational limits of subalpine trees. Some models predict a future gradual creeping to higher elevations of subalpine vegetation.

Fossil evidence shows that until about 12 million years ago, the widespread Subalpine Forest was composed of many species. It was a diversified forest that has gradually become impoverished. Today in the eastern Great Basin, there may be up to four species of subalpine trees in a mixed forest. This is where summer precipitation is still significant. In California, however, an area of Subalpine Forest seldom contains more than two dominant species. Single-species stands are most common. Furthermore, in California each of the four species of subalpine five-needle pines has a different, barely overlapping geographic distribution. Perhaps extensive Pleistocene glaciation in California reduced the former forest to localized pockets of trees in single-species stands. Different species must have been isolated in different localities. Reinvasion of specialized high-country habitats from low-elevation refugia occurred opportunistically by whatever species was nearby.

Refugia may have been extensive. Fossil evidence shows that during the Pleistocene, Foxtail and Bristlecone Pines occurred at considerably lower elevations. Fossil Foxtail cones are found in Pleistocene deposits at Clear Lake, in the northern Coast Range. This locality is 5000 ft (1600 m) below and 65 mi (100 km) south of the nearest presently growing population in the Klamath Mountains. All over the Great Basin, there is evidence of Bristlecone Pines that grew more than 2000 ft (700 m) lower than they do today. It must have been a species with widespread distribution during the Pleistocene.

It also could be argued that each land of subalpine climate is uniquely harsh, and that from a refugium of several species only one had the ability to tolerate the unique combination of ecological circumstances in a particular local-ity. In effect, each subalpine species today occurs in its locality by default; nothing else can tolerate the specific mix of conditions. This hypothesis seems simpler and therefore more believable, but it leads to the search for a suite of similarities and tolerances to explain apparent disjunctions in distribution. For example, it is difficult to imagine what elements of the climate could be similar between the southern Sierra Nevada and the Klamath Mountains of the Pacific Northwest. Foxtail Pines grow in both places, but nowhere in between.

Changes in montane vegetation during and since the Pleistocene can be analyzed by studying pollen, needle, and cone assemblages found in lake sediments. Similar climatic changes from the deserts can be analyzed by studying organic material fossilized in woodrat middens, which are piles of vegetational debris preserved by rat urine in rock crevices and other dry places. For example, we can infer that a coniferous forest of Rocky Mountain affinities was once more extensive in what is now the eastern Mojave Desert because a woodrat midden on Clark Mountain revealed that Rocky Mountain White Fir, *Abies concolor concolor*, has existed there continuously for 28,000 years. Distribution of Rocky Mountain White Fir in the area today is restricted to north-facing slopes above 6000 ft (1900 m) in the Clark Mountains, Kingston Mountains, New York Mountains, and Charleston Mountains north of Las Vegas. Species analyses in the Sierra Nevada suggest a drying trend following the retreat of glaciers about 10,000 years ago. Then, about 6000 years ago, a moist period including summer precipitation became apparent. At this time, there was an increase of Mountain Hemlock, firs, and Limber Pine in the Sierra Nevada. At the same time in the Great Basin, there was an increase in the two-needle version of Pinyon Pine, indicating a similar trend. This was followed by another dry period, interrupted by a short glacial episode about 600 years ago. Apparently, the glaciers that remain in California today date back to that "little ice age."

FIGURE 5.10 Clark's Nutcracker, *Nucifraga columbiana.*

Animals of the Subalpine Zone

In general, mammals of the subalpine zone are shared with the Montane Forest or the alpine zone. Unlike plants of the subalpine zone, endemism or restricted distribution of subalpine mammals seems not to occur frequently. Golden-mantled Ground Squirrels (plate 18D) are perhaps the most conspicuous. They feed on pine seeds and the cones of Whitebark Pines, and in the spring they eat various fungi, such as truffles. Chipmunks, such as the Alpine Chipmunk, *Tamias alpinus,* also occur here, feeding on seeds, buds, and insects.

Certain birds of the subalpine zone show some interesting behavior. Perhaps the most conspicuous member of the Subalpine Forest is Clark's Nutcracker, *Nucifraga columbiana* (figure 5.10), the high-elevation member of the jay family (Corvidae). Resembling the Gray Jay, *Perisoreus canadensis,* of the Rockies, it is a large gray bird with black wings and white wing bars. These birds are uncommonly noisy. Their raucous sounds attract attention throughout the subalpine zone.

In winter, Clark's Nutcrackers migrate to lower elevations, usually to the Pinyon Pine zone. Nutcrackers gather seeds of Whitebark, Limber, Bristlecone, Foxtail, Pinyon, and Jeffrey Pines, and they store them in caches for the following winter and spring. Research shows that one Clark's Nutcracker may store up to 38,000 pine seeds in a single season, and it has been demonstrated that they locate these caches by memory. Birds are known to lack a well-developed sense of smell. Nutcrackers remember patterns of rocks and trees in the vicinity where they bury their seeds, and relocation of rocks can cause them to dig for them in the wrong place. Many stored seeds are never recovered and subsequently sprout. The clumping pattern commonly associated with Whitebark Pines has been attributed to this phenomenon, as has clumping of Limber and Bristlecone pines. These pines do not have winged seeds; hence, Clark's Nutcracker is probably the major mode of dispersal.

Nutcrackers store seeds in two ways. They disperse them in forested areas and at lower elevations, particularly in the Pinyon Pine zone. Seeds are also concentrated in a communal cache on a slope of at least 20° to 30°. Nutcrackers have a pouch in the throat—the crop—where they can carry many seeds. It has been demonstrated that they will bury up to 15 seeds in a cache. Do Clark's Nutcrackers know how to count, or is the number of seeds dictated by the size of the crop? Although the answer has not been determined, the strategy of burying seeds in small caches is to make it energetically foolish for other animals to search randomly for them; more than 15 seeds per cache would be a reward worth searching for. Likewise, the energy expended in searching for seeds on a steep slope would not be rewarded unless the animal knew where to look.

In late August and September, Clark's Nutcrackers become concentrated in the Foxtail Pines and Whitebark Pines of the southern Sierra. The young have fledged and are able to fly, but they are still being fed by the adults. Much of the racket associated with groups of Clark's Nutcrackers is from young birds begging for food. As winter sets in, they move downslope, particularly on the east side, to overwinter at lower elevations in the Pinyon Pines.

About 30 species of birds can be identified as typical of the subalpine zone. Most of them migrate to escape the rigors of winter. Some of them, such as the Hermit Thrush (figure 4.44), go to Mexico. Others, such as American Robins, move into the western foothills. Clark's Nutcrackers and White-crowned Sparrows (figure 4.38) move down the east side of the Sierra. Mountain Bluebirds, *Sialia currucoides*, congregate in large flocks and move to the Central Valley or to the Great Basin for the winter.

Mountain Bluebirds are conspicuous cavity-nesting birds of the subalpine zone. They nest in abandoned woodpecker holes and feed in nearby meadows. They are conspicuous by virtue of their bright blue feathers and their habit of hovering in open space while searching for insects on the ground. They are frequently observed perching on a high open snag, from which they are able to dart out and grab insects on the fly.

The reproductive behavior of these birds has been used as an example of how natural selection favors species that produce the greatest number of offspring. Once a pair-bond is established between a male and a female, they take turns tending the nest. If a male returns to the nest and finds another male attentive to his mate, he will drive away the intruder. He will drive away the mate as well, but not until he has pulled out some of her feathers. If, however, the female already has laid a clutch of eggs, for which he is the father, he will pay little attention to either the intruding male or his mate. The implication is that the important thing is for the bird to pass on his genetic material to the next generation. If he already has mated and his mate is tending a clutch of eggs, it would be counterproductive to drive her away, because he would be left to attend the nest himself. On the other hand, if he finds that another male has been attentive to his mate before the eggs are laid, it is possible that he would be expending energy to rear another male's offspring. To prevent that, he drives away both the male and female, maims her to make her less attractive, and proceeds to court and mate with another female. In this way, he optimizes the production of his own offspring.

Pinyon-Juniper Woodland

Pinyon-Juniper Woodland is the lowest of the mountain communities, or it may be considered the uppermost desert community. Because the community often becomes isolated as ecologic islands on the tops of desert ranges, it will be discussed here.

Pinyon pines occur in the most hostile of habitats, experiencing snow and cold in the winter, followed by heat and drought in summer. The trees grow on rocky, exposed ridges where wind is often intense. Nevertheless, this is one of the most attractive landscapes of desert mountains. Pinyon pines are the nut pines. Their large seeds, known as piñones (pine nuts), which were a staple for Native Americans, still may be purchased in specialty stores.

The Singleleaf Pinyon Pine, *Pinus monophylla* (figure 5.11), occurs at upper-elevation localities on desert mountain ranges, and along the eastern side of Sierra Nevada and Transverse and Peninsular Ranges.

There are two other kinds of Pinyon Pines in California. A two-needle form, the Colorado Pinyon Pine, *P. edulis*, occurs at high elevation in the eastern Mojave, in the New York Mountains and the Mid Hills area to the southeast. The Four-needle or Parry Pinyon Pine, *P. quadrifolia*, occurs in patches in the Peninsular Ranges from Garner Valley in the Santa Rosa Mountains south to the northern part of Baja California. At some localities, Singleleaf Pinyon Pine overlaps with one of the other types. Hybridization often occurs in these overlap areas. Classification of these pines has been controversial, and studies of hybridization are ongoing.

Differences in appearance and microhabitat are apparent among these different kinds of Pinyon Pine. In the New York Mountains where the two- and single-needle types occur together, the two-needle form occurs at higher elevation. At the summit, it occurs with a few Rocky Mountain White Firs. Both white firs and Pinyon Pine enjoy widespread distribution throughout the Great Basin eastward to the

FIGURE 5.11 Singleleaf Pinyon Pine, *Pinus monophylla.*

Rocky Mountains. The Two-needle Pinyon Pine's occurrence in the New York Mountains represents its westernmost extreme in distribution. Ongoing studies of the two types of pinyon pine here seem to indicate that there is a broad range of intergradation between the upper-elevation two-needle form and the lower-elevation one-needle form. Intergrade characteristics include variations in color, anatomy, and number of needles, number of resin ducts in needles, as well as differences in the cones. Some authorities still insist that the southern populations of pinyon pines warrant a separate name, *californiarum,* as either a separate species or a subspecies.

Where the Singleleaf Pinyon Pine overlaps with the four-needle species in the Peninsular Ranges, the four-needle type occurs farther from the desert. Its habitat is on gravel flats and slopes associated with Great Basin Sagebrush and Jeffrey Pine. The Singleleaf Pinyon occurs to the east, on ridges bordering the desert. In an area east of Tecate, just south of the border in Baja California, there appears to be a hybrid swarm between the two species (figure 5.12). Pure Four-needle Pinyon Pines tend to be broadly pyramidal in shape, and the foliage is blue green. Pure Singleleaf Pinyon Pines have a sprawling, rounded shape, and the needles are yellow green. The apparent hybrids show both color and shape variations, and the needles occur in clusters of one, two, three, four, and five!

These four-needle Pines were probably carried to their present latitude by drift along the San Andreas fault. When the Baja California peninsula was farther south, these trees would have been adjacent to the range of the three-needle Mexican Pinyon Pine, *P. cembroides.* It is possible that the four-needle type was derived from the three-needle species since Baja California began to move away from the mainland.

Throughout their range, Singleleaf Pinyon Pines are associated with various species of juniper. On the eastern side of the Sierra north to the Nevada state line, pinyon pines are associated with Sierra Juniper, *Juniperus grandis* (figure 4.41). Eastward, throughout the Great Basin and Mojave, Pinyon Pines are associated with Utah Juniper, *J. osteosperma.* In southern California and northward to the southern Sierra Nevada, Singleleaf Pinyon Pine is associated with California Juniper, *J. californica.* They also occur together in the Granite Mountains of the Mojave Desert. Ranges of California Juniper and Utah Juniper overlap on the north side of the Transverse Ranges, where Utah Juniper has spotty distribution. All three species of juniper are found on the north side of the San Bernardino Mountains.

FIGURE 5.12 Pinyon Pine hybridization near Tecate, Baja California. Singleleaf Pinyon Pine, *Pinus monophylla*, to the left; Parry Pinyon Pine, *Pinus quadrifolia*, to the right; and a hybrid in between.

Pinyon pines tend to occur on rocky slopes and ridges. Junipers tend to occur on coarse soils of the upper slopes. Because of these habitat preferences, it is not uncommon for junipers and pinyon pines to occur in the absence of each other. California Juniper occurs to the exclusion of Singleleaf Pinyon Pine on the desert side of the Transverse Ranges and in southern California in the Elsinore Valley near Perris, between the Santa Ana and San Jacinto Ranges. It also occurs in similar habitat on the eastern side of the Coast Ranges, bordering the Great Central Valley.

On the west side of the Kern River along lower Peppermint Creek at 5200 ft (1600 m), Pinyon Pines occur on open granite outcrops along with Jeffrey Pines and Gray Pines. In the vicinity of Kennedy Meadows, from 5500 to 6500 ft (1700–2200 m), there is an extensive stand of Pinyon Pines with Sierra Junipers scattered throughout. At higher elevations and along drainage, this woodland is joined by Jeffrey Pines and Canyon Live Oak. Large fires in the Kennedy Meadows area in 2000 and 2002 burned nearly 200,400 acres (100,000 ha), much of which was Pinyon Pine Woodland. Pinyon-Juniper Woodlands are not fire adapted, and recovery is on the order of hundreds of years. Seed dispersal by animals such as seed-caching rodents and Clark's Nutcracker, along with favorable climate for seedlings should eventually result in reestablishment of pines and junipers, but not within our lifetimes.

Other associated plants of Pinyon-Juniper Woodland include various members of other drought- and snow-tolerant communities, such as Sagebrush Scrub, Blackbrush Scrub, and Desert Chaparral. Included among these species are various manzanitas (*Arctostaphylos* spp.), mountain mahoganies (*Cercocarpus* spp.), and scrub oaks (*Quercus* spp.). Surprisingly, two snow-tolerant species of prickly pear cactus (*Opuntia* spp.) reach the upper-elevation limit of their distribution in the Pinyon-Juniper Woodland. These species are characterized by rather long spines, which may form a sort of fur coat that helps insulate the cacti.

Animals of the Pinyon-Juniper Woodland

Animals of the Pinyon-Juniper Woodland tend to be transient species. Food for these animals is surprisingly abundant and includes pine seeds (piñones), acorns, and juniper berries. Browse is available in the form of shoots and young twigs on Antelope Brush and mountain mahoganies. Native Americans also found abundant food in the pinyon zone, spending summer and autumn months at higher elevation and winter in the desert or at the coast.

During summer months, desert animal species move up into the Pinyon-Juniper Woodland

TABLE 5.1
A Summary of Alpine Communities

Community	Physical Indicators	Plant Indicators	Animal Indicators
Alpine pioneer zones	Rock faces, talus slopes; little or no soil	Crustose lichens, Sky Pilot, Alpine Gold	Pika, Yellow-bellied Marmot
Alpine fellfields	Rocky gravel flats and slopes, patterned ground; rapid percolation, strong wind, needle ice	Cushion plants	White-tailed Jackrabbit, Sierra Nevada Bighorn Sheep
Alpine heath	Open shrubland; upper edges of meadows and lake shores	Heathers, willows	Sparrows, warblers
Dry meadows	Well-drained acidic soil; higher parts of meadows below the shrub zone	Mat-forming grasses, bunchgrasses, sedges, wildflowers	Gophers, voles, Mule Deer
Wet meadows	Organic waterlogged soil, hummocks, peat; lower parts of meadows; may border streams, lakes, or bogs	Sod-forming sedges and grasses	
Snowfields	Areas where snow remains most of year; north-facing slopes, protected ridges	"Snow algae"	Nematodes, rotifers, ice crawlers

to escape the heat. This group includes a variety of birds and large Mammals, such as Desert Bighorn Sheep, *Ovis canadensis nelsoni*. On the western borders of the desert where the mountains are high, numerous alpine and subalpine species migrate down the mountain to the Pinyon-Juniper Woodland in winter. These also include the Sierra Nevada Bighorn Sheep, *O. c. sierrae*, and various bird species, such as Clark's Nutcracker and Steller's Jay.

One bird that is identified particularly with Pinyon-Juniper Woodland is the Pinyon Jay, *Gymnorhinus cyanocephalus*. This jay differs from others by its uniform blue color and lack of a crest. These jays feed in large, noisy flocks that move incessantly from tree to tree. In the Peninsular Ranges, Pinyon Jays are also associated with Jeffrey Pines.

The importance of these birds for seed dispersal has been well documented. Clark's Nutcrackers and Pinyon Jays have been observed to cache seeds as far as 14 mi (22 km) from the parent trees. One study documented a flock of 150 Clark's Nutcrackers that succeeded in caching between 3.3 and 5 million pinyon seeds in a single season.

Alpine Zone

The alpine zone is the land above the trees. It was named in honor of the European Alps, but C. Hart Merriam called it the Arctic-Alpine zone, referring to the similarity of Alpine Tundra to Arctic Tundra. It is common for various authors to refer to the alpine zone as tundra, as if there were only one kind of community above tree line. At the other extreme are authors who find up to 31 different categories of alpine vegetation. This book will use six categories, broadly delineated according to soil characteristics and available moisture (table 5.1).

In a general sense, the alpine climate is one of intense light, temperature extremes, extreme wind, and short growing season. The important

154   MOUNTAINTOPS

difference between arctic and alpine climates is that in the Arctic, day length in the summer is very long, and the soil may be permanently frozen (permafrost). One of the critical limiting factors for plants in the alpine zone is cold nights.

In the alpine zone of the Sierra Nevada, the precipitation of 35–40 in (87–100 cm) per year is virtually all snow. Summer thunderstorms contribute patchy precipitation that may be locally heavy. To the north, total annual precipitation is greater, but due to the lower elevations, alpine regions are more restricted there.

The greatest expanse of alpine terrain occurs from Lake Tahoe southward. From Sonora Pass south to the Cottonwood Basin, it forms a discontinuous belt, broken broadly by the headwaters of the San Joaquin River near Mammoth Lakes. Near Lake Tahoe, alpine zones occur above 9900 ft (3000 m). To the south, the elevation of alpine terrain gradually increases to 11,000 ft (3300 m). In the southern Sierra Nevada, alpine communities are divided by the Kern River gorge. Along the western divide, they extend southward to the Mineral King area. The Cottonwood Lakes area is the southernmost alpine region of any size along the main divide. Southward, distribution is patchy.

In southern California, due to drier conditions, the alpine zone occurs above 10,000 ft (3050 m). It is well represented only on three peaks: Mount San Antonio (Baldy), Mount San Gorgonio, and Mount San Jacinto. Other high peaks such as Mount Baden-Powell and Cucamonga Peak have small areas of alpine vegetation. Some authors contend that these southern California peaks do not have true alpine vegetation, that the alpine appearance is a consequence of strong winds and poor soil. There is a strong resemblance to the alpine vegetation of the Great Basin.

North of the Sierra Nevada, alpine vegetation occurs on Mount Lassen, Mount Shasta, and several localities in the Klamath Mountains, where the elevation of the alpine zone drops to about 8000 ft (2400 m). The alpine zone of the Pacific Northwest is a good deal wetter than it is in the Sierra. In this respect, it resembles what one might find farther north in Oregon and Washington. Inclusion of some Sierran and Great Basin elements makes it unique.

In the California portion of the Great Basin, alpine vegetation is well established only in the White Mountains above 11,400 ft (3400 m), where it is divisible based on several distinct soil types. Similarly, in the Inyo Mountains to the south, there is a strong edaphic influence. The only other area of the Great Basin in California that supports alpine vegetation is Eagle Peak above 9000 ft (2700 m), in the Warner Mountains. Here the terrain is volcanic. What makes the alpine zone of the Great Basin unique is its intense solar radiation. It has been claimed that the solar radiation in the White Mountains is more intense, for this latitude, than anywhere else on the planet.

In California, the alpine zone is notable for occurring in a region of Mediterranean climate. Throughout the rest of North America, high mountains occur in regions where there is abundant summer precipitation. In the Rocky Mountains or in mountain ranges to the north of California, there is commonly cloud cover and frequent rain in the alpine zone during summer months. In California, with the possible exception of the Pacific Northwest, summer days are clear.

Fourteen percent of the plants in the Sierra Nevada are unique to the range, but only about 8% of alpine plants are endemic. Above 10,800 ft (3292 m), there are a total of 570 species, 47 of which are endemic. Most alpine plants are drought-adapted perennials. It is interesting that many alpine plants have near relatives that are desert plants, including buckwheats, lupines, grasses, and many members of the sunflower family. The most common plants of alpine regions are erect perennials in the Sunflower Family (Asteraceae), Mustard Family (Brassicaceae), Legume Family (Fabaceae), and the Buckwheat Family (Polygonaceae). Other common growth forms are cushion plants, and grasslike members of the Sedge (Cyperaceae),

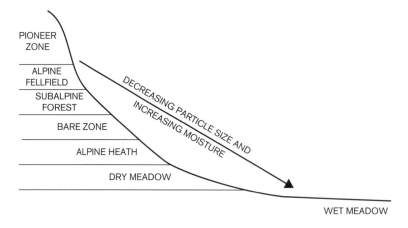

FIGURE 5.13 Generalized distribution of mountaintop vegetation.

Rush (Juncaceae), and Grass (Poaceae) families. The Sierra Nevada is further unique by having alpine plants that complete their entire life cycle in one season. These are known as ephemeral or annual plants. During a single growing season, they germinate from seeds and complete their growth and flowering in time to set seed, which is also a strategy of many desert plants. Accomplishing an entire life cycle in less than 1 year requires abundant sunlight. No other mountain range has a significant number of annual plants because no other mountain range has enough light in the summer. There is only one ephemeral species in the Rocky Mountains, for example.

The close relationship between Sierran alpine plants and desert plants might be expected for two reasons. First, the climate is similar: both are regions of drought and high light intensity. Second, the desert is nearby, which means that winds, birds, or other dispersal agents can carry seeds from one location to the other in a short period of time. Alpine and desert areas could share a reservoir of similar species.

In the Sierra Nevada, alpine communities are found south of Lake Tahoe. Here soil has a strong influence on the distribution of plant communities. Soil particle size influences water-holding capacity and drainage. Soil is also important because of the specific chemical composition associated with the rock from which it was weathered. North of Bishop, there is a good deal of metamorphic rock in the high country. Sandstone, limestone, volcanic rocks, and schist weather into soils of different consistency and chemistry. In the Carson Pass area, a thorough study of alpine vegetation found that distribution of vegetation types was related positively to parameters such as snow depth, soil moisture, and parent rock material, with the last factor judged to be highly significant. Likewise, in the White Mountains, where many rock types are present, the alpine vegetation has been classified according to soil type based on parent rock material.

In the southern Sierra Nevada, most parent rock material is granitic. Here the distribution of vegetation is clearly a function of water availability based on particle size of the soil. Progressing from steep cliff faces, where there is virtually no soil, to the center of meadows, soil particle size gradually gets smaller, and available soil moisture gradually increases. Along this gradient are distributed most of the alpine communities (figure 5.13). Exposure to wind and sun and duration of snowpack are other locally important variables.

Alpine Pioneer Zones

Pioneer vegetation is the first to inhabit a bare area. Pioneer zones occur on cliff faces and talus slopes, where soil is so scarce that there is not enough substratum for typical plants to take root. Most of the vegetation here is made up of crustose lichens that occur in colorful patches on bare rock surfaces. These organisms derive their moisture from the air or from that which condenses on the rock. The fungal component of the lichen anchors it to the rock, and the algal component provides the photosynthesis. The weak acid produced by the fungus as it metabolizes is a primary force that gradually breaks up the rock into particles of soil.

Angular boulders break off of cliff faces to form rubble piles known as talus slopes. Lichens continue to adhere to these boulders where possible. Soil accumulates in cracks and low places between these boulders. In this habitat, some of the most spectacular alpine wildflowers are found. Many of these plants grow on sites that extended above glaciers during the Pleistocene Epoch. Outcrops that extend above glaciers are called *nunataks*, and they may be recognized by the absence of glacial striations or polish on the rocks. The summits of many high mountains and some of the high plateaus such as Dana Plateau escaped glaciation and became refugia for high-elevation wildflowers during Pleistocene.

Sky Pilot, *Polemonium eximium*, grows in rock crevices at higher elevation than any other Sierra plant. It can be found near the summit of Mount Whitney. This small plant, with bright blue flowers occurring in a globular head, is a source of mixed fragrances. The sweet odor of the flowers is extremely pleasant, but the resin-covered foliage has a very unappealing odor, something like that of urine. Here we see various adaptations to this harsh climate. First, the air is very dry at high elevations, and interpretation of odor requires moisture. Therefore, most alpine plants rely on large, showy flowers rather than odor to attract pollinators. Sky Pilot compensates for the dry air by producing a very strong odor, and it is also brightly colored. However, the foliage has a noxious odor, perhaps to discourage herbivores. The resin coating of the foliage and the tight clustering of leaves about the stem are also desert adaptations that inhibit evaporation.

Perhaps the most striking odor produced by an alpine plant is the on the foliage of Mountain Pennyroyal (Coyote-mint), *Monardella odoratissima*. This plant with many square stems and tiny leaves has dense, rounded heads of many small violet flowers. It is one of the most colorful of the plants along trails through rocky areas. A member of the Mint Family (Lamiaceae), it produces a pungent odor that presumably would repel an herbivorous animal, but delights a hiker that might inadvertently brush against the plant. It apparently also can be used to make a flavorful tea.

Another typical plant of these rocky crevices is Alpine Gold, *Hulsea algida*. This member of the sunflower family (Asteraceae) shows typical alpine adaptations. The flower head is very large, up to 2 in (5 cm) across, whereas the whole plant is seldom taller than 8 in (20 cm). The blossom is brilliant yellow and has no odor. As is true of many members of the sunflower family, the blossom faces the sun at all times of day. This action, known as sun tracking, enables direct rays of sunlight to warm reproductive parts, which accelerates their rates of development. The foliage is covered with a myriad of fine silvery hairs that reflect light and trap air, which acts as an insulation. Light-colored, hairy foliage is a common characteristic of alpine and desert plants.

There are two common species of columbine in the Sierra. The white-flowered Alpine Columbine, *Aquilegia pubescens* (plate 20F), is common among the rocks of these talus slopes. Flowers are large, consisting of five petals, each of which is equipped with a nectar spur. The nectar spurs of this species are oriented downward with the openings upward. Close inspection will reveal that the spurs, at their very tips, are translucent providing a light spot or window that is probably a target to improve the aim of the tongue of the pollinator. Similar to many white-colored flowers, they are pollinated by Sphinx Moths. An interesting high-elevation sphinx moth of the Sierra Nevada is the One-eyed Sphinx Moth,

FIGURE 5.14 Hybrid between Alpine and Red Columbine.

Smerinthus cerisyi. Growing nearby, but more commonly along streams is the Red Columbine, *A. formosa* (plate 20E). This is a common columbine of the western United States from Baja California to Alaska. Flowers are smaller than those of the Alpine columbine, and the openings of the nectar spurs aim downward. Typical of red-colored tubular flowers, they are mostly pollinated by hummingbirds, although White-lined Sphinx Moths, *Hyles lineata* (figure 9.38) also pollinate tubular flowers. Hummingbirds or Sphinx Moths have no problem hovering below the flowers and projecting their tongues upward toward the bright spot at the end of the spur where the nectar is located. Where the two species occur together, they readily hybridize, producing flowers with a variety of colors from red to pink to purple to yellow to white (figure 5.14). Some have spurs aiming upward and some downward, no doubt confusing pollinators and enhancing hybridization.

In addition to anatomical adaptations, certain physiological processes are commonly associated with these plants. Most of them are long-lived perennials, the age of which can be determined by examining growth rings in the roots. Two notable examples in the Phlox Family (Polemoniaceae) include look-alike, low-growing shrubs, with abundant white flowers. Granite Gilia, *Linanthus* (= *Leptodactylon*) *pungens*, can reach 120 years of age and Spreading Phlox, *Phlox diffusa*, has been dated at 40 years of age. This means that they grow for many years, during which time they establish an elaborate root system. The bulk of most alpine plants is in the roots. Also, because the growing season is short, it is common for plants to produce flower buds during the year prior to their blooming. Energy of the plant is thus conserved, and it will not break its dormancy too early. Most of these plants are capable of photosynthesis at remarkably low temperatures, but they will not begin to function until soil temperature is high enough and day length long enough. If they came out of dormancy too early, they could be damaged by an unexpected cold snap or snowfall. Some flowers are more tolerant of desiccation. For example, in the Plantain Family, Plantaginaceae (= Scrophulariaceae), Mountain Pride, *Penstemon newberryi*, is a form that is common among dry locations in the rocks. It has conspicuous magenta, tubular flowers that are hummingbird pollinated. It is able to minimize water loss by partially closing its stomates on warm days, while a related species, Davidson's Penstemon, *P. davidsonii*, a purple-flowered form, has to restrict its growth to patches of wetter soil. Where they occur in close proximity, interesting hybrids between these two species possess a variety of colors and tube shapes. Tubular flowers such as these provide an interesting refuge for insect pollinators. Pollinators such as bees are known to crawl inside the flowers and spend the night, in sort of a floral sleeping bag, in order to become insulated from cold nighttime temperatures.

Close inspection could reveal that some of these talus slopes are actually rock glaciers. They have an ice center, and like a glacier they slowly slide down hill. As the ice melts, water accumulates at the downhill side of the rocks and forms a wetland that supports a large number of species. Particularly conspicuous, forming mats around borders rocks, would be the bright pink clusters of Sierra Primrose, *Primula suffrutescens*, or Rock Fringe, *Epilobium obcordatum*.

Alpine Fellfields

The term fell refers to a stony field. A fellfield is an area, usually flat or slightly sloped, that is composed mostly of gravel and rocks (figure 5.15). Water percolates rapidly through fellfield soils, so they do not retain moisture for long. Strong winds sweep across these flat areas and carry away many fine soil particles, so these areas remain coarse for many years. Many botanists who classify alpine vegetation object to calling these communities fellfields, apparently because the term describes the rocks and not the plants. The expression was originally used to describe European rocky tundra, and there is some concern that it is inappropriate to use it for a North American alpine zone composed of non-European plant species. Some botanists refer to this and the Pioneer zone as Alpine Steppe, and others classify each region according to dominant plant species. The term Fellfield will be used here because it is widely used in the literature, and it aptly describes the high, flat, rocky areas of the California high country.

Many plants of the Fellfield are known as cushion plants. These are long-lived perennials with a deep taproot. All the foliage grows close to the ground in a small mound about the size of a pincushion. By growing close to the ground, the foliage remains out of the wind and takes advantage of the warmer temperatures near the ground. Another advantage of a low, cushion-like growth form is that the plant traps its own soil and organic matter. As wind blows over the cushion, friction causes the wind to lose some of its energy. Fine dust particles, bits of leaf debris, and other matter fall directly into the cushion. There is far more soil and organic matter within the leaves of the cushion than out in the open between plants.

Alpine plants in general have leaves that are only 10% as large as lowland counterparts. Because of intense sunlight, large leaves are unnecessary, and small leaves reduce evapora-

FIGURE 5.15 Alpine Fellfield, New Army Pass, Sequoia National Park.

tive water loss. Plants that form cushions usually have foliage covered with silvery hairs, an adaptation that helps to intercept harmful ultraviolet (UV) radiation. There are two wavelengths of UV light. Ultraviolet A (UVA) is a longer wavelength and lower energy. Ultraviolet B (UVB) has higher penetrating power, and many alpine plants have silvery hairs or waxy surfaces that repel the harmful rays. Other plants have enzymes or chemicals that harmlessly absorb or dissipate UVB. Of particular importance are flavonoids such as red to blue pigments common in the petals of many flowers. Also of significance are pigments in petals and pollen, which are invisible to the human eye, but reflect UVA light visible to insects and birds. These pigments form patterns in petals that lead pollinators to the center of flowers where the nectar is. Some yellow pollen absorbs UV light, which causes anthers to appear as dark dots that stand out in visual relief to insects. Research on the vivid yellow color of Buttercups (*Ranunculus* spp.) has also revealed that surface cells reflect light as in a mirror, and deeper cells reflect a more diffuse light. The combination of these two forms of reflection is apparently very attractive to bees.

Unlike plants in rocky crevices, cushion plants typically have many small flowers. The entire upper surface of the plant becomes covered with blooms, the strategy being to attract pollinators by sheer abundance rather than by size. Most common among the cushion plants are lupines and buckwheats. Lupines are in the legume family (Fabaceae) and have nitrogen-fixing bacteria in their roots, which helps to fertilize the entire ecosystem. When they blossom, the entire upper surface becomes covered with tiny blue flowers.

The buckwheat family (Polygonaceae) has many members in mountain and desert environments. Buckwheat cushion plants produce a myriad of yellow blossoms. As the flowers become older, the petals become reddish. This color change ensures that pollinators continue to visit flowers that are unpollinated. Pollinators are attracted to the yellow color but tend not to visit the red ones. In this way, foraging energy is optimized for both the pollinator and the plant. By retaining the older red flowers, however, the plant apparently maximizes its total color display, which is helpful for long-distance attraction of pollinators.

Another characteristic of Fellfields is that the rocks may form conspicuous patterns indicative of stream channels, even though water does not flow among them today. These patterns were formed many years ago during times of heavy precipitation, apparently when rocks were forced to the surface by freezing of water-logged soil. The rocks then slid or were blown by the wind to the lowest adjacent area, which must have been a small stream channel. Later, the wind blew away the soil between the rocks, so that they remain elevated in the pattern taken by the former stream channel. This process takes thousands of years. Because the Sierra is a young mountain range, such patterned

ground is not as common as it is in the Rockies. Furthermore, soil in the Rockies is more likely to remain waterlogged for long periods of time.

In the western Sierra foothills, at about 400 ft (130 m) elevation, are regions of "rock stripes." These are lines of fragmented rocks, 60–100 ft (20–30 m) apart, separated by alluvial soils. This phenomenon has been interpreted as a form of patterned ground that apparently dates back to the time of maximum glaciation during the Pleistocene. Intense frost wedging broke up the rocks of uphill outcrops. Aided by snow and ice, the rocks were carried to their present positions in rows. Rock stripes are further evidence of the profound impact of glaciation on the topography of the Sierra Nevada.

In some areas of the alpine zone are gravel flats of rather uniform particle size. These areas are noteworthy because they support very limited plant cover. These barren zones may be 10–20 ft (3–6 m) wide at the edge of a meadow, below the tree zone, or they can cover many acres. One such large bare area, known as Guyot Flat, lies on a plateau at 11,000 ft (3300 m) on the eastern side of the Kern River gorge. This plateau probably formed during the same erosional interval like many of the other sandy plateaus, such as Siberian Flat, Sandy Meadow, and Bighorn Plateau, well-known landmarks along the Pacific Crest Trail. The surface was flattened by erosion and subsequently covered by deposits of gravel. Guyot Flat gravels have been measured to a depth of over 100 ft (30 m). The reason these zones are bare is not certain. One hypothesis is that needle ice forms during winter and kills the roots of trees. Another suggestion relates to the water-holding capacity of the soil during the growing season. When snow melts, water sinks rapidly through this gravel and is available to plants for such a short period of time that nearly nothing is able to grow.

One conspicuous plant ekes out a living on these gravel benches and bare zones, particularly in the southern Sierra. This is the Woody-fruited Evening Primrose, *Oenothera xylocarpa* (figure 5.16). Its flowers have four large, bright yellow petals, up to 1.5 in (3.8 cm) long. Similar

FIGURE 5.16 Woody-fruited Evening Primrose, *Oenothera xylocarpa*.

to flowers of rock crevices, the blossom is disproportionately large. There may be one to four blossoms on a plant, clearly exceeding the leafy portion in size. The petals arc upward to form a large disc that captures light in the same way a satellite disc captures radio waves. Capturing light in this way increases the temperature of the reproductive structures and accelerates development. Insects have discovered that it is warm in these flowers, and many of them rest there for long periods of time, particularly in the morning. The light color and large blossoms of this plant are useful for attracting nocturnal pollinators such as sphinx moths (Sphingidae).

The leaves on this plant show another alpine and desert adaptation. They occur in a circular rosette that lies flat on the ground, a growth habit that makes the most efficient use of heat and light while keeping the foliage out of the wind. When the snow melts, new leaves are produced. Leaves and flowers grow rapidly in order to accomplish their life cycle during the short growing season. The root of this plant is a thick, vertical, carrot-like structure that stores carbohydrate. The root probably is a succulent water source that enables the plant to continue growing after soil water has percolated away. As is true of many high Sierra plants, the nearest

relatives of this alpine species are desert plants, and have no odor.

Pussy Paws, *Calyptridium umbellatum* (figure 5.17), is a common, widespread plant that grows on bare, sandy flats from the Yellow Pine Forest to the alpine zone. This is another desert-related species. Similar to the Evening Primrose, its leaves lie prostrate on the ground in a basal rosette. Unlike the Evening Primrose, the flowers are small and borne in fuzzy clusters on the ends of long, wiry stems. During most of the day, these flower stalks lie on the ground, gathering heat and light. In the early morning, however, when the sun is low in the sky and the ground is cold, these stalks curl upward, lifting the flowers 2–3 in (5–8 cm) above the ground (figure 5.17). This enables solar radiation to warm the flower parts, accelerating their development. This form of sun tracking is the way Pussy Paws tries to cope with cold alpine nights.

Alpine Heath

In Great Britain, there are many acres of low, open brushland known as heaths. Many of the shrubs there are also known as heaths or heathers. In North American mountains, there is an alpine shrubland also known as a Heath. Many

FIGURE 5.17 Pussy Paws, *Calyptridium umbellatum*, sun tracking.

of the shrubs in this community are either in the heath family (Ericaceae) or in the willow family (Salicaceae).

In the Sierra Nevada, members of the heath family are common in areas where snow provides most of the precipitation and soils tend to be acidic. These plants range upward in the mountains from Upper or Cold Chaparral, through conifer forests, to the Alpine Heath. Manzanitas of Upper Chaparral are in this family, as are azaleas and rhododendrons of the forest communities. In the alpine zone is Red Mountain Heather, *Phyllodoce breweri*; White Heather, *Cassiope mertensiana* (figure 5.18); Alpine Laurel, *Kalmia polifolia*; Western Blueberry, *Vaccinium occidentale*; and Dwarf Bilberry, *V. cespitosum*. Flowers of the heath family are colorful and urn-shaped. On manzanitas, bilberries, and White Heather, the open end of the urn is tipped downward; on others, the flower opens upward, in a cup shape.

Water seems to be an important limiting factor for distribution of Alpine Heath. Quite often, the shrubs are found ringing a meadow, a lake, or bordering a stream. Of equal importance, however, is well-aerated soil. This combination of moisture and soil conditions is not widespread in the Sierra. Consequently, Heath zones are not as conspicuous or as widespread

FIGURE 5.18 White Heather, *Cassiope mertensiana*.

as they are in the Rockies or the mountains of Oregon and Washington. Heath zones cover certain areas in the northern Sierra, but in California one must travel to the Klamath

FIGURE 5.19 Zonation of Horseshoe Meadow, a southern Sierran meadow, Inyo National Forest.

Mountains to see the Heath community at its best. Even there, its distribution is more localized than in areas that receive abundant summer precipitation.

In the southern Sierra, Heath shrubland commonly forms a fringe around meadows, but the shrubs do not occur in the wettest soil in the center of the meadows (figure 5.19). Apparently, long roots of the shrubs are able to reach water deep in the ground, and more drought-tolerant vegetation may occur between the shrubs and the wet part of the meadows. It is common for Heath shrubs to form a transition between the bare zone of coarse soil and the actual meadow.

Alpine Meadows

There are two kinds of meadows in the Sierra: Dry Meadows and Wet Meadows. Most meadows of the Sierra Nevada would be described best as Dry Meadows, although available moisture occurs in a downslope gradient so that in the center of many meadows, at the lowest point, where soil is finest, there is true Wet Meadow.

Dry Meadow is characterized by mat-forming grasses (*Calamagrostis* spp. and *Poa* spp.), Bunchgrasses (*Festuca* spp.), and various sedges (*Carex* spp.) that resemble grasses. Close inspection of the stems of sedges will reveal that they are triangular in shape. Various species of wildflowers are also common in these meadows. Conspicuous among them are various phloxes (*Phlox* spp.), paintbrushes (*Castilleja* spp.; plate 20A), lupines (*Lupinus* spp.; plate 8A), Elephant's heads (*Pedicularis attollens*), and a number of members of the sunflower family (Asteraceae).

This is also an area where annual wildflowers may occur. Many of these flowers are quite small, usually less than half inch (1.0 cm) in diameter, and sometimes occur in great abundance. These may be so small that a person who wants to inspect one will probably get down on his stomach for a close look; hence, some people call them "belly flowers." Among these species is a diminutive member of the Phlox Family (Polemoniaceae) that goes by several names, Whisker Brush, Mustang Clover, or Yellow-throated Gilia, *Leptosiphon ciliatus*. The latter common name refers to the yellow throat in the center of its pink tubular flowers. Another in the same family that is about the same size and color is Skunky Monkeyflower, *Mimulus nanus* var. *mephiticus*. The yellow pattern in the center of this flower is so arranged that each flower including its petals resembles the face of a monkey. There are some 39 species of monkeyflower in the Sierra Nevada, making this one of the most common genera in the mountain range.

Another interesting wildflower of these meadows is Scarlet Gilia, *Ipomopsis aggregata* (plate 20B). In early summer, those plants have many brilliant red, tubular flowers. When they are abundant, it is truly a spectacular sight. They are fairly tall for wildflowers, up to 4 ft (1.3 m) high. In early summer, they are pollinated primarily by hummingbirds, who favor red, tubular flowers. Later in summer, however, hummingbirds tend to migrate to lower elevations. At this time, flower colors change to pink or white, attracting sphinx moths (Sphingidae), which pollinate at night. The shift in flower production from red to light colors is clearly an adaptation that makes these flowers more visible at night. No one is certain, however, how the plants "know" when to change color. It also has been demonstrated that during years when pollinators are scarce, the plants produce auxiliary vegetative growth that can produce flowers in a subsequent year.

Scarlet Gilia is often eaten by Mule Deer. An interesting study on Scarlet Gilia in the Rocky Mountains showed that once plants had been nipped off, they responded by growing larger roots, more stems, more flowers, and more fruits. This study showed that partial browsing actually increased the vigor of the plants. Therefore, both Mule Deer and Scarlet Gilia benefit by partial browsing.

In wetter parts of these meadows and along watercourses, there are a number of conspicuous wildflowers. Among these are the shooting stars, which are easily recognized by their pink, bent-back petals and long strap-like leaves. There are two species. Scented Shooting Star, *Primula fragrans* (= *Dodecatheon redolens*) has five petals and occurs in slightly drier localities. Sierra Shooting Star, *P. jeffreyi* (= *D. jeffreyi*) is a taller plant and its flowers usually have only four petals. When the flowers first open, the petals extend outward without being bent back. Apparently the bent-back position occurs after it is pollinated. Polli-nation occurs by bees, particularly Bumble Bees (*Bombus* sp.) in a unique process called *buzz pollination*. The female Bumble Bee curls her body around the pollen sacs that are clustered against the stigma and vibrates her body causing the pollen to be released. When she visits the next flower and repeats the process, some of the pollen she is already carrying is transferred to the stigma ensuring cross-pollination. Heavy-bodied Bumble Bees are especially well suited for flying close to the ground in windy habitats such as open meadows.

Honey Bees, *Apis mellifera*, are not common in alpine areas, but they can be observed there from time to time. Native to the Old World, they were imported to pollinate agricultural crops and now have spread everywhere. Wild hives are found in the foothills and the bees spread out from there in warm weather, but they are unable to overwinter at higher elevations.

Pollinators may not always be common, which has led to an interesting reproductive adaptation that occurs in some members of the Mustard Family (Brassicaceae). Some of the Drabas (*Draba* spp.) and the Rockcresses (*Boechera* spp.) are able to reproduce asexually by producing seed without pollination.

Soil in Dry Meadows tends to be thin and sandy. Duration of snowpack varies with exposure and season. At its upper edge, a Dry Meadow grades into a gravel flat or Alpine Heath zone. At its lower end, it grades into a Wet Meadow. As snow melts, soil becomes exposed to light. In some areas, snow becomes deflated, or is blown away by wind, exposing soil to sunlight during winter months. If plants began to sprout at that time, they would surely be killed by cold winter temperatures. Breaking of dormancy requires exposure of soil by melting of snow or wind deflation followed by warm soil or proper summer photoperiod. Combinations of proper moisture and appropriate temperature cause flowers to appear in a mosaic in the alpine zone. A common pattern is for flowers to appear first on the upper edge of a meadow, and as the season progresses, a belt of blooming plants moves toward the center. The

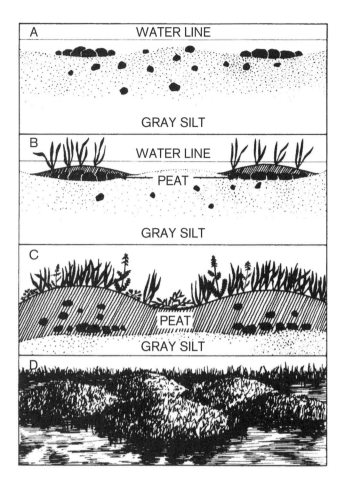

FIGURE 5.20 Formation of peat hummocks (from Whitney 1979; reprinted by permission).

A. During the winter, freezing churns the soil, heaving buried rocks to the surface, where they roll to the perimeters of the raised mounds to form rock polygons.

B. Continued churning prevents plants from invading the centers of the polygons, but moisture-loving sedges may become established among the submerged rocks of the perimeter.

C. The sedges contribute organic matter, initiating the formation of peat hummocks, which eventually protrude above the water surface. As the mounds increase in height, they become somewhat less soggy so that Wet Meadow plants such as elephant's heads and Alpine Willow can become established.

D. Mature peat hummocks appear as groups of low, grassy mounds in moist meadows.

outer edge of the meadow, where soil is coarsest, drains and warms first.

Wet Meadows are characterized by sodforming sedges and grasses. These plants tend to reproduce by means of underground stems, or rhizomes. Soil tends to be fine, acidic, and rich in organic matter. The thick mat of rhizomes and black soil is known as sod. At its extreme, it decays to carbon-rich organic material known as peat. Where water never drains, this community is commonly known as a Peat Bog.

Freezing of waterlogged soil can damage plants. The most conspicuous effect of freezing is that expansion of the ice causes the earth to heave up in mounds, or hummocks (figure 5.20). These hummocks cause the center or wet part of a meadow to have a very lumpy surface that is difficult to walk on. Sometimes these hummocks are bare on top. What happens is that the earth, in rising, uproots the vegetation, which usually consists of sedges. When the uprooted plants die, wind scours the top of the hummock, blowing away dead parts of the plant and much of the soil. The longterm effect is that the sedges appear to grow in circles. The appearance of these circular growths is particularly striking in those areas that are waterlogged during winter but drain rapidly in summer, so that very little else seems to grow there.

Where waterlogged soil is present on slopes, sliding may occur. This is known as solifluction (figure 5.21). As a result, the soil may pull the sod apart in a corduroy pattern, leaving alternating strips of vegetation and bare ground.

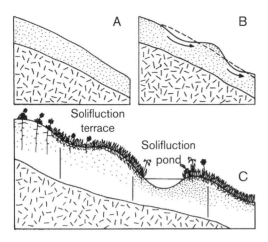

FIGURE 5.21 The effect of waterlogged soil on steep slopes (from Whitney 1979; reprinted by permission).

A. The soil begins to creep downslope in response to gravity because the particles are lubricated by abundant moisture and no longer adhere to one another.

B. Small terraces form where the soil piles up.

C. Eventually, moisture-loving sedges and grasses may invade the slope, inhibiting further creep. Tiny ponds impounded behind solifluction terraces are common features of moist alpine meadows.

Some authors consider Wet Meadows to be the only true alpine meadows. To them, alpine means a Wet Meadow, and nothing else. This interpretation is based on the similarity of Alpine Wet Meadows to Arctic Tundra, zones with many plant species in common. Plants of the Arctic Tundra are said to be circumpolar in distribution. In the Rocky Mountains, about 40% of the alpine plants are circumpolar, that is, shared with the Arctic. In the Sierra, however, only about 20% of the plants are circumpolar in distribution. This is a testimony to the limited distribution there of Wet Meadow and the scarcity of summer rain.

Due to the lower elevation north of Lake Tahoe, there is limited distribution of Alpine Wet Meadow vegetation in the Sierra Nevada. Likewise, on Mount Shasta and Mount Lassen, coarse volcanic soil limits alpine vegetation to the Pioneer and Fellfield categories. In the Klamath Mountains, however, there are Wet Meadows more closely resembling those of the north. Although Rocky Mountain elements are present, the bulk of the plants in the meadows of the Klamath region are of northwestern affinities. In southern California, the alpine zone consists primarily of gravel and Fellfields, and the plants are associated primarily with the Great Basin. These communities are probably associated primarily with soil and wind as overriding ecological factors.

So far, no attempt has been made to distinguish meadows at different elevations. At lower elevations, meadows occur at low spots or basins in forest communities. As meadows, these communities possess more similarities than differences. They are open, windblown habitats with low-lying vegetation, consisting mostly of turf-forming sedges and grasses. They are subject to cold-air drainage; frost can occur any time of year. In general, these are habitats where moisture is abundant in the top few inches of soil during at least part of each year. It is this persistence of moisture that seems to exclude the invasion of shrubs or trees even within forest belts.

In wetter sites, a common meadow plant is the Corn Lily, *Veratrum californicum*. It is sometimes called Skunk Cabbage, which it resembles only by having large leaf blades. On the other hand, it is suspected to share with Skunk Cabbage the ability to produce heat and cause early melting of snow. A common sight in early spring is young shoots of Corn Lily growing up through holes in the snow. Corn Lily is further notable because it is avoided by deer as a browse. It is one of the most poisonous plants in the Sierra, and Native Americans reportedly ate the root in order to commit suicide.

Meadows occur today where formerly there were lakes. These were low spots in glacially scoured terrain. As sediment accumulated in the bottom of the lakes, soil accumulated around the edges, and meadows gradually replaced the lakes (figure 5.22). Many meadows still have a lake in the center. At the edges of the meadows, where soil is deeper and better drained, the shrubs of the Heath community begin to encroach. In areas below timberline, trees move into the areas behind the shrubs, most commonly Lodgepole Pines. Transpira-

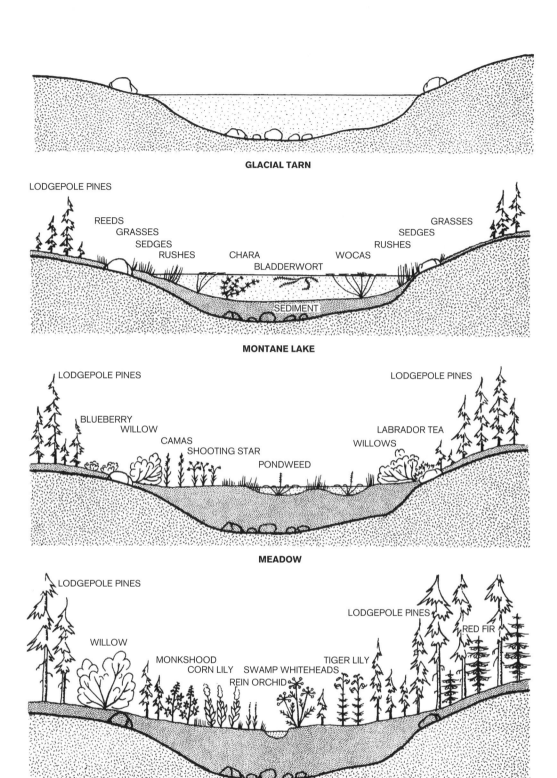

GLACIAL TARN

LODGEPOLE PINES
REEDS
GRASSES
SEDGES
RUSHES
CHARA
BLADDERWORT
WOCAS
GRASSES
SEDGES
RUSHES
SEDIMENT

MONTANE LAKE

LODGEPOLE PINES
LODGEPOLE PINES
BLUEBERRY
WILLOW
CAMAS
SHOOTING STAR
PONDWEED
LABRADOR TEA
WILLOWS

MEADOW

LODGEPOLE PINES
LODGEPOLE PINES
RED FIR
WILLOW
MONKSHOOD
CORN LILY
SWAMP WHITEHEADS
REIN ORCHID
TIGER LILY

SWALE

FIGURE 5.22 Stages in succession from lake to Montane Forest (from Bakker 1984).

FIGURE 5.23 Big Whitney Meadow, Inyo National Forest.

where glaciation has scoured vast flat areas that are now accumulating soil. Very few large meadows occur on the east side of the Sierra because not enough time has passed in this area of reduced precipitation. It takes thousands of years for the soil accumulation required for meadow succession.

In the southern Sierra, on the Kern Plateau, there is another large grouping of meadows. In this area, extensive glaciation was avoided, and longer periods of time have elapsed since major scouring of the landscape occurred. Some of these areas, such as Kennedy Meadows, are not true high-country meadows but are composed of Sagebrush Scrub and annual grasses, which are more typical of regions at lower elevation. In the Golden Trout Wilderness, there are some true mountain meadows (figure 5.23), but most of the research on high-country meadows has been to the north, in locations such as Carson Pass, Yosemite National Park, Sequoia National Park, Humphreys Basin, and Convict Creek Basin. Very little is known ecologically about the meadows of the Kern Plateau, perhaps because they have been disturbed by cattle and sheep grazing since the 1800s. What research has been done on these meadows relates to the impact of grazing. It has been determined that the trampling alters the wet zone near the streams and encourages encroachment by sagebrush. Apparently, in Templeton and Ramshaw Meadows at least half of the sagebrush zone used to be occupied by perennial sod-forming vegetation.

The Kern Plateau is home to numerous plants and animals with limited, endemic distribution. Foxtail Pine was discussed earlier. The Kern Plateau is also the home of the state fish, the Golden Trout, *Oncorhynchus mykiss aguabonita*, as well as several species of sedentary slender salamanders, *Batrachoseps* spp. It is also the only Sierra Nevada location for the Western Skink *Plestiodon (= Eumeces) skiltonianus*, and two species of whipsnakes, *Masticophis* spp., and a racer, *Coluber (= Masticophis)*.

Why has the Kern Plateau been the site of evolution for several endemic animals and plants? Presumably it was because ecological

tion of the trees helps to dry the meadow, and this drying makes the habitat appropriate for the Alpine Heath community; therefore, the shrubs often grow in rings around the bases of Lodgepoles.

At lower elevations, meadows are destined to be eventually replaced by forests. At upper elevations, meadows may avoid forest encroachment for thousands of years. As they accumulate soil, they become more like tundra of the north. Deeper soils remain wet for longer periods of time, and Wet Meadows become more common.

In the Sierra today, the largest meadows are fringed by Subalpine Forest. Tuolumne Meadows in Yosemite National Park is the largest complex of meadows in the Sierra, and it is ringed by Lodgepole forest. Nearby are other large meadow areas, such as Dana Meadows near Tioga Pass, which has Lodgepole and Whitebark pines on its border. Most of the large Sierran meadows are in the Yosemite area, islands of favorable habitat endured throughout the Ice Age, during which time relicts were preserved and specializations evolved. Furthermore, the near absence of organic material in the soil indicates that the Kern Plateau has been an arid region for a long time.

Snowfields

In the Sierra Nevada, snowfields persist on high passes, mountaintops, and north-facing slopes throughout most of the year. Likewise, snowfields persist throughout the year on Mount Shasta, Mount Lassen, and protected locations in the Klamath Mountains. In southern California, snowfields rarely persist on a yearly basis.

Snowfields have a surprising number of organisms associated with them. Food for the entire system is produced by microorganisms variously called snow algae, snow protozoa, and snow protists. For the most part these are single-celled, photosynthetic organisms (*Chlamydomonas* spp. and *Chloromonas* spp.) that bear a pair of flagella. They are autotrophic and motile (i.e., bearing a combination of plant-like and animallike characteristics), which has been the problem with giving them a common name. Botanists still think of them as algae, but zoologists think of them as protozoa. The debate over their classification is pointless.

These microorganisms are circumpolar in distribution. They are distributed mostly by the wind. They overwinter in spore form. In the spring, when the surface layer of snow melts, they change to their motile, flagellated form and are able to swim in the few millimeters of water on the surface of a snowfield. They absorb windblown nutrients associated with dust particles and organic debris. As the summer progresses, they become more abundant and are further concentrated as the snowfields decrease in size due to melting. Each of these tiny organisms contains a small amount of red pigment in addition to its chlorophyll. By the end of summer, they are so concentrated that the snow takes on a reddish hue. Stepping on the snow concentrates the organisms even more, and each footprint may be decidedly red in appearance.

Feeding upon the photosynthetic microorganisms are various small herbivores, such as nematode worms, protozoa, and rotifers, common semiaquatic invertebrates in many habitats. The carnivores in this ecosystem are mostly insects, including ants, springtails, and ladybird beetles. Spiders feed on the insects and birds feed on all organisms that are large enough to see.

Most remarkable of the insects are cricketlike forms known as Ice Crawlers, *Grylloblatta* spp. These are wingless, nocturnal, high-elevation forms that are most active during the cold seasons. Their optimal temperature range is around freezing, and they die quickly above 59°F (15°C). They live deep in crevasses under snow or glaciers and feed on soft-bodied insects or plant materials. They seldom venture out onto the snow except at night. However, when they do, these remarkable arthropods are snapped up by birds such as Gray-crowned Rosy-finches.

The mechanisms by which insects tolerate this cold climate are fascinating. Apparently, many arctic and alpine insects can tolerate solid freezing. One insect survival mechanism is to avoid freezing by producing natural antifreeze such as glycerol or even ethylene glycol, the same stuff used in automobile antifreeze. Other insects actually are freeze tolerant. They reduce the proportion of body water locked up in ice by using a chemical such as glycerol or even glucose as a "cryoprotectant." Similar to plants, their key to survival is to prevent dehydration and ice formation inside cells while ice slowly fills the rest of the body.

Another problem with high elevation is high-intensity light. Insects and spiders living at these elevations are known to have high densities of wax on their outer surface. This wax probably helps to intercept damaging UV light. It also may serve to repel water in this environment, which is frequently wet.

Snowfields may remain in place throughout the summer on north-facing slopes and in

FIGURE 5.24 Suncups on Mono Pass, Inyo National Forest.

FIGURE 5.25 Gray-crowned Rosy-finch, *Leucosticte tephrocotis.*

depressions that are free of wind. Their surfaces often become pitted with "suncups" (figure 5.24), pits formed by small particles of windblown debris. A combination of gravity and heat absorption by these dark particles causes them to sink gradually into the snow until the surface looks like an enlarged egg crate. Each pit serves as a trap to capture more windblown debris and nutrients.

The toe of a snowfield melts, or retreats, at a surprising rate. To an unobservant visitor, the snowfields seem not to change much, but on a typical summer day the edge of the snow will retreat 3–6 ft (1–2 m). This melting creates new snow-free habitat every day, where insects hatch and seeds germinate. This habitat at the edge of the snow is a rich feeding ground for alpine birds such as the Gray-crowned Rosy-finch, *Leucosticte tephrocotis* (figure 5.25).

Climate change, with its associated increase in temperature and drought, seems already to have an influence on the distribution of alpine and subalpine plants and animals. At the lower reaches of each zone, plants are dying out and some animals are disappearing. Plants and animals adapted to the harsh extremes of high elevation may not be able to compete with more adaptable species from below. We may see a replacement of species occurring, or we may see plants and animals colonizing the upper zones as temperatures become more moderate. We also may see a northward emigration of certain species. No one really knows what may happen if there is a change in the precipitation regime in which drier, warmer winters are coupled with wetter spring months.

Animals of the Alpine Zone

Most of the conspicuous animals of the alpine zone are birds and mammals that occur in other montane communities as well. On a yearly basis, the alpine zone becomes too cold for ectothermic species, and there is a coincident decrease in the numbers of species at higher elevations. There are not many reptiles, but there are some remarkable salamanders, as will be discussed later.

PLATES

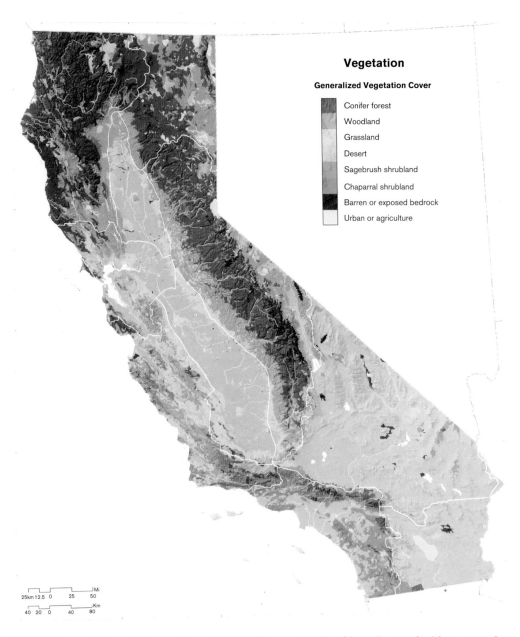

Vegetation

Generalized Vegetation Cover

- Conifer forest
- Woodland
- Grassland
- Desert
- Sagebrush shrubland
- Chaparral shrubland
- Barren or exposed bedrock
- Urban or agriculture

| | | | Mi |
| 25km 12.5 | 0 | 25 | 50 |

| | | | Km |
| 40 20 | 0 | 40 | 80 |

California's generalized vegetation cover (from Keeler-Wolf, T. 2003. *An Atlas of the Biodiversity of California*. State of California, The Resources Agency, Department of Fish and Game, p.19).

PLATE 1

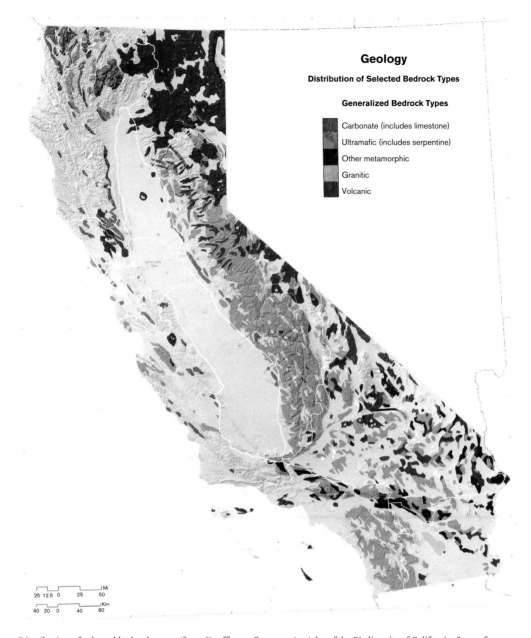

Geology

Distribution of Selected Bedrock Types

Generalized Bedrock Types

- Carbonate (includes limestone)
- Ultramafic (includes serpentine)
- Other metamorphic
- Granitic
- Volcanic

Distribution of selected bedrock types (from Kauffman, E. 2003. *An Atlas of the Biodiversity of California*. State of California, the Resources Agency, Department of Fish and Game, p.17).

PLATE 2

3A   Yosemite National Park, Mixed Coniferous Forest in winter.

3B   Emerald Bay State Park, Lake Tahoe.

PLATE 3

4A Joshua Tree Woodland and granitic boulders, Joshua Tree National Park.

4B Cactus Scrub featuring Ocotillo, Barrel Cacti, and Desert Brittlebush, Anza-Borrego Desert State Park.

PLATE 4

General Sherman, the largest tree in the world, in winter, Sequoia National Park.

PLATE 5

6A Mixed Coniferous Forest featuring Jeffrey Pine and Wilson Butte, a volcanic dome, north of Mammoth Lakes.

6B Montane Forest featuring Lodgepole Pines and Mountain Hemlock to left of center against horizon, McCabe Lake, Yosemite National Park.

PLATE 6

7A Subalpine Forest featuring Foxtail Pines, Sequoia National Park.

7B Old Bristlecone Pine in the Patriarch Grove, White Mountains.

PLATE 7

8A Alpine Meadow in Humphreys Basin, Inyo National Forest, featuring lupines in bloom and Mount Humphreys in background.

8B Autumn color, featuring Quaking Aspens in a Riparian Woodland, North Lake on Bishop Creek.

PLATE 8

9A Dry Mixed Coniferous Forest dominated by Jeffrey Pines and Manzanitas. Mount Lassen as viewed from the northeast.

9B Old-growth Douglas fir, Siskiyou Mountains.

9C Moist subalpine community of Salmon Mountains at Cuddihy Lakes.

PLATE 9

10A King Range at the Lost Coast north of Shelter Cove, Humboldt County.

10B Mixed Evergreen Forest, northern Coast Ranges.

PLATE 10

11A Eel River and Coast Redwoods in Humboldt Redwoods State Park.

PLATE 11

11B Oak Woodland featuring Blue Oaks and Gray Pines, Del Valle Regional Park.

PLATE 11C California Poppies, *Eschscholzia californica*, the California State Flower in the Antelope Valley California Poppy Reserve, Los Angeles County.

12A Coastline at Big Sur.

12B Julia Pfeiffer Burns State Park. McWay Falls drops directly into the ocean.

PLATE 12

12C Bristlecone Fir, *Abies bracteata*, endemic to the Santa Lucia Mountains in Monterey County.

13A Coastal Sage Scrub in autumn, showing diversity of species represented by various colors.

13B Oak Woodland on Santa Catalina Island featuring tree-sized Island Scrub Oaks, *Quercus pacifica*.

PLATE 13

14A Mixed Coniferous Forest on the north side of the San Bernardino
Mountains featuring Sugar Pines and White Firs.

14B Wildflowers after a fire.

PLATE 14

15A Great Basin Sagebrush in Owens Valley. Sierra Nevada in background.

15B Creosote Bush Scrub, the dominant plant community of the warm deserts in California.

PLATE 15

16A Cactus Scrub in Anza-Borrego Desert State Park.

16B Thousand Palms Oasis, featuring California Fan Palms, *Washingtonia filifera*.

PLATE 16 DESERTS

17A Vernal pool surrounded by a ring of flowers, endemic *Downingia* sp.

17B Paternoster of cirque lakes. View southward toward Sawtooth Peak from Black Rock Pass, Sequoia National Park.

17C Thousand Island Lake, a large tarn in a glacial scour. Mount Banner and Mount Ritter with small glaciers in background.

PLATE 17 INLAND WATERS

18A California Ground Squirrel, *Otospermophilus beecheyi*.

18B Chickaree or Douglas's Squirrel, *Tamaisciurus douglasii*.

18C Chickaree or Douglas's Squirrel, *Tamiasciurus douglasii*.

18D Golden-mantled Ground Squirrel, *Callospermophilus lateralis*.

18E White-tailed Antelope Squirrel, *Ammospermophilus leucurus*.

18F Yellow-pine Chipmunk, *Tamias amoenus*.

18G Belding's Ground Squirrel, *Urocitellus beldingi*.

18H Desert Woodrat, *Neotoma lepida*. (photograph by Greg Stewart).

PLATE 18 RODENTS

19A California Quail, *Callipepla californica.*

19B Hooded Oriole, *Icterus cucullatus.*

19C Wood Duck, *Aix sponsa.*

19D Western Bluebird, *Sialia mexicana.*

19E Red-winged Blackbird, *Agelaius phoeniceus.*

19F Western Scrub-jay, *Aphelocoma californica.*

19G Western Tanager, *Piranga ludoviciana.*

19H Yellow-billed Magpie, *Pica nuttalli.*

PLATE 19 BIRDS

20A Paintbrush, *Castilleja* sp.

20B Scarlet Gilia, *Ipomopsis aggregata.*

20C Harlequin Lupine, *Lupinus stiversii.*

20D Leopard Lily, *Lilium pardalinum.*

20E Red Columbine, *Aquilegia formosa.*

20F Alpine Columbine, *Aquilegia pubescens.*

20G Broad-leaved Helleborine,
Epipactis helleborine.

20H Purple Owl's Clover, *Castilleja exserta.*

PLATE 20 MOUNTAIN WILDFLOWERS

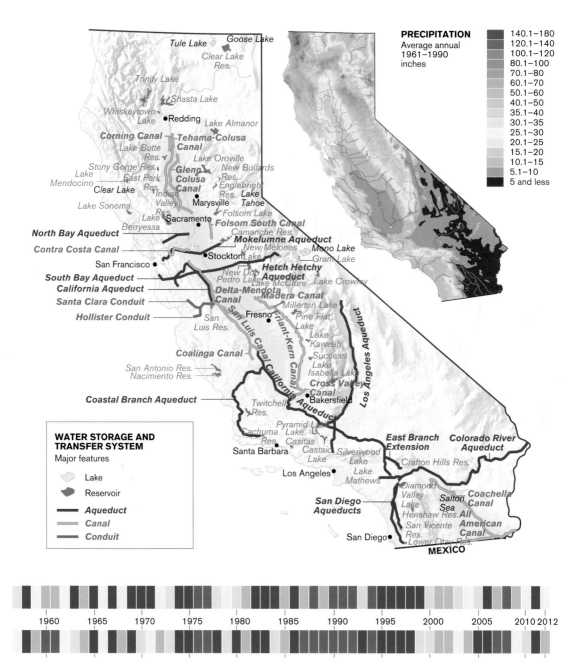

FIGURE 11.6 California has the most intricate aqueduct system in the world. The aqueducts shown here include state projects such as the California Aqueduct and Federal Projects such as the Coachella Canal and the canals of the Central Valley. The Colorado River Aqueduct is part of the Metropolitan Water District, the Los Angeles Aqueduct belongs to the Los Angeles Department of Water and Power, and the Hetch Hetchy Aqueduct was built for the City of San Francisco.

BUTTERFLIES One of the interesting phenomena documented for mountaintops in general is that they seem to attract certain kinds of insects. Monarch Butterflies, *Danaus plexippus* (figure 7.25A), which will be discussed in the chapter on the Coast Ranges, are migratory. Thus, they commonly occur in the high country in the course of their migration. What is difficult to understand is why so many other species of insects become concentrated on mountaintops, a phenomenon known as hilltopping. Swarms on the summits of ridges, hills, and mountains have been documented many times. One study in San Diego County documented 45 species of hilltopping butterflies alone. Several alternative hypotheses have been proposed to explain it, but it seems to be related to mating behavior. Most hilltopping butterflies are males, and females seem not to remain there long. Furthermore, the males on the hilltops engage in territorial patrolling behavior.

The California Tortoiseshell, *Nymphalis californica*, is an example of a butterfly that undergoes mass emigration in association with population outbreaks. These mass movements may be observed in early summer in the Cascades, the Sierra Nevada, the Santa Cruz Mountains of the southern Coast Ranges, and the San Bernardino Mountains of the Transverse Ranges. The species is particularly renowned for its massive outbreaks on Mount Shasta, where millions may appear in favorable years. Adults fly in such numbers that one collector gathered 50 of them with a single sweep of his net. Great numbers of dead Tortoiseshells may be seen on the snowfields in some years. The adult looks like a large anglewing, but it is less tattered in appearance. It has a 2 in (5 cm) wingspan, and the centers of the wings are orange. Larvae of these butterflies are velvet black with a broken yellow line along the back. Each segment bears five branching spines supported on blue tubercles. The caterpillars feed on various species of *Ceanothus*, which they readily defoliate during their unpredictable population explosions.

There are three common but inconspicuous butterflies that characterize the alpine region of the Sierra Nevada. Ridings' Satyr, *Neominois ridingsii*, occurs from Inyo County north to Carson Pass, and across the Owens Valley to the White Mountains. This "waif of the mountains" is gray with an irregular light band across the wings and two black spots. It is mothlike in appearance, relying on its coloration to conceal it when at rest. It flushes from underfoot when approached. It is most common in even-numbered years.

The California Ivalida, *Oeneis ivalida*, occurs from Inyo County north to Donner Pass, where it flits about cliffs and summits. Its dull yellow-brown color enables it to blend with the granite rocks upon which it rests with open wings, perhaps to increase the absorption of solar radiation.

The Small Apollo or Phoebus Parnassian, *Parnassius phoebus*, is in the same family as the swallowtails (Papilionidae), although it lacks tails. It occurs in the alpine zone of the Siskiyou Mountains and the Sierra Nevada from the Mount Whitney area north to Plumas County. This butterfly is cream to snow white with black and red spots on its wings. It occurs on rocky summits and has been observed flying in snowstorms. Larvae feed on various stonecrops (*Sedum* spp.), small succulent plants that occur on exposed rock outcrops. A related butterfly the Clodius Parnassian, *P. clodius*, also feeds on stonecrops from Carson Pass southward.

AMPHIBIANS AND REPTILES One of the strangest examples of an alpine animal is the Mount Lyell Salamander, *Hydromantes platycephalus* (figure 5.26). Its scientific name, *platycephalus*, means "flat-head," in reference to its body form, which is specialized for living in cracks and under rock flakes. This is another one of the lungless salamanders (Plethodontidae). The origin of lunglessness is an interesting theory. It is believed to be an adaption for living underwater in mountain brooks. Whereas air-filled lungs would cause an animal to float, lungless species can walk on the bottom and absorb oxygen directly through their skin. This mode of breathing functions particularly well in cool,

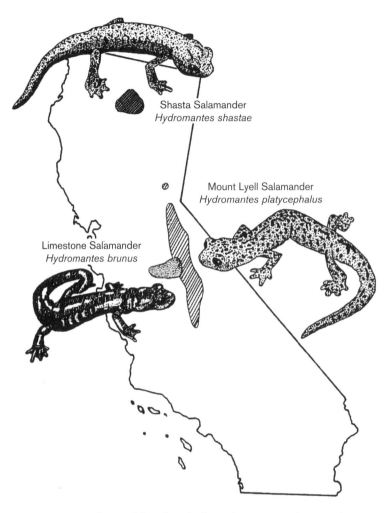

FIGURE 5.26 Distribution of the web-toed salamanders (genus *Hydromantes*) in California (illustration by Philip R. Brown, from Gorman, J. 1988. The Effects of the Evolution and Ecology of *Hydromantes* on their Conservation. Pp. 39–42. In De Lisle, H. F., Brown, P. R., Kaufman, B., and McGutty, G. M., eds. *Proceedings of the Conference on California Herpetology.* Southwestern Herpetologists Society Special Publication no. 4. Reprinted by permission).

well-aerated stream water. This is the type of habitat where lungless salamanders are found today in Appalachia, which is believed to be the center of origin for the family. The fact that western members of the family are nearly all terrestrial is incidental. As long as they remain cool and moist, they can absorb all the oxygen they need through their skin, as well as mouth and throat membranes.

In general, ectothermic animals are not typical of cold climates, but this salamander seems to be associated with open rock habitats that are wet by water from snowmelt or spray from waterfalls. Its distribution in the Sierra was originally known primarily in high-elevation bare habitats from Silliman Pass, in Sequoia National Park, to Sonora Pass. It was discovered in recent years at Smith Lake, west of Lake Tahoe. It was originally discovered on the east side of Mount Lyell at 10,800 ft (3214 m), and it also occurs in cracks near the large waterfalls in Yosemite Valley. Isolated populations also have been discovered at relatively low elevation, in deep canyons just above the floor of Owens Valley. The greatest number of observations for this salamander has been recorded on the top of Half Dome at 8852 ft (2634 m), in Yosemite National Park. This is a granite surface with no humus; the only soil is decomposed granite. Here the salamander spends daylight hours under rocks or in cracks, particularly in the areas of snowmelt, and forages at night for spiders and insects.

In late summer, when habitats become dry, these salamanders slip into fissures and cracks, where they become dormant. This state of estivation grades into hibernation with the onset of winter. They remain dormant, living off of stored fat until the snow melts the following spring. Lifespan for these interesting salamanders averages 8–9 years.

The distribution of this salamander and its relatives is an enigma. Not only is it strange for a delicate animal such as this to be localized in harsh, isolated places, but its nearest relatives are also part of a strange pattern. These are the web-toed salamanders (Plethodontidae), a group with five recognized species, although there is some controversy over whether each member deserves species status. Most of the species are restricted to limestone. Two are found in limestone caves in western Europe, and two others are found on limestone in California. One of these, the Limestone Salamander, *H. brunus*, occurs only along the Merced River in the foothills of the Sierra at about 2000 ft (600 m). The other, the Shasta Salamander, *H. shastae*, occurs in at least nine localities in Shasta County, mostly on limestone or metavolcanic substrates in the vicinity of coniferous or Canyon Live Oak habitats. The highest elevation at which it was recorded was on limestone on Tombstone Mountain at 5450 ft (1661 m) in the upper McCloud River drainage.

It is difficult to reconstruct a history to explain this pattern of distribution—western Europe and western North America. These are sedentary animals; over their entire lives, they probably move no more than a few meters. It defies logic to think of them walking long distances in order to become established at such isolated locations. To consider them relicts of a former widespread distribution is more logical, but this means they would have had to be distributed nearly worldwide at one time, perhaps as long ago as the Cretaceous Period (80–90 million years ago). It is thought that at this time, salamander evolution was explosive, occurring very rapidly. The center of evolution for the family (Plethodontidae) was in Appalachia, in eastern North America. At the time of that evolution, North America was still attached to Europe in a giant continent known as Laurasia; the Atlantic Ocean was not yet formed. If the group is old enough, it could have occupied a wide range of distribution from eastern North America to western Europe. As Europe and North America separated and the Atlantic Ocean widened, the salamanders simply remained in place and slid with the continents as if they were passengers on a giant, slowly moving ferry.

Getting the salamanders from eastern North America to western North America is another problem. Web-toed salamanders no longer occur in eastern North America, but more evolved relatives do. One hypothesis has the web-toed salamanders dispersing westward from Appalachia through belts of coniferous forest to the newly uplifted Rocky Mountains. This migration would have had to have taken place after the retreat of an interior Cretaceous seaway that bisected North America into eastern and western halves, but it would have had to occur before drying of the interior of the continent took place. Once the salamanders reached the west, they were free to occupy new habitat as land above sea level grew westward. Presumably, salamanders were well established in the west at the time of initial uplift of the proto-Sierra, approximately 80 million years ago. As the old ocean-floor sediments were eroded away, the two foothill species survived in stable microhabitats, possibly limestone caves. The Mount Lyell Salamander, after all prebatholithic rock was stripped away, became associated with fissures and cracks in granite. Elevation of the Sierra to its present height took place during the last 1–3 million years. Several periods of glaciation scoured the high country, destroying salamander habitat. Mount Lyell, Half Dome, and other high peaks remained above the glaciers, and it is believed that the salamanders survived in places above the glaciers and exist there today as relicts. Dispersal to lower-elevation sites could have occurred by various gravity-aided means, such as sliding on ice or being carried by water or rocks.

There are sophisticated means by which scientists can determine how many years ago closely related species diverged. Application of these techniques to the five species of web-toed salamanders puts their age of divergence no earlier than Oligocene, 37 million years ago. If that is true, it would not have been possible for them to have occurred continuously from Appalachia to Europe. By that time, the continents had spread well apart, and the Atlantic was a sizable barrier. In this case, it would have been necessary for salamanders to walk from western North America across the Bering Strait to eastern Asia. From there they would have made their way westward to Europe without leaving any descendants behind in Asia.

Because all this walking seems unlikely for a sedentary salamander, scientists have been searching for an alternative explanation for such a peculiar pattern of distribution. One hypothesis is associated with plate tectonics. The salamanders rode, as if they were cargo, on moving pieces of land known as microplates or terranes. To explain the distribution of web-toed salamanders using this theory, we must first assume that their lineage is older than 37 million years. The disjunction between Appalachia and Europe could be explained by a population that extended across the area before the separation of Europe and North America. Distribution to western North America took place by a surprisingly devious route. Evidence supports the contention that the southern part of Mexico, a region known as the Maya Terrane, was formerly adjacent to southern Appalachia.

This piece of land could have drifted away from eastern North America, carrying with it a cargo of salamanders. After that, pieces of western Mexico were carried northward and, during subduction, become attached to the western border of the continent in the vicinity of the developing proto-Sierra.

This is an attractive hypothesis because it provides a mechanism for delivering animals and plants to their present locations without involving them in some form of long-distance dispersal. The hypothesis is also supported by similar distributions of other plants and animals with poor dispersal ability. For example, the slender salamanders mentioned earlier also fit this pattern. This distributional range includes eastern North America, western North America, Mexico, and Central America. Furthermore, long-distance transport of the accreted terranes of western North America has been substantiated.

A weakness to this hypothesis is that to support proper "hitchhiking," the terranes would have to remain above sea level, and they would have to include portions of land that were translocated from one part of the continental margin to another. Most of the terranes are clearly of seafloor origin or are island-arc volcanics formed by eruption out at sea. A variation of the hypothesis proposes that the organisms were carried by some form of rafting, such as floating on trees, from mainland Mexico to distant islands, which were then translocated northward and accreted to the western border of what is now the northern part of North America. Whatever explanation finally provides the solution to this peculiar distribution, it merely will add more interest to this remarkable group of salamanders.

Another amphibian of alpine meadows is known as the Yosemite Toad, *Anaxyrus (= Bufo) canorus*. This threatened species is most active in the daytime and is characterized by males and females which do not resemble each other as if they were different species. The Yosemite Toad and its behavior is discussed in Chapter 11.

BIRDS Very few animals stay in the alpine zone year-round. Of the birds that occur there on occasion, none is more truly alpine than the Gray-crowned Rosy-finch, *Leucosticte tephrocotis* (figure 5.25). It is the only bird known to nest above timberline. Usually it makes its nest among rocks, out of the wind, but it also has been reported to nest in the bergschrund at the top of a glacier. As is often the case in harsh habitats, these birds cannot afford to be too specialized in their food habits. They eat a wide variety of seeds and insects, commonly feeding on or near the edge of snowfields. When insects such as mayflies emerge from mountain lakes, the birds feed voraciously in flocks, catching them on the water and in midair. On the tops of popular peaks such as Mount Whitney or Half Dome, these birds have become quite tame. They approach closely to feed on whatever scraps or handouts are offered. During winter, these birds remain in the high country unless the weather is unduly harsh, in which case they will fly down the east side of the Sierra to feed as low in elevation as the sagebrush belt.

MAMMALS There are three ecological rules or principles, based on surface-volume ratio and color that relate to temperature regulation, body form, and distribution among endothermic animals. These rules are based on minimizing heat lost or gained by radiation. Bergmann's rule states that within a given species, body size tends to increase in colder climates. Allen's rule states that within a given species, limbs tend to become longer in warmer climates. Gloger's rule states that within a given species, organisms tend to be darker in tropical (warm, moist) climates. In accordance with these rules, mammals of alpine zones should be expected to have large body mass, short limbs, and to be light colored, especially if they overwinter in the high mountains.

Among mammals of the alpine zone, there are three basic strategies for dealing with harsh winter. Animals can migrate, hibernate, or remain active by storing food or somehow making use of available food.

Mountain or Sierra Nevada Bighorn Sheep, *Ovis Canadensis sierrae*, spend summers in the high country. This is the subspecies that occurs in the Sierra Nevada and the Warner Mountains. During summer in the mountains, males gather in herds separated from ewes and lambs. In years past, the Sierra Nevada herds would migrate down the east side of the range to overwinter in the pinyon pine and sagebrush belt of the Sierra Nevada, Owens Valley, and the White Mountains. They have declined in number seriously. At its lowest, it was estimated that there were only about 100 members of this high-mountain subspecies left in the state. Their decline is due to many factors, including diseases, hunting, and competition from cattle and domestic sheep. Some of the diseases were introduced from the domesticated animals. A particularly serious disease for the Bighorns is a virulent pneumonia. In 1988, for example, 11 out of a herd of 49 were found dead in the Warner Mountains. These outbreaks can be very serious. In Oregon in 1987, one outbreak of pneumonia killed 85 Rocky Mountain Bighorn Sheep out of a herd of 120.

Another serious problem for the herds in the Sierra Nevada was their inability to cross Owens Valley to the White Mountains because of the highway and the Los Angeles Aqueduct. Restricted to meager winter habitat on the steep east face of the Sierra, their numbers declined steadily. Various government agencies then stepped in to help them out. Areas in the vicinity of Mount Williamson and Baxter Pass were closed to entry during late summer and early winter. It was hoped that by restricting human entry to the high country during the time that sheep band together to breed, the disturbance would be reduced to a minimum.

Apparently, learned behavior is very important for sheep. Younger individuals must follow older sheep in order to learn where to step as they traverse steep cliff faces. Trying to cross these routes in panic causes them to fall. Although adults have been observed with well-healed broken limbs, it has been reported that a major cause of death of young sheep is starvation after suffering a broken leg caused by an untimely fall. As long as sheep remain away from well-traveled routes in the summer, frightening by humans should be minimal.

Since adoption of the closures, the Baxter herd has done well and is maintained at about 200 sheep. For some reason, the Williamson herd has not fared as well; its number remains at around 30. Expansion of the Baxter herd has provided a source of sheep to be transported to other localities of the Sierra within their former range. Sheep have been transferred to the Mount Langley area, to Wheeler Crest, and to Lee Vining Canyon, in Yosemite National Park. At one point, in order to protect animals from predation by Mountain Lions, it was even suggested to translocate a small population to Paoha Island in Mono Lake. A major problem with these translocations is that sheep apparently also learn their migratory routes and, when translocated to a new area, they must learn these routes without help from sheep that have traversed the area before. Translocations therefore must include a few old sheep that remember to migrate. Apparently, these old-timers lead the way as a new route is learned. It has been reported that without the old sheep to lead them, a group of young translocated sheep will attempt to overwinter in the high country, much to their discomfort. Since 1999, following listing of the Sierra Bighorn as endangered, under the Endangered Species Act, the population has steadily increased. As of 2012, it was estimated that the population was over 500 animals in five separate herds.

The Peninsular Bighorn Sheep, originally called *O. c. cremnobates*, recently was described as a separate population of the Desert Bighorn, *O. c. nelsoni*. The Peninsular Bighorn occupies the Peninsular Ranges from the San Jacinto Mountains southward (figure 5.27). Like the California Bighorn Sheep, it was listed as endangered in 1998 and may not be hunted legally. Legal hunting of all types of Bighorn Sheep was halted in 1883. There were about 400 animals in 2000, and since then the population gradually increased to about 955 in 2010.

FIGURE 5.27 Peninsular Bighorn Sheep, *Ovis canadensis.* Deep Canyon, Santa Rosa Mountains, Riverside County.

One of the reasons for the establishment of Anza-Borrego Desert State Park was protection of the Peninsular Bighorn Sheep. (Borrego is the Spanish word for the sheep.) Another refuge, classified as critical habitat, occurs on the slopes of the Santa Rosa Mountains in the headwaters of Deep Canyon.

The Desert Bighorn Sheep has responded to management better than the other populations. These sheep occur in ranges throughout the Mojave Desert, and they also are found in the San Gabriel and San Bernardino Mountains. In 1987, the first legal hunt in 103 years was allowed. It was determined by the Fish and Game Department that the number of Desert Bighorns had reached about 3700, and some of the older sheep should be removed. The state legislature therefore passed a bill to allow hunters to take 15% of the adult rams from two herds in the eastern Mojave Desert. Thinning of a herd can have beneficial effects on the remaining sheep. It allows better utilization of resources and helps to retard epidemic diseases. The biggest threat to the existence of Desert Bighorn Sheep is the presence of feral Burros in many parts of the desert, a problem that will be discussed in Chapter 9.

Of the mammals that avoid the high-country winter through hibernation, none has been studied more thoroughly than Belding's Ground Squirrel, *Urocitellus (= Spermophilus) beldingi* (figure 5.28 and plate 18G). These little squirrels, about 10 in (25 cm) long, are gray on the back and buff on the sides and belly. They sit up straight on their hind legs and watch for danger, a posture that has earned them the common name of "Picket Pins" in reference to the stakes packers use to tether their livestock. These ground squirrels occur in colonies in Sierran meadows from about 6500 to 11,800 ft (2000–3700 m) in elevation, between Kings Canyon and Lake Tahoe. They are restricted to meadows where soil is abundant enough to construct burrows but does not remain waterlogged. In particular, they may be observed easily in any of the large meadows along the Tioga Pass Road in Yosemite National Park. In one of these, Dana Meadows, just east of Tioga Pass, researchers constructed small platforms atop tripods about 15 ft (4 m) above the ground. Here they sat, hour after hour, with binoculars, observing the behavior of Belding's Ground Squirrels.

These squirrels spend about eight months in hibernation. During summer months, they fatten up by eating seeds and insects. They double their body weight by increasing their fat reserves up to 15-fold. Males emerge first, in late April, when snow is still abundant. Females do not emerge until the snow has melted from the tops of their burrows. A few days after emerging, females become sexually receptive and remain so for only 3–6 hr on 1 d. Receptive females are followed by males, who attempt to mate with them and who chase and fight each other in order to be chosen. In actuality, a few large, old males do most of the mating. One study found that 71% of copulations were accomplished by 3–12 males. Females generally mate four to five times, with about three different males, and give birth to as many as eight offspring.

Gestation lasts 23–25 d. Females either dig their own burrows or inhabit abandoned gopher burrows. Each burrow has at least two entrances.

FIGURE 5.28 Belding's Ground Squirrel, *Urocitellus beldingi.*

Burrows often must be changed, as they may become flooded or dug up by predators such as Badgers, weasels, or Martens. The young spend their early days in the burrow, in a nest made of dry grass. After nearly a month of nursing, they appear aboveground, in late July and August. By that time, some adults already have entered hibernation. Juvenile males disperse, but juvenile females remain near the nest.

Studies show strong kin recognition in ground squirrel families. They recognize each other by odor. Males and females raised together are nonaggressive toward each other. Female siblings separated experimentally at birth are able to recognize each other, but sibling males reared apart are aggressive. Females remain for many generations in the same area where they were born.

Juveniles enter hibernation last and these yearlings are the last to emerge the following spring. Yearling males are highly carnivorous. They must have a good protein source in order to overwinter and to be ready to reproduce in their second year of life. Protein is provided by insects, carrion, small mice, and young nursling ground squirrels. Infanticide is practiced by yearling males presumably as a protein source. It is also practiced by adult females, who prey upon the nurslings of other females. These nurslings are not eaten, but merely dragged from the nest and killed. This practice is assumed to increase the contribution of genetic material from the female that does the killing, by reducing competition from offspring other than her own. Neither yearling males nor females kill the offspring of relatives.

Females vigorously defend their nests. After juveniles emerge, young females cooperate to enlarge the defended territory. Juvenile females will hibernate together, and the following year, they will construct their own nest near their original home. This means that females living near each other are related. Nonrelated males and females are kept from entering the territory.

When predators approach, some ground squirrels give alarm calls that cause others in the family to seek shelter. There are two kinds of alarm calls. A series of five short rapid squeaks means there is a predator, usually terrestrial nearby. A loud shrill whistle means immediate danger, usually a predatory bird. A similar behavior was described for California Ground Squirrels in the foothills. In the case of Belding's Ground Squirrel, the callers usually are large, old females. Because calling may attract the predator, callers are more frequently killed, but these old females are the most expendable of the group. This form of behavior, where one animal sacrifices itself to protect its descendants, is known as altruism. By her activity, which may lead to death, the female enhances her contribution of genetic material to future generations.

Of the forms of behavior that have been observed in Belding's Ground Squirrel, some have clear survival value or increase the probability of genes surviving in offspring. It might also be argued that many behavioral features are associated with density-dependent regulation of population size. This means that mechanisms in the population function to keep the population from becoming so large that it outstrips the available resources. Consider infanticide. If food is abundant, females are able to forage near their nests and defend it from invasions by other squirrels. Abundant food therefore helps to ensure survival of nurslings. Likewise, if insects are abundant, yearling males may not need to prey on nurslings in order to receive adequate protein. The reciprocal of this is that if yearling males do not receive adequate protein, they may perish during hibernation, or they may survive in a weakened condition and not be part of the breeding pool the following year. They may not become reproductive due to poor nourishment, or they simply will be too weak to win the battles necessary to mate with a female.

Differential lifespan can also be linked to density-dependent regulation of population size. Females typically live 4–6 years, males only 3–4 years. Some individuals live considerably longer, however. Females up to 11 and males up to 6 years of age have been reported. Differential mortality means that the sex ratio becomes skewed in favor of females. Because so few males actually succeed in mating, they are more expendable. When food is scarce, more females survive. The few males that survive are the strongest, and they become the parents the following year. It actually takes very few males to make all the females pregnant. The extra males are a luxury; they provide variation for the gene pool so that all the offspring are not sired by only a few males. This luxury, however, can be postponed until food is abundant. The fact that a female mates with several males is another mechanism to ensure that all babies are not fathered by the same male.

Of the major causes of death, two may be related to density-dependent regulation. In one 4 year period, infanticide accounted for 8% of the deaths. The regulatory nature of infanticide was discussed earlier. Another study determined that 4–11% of the study animals were killed by predators such as Badgers, Coyotes, weasels, and hawks. These predators are opportunistic, feeding on whatever is abundant. When squirrels are common, they feed on them more heavily, which also serves to reduce the size of the squirrel population when it is densest.

Apparently, the primary cause of death for these squirrels is the weather. Depending on the severity of the winter, 54–93% of juveniles and 23–68% of adults perish during hibernation. This pattern selects for large, well-fed adults that are able to endure a long, harsh winter. If the winter is mild, those squirrels that emerge early have an advantage: their offspring have a longer time to feed and develop and they become large and fat. A critical balance between opposing forces is evident. If squirrels emerge too early, they may be caught by a late snowstorm, so this favors those squirrels that emerge late. Conversely, if squirrels emerge late, their offspring will not have as long a time to grow. The critical balance between early and late emergence dictates which squirrels will survive to pass on their genes. Variation is also important. If all squirrels emerged at the same time, a population could be obliterated by the weather. Weather-related mortality is a classic example of density-independent regulation: the initial size of the population has very little influence on the number of offspring that are killed.

Hibernation in the Golden-mantled Ground Squirrel also has been studied. The interesting thing that has been demonstrated with these squirrels is that they maintain persistent circadian body temperature rhythms during their period of torpor, although the rhythms were of low amplitude compared to the rhythms during periods of activity. Circadian rhythms, normally set by daylight hours, have been demonstrated in total darkness in other animals, including humans, as well.

The most conspicuous mammal of the high country is the Yellow-bellied Marmot, *Marmota flaviventris* (figure 5.29). Also known as a Rock Chuck, this large rodent occurs in meadows and rock piles from about 6200 ft (1900 m) to the tops of peaks. Although it lacks cheek pouches, it is correctly considered to be a large squirrel. Related to the Hoary Marmot (*M. caligata*) of the Rockies and the Woodchuck

FIGURE 5.29 Yellow-bellied Marmot, *Marmota flaviventris.*

(*M. monax*) of the eastern United States, the Marmot is the largest of the Sierran squirrels. It reaches a body length of 20 in (45 cm) and a weight of about 10 lb (4.5 kg). By late summer, about half of its body weight is fat, because, like the other high-country squirrels, it is a hibernator. During hibernation, its body temperature drops from 97°F (36°C) to about 40°F (4.5°C). Its heart rate drops from about 100 beats per minute to about 4 beats per minute. It inhales about once every 6 min.

Marmots are common where there are rocks, and burrows are common under large boulders at the edges of meadows. They sometimes construct burrows among the roots of trees, and where little soil is present, they construct dens in talus slopes. A den or burrow system usually has several entrances, and there is one sleeping chamber lined with grasses. Colonies may contain 10–20 individuals.

Mating occurs upon the marmots' emergence from hibernation. Females dig new burrows, and males wander from burrow to burrow seeking a mate. Males mate with and defend one or more females. Unlike Belding's Ground Squirrel, mating takes place underground, so little is known about mating or territorial behavior. Presumably, a nonreceptive female drives males from her burrow until she is ready to mate, at which time the male joins her in the burrow. Gestation takes about 30 d, after which the male is evicted and offspring are cared for by the female for about six weeks.

Marmots spend a great deal of time basking in the sun. Apparently this is an energy-conserving device that enables them to maintain their body temperature. They will need all the energy they can store to get through the winter. It may seem that they are unduly exposed to predators while they bask, but, like other communal squirrels, they have an acoustic warning system. They have three kinds of calls, chuck, whistle, and trill. A single loud whistle or a trill from one member of the group causes others to dive for cover. To many back-country visitors, this is the most familiar animal sound. Looking in the direction of the sound, a person often sees just the head of a marmot peering from the top of a boulder. There may be a dominance hierarchy established among marmots. Each seems to have its own high perch, and the largest animal is usually on the highest perch. Concentrations of feces and urine are often found on the tops of rocks and may be a territorial marking system.

Marmots spend about 60% of their time underground; 80% of that time during a year is in hibernation. The onset of hibernation in mammals has long been a mystery. Recent studies show that lowering of blood temperature causes the release of natural opium-like chemicals (opiates) from a region of the brain known as the hypothalamus. This region is known to be associated with many other important physiological urges, such as thirst, hunger, and sleep. Extracts of hibernating Woodchuck blood can drive nonhibernators such as monkeys into a torpid condition. Furthermore, hibernating ground squirrels seem to show low sensitivity to morphine, another opiate. Appar-

FIGURE 5.30 American Pika, *Ochotona princeps*.

ently, natural opiates produce the suite of characteristics we call hibernation by activating certain receptor sites in the brain. Opiates such as morphine have little effect at this time because the receptor sites are already occupied by natural opiates.

A habit of Marmots that is particularly disturbing to hikers and backpackers is that they tend to chew holes into packs that may contain the odors of food or sweaty clothes. Worse than that, in some parking areas, in particular the trailhead at Mineral King in Sequoia National Park, the Marmots have taken to chewing on the rubber, flexible parts of brake lines, and water hoses. Presumably this behavior is because typical brake fluid and antifreeze are composed of glycol compounds, which have a sweet flavor. Obviously brake failure while descending a steep mountain road is a serious problem, and overheating can seriously damage an engine. Chicken wire barriers are now often placed around cars at Mineral King.

Another common mammal of the alpine zone is the American Pika or Cony, *Ochotona princeps* (figure 5.30). These animals employ a third strategy available to alpine animals: they remain active throughout the winter. They do not migrate, and they do not hibernate. They live on stored piles of sedges and grasses (hay piles) stashed among rocks in talus slopes. Apparently they are able to get all the water they need from the food they eat. They seldom venture from the rock piles except to gather food. Although they are rodent-like in appearance, American Pikas are more closely related to rabbits and hares; they are lagomorphs. Similar to other lagomorphs, American Pikas eat their own feces in order to get full digestion and absorption of nutrients. Feces deposited in the landscape during the daytime have been two times through the digestive tract. Also similar to other lagomorphs, their distribution is seriously limited by the potential for overheating. They have a relatively high body temperature for a mammal, 104°F (40°C), and they can tolerate only a few degrees of warming before they die. The large ears of lagomorphs are renowned for their ability to dissipate body heat by radiation, and the ears of American Pikas are proportionately larger than those of other mammals of the alpine zone. However, as predicted by the rules proposed by Bergmann and Allen, relating to body size and limb length, American Pikas have the smallest ears of the hares and their allies.

Studies in rock piles at various elevations show that the ability of American Pikas to colonize new habitat depends on their ability to run without overheating from one pile to the next. At high elevation, where it is cooler, this is less of a problem, and American Pikas seem to inhabit nearly every rock pile and talus slope. At the lower extremes of their range, however, the rock piles are like cool ecological islands separated from each other on the basis of heat. Rock piles too far from a source of colonizers have no American Pikas, whereas rock piles that may be reached during periods of cool weather maintain steady population size, with the number of colonizers balanced by the number that die. As might be expected in this situation, the apparent downhill dispersal of American Pikas is limited by temperature, not by the abundance of rock piles or food. Recent studies of American Pika colonies reveal that warming, apparently associated with climate change, is causing American Pikas to disappear from lower elevation sites. Apparently they can die within a few hours at air temperatures of 78°F (25.5°C) or higher. Some scientists are alarmed that extirpation of American Pikas could occur in marginal habitats, particularly if there are no suitable uphill refugia nearby. Disappearance of American Pika colonies already has been documented in the White Mountains and Yosemite National Park. Studies in the Great Basin indicate an unprecedented average of 476 ft (145 m) uphill displacement since 1999.

Vocalizations of American Pikas are also familiar to back-country hikers. They utter loud, short squeaks as a territorial signal. The sound is distinctly not whistle-like and is not likely to be confused with the sound of the marmot. As in the case of the marmot, a person searching the talus in the direction of the sound may locate the American Pika peering from the top of a rock. Studies of American Pikas in the Rocky Mountains have revealed a great deal of diversity in their vocalizations. They have long songs vocalized by males and females. There are also calls emitted during courtship, and there are wailing calls that signal the end of courtship. There are distress calls, aggressive calls, and territorial calls. Apparently, these territorial calls differ enough among individuals in a population that American Pikas are able to recognize each other by sound alone.

Another aspect of the territorial call is that it is uttered only when the animal has a retreat in the talus. If a American Pikas is out on a meadow gathering food when a predator approaches, it utters no sound. It waits motionless, hoping for the danger to pass. When in the talus it utters its call, and if the predator approaches too closely, the American Pika drops out of sight into the labyrinth of tunnels and cavities among the rocks. No call is uttered if the predator is a Marten or a Long-tailed Weasel, *Mustela frenata* (figure 5.31); the American Pika simply drops out of sight without warning the others. Apparently, the sound would attract the attention of the only predators able to follow the American Pika into its retreat.

In the spring, American Pikas establish territories in the talus in an alternating male-female pattern. Males court the females in areas where the territories overlap. If mating occurs, two broods of young are produced.

FIGURE 5.31 Long-tailed Weasel, *Mustela frenata.*

After the first mating, about three young are produced within about 30 d. Mating occurs again, and the second brood is produced about 30 d after that. After the second mating, the male is driven from the territory, and the young are cared for by the female only.

During late summer, males and females venture into meadows to gather sedges and grasses, which are stored in the talus as "haystacks." After the first snows of the winter, American Pikas continue to feed by burrowing in the snow or venturing out upon it to feed on nearby shrubs. They also eat crustose lichens, which they scrape off rocks. It is not until the snow becomes so deep that it would be too energetically expensive to dig that the American Pikas remain in the rock piles feeding upon their stored hay.

American Pikas are so efficient at gathering vegetation that they produce a gradient of grazing pressure in the vicinity of their talus piles. Preferred vegetation becomes depleted near the talus. If there are too many American Pikas in the talus pile, they are forced to venture farther in search of food. If this happens, they are more likely to overheat or be caught by a predator. This is another example of density-dependent regulation of population size.

A variation of the two preceding strategies is illustrated by the North American Deermouse, *Peromyscus maniculatus* (figure 8.48). This small mouse is notable for enjoying a wide range of distribution from below sea level in Death Valley to the alpine zone above 11,000 ft (3300 m). It has been collected as high as the summit of the White Mountains, at 14,250 ft (4240 m). On the one hand, this mouse stores food in its nest to feed upon during winter months. On the other hand, its body temperature drops every day when it goes to sleep. It is as if the animal hibernates every day, which can be a very important energy-conserving technique, particularly for such a small mammal with a high surface-volume ratio. They are highly reproductive, having several litters of young each year. Most mice live only a year, but many have been known to live for 5 years or more.

A number of studies have been conducted on these mice in an attempt to determine how animals can live at high elevation. By raising mice at different elevations and then moving them up and down the mountain, a number of generalizations have been established. At high elevations, animals tend to develop larger hearts, increased lung volume, and a greater number of red blood cells. In accordance with Bergmann's rule, body weight increased with elevation, but only to about 10,000 ft (3000 m). Above that, body weight decreased, presumably because lack of oxygen restricted growth. Animals also exhibit very high metabolic rates at high elevation. Other sampling at various localities in the Sierra illustrated that up to about 10,000 ft (3000 m), limb length, ear length, and tail length showed a decrease, in accordance with Allen's rule.

Breeding experiments at various elevations and subsequent transport of animals to different elevations illustrated that these changes were a function of development and acclimatization rather than hereditary. An interesting genetic component did become clear, however. Hemoglobin of mice born at high elevation has a greater attraction for oxygen than it does in mice at low elevation. Further studies showed that Deer Mice have polymorphic hemoglobins, which means that they are genetically capable of producing several kinds of hemoglobins. The one they produce as an adult is determined during development in the uterus. They produce the type of hemoglobin that functions best at the elevation of early development.

Experiments at different temperatures showed that mice exposed to cold developed a greater amount of brown fat—the high-calorie, heat-generating tissue. Their metabolic rate also went up, but this trend also was limited by reduced oxygen at high elevation. More brown fat was produced, but the limit in oxygen caused the metabolic rate to decline, so that heat production was actually reduced in spite of the greater mass of brown fat. These animals were probably living at the very fringe of existence.

Recently it has been determined that Deer Mice are vectors for hantaviruses and Lyme disease. Hantavirus is an influenza-like disease caused by the Sin Nombre virus. It is usually picked up when a victim inhales dust associated with droppings from the mouse. The disease unfortunately is lethal about 50% of the time. A typical case of hantavirus is picked up when a person cleans up a cabin after it has been closed up for a time. There have been several lethal cases in the Owens Valley on the east side of the Sierra, and at least 10 cases have been reported for people who had visited Yosemite Valley.

Lyme disease is caused by a spirochete bacterium that is transmitted from mammalian hosts by the Deer Tick, *Ixodes pacificus*. Persons with the infection could include influenza-like symptoms, skin rash, and recurring arthritis. The disease is usually treated with

FIGURE 5.32 Bushy-tailed Woodrat, *Neotoma cinerea* (photo by Mick Bondello).

antibiotics. Lyme disease is not common in California, but people always should take care not to pick up ticks when they are hiking.

Sylvatic plague is another disease occasionally carried by rodents in the Sierra Nevada. The vectors are fleas that parasitize various squirrels and chipmunks. In California, there have been 42 cases of human plague since 1970. Nine were fatal. In 2014 plague was detected in animals from seven counties, including El Dorado, Mariposa, Modoc, Plumas, San Diego, Santa Barbara, and Sierra. In August 2015, Crane Flat and Tuolumne Campgrounds in Yosemite were closed because Golden-mantled Ground Squirrels had tested positive for plague. A child who was camping at Crane Flat contacted the disease, which apparently was the first case reported in Yosemite since 1959.

The Bushy-tailed Woodrat, *Neotoma cinerea*, is not exclusively an alpine animal, but it commonly inhabits talus piles along with American Pikas (figure 5.32). It is larger than the Dusky-footed Woodrat, *N. fuscipes*, common in Oak Woodland and Chaparral areas, and it has a long bush tail. On the east side of the Sierra, it ranges down into the Pinyon-Juniper Woodland and upper reaches of the Mixed Coniferous Forest. Like other woodrats or trade rats, it gathers bits of debris or shiny objects and carry them back to its nest, which is usually a stick pile among the rocks. A particular annoying habit exhibited by these nocturnal rodents is their habit of climbing into the engine compartments of parked cars, chewing on wires,

FIGURE 5.33 Pocket Gopher, *Thomomys* sp.

FIGURE 5.34 White-tailed Jackrabbit, *Lepus townsendii,* on Mount Langley, Sequoia National Park.

and sometimes building a stick nest on the engine. Imagine returning from a backpacking trip to discover that your car won't start, or you may drive a distance before you discover that the heat of your engine has set fire to the nest. As discussed above regarding Marmots, rodents in general have taken a toll on automobiles parked in mountain areas from the Rockies to the Sierra.

Various species of pocket gophers (*Thomomys* spp.; figure 5.33) range from the Great Central Valley to the alpine zone of the Sierra. Unlike the North American Deermouse, different species occur at different localities, but their habits are similar. In meadows, they construct lateral burrows extending to the surface, through which excavated earth is dumped to the outside. Gophers usually feed on roots and underground parts of plants, such as bulbs and stems (rhizomes). At night, however, they venture from their burrows to feed on grasses and leaves. During winter, they leave the ground and maintain burrows under the snow at the surface of the ground. There they are able to move about, feeding on plants under the snow. In spring, when the ground thaws, they excavate underground burrows once again, filling the snow burrows with excavated earth. After the snow melts, ridges of earth indicate the outlines of their snow burrows on the surface of meadows. Like other burrowing forms, gophers make a nest beneath the ground and line it with shredded plant parts. Six to eight babies are produced in the spring in these nests. Gophers are

victimized by the usual predators of alpine animals, including the Great Gray Owl and Pine Marten, as described previously.

Jackrabbits are technically classified as hares. The White-tailed Jackrabbit, *Lepus townsendii* (figure 5.34), is an interesting mammal of the alpine zone that has been poorly studied. This high-elevation form of the common White-tailed Jackrabbit is widely distributed throughout the Great Basin. It is distributed in the southern Sierra, mostly in the subalpine zone, but it ranges widely into the alpine zone. It has been observed as high as 13,000 ft (4100 m) on Mount Langley. Compared to the desert members of this species, this jackrabbit clearly shows a trend toward enlarged body size and reduced size of ears and legs, as predicted by Bergmann's and Allen's rules. Furthermore, as predicted by Gloger's rule, during winter it grows white fur to protect it from predation and to increase its thermal efficiency. In cold weather, it is an advantage for an endothermic animal to be white, because it loses less heat by radiation than a dark animal. Furthermore, because white hair has microscopic air spaces, it is a better insulator than pigmented hair. The switch to a dark color in the summer is likewise more efficient for heat loss. As mentioned earlier, the distribution of hares and rabbits is largely controlled by temperature. In the winter, this jackrabbit grows hair on its feet so that it may travel in snowshoe fashion across the snow. Food for the White-tailed Jackrabbit during winter has not been identified. Presumably, it feeds on leaves, twigs, and bark, digging down to bush tops through deep snow.

The Snowshoe Hare, *L. americanus*, occurs widely through northern forests up to 8000 ft (2500 m), but it is not commonly seen due to its tendency to remain concealed. Although it is most common in the northern Sierra Nevada, the Snowshoe Hare reaches it southernmost distribution in Tulare County. It relies on protective coloration for concealment. Its preferred habitat seems to be dense creekside thickets, particularly where alders and willows abound. The Snowshoe Hare also occupies dense stands of manzanita that grow after a fire. Unlike the White-tailed Jackrabbit, it seldom ventures into the alpine zone. Where it coexists with the White-tailed Jackrabbit, the Snowshoe Hare can be distinguished by its short ears, scarcely longer than its head, and its large "snowshoe" feet and long hind legs. Like the White-tailed Jackrabbit, the Snowshoe Hare changes color with the seasons.

6

Pacific Northwest

FIGURE 6.1 The Pacific Northwest, where the forest meets the sea.

CLIMATE

Precipitation in the Pacific Northwest is the highest in the state. Average annual precipitation in the redwood belt along the coast exceeds 80 in (200 cm). In the wettest part of the Siskiyou Mountains, in Douglas-fir forest (plate 9B), annual precipitation is reported to reach 140 in (350 cm). Most of this precipitation occurs in the winter, but temperatures are not particularly cold. Soil never freezes, even in the subalpine zone, where snowfall may reach 10 ft (3 m) and remain on the ground until July. This means that, unlike most regions characterized by conifers, most photosynthesis occurs during the winter months. This soil property also selects against winter-deciduous hardwoods. The driest month is July, but summer precipitation is common. The tail ends of Alaskan storms produce more summer rain in the Klamath Mountains than is found anywhere else in the state. It is the predictability of this precipitation that favors the growth and maintenance of relict forest species that occur nowhere else in the state. Many plant species of the Pacific Northwest cannot exist without summer rain. Precipitation remains high on slopes that face the coast, but to the interior, in the rain shadow, precipitation drops to about 20 in (50 cm) per year.

The northern Cascade Range also lies in the rain shadow of the Klamath Mountains, but the southern part of the range is east of the Coast Ranges where rain shadow is not profound. The town of Mount Shasta at 3500 ft (1067 m) elevation to the north averages around 40 in (102 cm) of precipitation per year. To the south, precipitation is higher. The town of Shasta Lake gets around 66 in (168 cm) of precipitation per year. Mount Lassen (figure 6.1) at the southern end of the Cascades, where rain shadow is diminished, is one of the snowiest places in California. At 6700 ft (2040 m), snowfall averages 430 in (10.9 m) and at Lake Helen with an elevation of 8200 ft (2500 m) 700 in (17.8 m) of snow is common, the heaviest in the Cascades.

The cool, moist climate of the Pacific Northwest not only encourages vigorous forest growth, but also promotes conditions that discourage large forest fires, particularly in old-growth forests where there is a diversity of size and age among the trees. Fire intervals range from 140 to 10 years, depending on the site, with more fires toward the east and in former logged areas. Most fires are started by lightning, but human-caused fires have increased in association with access on logging roads. Fires in logged areas are more severe due to abundant slash and debris that remains on the ground, and fire is carried more readily in even aged stands of tree plantations. During drought years, however, particularly when long periods of fire suppression have allowed an accumulation of downed wood or other fuel, conditions favorable to large fires can occur. Research on fire scars and tree-ring analysis shows large fires during the droughts of 1918, 1956, 1977, 1987, 1994, and 2008. The fire in 2008 burned 105,855 acres (42.84 ha) of wilderness of the Trinity Alps. In 1987, a drought year, and the worst year on record, summer thunderstorms started an unprecedented number of fires throughout the northern part of the state. An estimated 1.1 million acres (0.5 million ha) of forest burned in northern California and southern Oregon. The pall of smoke in the Klamath River Canyon was so thick that during September of that year, high temperatures averaged 27°F (15°C) below normal. In a scenario supportive of the hypothesis of a "nuclear winter," high temperatures in the area hovered in the mid-50°F (13°C) range, while similar areas not covered with smoke averaged temperatures in the low 80°F (28°C) range.

KLAMATH MOUNTAINS

In the northwest corner of California is one of the least accessible and least known mountain regions in the state. This region is a rocky upland about 5000–7000 ft (1500–2100 m) in elevation. It has been cut deeply by three river systems, the Smith, Klamath, and Trinity Rivers. The highest peak is Thompson Peak at 9001 ft (2744 m) in the trinity Alps. Heavy precipitation occurs here. In the wettest part of the area, the

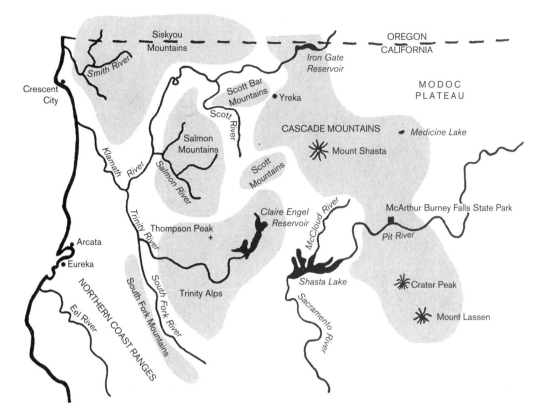

FIGURE 6.2 The Pacific Northwest (illustration by Phil Lingle).

Douglas-fir forest of the Siskiyou Mountains, average precipitation is reported to be about 140 in (450 cm) per year. Runoff from this rain is the cause of the large river systems.

Most of this mountainous region is not accessible to the general public because there are large tracts of wilderness. Of the roads that have been built, most are logging roads that are closed off. The few paved roads that traverse the area follow the coast or major rivers. The average person traveling by automobile in the area would see no difference between this forested region and the northern Coast Ranges.

A mode of entry that may become popular in the future is along abandoned railroads. Many of these are now being developed in northern California, usually on the bed of an abandoned logging or mining railroad. The advantage to these conversions is that they are seldom steep. They may be covered with planks to allow access by mountain bicycles. For example, a 25 mi trail

in Lassen National Forest occupies the bed of the old Fernley and Lassen Railroad.

Subunits of the Klamath Mountains are formed by deep incisions that have been made by the large rivers (figure 6.2). The northernmost region, which lies astride the California-Oregon border, is known as the Siskiyou Mountains. The main drainage to the west of the Siskiyous is the Smith River. To the south, the main gorge of the Klamath River separates the Siskiyous from the Salmon and Marble Mountains. These are drained by the Salmon River, a tributary of the Klamath. South of the Salmon River, the highest peaks of the area rise in the Trinity Alps. The Trinity River, which drains the Trinity Alps, is also quite large. It enters the Klamath just before that river dumps into the Pacific Ocean. Although the uppermost reaches of the Klamath and Trinity Rivers have been dammed, the lower reaches still flow freely. These rivers of the Pacific Northwest are

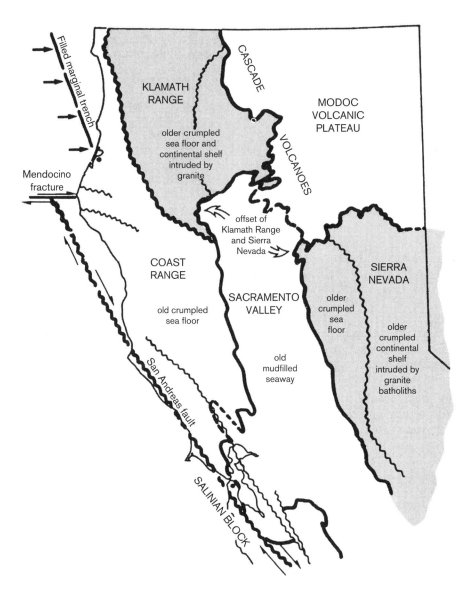

FIGURE 6.3 Geology of the Pacific Northwest (from Alt and Hyndman 1975; reprinted with permission).

the last such free-flowing rivers in California. Because of this, they are among the last places in California where salmon and steelhead are able to swim unimpeded upstream to their ancestral spawning grounds. Recognizing the economic and esthetic value of these rivers, the federal government has assigned wild and scenic status to them. This designation means that no human activities will be allowed to reduce their flow. In California, the wild and scenic rivers include the Smith, Klamath, and Eel Rivers of the Pacific Northwest as well as parts of the American, Tuolumne, Kings, and Kern Rivers of the western Sierra Nevada.

Geology of the Klamath Mountains

The structural framework of the Klamath Mountains is the same as that of the northern Sierra Nevada (figure 6.3). The batholithic rock (granitic) of the Klamath region was formed by subduction at the same time as the Sierra batholith. This batholithic material intruded into a series of prebatholithic metamorphic rocks that occur in long belts to the west of the plutonic rocks.

The metamorphic rocks were added to the western border of North America by accretion. The youngest rocks, the last to be added, are those farthest to the west. There is a near-perfect matchup of these belts on the western side of the Klamaths and the Sierra Nevada. Fossils of animals that lived along the shoreline during Cretaceous time, about 130 million years ago, are found in some places along the western slope of the Klamaths and the Sierra Nevada. In many areas, former trench rocks have been uplifted and become accreted to the edge of the continent. Soils derived from these rocks, including serpentines, are known collectively as ophiolitic or ultramafic soils.

Apparently, about 130 million years ago, the two ranges became dismembered. The Klamath Mountains began to slide westward in a manner that geologists call extension. They are now about 60 mi (100 km) west of their point of origin, in line with the northern Coast Ranges (figure 6.3).

The intrusion of plutonic rocks into these ranges occurred at about the same time as material was accumulating by accretion in the Coast Ranges. The presence of granite and gabbro in the Klamaths and the Sierra, and the degree of metamorphism caused by the intrusion are the basic differences between these mountains and the northern Coast Ranges.

During the end of the Mesozoic, about 65 million years ago, the Klamath Mountains stood as an island. The northern Coast Ranges were still accumulating displaced terranes, but seawater covered what are now the Cascades and the Modoc Plateau.

About 65 million years ago, at the beginning of the Cenozoic, a long period of erosion and tectonic quiescence began. Materials eroded off the Klamaths and the Sierra Nevada and became deposited westward. The climate was relatively warm and humid. Lateritic soils (oxisols) formed along the western borders of the Sierra and Klamath provinces. These red clay soils are still present in many areas.

About 25 million years ago, at the beginning of Miocene, tectonic activity began again. The San Andreas Transform System began its march northward, and the southern Coast Ranges began their uplift. At this time, volcanic activity began in northeastern California, and black basaltic lava began to pour out over the Modoc Plateau, eventually filling the gap between the Klamath Mountains and the newly rising Sierra Nevada. As these volcanic rocks accumulated, the sea was displaced. Extension and faulting, associated with the Basin-Range Province, began. This activity culminated with uplift of the Sierra Nevada and the mountains of the Great Basin. Erosion of materials off the top of the Sierra batholith exposed masses of granitics to weathering. These sediments were carried westward to fill the Great Central Valley, pushing back the ocean even farther. The reason the Klamaths look different from the Sierra today, in spite of their common origin and structural similarity, is that the Klamaths did not experience this secondary uplift and erosion. The Klamath Province is the product of millions of years of erosion under a dense forest cover.

Soon after the discovery of gold in the Sierra Nevada, prospectors found rich placer and lode gold deposits in the Klamath River area. The swiftness of the river made navigation nearly impossible, and the Native Americans in the area were hostile. Nevertheless, miners diverted the waters of the major rivers and mined the river gravels. After that, several hydraulic mining operations were undertaken, with incredible environmental degradation. On the South Fork of the Scott River from Etna Mills to Callahan, there are several miles of gravel piles left over from that period.

The threat of environmental degradation from mining operations is not over. A gold-mining process known as cyanide heap leaching has been developed. In this process, low-grade ore is piled up, and a cyanide solution is sprayed on top. The leachate contains the gold, which can be recovered by passing the solution

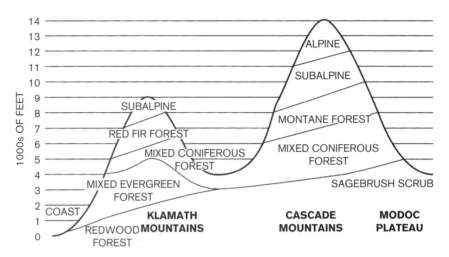

FIGURE 6.4 Biotic zonation of the Pacific Northwest Mountains.

through carbon-filled columns. Obviously, there is concern that the highly toxic cyanide solution may escape into groundwater or streams, but miners assure all those concerned that the process is safe and efficient. Environmentalists are concerned because over 500 mammals and nearly 7000 birds were reported killed at cyanide extraction gold mines in Arizona, California, and Nevada between 1980 and 1989. Of the mammals that died, one-third were bats, many of which were classified as endangered, threatened, rare, or protected by government agencies. It remains to be seen how this procedure in the quest for gold fares with the environment.

In addition, several mining companies would like to establish viable strip mines in the Gasquet Mountain area of the Siskiyou Mountains. These companies propose to mine the laterite soils and extract cobalt, nickel, and chromium. The largest of these proposed mines would remove about 3.3 million tons (3 million t) of dirt annually in order to yield about 50 lb of metals per ton (22 kg/t). Strip-mining laws require that spoils would have to be mulched, fertilized, and revegetated.

Obviously, this is an expensive operation. At world market prices, it would not be economically feasible to operate these mines. The mine owners are hoping that the federal government will subsidize the operation because these are "strategic minerals," components of hardened alloys used in the aerospace industry.

Environmentalists are opposed to the plan because the mining operations are in the Smith River drainage, a wild and scenic river. Furthermore, the area has been proposed for inclusion in a new national park. Leach water from the mine could contaminate the water. Dams to supply water for the mining operation would become the only dams in the drainage. Severe air pollution from sulfur dioxide and nitrogen oxides, the major pollutants associated with acid rain, could occur. The mining operation could thus contribute to the formation of acid rain, which would fall on nearby wilderness areas.

Biotic Zonation of the Klamath Province

Biotic zonation of the Klamath Province is shown in figure 6.4. Along the coast (figure 6.1), plant communities on the terraces are quite similar to those that will be discussed in association with the northern Coast Ranges. Shore Pines, *Pinus contorta contorta*, grow on stabilized sand dunes. On the first terrace is a well-established Coastal Prairie, which has been altered for agriculture, pastures, and urbanization. Moist pockets contain a forest of Alaskan affinities, including such species as Grand Fir, *Abies grandis*; Sitka Spruce, *Picea sitchensis*; and Western Hemlock, *Tsuga heterophylla*. On dry sites in this area, there is a closed-cone pine, the Knobcone Pine, *P. attenuata* (figures A and B, p. 348). It often grows on poor soil that is either well drained or serpentine, and it usually gets its water from fog drip.

A belt of Coast Redwoods, *Sequoia sempervirens*, is extensive, particularly along Highway 101. The Redwood Forest will be discussed in detail in Chapter 7 on the Coast Ranges. Redwood National Park and Prairie Creek Redwoods State Park are in this area. Coast Redwoods claim the title as the tallest trees in the world, and most of the tall trees are in these parks. Formerly the tallest tree was 365 ft (111 m) tall. In recent years, at least four trees have been discovered that were taller. At the present time, at 379.3 ft (115.6 m), the Hyperion Tree in Redwood National Park holds the record. Most of the forest along Highway 101 in this area is dominated by Douglas-fir, *Pseudotsuga menziesii* (plate 9B). Interestingly, there is some evidence that Douglas-firs may be taller than Coast Redwoods. In Mineral, Washington, a giant Douglas-fir was felled. Had it remained standing it would have been 393 ft (119.8 m) tall. Of primary economic importance are dense stands of Douglas-fir, the major source of construction lumber. The little access that exists in the Klamath Mountains is on logging roads that have been cut into steep slopes to carry out timber.

Above about 2000 ft (600 m) elevation, the montane forest of the Klamath Mountains takes on a different character. To see that, a person has to hike, ride a horse, or gain access to a logging road.

These forests have attracted attention among botanists because there are many endemics and relicts. Vegetation patterns are highly complex, and there is a great deal of species diversity. For example, there are over 20 species of conifers in the area, more than in any other part of California. Studies of fossils throughout the western United States suggest that the forests of the Klamath Mountains most closely resemble those that covered western North America until a million years ago. In this respect, Klamath vegetation represents living fossil forests.

At higher elevations are a few species, such as Subalpine Fir, *A. lasiocarpa* var. *lasiocarpa*, that are also found in the Rocky Mountains. The presence here of Foxtail Pine, *P. balfouriana*, which grows primarily in the southern Sierra Nevada at about 11,000 ft (3300 m) in elevation, is an interesting enigma. Foxtail Pines are closely related to Bristlecone Pines, which grow on mountaintops throughout the Great Basin.

Mixed Evergreen Forest

Above about 1000 ft (300 m) elevation, the Mixed Evergreen Forest dominates (plate 10B). This forest is a mixture of broad-leaved evergreen species and conifers. Unlike the Coast Ranges, in the Klamath Mountains this forest is often dominated by Douglas-fir, *Pseudotsuga menziesii* (plate 9B).

Various authors have divided Mixed Evergreen Forest into subcategories based on different dominant species. These subdivisions are, for the most part, an expression of moisture gradient or soil type. One author recognized 12 forest types on sedimentary, metasedimentary, volcanic, granitic, and ultramafic (ophiolitic) parent materials.

A simpler way is to divide the forest into four categories, roughly associated with soil type, exposure, and moisture. On gentle slopes, north-facing slopes, ridges with deep soil, or river terraces with deep sediment, there is a luxuriant forest dominated by Douglas-fir. Commonly, these are soils weathered from sedimentary rocks. This forest is recognized by many authors as Douglas-fir Forest. Nearly 70% of forest cover is occupied by Douglas-fir, huge trees that may rival Coast Redwoods in size and growth rate. Of secondary importance in this forest are components of moist Mixed Coniferous Forest, including Sugar Pine, *Pinus lambertiana* (figure 4.30); Ponderosa Pine, *P. ponderosa*; and White Fir, *Abies concolor* (figure 4.29). In this forest, broad-leaved evergreens play a minor role.

Of great importance in coniferous forests are numerous fungi, including the masses of underground filaments known as mycorrhizae, which are discussed in Chapter 4. One of the most important of these fungi is the Oregon White Truffle, *Tuber oregonense*. Its range extends from British Columbia to northern California. Although it is most often associated with roots of Douglas-fir, it also is found in soil under various species of oaks. This is a convoluted mushroom that grows entirely beneath the ground. It is a favored food for rodents, particularly the Golden-mantled Ground Squirrel (plate 18D), which digs up truffles and disperses the spores through its digestive tract. Also a favored food for humans, the flavor of this species of truffle has been likened to the European truffles considered gourmet delicacies. In fact, it has been the human quest for these underground mushrooms that has also revealed their presence in the Oak Woodlands of the northern Coast Ranges.

On granitic soils, typically at lower elevation than the Douglas-fir Forest, dominant trees are similar, but broad-leaved evergreens are more common. Particular in canyons, Tanoak, *Notholithocarpus (= Lithocarpus) densiflorus* (figure 7.9), and Canyon Live Oak, *Quercus chrysolepis*, form forests of broad-leaved trees. An interesting species in this community is Giant Chinquapin, *Chrysolepis chrysophylla*. This member of the beech-oak family (Fagaceae) occurs in tree form at widely scattered localities in Redwood and Mixed Evergreen forests. On dry sites, associated with chaparral species, it occurs in shrub form. In western Siskiyou County, distribution of the shrub form overlaps that of Bush Chinquapin, *C. sempervirens* (figure 4.42); hybridization of the two is believed to occur. Bush Chinquapin was discussed as an under-

story shrub in Chapter 4, in the section on Montane Forest.

On steep slopes where soil is coarse and well drained, dominance shifts to Canyon Live Oak, also known as Golden-Cup Oak. Douglas-fir is of minor importance in this community, representing a dry extreme of Mixed Evergreen Forest. In this form, the forest is quite similar to Canyon Live Oak woodlands that occur in canyons throughout the state. Canyon Live Oak is the most widely distributed oak in California.

The fourth type of forest is associated with the ophiolitic soils, such as those derived from serpentine or gabbro. Here we find strikingly different patterns of vegetation depending on exposure and mineral composition of the soil. In ravines are moist forests dominated by Port Orford Cedar, *Chamaecyparis lawsoniana*; Western White Pine or Silver Pine, *P. monticola* (figure 4.26); and, to a lesser degree, Douglas-fir. Port Orford Cedar, a member of the cypress family (Cupressaceae), occurs at patchy localities for about 200 mi (320 km) along the coast from Coos Bay, Oregon, to Humboldt County, California. It is often described as a serpentine endemic, but it also occurs on other soil types, in moist areas within Douglas-fir Forest. The fine wood of this forest tree is treasured for beautiful interior finish work, particularly in shipbuilding. It also has been planted extensively in the Pacific Northwest as a horticultural species known as Lawson Cypress. Western White Pine in this forest is the same species that occurs on dry exposed sites at middle elevations in the Sierra Nevada, most often in association with Lodgepole Pine.

Studies of gradients between diorite, serpentine, and gabbro show a decreasing abundance and diversity on gabbro. California Bay Laurel or Bay Laurel, *Umbellularia californica*, California Huckleberry, *Vaccinium ovatum*, California Coffee Berry, *Frangula (= Rhamnus) californica*, and Sonoma Canescent Manzanita, *Arctostaphylos canescens* subsp. *sonomensis*, all occur more commonly on gabbro soils. The beautiful Stream Orchid, *Epipactis gigantea*, only occurs on gabbro in this area. A fire in

2002 completely burned the gabbro study area. Recovery has yet to be reported.

In more exposed localities on ophiolitic soils, or on steep slopes, a forest more similar to dry Mixed Coniferous Forest occurs. Trees here include many that have been discussed previously, such as Incense Cedar, *Calocedrus decurrens* (figure 4.25); Jeffrey Pine, *P. jeffreyi* (figure 4.27); and Knobcone Pine, *P. attenuata* (p.348). Also possibly found here are Douglas-fir, Western White Pine, and Sugar Pine (figure 4.26). This forest has a completely different appearance than other parts of Mixed Evergreen Forest. Trees are farther apart, occurring in rather open stands, sometimes single-species stands.

On ophiolitic soils in warm, dry localities, particularly on south-facing slopes, the Mixed Evergreen Forest gives way to Oak Woodland and Chaparral communities. Among the oaks found here is a form of white oak known as Oregon Oak, *Q. garryana*. Some authorities refer to a community dominated by Oregon Oak as Northern Oak Woodland. They consider this to be a form of Foothill Woodland in which there is a change in the dominant white oak from Blue Oak to Oregon Oak. This tree resembles Blue Oak, often growing in savannas, and its associated species are similar to those found with Blue Oak. An example of ecological separation between Oregon Oak and Blue Oak can be seen near Browns Creek, in Trinity County. Here Oregon Oaks occur in open, rocky locations on ophiolitic soils within Mixed Evergreen Forest, whereas the Blue Oaks occur on disjunct parches of clay loam derived from shale.

Oregon Oak is a northern oak species, occurring all the way to British Columbia, that reaches its southern limits in California. This is the only California oak that has widespread distribution beyond the state. Farther south, in the Coast Ranges, Oregon Oak becomes a species of north-facing slopes, or those slopes that receive more precipitation than the surrounding areas, which may be dominated by Blue Oak. Also in the northern Coast Ranges, Oregon Oak dominates on ridgetops up to 5000 ft (1600 cm) in elevation. These ridgetop communities, known locally as balds or bald hills, are dominated by grasses or by Oregon Oaks, depending on the type of soil.

As soils become increasingly exposed and dry, tree species drop out. A rather continuous layer of shrubs may dominate, forming a chaparral-like community. Several species of shrubby oaks may coexist, including a shrub variety of Oregon Oak. A scrub oak known as Huckleberry Oak, *Q. vacciniifolia*, occurs here along with Deer Oak, *Q. sadleriana*. These species of scrub oak do not have spiny margins on the leaves. Huckleberry Oak leaves are oblong, resembling those of a Huckleberry plant. The distribution of Huckleberry Oak is typically northwestern Californian, although its total range is from southern Oregon, through the Cascades, down the western Sierra foothills to the region of Yosemite National Park. It is the dominant scrub oak at Glacier Point, one of the most popular view-points in Yosemite. The shrub form of Oregon Oak also occurs in a disjunct pattern along the foothills of the Sierra, but it extends as far south as the western Transverse Ranges.

Deer Oak is also known as Sadler Oak. Its distribution resembles that of Huckleberry Oak, but it does not occur eastward into the Sierra foothills. Leaves of Deer Oak resemble those of an alder or an elm. They are 2–4 in (5–10 cm) long, and they are serrated but not spiny. Leaves are pale gray underneath and shiny on top.

Also in these shrub-dominated communities are shrub varieties of other common tree species of the northwest, such as Canyon Live Oak, Tanoak, and California Bay Laurel. True chaparral species such as California Coffee Berry, *F. (= R.) californica*, Buckbrush, *Ceanothus cuneatus* var. *cuneatus*, and Hoary Manzanita, *A. canescens* subsp. *Canescens*, also occur here.

Overall, the patterns of community distribution just described have become complicated by a diversified history of fires and logging. Succession following a disturbance may vary depending on the intensity of disturbance. For example, Douglas-fir does not sprout; if an intense fire were to kill all the Douglas-fir in an area, or if clear-cutting were extensive, there would be no source of seeds. In this case, sprouting species such as Tanoak or Pacific Madrone would dominate for many years. Coast Redwoods also regenerate after a severe fire because they too are sprouters. If a fire is mild, or if many acres have not been logged, there could be a nearby source of Douglas-fir seeds. Seedlings of Douglas-fir could start in the partial shade of sprouting species, but eventually it would overtake the other forms. Douglas-fir is a long-lived species, reaching an age of 400–600 years. In old-growth situations, Douglas-fir is certain to dominate.

Moist Mixed Coniferous Forest

Some authorities use the term Montane Forest to include two communities that have been discussed previously: Mixed Coniferous Forest and Montane Forest. On western slopes of the Klamath Mountains, these communities are present in their moist form.

From about 2000–4500 ft (600–1400 m), the moist Mixed Coniferous Forest is dominated by White Fir, *Abies concolor* (figure 4.29). Douglas-fir may be common here as well. Subdominant species include Sugar Pine and Canyon Live Oak.

An interesting understory tree in this moist forest is Pacific Yew, *Taxus brevifolia*, a slow-growing, long-lived conifer. Pacific Yew is widespread in the west, ranging from Alaska to California and into the Rockies of Idaho and Montana. It is not common anywhere, usually occurring in isolation or in small stands in old-growth forest. In California, it occurs in the northern Coast Ranges, Klamath Mountains, the Cascades, and the northern Sierra as far south as Calaveras Big Trees State Park.

Pacific Yew is a small tree, seldom over 25 ft (8 m) in height. On drier soil, it becomes shrub-like. Because it grows in deep shade, its trunk is usually twisted and it has sprawling branches. Its needles resemble those of a fir, but each is prickly tipped. The bark has purple,

papery scales under which there is a clear reddish hue. These trees differ from other conifers in that male and female reproductive structures are borne on separate trees. The female part, the cone, is red and fleshy, resembling a large berry with a hole in it. This sweetish, fleshy fruit is an important food for some birds, but toxic to humans. Pacific Yew has heavy, rough wood that is surprisingly resistant to decay. For that reason, it has been used for gateposts. The wood also has a good deal of resilience and has been used for canoe paddles and bows for archery.

In 1962, a program conducted by the National Cancer Institute identified chemicals from the bark and roots of Pacific Yew known as taxanes, including taxol, as having the ability to inhibit tumors in mice. The National Cancer Institute initially contracted for 60,000 lb (27,000 kg) of yew bark, which would require cutting down about 12,000 trees. By 1991, it was estimated that 38,000 trees would have to be harvested to satisfy demand for the drug during that year. Because the Pacific Yew is so sparsely distributed and grows so slowly, it may be unrealistic to expect such a quantity of bark ever to become available even with the use of cultivars. A synthetic substitute extracted from yew needles may solve the problem.

Pacific Yew is California's only true yew. It should not be confused with another member of this family (Taxaceae), which is known as California-nutmeg, *Torreya californica*. California-nutmeg occurs in the shade of canyons in the northern Coast Ranges and foothills of the Sierra Nevada, and it does not occur in the Klamath Mountains, where Pacific Yew is most common. The trees resemble each other, but *Torreya* has even stiffer prickly needles, and its fruit is blue green with no hole. When its foliage is broken, it gives off a foul odor. In the Sierra Nevada, it occurs at lower elevations and farther south than the Pacific Yew.

Dry Mixed Coniferous Forest

East of the main crest of the Klamath Mountains, vegetation zones follow a pattern similar

FIGURE 6.5 Stanshaw Meadow, Marble Mountain Wilderness, Salmon Mountains.

to that of the western slopes, but the elevation of each zone is deflected upward about 600 ft (200 m). There is less precipitation on the east side of the mountains, so the forest takes on a more open appearance, and the moisture-loving species either drop out or become common only locally. Below about 4000 ft (1300 m), Ponderosa Pine, Sugar Pine, and California Black Oak, *Quercus kelloggii* (figure 4.39), become important components of the forest, which more closely resembles the Mixed Coniferous Forest of the Sierra Nevada.

On ophiolitic soils in the western Klamath Province, the forest is also typical of dry Mixed Coniferous Forest, but the dominant pine is usually Jeffrey Pine. On the eastern slopes of the Klamath Mountains, the dry Mixed Coniferous Forest becomes interspersed with chaparral species. Of particular interest is the occurrence of Leather Oak, *Q. durata*, a serpentine endemic. The distributions of these serpentine-associated species are discussed in more detail in Chapters 4 and 7.

Montane Forest

Elevational gradients in the western Klamath Mountains are similar to those in the Sierra Nevada. Above Mixed Coniferous Forest, there is a zone dominated by Red Fir, *Abies magnifica*. Occurring from about 4500 to 6000 ft (1400–1950 m) in elevation, this forest often borders meadows (figure 6.5). Associated species include Lodgepole Pine, *Pinus contorta* subsp. *murrayana*; Western White Pine, *P. monticola*, and Mountain Hemlock, *Tsuga mertensiana* (figure 5.3). At higher elevations, Mountain Hemlock (figure 6.5) becomes the dominant species. Above 6000 ft (1950 m), Mountain Hemlock is joined by Whitebark Pine, *P. albicaulis* (figure 5.5), in a typical subalpine community. On ophiolitic soils and on south-facing slopes, Foxtail Pine, *P. balfouriana* (figure 5.8), becomes the dominant subalpine species. The subalpine zone on the Klamath and Cascade Mountains is discussed in Chapter 5.

Forests of the Klamath Mountains are noted for the presence of relict conifers. In particular, in the Russian Wilderness of the Salmon Mountains, 18 different species of conifers have been recorded. Typically, there are species of northern affinities that reach their southernmost distribution in northern California. One exception is Foxtail Pine, which also is found in the southern Sierra Nevada. Its enigmatic distribution is discussed in Chapter 5. The usual summer precipitation from the tails of Alaskan storms is deemed particularly important for these relics.

Of special interest in the Siskiyou and Marble Mountains are two large fir species that resemble Red Fir. The Pacific Silver Fir, *A. amabilis*, primarily a subalpine species, is common in the Olympic and Cascade Mountains. It is scarce south of Crater Lake, Oregon. Young trees of this species may be sold as an expensive Christmas tree under the name "Silver Tip," although, as explained earlier, any of the true firs might be sold under that name. Noble Fir, *A. procera*, occurs in association with Pacific Silver Fir in the Cascades of Oregon and Washington. In California, it occurs along with Red Fir on the western side of the Siskiyous. It is so similar to Red Fir with which it may hybridize that only an expert can tell the two species apart, and then only by closely examining a mature cone.

Another fir with restricted distribution occurs in the Red Fir Forest of the Salmon Mountains. This is the Subalpine Fir, *A. lasiocarpa*. It grows on the edges of meadows on wet slopes, and its distribution is apparently expanding into the meadows. This is the most widely distributed fir in North America, occurring from Alaska to Arizona. Throughout most of its range, it is a subalpine species, but on the coast of California it occurs lower. It is believed to have survived above the glaciers during the Pleistocene and to have extended its range down the mountain during the last 10,000 years or so.

Port Orford Cedar was mentioned previously as a relict occurring in moist Mixed Coniferous Forest, often on serpentine. A close relative of this species, Alaska Cedar, *Callitropsis nootkatensis*, occurs at higher elevations, in wet meadows and seeps. Where it grows in moist forests to the north, this is a large, stately species. At the few localities where it is found in the Klamath Mountains, it is low growing and shrubby. Apparently, it reproduces vegetatively.

Spruces are not common in California. There are three species, all in the north-west. They resemble firs, but their needles are stiff,

FIGURE 6.6 Brewer Spruce (Weeping Spruce), *Picea breweriana*.

encircling the branches, and their cones hang down. Fir cones fall apart when mature, and they point upward on the stem. Spruces most nearly resemble Douglas-fir. Sitka Spruce, *Picea sitchensis*, mentioned earlier, is a component of the wet coastal forest in association with sand dunes. Engelmann Spruce, *P. engelmannii*, is widespread in mountains of western North America. In California, it occurs primarily on the east side of Russian Peak in the Salmon Mountains. This spruce is concentrated in valley bottoms in the Montane and Mixed Coniferous Forests, and its upper limit is about 6500 ft (2150 m). The third spruce is Brewer Spruce (Weeping Spruce), *P. breweriana* (figure 6.6). This spruce is endemic to the Klamath Mountains of California and southern Oregon, and it gets its name from its long, down-drooping branches. It is a very beautiful tree, often standing by itself on a rocky outcrop. It does not grow in large stands but is associated with a wide variety of species at a wide range of elevations. It is probably fire sensitive and cannot tolerate saturated soil. It is remarkably drought tolerant, withstanding moisture stresses below −20 bars.

Riparian Woodland

The contrast between Riparian Woodland and surrounding communities is less dramatic in the Klamath Province than in parts of the state where there is a pronounced period of summer drought. At lower elevations along the coast, the forest understory contains many riparian species, including broad-leaved evergreens such as Pacific Madrone, *Arbutus menziesii* (figures 6.7; Tanoak, *Notholithocarpus (= Lithocarpus) densiflorus* (figure 7.9); and California Bay Laurel, *Umbellularia californica* (figure 6.8). Among the component deciduous species of the surrounding forest are Mountain Dogwood, *Cornus nuttallii* (figure 4.49); Big-leaf Maple, *Acer macrophyllum* (figure 6.9); and Oregon Ash, *Fraxinus latifolia*. In northern forests, Oregon Ash is one of the hardwood species with some commercial value. The yellowish wood is used for furniture, drawer sides, and cabinets.

Also typical of these coastal drainages are willows that are common in similar habitat throughout much of California, such as Pacific Willow, *Salix lasiandra* var. *lasiandra*, Narrow-leaved Willow, *S. exigua* var. *exigua*, and Arroyo Willow, *S. lasiolepis*. Below 500 ft (150 m) in these coastal drainages are at least six species of willow, some with localized distribution, most of which are northwestern in distribution. At higher elevations, usually above 5000 ft (1500 m), there are at least four species of shrubby willows. Lemon's Willow, *S. lemmonii*, which occurs statewide, is the most common species.

Upstream there is White Alder, *Alnus rhombifolia* (figure 4.48), also common throughout the state. In the Pacific Northwest, three other species of alder occur. Sitka Alder, *A. viridis* subsp. *sinuata*, occurs in marshes and stream bottoms in the wet coastal forest of the Klamath Mountains. It is a nitrogen-fixing, pio-

FIGURE 6.7 Pacific Madrone, *Arbutus menziesii*: leaves.

neer species in that area. Red Alder, *A. rubra*, grows in slightly drier areas of coastal forest, usually associated with Douglas-fir. Red Alder is a large species, reaching 80 ft (25 m) in height. It grows along the streams at lower elevations than White Alder. Its range overlaps both Sitka Alder and White Alder, but the three species rarely grow together. Red Alder can be a forest species, too.

Along with willows, Mountain Alder, *A. incana* subsp. *tenuifolia*, forms a typical montane type of riparian community resembling that of the Sierra Nevada. Mountain Alder is widely distributed along mountain lakes and streams of the eastern Klamath Mountains and the Cascades. The importance of alders as nitrogen fixers may be illustrated by a study conducted at Castle Lake, a nutrient-poor (oligotrophic) lake in the eastern Trinity Mountains. One-third of the lake's nitrogen is supplied by Mountain Alders that grow along

FIGURE 6.8 Bay Laurel with leaves and fruit.

FIGURE 6.9 Big-leaf Maple, *Acer macrophyllum*, with winged fruit (samaras).

the east side. Sampling of soil, spring water, and lake sediment from both sides of the lake revealed that the east side had far higher concentrations of water-soluble nitrogen compounds.

At higher elevations and on the interior slopes of the Klamath Mountains, the Riparian Woodland is more Sierran in appearance. This type of riparian habitat is discussed in Chapter 4. Of special interest, however, is the presence of Water Birch, *Betula occidentalis*. Its distribution is patchy. In California, this Great Basin species occurs throughout the Klamath Mountains, in the interior of the northern Coast Ranges, and on Mount Shasta. Southward it reappears on the eastern slope of the Sierra Nevada from about 5000 to 9000 ft (1500–2700 m). This sort of disjunct distribution is similar to other plants such as the Foxtail Pine and Oregon Oak, discussed in Chapters 4 and 5.

CASCADE MOUNTAINS

To the east of the Klamath Mountains is the southernmost extension of the Cascade Range. This is a volcanic region consisting mostly of rolling, forested terrain about 4000 ft (1200 m)

FIGURE 6.10 Bumpass Hell, Lassen Volcanic National Park. This region of geothermal activity is typical of areas associated with recent volcanism.

in elevation. The southernmost of the Cascade volcanoes is actually represented by the Sutter Buttes, which project 2132 ft (656 m) from the sediments of the Sacramento Valley. In California, two well-known volcanic peaks, Lassen and Shasta, project boldly above the surrounding terrain. From a biological perspective, the southernmost extent of the province is marked by Lassen Peak (figure 6.17), which reaches 10,457 ft (3187 m) in elevation. It and the surrounding region are included in Lassen Volcanic National Park. This is a region of open forests, hot springs, and fumaroles (figure 6.10).

To the north, Mount Shasta towers above the surrounding landscape. Reaching 14,162 ft (4317 m), it is visible for many miles in any direction (figure 6.17). Lassen and Shasta in California are but 2 of the 12 major peaks of the Cascades, which extend in a north-south line northward to Washington. The northernmost peak, Mount Baker, is about 500 mi (800 km) from Lassen.

In general, the Cascade Range lies west of the volcanic tableland known as the Columbia Plateau. It extends from southern British Columbia to California. The California portion of the tableland, known as the Modoc Plateau,

lies in the rain shadow of the Klamath and Cascade Mountains. It is discussed in Chapter 9, in the section on the Great Basin Desert.

Mount Lassen and Mount Shasta loom spectacularly above the surrounding countryside (plate 9A). Each of these large, treeless peaks can be seen from the other, as is typical in most of the Cascade topography. The peaks are surrounded by a coniferous forest dominated by Ponderosa Pine, which is of prime importance for wood products. This forest is distributed widely over rounded foothills of lower aspect. These hills represent numerous episodes of volcanism.

Precipitation on the western side of the California portion of the Cascades averages about 80 in (200 cm) per year. Abundant snow often caps Lassen and Shasta all year long. Mount Shasta is nearly always white-topped, and its glacier field is perhaps the most spectacular in the state. Interestingly, scientists tell us that the glacier field on Mount Shasta is the only one in California that continues to increase in size, in spite of climate change. The Cascades are drained by the Pit River, which comes from the Modoc Plateau to the east. The Pit River formerly ran into the Sacramento River, but it now drains into Shasta Reservoir, a part of the

FIGURE 6.11 View of Mount Shasta from the crater atop Mount Lassen.

massive State Water Project that includes the California Aqueduct.

Geology of the Cascade Mountains

From a geological point of view, the entire Cascade Range is young. Its oldest portion may be no older than about 5 million years. The process that formed the range began several million years ago and was responsible for uplifting volcanic peaks in excess of 14,000 ft (4300 m). Older volcanics, of the Tuscan Formation to the west, have been dated at about 5 million years of age. Apparently, the interval between eruptions is much greater than it was centuries ago, and there are few earthquakes in the area today. This reduction of activity indicates that the forces responsible are subsiding. Recent eruptions are from magma that formed a million years ago. Many of its peaks are much younger: Mount Lassen probably appeared only 10,000 years ago. Recent eruptions of Mount Saint Helens in Washington attest to the fact that some of the Cascade volcanoes are still capable of significant activity.

The California coast south of Cape Mendocino is an example of a rifted borderland (figure 3.3). This is the area associated with the San Andreas fault. North of Cape Mendocino, Cali-

fornia's most westerly projection, the coastline changes to the Andean type. Subduction is recent and may still be going on. Offshore there is a deep marine trench where the ocean floor slides beneath the continent. To the interior is an arc of volcanoes, above the region of magma formation. This volcanic arc is the Cascade Range. The higher Cascades, to the east, are younger than 5 million years, indicating that the leading edge of the subducted plate moved eastward under the North American continent. It is believed that the leading edge is now under Montana and is responsible for the volcanic and geothermal activity in the Yellowstone area. Mount Shasta last erupted in 1786, and Mount Lassen erupted sporadically from 1914 to 1917. Geothermal activity in the area today proves that the region is still hot (figure 6.10).

The primary type of volcanic rock in the Cascades is andesite. This type of rock is consistent with a subduction-volcanic arc borderland and is similar to what occurred to form the Andes of South America, which is why the rock is called andesite. Mount Lassen is composed of andesite and a related rock called dacite. Mount Shasta has two cones. The larger, older cone is composed of dacite. The smaller cone, known as Shastina, is composed of rhyolite, a light-

FIGURE 6.12 Moist Mixed Coniferous Forest in McArthur-Burney Falls State Park.

colored volcanic rock that erupts in a more viscous state. Shasta, the higher peak, shows evidence of multiple glaciation, which indicates that it is quite young. An active glacier is still present on the top of Shasta.

Biotic Zonation of the Cascade Mountains

East of the Klamath Province, the volcanic soils of the Cascades make their appearance. In lowland areas, the forest is typically moist Mixed Coniferous Forest dominated by Ponderosa Pine, Sugar Pine, White Fir, and Incense Cedar. Elevational relationships in the Cascades are similar to those in the northern Sierra Nevada. Forests on the west side are moist, and those on the east are drier. On the western side of the Cascades, Mixed Coniferous Forest occurs in a belt that covers about 1500 ft (500 m) in elevation. In some areas on the west side of the Cascades, the moist Mixed Coniferous Forest is particularly luxuriant. One such area lies in McArthur-Burney Falls State Park. One of the main attractions in this park is Burney Falls itself, a two-tiered waterfall (figure 6.12). The upper part of the fall occurs where the water flows over the top of a cliff, and the lower part occurs at an interface between volcanic rock and old sedimentary rock. Water percolates

downward through the volcanic rock and then flows horizontally along the sedimentary rock until it emerges halfway up the cliff to complete the waterfall.

On the east side of the Cascades, understory plants in the Mixed Coniferous Forest are typically those of Great Basin affinities. Trees in the forests of Ponderosa and Jeffrey pines are widely spaced, with abundant shrubs in between. Ponderosa Pine is more common to the north. Western Juniper, *Juniperus occidentalis* (figure 6.13), is an important tree species occurring with Jeffrey Pine to the south and east. Shrubs such as Great Basin Sagebrush, *Artemisia tridentata* (figure 9.20A), and Bitterbrush, *Purshia tridentata* (figure 9.20B), form the understory. This forest grades off to the east to the Modoc Plateau.

An interesting conifer known as Baker Cypress, *Hesperocyparis (= Cupressus) bakeri*, grows on the volcanic soils of the east side of the Cascades in California. There are 10 species of cypress in California, and all of them have restricted ranges, usually in association with specialized soils such as serpentine. Baker Cypress grows primarily on volcanic soils, but it also grows in the Siskiyou Mountains on serpentine soils. Distribution of the species is patchy, with groves separated by many miles. It grows at

FIGURE 6.13 Western Juniper, *Juniperus occidentalis*, mixed with Great Basin Sagebrush, *Artemisia tridentata*. A community typical of the east side of the Cascade Range.

only eight locations in California and one in Oregon, 25 mi (40 km) south of Crater Lake.

Baker Cypress grows straight and tall. It is an attractive tree, often having cherry red bark, particularly on young specimens. This is a unique species of cypress by virtue of its tolerance for snow and cold. Throughout most of its range, it is associated with Ponderosa Pine, Knobcone Pine, Incense Cedar, Western Juniper, Oregon Oak, and California Black Oak. In Plumas County, however, among Red Firs of the northern Sierra Nevada, Baker Cypress grows at an elevation of 7200 ft (2300 m). This is the highest elevation at which cypresses are known to grow in California.

At higher elevations, zonation of the Cascade Province is typically Sierran. Volcanic soils are coarse, so trees are well spaced. A variety of Red Fir, known as Shasta Red Fir, *Abies magnifica var. shastensis*, occurs here, primarily on the west side. This variety is now known to be a hybrid of Red Fir and Noble Fir, *A. procera*. Recent studies in the Pacific Northwest are showing significant mortality in Shasta and Red Fir, presumably in association with decreased snowpack and warmer temperatures that increase the numbers of native pathogens. The subalpine and alpine communities on the peaks are discussed in Chapter 5, on mountaintops.

ANIMALS OF THE NORTHWEST FORESTS

Animals of Oak Woodland, Chaparral, and Yellow Pine Forest are similar throughout the state, but in the moist forests along the coast are a number of animals unique to California. Mostly, these are animals that reach the southernmost portions of their range here, and many of them enjoy widespread distribution along the coast farther north. However, conspicuous species are not numerous.

In an environment as productive as these moist forests, a considerable biomass of animal life can be sustained. One of the reasons that the forest seems to lack animal life is that there is a large community of detritus feeders and decomposers. There are fungi, slime molds, termites, and many kinds of insects. Banana Slugs, *Ariolimax* spp. (figure 7.24), are conspicuous detritus feeders. These large, shell-less mollusks, up to 6 in (16 cm) in length, creep about on the forest floor consuming litter and debris. Banana slugs will be discussed in more detail in the next chapter.

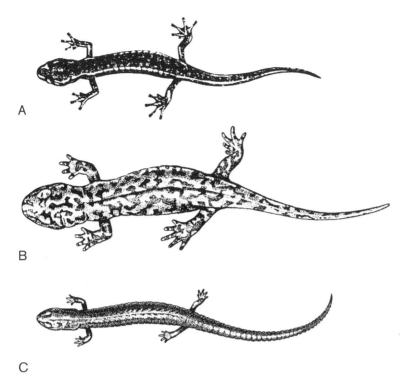

A

B

C

FIGURE 6.14 Some salamanders of the Pacific Northwest. A. Clouded Salamander, *Aneides ferreus*. B. California Giant Salamander, *Dicamptodon ensatus*. C. Del Norte Salamander, *Plethodon elongatus* (from Stebbins, R. C. 1951. *Amphibians of Western North America*. Berkeley: University of California Press).

Salamanders

The Pacific Northwest is the land of the salamander. Here in the northern forests there are at least 15 species, most of which reach the southerly limits of their distribution in the area.

Some of the salamanders are discussed in other chapters. The rassenkreis, or circle of races, of *Ensatina* (figure 8.32) will be discussed in the chapter on Cismontane Southern California. In the northern part of the state, the spotted Sierra form, *E. eschscholtzii platensis* (figure 4.34), merges with a mostly solid-colored coastal form, *E. e. oregonensis*. Furthermore, a dwarfed form known as the Painted Ensatina, *E. e. picta* (figure 4.34), from the northern Coast Ranges, reaches its northernmost distribution here. Intergradation between the different subspecies occurs in the regions where the ranges overlap. In the Redwood For-

est, along the coast, two species of newt (*Taricha* spp.) overlap their ranges (figure 7.26).

Among the lungless salamanders (Plethodontidae), three species of climbing salamanders (*Aneides* spp.) overlap their ranges in these northern forests. The Arboreal Salamander, *A. lugubris*, reaches its northern distribution in Humboldt County and is associated primarily with oaks. The Black Salamander, *A. flavipunctatus*, ranges from Santa Cruz County to the Klamath Mountains, occurring mostly in rock talus along streams. The Clouded Salamander, *A. ferreus* (figure 6.14A), the most arboreal of the three, is associated with Coast Redwoods, Douglas-fir, and Port Orford Cedar from Mendocino County to the Columbia River.

Other lungless salamanders are the Woodland Salamanders (*Plethodon* spp.). These salamanders occur only in the extreme northwestern corner of the state. The Del Norte Salamander,

P. elongatus (figure 6.14C), occurs in wet, mossy rock slides and talus from Humboldt County to just north of the Oregon border. A close relative of the Del Norte Salamander is the Siskiyou Mountains Salamander, *P. stormi*. An endangered species, it occurs only in the Rogue River area of the Siskiyou Mountains. Its range barely extends into California just south of the Oregon border. Dunn's Salamander, *P. dunni*, occurs together with the Del Norte Salamander in talus along the Rogue River. It has not been collected in California, but its range occurs so near to the border that it probably will be collected here sooner or later.

The Shasta Salamander, *Hydromantes shastae*, is a cave-dwelling species that is endemic to limestone. This species is also endangered. Its range is restricted to the McCloud River area of Brock Mountain, north of Shasta Reservoir. Closely related to the Mount Lyell Salamander, its distribution is discussed in Chapter 5 (figure 5.26).

The mole salamanders (Ambystomatidae) differ from the lungless salamanders in that adults are terrestrial and larvae are aquatic. In this respect, they are similar to newts. Unlike newts, the adults are seldom seen, spending most of their lives underground. Also unlike newts, they seldom venture far from water.

Of special interest among the mole salamanders is the California Giant Salamander, *Dicamptodon ensatus* (figure 6.14B). Adults of this species are known to grow up to 12 in (30 cm) in length, but may remain in their larval form with external gills all of their lives. The adults that become terrestrial are among the largest terrestrial salamanders in the world. Adults are rarely seen, but they are distinctive, with a light brownish color and a marbling of brown to black blotches on their backs. More of a Coast Range form, they range from Santa Cruz County to Mendocino County. In California, they are associated with streams in the Redwood country. They are known to burrow in stream gravels, particularly at the heads of streams. They are voracious feeders eating everything from the larvae of aquatic insects to fish, frogs, and other salamanders. When molested, they make a noise that has been described variously as a scream or a bark. This is the reason they are sometimes called barking salamanders. A smaller form that is more common in Oregon and Washington is the Coastal Giant Salamander, *D. tenebrosus*. Its range is north of the California giant Salamander where it occurs along the coast and inland to the vicinity of Shasta Lake.

Two other mole salamanders are closely related. The Northwestern Salamander, *Ambystoma gracile*, occurs in wet forests from southeast Alaska to Sonoma County. Adults are a uniform brownish to black, but they are seldom seen. They breed in quiet water. The Long-toed Salamander, *A. macrodactylum*, occurs in the Cascades, the eastern Klamath Mountains, and the northwestern parts of the Sierra Nevada. It ranges up to about 9000 ft (2900 m) in elevation, almost always near lakes. Sometimes adults are found under rocks and logs at the edges of lakes. It differs in appearance from the Northwestern Salamander by having an undulating light stripe down its back.

The Southern Torrent Salamander, *Rhyacotriton variegatus*, is small species associated with rapidly flowing mountain streams in the coastal forest from Point Arena in Mendocino County to Oregon. It ranges inland to the upper reaches of the McCloud River in Siskiyou County. It may be found in the water or under rocks on the water's edge. It is chocolate brown with black flecks on its back.

In similar habitats with the Southern Torrent Salamander can be found the Coastal Tailed Frog, *Ascaphus truei*, which isn't tailed at all. This little frog which resembles a Northern Pacific Treefrog or Chorus Frog, *Pseudacris* (= *Hyla*) *regilla*, has a taillike copulatory organ with a vent opening at its tip. It ranges through Mixed Evergreen Forest on the coastal side of the Cascades from British Colombia to Anchor Bay in Mendocino County. Its range extends eastward from the coast in the vicinity of Big Bend in Shasta County. A rocky streambed is important for larvae and eggs to attach. Tadpoles have a large sucker-like mouth, which they use to scrape algae and cling to rocks in rapids and waterfalls. Interestingly, this frog is the only member of its family (Hylidae) outside of New Zealand, where its relatives lack the "tail."

Birds

One would expect birds to be abundant in a forest. Most of the birds that are present in the northwest are similar to those that occur throughout the forests of California, but they are not particularly abundant. Most authorities recognize only seven kinds of birds nesting in the coastal coniferous forests of northern California. Two of them are hummingbirds—the Rufous, *Selasphorus rufus*, and Allen's, *S. sasin*—both of which are also found in the Sierra Nevada. These two species are so similar in appearance that it is difficult to tell them apart. These are among the most beautiful of California's hummingbirds. They are mostly bronze with a green back.

Two grouses occur in the area, but only the Sooty Grouse, *Dendragapus fuliginosus* (figure 4.45), nests here. This species is discussed in the chapter on the Sierra Nevada. The other grouse is the Ruffed Grouse, *Bonasa umbellus*. It is a brownish bird and is more pheasant-like than the Sooty Grouse. It reaches its southernmost distribution in the Klamath Mountains.

Among gleaners, there are creepers, nuthatches, and warblers. The only nesting species is the Chestnut-backed Chickadee, *Poecile (= Parus) rufescens*. This chickadee lacks the lyric call of its relative, the Mountain Chickadee, and differs in appearance by having a brownish-colored back.

A conspicuous bird of the forest is the Varied Thrush, *Ixoreus naevius*. This bird is similar in size and appearance to the American Robin, also in the Thrush family (Turdidae). The Varied Thrush has a black band across its rusty-colored chest. It nests in the forests of the Pacific Northwest and winters southward in the Coast Ranges or out in the Great Basin. Similar to the Hermit Thrush, its clear whistle is a sound commonly associated with the forest.

Swallows and swifts are acrobatic birds that catch insects on the wing. Vaux's Swift, *Chaetura vauxi*, is a small dingy-looking swift. It looks like a swallow with no tail. The Black Swift, *Cypseloides niger*, is the largest of this group, about 8 in (12 cm) in length. It is not a common bird in California, but one of the few places where it may be seen is over the cliffs in McArthur-Burney Falls State Park. Black Swifts make nests on cliffs by packing algae and moss into moist cracks. They often are solitary and may fly along with other species of swifts and swallows. One biologist described the flight of the Black Swift as looking like a great winged spider gone mad.

Large birds that are considered rare may occur here. During salmon or steelhead runs, Bald Eagles, *Haliaeetus leucocephalus* (figure 6.15), may be observed fishing in the larger rivers. The largest winter concentration of Bald Eagles in the lower 48 states occurs in the Klamath Basin. Approximately 500 may occur at one time in the Tule Lake National Wildlife Refuge, just south of the Oregon border. The Bald Eagle formerly occurred over much of the state, but because of hunting and widespread use of DDT (dichlorodiphenyltrichloroethane), the number of nesting pairs by the late 1960s was reduced to 30. Protective measures were initiated in association with the Endangered Species Act. The Bald Eagle was added to the federal endangered list in 1967 and the state list in 1971, and the use of DDT was discontinued in 1972. Since that time, the number of Bald Eagles has increased markedly. As of 1989, there were 83 known pairs of nesting birds in the state, and by 2015 the number of winter visitors had risen to over 1000. Most of the overwintering birds were in the Klamath Basin of northern California, but about 30 made it as far south as Big Bear Lake in the San Bernardino Mountains, Lake Mathews in Riverside County, and Lake Henshaw in San Diego County. Successful reintroduction of Bald

FIGURE 6.15 Bald Eagle, *Haliaeetus leucocephalus.*

FIGURE 6.16 Spotted Owl, *Strix occidentalis* (photo by William B. Cella).

Eagles to Santa Catalina and Santa Cruz Islands has bolstered the southern California populations. The Bald Eagle was removed from the list of endangered species in 2004.

Pileated Woodpeckers, *Dryocopus pileatus* (figure 4.37), also reach their southern, coastal distribution limits in the Klamath Mountains, although they extend southward in the Sierra Nevada at least to Yosemite Valley. The secretive Spotted Owl, *Strix occidentalis* (figure 6.16), might also be observed in its daytime haunt, resting on a branch close to the trunk, where it is well camouflaged. Spotted Owls have been particular victims of habitat destruction. As inhabitants of old-growth forests, they have become a symbol of environmentalists versus loggers. They are threatened by clear-cut logging, and salvage logging after fires. Furthermore, in the summer of 1990 the Northern Spotted Owl (*S. o. caurina*) was officially listed by the federal government as threatened, promoting the appearance of bumper stickers claiming, "Spotted Owl Tastes Like Chicken." Range of the Northern Spotted Owl extends from British Columbia to Marin County. The California Spotted Owl (*S. o.*

occidentalis) ranges from the southern Cascades down the western side of the Sierra Nevada to Fresno County. It also occurs at various localities in the Coast Ranges and southern California to Baja California. Where the ranges of the two subspecies overlap, hybridization may occur. The California Spotted Owl is considered a "Species of Special Concern" by the state, and it may become listed by the federal government as threatened.

A coincident threat to Spotted Owls has come with an invasion from the east of Barred Owls, *S. varia.* At first, Barred Owls seemed to prefer logged areas, but they rapidly moved into old-growth forests where aggression, competition, and hybridization threatened the existence of Spotted Owls. Numbers of Spotted Owls continued declining, which in 2013 spurred trained hunters from the US Fish and Wildlife Service to begin killing Barred Owls in a study area on the Hoopa reservation in Humboldt County. Future activity and success of this program remains to be seen.

One of the strangest birds that a person can expect to find in moist coastal forests is a sea-going bird known as the Marbled Murrelet, *Brachyramphus marmoratus*. This species, which spends most of its life out on the Pacific Ocean, nests in the tops of tall trees up to several miles from the sea. This is a robin-sized, stout-bodied bird with short wings. It is quite agile in flight, but it apparently experiences some difficulty getting airborne from a perch. When it takes off, it is said to require the height of a tall tree to build enough momentum for flight. The first Marbled Murrelet nest discovered in California was found in 1973 in the Santa Cruz Mountains. It was 135 ft (42 m) up in a large Douglas-fir. As another symbol of old-growth forests, this bird was listed as threatened in 1992 by the US Fish and Wildlife Service.

Mammals

Among the mammals of the coastal forest, squirrels are the most conspicuous. Three species overlap their ranges here. The Northern Flying Squirrel, *Glaucomys sabrinus*, is the nocturnal member of the group. The Western Gray Squirrel, *Sciurus griseus* (plate 18B), occurs where there are acorns, and the Chickaree or Douglas's Squirrel, *Tamiasciurus douglasii* (plate 18C), feeds mostly on cones of various conifers. There are three species of chipmunk. The Sonoma Chipmunk, *Neotamias (= Tamias) sonomae*, occurs in open places in the forest and in chaparral. The chipmunk of the forest floor is Townsend's Chipmunk, *T. (= N.) townsendii*. This is a wide-ranging species of northern forests that occurs as far south as Marin County and the central Sierra Nevada. This is the darkest species of chipmunk and is often used to illustrate Gloger's rule, that dark coloration is associated with moist habitats. One authority proposed to split this form into four separate species based on anatomy of the penis bone (baculum). The Siskiyou Chipmunk, *N. (= T.) siskiyou*, is one such species that occurs from the Klamath River north to southern Oregon.

Another ecological principle illustrated by rodents of the coastal forest is parallel evolu-

FIGURE 6.17 California Vole, *Microtus californicus*.

tion. Mice of the subfamily Microtinae are known as microtine mice. They look very much alike, and it often requires inspection of teeth and skeleton to tell them apart. Included here are mice from three genera. They all have relatively small ears that may be nearly covered by long hair and they also tend to have short tails. They are separated by habitat and behavior, however. Meadow mice or Voles (*Microtus* spp.; figure 6.17) often occur in meadows and along streams. There are at least four species that may occur in these forests. Good swimmers, they often have been observed to swim underwater with their front legs folded back, propelling themselves with their hind feet. Many of them make tunnels through the grass, which they use to conceal themselves from predators.

There are two species of tree voles. The White-footed Vole, *Arborimus albipes*, lives on the ground and in trees. It ranges from northern California into Oregon. It is distinguished from other microtine mice by being the only species with white feet and a nearly hairless tail. It is a rare mouse, and very little is known about its life history. The Red Tree Vole, *A. longicaudus*, can be distinguished from the others by its reddish color and long tail, which is usually about 65% of its total length. These are truly mice of the moist forests west of the Cascades. They live high in the trees, seldom coming to the ground. Their diet consists of needles of fir, spruce, and hemlock. They make large nests of the inedible parts of the needles. Of particular interest to physiologists are the rather long gestation periods and small litters of these mice. There are usually only one to three offspring per litter, and the mother carries them up to 48 d. This is a low reproductive capacity for a mouse. It may be related to the low-calorie diet of fir needles, or it may mean that these rodents are so well adapted to their environment that they need not produce many young. An investigation has discovered that Red Tree Voles south of the Smith River have a different chromosome number than those north of the river. Classification is still in flux, and the two forms may be assigned separate species names.

The Western Red-backed Vole, *Myodes (= Clethrionomys) californicus*, is another mouse of the deep forest. It usually lives on the forest floor, making a nest out of lichens under logs, but sometimes it occurs in wet meadows. These mice can be distinguished from Red Tree Voles by having a shorter, bicolored tail and grayish underparts. The back is not as rusty in appearance as that of the Red Tree Vole; it is more of a chestnut brown. These mice feed mostly on leaves and other green vegetation.

A unique rodent of moist northwestern forest is the Mountain Beaver, *Aplodontia rufa*. Also known as a Boomer, Ground Bear, or giant Mole, this large rodent is not really a mountain dweller, nor is it a beaver. It is a primitive squirrel-like creature that looks like a large gopher. Mountain Beavers grow up to 14 in (35 cm) in length and have large claws, small ears, and a stubby tail. They spend most of the time underground in burrows that they may share with California Giant Salamanders. When they emerge from their burrow, it usually is at night. They make their burrows in thickets of wild Blackberries, Salmonberries, and Thimbleberries, on which they feed. They also feed on a wide variety of green vegetation. During winter, at higher elevations, they burrow under the snow in a manner similar to a gopher. They often strip the bark from young White Firs during winter. There is only one species in the

FIGURE 6.18 North American Porcupine, *Erethizon dorsatum* (photo by Sreven D. Cain).

Mountain Beaver family (Aplodontidae). Its distribution is limited entirely to the forests of the Pacific Northwest and the northern Sierra Nevada.

Of the larger mammals, the most conspicuous is the North American Porcupine, *Erethizon dorsatum* (figure 6.18). North American Porcupines are seldom reported in humid coastal forests, but east of the crest of the Klamath Mountains and throughout the Cascades, they are commonly observed as road kills. North American Porcupines are mostly nocturnal, but they are occasionally encountered during daylight hours sauntering along the trail, humming to themselves. During the daytime, they usually conceal themselves in the foliage of a tree. In the summer, they eat leaves and herbs, or they may climb into treetops and eat buds, an activity that may cause forking of treetops. During winter, they primarily gnaw on the bark of young trees, and it is this habit that makes them the enemy of loggers. Actually, they may provide a beneficial thinning effect, because most of the trees they damage are saplings that do not grow well under the forest canopy. Their habit of erecting their spines and slapping with their spiny tail when molested protects them from most predators, although American Martens have been reported to burrow under snow and attack North American Porcupines from beneath.

Another nocturnal mammal that is sometimes conspicuous is the Northern Raccoon, *Procyon lotor*. This animal is widespread throughout northern forests, usually occurring along watercourses and lakes, where it feeds on nearly everything edible. The reason they are conspicuous is that they are apparently very curious. They also are clever and brazen. They have learned that campers are careless with food and that for a simple glimpse of a "wild" animal, humans will leave tempting morsels of food in easy-to-reach places. Many campers in public campgrounds are delighted by the visits of Northern Raccoons to the campfire, and dismayed when they raid the larder. They occur statewide and have become urbanized in many regions where they eat pet food, garbage, and dig up suburban gardens in search of grubs. The reason that raccoons appear to wash their food before they eat it has never been resolved. Perhaps it makes the food easier to swallow.

Of the predatory mammals in the northwest forests, most important are members of the weasel family (Mustelidae). Every Pacific representative of this family can be found somewhere in the northwestern mountains, each in a different habitat. This list includes the Long-tailed Weasel, *Mustela frenata* (figure 5.31), American Marten, *Martes americana*, Fisher, *M. (= Pekania) pennanti*, Ermine, *M. erminea*, American Mink, *Vison (= Mustela) vison*, Wolverine, *Gulo gulo*, American Badger, *Taxidea taxus*, Western Spotted Skunk, *Spilogale gracilis*, Striped Skunk, *Mephitis mephitis*, and Northern River Otter, *Lontra canadensis*.

These mammals have a number of features in common, including highly developed scent glands, long bodies with short legs, and fine-quality fur. The rarest of these mammals are the Fisher and the Wolverine. The Fisher is now rare because its prime pelt has commanded high prices, and it has been hunted to the brink of extinction. Logging has also seriously reduced its available habitat. Fishers still live in the Trinity Mountains, Siskiyou Mountains, and inaccessible parts of the Sierra Nevada. They are probably misnamed, for they principally eat small mammals and birds.

The Wolverine was never common, and now it is listed as an endangered species. The Wolverine is largely a subalpine mammal. It is the largest of the terrestrial mustelids, weighting up to 35 lb (16 kg). Their body form resembles a large, dark American Badger. Instead of having white lines on the face, as an American Badger does, it has a broad gray forehead. Its food consists of a variety of small mammals, including North American Porcupines, and sometimes it feasts on the remains of large animals. No one knows why the species is disappearing. Because it was always scarce, it was not the object of serious hunting. Its fur, however, is reputed to be of high quality. Apparently, Wolverine fur has been used to line hoods of parkas because frosty breath will not adhere to it. Wolverines today are reported occasionally at timberline in the Klamath Mountains and the central Sierra Nevada.

The Northern River Otter is another protected mammal. Its short, thick fur was highly prized by trappers and was used for ceremonial purposes by Native Americans in northwest California. They still occur in small numbers in larger mountain streams and lakes in the Klamath Mountains, northern Coast Ranges, the Sacramento-San Joaquin Delta, and the northern Sierra Nevada. Interestingly they apparently have followed the large reservoirs of the Stanislaus drainage up and over the summit of the Sierra and now occur on the east side of the Sierra in the Carson River in Nevada. They do not spend a considerable time on land, but they live in burrows of other mammals some distance from water. One of their interesting

FIGURE 6.19 Roosevelt Elk, *Cervus canadensis roosevelti (= C. elaphus roosevelti)*, Prairie Creek Redwoods State Park.

habits is to build a slide, which may be 25 ft (8 m) long, on the bank of a stream or lake. They repeatedly climb to the top and slide into the water, as if playing. Their food is primarily aquatic, consisting of fish, frogs, turtles, crayfish, and aquatic insects.

The Gray Wolf, *Canis lupus*, is a remarkable story of the reappearance of an extirpated animal. It had been nearly 100 years since wild wolves lived in California. Driven out by a federal bounty on large predators, the last time a wolf was observed in California was 1924. Hopes for a recovery arose in December 2011 when a lone Gray Wolf wearing a radio collar wandered into northern California from Oregon. However, it was not long before it wandered back. Yet, in the aftermath the state voted to list the Gray Wolf as an endangered species. The federal listing covering the entire United States has been controversial. Since the recovery of wolf populations in Yellowstone National Park, there have been efforts to delist the species. Nevertheless, in California, in May and June 2015, by camera traps set up by the California Department of Fish and Wildlife, a single wolf was photographed, which prompted the department to set up more cameras. Then

in August 2015, a mated pair and five cubs were photographed on public land in southern Siskiyou County, raising hopes among the environmental community that wild Gray Wolves may have returned to the state.

The largest mammal of this region is the Roosevelt Elk, *Cervus canadensis roosevelti (= C. elaphis roosevelti;* figure 6.19). Not to be confused with the somewhat smaller Rocky Mountain Elk, *C. c. nelsoni (= C. e. nelsoni)*, which occurs near Shasta Lake in a small part of its former range in northeastern California, Roosevelt Elk is the subspecies that occurs in moist forests along the coast, as far north as Vancouver Island, British Columbia. In Europe, members of this species are called Red Deer. Usually they occur in herds in the vicinity of meadows. A herd is generally composed of a single dominant bull and many cows. Sometimes young bulls may remain with the herd.

Mature Roosevelt Elk bulls, usually at least 5 years of age, engage in fierce fights to establish dominance. These fights occur in autumn, the rutting season. Bulls sound a challenge by bugling. If the challenge is accepted, two bulls will charge each other head on. They may lock antlers, or they may charge several times. The winner is the stronger, the one that is able to push an intruder away from his territory.

One of the marvels of these fights is that bulls rarely hurt each other. Innate signals of aggression and appeasement ensure that these fights remain rituals and stop short of death or serious harm. It is not necessary to kill the loser in a fight to keep him from mating. Usually, the dominant bull succeeds in keeping his herd of females together and chasing away subdominant males. However, subdominant males occasionally sneak into a herd and successfully copulate.

Food of these elks includes grasses and leaves of trees. They feed a great deal, for an elk may weigh 1000 lb (450 kg). They are not easily confused with deer, which are much smaller. Antlers of a bull elk also have a different branching pattern than those of a deer. The only native deer in California, the Mule Deer, *Odocoileus hemionus*, has antlers that fork in pairs. Elk antlers have one main beam with all of the branches aiming forward. Furthermore, the subspecies of deer, *O. h. columbiana*, that occurs in the northern Coast Ranges and Klamath Province is smaller than other deer in California. This subspecies, known as the Black-tailed Deer, ranges along the coast from California to Canada.

Recovery of the Roosevelt Elk population is a conservation success story in California. Habitat destruction might have reduced the population to a few pairs, but now they are thousands, although in California, for inexplicable reasons the herds in recent years seem to be getting smaller. They are fully protected in essentially a pasture situation in Redwood National Park and Prairie Creek Redwoods State Park, where there are five different herds, of up to 30 individuals. They can be viewed by nearly everyone in a large meadow next to Highway 101 in Prairie Creek Redwoods State Park. This herd is so tame that thousands of people every year watch and photograph these stately animals.

7

Coast Ranges

FIGURE 7.1 Aerial view of northern Coast Ranges near Elk, Mendocino County, 15 mi (24 km) north of Point Arena. Note coastal terraces, north-south trending ridges and valleys, and vegetation mosaic with bald hilltops (from Bailey, E. H., W. P. Irwin, and D. L. Jones. 1964. *Franciscan and Related Rocks, and Their Significance in the Geology of Western California*. Bulletin 183. California Division of Mines and Geology).

CALIFORNIA'S COAST RANGES extend along the coast for about two-thirds of the length of the state. They run about 550 mi (880 km) from the South Fork Mountains of the Klamath region southward to the Santa Barbara area. Here they meet the Santa Ynez Mountains of the Transverse Ranges. The Coast Ranges rise abruptly from the sea to nearly 6000 ft (2000 m) in elevation. On the east, they dip beneath the sediments of the Great Central Valley. In general, they consist of a series of northwest-to-southeast-trending ridges and valleys associated with faulting and folding. Many of the rivers of the Coast Ranges flow northward to empty into the Pacific Ocean many miles from their source. In the northern Coast Ranges, the Eel River dumps into the ocean south of Eureka, approximately 150 mi (240 km) north of its source. In the southern Coast Ranges, the Salinas River does likewise to empty into Monterey Bay. This river follows the long valley made famous by the writings of John Steinbeck. A major river of the northern Coast Ranges is the Russian River. It flows for about 40 mi (64 km) southward along the structural grain of the Coast Ranges. Near Healdsburg, however, it turns abruptly westward to cut through the range in a gorge nearly 1000 ft (300 m) deep, exiting on the coast near Jenner. This east-west orientation permits marine air to flow a considerable distance inland along the Russian River.

A significant portion of the California coastline has been set aside as public land in order to protect its resources and scenic beauty. The California Costal National Monument mostly is associated with offshore rocks, islands, and reefs that are exposed above mean high tide out to a distance of 12 nautical miles, but it also includes a mainland component south of Eureka in Mendocino County, which is known as the Point Arena-Stornetta Unit. Other notable federal land includes Redwood National Park, the Point Reyes National Seashore, the King Range National Conservation Area of the Lost Coast, and the Santa Lucia Mountains of the Los Padres National Forest. Numerous state

parks occur along the coast as well. Notable among them are the redwood state parks, the Sonoma Coast State Park north of Bodega Head, and the Sinkyone Wilderness State Park at the southern end of the Lost Coast in northern Mendocino County.

An interesting feature of the coastal side of these ranges is the presence of coastal terraces, old wave-cut benches that have been elevated to their present positions by rising of the land. A combination of variable sea level and tectonic forces during the Pleistocene has produced the steplike appearance of the land. Seventeen such terraces are visible in the Ventura area. Vegetation on these terraces varies in accordance with age and differences in climate.

On the coastal side of the ranges, the climate is heavily influenced by the presence of comparatively cold water offshore. As a result, heavy fog often cloaks the coastal slopes. An interesting assemblage of plants that water themselves with fog drip is characteristic of this region. To the north, particularly on the second terrace above the sea, the towering Coast Redwoods depend on this fog drip. Unlike Sierra Redwoods, Coast Redwoods are an important lumber crop. Only about 5% of the original Coast Redwoods remain, but fortunately, large tracts of these iconic trees have been preserved within a series of state parks and a national park that are showcases of natural beauty in northern California (figure 7.17).

The northern and southern Coast Ranges are separated by San Francisco Bay. In some of the classifications of California's bioregions or ecoregions, the northern and southern Coast Ranges are split by a region called Bay/Delta. That region will be discussed in Chapter 11 along with estuaries.

The northern Coast Ranges (figure 7.2A) are generally higher than the southern Coast Ranges (figure 7.2B). Just south of the Klamath Province, in the southwest corner of Trinity County, Solomon Peak reaches 7581 ft (2312 m). The longest primitive stretch of coastline in the state is found along the northern Coast Range about 70 mi (110 km) south of Eureka. This

23 mi (37 km) stretch of beach is south of the Mattole River along the King Mountain Range. In this area, the crest rises over 4000 ft (1200 m) in view of the ocean. Highway 1 cuts inland to avoid the range, which has left this "lost coast" largely unvisited and unexploited (plate [10A]). A single paved road to Shelter Cove penetrates this remote area, which is administered largely by the Bureau of Land Management. Most of the area is protected as the Sinkyone Wilderness State Park and the King Range National Conservation Area.

In the south Coast Ranges (figure 7.2B), the highest peaks are at the southern end, where the general orientation of the mountains is east-west, similar to the Transverse Ranges. North of Santa Barbara, Big Pine Mountain, at 6828 ft (2083 m), is the highest in the San Rafael Range. In the southeastern corner of the Coast Ranges, in a terrane that may be more appropriately a part of the Transverse Ranges, Mount Pinos reaches 8826 ft (2758 m). Mount Pinos lies just west of the juncture of the San Andreas and Garlock faults, about 40 mi (64 km) inland.

The beauty of the southern coast has been celebrated by poets, photographers, and naturalists. Robinson Jeffers, who lived at Big Sur, called the area "the greatest meeting of land and water in the world." Another Big Sur resident, novelist Henry Miller, declared, "This is the face of the earth as the Creator intended it to look." Highway 1, constructed with convict labor between 1924 and 1937, runs 92 mi (147 km) from Cambria to Carmel. It is considered by many to be one of the most scenic highways in the world.

The crown jewel of the southern Coast Ranges is the Santa Lucia Range, which rises abruptly from the Big Sur coast to a height of 5868 ft (1893 m) at Junipero Serra Peak. In this area, there are at least 57 endemic plant species and over 220 total species, including Coast Redwood, which reaches its southernmost distribution here. Average precipitation at Big Sur is nearly 60 in (150 cm) a year, but at higher elevations the annual total can exceed 100 in

(250 cm). The area's most impressive figure, however, is the highest official annual precipitation ever recorded in California, 161 in (403 cm), which fell during the 1940–1941 rainy season.

CLIMATE

On the coastal side of the Coast Ranges, maritime climate predominates. The temperature of the ocean along the California coast is generally cold because the water flows southward from Alaska and heats slowly. The cold temperature causes condensation of water vapor in the air near the ocean surface. The result is frequent fog along the coast, which is held there by the Coast Ranges. This fog is very significant to the vegetation because it reduces water lost by transpiration, and it contributes significantly to soil moisture through fog drip. It has been determined that fog drip in the Berkeley Hills accounts for an equivalent of 10 in (25 cm) of precipitation. In the northern Coast Ranges, the fog drip is responsible for the belt of Coast Redwoods, which to many people typifies northern California.

To the interior of the Coast Ranges, drying occurs because of the rain-shadow effect. At its southern end, the rain shadow is so effective that the vicinity of the Great Central Valley near Taft is truly a desert. Much of the Coast Ranges vegetation is Oak Woodland or Foothill Woodland, similar to that of the western Sierra Nevada. In the southern Coast Ranges, however, Coastal Sage Scrub and lower Chaparral are important, particularly on south-facing slopes. Slope effect on the south coast is as distinct as will be seen anywhere in the world. Accentuated by Mediterranean climate, south-facing slopes suffer the extremes of drought. These slopes are covered with Chaparral that appears dry and brown by midsummer. In contrast, north-facing slopes are cloaked with evergreen oaks whose roots reach deeply to permanent water. The patterns in color form a visual contrast that bears witness to the profound influence of slope exposure.

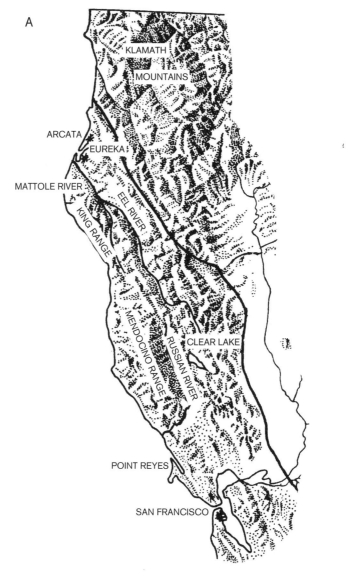

FIGURE 7.2 A. The Northern Coast Ranges.

GEOLOGY

The Coast Ranges are folded and faulted on an axis that parallels the coastline. Formation of these ranges is generally attributed to events associated with subduction of the Pacific plate beneath the western border of North America. To the north, marine sedimentary and meta-sedimentary rocks have been uplifted by action of the Mendocino Triple Junction. Island-arc volcanics and materials deposited in a deep submarine trench apparently became accreted to the western edge of California as the ocean floor slid beneath the continental margin. Most of these rocks are late Mesozoic in age (150–65 million years old). Eastward-directed pressure folded and cracked the leading edge of the continent to form the parallel series of ridges and valleys that make up the present ranges (figure 7.1).

Long-distance northward transport of rocks in Alaska indicates that a transform fault sys-

B

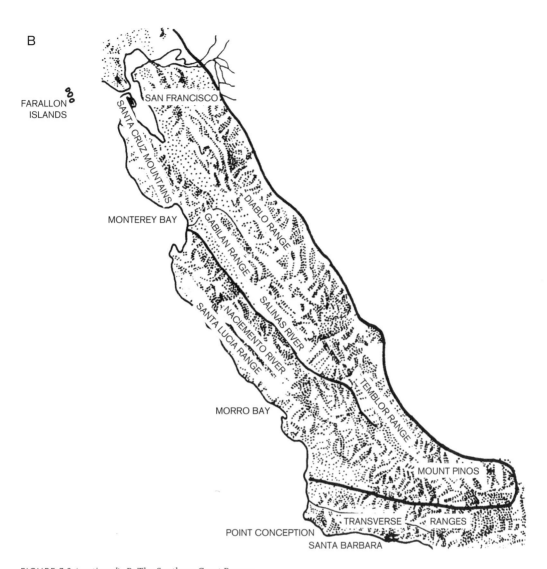

FARALLON ISLANDS

SAN FRANCISCO

SANTA CRUZ MOUNTAINS

MONTEREY BAY

DIABLO RANGE

GABILAN RANGE

SALINAS RIVER

SANTA LUCIA RANGE

NACIEMENTO RIVER

MORRO BAY

TEMBLOR RANGE

MOUNT PINOS

TRANSVERSE   RANGES

POINT CONCEPTION

SANTA BARBARA

FIGURE 7.2 (*continued*) B. The Southern Coast Ranges.

tem must have existed along the western margin of North America until about 100 million years ago. Many of the rock units of the Coast Ranges had their origin much farther south than where they lie today. The southern Coast Range has a granitic core that extends northward to terminate at the Farallon Islands. Formed at the same time as the Sierra and Southern California batholiths, this granitic material originally lay in place south of the present Sierra Nevada. It has since been carried

to its present position by northward displacement along the San Andreas fault and is now bounded on the east by the San Andreas fault and on the west by the Nacimiento fault.

In the southern Coast Ranges, batholithic rocks (granitics) form a major component of what is known as the Salinian block. These plutonic rocks were probably formed by subduction south of the present Sierra Nevada and carried to their present position west of the San Andreas fault during the last 25 million years by right-lateral slip. Uplift and emplacement of the southern Coast Ranges in its present position took place very recently compared with the formation of the northern Coast Ranges.

North of Red Bluff, the boundary between the northern Coast Ranges and the Klamath Mountains generally lies along the South Fork Mountain fault. This is a thrust fault in which rocks of the Klamath Mountains, resembling those of the northern Sierra Nevada, have been thrust over the sedimentary rocks of the Franciscan Formation. Most of the northern Coast Ranges is an assemblage of Franciscan rocks that are late Mesozoic in age (figure 7.3). These rocks are associated with the offshore trench that was formed during subduction of the ocean floor and were named for deposits near the city of San Francisco. Franciscan rocks are dominated by shales and sandstones that were formed by rapid erosion of a volcanic highland and deposited in deep marine basins. They are gray green in color and known as graywackes. About 90% of Franciscan rocks are of this type. It has been estimated that these deposits are about 25,000 ft (7600 m) thick. Graywackes are interbedded with lesser amounts of dark-colored shale, occasional limestone, and reddish silica-rich cherts (figure 7.4). East of the Franciscan rocks is a belt of sedimentary rocks that formed along the continental shelf of the Mesozoic ocean. These rocks are known as the Great Valley sequence.

Franciscan sedimentary rocks have become intruded by igneous rocks that have been metamorphosed, probably by heated seawater, to become the waxy green rock known as serpentine or serpentinite. This attractive, hydrothermally altered rock has been designated the state rock. Mixed in with this assemblage are other igneous rocks, believed to be associated with ocean floor deposition, collectively known as ophiolites. These rocks are composed of a great variety of different minerals, and it is their weathering that has caused the enormous variety of soil types scattered like islands throughout the Coast Ranges. These soils, which may be called ophiolitic, serpentine, ultramafic, or ultrabasic, have in common an odd assortment of minerals. They are unusually rich in magnesium and iron, but they are deficient in calcium, sodium, and potassium. They also may contain large amounts of nickel, cobalt, and chromium. Many unique plants grow on the special soils, forming the edaphic ecologic islands that are numerous in the Coast Ranges and the Klamath Mountains. They also occur in the western foothills of the Sierra Nevada and sparsely in southern California. This unique combination of minerals degrades into a specialized impoverished soil. In turn, this soil supports a unique assortment of plant species. These *serpentine endemics* are an important part of California's species diversity.

North of San Francisco, in an area of 30 mi^2, is the world's largest geothermal field, which bears testimony to the fact that tectonic pressures still exist in the area. In an area along the Sonoma County line, numerous hot springs, or fumaroles, occur. The area is generally known as "the Geysers." The source of the heat is apparently a large magma pool over 4 mi (6.4 km) below the surface. The steam produced here naturally by the earth's heat comes from a greywacke sandstone reservoir and is used to generate electricity. The first power plant was opened in 1921. Today there is a complex of 22 power plants, drawing steam from more than 350 wells, that provides about 60% of the power used by coastal California between the Golden Gate and the Oregon border. By 1999, the natural source of steam became depleted. However, since 1997 the steam field has been recharged by the injection of treated sewerage effluent. The effluent comes from the Lake County Sanitation treatment plant and piped up to 50 mi (80 km) from its source. The project uses about 85% of the effluent produced in the area and helps to protect Clear Lake, which formerly received the treated effluent. A downside to the injection of waste water is that it has increased the number of magnitude 4.0 earthquakes in the region. Initially, an oil company owned the mineral rights to the steam and apparently the price of the steam was linked to the price of oil.

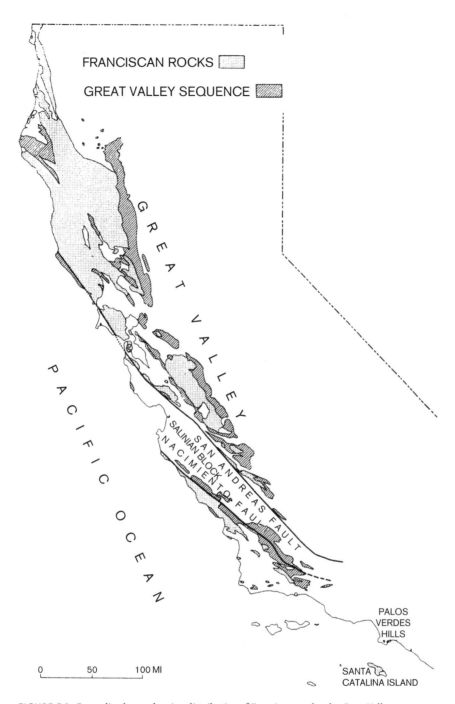

FRANCISCAN ROCKS

GREAT VALLEY SEQUENCE

GREAT VALLEY

PACIFIC OCEAN

SAN ANDREAS FAULT

SALINIAN BLOCK

NACIMIENTO FAULT

PALOS
VERDES
HILLS

SANTA
CATALINA ISLAND

0   50   100 MI

FIGURE 7.3 Generalized map showing distribution of Franciscan rocks, the Great Valley sequence, and the Salinian block (after Bailey, E. H., W. P. Irwin, and D. L. Jones. 1964. *Franciscan and Related Rocks, and Their Significance in the Geology of Western California*. Bulletin 183. California Division of Mines and Geology).

FIGURE 7.4 Franciscan rock, thinly bedded chert in the Coast Range northwest of the Golden Gate Bridge.

major faults on either side. If the Salinian block is aligned with the southern Sierra, the Nacimiento fault lines up south of the Mother Lode fault. This is the fault on the western side of the Sierra that separates the granitic rocks from the western metamorphic complex, which also includes Franciscan rocks. To the south, the Nacimiento fault more or less coincides with the Newport-Inglewood fault, which parallels the Los Angeles Basin along the coastline. The granitics of the Peninsular Ranges formerly were in line with the Salinian block at its southern end.

The Gabilan Range lies east of the Salinas Valley and west of the San Andreas Fault. The basement core of granitic rocks, 78–100 million years old, is overlain by a volcanic breccia that is about 23 million years old. This breccia is the basis for the spectacular cliffs, caves, and spires of Pinnacles National Park. First established as a National Monument in 1906 by Theodore Roosevelt, it was dedicated as a National Park in 2013. Composed mostly of rhyolite, these volcanic rocks are correlated with the Neenach volcanic rocks in the western Antelope Valley on the other side of the San Andreas Fault about 180 mi farther south.

The southern Coast Ranges are mainly composed of sedimentary rocks much younger than the Franciscan or granitic rocks. In early Miocene, over 20 million years ago, the sea covered much of the southern Coast Ranges, forming numerous bays, straits, islands, and inlets. In a setting similar to the modern Gulf of California, deep-water sediments such as the 700 ft thick diatomaceous shales near Lompoc were deposited in deep, elongate basins. These and other sediments began to be uplifted only in the last 10 million years or so, making the southern Coast Ranges quite young. By the end of the Pliocene, 1–2 million years ago, the southern Coast Ranges were above sea level. These sedimentary rocks are reasonably soft and unstable, and they are easily eroded by surf, particularly during heavy winter storms. A series of wet winters since 1978 has greatly exacerbated the problem. All along the coast, people have built

It seemed that if the price of steam were kept artificially high, geothermally generated electricity would not undermine the demand for fossil fuel.

In the southern Coast Ranges, Franciscan rocks lie east of the San Andreas fault. West of the San Andreas is a zone of granitics and metamorphics, known as the Salinian block (figures 7.3 and 7.5). Mount Pinos lies at the southern tip of this block. The Salinian block represents a rock assortment similar to the Sierra Nevada and Transverse Ranges. It was formed south of the Sierra Nevada and carried northwest about 190 mi (300 km) by motion on the San Andreas fault. West of the Salinian block is the Nacimiento fault, which separates the granitics from more Franciscan rocks to the west. In cross section, the southern Coast Ranges have a central core of granitic rocks separated from Franciscan rocks by

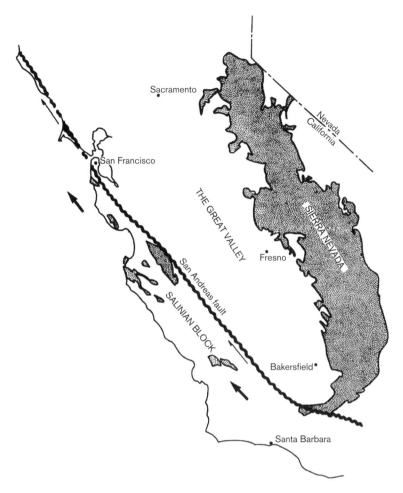

FIGURE 7.5 Granitics of the Salinian block, originally formed by subduction south of the Sierra Nevada. They were carried to their present position by northwest displacement along the San Andreas fault (from Alt and Hyndman 1975; reprinted with permission).

homes too close to the edge. In the Santa Cruz area, retreat of the cliffs away from the sea averages about 1 ft (30 cm) per year, and loss of 75 ft (24 m) has been recorded in a single year.

Furthermore, these sedimentary beds are not always horizontal; tectonic activity has tilted many of them. When these layers become wet, they slip on each other, causing landslides. The heavy rains of 1978 and 1983 have contributed to this wetting, but the condition is made even worse by homeowners on the cliffs who insist on planting landscapes that demand lots of water. Erosion by surf or landslides associated with overwatering have claimed many multimillion dollar homes on coastal cliffs.

Coastal terraces are formed by the action of surf on a rocky headland. Surf, with its load of sand, can be highly abrasive. The action of the surf is to cut at right angles into the rock along the shore. The outcome is a relatively flat, wave-cut bench or reef with a steep cliff on the shoreward side. Because the west coast is continually rising by tectonic activity, old wave-cut terraces have become elevated above present sea level. Superimposed on this uplift are intervals of changing sea level, which coincide with episodes of glaciation associated with the Ice Age (Pleistocene). During periods of glaciation, sea level was lowered as a great deal of the earth's water was bound up in polar ice caps and continental

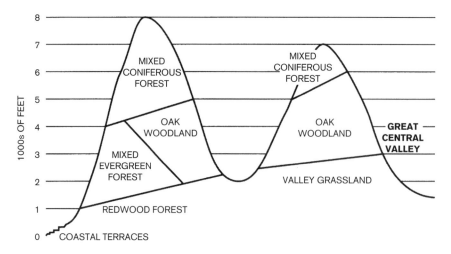

FIGURE 7.6 Biotic zonation of the northern Coast Ranges.

ice sheets. During interglacial intervals, sea level rose as the ice caps melted. Furthermore, expansion and contraction of the water itself is a temperature-related phenomenon. Warm water, like warm air, expands and takes up more space. Sea level today apparently is as high as it has been in the last 125 million years.

These flat terraces or platforms, rising like staircases on hills that border the ocean, are particularly conspicuous where they have been formed on rocky headlands. In the Palos Verdes area of Los Angeles County, 13 terraces are visible. In the Point Lorna area near San Diego, there are 11 terraces. Perhaps the most terraces, 17, are visible on the southern coast near Ventura. Up and down the Coast Ranges, the number of visible terraces varies, although 5–7 are typical (figure 7.1).

BIOTIC ZONATION

Oak Woodland

As illustrated in figure 7.6, throughout most of the Coast Ranges the predominant community is Oak Woodland (figure 7.7). The significance of oaks and pines to the diversity of species in California is discussed in Chapter 4. For all practical purposes, Oak Woodland in the Coast Ranges is the same community as Foothill Woodland of the Sierra Nevada. The community encircles the Great Central Valley. For the purposes of this book, the two terms will be considered equivalent. This is a mixed community of trees and grasses. Many authorities refer to it as a savannah. Dominant tree species include Blue Oak, *Quercus douglasii* (figure 4.20), and Gray Pine, *Pinus sabiniana* (plate 11B). At slightly higher elevations and on north-facing slopes, particularly in the northern Coast Ranges, California Buckeye, *Aesculus californica*, is locally common. Western Redbud, *Cercis occidentalis*, may be locally common in the northern Coast Ranges as well.

A community known by some authors as Northern Oak Woodland dominates ridgetops up to 5000 ft (1600 m) in the northern Coast Ranges. This community, dominated by Oregon Oak (*Q. garryana*) rather than Blue Oak, is also discussed in Chapter 6. In Humboldt and Mendocino Counties, grassy "balds" and open woodlands of Oregon Oak occur in patches on ridges in the Mixed Evergreen Forest. These bald hills occur in patterns caused by different soil types. This grass-tree mosaic reflects the soil mosaic associated with the diverse geologic nature of the region.

FIGURE 7.7 Oak Woodland of the northern Coast Ranges. Note Lace Lichen hanging from the tree in the left foreground.

Other oaks include Coast Live Oak, *Q. agrifolia*, Interior Live Oak, *Q. wislizeni*, and Valley Oak, *Q. lobata* (figure 10.10). To the south of the Coast Ranges, Blue Oak is ultimately replaced by Coast Live Oak. Where the two species occur together, in the southern Coast Ranges, Blue Oak tends to grow on south-facing slopes, and Coast Live Oak grows on north-facing slopes. Interior Live Oak and Valley Oak are more common toward the interior of the Coast Ranges where Oak Woodland grades into Valley Grassland. Valley Oak dominates in valleys and on gentle upper slopes with deep soils. These deep soils are the result of reduced erosion rates. One author describes these open woodlands of Valley Oak as montane savannahs that are ecologically equivalent to the open woodlands of Oregon Oak. Valley Oak grows as far south as the Tehachapi Mountains, where it occurs on gentle, sloping ridges up to 6000 ft (1800 m) in elevation.

Old gnarled oaks can be very picturesque. Large specimens are spectacular. The largest Blue Oak, located in Alameda County, is over 6 ft (2 m) in diameter and stands 94 ft (29.4 m) high. Its crown spread is 48 ft (15 m). The largest Coast Live Oak, found near Gilroy, is even more impressive. With a diameter of over 9 ft (3 m), it stands 85 ft (26.6 m) high and has a crown spread of a whopping 127 ft (40 m). The largest Valley Oak is found in Butte County in the Great Central Valley. It is also an impressive specimen, standing 120 ft (37.5 m) high with a crown spread of 103 ft (32 m). Its trunk is also nearly 9 ft (3 m) in diameter.

In canyons, particularly on north-facing slopes, there is Canyon Live Oak, *Q. chrysolepis*. The most widespread oak in California, Canyon Live Oak, resembles Coast Live Oak in size and shape but has small whitish hairs on the undersides of the leaves. In addition, the leaf margins of Canyon Live Oak are highly variable, ranging from smooth to spiny. Canyon Live Oak therefore may be identified at a glance by noting the undersides of the leaves and the variety of leaf shapes on a single tree. The largest Canyon Live Oak is a massive tree found in the Santa Ana Mountains of southern California. Its trunk diameter is nearly 11 ft (3.5 m), but it stands only 72 ft (22.5 m) in height. Its crown width is 80 ft (25 m). Another 70 ft (22 m) Canyon Live Oak, located at the foot of Duckwall Mountain in Tuolumne County, was lost in 1965 when it split apart and toppled during a snowstorm.

North of San Francisco, the ranges of Coast Live Oak and Interior Live Oak overlap. Where this occurs, microclimatic preferences between the two species become apparent. Coast Live Oak tends to occur more commonly on the coast-facing slopes, where there is more soil moisture. Interior Live Oak prefers slopes that face away from the coast and becomes more common toward the interior of the Coast Ranges. Where the two species occur together, they may hybridize, which indicates that they have not been separate species for long. Identification of each species where their ranges overlap is complicated by hybridization, but in their pure forms, they can be identified by the color of their leaves. Interior Live Oak is bright green and shiny on both leaf surfaces. The leaf of the Coast Live Oak is a darker green and shiny only on the top surface. Coast Live Oak also has small tufts of hairs where leaf veins intersect on the lower surface.

Where fog is common in the northern Coast Range, the vegetation is coated with a reticulated, filamentous lichen known as Lace Lichen, *Ramalina menziesii* (figure 7.7). This yellow-green growth cloaks and dangles pendulously from everything that is bathed by the fog. The apparent function of the netlike branching is to increase the surface area available for condensation of water. Condensation not only supplies additional water to the lichen and promotes fog drip, but also provides a warming effect that could enhance photosynthesis. The warming of a surface by condensation is the opposite of evaporative cooling, which is so important to organisms in hot climates.

Oak Woodland may be a threatened community. For a variety of reasons, it seems that the number of hawks, owls, Coyotes, and Bobcats is decreasing. This change, if nothing else, has allowed an increase of rabbits, ground squirrels, gophers, mice, and deer. In addition, introduced annual weeds that produce a large number of seeds have led to an increase in seed-eating mammals. All these different herbivores have interfered with regeneration of Blue Oaks. In many areas, there are no Blue

Oaks younger than 50 years of age. Mule Deer, rabbits, and mice browse on young oaks from above while gophers attack the roots. Species composition in the Oak Woodland seems to be undergoing an important change. Coast Live Oaks seem not to suffer as much damage, so they are on the increase while Blue Oaks are disappearing.

Furthermore, Wild Pigs, *Sus scrofa*, are also becoming more common. The impact of wild pigs has yet to be addressed fully. These pigs are either escaped domestic pigs or descendants of European wild pigs (also *S. scrofa*), which were introduced to Monterey County in 1923. They are now distributed statewide. Feral pigs uproot acorns and seedlings, and the abundance of acorns is an important limiting factor for many animals, including Mule Deer. At equal densities, wild pigs eat three times as many acorns as deer, which, if nothing else, could severely curtail the size of native deer herds. In addition, at the Pepperwood Preserve in Sonoma County, pigs gnaw the bark off Knobcone Pines, ultimately killing them. Pigs also consume a wide variety of underground plant parts, including bulbs, corms, tubers, and fungi. It has been suggested that attempts to hunt wild pigs be intensified. The California Department of Fish and Wildlife has authorized bag limits for pigs at one per hunter per day, and special depredation permits may be issued for private land owners if pig damage is extensive. Since 1992, hunters have been required to possess wild pig license tags. It has been recommended that no more than six pigs per 40 acres (16 ha) be allowed to persist. Unfortunately, wildlife corridors that have helped large animals such as Mountain Lions escape habitat destruction have also provided an avenue for dispersal of wild pigs.

In recent years, *sudden oak death* has taken a devastating toll on several species of oaks in the Coast Ranges, from Big Sur to Humboldt County. Particularly susceptible are Tanoak, Coast Live Oak, California Black Oak, and Blue Oak. The culprit is a fungus, *Phytophthora ramorum*, which is carried from tree to tree by

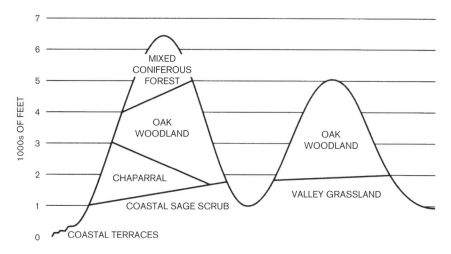

FIGURE 7.8 Biotic zonation of the southern Coast Ranges.

Ambrosia Beetles, *Monarthrum scutellare*, and Bark Beetles, *Pseudopityophthorus pubipennis*, among others. Beetles bore into the bark and inoculate the tree with the fungus that ultimately produces bleeding cankers of the tree trunk and dieback of the foliage. The most susceptible trees are those that seem to be stressed, particularly by drought. The disease also affects other forest trees, but some seem to harbor the fungus without being affected. In particular, California Laurel is known to be a host, and one of the mechanisms of reducing the disease in oaks is to remove California Laurels. Early treatment of infected trees with a phosphate fungicide has been mildly successful. The oak disease in southern California is also carried by an Ambrosia Beetle called the Polyphagous Shot Hole Borer, *Euwallacea* spp., which introduces a fungus, *Botryosphaeria fusarium*, that grows under the bark and feeds the larvae of the beetle, producing galleries that ultimately kill the tree. This disease and its implications will be discussed in more detail in the chapter on Cismontane Southern California (Chapter 8).

In most coastal valleys, oak woodlands, native marshes, and grassland have been replaced by pastures or agriculture. In the south, these valleys produce a wide variety of crops, including cotton, corn, tomatoes, beans, and sugar beets. Castroville, near the mouth of the Salinas Valley, is known as the artichoke capitol of the world. Valleys of the northern Coast Ranges are renowned for grapes, producing some of the finest wines in the world. Many famous-name vineyards are in Napa Valley, northeast of San Francisco.

Maritime Chaparral

As figure 7.8 shows, Chaparral-covered slopes are common in the southern Coast Ranges. Chaparral predominates on south-facing slopes and becomes more common toward the south. This community will be discussed in more detail in the chapter on Cismontane Southern California (Chapter 8).

Maritime Chaparral is a unique kind of Chaparral that occurs in patches on the coastal side of the southern Coast Ranges. In general, the shrubs form low, conspicuous mounds in open areas among the woodlands of oaks or closed-cone pines. In the Monterey area, the community occurs on sandy soil, the remnant of Pleistocene sand dunes that occurred along the coast. The community seldom grows on shale.

The significant feature that differentiates this form of Chaparral from that in the southern part of the state is the presence of abundant summer fog. Data gathered from the Monterey Bay area indicated an average of 135 foggy days per year. The average distribution of fog during summer months was as follows: June, 11 d (9 hr/d); July, 20 d (13 hr/d); August, 22 d (14 hr/d); and September, 17 d (12 hr/d). It is this foggy weather that also dictates where other ecologic-island species such as Coast Redwood and various closed-cone pines will grow.

Maritime Chaparral is dominated by a number of endemic manzanitas (*Arctostaphylos* spp.) and California lilacs (*Ceanothus* spp.). The community also includes Chamise, *Adenostoma fasciculatum* (figure 8.19), the most common species in many chaparral communities. The amount of endemism in the Monterey area is remarkable and will be noted frequently in this chapter. The endemic shrubs that occur in Maritime Chaparral in the Monterey area include Hooker's Manzanita, *A. hookeri* subsp. *Hookeri*; Toro Manzanita, *A. montereyensis*; Pajaro Manzanita, *A. pajaroensis*; Sandmat Manzanita, *A. pumila*; Cropleaf Ceanothus, *C. dentatus*; Monterey Ceanothus, *C. rigidis*; and Eastwood's Goldenbush, *Ericameria fasciculata*. However, habitat destruction has nearly wiped out this community. Nearly every one of the endemic species mentioned is classified as threatened or endangered.

Coastal Scrub

This community, dominated by California Sagebrush, *Artemisia californica*, and Black Sage, *Salvia mellifera*, is the northern extension of Coastal Sage Scrub, which will be discussed in detail in Chapter 8 on southern California. It differs from Coastal Sage Scrub by lacking a large component of succulent species such as cacti and it has a greater proportion of Coyote Brush, *Baccharis pilularis*, and shrubby lupines, *Lupinus* spp. This community is particularly important in the southern Coast Ranges where

it dominates on coastal bluffs and on low coastal terraces. South of San Francisco, scenic Highway 1 winds through this community throughout most of its length.

Mixed Evergreen Forest

In the northern part of the Coast Ranges, in Mendocino and Humboldt Counties, precipitation increases markedly. On cooler, moister sites, particularly canyon bottoms, Oak Woodland merges with Mixed Evergreen Forest composed of conifers and broad-leaved evergreen trees (plate [10B]). Douglas Fir is an important component of this forest to the north. In the Klamath region, as discussed in the preceding chapter, this community occurs in an even moister form.

Tanoak or Tanbark Oak, *Notholithocarpus* (= *Lithocarpus*) *densiflorus* (figure 7.9), is an evergreen species of the beech-oak family (Fagaceae) that is more closely allied to the oaks of southeast Asia. The tree is called Tanoak because it was once a major source of tannin, the substance used to preserve or tan leather. Except for the acorns, a person would not guess that this is an oak. The leaves are large, oblong, and leathery, up to 5 in (13 cm) in length. They have a wavy, toothed margin, and the undersides are hairy. This is a relict, distributed in moist locations from southern Oregon to southern California. Along with Pacific Madrone, California Laurel, and Coast Redwood, it is a remnant of a forest that once enjoyed a much larger distribution. Native Americans found the fairly large acorns to be particularly palatable after the tannins were leached.

Pacific Madrone, *Arbutus menziesii*, looks like a large manzanita. It is an evergreen tree with large, oblong leaves similar in shape to those of the Tanoak. The leaves differ from those of the Tanoak by lacking the hairy underside and, usually, the toothed margin. Pacific Madrone grows as far north as Canada; as such, it is the most northerly occurring evergreen hardwood tree. In California its distribution is similar to that of Tanoak, and the two

FIGURE 7.9 Tanoak, *Notholithocarpus densiflorus*.

species usually occur together. There should be no mistaking the two species, however. Tanoak has brown, deeply furrowed bark. Madrone has smooth, red bark that peels off in papery sheets on the branches. Flowers on Pacific Madrone are similar to manzanita flowers. They are white and urn-shaped, and they occur in clusters. The flowers are typical of the heath family (Ericaceae), to which manzanitas and Pacific Madrone belong. Also similar to manzanita, it produces bright orange to red berries in autumn. Pacific Madrone is a handsome tree that does well in residential landscapes when it gets adequate water. It is particularly attractive where it occurs among Coast Redwoods. The towering redwoods create so much shade that the Pacific Madrone, in its quest for light, commonly possesses a tortuously twisted trunk and is cloaked with bright green lichens.

California Bay Laurel, *Umbellularia californica* (figure 6.8), is an evergreen with dark-green, lance-shaped leaves, up to 5 in (13 cm) in length. The most distinctive feature of the leaves is a peppery, aromatic odor. They are shaped like bay leaves, and they may be used in cooking; therefore, this tree sometimes is referred to as California Bay or Bay Laurel. The odor is quite pungent when the leaves are bro-

ken, and it is not wise to sniff it for too long because it is known to cause headaches. In small doses, however, it is very pleasant.

California Bay Laurel is more common in southern California than the other evergreen species mentioned here. It is a frequent component of riparian communities there. This tropical relict belongs to the laurel family (Lauraceae), which includes several trees of economic importance, such as avocado, cinnamon, and camphor. California Laurel is the only native member of this tropical family in California. The wood is very beautiful and is commonly used for carved figurines, gun stocks, trays, and bowls. Along the southern coast of Oregon, the wood is known as myrtlewood, and the tree is called Oregon Myrtle. Carvings made of myrtlewood are a popular tourist attraction in Oregon, and they often carry a sticker erroneously stating that this tree is unique to the Oregon coast. How strange that they would make such a claim when the scientific name refers to the state of California.

In the northern Coast Ranges, California Bay Laurel is commonly called Pepperwood. In the northern Coast Ranges alone, there are at least 10 localities called Pepperwood. In areas associated with serpentine-derived soils, particularly where springs keep the soil wet,

California Laurel may form shrubby thickets. In contrast to the Shrubby forms found on serpentine, an extremely interesting locality in Marin County, west of Novato, is occupied by a California Laurel savannah with trees up to 80 ft (24 m) in height.

In the southern Coast Ranges, Mixed Evergreen Forest is restricted to north-facing slopes at higher elevation. These sites, characterized by Tanoak and Pacific Madrone, may be highly localized. On disturbed sites, or where the soil is derived from granitics or schist, Coulter Pine, *Pinus coulteri* (figure 8.27), is an important species. Coulter Pine is the southern California equivalent of Gray Pine. It is the most drought-adapted member of the yellow pine group, which includes Ponderosa and Jeffrey pines. It can be distinguished by its very long needles, often over 10 in (25 cm) in length. Its volleyball-sized cone is the largest of all pine species. At the few high-elevation localities of the southern Coast Ranges where true Mixed Coniferous Forest occurs, Coulter Pine may coexist with Jeffrey Pine, with which it may hybridize. In the New Idria Barrens discussed later with edaphic communities, Jeffrey Pine, Coulter Pine, and Gray Pine occur together at the same site. The three species can be distinguished by the size and shape of the cone. Coulter Pine will be discussed in more detail in Chapter 8, in the section on southern California Chaparral communities.

Riparian Woodland

In the northern Coast Ranges, where annual precipitation is high, the Riparian Woodland is similar to that of the Klamath Mountains. Differences are not dramatic between the Riparian Woodland and the adjacent forest areas. Conifers such as Coast Redwood, Douglas Fir, and California-nutmeg, *Torreya californica*, are forest species that may be included in the Riparian Woodland. Broad-leaved evergreen species such as Pacific Madrone, Tanoak, and California Laurel may occur along the streams as well as in the adjacent forest.

Among the winter-deciduous trees that occur in the Riparian Woodland as well as the bordering forest are Big-leaf Maple, *Acer macrophyllum* (figure 9.6); Mountain Dogwood, *Cornus nuttallii* (figure 4.50); and Red Alder, *Alnus rubra*. Upstream, White Alder, *A. rhombifolia* (figure 4.49), replaces Red Alder. In Oak Woodland areas, riparian species frequently are joined by Canyon Live Oak, *Quercus chrysolepis*. The most common willows of the Coast Ranges are Red Willow, *Salix laevigata*, and Pacific Willow, *S. lasiandra* var. *lasiandra*.

Shrubs associated with Riparian Woodland in the northern Coast Ranges are similar to those that occur in the Klamath Mountains. The common willow species are discussed in Chapter 6. Among the shrubby dogwoods that occur in moist localities in the Coast Ranges is Brown Dogwood, *C. glabrata*. This species resembles American Dogwood, *C. sericea*, with which it may occur, but the leaves are smaller, seldom over 2 in (5 cm) in length, and the bark is brownish rather than bright red. Brown Dogwood is truly a California species. Its range barely extends into southern Oregon, in the Siskiyou Mountains. It also has been reported at patchy localities in moist areas in the foothills of the western Sierra Nevada and in southern California.

A very attractive spreading shrub in the dogwood family (Cornaceae) is known as Bunchberry, *C. canadensis*. This little shrub grows only 3–9 in (7–22 cm) high and has a whorl of four to six dogwood-like leaves. It is most spectacular when it is in fruit. A cluster of large, bright red berries is framed by the whorl of leaves. These plants grow as ground cover in swamps and moist places along the coast from Mendocino County northward.

In the moist northern forests from Monterey County to British Columbia grows a beautiful flowering shrub known as California Rhododendron or Western Rose-bay, *Rhododendron macrophyllum*. Related to California Azalea, *R. occidentale*, with which it may occur, California Rhododendron is an evergreen species with dark green, leathery leaves, up to 8 in (20 cm) in length. Its leaves resemble those of a madrone, but this species is darker green, and it is always a shrub. Its large, funnel-shaped, rose-colored flowers, however, could not be confused with anything else. Sometimes these shrubs are cultivated in yards of the Pacific Northwest.

Farther south, where summer drought becomes an important climatic factor, species that are shared with the adjacent forest in the north become strictly riparian. Pacific Madrone, California Laurel, Big-leaf Maple, Pacific Dogwood, White Alder, and Canyon Live Oak range southward all the way to the Peninsular Ranges, where they occur along streams or strictly on north-facing slopes.

In the southern Coast Ranges, the Riparian Woodland becomes more southern Californian in appearance. Species such as Coast Redwood, Douglas Fir, Red Alder, and California Azalea drop out. Western Sycamore, *Platanus racemosa*, and Fremont Cottonwood, *Populus fremontii* subsp. *fremontii*, become dominant species along with White Alder, Red Willow, and Pacific Willow.

Mixed Coniferous Forest

True Mixed Coniferous Forest occurs in the northern Coast Ranges at high elevations and on dry sites associated with serpentine or other ophiolitic soils. In the southern Coast Ranges, this community occurs only on a few high peaks. Where it occurs, such as on Figueroa Mountain in Santa Barbara County, the forest is dominated by Jeffrey Pine. Also in the area may be found Coulter Pine, Gray Pine, Big-cone Douglas Fir, Canyon Live Oak, Blue Oak, Coast Live Oak, and Interior Live Oak. Such an interesting mix of species should provide ecologists many hours of study on microclimate preferences.

Edaphic Communities

Edaphic communities are those controlled by specialized soils. The significance of these soils to the distribution of plants has been emphasized already in the section on soils in Chapter 3 and in the chapters on the Sierra Nevada and the Pacific Northwest. In particular, mention was made of the degree of specialization associated with serpentine soils. These soils occur in patches in the northwestern foothills of the Sierra Nevada, the Klamath Mountains, and the Coast Ranges (figure 7.10). In the northern Coast Ranges, serpentine outcrops can usually be recognized by the presence of chaparral species such as Chamise, although on the coastal side, where moisture is more abundant, closed-cone pines or open stands of Jeffrey Pine and Incense Cedar may occur there. The bald hills of the northern Coast Ranges are also associated with special clay soils derived from shales.

There are also special soils derived from sandstone. For example, in western Fresno County near the town of Coalinga, there is an interesting area known as Anticline Ridge. Soft sandstone in this area weathers to sandy mounds that harbor a number of desert species, such as Interior Goldenbush, *Ericameria linearifolia*, and Mojave Sand Verbena, *Abronia pogonantha*. Also found in this area is California Juniper, *Juniperus californica*, and some of its associates. Of particular interest is an abundance of Desert Tea or Mormon Tea, *Ephedra californica*, a leafless shrub sometimes harvested for the purpose of brewing a rich tea. This plant contains tannins and small amount of a stimulant known as ephedrine, a component of some commercial decongestants. Commercial ephedrine, however, is either derived synthetically or obtained from an Asian member of this genus. The desertlike nature of this area is not only a product of its soils. Lying in the rain shadow of the southern Coast Ranges, this region experiences a pattern of precipitation characteristic of a desert. Desert plants of various types occur along the inner side of the southern Coast Ranges.

Biologists seem to emphasize the edaphic communities associated with serpentine. In the Coast Ranges, indicator species for serpentine outcrops include Leather Oak, *Quercus durata*; Musk Brush, *Ceanothus jepsonii*; and Interior Silktassel, *Garrya congdonii*. Serpentine soils

FIGURE 7.10 Distribution of serpentine in California (courtesy of California Division of Mines and Geology).

of the inner Coast Ranges from Napa County northward are characterized by Mcnab Cypress, *Hesperocyparis (= Cupressus) macnabiana*. Sargent Cypress, *H. (= Cu.) sargentii*, occurs on serpentine soils of the outer Coast Ranges from Mendocino County southward. Among herbaceous species, there are also many serpentine endemics. One group that has been studied extensively is *Streptanthus* spp., a highly diversified group in the mustard family (Brassicaceae).

Perhaps the best showcase for Sargent Cypress is a region in the headwaters of the Russian River in northwestern Sonoma County. In honor of the cypress trees, the region is known as The Cedars. Most of the property is in private hands, but the portion administered by the Bureau of Land Management is classified as an Area of Critical Environmental Concern (ACEC). Geologically the area is marked by serpentine barrens and picturesque travertine waterfalls. The area is also home to at least nine endemic species or subspecies of plants, including the Cedars Manzanita, *Arctostaphylos bakeri* subsp. *sublaevis*.

One interesting area characterized by serpentine soils is known as the New Idria Barrens (figure 7.11). This region of Franciscan rocks occurs in the Clear Creek Management Area in the Diablo Range of the inner Coast Ranges in San Benito County. This is a visually striking area characterized by yellowish soil nearly devoid of vegetation. Because of its apparent barren nature, the Bureau of Land Management has authorized this area for open use by off-road vehicles, an interesting use of land considering the dust from serpentine soils often includes asbestos. Areas characterized by endemic plants have been fenced. Included here is a depauperate chaparral community dominated by Bigberry Manzanita, *A. glauca*. The area is also unique because of the occurrence together of Gray Pines, Jeffrey Pines, and Coulter Pines. Coulter Pines occur here at higher elevation and on south-facing slopes. The Jeffrey Pines are scarce and seem to be disappearing. The slow growth of plants associated with serpentine soils was dramatically illustrated by paired photos taken in 1932 and again in 1960. There was no apparent change in the size or abundance of vegetation during the 28 year interval.

FIGURE 7.11 New Idria Barrens, San Benito County. This is an area characterized by serpentine soils.

Strategies of plants for tolerating serpentine soils include drought tolerance and the ability to deal with the peculiar minerals. Some species have acquired ways to exclude the undesirable elements, such as nickel and chromium. At the opposite extreme are species called hyperaccumulators. For example, the serpentine endemic known as Milkwort Jewelflower, *S. polygaloides*, can take up amounts of nickel in excess of 1000 parts per million.

As might be expected, where there are endemic plants, there are also endemic animals, particularly insects that feed or breed on specialized plants. One study documented nine species and subspecies of butterflies associated with the inner northern Coast Ranges alone.

FIGURE 7.12 Some plants of the Coastal Strand.

A. Crystalline Iceplant, *Mesembryanthemum crystallinum*.

B. Beach Morning Glory, *Calystegia soldanella*, mixed with Freeway Iceplant, *Carpobrotus edulis*.

For example, the Muir's Hairstreak, *Callophrys muiri*, occurs only on Sargent Cypress, and the Leather Oak Dusky-wing, *Erynnis brizo lacustra*, occurs only on Leather Oak. Similarly, certain butterflies are associated with the mustard genus *Streptanthus*. The California White, *Pontia (= Pieris) sisymbrii*, is very nearly endemic on serpentine because it lays eggs mostly on the endemic members of the *Streptanthus* group. An unusual Edith's Checkerspot, *Euphydryas editha luestherae*, is restricted to serpentine even though its two host plants, a paintbrush (*Castilleja* sp.) and a lousewort (*Pedicularis* sp.), also occur off of serpentine. In this case, there may be something special about

the temperature and moisture regimes of serpentine that keep the butterfly there, or there may be something about the timing of growth periods and flowering that is unique to the plants growing on serpentine.

Coastal Strand

Coastal Strand is a terrestrial community of sandy beaches and dunes, the pioneer community of the coastline. It occurs where there is no sea cliff. Here, the ever-present winds stack up sand carried to the area by a nearby river. Water-holding capacity of the sand is low, and accumulation of sea salts (sodium chloride) is high. This

FIGURE 7.12 *(continued)*
C. Pholisma, *Pholisma arenarium.*
D. Beach Evening-primrose, *Camissoniopsis cheirianthifolia.*

is a stressful environment for plants. Those that become established here have long taproots and are commonly prostrate and succulent. The long taproot enables them to reach water deep in the soil. Succulence enables them to cope with the salty soil in a manner similar to that used by plants of the Salt Marsh. By storing water in their tissues, plants are able to dilute the concentration of salts. Many common plants here are related to desert plants. Among these are Coastal Sagewort, *Artemisia pycnocephala* (a sagebrush); Seacliff Wild Buckwheat or Dune Buckwheat, *Eriogonum parvifolium*; and Beach Ragweed or Beach Bur-sage, *Ambrosia chamissonis*, which is related to the desert Bur-sage. These low shrubs of the dunes have silvery herbage and grow in loose mats up to 10 ft (3 m) across.

Members of the iceplant family (Aizoaceae) occur on the beaches of the southern Coast Ranges and down to Baja California. These succulent plants are native to Africa, but they have colonized beaches here to the degree that they are among the most common species. One species is the familiar succulent ground cover often called Hottentot Fig or Freeway Ice Plant, *Carpobrotus (= Mesembryanthemum) edulis* (figure 7.12B). It has been used to stabilize slopes in all forms of landscaping. The other iceplant is a succulent annual with tiny water-filled blisters on its leaves. Its scientific name, *M. crystallinum*, refers to the crystal-like appearance of the blisters, which apparently are related to the plant's attempt to excrete salts (figure 7.12A). Also found on these southern sand dunes are plants related to those found on desert dunes, such as the Red Sand-verbena, *Abronia maritima*. Similar to the desert Sand Verbena, this viny species covers many dune areas. During spring, many clusters of reddish violet flowers form a colorful carpet on the dunes. Another species with desert relatives, the Beach Evening-primrose, *Camissoniopsis cheiranthifolia* (figure 7.12D), is a prostrate evening primrose that bears bright yellow, four-petaled flowers. The flowers grow on long wiry stems that radiate from a central rosette. It is a conspicuous species all along the California coast.

Many of the coastal species differ from their desert relatives by being perennial as opposed to ephemeral. The constant overcast along the coast limits light intensity so much that it is nearly impossible for annual plants to complete a life cycle in a single year. A true desert plant that occurs on dunes is a root parasite, *Pholisma arenarium* (figure 7.12C). This low-growing species is a relative of the plant known as Sand Food, which is discussed along with desert sand dunes. Unlike Sand Food, which resembles a small bagel, this sand plant is more like a mushroom in appearance. Close inspection will show that it bears many small flowers on its rounded upper surface.

Perhaps the most spectacular flower of the beach dunes is Beach Morning Glory, *Calystegia soldanella* (figure 7.12B). This species is a geophyte, returning each year from deep-seated, fleshy rootstock. It is not an annual, even though it may appear so. It produces a large rose to purple flower that looks like a typical morning glory, but it is not a vine. It is common on beaches from Washington to San Diego. Silver Bush Lupine, *Lupinus albifrons*, is an important stabilizing plant on the Channel Islands.

American Dune Grass, *Elymus mollis* subsp. *Mollis*, was formerly common from Monterey County north to Alaska. Unfortunately, it is being replaced by an introduced grass known as European Beachgrass, *Ammophila arenaria*. The native grass reproduces vegetatively and by seed. The European grass reproduces mainly by vegetative means, but more rapidly than the native form. It spreads by means of underground stems (rhizomes) that can be broken up by ocean waves and dispersed to new localities. It grows so fast that it is more effective at stabilizing sand than is the native species. The result is a wall of grass tussocks near the shoreline that blocks the introduction of new sand. Starving a dune field of new sand leads to rapid degradation of the entire ecosystem.

A dune field is a food-poor ecosystem. Consequently, the most common animals are insects. Of the vertebrate animals, there are a few lizards and small birds that visit the area. Most of the predators are birds such as Kestrels (Sparrow Hawks) and Red-tailed Hawks. The sandspit at Morro Bay is one of the few places in the state where a person can still see Peregrine Falcons. They nest on nearby Morro Rock.

The endangered Least Tern, *Sternula antillarum* (figure 7.13), is a member of the gull family (Laridae). It nests on open sand, just below the zone where dune vegetation occurs. In order to protect the Least Tern, government employees have erected small fences around the bird's nesting areas. At Bolsa Chica, near Huntington Beach, a wetland has been reconstructed. The area has been off limits to development, largely to protect a breeding colony of Least Terns.

An interesting irony at Bolsa Chica involves Red Foxes, *Vulpes vulpes*, which were introduced at the nearby Naval Weapons Station Seal Beach. They have found refuge at Bolsa Chica and have been implicated in the demise of a colony of Light-footed Rails at Seal Beach and a group of fledgling Least Terns at Bolsa Chica. Under criticism from environmentalists, the Navy began a trapping program to remove the nonnative foxes. Between 1986 and 1989, 250 of the animals were trapped and removed. As is often the case, the Navy experienced difficulty

FIGURE 7.13 The Least Tern, *Sternula antillarum*, an endangered species that nests on the Coastal Strand.

finding new homes for these animals and aroused the ire of animal rights activists at the same time. In 1989, a permit to shoot the foxes was granted to the Navy. Another interesting situation with Red Foxes occurred at Point Mugu. Coyotes, *Canis latrans*, were reintroduced in the 1980s and seemed to chase out the Red Foxes. (Generally, Coyotes do not prey on birds.) After 3 years, not a single Red Fox could be located, and 100 Least Tern chicks fledged in a single year.

The Snowy Plover, *Charadrius nivosus*, is another bird that nests on the dunes, and it too has experienced a serious decline in numbers. It also nests to a small degree on the shores of alkaline lakes such as Mono Lake, and some have adapted to nesting on dikes constructed for salt evaporation ponds. This change in behavior may be its salvation.

Interesting symbiotic relationships between plants and their pollinators occur in dunes. For example, at Morro Bay, the Beach Primrose is pollinated only by bumblebees (*Bombus* spp.). Consistent, strong winds pose a problem for pollinators. It takes a great deal of energy to fly in such an environment. Large, powerful bumblebees apparently are able to cope with this problem by flying into the wind as they approach each flower. Because the wiry stems of Beach Primrose are flexible, the wind causes flowers to be oriented downwind. Also, because the flowers are so close to the ground, the bum-blebees are able to fly low where wind intensity is reduced. Here the Beach Primrose and bumblebee are able to thrive and reproduce by mutual cooperation.

Tiny blue butterflies (*Euphilotes* spp.) are found in many dune areas. These are diminutive but beautiful butterflies that occur in association with beach buckwheats. Buckwheats are their preferred nectar sources, and the larvae feed inside the flowers. Different species of these butterflies are associated with buckwheats in other habitats, but along the coast, and in association with extreme habitat destruction, several subspecies occur today at only a few localities. Five of California's federally listed endangered species are blue butterflies. The most restricted is the El Segundo Blue Butterfly, *E. battoides allyni*, which occurs only in the dunes west of the Los Angeles International Airport. The reason this area has not been developed is that a clear space is required for overflying aircraft. The dunes formerly covered the Coastal Strand from Playa Del Rey to San Pedro. All that is left are about 300 acres (120 ha) that the Los Angeles World Airports wanted to convert to a 27-hole golf course. The Palos Verdes Blue Butterfly, *Glaucopsyche lygdamus palosverdesensis*, was lost forever when its habitat was bulldozed for a baseball diamond.

Sand dunes are one of California's most disturbed communities. There are few natural dune systems left. All have been invaded by introduced plants, and most have been altered by sand mining, highway construction, military and recreation activities, and urbanization. Probably the least disturbed area is the Lanphere-Christensen Dunes in Humboldt County. Here the Samoa Peninsula forms a barrier between the Pacific Ocean and Humboldt Bay. These dunes were protected by the Lanphere and Christensen families until 1974, when the land was donated to The Nature Conservancy. Since then, more land has been purchased so that over 200 acres (80 ha) are now administered jointly by The Nature Conservancy and Humboldt State University. This is one of the few places in California where a dune field ecosystem can be studied in a nearly natural state, although there are some relatively undisturbed dune fields on the Channel Islands.

Another, much smaller natural dune field may be observed at Morro Strand State Beach, where a sandspit forms a barrier enclosing Morro Bay, the last natural estuary in southern California. This sandspit, about 4 mi (7 km) long, is the only place in southern California where a dune field can be observed in a semi-natural state, and where the forces of nature are allowed to continue shaping the land.

The largest field of coastal dunes occurs south of Pismo Beach. Here, near the small town of Guadalupe, there is an 18 mi² (32 km²) ecosystem of dunes and lakes. The area, known as the Guadalupe-Nipomo Dunes National Wildlife Refuge, was designated a National Natural Landmark in 1980 by the secretary of the Department of the Interior. Huge dunes along the coast have impounded 10 freshwater lakes. This type of landform was formerly a bay separated from the ocean by a sandspit, much like the Morro Bay sandspit today. The area is a treasure of restricted species; the California Native Plant Society lists at least 18 species as rare or endangered, or of very limited distribution, in these dunes. One of the species, the Nipomo Mesa Lupine, *L. nipomensis*, had been reduced to six known plants.

Up until 1982, when the state park system took stewardship of 3000 acres (1200 ha), hundreds if not thousands of people on any weekend would drive over the dunes in various forms of all-terrain vehicles. The Nature Conservancy manages a 567 acre (229 ha) parcel just south of the Santa Maria River that protects the rare La Graciosa Thistle, *Cirsium scariosum* var. *loncholepis*; Surf Thistle, *C. rhothophilum*; and Dune Indian Paintbrush, *Castilleja affinis* subsp. *affinis*. Of importance to the dunes is that these species, particularly Surf Thistle, help to stabilize the system, preventing loss of sand. Off-road vehicles disrupt the sand, causing it to move out from the root systems of the plants. This in turn allows the sand to be blown away. Areas occupied by sensitive species were marked and fenced, but a few inconsiderate people showed no respect for these organisms and drove wantonly through the restricted areas. Since state park acquisition in 1982, the dunes at Oso Flaco Lake have been closed to off-road vehicles, and camping along the beach has been prohibited. Off-road vehicle activity has been brought partially under control by better enforcement and fencing. An agenda for restoration of damaged areas has been implemented, including an ambitious plan to rid the area of European Beachgrass.

In Orange County, a remnant of a large dune field is still present on the Newport Peninsula near the town of Balboa. Formerly, the entire peninsula was a large sandspit similar to that at Morro Bay. In this region, development is set well back from the surf. Between the houses and the ocean is a small dune field occupied mostly by iceplants. But native species, such as Beach Primrose, Beach Morning Glory, and Beach Ragweed, may still be found among the introduced species.

In San Diego County, the best examples of dune and beach vegetation occur at the mouth of the Santa Margarita River at Camp Pendleton, although vestiges of the beach communities also occur along San Diego Bay and the Tijuana Estuary.

On the north coast, the interior portion of the dune fields is occupied by an interesting forest of conifers. Where the sand has become somewhat stabilized, Shore Pine, *Pinus contorta* subsp. *Contorta*, becomes established (figure 7.14). This is a pioneer species that is notable as

FIGURE 7.14 Shore Pine, *Pinus contorta* subsp. *contorta*.

a close relative of the pioneer in mountain mead-ows, the Lodgepole Pine, *P. contorta* subsp. *mur-ryana*. Similar to the Lodgepole Pine, needles occur in clumps of two. Although the cones resemble those of the Lodgepole in size and appearance, they often remain closed and on the tree for many years. Also, they tend to be visibly asymmetrical, curling backward along the stem.

These Shore Pines are distributed as far south as Mendocino County. To the north, how-ever, they are associated with coastal communi-ties such as dunes and bogs as far north as Alaska. In association with the pines are other species of northern affinity such as Sitka Spruce, *Picea sitchensis*; Grand Fir, *Abies gran-dis*; and Western Hemlock, *Tsuga heterophylla*. They reach their southernmost distribution along the north coast of California, where they depend on the moisture from fog drip. One of the best examples of this assemblage of species is preserved in the Lanphere-Christensen Dunes of Humboldt County.

The Ecological Staircase of Mendocino County

Perhaps the most interesting example of biotic zonation occurs on the coastal side of the mountains in Mendocino County. On the coastal terraces near Fort Bragg, there is an eco-logical staircase where each terrace possesses a distinctive vegetation type (figure 7.15). The cause of the zonation is a combination of fac-tors including climatic differences associated with elevation, as well as soil differences associ-ated with age. These terraces illustrate the suc-cession of changes over time. The oldest terrace is at the top, about 650 ft (200 m) above sea level. There is a guided nature trail to the stair-case in Van Damme State Park. A walk up the ecological staircase is, in fact, a walk back in time. It is also possible to think of it as a walk into the future, because it demonstrates what eventually will become of the lower terraces.

The Lowest Terrace: Coastal Prairie

The lowest terrace is the youngest. It was for-merly the beach and had its association of tide pools and beach dunes. The native community of this terrace today is a grassland known as North Coast Grassland or Coastal Prairie. It is characterized by grasses and spring wildflow-ers such as California Poppy, *Eschscholzia cali-fornica*; Johnny-tuck, *Triphysaria (= Orthocar-pus) eriantha*; and Beach Strawberry, *Fragaria chiloensis*. There are no large tree or shrub

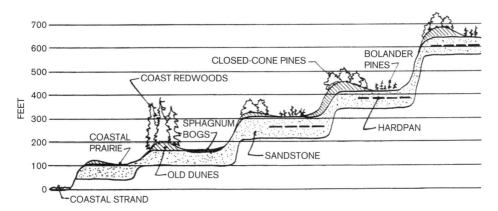

FIGURE 7.15 The ecological staircase of Mendocino County (illustration by Geoff Smith, after Fox, W.W. 1976. Pygmy Forest: An Ecological Staircase. *California Geology* 29(1):4–7).

species because heavy winds and salt spray produce a drying effect that perpetuates the coastal prairie. Hundreds of thousands of years of accumulation of organic material from the grasses and wildflowers have produced a rich, dark prairie soil (mollisol) highly prized for vegetable farming. Natural percolation of water from the next terrace above keeps the soil on this terrace wet enough that the vegetables usually do not have to be irrigated during the summer dry season. Furthermore, grasses remain green all summer, so the Coastal Prairie is also prize terrain for dairy cattle. A consequence of the economic importance of the Coastal Prairie is that most of it has been converted to agricultural or pasture land. As the population of California continues to grow, we shall probably see the phenomenon of urban sprawl creep over the north coast just as it has in the south.

Where percolation from upper terraces does not occur, the Coastal Prairie is subject to typical seasonal growth spurts. Primary productivity in such cases is low. For example, studies at Bodega Head revealed that during a 289 d growing season, photosynthesis accounted for about 4600 lb of vegetation per acre (527 g/m²). Although this is more productive than a typical desert, it is only half as productive as most chaparral communities. Even though the Coastal Prairies of California have been used as pastures since the days of the missions, this means that on a steady basis, Coastal Prairie is not capable of supporting much animal life.

Where there is some protection from salt spray, Coastal Prairie grades into a community of herbs and evergreen shrubs known as Northern Coastal Scrub. This community includes a number of species associated with Coastal Sage Scrub in southern California, such as Sticky or Orange Bush Monkeyflower, *Mimulus (= Diplacus) aurantiacus* (figure 8.12), and Coyote Brush, *Baccharis pilularis*. Northern Coastal Scrub differs from Coastal Sage Scrub by lacking a large number of drought-deciduous species and by possessing a large component of herbaceous species, many of which are components of the Coastal Prairie. In some areas, the dominant plants are perennial lupines, such as Varicolored Lupine, *Lupinus variicolor*, and Yellow Bush Lupine, *L. arboreus* (figure 7.16). The former is a low-growing shrub about 1.5 ft (30 cm) high. Its flowers may be yellow, pink, white, purple, or blue—hence the name "varicolored." Bush Lupine grows up to 9 ft (2.5 m) in height and has yellow flowers. Other conspicuous species include the invasive climber, California Blackberry, *Rubus ursinus*, and Salal, *Gaultheria shallon*, a glossy-leaved evergreen

FIGURE 7.16 Yellow Bush Lupine, *Lupinus arboreus*, in Northern Coastal Scrub.

often used for Christmas decorations. Its berry-like fruit is an important food for birds and other wildlife.

The Second Terrace: Redwood Forest

As time passes, these perpetually wet soils undergo podzolization, accumulating decay products and become increasingly acidic. This process leads ultimately to a fine-grained, ash gray soil with an acidic pH. The upper terraces on the Mendocino coast are characterized by soils of this type and have been measured as the most acidic in the world. The soil here has been measured at pH 2.8.

The second terrace is about 200 ft (65 m) above sea level and is believed to be about 100,000 years of age. It is nearly a mile (1.6 km) back from the sea cliffs. Salt spray is much less of a factor here, and the increase in elevation is enough to cause an increase in pre-cipitation, which measures 40–60 in (100–180 cm) per year. It is here that the "towering rain forest" occurs, characterized by Coast Redwoods, and Douglas Fir. Western Hemlock, Tanoak, and Pacific Madrone are also important in this community.

This forest is best established on the forward portion of the terrace upon old sand dunes. Soil has become weathered to a light brown, and pH is about 5.0. This is a rich soil with abundant organic material and ample nitrogen. A constant shower of needles and leaves from the forest adds to the decay products.

The distinction between Sierra Redwood and Coast Redwood is explained in Chapter 4 on the Sierra Nevada. The California legislature has declared that both trees are the state tree. Sierra Redwoods are the largest living things and the Coast Redwoods are the tallest. While there is some controversy over the possibility that there are Douglas Firs or even Eucalyptus in Australia that are taller, at the present time the Hyperion Tree at 379.3 ft (115.6 m) is the record holder for tallest in the world. The widest Coast Redwood is 29 ft (10 m), whereas the width of many sierra redwoods is greater. General Sherman, the largest tree in the world is 36.5 ft (11.1 m) across, but only 273 ft (83 m) tall. Sierra Redwoods also tend to live longer, with many specimens older than 2000 years. The oldest Coast Redwood is 2200 years of age.

The redwood forest of the north coast is one of California's treasures. On the Mendocino Coast, the forest has been logged, but the second growth itself is impressive. In order to see pristine redwood forest, one must travel to Redwood National Park or one of the series of state parks scattered along the coast from near the Oregon border to as far south as Big Sur.

Because of its dependence on high humidity, provided by a combination of precipitation and fog drip, redwood forest does not occur far inland. Where it does occur inland, the forest is associated with river bottoms. Some of the most beautiful groves occur along the "Avenue of the Giants" in Humboldt County (figure 7.17). Old Highway 101 in this area follows the at about 1500 tons per acre (346 kg/m²), is greater than in any other community in California. Douglas Fir (plate 9B[4B]) is the most important source of construction lumber used to build the ever-increasing number of buildings required to accommodate the human population.

Distribution of redwood groves is patchy. This is an ecological-island situation associated with a complex set of factors including temperature, moisture, soil, and exposure. For example, it has been reported that Coast Redwoods prefer temperatures around 65°F (18°C). Moisture, however, seems to be the most important factor. The redwood forest is the southern extreme of the temperate rain forest that occurs along the coast from Alaska to California. In the wettest parts of its range, as in the Klamath region, the forest is composed of Sitka Spruce, Grand Fir, Western Hemlock, and Douglas Fir. Some authorities propose that Coast Redwoods are actually a successional species, and that if precipitation in California were heavier, they would be replaced except in disturbed areas by more moisture-loving species.

Coast Redwoods grow inland beyond the reach of salt spray, but not beyond the influence of maritime climate. A combination of precipitation and fog drip keeps them watered year-round. The importance of fog drip should not be underestimated. In addition to about 60 in (180 cm) of moisture contributed by rainfall, it has been determined that up to 12 in (30 cm) is provided by fog drip. Coast Redwoods occur inland along river bottoms only where the canyon mouths face westward, so that interior heating can draw moist maritime air some distance inland. Along the Russian River near Cloverdale, Coast Redwoods occur as far as 23 mi (37 km) inland. Because most river valleys in the Coast Ranges run north to south, parallel to the coastline, Coast Redwoods seldom occur far from the coast. The reason they occur so far inland on the Russian River is that this is one of the few rivers that has an extensive westward run to the sea. It is also important to note that, in the interior groves, Coast Redwoods occur only

FIGURE 7.17 Coast Redwood, *Sequoia sempervirens*, along the Avenue of the Giants, Humboldt County.

course of the Eel River, winding its way among some of the most impressive trees in the world. In this area, redwoods extend inland about 10 mi (16 km) from the coast (plate 11A).

The ecology of redwood forests has been thoroughly studied. This is such an important lumber crop that logging companies have been willing to spend large amounts of money on research, particularly to determine why they grow where they do and what complex of factors makes them grow well. Coast Redwood and Douglas Fir are the two most important timber crops in the state. Redwood is valued for its rich color and resistance to termites and decay. Coast Redwoods are also among the fastest growing. Redwoods accumulate more mass of wood per year than any other forest type. It has been estimated that a hectare (2.5 acres) of forest accumulates about 42 m³ wood per year, or about 4000 lb (1815 kg) per acre. Furthermore, in a redwood forest, the aboveground biomass, on north-facing slopes. In these regions, the south-facing slopes are occupied by Douglas Fir.

The influence of wind patterns on the distribution of Coast Redwoods is illustrated by the Mattole River basin, where average precipitation is the highest in the area, and there is a large gap in redwood distribution. During summer, as the prevailing wind sweeps past the King Range, an eddy occurs, causing the wind to blow offshore instead of onshore. Dry downdrafts blow out to sea, and there is very little fog in the Mattole Valley. The area is characterized by Valley Grassland and Oak Woodland.

The influence of soil on Coast Redwood distribution is also profound. Redwoods grow on sandy soils such as old sand dunes, alluvial deposits along riverbanks, or soils derived from weathering of sandstones. Abrupt boundaries in redwood distribution occur where soil is derived from rocks such as serpentine or other ophiolites. On these soils, with their strange mineral composition, there are no Coast Redwoods. Instead, these areas are occupied by Mixed Coniferous Forest species such as Jeffrey Pine, Ponderosa Pine, Sugar Pine, and Incense Cedar. At drier localities, these ophiolitic soils are occupied by various closed-cone pines, which will be discussed later.

The boundary of a redwood forest is often abrupt. Where this boundary is associated with a change in soil, there may be a complete change of species within a few hundred yards (meters). One such area is at Red Mountain, in northern Mendocino County. A highly visible example is in Jedediah Smith Redwoods State Park.

Fire and flooding also play important roles in the dynamics of a redwood forest. Redwoods will stump-sprout. The well-known redwood burls sold throughout the northwest are examples of these buds, which will begin to grow if provided with ample water and light. After logging, or after a fire, the redwood trees will rejuvenate rapidly from burls at the base of the trunk. The importance of fire to the redwood forest was discussed in reference to Sierra Redwoods. Fire removes competing species, clears downed wood, recycles minerals, opens the forest canopy, and stimulates seedfall. It is ironic that the California State Parks system still believes in fire suppression.

On occasion albino Coast Redwoods are found among the trees with normal amounts of chlorophyll in their needles. Apparently, these albinos, incapable of photosynthesis, tap into the root systems of nearby trees and behave as parasites, deriving their nutrition for the roots of the green trees.

Along the rivers of the northwest, periodic flooding is a natural phenomenon. Each time a flood occurs, a new layer of soil is deposited on the alluvial flat. This deposition tends to clear out undergrowth and provides new fertile soil for the forest. Redwood trees grow a new set of surface roots that penetrate the new soil. The result is that the root systems of old redwoods penetrate deeply into the ground. They have multiple layers of roots, each corresponding to a former flood (figure 7.18). At Bull Creek Flat, one researcher's crew dug down about 28 ft (9 m) into the alluvium. He found evidence of 15 different floods, at intervals of 30–60 years, dating back about 1000 years.

Each new deposit of sediment also creates ideal conditions for seed germination and seedling survival. In open areas, where light is abundant, the seedlings grow rapidly. At Stephens Grove, in Humboldt Redwood State Park, there was a flood in 1861. Seedlings that germinated that year, now over 100 years old, can be tracked exactly. In areas at the margin of the grove, where there has been abundant light, those trees are now over 180 ft (60 m) tall. In the interior of the grove, where light is not abundant, tree seedlings dating to the same flood are only about 30 ft (10 m) tall. It should be obvious that disturbance is important to the health of a redwood forest. A combination of fog, fires to open the canopy, and floods to stimulate proper germination represent the best of all possible worlds for a redwood forest.

Sphagnum Bogs

On the second coastal terrace of northern California, the redwood forest occurs primarily on

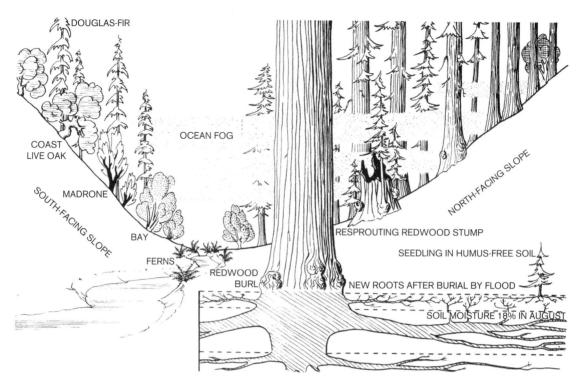

FIGURE 7.18 Slope effect in a grove of Coast Redwoods and diagrammatic representation of a Coast Redwood root system on a flood plain (from Bakker 1984).

the outer edge, in the vicinity occupied by former sand dunes. To the interior of the forest, in the center of the terrace, the soil is derived differently. The area is low-lying and boggy, fed by seeps and springs. In some areas, these bogs dry out every summer. In other areas, they remain moist and are invaded by thick layers of a moss called sphagnum. Water in sphagnum bogs is very acidic, from the accumulation of decay products. At this acidic pH, water-soluble nitrate (NO_3) cannot exist, which leads to a deficiency of nitrogen for bog plants. It is for this reason that carnivorous plants are associated with bogs. Carnivorous plants get their nitrogen from the protein in the insects they consume. They do not depend entirely on insects for survival. They do contain chlorophyll and can produce their own food.

The largest of these carnivorous plants is the California Pitcher Plant or Cobra Lily, *Darlingtonia californica* (figure 7.19). Its distribution is confined mostly to areas where a source of cold water flows through a serpentine substratum. Apparently, they seldom produce flowers, reproducing by underground runners. Most populations are clones, but they can produce maroon-colored flowers.

Pitcher Plants are yellow green with conspicuous veins. Their hollow stem is a modified leaf that is gradually enlarged upward into a rounded, dotted hood with an opening underneath. A two-lobed appendage under the hood gives the plant its cobra-like appearance. Plants may be 1.5 ft (60 cm) tall. In the hollow part, or pitcher, is a liquid that contains digestive enzymes. Insects are attracted under the hood by light shining through transparent "windows," after which they fall into the digestive fluid. A series of small, downward-pointing hairs prevents the insects from climbing out once they fall in.

A less conspicuous carnivorous plant is the Round-leaved Sundew, *Drosera rotundifolia*. This small plant may be 4 in (10 cm) across. It

FIGURE 7.19 California Pitcher Plant, *Darlingtonia californica*. Carnivorous plants grow on nitrogen-poor soil. They obtain nitrogen from the proteins of trapped insects.

FIGURE 7.20 Bishop Pine, *Pinus muricata*.

grows flat on the ground in a basal rosette. Its leaf blades are covered with tentacle-like reddish hairs that secrete a sticky fluid that traps insects. Entrapment is semiactive on the part of the plant; after an insect is trapped, the leaves are able to bend, which brings the prey in contact with more glands and sticky fluid. Sundews have a wider distribution than the Pitcher Plant. They may also occur in the Sierra Nevada as high as 8000 ft (2600 m) in elevation.

Upper Terraces: Closed-Cone Pine Forests

On the ecological staircase of Mendocino County, closed-cone pines begin to appear on the interior of the second terrace, and they are the dominant form of vegetation on the outer margins of the third, fourth, and fifth terraces. Soil here has become more podzolized. It is shallow, nutrient poor, and acidic. The closed-cone pine in this area is Bishop Pine, *Pinus muricata* (figure 7.20).

In its best-developed stands, it forms a nice forest of trees up to 40 ft (13 m) in height. In more impoverished conditions, the trees are short and scrubby, occurring in a Chaparral-like community, particularly in association with manzanitas. Whorls of firmly attached cones grow on the main trunks each year. These cones may remain closed for many years. If a fire occurs, the heat opens the cones, and seeds are dispersed. Because of this association with fire, most stands of Bishop Pines are even-aged.

Distribution of Bishop Pines is patchy, in the typical ecological-island pattern. The trees occur in groves from Humboldt to Santa Barbara Counties. They also occur on Santa Cruz and Santa Rosa islands, where some authorities recognize another, closely related species—the Santa Cruz Island Pine, *P. remorata*. Bishop Pines usually grow on fine-grained, nutrient-poor, shallow, poorly drained soil with an acid pH. In the Lompoc area, they occur on fine-grained diatomite, or diatomaceous shale. In areas where soil is dry, they water themselves with fog drip. They never grow far from the influence of the ocean.

The species is not limited to California, but it is nearly so. It also occurs in Baja California in one area near San Vicente. The closed-cone pine on Cedros Island is also assigned to this species by some authorities. This peculiar distribution is usually explained as remnants from a formerly widespread coastal distribution in the late Tertiary, several million years ago. Changing climatic conditions during the Pleistocene (the Ice Age) caused them to become isolated as relicts in their present localities.

The pattern is similar for the other closed-cone pines. Knobcone Pine, *P. attenuata* (figures A and B, p. 348), enjoys the largest distribution of the closed-cone species, and it also has rather specific requirements for fog, fire, and fine-grained soils. In Big Basin Redwoods State Park in the Santa Cruz Mountains, the location of Knobcone Pines on the outer margins of upper terraces is analogous to locations of Bishop Pines on the Mendocino coast. Knobcones replace redwoods to the interior and on the ridges. Isolated Knobcone Pine populations also occur in southern California and Baja California. Those populations will be discussed in the chapter on Cismontane Southern California.

Another closed-cone pine with similar requirements is Monterey Pine, *P. radiata.* There is a widespread fossil record for Monterey Pine in coastal California, but now its natural distribution is limited to three widely spaced regions extending over 100 mi (160 km) on the central California coast. Far to the south, however, on Guadalupe Island, about 460 mi (730 km) from the southernmost population near Cambria, there is a population of closed-cone pines that have been designated as either Bishop Pines or Monterey Pines (figure 7.21). Different authorities have different opinions. As of this writing, the closed-cone pines of Cedros and Guadalupe islands are called Monterey Pines. They appear to be intermedi-

FIGURE 7.21 Monterey Pines, *Pinus radiata*, on a fog-shrouded cliff above the sea on Guadalupe Island.

ate between the two species. Whatever the case, the distribution of the closed-cone pines is a fascinating case of biogeography and isolation.

Monterey Pine is a large, spreading tree under ideal conditions, but in its native range it often takes on a scrubby form, or it is flagged by salt spray. Monterey Pine has become one of the most commercially valuable timber crops in the world. It has been planted widely in the southern hemisphere, where, under optimal conditions, the trees grow straight and tall and reach harvestable size in 25–30 years.

The three species of closed-cone pines seldom grow together, although their ecological requirements are similar. The northern locality of Monterey Pine in the Año Nuevo area, about 30 mi (48 km) north of Monterey, is a region where Monterey Pine and Knobcone Pine occur in close proximity and hybridize. On the other hand, on the Monterey Peninsula on Huckleberry Hill, there is a small stand of Bishop Pines surrounded by Monterey Pines, and there is little or no hybridization.

FIGURE 7.22 Monterey Cypress, *Hesperocyparis (= Cupressus) macrocarpa*. This specimen, along the 17-Mile Drive on the Monterey Peninsula, has been shaped by wind and salt spray.

If the three species were to occur together commonly, they would not be difficult to tell apart. Bishop Pine has needles in clusters of two, and the others have needles in clusters of three. The cones, which are persistent on the trees, are easy to tell apart, if by no other means than their size. Bishop Pines have the smallest cones; they are about 2 in (5 cm) long, and they have prickly spines. Monterey Pine cones are about 3 in (7.5 cm) long, they are distinctly rounder, and they have no spines. Knobcone Pines have cones that are 4–5 in (10–12 cm) in length, and they are distinctly asymmetrical, with large, pointed cusps on each scale.

The possible reasons for different cone size are often debated. Why do some pine trees have such large cones? The answer is not certain, but it appears that pines in hot, dry climates have larger cones. If we look at the localities where the three closed-cone pines occur, this seems to be the case. Bishop Pine has the smallest cones, and Knobcone Pine, the most interior in distribution, has the largest cones. Large size may protect the seeds, or larger seeds may have more stored energy to help them germinate and endure the harsher climate.

Another group of conifers with localized distribution along the coast are the cypresses. There are 10 species of cypress, *Hesperocyparis* (= *Cupressus* spp.), in the state. They are characterized by gray, fleshy cones and scalelike leaves. Each is rather localized in distribution, usually associated with a specialized soil. Eight of the species are endemic, and the other two have limited distribution outside the state. Cedars, junipers, and cypresses are all in the cypress family (Cupressaceae). These common names cause a great deal of confusion.

Two species of cypress occur in the southern California area. Three others are localized foothill species found in the Sierra Nevada and the Cascades. The remaining five species are coastal forms with requirements similar to the closed-cone pines. Perhaps the best known species is Monterey Cypress, *H. (= C.) macrocarpa*. Its natural distribution is on granitic headlands with thin marine deposits in the vicinity of Monterey. One population occurs in Point Lobos State Natural Reserve; the other occurs along the famous 17-Mile Drive near Carmel. It is a picturesque species, occurring in gnarled form where it is exposed to wind and salt spray (figure 7.22).

Monterey Cypress in more favorable localities develops into a large, well-shaped tree. Near the Pebble Beach golf course is an old stand of large Monterey Cypress growing on a rich organic soil. Scattered among these trees, which are estimated at about 300 years of age, are old Indian middens. About 90% of the native population of Monterey Cypress grows in the Carmel area, but the species has been widely planted in landscapes, particularly along the coast. The phenomena of ecologic islands, specialization, endemism, and relicts are well illustrated by the pines and cypresses of the coast of California.

The Pygmy Forest

The pygmy forest occurs on the most impoverished soils of the Mendocino County ecological staircase. On the upper terraces to the interior of the Bishop Pine forest is a region of extreme podzolization. These soils were formerly bogs. Hundreds of thousands of years of leaching by the acidic water carried minerals into deeper soil, where they precipitated, forming a hardpan. Once a hardpan formed, percolation of water into deeper layers was prevented, and the acidification process was accentuated. The soil known as a spodosol or podzol is white, it has a peculiar mineral consistency, it is very acidic, it is oxygen poor, and it has poor drainage. Only specialized plants can survive here. Most characteristic of the species on the Mendocino coast that survive these conditions are dwarfed conifers that occur nowhere else in the world. A unique manzanita, known as Fort Bragg Manzanita, Arctostaphylos nummularia ssp. mendocinensis, also occurs here on this specialized soil.

Of the two conifer species that inhabit the pygmy forest, one is a cypress, and the other is a pine. The cypress is known as Pygmy Cypress or Mendocino Cypress, *Hesperocyparis pygmaea* (= *Cupressus goveniana pygmaea*), and it grows as a cane-like dwarf. It may be centuries old, but it seldom stands more than 10 ft (3 m) tall. Its dwarf status is caused not by a lack of water, for the ground is frequently saturated, but by a lack of essential nutrients. Perhaps it is lack of oxygen, which could severely

FIGURE 7.23 Bolander's Beach Pine, *Pinus contorta* subsp. *bolanderi*. This dwarfed type of Lodgepole Pine grows on impoverished, acidic soil of the upper coastal terraces in Mendocino County.

restrict metabolism in the root system. Ironically, when this species is planted in a favorable locality, it attains great size. In fact, a Pygmy Cypress nearly 100 ft (30 m) in height is one of the tallest specimens of cypress ever recorded.

The pine that grows in the pygmy forest is a variety of Lodgepole or Shore Pine. Known as Bolander's Beach Pine, *Pinus contorta* subsp. *bolanderi* (figure 7.23), it is sometimes considered a distinct species, but it is probably best to consider it a subspecies of Lodgepole Pine. Whether it grows on impoverished soils or in more favorable conditions, the cones remain closed. In the pygmy forest, growth is very slow. Bolander's Beach Pine 4–5 ft (1.3–1.6 m) high were found to be about 60 years of age. Thickets of stunted trees often are found to be the same age, a reflection of their closed-cone habit and their fire-stimulated origin.

FIGURE 7.24 Banana Slug, *Ariolimax columbianus.*

Other Upper-Terrace Communities

Huckleberry Hill on the Monterey Peninsula also has a pygmy forest. The upper terrace where this occurs is only 400 ft (130 m) in elevation. It is a terrace occupied by an acidified, claypan soil similar to that in Mendocino County. The dwarfed species here are Bishop Pine and Gowen Cypress, *Hesperocyparis* (=*Cupressus) goveniana.* In this area, on better soil, the pygmy forest is surrounded by a forest of Monterey Pine. This is the only place where these two closed-cone pines have survived together.

Of special interest on the upper terraces of the Santa Lucia Range in Monterey County is an endemic fir, the Bristlecone Fir or Santa Lucia Fir, *Abies bracteata* (plate [12C]). It is known as Bristlecone Fir because its small, upright cones have spinelike tips on the scales. These firs are found only in a strip about 13 mi (20 km) wide and 55 mi (90 km) long. Its distribution is limited to steep, rocky slopes and ridges, where it occurs with Canyon Live Oak. More than any other factor, however, its distribution is limited by fire. Unlike the closed-cone pines and cypresses, this fir cannot tolerate fire. Its distribution is limited to the steep slopes and ridges with little fuel to carry a fire. When it becomes established in adjacent Oak Woodland, it is soon eliminated by fire.

ANIMALS OF THE COAST RANGES

Most animals of the Coast Ranges are not particularly distinctive, although the unique insects associated with serpentine were mentioned earlier. In the southern Coast Ranges, most animals are typical species of Foothill Woodland and Chaparral. They are discussed in the sections devoted to those communities. In the northern Coast Ranges, Foothill Woodland and forest species are similar to those discussed in Chapter 4. The animals of the moist forests were discussed in Chapter 6.

In the moist parts of the Coast Ranges, there are many moisture-loving animals, such as insects, mollusks, and amphibians. Banana Slugs (*Ariolimax* spp.) are bright yellow detritus feeders that are conspicuous on the forest floor. There are four species native to California. They are very important as processors of organic material. In 1988, the state legislature voted that one species, *A. columbianus* (figure 7.24), be elevated to the status of state mollusk. George Deukmejian, the governor at the time, vetoed the bill, believing erroneously that the species was not native to California. Slugs do not have shells, but they are in the same group of mollusks (Gastropoda) as the snails. Of particular note is that slugs are hermaphroditic and have internal fertilization, mediated by a penis.

When they mate, they impregnate each other. *A. dolicophallus* has a penis longer than its body. Nevertheless, the charismatic Banana Slug has been named the school mascot at the University of California Santa Cruz; there is an annual Russian River Banana Slug Festival as well.

Butterflies

Among insects of the Coast Ranges, perhaps none has captured people's hearts more than the Monarch Butterfly, *Danaus plexippus* (figure 7.25A). This is the only insect known to make a long-distance migratory round trip. On the eastern side of the Rocky Mountains, Monarchs migrate between Canada and the Sierra Madre of central Mexico. West of the Rockies, some 5 million Monarchs spend the winter months along the foggy California coast from Mendocino County to south of San Diego. The greatest concentration occurs along the central coast. They are particularly visible in Pacific Grove, where thousands roost in "butterfly trees," which are usually Monterey Pines or eucalyptus. Each October, to celebrate the return of the Monarchs, Pacific Grove has a Butterfly Festival and parade sponsored by the Chamber of Commerce.

The same butterflies do not make the entire round trip, but flocks do return to the same trees every year. During migration, they glide on the wind to save energy. They may travel 3000 mi (4800 km) and reach altitudes of 10,000 ft (3000 m). On their southerly migration, they often follow the Sierra Nevada, and they are frequently spotted on the high peaks. The adults feed on nectar during migration. At their winter sites, they search for nectar during the day and bunch up at night in communal roosts. In the spring, they leave the coast and fly northward and inland, lay their eggs on milkweeds (*Asclepias* spp.), and then die. Caterpillars (larvae; figure7.25B), which feed on the milkweed, hatch from the eggs in 2–12 d, depending on the temperature. They are greenish with black and yellow bands, and they have a pair of long filaments. The caterpillar contin-

FIGURE 7.25 A. Monarch Butterfly, *Danaus plexippus*. B. Monarch caterpillar.

ues to grow, shedding its skin (exoskeleton) about five times before weaving a bright green cocoon (pupa). In about two weeks, the metamorphosis to an adult butterfly takes place. The adults migrate farther north or farther east and repeat the process.

Researchers have marked thousands of adult Monarchs in attempts to track their migratory routes. East of the Rockies, a tagged

Monarch flew from Toronto, Canada, to southern Mexico, a distance of over 2000 mi (3200 km). To date, the longest distance a California Monarch has been known to travel is 660 mi (1056 km). Between November 7, 1987, and April 9, 1988, this specimen traveled from Ellwood, in Santa Barbara County, to southeastern Arizona. Apparently, a California Monarch reproduces several times before returning to its winter roost. It is believed to take about four generations to make the round trip, although the southern leg of the trip is accomplished by a single butterfly.

The mechanism by which Monarchs are able to navigate has also been the subject of some research. Because the same butterflies do not make the full round trip, learning has very little to do with it. The entire procedure must be instinctive. Furthermore, it has been demonstrated that Monarchs are able to find their direction by using sun orientation, an ability exhibited by nearly all migratory animals, including birds, whales, and salmon.

Monarchs are Milkweed Butterflies (Nymphalidae). The fact that Monarch caterpillars feed on poisonous milkweeds has been known for a long time. The toxins, known as cardiac glycosides, are capable of killing most animals that would forage on the plant; that is the plant's way of defending itself. The butterfly, however, has become immune to the poison and uses it as a mechanism to defend itself from predation. This is known as coevolution. Similar patterns have been observed for other plants and insects. The case of willows and their associated beetles was mentioned in the section on Riparian Woodland in the Sierra Nevada.

Because potential predators, particularly birds, have learned that Monarchs are distasteful, if nothing else, they avoid them. This mechanism also has inadvertently protected other species of butterfly that resemble Monarchs. In time, the nonpoisonous butterflies become more and more similar to Monarchs in appearance, an evolutionary phenomenon known as mimicry.

Mimics need not be identical to be protected. In fact, it would be a disadvantage to be identical, because it would cause confusion between the species during mating. Birds seem to avoid nearly all orange butterflies as if they were poisonous. In California, this means that common butterflies such as silverspots or fritillaries, *Speyeria* spp., also receive a certain amount of protection from predation. The best known mimic of the Monarch is the Viceroy, *Limenitis archippus*, a common butterfly throughout the range of the Monarch in the eastern United States. In California, however, the Viceroy enters the range of the Monarch only along the southeastern border of the state.

In southern California, a butterfly known as the Queen, *D. gilippus*, is another poisonous milkweed butterfly. It is closely related to the Monarch, which it resembles. This is also a form of mimicry, because the two species of poisonous butterfly, by resembling each other, derive cooperative protection. The resemblance reinforces the protective nature of the coloration. Again, it would not be advantageous for the two species to be identical, because confusion between mates could lead to considerable waste of reproductive effort.

Amphibians

Of special interest in the Coast Ranges are sedentary animals such as certain amphibians and reptiles. There are several cases of breaks in distribution that correspond to geologic breaks. A case in point is represented by slender salamanders (*Batrachoseps* spp.), in the family of lungless salamanders (Plethodontidae). These salamanders are distinguished by their slender, worm-like shape and the fact they have only four toes on their hind feet. It was indicated previously that distribution and diversity within this sedentary group of salamanders could be correlated with specific geologic terranes that experienced long-distance transport before becoming attached to the western border of North America (figure 7.26). Breaks in distribution between species or subspecies occur along major fault

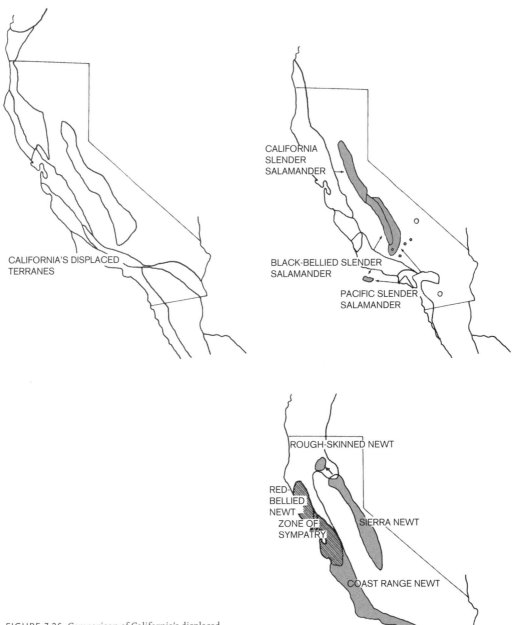

FIGURE 7.26 Comparison of California's displaced terranes with distribution of slender salamanders and newts. Arrows indicate discontinuous distributions (displaced terrane map after Hendrickson, D. A. 1986. Congruence of Bolitoglossine Biogeography and Phylogeny with Geologic History: Paleotransport on Displaced Suspect Terranes? *Cladistics* 2(2):113–129).

lines or in association with major rock units with distinctive geologic histories. The separation of animals into species or subspecies populations usually requires spatial isolation over long periods of time. The implication is that fragments of land, like massive ferries, moved from positions to the south, carrying with them cargos of organisms. Once these ferries became docked to the North American continent, the animals were free to move onto the mainland. Many of these salamanders, however, are so sedentary that they apparently chose not to move and have remained in place on the displaced terranes. As evidence of their sedentary nature, one researcher found that over an entire lifespan the farthest one of these salamanders moved was about 4 ft (1.5 m). When populations of slender salamanders are compared to each other on opposite sides of the San Andreas fault, it was discovered that salamanders displaced by right lateral drift on the west side of the fault are more closely related to populations farther south on the other side of the fault than they are to populations directly opposite to them across the fault.

To review the basic distribution of these salamanders in California, they are found in the western foothills of the Sierra Nevada, the Kern Plateau, the Inyo Mountains, and along the coast from southern Oregon to Baja California. They are associated primarily with various Oak Woodland and forest communities. At present, there are 20 species in California and one disjunct population in the Douglas Fir Forest in Oregon's northern Cascades. There are two groups of these salamanders. A robust body type is associated with three species of ancient lineage whose ancestors date to about 10 million years ago. These three, presently inhabiting disjunct localities, are the Oregon Slender Salamander, *B. wrighti* (= *B. wrightorum*), the Inyo Mountains Slender Salamander, *B. campi*, and the Kern Plateau Slender Salamander, *B. robustus*. The Inyo Mountains form lives in springs and riparian areas in canyons on both sides of the mountain range. On the Kern Plateau, they may be associated with conifers, as is discussed in Chapter 6. The rest of the species are so similar in appearance that they are difficult to tell apart, and this is further complicated by the fact that their distributions overlap and in several localities they coexist. Scientists call these *cryptic species*. Externally they are barely distinguishable from each other, but they have been assigned to different species on the basis of DNA analysis. For example, in the central part of the southern Coast Ranges in the vicinity of the Gabilan and Santa Lucia Mountains, the ranges of six different species overlap. The widely distributed California Slender Salamander, *B. attenuatus*, occupies oak woodland, forests, and suburban gardens in the northern Coast Ranges and the northern foothills of the Sierra. In both regions, its distribution is associated with rocks of the Franciscan Formation, 70–150 million years old, but the ranges in the two areas are separated by the Cascade volcanics. Its distribution overlaps that of other species in Santa Cruz and Calaveras Counties in the southern portions of its range. The Gregarious Slender Salamander, *B. gregarius*, occupies foothill habit in the southern Sierra from Yosemite to the Kern River. Its distribution overlaps at least three other species, including the Kern Canyon Slender Salamander, *B. simatus*, which is still being studied and may be split into at least two species. In southern California, there are two major species, the Garden Slender Salamander, *B. major*, and the Black-bellied Slender Salamander, *B. nigriventris*. The former is a lowland form associated with suburban gardens, and the latter is primarily associated with Southern Oak Woodland occurring in the foothills at higher elevations than the Garden Slender Salamander. A third species, the San Gabriel Mountains Slender Salamander, *B. gabrieli*, occurs on the coastal side of the San Gabriel Mountains between San Gabriel Canyon and Waterman Canyon on the western end of the San Bernardino Mountains. An isolated form of the Garden Slender Salamander has been described as the Desert Slender Salamander, *B. major aridus*. It occurs in a California Fan Palm oasis high in Deep Canyon on the desert side of the Santa Rosa Mountains. In the southern

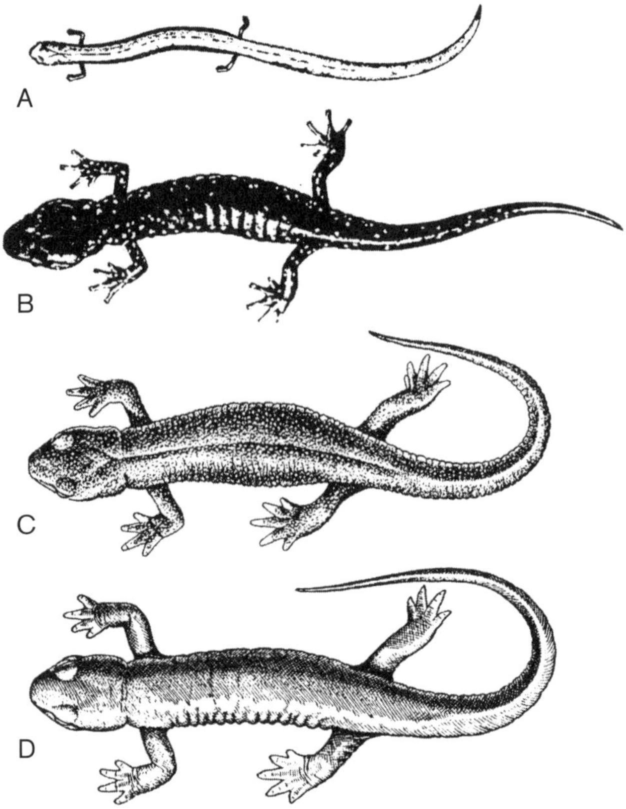

FIGURE 7.27 Some salamanders of the Coast Ranges.

A. California Slender Salamander, *Batrachoseps attenuatus.*

B. Arboreal Salamander, *Aneides lugubris.*

C, D. California Newt, *Taricha torosa* (terrestrial and aquatic stages). (From Stebbins, R. C. 1951. *Amphibians of Western North America.* Berkeley: University of California Press.)

Coast Ranges, the Black-bellied Salamander is associated with Oak Woodland, but primarily in the area of sedimentary rocks that have been high and dry only since the Pliocene, some 2–3 million years ago. It coexists with seven of the other species from the foothills of the Sierra and the central Coast Ranges to southern California where it occurs in association with the San Gabriel Mountains Slender Salamander and many locations with the Garden Slender Salamander, including sites on Santa Cruz Island. Interestingly, the two salamanders tend to occur on opposite sides of the Central Valley of Santa Cruz Island, which marks the contact between terranes with different geological histories.

North of the sedimentary rock units occupied by the Black-bellied Salamander are two interesting units of rock associated with the Salinian block. The Santa Lucia Mountains occur along the coast south of Monterey and are occupied by an endemic species related to the Garden Slender Salamander. Directly east of the Santa Lucia Range, on the other side of the Salinas Valley, is a large block of granitic rocks known as the Gabilan Range, where there is another endemic species. The Gabilan and Santa Lucia Ranges are comparatively high, remaining above sea level and forming islands during times of maximum inundation by seawater. These rocks are at least 100 million years of age. That, plus their separate origins from positions hundreds of miles to the south, are responsible for the isolation and subsequent derivation of at least two kinds of slender salamander. The two ranges stand today as examples of ecologic islands separated by a region of low relief that is now occupied by Valley Grassland and agriculture.

Another salamander that shows differentiation in association with fault lines is the Arboreal Salamander, *Aneides lugubris* (figure 7.27).

Another lungless salamander, it is commonly associated with hollows in large oaks; however, in spite of its name, it also is commonly found on the ground under rocks and logs. Its distribution parallels that of the slender salamanders, although it is seldom found away from oak trees. Although separate subspecies have not been named, different patterns of spotting are correlated with different geologic units. In the Gabilan Range, there are forms with spots larger in diameter than 1 mm. In the Santa Lucia Range and elsewhere, spots are smaller than 1 mm. These salamanders occur generally above 300 ft (100 m) in elevation, and they do not occur below in the Salinas Valley. This appears to be an ecologic-island situation comparable to that described for slender salamanders. It is also noteworthy that the only other population of Arboreal Salamander with large spots is on South Farallon Island, a granitic outcrop that is also part of the Salinian block (figure 7.5), but 90 mi (150 km) to the northwest of the Gabilan Range. This island is about 30 mi (50 km) west of San Francisco. A 5-million-year-old fossil of this salamander was found in the foothills of the Sierra.

Breaks in the distribution of subspecies of the salamander genus *Ensatina* also correspond to similar geologic units (figure 8.32). Distribution of *Ensatina* has been discussed already, and it will be analyzed thoroughly in the following chapter. These are the lungless salamanders that occur as spotted forms in the Sierra Nevada (figure 4.34) and as solid-colored forms in the Coast Ranges. The total range of the species encircles the Great Central Valley, and at the southern end (figure 8.33 and 8.34), where the ranges overlap, they do not hybridize. The pattern of distribution of the solid-colored forms parallels in a remarkable way that of the arboreal and slender salamanders of the Coast Ranges and Sierra foothills. The breaks in subspecies distribution parallel closely those of the slender salamanders.

Another family of salamanders is known as newts (Salamandridae). The most conspicuous salamanders in California, they are brownish on the back, and the underside is bright yellow, orange, or red, depending on the species. Apparently not fearful of predators, they are commonly active during daylight hours. Their skin secretions are toxic to most animals. When molested, they arch their backs and expose the bright color of the belly. Bright colors are often warnings, as has been discussed in other chapters.

These are not sedentary salamanders. They spend summer months estivating under rocks and logs or in cracks. With the first rains of autumn, they migrate to water in large numbers. They reproduce in the water and remain aquatic throughout the breeding season. During this time, their skin becomes smooth, and the tail develops into an oar-like fin. During their migrations, which often occur in daylight hours, they may be observed in great numbers. In the Berkeley area, there are signs along roadways that say "Watch for Newts Crossing."

This migratory habit enables these salamanders, more than any other, to colonize new habitats. For this reason, the breaks between populations are not as obvious as those for the lungless salamanders. Nevertheless, inspection of the ranges of the three California Newt species will show that ancestral geographic barriers are the same as they were for the others (figure 7.26).

Overall, distribution of the three species resembles that of the *Ensatina* group (figure 7.26). They occur along the coast and around the north end of the Great Central Valley. In the Sierra Nevada they occur as far south as Kern County. Unlike *Ensatina*, the two arms of distribution do not overlap at the southern end.

The northernmost species is the Rough-skinned Newt, *Taricha granulosa*. It occurs in moist forests from southeast Alaska to Santa Cruz County, California. The southern boundary of its distribution coincides generally with the boundary between Franciscan rocks and the younger sedimentary rocks of the southern Coast Ranges. The California Newt, *T. torosa* (figure 7.27), occurs in canyons and along streams from the southern Coast Ranges to southern California. A closely related subspecies, the Sierra Newt,

T. sierrae, occurs across the Great Central Valley in canyons and streams of the Foothill Woodland. To the north, the ranges of these two subspecies overlap that of the Rough-skinned Newt, but hybridization is evidently not common. The California Newt and the Rough-skinned Newt actually inhabit the same aquatic and terrestrial habitats where their ranges overlap. The third species, the Red-bellied Newt, *T. rivularis*, occurs in association with flowing water in the Coast Redwood Forest. Its range overlaps that of the Rough-skinned Newt with which it occasionally hybridizes. At its southern end, its range overlaps that of the Coast Range Newt. The three species could occur together, but they seldom do. Presently the three species are separated ecologically, but they were likely separated geographically in former times. Their migratory habit apparently has brought them together. In particular, it seems that the Coast Range Newt has moved northward for some considerable distance, so that its range now overlaps those of the other two.

Genetic analysis of the newts has been undertaken. Because of the great difference between the Rough-skinned Newt and the Coast Range Newt, it is estimated that they were geographically separated about 9 million years ago. This time corresponds well with the period of inundation of the southern Coast Ranges. A comparison of northern and southern populations of the Coast Range newts indicated that their time of divergence occurred about a million years ago. This coincides with uplift of the southern Coast Ranges, indicating that they probably migrated northward during that time.

Among the mole salamanders discussed in Chapter 6 is the California Tiger Salamander, *Ambystoma californiense*. This salamander, which usually has large yellow blotches across its back, is seriously threatened within its native distribution in the central Coast Ranges and Sierra foothills. It has been listed as threatened by the federal and state governments. The larval forms, without yellow bands, and with distinctive external gills are aquatic, and the adults are terrestrial. A closely related form the Western Tiger Salamander, *A. mavortium*, may remain in larval form for many years. These larvae, known as Axolotls or Water Dogs, have been imported to California to be used as bait for game fish. They have escaped into nature in California and are known to hybridize with the California Tiger Salamander.

Salamanders are not the only animals that show this pattern of diversification in association with geologic units. The Foothill Yellow-legged Frog, *Rana boylii*, and the Southern Mountain Yellow-legged Frog, *R. muscosa* (figure 11.30), have the same two-armed distribution around the Great Central Valley. These two species formerly occurred together in southern California in the San Gabriel Mountains. In particular, they once were known to inhabit the same localities in the North Fork of the San Gabriel River and in Bear Creek. Distribution of these frogs is discussed in Chapter 11.

Reptiles

The Western Pond Turtle, *Actinemys (= Emys = Clemmys) marmorata* (figure 11.33), occurs along the coast and in the Lake Tahoe area of the Sierra Nevada. There are two subspecies in the Coast Ranges, a northern and a southern form. Their ranges overlap in the central area north of Monterey. Pond turtles live up to 50 years. Nevertheless they reproduce slowly, and their populations are in decline throughout the state. Their status in southern California is particularly threatened and will be discussed in Chapter 11.

Among lizards, the alligator lizards and the skinks are noted for their sedentary habits. Like salamanders, they spend a great deal of time under rocks, logs, and other debris. Both of these lizards are noted for their cigar-shaped (fusiform) bodies and small legs. They move in a snakelike fashion. Alligator lizards have conspicuous squarish scales on their backs. Skinks are very sleek and shiny and are noted for their bright blue or red tails.

Ranges of the Northwestern Alligator Lizard, *Elgaria coerulea principis* (= *Gerrhonotus*

FIGURE 7.28 Southern Alligator Lizard, *Elgaria multicarinata*.

coeruleus), and the Southern Alligator Lizard, *E. multicarinata* (= *G. multicarinatus*; figure 7.28), overlap in a manner similar to that of the newts, and subdivisions into subspecies occur in approximately the same places as in *Ensatina*. Similarly, ranges of the Western Skink, *Plestiodon (= Eumeces) skiltonianus*, and Gilbert's Skink, *P. gilberti*, follow a similar pattern. The Western Skink ranges along the coast from Canada to Baja, and it may coexist with Gilbert's Skink in the inner Coast Ranges from Monterey south to Baja. Subspecies of Gilbert's Skink are also fragmented along similar boundaries. Alligator lizards and skinks are also found on the Kern Plateau and in southern California in association with Southern Oak Woodland.

The Ring-necked Snake, *Diadophis punctatus*, is a sedentary species found under a variety of objects throughout its range. It occurs mostly in moist habitats such as forests and Oak Woodland, and it commonly feeds on slender salamanders. It is a small snake, seldom over 12 in (30 cm) in length. It has a dark-colored head with a yellow to red ring around its neck. The belly is also bright yellow to red. When a Ring-necked Snake is disturbed, it coils its tail like a corkscrew, which reveals the bright color. The snake waves the tail, presumably as a dis-traction. A Ring-necked Snake is not able to shed its tail like a skink, but if a predator attacks the tail, more vulnerable parts of the body are likely to escape damage. The distribution of Ring-necked Snakes is notable because it is very much like that of *Ensatina*. Its circle of subspecies is divided in much the same way, and the two arms of distribution overlap in southern California.

Distribution of all these species of amphibians and reptiles, taken as a group, illustrates a remarkable pattern of biologic differentiation caused by the interaction of climatic and geologic history. These are but more examples of how California's diversity is the product of an evolutionary laboratory.

Birds

Most birds and mammals of the Coast Ranges are discussed elsewhere in this book. Those of the northern forests are discussed in Chapter 6. Those of Mixed Coniferous Forest are discussed in Chapter 4. Oak Woodland species are discussed in association with Foothill Woodland, Southern Oak Woodland, or Chaparral. The one true inhabitant endemic to Oak Woodlands in the Coast Ranges is the Yellow-billed Magpie, *Pica nuttalli* (plate 19H). California

FIGURE 7.29 Acorn Woodpecker, *Melanerpes formicivorus*.

has only two endemic birds. This magpie is the only bird endemic to California mainland. The other endemic is the Island Scrub-jay, *Aphelocoma insularis*, which will be discussed in the section on the Channel Islands. The Yellow-billed Magpie is fairly common from Humboldt County south to Santa Barbara County. A spot where many motorists can watch these birds beg for food is the rest stop on Highway 101 near Paso Robles.

Among the birds, those associated with oaks are the most conspicuous. This includes the California or Western Scrub-jay, *A. californica* (plate 19F), the Oak Titmouse (*Baeolophus inornatus*), and the Acorn Woodpecker, *Melanerpes formicivorus* (figure 7.29). The Scrub Jay is very important to Oak Woodland because it buries hundreds of acorns for future consumption, but it does not retrieve them all. This is a major mechanism for planting and uphill dispersal of oak trees.

The habit of Acorn Woodpeckers of drilling holes and storing acorns and the seeds of Gray Pines has been mentioned in Chapter 4. Perhaps the most interesting thing about them, however, is their social behavior. Because they are typically associated with oaks, and because

much of their behavior has been studied in the southern Coast Ranges, they will be discussed here.

Acorn Woodpeckers are conspicuous, gregarious birds that are easy to watch. They are dark birds with a white chin and a small red cap. When they fly, their white rump and white wing patches are conspicuous. They also attract attention because they make considerable noise. Their call is an easily recognized "wack-a, wack-a." While we associate these woodpeckers with storing acorns for food, they also eat insects, particularly in spring and summer, which the catch on the fly, in the same manner as a flycatcher. It seems they are particularly fond of flying ants.

Apparently, the urge to fill a hole with an acorn is instinctive. If the hole happens to be in a cabin wall, Acorn Woodpeckers will continue to push acorns through it until one sticks or until the space in the wall becomes filled up to the hole. When the woodpecker sees an acorn in the hole, it loses the urge to stuff another one in there. According to one story, when a cabin in the Santa Ana Mountains was torn down, enough acorns were taken from the walls to fill two large trash barrels.

Acorn Woodpeckers are unique among birds in north temperate regions because they feed and breed in cooperative groups that may include 12–15 members. These social groups defend a territory, which consists of a tree full of holes into which acorns have been stuffed. These trees are known as granaries. In regions such as southern Arizona, where acorns are not abundant, these woodpeckers do not defend granaries, and they mate in single pairs. In California, where acorns are commonly abundant enough to provide food for the entire winter, these social groups have evolved.

Males in the social group are commonly siblings. Females are also commonly siblings but may be unrelated to the males. A social group may include one to three breeding females and up to seven breeding males. The remainder of the birds are nonbreeding helpers, usually youngsters. Mate sharing is common among the breeders. Males mate with multiple females and vice versa. The helpers gather food, defend the granary, and feed the young. When the granary is full, helpers drill more holes. If the granary is not full, they spend most of their time searching for food. Young birds are fed insects, not acorns. The acorns are food for the birds that gather insects to feed the young. Nesting takes place in communal cavities. Females and males incubate eggs and carry out fecal packages to keep the nest clean.

If mating birds die or disappear, new breeders are not recruited from the helpers. Rather, if all mating birds of one sex are gone, a mate vacancy appears. Groups of woodpeckers from nearby colonies will then appear at the granary, and a great deal of commotion ensues as mature birds compete, chase, and display for the privilege of filling the vacancy. When the vacancy is filled, it is by new sibling breeders, which immediately begin to destroy eggs and offspring belonging to the former breeders. This behavior ensures that new breeders will contribute genes to the next generation.

Competition between females takes place over which of the sisters will lay the first egg. Most females lay about five eggs. If one breed-ing female begins to lay eggs before another, the second breeder will remove the egg and carry it to another tree, where it is broken and eaten by the group. This procedure is continued until the second and third females begin to lay eggs. When two females are laying eggs in the same nest, the eggs cannot be told apart, and none of them is destroyed. This process ensures that more than one female contributes genetic material to the offspring, and the bird that removed the eggs will contribute more genetic material than the first to lay.

Sometimes, whole new colonies are formed by a merging of sibling coalitions—females from one group and males from another. Formation of a new colony requires an abundance of acorns and, more important, a granary. Each bird is capable of drilling only a few holes per year, and a granary has anywhere from a few hundred to a thousand storage holes. Marginal habitats are scarce, and it takes the work of many birds to create a new granary. New granaries are produced only when all existing holes are filled and acorns are still abundant.

The California Condor, *Gymnogyps californianus* (figure 7.30), is primarily a species of the southern Coast Ranges, although its official range also includes the Tehachapi Mountains and southern Sierra foothills. As of 1988, however, there were no animals remaining in the wild. The last natural home of the California Condor was the Los Padres National Forest, east of Ventura. A captive-breeding program was begun, and in 1992, when sufficient numbers of these birds were living again, captive condors began to be returned to their former range. One of the captives added to the breeding program in 1985 ultimately sired 29 chicks. At the age of 35, it was finally released into the wild at the Bitter Creek National Wildlife Refuge, the same location where it was captured. That is considered middle-aged for this long-lived species.

The California Condor is the largest flying bird in North America. It weighs about 20 lb (8 kg) and has a wingspread of around 10 ft (3 m). A large vulture, it is black and has a naked head. It soars on rising currents of air,

FIGURE 7.30 California Condor, *Gymnogyps californianus*, sunning.

called thermals. It covers hundreds of miles in a day searching for carrion, which, similar to the Turkey Vulture, *Cathartes aura*, it probably locates by sight and odor.

The California Condor roosts and nests in tall trees and on cliffs, usually in remote areas. It flies during the day over pastureland and other open habitat, searching for food. A California Condor would typically soar over 120 linear miles (190 km) in a day. Because of its large size, it is not able to become airborne merely by flapping its wings. It must wait for the thermals and then launch itself from a high place into the rising air. In the morning, similar to Turkey Vultures, it may be seen sunning itself with wings spread (figure 7.30). When it does land on the ground, it experiences great difficulty becoming airborne again, particularly if its stomach is filled with food. In this circumstance, birds have been observed to walk up a hill and then run back down, flapping their wings until they reach sufficient speed to take off.

Activities of humans drove the California Condor to the brink of extinction. In the old days, a large number simply were shot. Since the 1800s, loss of habitat has been very impor-

tant. Conversion of Valley Grassland to farmland took away the feeding grounds. The prey as well as the predators were driven out. Carcasses of Tule Elk and Pronghorn were no longer available for food. In recent years, the birds fed primarily on dead livestock and on Mule Deer that were shot and not located by hunters. Although deer carcasses, the remains of mountain lion kills, were still available, these were in Oak Woodland and Chaparral areas. After the turn of the century, following the introduction of a policy of fire suppression, brushlands became overgrown, resulting in a lack of open space for landing and feeding of California Condors. Deer moved into woodlands, where more food was available. The California Condor gradually disappeared.

Extirpation of the California Condor is not restricted to California. Fossil remains indicate that during the Pleistocene, about a million years ago, they occurred widely over North America. There is evidence of their existence from Canada to Florida. Their decline is associated with a general reduction in the number of bird and mammal species over the last few thousand years. This may be the result of climatic change, but a strong body of evidence relates the extinctions to a rise in human populations in North America. The La Brea Tar Pits include the remains of California Condors that lived up to 40,000 years ago. As recently as 2000 years ago, they still occurred across Arizona, New Mexico, and Texas. California Condor remains are found in caves all over the Southwest, often in association with Indian artifacts. Native Americans most certainly had an interest in these huge birds. Rituals involving sacrificial deaths of California Condors are known among San Diego Indians, and costumes and other artifacts made of California Condor skins or feathers are found in natural history museums. In fact, collecting of California Condors for display in zoos and museums has also taken its toll. By actual count, 177 California Condor specimens are known to exist in the world's natural history museums.

In the early 1800s, the California Condor still occurred along the Pacific coast. In 1806, the Lewis and Clark expedition recorded instances of California Condors feeding on dead salmon washed up on the banks of the Columbia River. In the early 1900s, they were eliminated from southern California. One of the early natural history studies of the California Condor took place in 1906 in Eaton Canyon, in the San Gabriel Mountains. Photographs taken with 5×7 in^2 glass plates are still among the best of nesting California Condor.

In the 1930s, it was estimated that there were 60–70 birds left in the wild. In the 1940s, an estimate of 60 birds was made. Some authorities claim that these estimates are not accurate, and that there must have been at least 150 birds as recently as the 1950s. The Sespe Condor Sanctuary, encompassing 35,000 acres (14,200 ha), was created in 1947. Then 18,000 acres (7300 ha) were added in 1951, and a 1200 acre (485 ha) parcel around Sisquoc Falls was declared a sanctuary in 1973. In 1986, the 11,000 acre (4450 ha) Hudson Ranch was purchased by the US Fish and Wildlife Service. Acquisition of this property, in the southeast corner of Kern County, put into public hands

most of the natural range of the California Condor. The wild population continued to decline, however, and in 1965 there were an estimated 40 birds left. In 1983, there were 20, in 1986 there were 6, and the last wild bird was trapped and taken to a zoo in 1987.

Much of the decline in recent years can be attributed to factors other than habitat loss. From the 1960s on, thinning of eggshells due to DDT (dichlorodiphenyltrichloroethane) was a factor. Some birds were poisoned as victims of coyote and rodent extermination programs, and many others were poisoned from consuming lead, presumably in the remains of deer or other animals that were wounded by a hunter and died later. Many authorities today consider lead poisoning to be the major cause of the California Condor's decline. Unfortunately, birds released to their former range along the central coast began feeding on dead marine mammals, particularly sea lions, and that caused reintroduction of DDT and PCB (polychlorinated biphenyl) residues into Condor tissues, and subsequent thinning of eggshells.

The California Condor is now the rarest bird in North America. Federal and state agencies, under the auspices of the 1973 Endangered Species Act, have begun a recovery program including the establishment of critical habitat, research on habits of the species, and a captive-breeding program. Regarding critical habitat, since 1969 over 4300 acres (1740 ha) have been acquired for management as reserves. As for research, observation posts were established where wild birds could be studied and photographed, and birds were fitted with tiny radio collars so their daily activities could be monitored. The most controversial part of the program has been the captive-breeding program.

In a captive-breeding program, birds in captivity produce and raise offspring that are destined to be released into the wild. Normally a California Condor lays a single egg every other year. This is a low reproductive rate, typical of long-lived, large animals that occur high on a food chain. If an egg is destroyed or removed, the California Condor responds by laying another egg, a phenomenon known as double-clutching. There is one known case of triple-clutching. In a captive-breeding program, eggs are removed from nests of wild birds and hatched in captivity. The wild birds then lay another egg, which doubles the reproduction rate. Similarly, mated captive birds are stimulated to double-clutch. In 1983, there were 20 birds in the wild and 8 in captivity. By 1986, there were 6 birds in the wild and 21 in captivity. In 1992 when there were a total of 52 live birds in captivity, the first captive-reared California Condors were released into the wild. They have since been introduced to southern Utah and western Arizona. They may be observed today in Grand Canyon and Zion National Parks. In California, there are 25 known in Pinnacles National Park. As of October 2014, there were 220 California Condors alive in the wild and a total of 435 in the wild or in captivity. Captive birds are housed in Los Angeles and San Diego zoos and wildlife parks.

Part of the controversy involved whether or not there should be any birds left to reproduce in the wild. Various studies have indicated the importance of genetic diversity to survival of the species. Diversity is maintained by reproduction among as many breeding pairs as possible. The proponents of captive breeding contended that to achieve as much diversity a possible, all birds ought to be kept in captivity until there were enough to establish a viable population in nature. Furthermore, hatching success is much higher in captivity, where adverse conditions such as predation by Common Ravens, *Corvus corax* (figure 9.49), can be eliminated.

Critics of the program said that if all birds were in captivity, their habitat would be doomed to destruction in their absence. Theoretically, there can be no critical habitat if there are no animals in the area to protect. Critics were also quick to point out that one California Condor death occurred from stress suffered during handling by a US Fish and Wildlife biologist. Critics also fear that birds raised in captivity will be too tame to make it in the wild.

Because of the precipitous decline in the numbers of these wild birds, an agreement was reached in 1986 to remove them all from the wild. Meanwhile, in order to get a better idea of ranges and feeding behavior, female Andean Condors, *Vultur gryphus*, with radio collars were released into the California Condor refuge. If lead poisoning is indeed the major threat to further existence of this bird, precautions will have to be made to ensure that newly released birds do not contact carrion containing lead pellets. The problem of lead poisoning will have to be addressed sooner or later anyway. Whatever may be the result of this outpouring of effort and money, it cannot be stated that California Condors were ignored. With luck, future generations will continue to see these magnificent birds in the wild.

The Golden Eagle, *Aquila chrysaetos* (figure 10.15), is one of the few birds that might be confused with a California Condor or a Turkey Vulture, *C. aura*, in flight. With a wingspan exceeding 7 ft (2.1 m), and soaring with slightly upturned wing tips, the Golden Eagle may be seen over open country throughout the state. Golden Eagles build huge stick nests in tall trees or on cliffs where they are afforded a good view of open country. With their extremely keen vision, Golden Eagles search for a variety of diurnal rodents and carrion. Despite their protected status, they are often shot by ranchers, who accuse them of preying on sheep or other livestock. Although Golden Eagles are not rare in California, they are not common either, and it is possible that these large birds could face a fate similar to that of the California Condor. Studies confirm that 30% of the Golden Eagles sampled in the southern Coast Ranges contained high concentrations of lead in their blood.

Another victim of eggshell thinning is the Peregrine Falcon (Peale's), *Falco peregrinus pealei*. Its numbers declined markedly in association with the widespread use of DDT. Before World War II, there were from 200 to 300 pairs in California. There were only five pairs left in California by 1970 and it was one of the first to be placed on California's list of endangered species. Under full protection and management, the number of breeding pairs had increased markedly by 2014, at which time the California subspecies was proposed for delisting. It was federally delisted in 1999. This is not specifically a Coast Ranges bird, nor is the species uniquely Californian. It is a species whose native range is in temperate regions throughout the world. This was the sovereign's bird of the Middle Ages, trained to the sport of falconry by noblemen and commoners. With a claim to fame as the "fastest bird in the world," it has been clocked at 217 mph (347 kph) in a dive.

It is appropriate to discuss the Peregrine Falcon here because Morro Bay is one of the places in California where these birds have persisted in nature. In 1970, the pair at Morro Bay was one of the few nesting pairs known in the state. Another was in the Coast Ranges in Sonoma County.

These raptors feed on other birds, which they catch in midair in a spectacular display of aerial acrobatics. In Yosemite, they are known to feed on swifts, acrobats in their own right. Their primary natural enemies are Great Horned Owls and Golden Eagles, which prey on chicks in nests. An "unnatural" enemy could be a human who steals the eggs or chicks. On the world scale, there apparently is a profitable black market for raptors.

In California, the species was hard hit by pesticides. All of the southern California populations and most of those in the Sierra Nevada were extirpated. The remainder of the state retained less than 10% of its historic population.

The Peregrine Falcon represents a success story for management practices. Since DDT was banned in the United States in 1972, the species has been making a comeback. Likewise, the threat of mass species extinction on a worldwide scale caused the abandoning, or at least curtailment, of DDT use all over the world, enabling Peregrine Falcons to make a global comeback.

In California, there is still some problem with thin eggshells. This problem is attributed to DDT, DDE (dichlorodiphenyltrichlo-roethane), and various other chlorinated hydrocarbons still present in nature. The source of these chemicals is not certain, but there are four possibilities. Three involve local conditions. Pre-1972 residues may still be in the food chain, the chemical may be being used illegally, or it could be present as "impurities" in related pesticides. The fourth possibility is that various migratory prey species for the falcons, particularly the swifts, are picking up the DDT when they migrate to Mexico, where the pesticide is still legally used.

In particular, management of the Peregrine Falcon represents a success story for captive breeding. Aided by falconers, captive-bred birds were released into their former haunts. Peregrines mate for life and annually breed in the same territory. Nests are built on cliffs such as Morro Rock and El Capitan in Yosemite. Breeding pairs in nature will double-clutch if eggs are removed from their nests. In the past, many of the eggs were thin-shelled and would not hatch in nature. Pressure or weight of the nesting bird during incubation would cause them to break. Eggs removed from the nests were replaced with artificial eggs during which time the thin eggs could be hatched in an incubator and the chicks returned to the nest after hatching.

Another technique involves what is known as hacking. After the eggs are removed, the adults often will lay more eggs, a process known as double-clutching. Eggs are hatched in an incubator and young birds are fledged in an area where there are no adult birds. A juvenile is placed in a box on a cliff or on a tower built for hacking, and it is fed indirectly while it practices flying and catching its own prey. After five weeks of training, it is left to the process of natural selection. The success rate has been quite high using this technique. About 80% of these birds become established on their own—a number that is particularly striking since the natural survival rate for birds in their first year of life is only 25%.

One of the most interesting success stories of Peregrine Falcons is the fact that they have adapted to urban life. Throughout the United States and Canada, these birds are nesting in cities, on the ledges of skyscrapers. As of 1988, there were seven birds living in Los Angeles and six in San Francisco. To the delight of the people who work and live in these cities, these magnificent birds may be watched as they hunt, feeding particularly on the pigeons (Rock Pigeons), *Columba livia*, so common in these areas. What an irony: the high-rise topography of cities that has displaced so many wild animals had become the home for one of the most endangered!

Each year the effect of these management techniques has become more apparent. In 1977, two young from the captive-breeding program were released; one survived. By 1983, however, over 200 had been released, either by hacking or by fostering of young that were hatched in captivity. The number had risen to 103 by 1989. The Peregrine Falcon is returning to California, largely because of the captive-breeding program. The population in 2015 has been estimated at 1650 pairs.

8

Cismontane Southern California

MAINLAND AND ISLANDS

FIGURE 8.1 Winter storm in the San Bernardino Mountains. View eastward from Rim of the World Highway.

THE TERM SOUTHERN CALIFORNIA, as it is used by many people, refers to the southwestern portion of the state, where rapidly proliferating urban sprawl is covering the coastal lowlands. The word *cismontane* means "this side of the mountain." Thus, the region covered by this chapter consists primarily of the coastal sides of the Transverse and Peninsular Ranges, an area that includes the coastal sides of mountains in Ventura, Santa Barbara, Los Angeles, Orange, and San Diego Counties, much of which is often referred to as the Los Angeles Basin. Technically, the Los Angeles Basin is part of the Peninsular Ranges, but the term specifically refers to an alluvial outwash that includes most of Los Angeles and Orange Counties, as well as western San Bernardino and Riverside Counties. In terms of area, this region encompasses less than one-sixth of the state, but it contains over half of the human population, most of which is concentrated on the coastal plain. Above 1000 ft (300 m) elevation, Cismontane Southern California is mostly uninhabited brushland.

Southern California is the part of the state characterized by scrub vegetation, commonly known as Chaparral. The word *chaparral* is derived from Spanish and originally referred to a thicket of shrubby evergreen oaks. Now it is used to denote habitats characterized by dense stands of brush. Chaparral is not restricted to Cismontane Southern California, but in this region it is the dominant form of vegetation. Statewide, chaparral types of vegetation also occur in the drier parts of the Sierra Nevada foothills and the Coast Ranges, particularly on south-facing slopes or in association with serpentine or other depauperate soils. In total, variations of Chaparral represent one of California's most common communities, covering about 12 million acres (29 million ha) of terrain. In this chapter, four brushy communities will be discussed, all of which have been referred to at one time or another as Chaparral.

Scrub vegetation occurs throughout the world in regions with a Mediterranean climate. This type of climate, characterized by long, hot summers and moderate winter precipitation including snow at upper elevations (figure 8.1), promotes native vegetation such as evergreen shrubs with relatively small leaves. Because the leaves of such shrubs are often covered with a waterproof coating of resinous or waxy material, some authorities prefer to call this sclerophyllous (hard-leaf) vegetation. Associated with the long summer drought, vegetation becomes very dry by autumn. At this time, it is susceptible to fire. Chaparral therefore has evolved in conjunction with fire for millions of years. In many respects, the chaparral communities require periodic burning for proper growth and vigor. The conflict between human habitation and periodic fires is never-ending.

The mountain ranges of Cismontane Southern California are shown in figure 8.2. North of the Los Angeles Basin lie the Transverse Ranges. These ranges are aligned on an east-west axis, an orientation quite different from other mountain ranges of the state. This alignment is caused by northward motion of the Pacific plate along the San Andreas fault. At the contact point between the Pacific plate and the westward-moving North American plate, there has been a good deal of rotation in subplates, which act as ball bearings between the two major plates. This rotation is reflected in the east-west orientation of the Transverse Ranges. From east to west, the three main ranges of this province are the San Bernardino Mountains, the San Gabriel Mountains, and the Santa Monica Mountains. The northern Channel Islands appear to be a westward extension of the Santa Monica Mountains where all but the mountaintops are covered by seawater. Cajon Pass, through which Interstate 15 passes, separates the San Gabriel and San Bernardino Mountains. San Fernando Pass, through which Interstate 5 runs, separates the San Gabriel and Santa Susana Mountains to the west. Two other lesser ranges lie north and west of the San Gabriel Mountains. The Sierra Pelona is north of Highway 14 and Santa Clarita, and the San Emigdio Mountains are farther north, west of Gorman along Interstate 5. Another prominent Trans-

FIGURE 8.2 Cismontane Southern California.

verse range is the Santa Ynez Range, which parallels the coast from Santa Barbara to Ventura.

The Peninsular Ranges are located east and south of the Los Angeles Basin. They are called the Peninsular Ranges because they are the northernmost ranges of the series of mountains that make up the Baja California peninsula. From north to south, the main ranges of this province are the San Jacinto Mountains, the Santa Rosa Mountains, and the Laguna Mountains. San Gorgonio Pass lies between the San Jacinto Mountains of the Peninsular Ranges and the San Bernardino Mountains of the Transverse Ranges. Interstate 10 lies in this gap. The Santa Ana Mountains are also part of the Peninsular Ranges, but they lie to the west of the San Jacinto and Santa Rosa Mountains, separated from them by a broad valley that includes the Hemet and Lake Elsinore areas. South of this basin are some smaller ranges such as the Agua Tibia Mountains that connect the Santa Ana Mountains to the Laguna Mountains. In the Agua Tibia Mountains, Palomar

Mountain, at 5202 ft (1586 m) is the home to the famous 200 in (450 cm) telescope that was built there to take advantage of clear skies and the absence of urban light pollution.

Peaks of the Transverse and Peninsular Ranges are remarkably high. Surrounding the Los Angeles Basin, which is slightly above sea level, three prominent peaks rise more than 2 vertical mi (3.0 km). San Antonio Peak (Mount Baldy), in the San Gabriel Mountains, rises to 10,064 ft (3068 m). Mount San Jacinto, at 10,805 ft (3293 m), is the highest in the Peninsular Ranges. Mount San Gorgonio (Greyback), in the San Bernardino Mountains, at 11,502 ft (3506 m), is the highest in southern California. The San Gorgonio Wilderness has two peaks above 11,000 ft (3250 m) and by some counts, 16 more peaks above 10,000 ft (3048 m), most along San Gorgonio Ridge. At 5,678 ft (1733 m) Santiago Peak is the highest in the Santa Ana Mountains. Including nearby Modjeska Peak and the low saddle in between, this mountain known locally as Saddleback, is visible from all over Los Angeles, Orange, western Riverside, and San Bernardino Counties.

The Los Angeles Basin is a flood plain surrounded by mountains. Deep beneath the flood plain, there is a series of north-south-trending faults, similar to the geologic picture in the Peninsular Ranges. This is the basis for including the basin as part of the Peninsular Ranges province. For millions of years floodwaters carried by the Los Angeles, San Gabriel, and Santa Ana Rivers deposited alluvial outwash from the mountains. This alluvium is up to 14,000 ft (4200 m) thick and extends some 75 mi (120 km) from the coastline to the base of the San Bernardino Mountains. The series of faults buried beneath the alluvium have been responsible for major earthquakes in the recent past. Trapped in pockets associated with these faults are several large petroleum deposits. For this reason, the Los Angeles Basin has been one of the major oil-producing regions of the United States. In fact, the deadly 1933 earthquake in Long Beach has been attributed to deep oil drilling and increased production prior to the quake.

CLIMATE

The climate of the southern California area is the product of cold ocean water and latitude, a combination of maritime and Mediterranean climates. The maritime influence usually prevails in the Los Angeles Basin, causing a persistent marine layer or temperature inversion layer. Consequently, the area is often hazy, foggy, or smoggy. Many winter storms reach their southernmost latitude in southern California. During summer, the region lies under a high-pressure zone associated with descending dry air from the upper atmosphere. This persistent high pressure generally prevents precipitation in summer, although tropical storms sporadically work their way into southern California, particularly in the Peninsular Ranges (figure 1.5). For example, in August 1981, a tropical storm dumped 11.5 in (290 mm) on Campo, in San Diego County, in only 80 min.

Four distinct seasons occur in southern California, but the primary growing season is winter. Winter rain is followed by spring fogs, which give way to summer haze and smog. Summer temperatures are often in the 80s and 90s. Autumn brings Santa Ana winds, which blow from the Mojave Desert toward the ocean. They push the marine layer out to sea and become heated by compression as they drop into the basin. This is the season when scrub vegetation is driest. Fires can be fanned into holocausts by this hot wind. Winter is characterized by alternating rainstorms and clear, sunny days. This is a time of renewal, when burned vegetation resprouts and ephemeral plants provide hillsides and valleys with a brilliant green hue followed by colorful wildflowers (plate 14B).

Precipitation at the Los Angeles civic center averages about 15 in (38 cm) per year. Closer to the coast, however, it is half that. On Mount Wilson, in the San Gabriel Mountains, it is twice that. This range of precipitation, from 8 in (20 cm) at the coast to 30 in (75 cm) in the mountains, is a clear example of the effect of elevation on precipitation. Adiabatic cooling causes lower average temperatures at higher elevations. This cooling increases precipitation in the mountains and also increases the proportion of precipitation that falls as snow. Also, along the coast, the extremes of temperature are reduced through the influence of maritime climate (Table 8.1). Close proximity of the ocean increases the amount of moisture in the air and tends to moderate temperature fluctuations.

The influence of rain-shadow effect may be illustrated by comparing records from two weather stations in the San Gabriel Mountains. Mount Wilson is directly above Los Angeles. This is the mountain with the television transmitters. Table Mountain, in the vicinity of Big Pines, is on the desert side of the San Gabriel Mountains. Both weather stations record the same average annual temperature, 56°F (13°C), but the average annual precipitation on Table Mountain is only 21 in (52 cm), about 10 in (25 cm) less than that on Mount Wilson.

TABLE 8.1
A Comparison of Average Temperatures at Three Points in the Los Angeles Area, Indicating Influence of Maritime Climate and Elevation

	Long Beach (coastal)	San Bernardino (interior basin)	Mount Wilson (San Gabriel Mts.)
Coldest month	53°F (12°C)	48°F (9°C)	32°F (0°C)
Warmest month	68°F (20°C)	80°F (27°C)	62°F (17°C)
Annual range	15°F (8°C)	32°F (18°C)	30°F (17°C)
Elevation	Sea level	400 ft (120 m)	5000 ft (1600 m)

Slope effect is superimposed upon the effects of temperature and precipitation. Mediterranean climate, with its long, hot summer, accentuates slope effect. South-facing slopes, with their greater degree of drought stress, are cloaked with drought-tolerant vegetation. This phenomenon is particularly evident in the Transverse Ranges because of their east-west orientation. The entire side of the ranges north of the Los Angeles Basin is south-facing. On that side, chaparral vegetation extends all the way up to about 5,000 ft (1600 m). At that elevation on the north side (the desert side) of the mountains, Mixed Coniferous Forest occurs even though actual precipitation there is lower (plate 14A[6A]). Water is more available because north-facing slopes experience less evaporation than south-facing slopes, and snow lasts longer before melting. In contrast, in the north-south-oriented Peninsular Ranges, vegetational zonation is more like that of the Sierra Nevada. As one would expect, Mixed Coniferous Forest occurs at a lower elevation on the coastal side of the mountain, where there is more precipitation.

GEOLOGY

Southern California has many faults (figure 8.3). Most of this region lies west of the San Andreas fault. About 25 million years ago (Miocene), right-lateral movement on this fault began carrying southern California to its present latitude from its former position adjacent to mainland Mexico. However, the area contains several other faults running in a parallel direction. From east to west, these are the San Jacinto fault, the Whittier-Elsinore fault, and the Newport-Inglewood fault. The Newport-Inglewood fault runs roughly parallel to the coast on the western edge of the Los Angeles Basin. It extends out to sea and remains offshore south of Corona Del Mar. At least two other major faults lie offshore. One lies between Santa Catalina Island and the mainland, and the other lies between Santa Catalina and San Clemente Islands.

Motion along these faults has been up-and-down (dip-slip) and/or sideways (strike-slip). Dip-slip motion has resulted in uplift of fault-block mountains. Authorities generally place most of the uplift during the last 2 million years. Strike-slip motion has been right-lateral, similar to that of the San Andreas fault. The overall picture is that west of the San Andreas fault, a series of slices of continental crust are sliding against each other as everything westward is carried toward the north.

The Transverse and Peninsular Ranges are all fault-block ranges. The San Jacinto and Santa Rosa Mountains have been uplifted along the east side of the San Jacinto fault. The Santa Ana and Laguna Mountains have been uplifted along the west side of the Whittier-Elsinore fault. The Palos Verdes Peninsula is an uplifted mountain lying in the western edge of the Los Angeles Basin just west of the Newport-Inglewood fault.

At many points along these faults, there are hot springs. Borrego, Gilman, Eden, and Soboba hot springs are located along the San

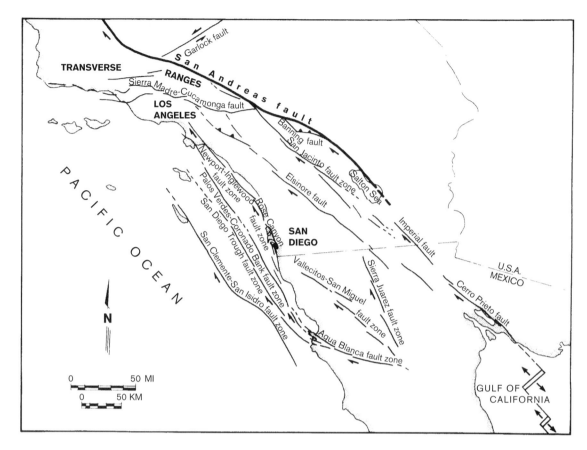

FIGURE 8.3 Some important southern California faults (reprinted with permission from Abbott, P. 1989. The Rose Canyon Fault. *Environment Southwest* 524 (Winter/Spring):12–16).

Jacinto fault. Agua Caliente, Warner, Murrieta, Elsinore, and Glen Ivy hot springs are on the Whittier-Elsinore fault. The therapeutic value of hot mineral baths is subject to debate; nevertheless, many of these springs are still operated in association with resorts. Some of these, such as San Jacinto and Elsinore, have had towns developed around them. In the Imperial Valley, to take advantage of this source of heat, geothermal electrical power plants have been constructed south of the Salton Sea.

There is a record of major earthquakes rocking the Los Angeles Basin as the earth slipped along these faults. The magnitude 6.3 earthquake near Long Beach in 1933 was on the Newport-Inglewood fault. This earthquake was the incentive to build homes that are more earthquake-proof. The well-known wooden-frame house with stucco exterior, considered in many parts of the United States to be a California-style house, was a response to a demand for a design that would flex and bend as the earth moved. Between 1968 and 1988, there were no fewer than five earthquakes of magnitude 6.0 or higher on these faults where they extend into the Colorado Desert east of the Peninsular Ranges.

The magnitude 6.6 earthquake in 1971 was on the San Fernando fault at the base of the San Gabriel Mountains. The 1987 magnitude 5.9 earthquake near Whittier was on a previously unknown blind-thrust fault. In 1994, at 6.7 the largest earthquake in modern times was on a previously unknown blind-thrust fault near Reseda. Now known as the Northridge earthquake, it resulted in serious damage to buildings and highways, and 60 people were killed.

The Transverse Ranges lie on the "big bend" of the San Andreas fault. Here, since the Miocene, the mountains have been rotated about 90° clockwise, so that they now lie on an east-west axis. The faults associated with them also run east-west. The San Fernando fault lies across the southern base of the San Gabriel Mountains, and the Santa Monica fault lies along the southern edge of the Santa Monica Mountains and the Channel Islands.

The position of the San Bernardino Mountains and two smaller ranges eastward represents a bit of an enigma. These members of the Transverse Ranges lie east of the San Andreas fault. The San Andreas passes between the San Bernardino and San Gabriel Mountains and continues along the northern base of the San Gabriel Mountains on a mostly east-west line before turning northward once again. This is the "big bend." Paleomagnetic data in volcanic rocks of Miocene age provide evidence that the San Gabriel Mountains, the Santa Monica Mountains, and the Channel Islands have rotated about 90° clockwise. The absence of such volcanic rocks in the San Bernardino Mountains leaves us without proof that this range also has been rotated, although from outward appearances it seems to lie on a similar east-west line. The issue is further complicated by the fact that, because they lie east of the San Andreas fault, the San Bernardino Mountains must not have experienced the northward translocation exhibited by the other Transverse Ranges. Apparently, as the San Andreas continues its relentless motion, the San Bernardino Mountains will be left behind as the western portion of the Transverse Ranges is carried northward.

Rocks of the Transverse Ranges are among the oldest in California (figure 8.4). The comparatively low precipitation in southern California has caused less erosion than in the Sierra Nevada. Consequently, many old rocks remain on top of the granitics of the Southern California batholith. A number of igneous and metamorphic rocks from the northwestern part of the San Gabriel Mountains apparently are older than a billion years. The oldest is a darkly banded gneiss dated at 1.7 billion years. This rock is visible on the Angeles Forest Highway near Mill Creek, where the highway passes through Singing Spring Tunnel (figure 8.5). Included among the old rocks of the San Gabriel Mountains is another gneiss dated at 1.05 billion years of age. An igneous rock known as anorthosite, dated at 1.02 billion years, is visible west of the road at Baughman Spring. Anorthosite is rare on this planet, limited to a few Precambrian localities, but interest was generated in this formation when the Apollo astronauts discovered that this rock constitutes the bright lunar highlands of the moon.

Correlation of old rocks with those in the Orocopia Mountains, 100 mi (160 km) southward, on the other side of the San Andreas fault, has been an important part of the evidence substantiating the magnitude of drift that has occurred. A green schist known as the Pelona Schist, which forms much of the eastern part of the San Gabriel Mountains, is found on the other side of the San Andreas fault in the San Bernardino Mountains, as well as in the Orocopia and Chocolate Mountains farther south. It is difficult to explain this distribution, but one hypothesis is that about 70 million years ago, as shallow subduction occurred, the schist, which has its origin to the west, was thrust on top of the eastern terrain. Later, about 25 million years ago, when the San Andreas fault began its activity, the schist west of the fault was dragged northward. The schist formerly of the San Bernardino Mountains was deposited north of the San Gabriel Mountains by the same events that deposited it in the Orocopia and Chocolate Mountains, but it now lies slightly to the south of the San Gabriels due to 100 mi (160 km) of displacement.

In addition to the very old rocks just mentioned, there are rocks of Paleozoic age (700–245 million years old), including schists, shales, sandstones, and limestones. These old rocks are exhibited in roadcuts that traverse the Transverse and Peninsular Ranges, especially

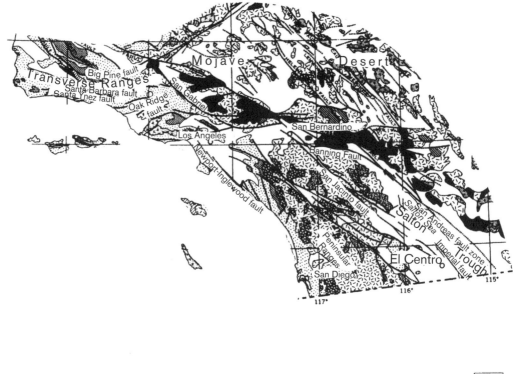

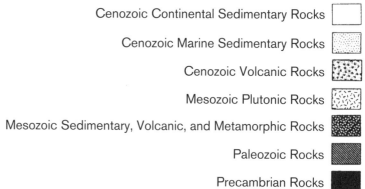

Cenozoic Continental Sedimentary Rocks ▢

Cenozoic Marine Sedimentary Rocks ▢

Cenozoic Volcanic Rocks ▢

Mesozoic Plutonic Rocks ▢

Mesozoic Sedimentary, Volcanic, and Metamorphic Rocks ▢

Paleozoic Rocks ▢

Precambrian Rocks ▢

FIGURE 8.4 Geologic map of Cismontane Southern California.

in the San Gabriel and San Bernardino Mountains. There are several cement plants where Paleozoic limestones are mined on the north side of the San Bernardino Mountains. For some reason, Paleozoic rocks are not common in the Peninsular Ranges. Those that are present are found mostly on the desert side of the San Jacinto and Santa Rosa Mountains. Some Paleozoic limestone has been mined for cement near Riverside.

Mesozoic rocks are best represented by the Southern California batholith. These are granitics and gabbros that became intruded into older rocks during the time of major subduction. Similar to the Sierra Nevada, all southern California ranges have a granitic component. In southern California, however, granite is often covered by older rocks. Where the older rocks have been stripped away by erosive processes, the granite is exposed and shows a

FIGURE 8.5 The oldest rock in the San Gabriel Mountains, 1.7 billion-year-old gneiss at Singing Spring Tunnel, Angeles Forest Highway.

weathering pattern similar to that of desert ranges. Lily Rock (Tahquitz) near Idyllwild is a good example of a granite dome. Where the Ortega Highway crosses the Santa Ana Mountains, near the summit is a well-exposed outcrop of the batholith. Similarly, where Interstate 8 crosses the Laguna Mountains between San Diego and El Centro, the granite is exposed.

Edaphic ecologic islands are found in the southern California area, too. In the Laguna Mountains are several gabbro outcrops that weather into reddish, iron-rich soil with peculiar chemical components. These gabbro soils resemble serpentine soils in that they have a high magnesium content. Gabbro soils support a large number of endemic plants found nowhere else in the world. The endemic Cuyamaca Cypress, *Hesperocyparis (= Cupressus) stephensonii*, is restricted to gabbro soil in upper King Creek on the west side of Cuyamaca Peak (figure D, p. 351). The only southern California location for serpentine is also one of the only locations for Knobcone Pine, *Pinus attenuata*, in southern California (figures A and B, p. 348).

Processes associated with intrusion of the batholith have resulted in many mineral deposits of economic value. Eighteen different gold-mining districts have been described from the Transverse Ranges alone. Some of the earliest gold-mining operations in the United States were in the northwest part of the San Gabriel Mountains. As early as 1834, placer operations in gravels near Castaic were under way; Placerita Canyon gets its name from these operations. In the Peninsular Ranges, many small gold-mining operations were in operation in the early 1900s, particularly in the Santa Ana Mountains. Important gold mines were located in the Laguna Mountains, in the area around Julian. Of special interest in the Julian area was one of the few nickel mines. Nickel was mined from important sulfide minerals associated with gabbro deposits. Other metals mined in southern California include lead, zinc, and some copper. Most of the ore deposits in the Santa Ana Mountains were associated with the Bedford Canyon Formation and were fairly small, but in 1878 a silver strike in Silverado Canyon triggered a rush that lasted several years.

Of special interest are gem mines along the San Luis Rey River on the southwestern side of the Agua Tibia Mountains. In this area, near the present town of Pala, transparent crystals of pink, green, and blue tourmaline are still being mined. Giant crystals from 5 to 20 in (12–50 cm) in diameter and up to 3 ft (1 m) long have been collected. Some gem-quality topaz and garnet also were collected from these areas. During World War II, pink lithium-bearing mica, called lepidolite, was also mined here. On a worldwide scale, such high concentrations of rare minerals are uncommon. The reason for their occurrence here is unknown.

Mesozoic rocks in southern California also include island-arc volcanics that became accreted to the continent during subduction. Remnants of these volcanics are found in the Peninsular Ranges near San Diego and in the Santa Ana Mountains, where they exist as roof pendants. The Santiago Peak (Saddleback) volcanic rocks have been estimated to be 2300 ft (700 m) thick in some places.

During the Mesozoic, much of southern California was under water; therefore, Mesozoic rocks include some interesting fossil-bearing marine sediments. In the Santa Ana Mountains near Silverado Canyon, and through the southern Peninsular Ranges in San Diego County, there is an interesting section of upper Cretaceous marine sediments, approximately 80 million years old, that have yielded large fossil ammonites. These ammonites, up to a meter in diameter, represent a group of organisms that were once very common but now are nearly extinct. Although these organisms had octopus-like soft parts, they bore a large coiled shell that is preserved in fossils. The modern-day chambered nautilus is the only remnant of that formerly abundant group. In 2002, in the Ladd Formation, dinosaur fossils were found. Toe bones and vertebrae were identified as part of a Hadrosaur skeleton, an animal that normally lived along the coastline. It is speculated that the animal probably died on land and was washed into the sea, after which the body sank and became covered by marine sediments.

In many areas of the southern California mountains, the Cenozoic story (the past 65 million years) begins as it does in the desert, with extensive layers of rock missing, indicative of some 40 million years of erosion. In association with such an unconformity, Miocene rocks are often found on top of much older Mesozoic rocks. Where early Cenozoic rocks are present, they are mostly marine sediments. The thickest section of these sediments is in the Ventura Basin, where they have accumulated to a depth of nearly 5,000 ft (15,000 m). Included in this record are about 2500 ft (760 m) of sediments of terrestrial origin. These are red siltstones dated at about 35 million years. These beds, known as the Sespe Formation, are also found farther south, in the Vasquez Rocks Natural Area Park of the northern San Gabriel Mountains. Sespe rocks are also found on both sides of the El Toro Valley, in Black Star Canyon and Laguna Canyon. The Los Angeles Basin includes about 14,000 ft (4200 m) of mostly marine sediments corresponding to the same time interval as those of the Ventura Basin.

Other old sedimentary beds scattered around southern California indicate that the area was under seawater for most of its history. There was an Andean type of shoreline until about 65 million years ago, when uplift of the coastline in conjunction with a tropical climate caused a great deal of erosion. Much of the eroded material was deposited offshore in a shallow marine environment. Shallow marshes and swamps occurred along the shoreline, and sediments from these terrestrial environments are found as siltstones and clay stones, many red in color. These can be found near the entrance to Silverado and Santiago Canyons on the western border of the Santa Ana Mountains (figure 8.6).

In the Cajon Pass area, on the eastern end of the San Gabriel Mountains, there is a region where the San Andreas and San Jacinto faults are only about 2 mi (3.2 km) apart. The San Jacinto fault passes through Lytle Creek Canyon, and the San Andreas passes through Lone Pine Canyon, a couple of miles to the north. In this area, some interesting sedimentary beds are exposed by road cuts and along the railroad. Among these are marine beds just northeast of the San Andreas fault. These are the youngest marine beds north of the San Gabriel Mountains. They are unique because of their age

FIGURE 8.6 Sespe Formation: sedimentary rocks of terrestrial origin near the junction of Silverado and Santiago Canyons, Orange County.

FIGURE 8.7 Mormon Rocks near Cajon Junction, San Bernardino County.

(40–65 million years old) and because they lie on the north side of the San Andreas fault. The distinctive tilted sandstone beds known locally as Mormon Rocks (figure 8.7) also lie north of the San Andreas in this area, but they are sandstones of terrestrial origin. They represent a river deposit that is perhaps only 5 million years of age.

About 25 million years ago (Miocene), the modern picture of coastal California, with its faulting and rifting, began to emerge. About this time, a large, mountainous island similar to, but much larger than, present-day Santa Catalina Island was uplifted on the western side of the Newport-Inglewood fault. This island was composed of metamorphic trench

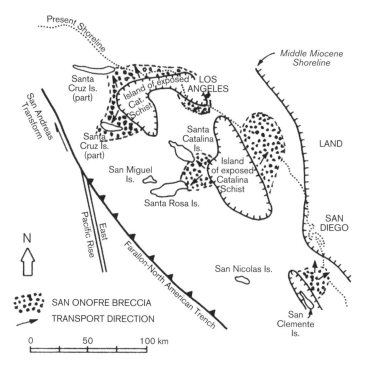

FIGURE 8.8 Reconstructed paleogeography of the southern California borderland during deposition of the San Onofre Breccia (Lower and Middle Miocene). Locations of islands are shown for reference; they were not all islands at the time (after Howell and others, 1974; from Rowland, S.M. 1984. Geology of Santa Catalina Island. *California Geology* 37(11):239–251).

rocks (Franciscan Formation) formed from the former subduction zone. To the interior of this island, where the present coastal plain is located, there was a region of deep marine water covering what is now the Los Angeles Basin.

The Upper Miocene shoreline is represented by Topanga sandstone, a yellowish rock that contains many windblown hollows and caverns. It is found around the edge of the Los Angeles Basin in places such as Topanga Canyon, the mouth of Silverado Canyon, and in the San Joaquin Hills along Laguna Canyon.

In Middle Miocene, 12–15 million years ago, landslides and alluvial wash carried rock debris eastward from the large island. This debris, originally deposited underwater, is now located along the present shoreline, where it has become exposed by more recent uplift. This cemented landslide material is known as the San Onofre Breccia (figure 8.8). It forms the headlands and reefs of the Orange County area, which look somewhat like crude, wave-washed concrete. It is particularly well exposed as the cliffed headland at Dana Point, and along the headlands in Laguna Beach.

Sediments from this deepwater episode include many small fossils of invertebrates such as sand dollars and pectens. These sediments are found up and down the coast from Monterey to San Diego, indicating an extensive, deep, but narrow marine basin, or several such basins. The nature of these sediments indicates that these basins were up to 15,000 ft (4300 m) deep. These sediments are now represented by thinly bedded shales, rich in diatoms. Where exposed, they often form whitish to buff deposits. The light-colored cliffs on both sides of upper Newport Bay are composed of this material, which is known as the Monterey Formation. As mentioned in the previous chapter, near Lompoc, north of Santa Barbara, there is a deposit about 700 ft (230 m) thick, which is mined and marketed as diatomaceous earth. This sediment was also rich in organic material and has yielded a significant amount of petroleum.

Miocene was also a time of volcanism along the coast. The El Modena area, east of the town of Orange, is made up of rocks known as El Modena volcanic rocks. They are exposed in a road cut on Chapman Avenue at the edge of Orange Hill. Miocene volcanic rocks are also found at the foot of the San Gabriel Mountains in the Glendora area. Volcanic rocks in the Crystal Cove area composed of andesite and may represent the core of an old volcano. Columnar basalt is visible on the headland south of the beach at Crystal Cove. The volcanic rocks here are about 10 million years old.

The last 15 million years was marked by shallowing of basins, uplift of fault-block mountains, and a generally emerging coastline. Most evidence, however, indicates that the bulk of the uplift of the Transverse and Peninsular Ranges has occurred in the last 2 million years, since the Pliocene, although some authorities would have most of the uplift occurring during the last million years. Because these mountains are young and because precipitation in the southern part of the state has not been as great as it has farther north, these mountains are steep and jagged. In the absence of severe erosion, many very old rocks are still in place.

Sediments exposed in the Capistrano, Puente, and Laguna areas have yielded fossils of shark's teeth and marine mammals in addition to many microfossils. This fossil evidence tells us what the marine environment was like at the time.

During the Pleistocene, when the Sierra Nevada and more northern mountain ranges were experiencing waves of glaciation, the Transverse and Peninsular Ranges were mostly free of the glaciers. Glaciers are one of the most powerful forces of erosion; hence, as mentioned earlier, many very old rocks are still exposed in these southern ranges. Mount San Gorgonio is the southernmost peak to show evidence of glaciation. Popular landmarks such as Dollar Lake, Dry Lake, and Poopout Hill are actually glacial landforms.

Also during the Pleistocene, there were changes in sea level associated with alternating periods of glaciation and global warming. When the polar ice caps melted, sea level rose and the Los Angeles Basin became flooded. At this time, high points such as Palos Verdes Hills, Baldwin Hills, and the Coyote Hills near Whittier became islands. During periods of maximal glaciation, sea level was so low that the northern Channel Islands near Ventura were fused and became one large island. It is now assumed that even at its lowest point, the depth of the water between the mainland and the Channel Islands was never less than 300 ft (100 m). The distance between the Channel Islands and the mainland was probably never less than about 4.5 mi (7 km).

Elevation of the shoreline has exposed a series of marine terraces that reflect ancient shorelines. These wave-cut benches are represented by flattened terraces all along the coast. Most coastal cities and the Pacific Coast Highway are laid out along the first terrace. In profile, hills along the coast show a series of flat regions representing the older terraces. Studies of fossil marine organisms on these terraces reveal that the water was alternately warm and cold, and different marine organisms existed along the coast at different temperatures. The periods of cold water are correlated with the periods of glaciation in the Sierra and the periods of maximal flooding in the deserts.

BIOTIC ZONATION

Scrub Communities

Elevational zonation of vegetation in southern California mountains (figure 8.9) is similar to that of the Sierra Nevada. Southern California differs, however, in its abundance of drought-adapted scrub vegetation. Plant communities

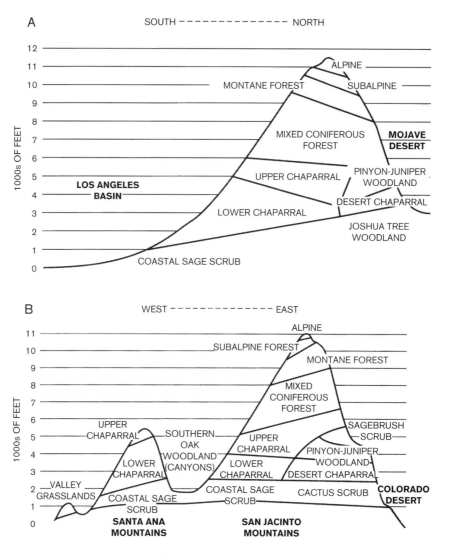

A. Biotic zonation in the Transverse Ranges.

B. Biotic zonation in the Peninsular Ranges.

FIGURE 8.9 Biotic zonation in southern California.

A. Biotic zonation in the Transverse Ranges.

B. Biotic zonation in the Peninsular Ranges.

of Cismontane Southern California consist primarily of different kinds of Chaparral. Chaparral communities dominate on the coastal side of the Peninsular Ranges, Transverse Ranges, and southernmost Coast Ranges (plate 12A). The community also predominates southward to the Ensenada area, in northern Baja California. Forest and mountaintop communities have been discussed in previous chapters. It is important, however, to point out that the southern California forest communities are composed mostly of elements that predominate in the drier Sierra Nevada forests. For example, the dominant conifer of the Mixed Coniferous Forest is Jeffrey Pine, *Pinus jeffreyi*, not Ponderosa Pine, *P. ponderosa*, as is often reported. Sierra Redwoods are not native to southern California today, although they occurred here in the past, as indicated by cones found in the La Brea Tar Pits.

Pronounced slope effect complicates an otherwise simple picture of chaparral communi-

TABLE 8.2
A Comparison of Southern California Scrub Communities

Community	Slope Aspect	Growth Habit	Examples
Coastal Sage Scrub	South-facing	Drought-deciduous; small leaves; phytotoxins	California Sagebrush, Coast Brittle-bush, monkeyflowers, true sages
		Succulent	Prickly Pears, Chollas, Live-forevers
	North-facing	Evergreen; large leaves	Toyon, Laurel Sumac, Lemonadeberry, Fuchsia-flowered Gooseberry
Lower Chaparral	South-facing	Evergreen; small, sclerophyllous leaves	Chamise, California lilacs
	North-facing	Evergreen; oval, spiny, sclerophyllous leaves	Scrub oaks, Holly-leaved Redberry, Holly-leaved Cherry
		Vines	Wild peas, honeysuckles, wild cucumbers
Upper Chaparral	South-facing	Evergreen; vertically oriented sclerophyllous leaves; sun tracking	Manzanitas, silk-tassel bushes, Western Mountain Mahogany
	North-facing	Evergreen conifers	Big-cone Douglas-fir, Coulter Pine
Desert Chaparral	South-facing	Drought-deciduous	Desert Apricot, Desert Almond, true sages
		Succulent	Prickly-pears, chollas, Mormon teas
	North-facing	Evergreen; large leaves	Sugar Bush, Desert Scrub Oak, Bigberry Manzanita, Jojoba

ties. Various attempts have been made to classify scrub vegetation into subgroups or communities. From 2 to over 10 subdivisions have been proposed. In this chapter, four communities will be recognized, based primarily on climatic considerations and easily recognizable generalizations in vegetative characteristics (Table 8.2). Within these four communities, there are pronounced differences based on slope exposure. The effect is that within each community, most groupings of related species appear on north-facing slopes at lower elevations and on south-facing slopes at the upper reaches of their distributions.

Coastal Sage Scrub

The Coastal Sage Scrub community has been referred to by some authorities as Soft Chaparral because many of the dominant plants bend easily and/or have soft, flexible leaves. Many of the shrubs are odoriferous and drought-deciduous. Such vegetation usually grows about knee-high (plate 13A), but even when taller shrubs are present, it is not difficult to walk through the community. This community occupies mostly alluvial soils at low elevation. It is not restricted to coastal regions but is found inland in suitable low-elevation valleys.

Coastal Sage Scrub does best under the influence of a maritime climate. Fog is important. Frost is also a critical factor; many of the component species do not do well where frost is common. The community is also remarkably drought adapted. Throughout much of its coastal distribution, precipitation averages 10 in (25 cm) per year or less. This precipitation regime qualifies Coastal Sage Scrub as a desert community. In some respects, it is not inaccurate to think of it as such, but the high humidity and moderate temperatures typical of maritime climate keep evaporation rates low, so that this community seldom experiences the drought stress typical of true desert vegetation. However, some of the community's plants and animals are also found in the desert, and several of the indicator species have close relatives in the desert. Aboveground productivity is low but is intermediate between average desert and average chaparral communities in other areas. The nearest desert analog to Coastal Sage Scrub is the Cactus Scrub community of the Colorado Desert. Both communities have many drought-deciduous and succulent components. This is particularly true farther south, toward Ensenada.

Also similar to desert vegetation, many species produce large, showy flowers. When in full bloom, the Coastal Sage Scrub community is very beautiful. It is surprising that more species from this community have not been adopted for suburban landscapes. They are hardy and attractive, and they often have pleasant odors.

There is a pronounced difference between the shrubs on south-facing slopes and those on north-facing slopes. On south-facing slopes and exposed flats, the vegetation typically has small leaves and is drought-deciduous. California Sagebrush, *Artemisia californica*, is one of the indicator species here (figure 8.10). It has finely divided, almost filamentous herbage, which is shed during drought. When winter rains arrive, this is commonly the first species to grow new leaves and shoots. In some areas, this shrub is so abundant that whole hillsides take on a blue-gray appearance. California

FIGURE 8.10 California Sagebrush, *Artemisia californica*, winter foliage.

Sagebrush has a distinctive odor, similar to that of Great Basin Sagebrush, its close relative. Unlike the sages of this community, California Sagebrush is in the sunflower family (Asteraceae). Its flowers are not conspicuous, however, nor do they resemble typical sunflowers.

A more typical sunflower of Coastal Sage Scrub is California Encelia or Coast Brittlebush, *Encelia californica*. This shrub resembles its relative in the Cactus Scrub community, Desert Brittlebush, *E. farinosa*. Desert Brittlebush has white leaves, whereas Coast Brittlebush has light green foliage. Both are drought-deciduous and display a spectacular canopy of bright yellow sunflowers that track the sun. In low passes such as San Gorgonio Pass, these two species hybridize. Hybrids are particularly common on disturbed sites such as road cuts.

True sages can be identified by several common characteristics. They are members of the mint family (Lamiaceae); as such, they usually have a pungent odor. They have two-lipped flowers with long anthers and stigma. They usually have square stems, which, if not visible, can be detected by rolling the stem between the thumb and forefinger. Finally, the leaves usually occur in pairs, opposite on the stem. Most shrub species have their leaves arranged singly on the stem, alternating on each side.

Several species of true sages also are drought-deciduous components of this community. Black Sage, *Salvia mellifera*, is perhaps most common. It gets its name from its black stems, which become particularly conspicuous when the leaves drop. Leaves of Black Sage are the darkest green of the sages, and hillsides predominantly of Black Sage can take on a blackish appearance. Leaves are narrow and lance-shaped, and their surface is rough and mealy to the touch. White to pale lavender flowers appear in spring. Typical of sage flowers, they are tubular and two-lipped, resembling flowers in the Lopseed family, Phrymaceae (= Scrophulariaceae). They differ from monkeyflowers by having long reproductive parts (anthers and stigma) that project for some distance beyond the petals.

In contrast to Black Sage, there is Purple Sage, *S. leucophylla* (figure 8.11), named for its purple flowers. Leaves of Purple Sage resemble those of Black Sage, except that they are whitish, and flowers are rose-lavender. This species is common from Orange County northward. Southward there is another species, *S. clevelandii*, that is sometimes called Purple Sage, but it is probably more correct to call it Fragrant Sage. It has flowers of a darker purple hue, but otherwise it resembles the northern Purple Sage.

White Sage, *S. apiana*, has white leaves. This sage is very common and also occurs on the desert side of the Peninsular Ranges. It differs from the others by having much larger, velvety leaves. Also, its flowers are borne on long, whiplike stalks with the leaves crowded at the base. It resembles, and may coexist with, its near relative of the Cactus Scrub community, Wand Sage, *S. vaseyi*, whose name refers to its long, wand-like flower stalks.

FIGURE 8.11 Purple Sage, *Salvia leucophylla*.

The value of odor to plants has been the subject of some debate. The odor of flowers helps to attract pollinators, and the odor of herbage apparently is one of the mechanisms plants use to discourage herbivory. Presumably, these odors have evolved in conjunction with resinous coatings that help inhibit evaporation. Some plants produce toxic chemicals but do not produce odors; animals learn to avoid these plants by previous experience or through heredity. Odors, however, advertise that the plant may not be palatable. Other evidence indicates that these chemicals inhibit germination and the detritus produced by leaf fall keeps the area clear of competing vegetation, a phenomenon known as *allelopathy*. Perhaps these chemicals also promote the invasion of grassland by shrub vegetation.

Humans have come to appreciate the odors and flavors of many plants, some of which require getting used to. A good example is the

FIGURE 8.12 Orange Bush Monkeyflower, *Mimulus aurantiacus.* Note that the stigma lobes are closed in the flower on the right.

flavor of juniper berries, which are what gives gin its distinctive flavor. There are also many examples of plants used as spices to flavor food, including the traditional parsley, sage, rosemary, and thyme. It is further interesting to note that the native habitats of many of these plants are associated with Mediterranean climates. The sage usually used in cookery is known as Garden Sage, *S. officinalis.* It is grown widely in the east, but originally it was imported from the Mediterranean region of Europe. Locally, *S. clevelandii* is an appropriate substitute for cooking. The leaves of the California Bay Laurel, *Umbellularia californica* (figure 6.8), a riparian tree in southern California, are a good substitute for commercial bay leaves, which come from a European shrub of Mediterranean origin known as Sweet Bay, *Laurus nobilis.*

Some of the most beautiful flowering shrubs in this community are the bush monkeyflowers. They are inconspicuous most of the year because they are small in size and drought-deciduous. In the spring, however, when they come into bloom, entire hillsides are colored by their tubular, two-lipped blossoms. The tip of the pistil, the stigma, is divided into two conspicuous lobes. When a pollinator, such as a bee, visits the flower, it touches the stigma lobes, causing them to fold together (figure 8.12). This change in appearance is a signal to the pollinator that the flower already has been visited, thus optimizing the efficiency of the pollinator and the plant.

There is a good deal of confusion about speciation in the bush monkeyflowers. As of this writing, there are six varieties of Sticky Bush Monkeyflower, *Mimulus aurantiacus* (figure 8.12). This is a wide-ranging species inhabiting rocky hillsides and canyons from the southern Coast Range and Sierra Nevada to Baja California and desert mountains. Flowers range in color from a whitish to pale yellow to orange, to red. The variety *parviflorus* (= *M. flemingii*) is a conspicuous member of Coastal Sage Scrub on San Clemente, Santa Cruz, Santa Rosa, and Anacapa Islands. The Red Bush Monkeyflower, variety *puniceus*, perhaps the most beautiful of all, with flowers ranging from brick red to orange, was found along the coast from Laguna Beach southward and on Santa Catalina Island.

An interesting feature of these drought-deciduous shrubs is that during the dry season they may not become entirely leafless. Rather, they exhibit what is known as seasonal leaf dimorphism. During summer, they may exhibit only small terminal leaves, which, although representing only a fraction of the total seasonal leaf area, are capable of maintaining low levels of photosynthesis during drought stress.

FIGURE 8.13 Coastal Sage Scrub showing differences between north- and south-facing slopes. San Joaquin Hills, Orange County.

Another common plant associated with this community is California Buckwheat, *Eriogonum fasciculatum*. This low evergreen herbaceous shrub is one of the most common representatives of a very large genus with over 75 different species in California. The buckwheat family (Polygonaceae) includes a large number of plants found in nearly every habitat in the state. The common California Buckwheat has small waxy leaves borne in clusters or fascicles. Elongated flower stalks persist as dried structures long after the flowers have set seed. Native Americans would gather the small black seeds by the millions and grind them into a flour—hence the name "buckwheat flour." This species grows naturally in sunny disturbed areas, such as road cuts, from the coast to the desert. Sometimes it is deliberately planted to help stabilize soil along roads.

On north-facing slopes, or where water is locally more abundant, the Coastal Sage Scrub vegetation is quite different (figure 8.13). In these areas, the vegetation is more Chaparral-like in that the shrubs are large and evergreen, and they possess large root systems that penetrate deeply. Furthermore, the leaves are large compared to other scrub species, and this distinguishes the plants from true chaparral species, which, even on north-facing slopes, usually have smaller leaves. Many scientists do not consider these plants as part of Coastal Sage Scrub. Some would call areas dominated by these shrubs a version of Chaparral. Others simply divide them into alliances in accordance with the dominant shrub. Perhaps the most conspicuous of these large-leaved shrubs is Toyon, also known as Christmas Berry or California Holly, *Heteromeles arbutifolia* (figure 8.14). The name Toyon is derived from an old Spanish word that means canyon, in reference to a common site where it grows. The long, lance-shaped leaves do not truly resemble holly; the resemblance is based on the presence of toothed margins. Near Christmas, the plant may be quite showy, with large clusters of bright red berries. One story claims that the famous city in the low hills north of Los Angeles was named Hollywood in honor of this shrub. This tale has been refuted by some authorities. Toyon is available in many nurseries, and it has been planted in many landscapes. Given abundant water, it grows quite large, resembling a many-stemmed tree. At Christmas time, many of the early settlers with European roots would adorn their homes with berry-laden branches of California Holly.

FIGURE 8.14 Toyon, *Heteromeles arbutifolia.*

Apparently, the practice was so widespread that in 1920 a state law was passed that made it illegal to harvest the plant. The berries are an important food for birds and Coyotes. Native Americans and early settlers also ate the berries, but they boiled them first in order to remove a bitter flavor. Early Spanish settlers baked the berries with sugar and used them for a filling in Toyon pie.

Associated with Toyon at many localities is Laurel Sumac, *Malosma (= Rhus) laurina* (figure 8.15), a member of a large tropical family (Anacardiaceae) that includes economically important products such as cashews and pistachio nuts. This large shrub has elongated leaves that tend to fold up like a taco shell when it gets dry. Unlike many members of the Coastal Sage Scrub community, it apparently seldom goes dormant, continuing to grow throughout the year. One of its distinctive features is that it is quite frost sensitive; frost causes it to turn yellow to brown and die back to the ground. However, when the weather warms in spring, new shoots rapidly grow back from the roots, as much as a full 3 ft (1 m) in a single year. After 2 years, a person can barely tell that the shrub was damaged. In years past, locating this shrub was important for farm-

ers searching for appropriate sites to plant citrus. Orange County was originally settled and inhabited in this way. Of course, most of the orchards are now being replaced by housing tracts and shopping malls. Here we see an interesting pattern of succession. Coastal Sage Scrub becomes replaced by citrus, which in turn becomes replaced by buildings.

Also in the cashew family is another associate on north-facing slopes, Lemonade Berry, *R. integrifolia* (figure 8.16). This shrub has oval, leathery leaves that are usually toothed. The plant's large, flattened red berries are covered with a sticky, acidic secretion that can be soaked in water to make a drink that tastes very much like lemonade. This plant flowers early, as soon as winter rains begin. The flowers are borne in clusters and are white to rose in color. This is another shrub that does well in coastal landscapes. Its flowers provide winter color, and the fruits are attractive, too. One feature that makes this a particularly useful shrub in coastal communities is that it is mildly tolerant to salt spray. It occurs naturally on cliffs next to the ocean.

In interior valleys, where it is warmer, Lemonade Berry is replaced by a close relative known as Sugar Bush, *R. ovata*. Leaves of Sugar

FIGURE 8.15 Laurel Sumac, *Malosma laurina*.

FIGURE 8.16 Lemonade Berry, *Rhus integrifolia*.

Bush are intermediate in appearance between Laurel Sumac and Lemonade Berry. They are leathery, similar to Lemonade Berry, but they are taco-shaped, similar to Laurel Sumac. The leaves are shorter and broader than those of Laurel Sumac, and the petioles (leaf stems) are usually red. Identification is sometimes difficult because where ranges overlap, the species sometimes hybridize.

Many plants in this family produce chemicals to discourage herbivores. Probably the most famous is Western Poison Oak, *Toxicodendron diversilobum*, which is most common in Southern Oak Woodland. The chemicals that cause the rash are phenolic compounds known as resorcinols and catechols. Many persons are sensitive to Western Poison Oak, and those that are especially sensitive will also get dermatitis from other members of the family.

Coyote Brush or Chaparral Broom, *Baccharis pilularis*, is another evergreen shrub that occurs in Coastal Sage Scrub. It is most common at low elevations, particularly where drainage increases soil moisture. This shrub has resinous oval leaves about 0.25–0.75 in (1.5–2.0 cm) in length, and there are usually five to nine teeth along the margins. The flowering period of this species is late summer through autumn, during which time the plants are covered with clusters of white flowers. Close inspection of the flower will reveal that Coyote Brush is in the sunflower family, but it lacks the ring of petal-bearing flowers characteristic of other sunflowers such as Coast Brittlebush. This attractive shrub is another component of Coastal Sage Scrub that is being used as a landscape plant. A dwarf or prostrate form, native to the coast from the Russian River south to Point Sur, is often used as a ground cover in domestic landscapes.

From Point Sur, Monterey County, north to southern Oregon, there is a version of Coastal Sage Scrub that many authors refer to as Coastal Scrub—an appropriate name because most of the sages go no farther north than the southern Coast Ranges. Furthermore, this community differs from Coastal Sage Scrub in that most of the shrubs are evergreen, and there is an important herbaceous element. Dominant shrubs include Coyote Brush; California Yerba Santa, *Eriodictyon californicum*; Salal, *Gaultheria shallon*; and Yellow Bush Lupine, *Lupinus arboreus* (figure 7.16). This community occurs in a narrow coastal strip that is often mixed with coastal prairie, a community discussed in the preceding chapter.

South-facing slopes throughout the range of coastal Sage scrub are occupied by drought-deciduous shrubs and succulents, representing two strategies to cope with drought. The shrubs drop their leaves and go dormant in response to drought, whereas the succulents store water when it is available and remain active during periods of drought. At the southern end of its distribution, where it is drier, Coastal Sage Scrub is characterized by numerous succulent species. This has led some authors to identify a separate community known as Maritime Desert Scrub, which in some respects resembles Cactus Scrub, a desert community. A number of elements of Coastal Sage Scrub as well as numerous succulents, such as cacti, stonecrops, agaves, and liveforevers, occur here. Among the cacti are several species of prickly pear and cholla, all of which are in the genus *Opuntia*.

There are at least three coastal species of prickly pear cactus. Where they coexist, they all hybridize, and in some areas there are three-way hybrid swarms. One such area is east of Orange in the El Modena area. A large hill that occurs here, known locally as Orange Hill, is composed of Miocene volcanics. Here, amid two restaurants and numerous houses, is a hybrid cactus swarm. Coastal Prickly Pear, *O. littoralis*, is the most common. It tends to have oval pads and sprawls out in prostrate swarms. Its flowers are yellow and the bright red fruit, called tunas by the Spanish, are important foods for birds, coyotes, and woodrats. A larger, more robust prickly pear with round pads is sometimes called Tall Prickly Pear, *O. oricola*. These plants may grow as high as 6 ft (2 m). They sometimes grow in association with Coastal Prickly Pear, but they are more common toward the south. Both of these species are common in Coastal Sage Scrub on the Channel Islands.

The introduction of cattle in the 1800s greatly altered the composition of native vegetation in some areas. Among the plants that cattle seldom choose to eat are prickly pear cacti. Therefore, in overgrazed areas, even where cattle are no longer present, prickly pear cacti may spread out to form thickets.

Chollas are also among the dominant species. Coast Cholla, *Cylindropuntia (= O.) prolifera*, is widespread along the coast. This thick-stemmed species occurs in dense stands on overgrazed sites, particularly on islands such as Santa Rosa, Anacapa, Santa Catalina, and San Clemente. In the interior valleys, the cholla of Coastal Sage Scrub is Cane Cholla, *C. californica* var. *parkeri*, a thin-stemmed, sparsely branched plant that grows up to 7 ft (2.4 m) tall. Coast Cholla and Coastal Prickly pear are important components of habitat for the Coastal Cactus Wren, *Campylorhynchus brunneicapillus sandiegense*.

Among the most handsome succulents of this community are various members of the stonecrop family (Crassulaceae), which are most common on rocky outcrops. Most are in the genus *Dudleya* and are known by that name, although they may be referred to as stonecrops or liveforevers.

Most of them have very localized distribution in specialized habitats, particularly cliff faces. There are at least 17 species in the southern California coastal area. There are seven species in the Santa Monica Mountains alone and another seven in San Diego County. Twelve kinds are restricted to the islands. They are not always easy to identify as many of the species hybridize. The largest and most spectacular is called Chalk Dudleya or Chalk Lettuce, *Dudleya pulverulenta* (figure 8.17). The rosette of leaves commonly reaches a width of 30 cm or more. It commonly grows on rocky slopes or cliff faces, including some desert sites. The most common species is Lanced-leaved Dudleya, *D. lanceolata*. The Laguna Beach Dudleya, *D. stolonifera*, only occurs on north-facing sandstone boulders in the Laguna Laurel Ecological

FIGURE 8.17 Chalk Dudleya, *Dudleya pulverulenta*.

Reserve, a state facility located within Laguna Coast Wilderness Park. The Channel Islands have at least 12 endemic taxa of Dudleyas.

They are significant because they often are endemic, localized in distribution, and may be classified as rare or endangered. Many-stemmed Dudleya, *D. multicaulis*, is a rare diminutive species associated with clay soils on coastal terraces. It disappears in the dry season and returns from underground stems after winter rains. Its presence can be a frustration to developers because many sites where it occurs are protected by law.

To the interior, particularly on outwashed flood plains and rocky slopes, there is another leaf succulent known as Chaparral Yucca or Our Lord's Candle, *Hesperoyucca (= Yucca) whipplei*. This species is a common component of another variation of Coastal Sage Scrub known as Alluvial Scrub (figure 8.18). Historically, this community was subjected to periodic flooding and consists mostly of the large, evergreen components of Coastal Sage Scrub in various stages of regeneration. Chaparral Yucca, similar to other yuccas, is pollinated by

FIGURE 8.18 Alluvial Scrub in Cajon Wash, featuring Chaparral Yucca, *Hesperoyucca whipplei.*

Yucca Moths (*Tegeticula* spp.). Unlike other yuccas, but similar to desert succulents known as agaves, it blooms once and then dies. This plant is recognized by its low-growing cluster of stiff, narrow, sharply pointed leaves, which give the appearance of a large pincushion. When it is time to blossom, plants send up a long stalk, up to 10 ft (3 m) in height, topped with a large cluster of white flowers. Hundreds of these plants may blossom at one time, providing a spectacular sight on flat flood plains.

In San Diego County and northern Baja California, more succulent species appear. This community formerly known as Diegan Sage Scrub is more commonly called Maritime Succulent Scrub. It can be seen at isolated localities in the San Joaquin Hills, Cabrillo National Monument, and Torrey Pines State Natural Reserve. It is also present on Santa Catalina and San Clemente Islands. The usual cactus species are joined by Golden-spined Cereus or Button Cactus, *Bergerocactus emoryi*, which forms clusters of upright stems with long golden spines. There is also a San Diego Barrel Cactus, *Ferocactus viridescens*, and a small desert Pincushion, *Mammillaria dioica*. Shaw Agave or Coastal Agave, *Agave shawii*, is a large leaf-succulent species that has spectacular flower stalks with large clusters of yellow flowers. They are so prolific at producing nectar that it is possible to drink the sweet nectar from plucked blossoms. Following habitat destruc-

tion, the species only occurred naturally in California at Point Loma and Border Field State Park, just north of the Mexican border, but it has been reintroduced to Cabrillo National Monument and Torrey Pines State Natural Reserve.

In the interior valleys, changes in the composition of Coastal Sage Scrub reflect generally warmer conditions. Components of the community here are more desertlike. Coast Cholla is replaced by Cane Cholla. Lemonade Berry is replaced by Sugar Bush. Coast Brittlebush is replaced by Desert Brittlebush. White Sage is joined by, and in some areas replaced by, Wand Sage. In the area around Santa Rosa Mountain, Black Sage becomes replaced by Desert (Santa Rosa) Sage, *S. eremostachya*. Chaparral Yucca becomes locally common on sandy soils. Jojoba, *Simmondsia chinensis* (figure 9.28), is a desert shrub with erect, semisucculent leaves that is also locally common on rocky soils.

In disturbed areas such as road cuts and abandoned fields, hybridization between coastal and interior species is common. This is particularly true among Sugar Bush, Laurel Sumac, and Lemonade Berry, as well as between the two brittlebushes, various sages, and various cacti. In these disturbed habitats, natural barriers to germination and cross-fertilization apparently are lost. Seedlings of related species may become established in close proximity, which facilitates pollen transfer.

Aerial photographs taken from the 1930s to today have been used to compare changes in vegetation in the Los Angeles Basin. These photographs show that the amount of area covered by grasslands has increased, while that covered by Coastal Sage Scrub has decreased. Coastal Sage Scrub is a threatened community. Indicative of its threatened status, one authority has cited 13 plant species from this community as rare, endangered, or threatened. Furthermore, the California Gnatcatcher, *Polioptila californica californica*, was listed as threatened by the state and federal governments, and the Coastal Cactus Wren, *C. brunneicapillus*, has been classified as a California State Species of Special concern. Formerly distributed widely from coastal bluffs to the foothills, Coastal Sage Scrub in most areas has been replaced by habitat fragmentation associated with grazing, agriculture, and urbanization. Nonnative grassland has invaded as a consequence of overgrazing and frequent fires. Fire frequency in grassland is high enough to prevent regeneration of Coastal Sage Scrub species.

Coastal Sage Scrub still exists in isolated pockets such as the Santa Monica Mountains parklands, the Marine Corps Base Camp Pendleton, and parklands in the San Joaquin Hills near Laguna Beach. Since the1980s, the Los Angeles Basin has been second only to New York City in population density per acre. At the present time, Coastal Sage Scrub is protected in the above locations, but if the situation were reversed and the entire habitat eliminated, California would lose one of its most attractive shrub communities. With it would go many native species of plants and animals, not to mention natural open space. Open space, free of the visual and acoustic trappings of civilization, has been determined to be very important for the psychological well-being of humans.

Lower Chaparral

Lower Chaparral is also referred to as Warm Chaparral, and for the most part is occupied by Chamise alliances. This community occurs on the lower slopes of mountains, below the region where snow is the major form of precipitation. Thus, Lower Chaparral is frost tolerant but not snow tolerant. A factor that separates this community from Coastal Sage Scrub is the significance of frost. Lower Chaparral, southern California's most common natural community of plants and animals from about San Francisco to Ensenada, occurs at elevations from about 1000 ft (300 m) to 5000 ft (1600 m), between the frostline and the snowline. It also is a significant component of south-facing western slopes of the southern Sierra Nevada.

As in other scrub communities, there is also a pronounced slope effect in Lower Chaparral. As is the case with Coastal Sage Scrub, south-facing slopes are characterized by shrubs with small leaves, and north-facing slopes have shrubs with larger leaves. Unlike Coastal Sage Scrub, however, plants of Lower Chaparral are seldom drought-deciduous. Most are evergreen, with resinous or waxy coatings on their leaves. These coatings, plus an increase of nondigestible fibers, make the leaves stiff to the touch. This consistency has led to use of the term sclerophyll ("hard-leaf") to describe chaparral vegetation. Chaparral plants also have woody stems and large root systems, and they grow to be much larger than most plants of Coastal Sage Scrub. A mature stand of Lower Chaparral may form an impenetrable thicket of tangled branches up to 10 ft (3 m) in height.

Chaparral plants are physiologically different from those of Coastal Sage Scrub. Drought-deciduous and succulent plants of Coastal Sage Scrub are adapted for longer periods of drought. Their relatively shallow root systems absorb water as soon as it rains, and they quickly grow new shoots and leaves. In contrast, evergreen shrubs of Chaparral have two-layered root systems that, over a long period of time, continue to absorb water from both shallow and deep regions of the soil. New growth of evergreen shrubs takes place about four months later than that of most Coastal Sage Scrub plants, and it continues later into the summer.

A difference between the two physiological strategies of these communities is also reflected in their photosynthetic rates. During winter, the photosynthetic rate of drought-deciduous species is twice that of the evergreens. Drought-deciduous species, however, are only about half as resistant to water loss. At the height of summer drought, Coastal Sage Scrub species have lost most of their leaves, and they experience severe water stress at this time. One sage species was measured with the low water potential of −64 bars. This is equivalent to 64 times the atmospheric pressure, or nearly 1000 lb/in² (160 kg/cm²). Effectively, they are dormant. In contrast, evergreen species at this time continue to photosynthesize, although at lower rates than during winter, and some species continue to photosynthesize at −85 bars. Evergreen species also apparently can carry on with photosynthesis over a wider range of temperatures, although there seems to be little difference in photosynthetic rate between the two communities in the range from 45°F to 87°F (7–30°C). This implies that the primary limiting factor on growth of evergreen chaparral shrubs is the short winter day length.

In Lower Chaparral, aboveground growth of evergreen shrubs takes place during about four to six months during the winter and spring. The rest of the year, growth goes into the roots. Overall, aboveground growth never seems great because, in order to discourage herbivores, much of the energy budget of evergreen plants goes into cellular maintenance and production of chemicals such as lignins and tannins. The gain provided by this effort is that leaves tend to stay on the plants for at least 2 years, whereas drought-deciduous species have to make a complete set of new leaves every year.

Lower Chaparral species have a number of other leaf adaptations that help them to deal with the stresses of their environment. Sclerophyllous leaves have reduced rates of water loss, and the hardened tissues and coatings make them difficult to digest—a discouragement to herbivores. In addition, it is notable that many species have serrated or spiny leaf margins. It is believed that the function of these spines is to increase surface area to help dissipate heat. A number of species roll their leaves, which traps air on the inside and reduces evaporation. Leaves of several species have fine hairs, which form a boundary layer that traps air. The hairs also provide a small amount of shade and help to dissipate heat. Other species have leaves that track the sun in order to optimize or minimize light intensity, and leaves of some plants are oriented vertically so that sun does not strike them directly.

Nutrient and water absorption in chaparral evergreens seems to be enhanced by mycorrhizal fungi. Chamise, *Adenostoma fasciculatum*. scrub oaks, *Quercus* spp., and manzanitas, *Arctostaphylos* spp., have been shown to possess these fine underground filaments. The absorption of phosphorus by mycorrhizae is believed to be one of the primary nutritional benefits. It also has been demonstrated that association of these fungi with the roots of seedlings helps prevent wilting during the first months after germination.

Lower Chaparral is the southern California community most likely to burn. The community has evolved over millions of years in association with fires, and in fact requires fire for proper health and vigor. Thus, it is not surprising that most chaparral plants exhibit adaptations enabling them to recover after a burn. Many species are sprouters; the aboveground parts may be killed, but new growth arises from roots or buds at the base of the stem in a region known as a root-crown burl. Other species have seeds that require fire in order to break dormancy; they will not germinate unless they have been heated, or bathed in water (leachate) which has been soaked in ashes. The cones of some chaparral conifers, such as Knobcone Pines (page 348) or various cypress species open only after they have been heated. Some herbaceous species will not germinate unless there is ash on the ground when it rains. Large oak trees have a thick, fireproof bark.

Not only do chaparral plants feature adaptations that help them recover after a fire, but also some characteristics of these plants, such as fibrous or ribbonlike shreds on the bark, seem to encourage fire. Other species contain volatile oils. In the absence of fire, a mature chaparral stand may become senile, in which case growth and reproduction are reduced. Without new growth and with fewer berries and seeds, there is less food for animals. The new growth of some shrubs contains about 14% protein. Browsers such as deer cannot survive without this kind of nutrition.

In the old days, prior to fire suppression, fires were seldom infernos because the slopes were a mosaic of burned and unburned terrain. Usually new fires were limited by recently burned regions with very little fuel. Dead wood and other fuels could not accumulate for long. When a fire started, it was usually due to lightning, and the fire would burn during conditions of high humidity. Old records show that a fire might start in early summer and burn for months, moving slowly along the ground, sometimes smoldering for a while before blazing again.

Of course, in the "old days," there was very little habitation to worry about. Now there are homes and villages all over in the Chaparral. When a fire breaks out, a tremendous amount of money and effort are expended to protect habitation areas. Unfortunately, most fires in southern California are caused by humans, through either arson or accidents. Arson fires are set during periods of Santa Ana conditions, when wind, low humidity, and heat are complicating factors. The accumulation of fuel and the great extent of unburned brush provide conditions ensuring that the fire will be severe, all of which is now complicated by climate change. In recent years the number and intensity of fires has increased and the fire season is longer. California's largest fire today, the Cedar Fire in San Diego County, burned 273,240 acres (110.579.5 ha), of predominantly Chaparral habitat.

These extremely hot, large fires have serious consequences. Burrowing animals that normally survive small fires are suffocated in their burrows. Animals that escape are forced to compete with those already established in the unburned areas. Thousands of acres may be burned all at once, drastically reducing total habitat. When the rains come, erosion is severe.

Dry creep or slow sliding is a characteristic of chaparral soils. Loose soil material is carried away with heavy rain. Furthermore, many chaparral plants produce waxy, nonwettable (hydrophobic) chemicals. These chemicals cause the soil to repel water, and this may inhibit germination of seeds under a healthy canopy. Water that supplies roots of mature plants gets there by trickling down branches and running into the soil at the base of the trunk. When a fire burns the litter, the chemicals are driven into the soil to form a nonwettable layer beneath the surface. After a fire, soil becomes saturated above the layer and washes away. Before the policy of fire suppression, when fires were small and cooler, this problem was not severe. Now large areas are denuded of vegetation, and run-off is increased. Tons of valuable soil are carried away, and homes below the burn area are flooded or damaged by mud slides.

Attempts to reduce the erosion-flood sequence usually are futile. Agencies concerned with management of chaparral habitats often reseed burn sites with nonnative grasses, particularly Perennial Ryegrass, *Lolium perenne*. Evidence has accumulated to indicate that this is not a good procedure. Unless the first rains are gentle, the grass will not germinate in time to do any good. If the first rains are gentle, the grass will take valuable water away from native species and retard their regrowth. Furthermore, grasses increase fire frequency.

A policy of outright fire suppression is disastrous for both plant and animal life. Government agencies are now undertaking programs to light fires deliberately in order to maintain natural habitats, a procedure known as prescribed burning. This requires careful analysis of fuel and weather conditions. When conditions of humidity allow the fire to burn slowly, it can be controlled. The answer to the problem is to allow Lower Chaparral to return to more natural conditions, where there is a mosaic of burned and unburned terrain. Where human habitation encroaches on Chaparral, homes should be built of nonflammable materials, such as masonry walls and tile roofs.

FIGURE 8.19 Chamise, *Adenostoma fasciculatum*. Mature plant on the west slope of the San Jacinto Mountains.

The most abundant plant in Lower Chaparral is Chamise, *A. fasciculatum* (figure 8.19). It is in the rose family (Rosaceae), but its white flowers are quite small, requiring close inspection to recognize the family characteristics. Chamise probably has the greatest biomass of any plant in Cismontane Southern California. It is characteristic of south-facing slopes, where it may occur in nearly pure stands. A textbook example of a chaparral shrub, it has small, drought-adapted leaves and a large root system, and it is fire-adapted.

Drought tolerance of this species is moderately high, although not as high as plants from Coastal Sage Scrub. Water potentials of –50 bars have been recorded, which places Chamise between Coastal Sage Scrub plants on south-facing slopes and those of north-facing slopes in terms of drought tolerance. This plant has tiny, resin-covered leaves that grow in clumps. The leaves are similar to those of the buckwheat mentioned previously but are often no longer than 0.25 in (7 mm). Following a fire, Chamise reestablishes itself primarily through new shoots that resprout from its root crown or burl. These shoots may grow a foot (30 cm) or more during the year following a fire.

Chamise also can regenerate by means of seeds, which are produced in enormous numbers. There are two kinds of seeds: those that require heat in order to germinate and those that do not. This variation, plus its ability to sprout, gives Chamise ample opportunity to become reestablished. Some experiments show that regeneration by seeds is more common following a high-intensity fire.

The relationship between fire and Chamise is illustrated by the plant's tendency to "encourage" burning. A thermometer was placed within a Chamise shrub as a fire approached, and the following changes were documented. At about 200°F (93°C), the plant began to wilt as its temperature approached the boiling point of water. At about 400°F (204°C), the plant began to emit combustible gases such as hydrogen, alcohol, and methane. At about 600°F (316°C), the shrub smoldered and began to turn black. At about 800°F (427°C), the plant burst into flames! This species must have evolved in association with frequent fires to have reached the point where it seems to encourage burning.

Another bit of evidence shows the evolution of these shrubs in association with fire. This is an example of coevolution. The color on the back of the Western Fence Lizard, *Sceloporus occidentalis*, closely matches the color of a burned Chamise stem. Data using a machine that measures reflected light (a reflectometer) showed that a Western Fence Lizard is nearly invisible when it rests on a burned Chamise

FIGURE 8.20 Chaparral Whitethorn, *Ceanothus leucodermis*.

stem. The surface that matches the lizard next best is granite.

Chamise may become the dominant plant on south-facing slopes in mature Chaparral by out-living the other species. If an area has gone unburned for 35 years or more, Chamise may seem to be the only species left, although it is not always completely healthy in these old stands. Although Chamise resprouts immediately after a fire, other shrubs dominate from 4 to 10 years after a burn. Common among these early succes-sional plants are several species in the buckthorn family (Rhamnaceae) known generally as Cali-fornia-lilacs (*Ceanothus* spp.). These include Hoaryleaf Ceanothus, *C. crassifolius*; Hairy Cean-othus, *C. oliganthus*; and Chaparral Whitethorn, *C. leucodermis* (figure 8.20). These species of *Ceanothus* do not sprout; they must return from seeds after a fire. They mature quickly and pro-duce a great number of seeds early in their lives. The seeds may remain dormant for hundreds of years, but heating stimulates them to germinate. The heat either melts or cracks the hard seed coat, allowing the uptake of water when it finally rains. Another characteristic of *Ceanothus* plants is that they throw their seeds. Seeds that bounce off a rock may be cracked by the impact, and cracked seeds can germinate without fire if they land in an appropriate open space. Of particular

importance to the chaparral community is that *Ceanothus* plants have nitrogen-fixing bacteria associated with their roots. Hence, new shoots and leaves are rich in protein, providing an important browse for deer and other herbivores. Guess why *C. cuneatus var. cuneatus* is known as Buckbrush.

On some hillsides, *Ceanothus* dominates in nearly pure stands, and some authors distin-guish this special community as *Ceanothus* Chaparral. Clusters of different *Ceanothus* spe-cies are recognized as different alliances. How-ever, in southern California this is usually a suc-cessional community that exists until the *Ceanothus* plants die of old age or become over-topped by Chamise. Depending on circum-stances, complete replacement by Chamise gen-erally takes place within 10–30 years after a fire.

In the mid-1980s, patches of *Ceanothus* with yellow leaves, a condition sometimes known as chlorotic decline, began to appear, particularly in the Transverse Ranges. At first, informed biolo-gists assumed it was a consequence of fire sup-pression and that the shrubs were dying of old age. As researchers began to study the problem and the number of dying plants increased, how-ever, other factors came to be suspected. The period from 1978 to 1983 was one of the wettest in history. It was followed by one of the worst droughts in history. Perhaps the additional precipitation stimulated so much growth that the plants were put under severe stress in attempting to maintain excessive foliage during the drought. Another complicating culprit may be dry fallout of nitrogen compounds from the polluted air. The excess nitrogen, particularly during periods of heavy rainfall, may also stimulate growth that cannot be maintained during drought.

Some herbaceous species or subshrubs have characteristics similar to *Ceanothus*. These species require heat or smoke for seed germination, and they have nitrogen-fixing bacteria in their roots. Among these are members of the Legume family (Fabaceae), including a variety of lupines (*Lupinus* spp.) and a small shrub known as Deerweed, *Acmispon glaber* (= *Lotus scoparius*). Guess why it is known as Deerweed.

Apparently, chaparral fires cause a tremendous loss of nitrogen from the ecosystem by the burning of nitrogenous compounds such as proteins in plant tissues and detritus. It has been determined that a direct loss of 133–178 lb of nitrogen per acre (150–200 kg/ha} could result from a moderate fire. During the first years after a fire, species equipped with nitrogen-fixing bacteria dominate because they grow quickly on the disturbed terrain, and they restore nitrogen to the system. Chaparral succession therefore is characterized in its early stages by herbs and suffrutescent (herbaceous) shrubs such as Deerweed.

The leaves of Chamise contain chemicals that are allelopathic, and inhibit germination of other plants. These chemicals are carried into the soil when it rains, and helps to explain why areas might appear to be a Chamise monoculture. After a fire, the chemicals dissipate, which allows germination of wildflowers or other herbaceous species, so for the first few years after a fire, Lower Chaparral is dominated by short-lived herbaceous plants, including many brightly colored wildflowers. Among these are the lupines and several members of the Water-leaf family (Boraginaceae), such as one of the Yerba Santas (*Eriodictyon* spp.) and heliotropes or Canterbury-bells (*Phacelia* spp.).

Among the well-known wildflowers associated with chaparral are the paintbrushes (*Castilleja* spp.; plate 20A[16A]). There are about 34 species of paintbrushes in California, all associated with shrubs. They occur in a variety of communities from the desert to the mountains, and their often-brilliant red blossoms add a striking accent to the landscape. Paintbrushes are hemiparasites; that is, they are partially parasitic. They are photosynthetic and produce their own food like any free-living plant. In addition, paintbrushes tap into the root systems of shrubs to enhance their absorption of water and minerals. Therefore, they are able to grow vigorously, even when surface layers of the soil become too dry for most small plants.

As might be expected, paintbrushes are not host specific. They appear to parasitize a number of common shrubs, such as Chamise and California Buckwheat. Although this type of parasitism does not severely affect the host plant, during times of drought, it has been shown that the host's growth and vigor are reduced. Another interesting study has shown that paintbrushes are also able to absorb phytotoxins such as alkaloids from host plants, thus parasitizing the chemicals that defend against herbivores.

Studies in the 1980s showed that compounds found in burned wood (charate or leachate) are capable of stimulating germination of a number of species. Among the wildflowers stimulated to germinate by charate are the phacelias. Also included in this group are common "fire followers," such as Whispering Bells, *Emmenanthe penduliflora*; Golden-yarrow, *Eriophyllum confertiflorum*; Fire Poppy, *Papaver californicum*, and two species of ear drops, *Dicentra* spp., also in the Poppy family (Papaveraceae). At the same time these herbs germinate, *Ceanothus* and Chamise are also making a comeback. *Ceanothus* seems to dominate for about 10 years. After that, Chamise dominates. The successional phases described before are apparently a function of growth rates and life span.

On north-facing slopes Bush Poppy, *Dendromecon rigida*, with striking bright yellow blos-

FIGURE 8.21 Scrub Oak, *Quercus berberidifolia.*

soms is a common fire follower. A closely related species the Channel Island Tree Poppy, *D. harfordii*, is taller and has larger flowers. This species is often used as an example of a phenomenon known as *island gigantism*. These poppies are unique among chaparral plants because their oil-rich seeds are carried by ants, in particular the native Harvester Ants, *Pogonomyrmex* spp., and Carpenter Ants, *Camponotus* spp. Like all poppies, these species contain chemicals that repel herbivores. Also on north-facing slopes, a common fire-following shrub is Chaparral or Bush Mallow, *Malacothamnus fasciculatus*. Various related forms of this mallow are also found as endemics on San Clemente, Santa Catalina, and Santa Cruz Islands. These poppies and mallows are fairly short-lived, but they may dominate the landscape for 3–4 years after a fire.

Other common associates of Lower Chaparral on south-facing slopes, particularly during early successional phases, are various members of the Coastal Sage Scrub community, such as Chaparral Yucca, California Buckwheat, White Sage, and Toyon. South-facing exposure provides the heat necessary for their proper growth.

On north-facing slopes in Lower Chaparral, there is a community dominated by shrubs with comparatively large leaves, many of which have serrated or spiny leaf margins. The most important of these is Scrub Oak, *Q. berberidifolia* (fig-

ure 8.21). The name *Q. dumosa*, formerly used to denote Scrub Oak, now refers to a separate species of scrub oak that grows in patches in canyon bottoms and on north-facing slopes in limited areas within Coastal Sage Scrub. Acorns from scrub oaks are among the most important food items for chaparral animals. Scrub Oaks can form a dense canopy up to 12 ft (4 m) in height, and their abundant leaf litter, up to 8 in (20 cm) deep, often precludes growth of understory plants. This is an important habitat for woodrats (*Neotoma* spp.), who build their nests of branches and litter on the ground and in the branches of the oaks. Scrub Oaks have dark green, thick, leathery leaves with soft, short hairs on the underside. The leaves have small lobes on the edge, but the margins are sometimes spiny. In some areas, there is another scrub oak with spiny leaf margins. This is the scrub version of the Interior Live Oak, *Q. wislizeni* var. *frutescens*, which is more common in the foothills of the Sierra Nevada and Coast Ranges. It can be distinguished from Scrub Oak by its leaves; the undersides are not hairy, and both sides are shiny green. The endemic Island Scrub Oak, *Quercus pacifica* (plate 13B) grows treelike in stature, another example of island gigantism.

More species are associated with Scrub Oak than with any other chaparral plant, and nearly all of them have comparatively large, oval leaves.

Among those that have leaves with serrated or spine-tipped margins are Holly-leafed Cherry, *Prunus ilicifolia* subsp. *ilicifolia*, and Hollyleaf Redberry, *Rhamnus ilicifolia*. Many of the species also have berries, such as California Coffee Berry, *Frangula (= R.) californica*, and a number of wild currants (*Ribes* spp.). The importance of berries as a food for animals and as a means for dispersal of seeds is obvious. In particular, Coyotes, woodrats, and birds are known to feed on these fruits. It has been demonstrated that the seeds of Holly-leaved Cherry require the acid treatment of the Coyote's digestive tract or they will not germinate. This is an interesting irony, because the seeds, if cracked, release cyanide, another example of a defensive mechanism to discourage animals from eating the seeds. Apparently, the Coyote must eat the fruit without cracking the seeds. Does the Coyote learn this trick, or is it innate? This appears to be another example of coevolution. Included among these species are two more examples of island gigantism. Catalina Cherry, *Prunus ilicifolia* subsp. *lyonii*, is a tree-sized subspecies endemic to Santa Catalina Island, and Island Redberry, *R. pirifolia*, is a small tree found only on Santa Catalina, San Clemente, Santa Cruz, and Santa Rosa Islands.

Also common in Lower Chaparral are species of woody vines, such as honeysuckles (*Lonicera* spp.), wild peas (*Lathyrus* spp.), and wild cucumbers, or man-roots (*Marah* spp.). Being vine-like enables these plants to grow over the top of the shrub canopy so that they are not shaded or crowded out. Honeysuckles and wild peas bear attractive, tubular, hummingbird-pollinated flowers. Many consider the flowers of the various species of wild pea to be the most beautiful of the entire community. The most common of the wild cucumbers is known as Chilicothe, *M. macrocarpa* (figure 8.22). This vine produces many white, star-shaped flowers about 0.25–0.50 in (8–13 mm) in diameter. Some people call the spiny, egg-shaped fruits "porcupine eggs." The plant returns each year from a large underground tuber and spreads widely over the vegetation. The name *man-root* refers to the huge tuber, which may reach the size of a man's body.

FIGURE 8.22 Chilicothe, *Marah macrocarpa*.

Many authorities consider Scrub Oak and its associates to be a distinct community known as Scrub Oak Chaparral, pointing out that it occurs on north-facing slopes over a wide range of elevations from within Coastal Sage Scrub to Upper Chaparral. Scrub Oak Chaparral requires more moisture than communities on adjacent south-facing slopes, and it is more snow tolerant than Chamise Chaparral. At upper parts of its range, it occurs on south-facing slopes, and in the Transverse Ranges it occurs on both north- and south-facing slopes above about 3000 ft (900 m) elevation. This community also tends to respond differently to fire than Chamise-dominated chaparral. Most species associated with Scrub Oak are sprouters that regrow rapidly after a fire. Even though herbaceous species are common for a year or two after a fire, it takes only a few years of vigorous growth to shift the dominance back to the common shrub and vine species.

FIGURE 8.23 Bigberry Manzanita, *Arctostaphylos glauca*.

Upper Chaparral

Upper Chaparral is also known as Cold Chaparral because much of the precipitation is provided by snow and fog drip. This community occurs on mountain slopes below the Mixed Coniferous Forest. Compared to Lower Chaparral, soil here is moderately organic, relatively deep, and usually acidic (pH 5.2–5.5). As in other chaparral communities, slope effect is pronounced. On south-facing slopes, there are evergreen shrubs with thick, oval leaves. On north-facing slopes, there are drought-tolerant conifers. This community is well established on the coastal side of the Transverse and Peninsular Ranges from about 4000 to 5000 ft (1300–1500 m) elevation.

On south-facing slopes, Upper Chaparral, throughout the state, is dominated by various species of manzanita. Some authors call this community Manzanita Chaparral. Particularly common are Eastwood Manzanita, *Arctostaphylos glandulosa*, and Bigberry Manzanita, *A. glauca* (figure 8.23). In some areas, manzanitas occur at lower elevations on north-facing slopes. At higher elevations, they are nearly always on south-facing slopes. Greenleaf Manzanita, *A. patula* (plate 9A), occurs this way, particularly on dry slopes in the Mixed Coniferous Forest (plate 9A). A manzanita is recognizable from a distance by its light green herbage.

Up close, the manzanita is distinctive because of its smooth, bright red bark. Leaves on manzanita have a mealy, waxy surface and are often oriented vertically, which serves to reduce the amount of light that directly strikes leaf surface. Manzanita is a member of the heath family (Ericaceae). Many members of this family are notable for their distribution in snow zones. Manzanitas represent the drought-tolerant, fire-adapted heaths, but their tolerance for snow is typical of the family. Flowers on manzanitas are urn-shaped, as is typical of many members of the heath family. The reproductive parts of the flowers (stamens and pistil) are concealed by the petals, with the opening on the underside, which makes it appear that pollination is special. As it turns out, this is another example of buzz pollination described for Shooting Stars in the High Sierra. The primary pollinators are bumblebees, which hang on the underside of the flowers and produce a buzzing sound, which turns out to be the musical pitch we call middle C. This vibration causes pollen grains to be released from pores at the ends of the anthers, dusting the underside of the bee with pollen, which then can be transferred to the stigma of the next flower that is visited. Release of pollen can be stimulated artificially by holding a middle C tuning fork against a manzanita flower.

FIGURE 8.24 Silk-tassel Bush, *Garrya fremontii.*

Manzanita is often associated with other cold-tolerant shrub species, such as Silk Tassel Bushes (*Garrya* spp.), which resemble manzanitas but whose bark is not red. Silk Tassel leaves are turned upward and arranged in pairs, opposite each other on the stem. The flowers are distinctive, hanging down in long chains like church bells (figure 8.24).

Another common associate in Upper Chaparral is Western or Birch-leaf Mountain-mahogany, *Cercocarpus betuloides* var. *betuloides* (figure 8.25). With its whitish gray bark, this shrub is distinctive. Its leaves are serrated, resembling a small birch leaf. Its fruits are dry, curled, featherlike structures. This shrub is in the rose family (Rosaceae), but it has nitrogen-fixing bacteria associated with its roots. It is therefore an important browse species, particularly in exposed, rocky locations, where it is important as food for Bighorn Sheep and Mule Deer.

An interesting characteristic of all three of these shrubs is that their leaves are oriented vertically, and to some degree they are able rotate in order to optimize light and heat for photosynthesis, a phenomenon known as *sun-tracking.*

Ribbon Wood, *Adenostoma sparsifolium*, also known as Red Shanks or Ribbon Bush, occurs in the Peninsular Ranges. This member of the rose family is a relative of Chamise that favors north-facing slopes at elevations above Lower Chaparral. It is an associate of manzanitas and Scrub Oak. On favored localities, particularly above 4000 ft (1300 m), it grows in nearly pure stands, and some authorities categorize this as a separate community known as Red Shanks or Ribbon Wood Chaparral. In the San Jacinto Mountains, east of Garner Valley and south to Anza, there is an area along the Pines to Palms Highway (Highway 74) where a stand of Ribbon Wood dominates for many miles. At the lower reaches of its distribution, Ribbon Wood is associated with Scrub Oak and Hoary Ceanothus. At upper elevations, it mixes with manzanitas. It appears to represent a transitional community between Lower and Upper Chaparral. Ribbon Wood has small, filamentous leaves and red shaggy bark, which increases its flammability. It is a sprouter, growing new shoots vigorously after a fire.

Upper Chaparral has abundant leaf litter, which promotes a soil reasonably rich in organic matter. When there is a fire, the abundance of

FIGURE 8.25 Birch-leaf Mountain-mahogany, *Cercocarpus betuloides*. Note the sun-tracking leaves oriented vertically.

litter contributes to a deep ash layer with alkaline pH (ca. 7.4) that tends to inhibit germination of seeds. Therefore, after a fire, the abundance of annuals associated with other types of Chaparral is not as evident. About half of the manzanita species are sprouters that return readily after a fire. The other half become reestablished by seeds. Seeds of manzanitas, along with those of California-lilacs and various herbs, must germinate in appropriate localities and compete for space. The abundance of seedlings of manzanitas and California-lilacs after a fire maintains their dominance in the community at these times. The forms of manzanita that resprout are apparently more common in areas where vegetation has greater ground cover. California-lilacs, however, are short-lived. Seedling California-lilacs have a high mortality within the first 2 years after a fire, and once seedlings become established, California-lilacs

die before manzanitas. For this reason, mature stands of Upper Chaparral may be composed entirely of manzanita.

A severe fire may burn the organic material in the soil and also burn dormant seeds. Recovery from such hot fires is very slow. In the Transverse Ranges, there have been several large fires since the early 1960s. In these areas, there seems to be no reestablishment of Bigberry Manzanita, one of the species that does not sprout but returns from seeds. In the San Gabriel Mountains in 1980, there was a large, hot fire in San Antonio Canyon and Ice House Canyon on the south side of Mount Baldy. As of 1988, no manzanita seedlings were apparent. In the San Bernardino Mountains along Highway 18 between Arrowhead and Running Springs, there is a site that burned in 1959 that still has not returned to a manzanita shrubland. The area is showing early successional stages of dry Mixed Coniferous Forest, including Bracken Fern, Incense Cedar, and California Black Oak. Some Coulter Pines are also becoming established. Seedlings of these species normally would be precluded by deep ash and competition with the manzanitas.

The mechanisms associated with seedling germination after a fire are not well understood for species of manzanita. For example, in some areas, sprouters occur with nonsprouters. Perhaps the most common crown-sprouting manzanita is Eastwood Manzanita. In an area east of San Diego, it occurs equally with Bigberry Manzanita, a nonsprouting species. Bigberry Manzanita produces about eight times as many seedlings after a fire, but examination of soil in nearby unburned areas revealed about 25 times more Eastwood Manzanita seeds. As its name implies, Bigberry Manzanita has larger fruit, which would seem to increase the competitive ability of seedlings because there would be more stored food. It might also be assumed that larger fruit and seeds would provide better protection from fire. In light of these generalizations, it is difficult to explain why over 99% of Bigberry Manzanita seeds do not germinate after a fire.

FIGURE 8.26 Bigcone Douglas-fir, *Pseudotsuga macrocarpa.*

FIGURE 8.27 Coulter Pine, *Pinus coulteri.*

Perhaps the size of the seeds is associated with some other adaptation, such as dispersal by animals. Here again, it would seem that larger fruit would attract more herbivores and therefore result in removal of a greater number of seeds from an area. This hypothesis was verified by predation studies. Presumably, Coyotes and ground-feeding birds removed the larger fruit from the area, which could explain why there are many more Eastwood Manzanita fruits remaining in the soil. But this does not explain why more Bigberry Manzanitas than Eastwood Manzanitas germinate. Perhaps the value of large fruits is to encourage dispersal of seeds. Further study may also show that passing of seeds through the digestive tract of an animal enhances its ability to germinate, as is the case with Holly-leaved Cherry.

On north-facing slopes in Upper Chaparral, shrub species are less common, and coniferous trees may dominate, leading many authorities not to consider this assemblage of species as a form of Chaparral. Although they may not occur together, most common among these conifers are Bigcone Douglas-fir, *Pseudotsuga macrocarpa* (figure 8.26), and Coulter Pine, *Pinus coulteri* (figure 8.27). Bigcone Douglas-fir or Bigcone Spruce is related to the Douglas-fir of northern forests, but the cone of this species is twice as large. Distribution of Bigcone Douglas-fir is entirely southern Californian, extending from Santa Barbara to San Diego Counties. Seeds germinate and grow well after a fire. It is notable that this conifer sprouts after a fire, but not from buds near the ground. These trees sprout from latent buds located at the bases of branches along the trunk. When a fire is too intense, the trees apparently fail to sprout, and regrowth must occur from seedlings.

Bigcone Douglas-fir is a moisture-dependent species commonly restricted to north-facing slopes and canyon bottoms above 3000 ft (980 m). It is notable that the only southern California occurrences of Pacific Madrone, *Arbutus menziesii* (figure 6.7), are with Bigcone Douglas-fir on north-facing slopes. These sites are in deep canyons such as the West Fork of the

San Gabriel River, Trabuco Canyon in the Santa Ana Mountains, and near Mount Palomar in the Agua Tibia Mountains. To the north, Pacific Madrone is normally associated with Coast Redwoods or other moisture-loving plants.

Coulter Pine resembles a short, stout Ponderosa Pine. Its growth habit tends to promote long branches that give the tree a large cone-shaped appearance. Its thick foliage consists of long needles, often greater than 10 in (25 cm) in length, which occur in clusters of three. One of these trees might be mistaken for a Ponderosa or Jeffrey Pine were it not for its enormous cones, which may be larger than a volleyball. Similar to Gray Pine, the cones have large curled hooks on each scale (figure 8.27). In the southern Coast Ranges, where the distribution of Gray Pine and Coulter Pine overlaps, the latter may be distinguished by its darker green needles and larger, light-colored cones. In some respects, Coulter Pine may be considered the ecologic equivalent of Gray Pine in southern California.

It is notable that large cones seem to be associated with hot, dry habitats. Perhaps larger cones are needed to protect the seeds in these climates. However, the seeds of these conifers are also quite large. Perhaps large seeds are necessary to store enough food to support extensive root growth after germination, and large cones are merely coincident with large seeds.

Coulter Pines often occur in pure stands, although in many locations they are often associated with Canyon Live Oaks, *Quercus chrysolepis*, the most widely distributed oak in the state. This association is similar to what a person might find on upper north-facing slopes in the southern Coast Ranges. California Ash, *Fraxinus dipetala*, is another northern associate on these north-facing slopes. These are large, winter-deciduous shrubs or small trees with pinnately compound leaves composed of three to seven leaflets. In the spring, the shrubs become covered with white flowers, making them conspicuous on these slopes. Other associates include shrubs of Upper Chaparral, such as manzanitas.

Coulter Pines seem to be less fire-adapted than other chaparral species, although they survive low-intensity ground fires quite well. Unlike Bigcone Douglas-fir, Knobcone Pine, and Jeffrey Pine, which may grow nearby, they show fewer specializations that may be called fire adaptations. It is important, however, that the cones of Coulter Pines open slowly. This causes seedfall to be delayed until December, after the fire season is generally over.

Coulter Pines may be less drought-tolerant than other conifers in southern California. Similar to the situation described previously for *Ceanothus*, large stands of Coulter Pines are dying, particularly at the lower extent of their range. Weakened by drought and smog, the trees succumb to bark beetle infestations, another phenomenon that promotes large fires.

Distribution of Coulter Pine is a bit of an enigma. In the southern Coast Ranges, it is an associate of Gray Pine, but it is absent from the Sierra Nevada in similar habitat. In southern California, its distribution is spotty in the Transverse and Peninsular Ranges to northern Baja California. It and Bigcone Douglas-fir are the most common conifers in the Santa Ana Mountains. There is some evidence to support the contention that in southern California its distribution is associated with granite. It seems to be absent, for example, along the front range of the San Gabriel Mountains, where soils are derived mostly from metamorphic rock. Where Coulter Pines are common, they may form an open woodland on shaded flats and north-facing slopes. Regions of great abundance occur in the Chilao-Charlton Flats area near Mount Waterman in the San Gabriel Mountains, and near Mountain Center in the San Jacinto Mountains. At higher elevations, Coulter Pines are replaced by typical Mixed Coniferous Forest, although on rocky ridges they may extend up to 6000 ft (1800 m) elevation. In the San Bernardino and San Jacinto Mountains, they may be found even higher.

Desert Chaparral

Desert Chaparral is found on the desert slopes of the Transverse and Peninsular Ranges (figure 8.28). It is also found on the inner slopes of

FIGURE 8.28 Desert Chaparral, Santa Rosa Mountains, Riverside County.

the southern Coast Ranges bordering the dry southern end of the San Joaquin Valley. Technically, this community is not in Cismontane Southern California, because it is located in the rain shadow of the mountain ranges. Its obvious relationship to other chaparral communities, however, dictates that it be discussed here. In southern California, this community usually occupies a zone below the Pinyon-Juniper Woodland but above the desert communities.

Desert Chaparral is an open type of community. Vegetation usually covers less than 50% of the terrain, but individual shrubs may be quite large, 10 ft (3 m) in height or more. Included here is a mixture of drought-adapted, snow-tolerant species found in other communities.

This community is less extensive on the desert slope of the Transverse Ranges than it is in the Peninsular Ranges. In the Transverse Ranges, it may include Chamise as a dominant species, or it may include one of the desert species of scrub oak, *Quercus cornelius-mulleri* or *Q. john-tuckeri*. It also may include Bush Poppy, *Dendromecon rigida*, which is often considered a fire follower. This shrub is also called Tree Poppy, which is what the name *Dendromecon* means. The name "Bush Poppy" seems to be more appropriate, however, because

it seldom grows in a treelike form, except on the Channel Islands. Bush Poppy is characterized by large, bright yellow flowers with four papery petals.

In the Desert Chaparral of the Transverse Ranges, Bigberry Manzanita and Cupleaf (Desert) Ceanothus, *Ceanothus perplexans* (= *C. greggii*), may also be included. Cupleaf (Desert) Ceanothus resembles Hoaryleaf Ceanothus of Lower Chaparral; it has coarsely toothed margins, but the underside of the leaf is not white. Desert Chaparral is also characterized by localized patches of the shrub form of Interior Live Oak. At higher elevations, Curl-leaf Mountain-mahogany, *Cercocarpus ledifolius*, may occur, a species more commonly associated with Pinyon-Juniper Woodland. This shrub is also a nitrogen fixer and therefore serves as an important browse species for Mule Deer and Bighorn Sheep.

An interesting species that occurs in Desert Chaparral and is also found in Chaparral of the western foothills of the Sierra Nevada is California Flannelbush, *Fremontodendron californicum* (figure 4.21), which recently has been moved to the Mallow family (Malvaceae). This large shrub has lobed leaves with a fine coating of hair. When it flowers in the spring, it becomes covered with large yellow blossoms.

The blossoms are formed by bracts and sepals; there are no petals. During the dry season, leaves are shed a few at a time, but the plant seldom becomes leafless. This beautiful shrub is being cultivated for locations where no supplemental watering is available. When it is in bloom, very few shrubs can rival it for beauty. The blossom of this shrub is used on the logo for the California Native Plant Society journal, *Fremontia*.

On desert slopes of the Peninsular Ranges, there is a more extensive belt of Desert Chaparral. The diversity here is extraordinary. In one study in upper Deep Canyon on the northeastern slope of Santa Rosa Mountain, over 50 species of perennial plants were identified, most of them shrubs. This diversity is a consequence of edge effect, a merging of communities. There is a mixture of species from Joshua Tree Woodland, Pinyon-Juniper Woodland, Cactus Scrub, Sagebrush Scrub, and Coastal Sage Scrub. Included are Sugar Bush, Cupleaf (Desert) Ceanothus, Western Mountain-mahogany, Bigberry Manzanita, Holly-leaved Cherry, Bitterbrush (*Purshia tridentata*), and Muller's Oak (*Q. cornelius-mulleri*). Muller's Oak differs from the Scrub Oak, *Q. berberidifolia*, by having small hairs on both surfaces. There are now three species of Desert Scrub Oak that were formerly known as *Q. turbinella*. *Q. turbinella* now refers to those populations that occur in the New York Mountains of the Eastern Mojave Desert and eastward into Colorado and Texas. Muller's Oak, *Q. cornelius-mulleri*, occurs on the desert side of the Peninsular Ranges from Baja California north to the San Bernardino Mountains. Tucker's Oak, *Q. john-tuckeri*, occurs from the north side of the San Gabriel Mountains to the interior side of the southern Coast Ranges. Bitterbrush is in the rose family (Rosaceae). It is spectacular in bloom, when it becomes covered with small cream-colored flowers. The variety *P. tridentata* var. *tridentata* (figure 9.20B) is an important component of the Sagebrush Scrub community of the Great Basin. A similar shrub, known also as Antelope Brush, *P. t.* var. *glandulosa* is the form found in the Desert Chaparral of the Peninsular Ranges.

There is some disagreement among authorities as to whether Desert Chaparral deserves recognition as a distinct entity. Because it contains plant species found in adjacent communities, some choose to consider it merely a transitional community. Others point out that the community contains many species from nonadjacent areas, and plants such as Sugar Bush, Desert Scrub Oak, and Cupleaf (Desert) Ceanothus are more common here than anywhere else. Furthermore, because most of the plants are shrubs with sclerophyllous leaves, the term *Chaparral* is appropriate. Finally, because most of the birds and large mammals are the same as those found in other chaparral communities, it seems best to refer to the community as a form of Chaparral.

A number of species in this community produce showy flowers. They do not always blossom at the same time, however, so a person can expect to find things blooming from early spring onward. Among the most spectacular species is a shrubby sunflower (Asteraceae) known as Narrowleaf or Interior Goldenbush, *Ericameria* (= *Haplopappus*) *linearifolia*, which is locally abundant on rocky or sandy slopes. It has fine, linear, resinous leaves. When it flowers, whole hillsides appear covered by bright yellow sunflowers.

Because this community merges with Pinyon-Juniper Woodland at its upper reaches, it is understandable that there are many important foods here for animals. Cahuilla Indians gathered acorns, pine nuts (piñones), juniper berries, cactus apples (prickly pears), and roots of yuccas while they took refuge from desert heat. In this community, there is a group of drought-deciduous shrubs that were also useful to Indians. Among these is another member of the cashew family (Anacardiaceae), so important in chaparral communities. This is Basket Bush or Squaw Bush, *Rhus trilobata*. Superficially, it appears to be a shrub-like version of Western Poison Oak, but it seldom causes dermatitis. Whereas Western Poison Oak has leaflets in groups of three, these leaves are each divided into three lobes of about the same size and shape. Western Poison Oak occurs in moister habitats, and should not be feared in Desert Chaparral. The two species do occur together, however, on the coastal side of the Transverse and Peninsular Ranges, where Scrub Oak Chaparral mixes with Southern Oak Woodland. Basket Bush was important to Native Americans because the stems are very flexible and were used in making baskets. This is also one of the plants that shows measle-like splotching in the presence of ozone. Therefore, it is being watched closely in some areas as an indicator of air pollution.

Other drought-deciduous species are two fruit-bearing members of the rose family: Desert Apricot, *Prunus fremontii* (figure 8.29), and Desert Almond, *P. fasciculata* (figure 8.30). In the spring, Desert Apricot is covered with clusters of white flowers resembling those of many cultivated fruit trees. The fruit has thin meat and a large seed but was eaten by Cahuilla Indians nevertheless. Desert Almond produces many hairy fruits that probably were eaten as well. This is the shrub that is commonly infested with tent caterpillars (*Malacosoma* spp.; figure 9.21).

Succulent members of this community include two yuccas, Chaparral Yucca and Mojave Yucca (*Yucca schidigera*), as well as several species of cacti. The Desert Century Plant, *Agave deserti* var. *deserti*, may also occur here, although it is more appropriately a member of the Cactus Scrub community. The prickly pear found here is the Brown-spined Prickly-pear, *Opuntia phaeacantha*, which is associated with Pinyon Pines throughout the deserts. Its fruit is another important food for animals. At least three species of cholla can be included as components of this community. *Cylindropuntia echinocarpa* may be known as either Silver or Golden Cholla, depending on the color of its spines. Buckhorn Cholla, *C. acanthocarpa* var. *acanthocarpa*, is a similar but less spinescent species that gets its name from the fact that its branching may resemble the antlers of a deer, and Cane Cholla, *C. californica* var. *parkeri*,

FIGURE 8.29 Desert Apricot, *Prunus fremontii.*

which is also a component of Interior Coastal Sage Scrub.

Productivity in Scrub Communities

Water is a limiting factor on photosynthesis in scrub communities. Primary production is therefore low. Figures reflecting food production in Chaparral vary, but Mediterranean climates produce an average biomass of about 5300 lb per acre (600 g/m²) per year. This is about six times greater than a typical desert, but compared to other terrestrial communities it is low. A biomass of 5300 lb will support about 530 lb of herbivores and only about 53 lb of carnivores per acre. That small amount of food production shifts the dominance to small, ectothermic animals in scrub communities.

Variation in primary production within scrub communities is also high. Communities dominated by species of California-lilacs (*Ceanothus* spp.) are nearly twice as productive as the others. This type of Chaparral produces a biomass of nearly 9000 lb per acre (1056 g/m²) per

FIGURE 8.30 Desert Almond, *Prunus fasciculata*.

year. Furthermore, because nitrogen-fixing bacteria are associated with the roots of California-lilacs, the food produced is rich in protein. About 80% of the production results in litterfall, contributing to a diverse community of insects that live in detritus. Many species of birds are ground dwellers that feed on these insects. It should be emphasized that California-lilacs are usually short-lived successional species, and that high productivity lasts from about 4 to 10 years after a fire.

Productivity in Coastal Sage Scrub is low, reflecting the desertlike amounts of precipitation. Typical Coastal Sage Scrub produces only about 3000 lb of food per acre (335 g/m²) per year. About 60% of that results in litterfall. Total aboveground biomass is small compared to chaparral communities. Aboveground biomass for Coastal Sage Scrub is about 10,430 lb per acre (1172 g/m²), and for *Ceanothus* Chaparral it is about 67,850 lb per acre (7624 g/m²). The total biomass for animals in Coastal Sage Scrub is correspondingly low.

Southern Oak Woodland

Southern Oak Woodland is the southern California equivalent of Foothill Woodland. Usually it is dominated by one of the large oak species. Three species of large oak trees occur in southern California: Coast Live Oak, *Quercus agrifolia*; Canyon Live Oak, *Q. chrysolepis*; and Engelmann Oak, *Q. engelmannii*. Because they frequently hybridize, these three species are not always easy to tell apart, particularly where they grow near each other. These oaks have enormous root systems that enable them to tap into water supplies deep in the soil. Thus, they are not forced into long periods of dormancy by summer drought. Their thick bark protects them from fire.

Coast Live Oak is most common along the southern Coast Ranges and at low elevations in southern California. It occurs at moist sites with deep soil, particularly canyon bottoms and north-facing slopes. Live oak woodlands exist as islands within Coastal Sage Scrub. In the southern Coast Ranges, Coast Live Oak occurs on the coastal side, usually on north-facing slopes. It resembles Interior Live Oak, *Q. wislizeni*, by having spine-tipped leaf margins, but it differs by having curled leaf margins. The leaves are more oval, and the underside has hairs at the junctions of the veins. Interior Live Oaks have no hair on the underside of the leaves.

Coast Live Oaks are also known as California Live Oaks, and in many ways they symbolize southern California country living. In areas such as the San Fernando Valley, communities have incorporated the native oaks into their plans for urbanization. Place names such as Sherman Oaks and Thousand Oaks bear witness to this trend. Unfortunately, many people don't know how to treat these trees and wind up killing them. One obvious mistake for a tree such as this is to cover its enormous root system with concrete. The other major mistake is to plant a lawn or put water-loving ornamental shrubs around the base of the tree. These forms of vegetation require summer watering, and water during the summer causes a fungus to attack the roots of the oak, which in time will certainly kill it. The best advice for a person who has one of these oaks is to leave it alone. It probably has lived at that spot for hundreds of years, surviving on natural precipitation. If further landscaping is desired, the best bet is a mixture of shrubs from north-facing slopes of Coastal Sage Scrub. Attractive shrubs that are common associates of Coast Live Oak include Toyon, Lemonade Berry, and Fuchsia-flowered Gooseberry, *Ribes speciosum*.

At least six species in four families of boring insects (bark beetles) are threatening our native oaks. The most serious seems to be the Goldspotted Oak Borer, *Agrilus auroguttatus*, in the family Buprestidae. In most cases, the larvae create channels as they feed on the phloem, the nutrient-rich tissue under the bark. This disturbance weakens trees causing crown thinning and dieback. Ultimately the trees will die. The most seriously infected trees are Coast Live Oaks and California Black Oaks. Insecticide treatments have not been very successful and natural enemies of the beetles such as parasitoid wasps have not become established. At this point, removal of infected trees has been the primary form of treatment.

In recent years, a fungus infection carried by the Polyphagous Shot Hole Borer (*Euwallacea* sp.) has also taken a toll on Coast Live Oaks. It was first found at Whittier Narrows in 2003.

The infection caused by *Botryosphaeria fusarium* (= *Fusarium euwallacea*) is inserted under the bark by the borer. Tiny females carry the fungus in their mandibles that is used to feed the larvae. The larvae bore galleries under the bark, interfering with the transport of nutrients for the tree. A wet spot may appear at the site of the entry hole. The fungus affects about 30 species of trees, most of which, such as Avocado, are nonnative, but the dramatic effect of iconic oaks dying is getting the attention of arborists and naturalists. On the other hand, at this time less than 10% of the infected oaks appear to be actually dying, but the condition appears to be exacerbated by drought. Dead and dying oak trees are becoming conspicuous throughout the area. Other native species of trees affected include Fremont Cottonwood, Red Willow, California Sycamore, and White Alder.

Perhaps the most common associate of Coast Live Oak is Western Poison Oak, *Toxicodendron* (= *Rhus) diversiloba*. In spite of its name, Western Poison Oak is not in the same family as the oaks. It is a member of the cashew family (Anacardiaceae), which includes several other important southern California shrubs, such as Laurel Sumac, Lemonade Berry, Sugar Bush, and Basket Bush. Western Poison Oak is winter-deciduous and is very beautiful in autumn, when its leaves turn bright red before they are shed. Western Poison Oak does not affect everyone equally. Apparently, Native Americans were immune to its toxin. Those who are sensitive may be affected by contact with leaves, stems, or even smoke from a brushfire. It is a serious mistake to use it for a campfire. When hiking in canyons, sensitive individuals should proceed with caution, particularly during winter, when there are no leaves. If a sensitive person must hike with a dog, it is advisable not to pet it or hug it, because the dog is certain to have brushed against the Western Poison Oak.

Another southern California tree that is commonly associated with Coast Live Oak is Southern California Black Walnut, *Juglans californica*. This is the state's native walnut. Although the nuts are not as large as the commercial walnut, they are considered by many to be every bit as tasty. Squirrels commonly eat the nuts and are believed to play an important role in the dispersal of this species. This tree is winter-deciduous. Leaves are pinnately compound, meaning that what appears to be many leaves arranged opposite each other on the stem is actually a large single leaf with many leaflets. Distribution of Southern California Black Walnut is patchy, from lower canyon mouths of the Santa Monica Mountains southward to about the Santa Ana River, where it becomes associated with Riparian Woodland. Many of these trees, particularly along the southern face of the Transverse Ranges, have been replaced by development. In the San Jose Hills, between Diamond Bar and Fullerton, is the state's last remaining extensive stand of Southern California Black Walnut intermixed with Coast Live Oak. Unfortunately, this site is trapped in the midst of urban sprawl, and it is hoped that plans to preserve this area as open space will not become sidetracked.

Canyon Live Oak occurs at higher elevations than Coast Live Oak. Also known as Goldencup Oak, the name of this species refers to the large yellow cups on each acorn. The most widely distributed oak tree in the state, it occurs in similar habitat throughout the Foothill Woodland of the Sierra Nevada, Klamath Mountains, and Coast Ranges. It also occurs in the Providence and New York Mountains of the Mojave Desert, where it associates with Pinyon Pines. Canyon Live Oak is associated with north-facing chaparral species such as Scrub Oak and Coulter Pine. Leaves of Canyon Live Oak differ from Coast Live Oak and Interior Live Oak by possessing white hairs on the underside. A bit of confusion may arise in relation to the leaves because, on any tree, some have smooth margins and some spiny margins. If acorns are present, there should be no confusion, for no other oak has such large acorns with a sulfur yellow cup.

Engelmann Oak is also known as Mesa Oak. Its typical habitat is flat tablelands of the Peninsular Ranges, where the trees are widely spaced and the understory is a mixture of grasses and herbs. This type of habitat is often called an oak savannah. Engelmann Oaks formerly grew along the base of the San Gabriel Mountains between Pasadena and Claremont, but no woodland of this type persists in the area today. The community was a victim of urbanization and grazing.

Engelmann Oak is in the white oak group. This is the southern California equivalent of the Blue Oak, *Q. douglasii*, which is characteristic of the Foothill Woodland surrounding the Great Central Valley. Engelmann Oak differs from Blue Oak by not having lobed leaves; its leaves are elongated and have smooth margins. Also, whereas Blue Oaks are winter-deciduous, Engelmann Oaks are semideciduous: they may lose many leaves, but they usually do not lose them all.

An interesting pattern of reproduction is emerging for these oak trees. Seedlings require shade, but this is a fairly open habitat. Seedlings and saplings therefore occur in groups near the dripline of the canopy of large old trees. These are known as nurse trees. Apparently, fog drip and shade are necessary for germination.

The woodland characterized by Engelmann Oaks is associated with deep, clay-rich soils. These soils are found on flatlands amidst Scrub Oak and Upper Chaparral. A significant characteristic of these soils is poor drainage. Also found within these areas are temporary ponds known as vernal pools, which will be discussed later as ecologic islands. Engelmann Oaks as well as several vernal pools are now protected in a large Nature Conservancy reserve on the Santa Rosa Plateau in the Santa Ana Mountains southwest of Elsinore.

Coniferous Forests

Composition of coniferous forests in southern California has been outlined in Chapters 4 and 5. The dominant species of Mixed Coniferous Forest include Jeffrey Pine, Incense Cedar, and California Black Oak; these are joined by Sugar

Pine and White Fir on moist sites. Ponderosa Pine, which is not common, reaches its southernmost distribution in the Laguna Mountains of San Diego County. Mixed Coniferous Forest is absent from the Santa Monica and Santa Ana Mountains, but it is well established in the Santa Ynez Mountains in the vicinity of Figueroa Mountain. Lodgepole Pines and Limber Pines are found on high peaks in the San Gabriel, San Bernardino, and San Jacinto Mountains. The southernmost Limber Pines are found on Toro Peak, the highest peak in the Santa Rosa Mountains. The rest of the species mentioned here reach their southernmost distributions in Baja California.

Riparian Woodland

Riparian Woodland of Cismontane Southern California resembles that of the southern Sierra Nevada and the southern Coast Ranges. Most of the riparian species are winter-deciduous; descriptions of these species are found in Chapter 4. These are the trees that provide the autumn color in southern California canyons. People from the east who yearn for autumn color can drive less than an hour from southern California's urban sprawl to any number of canyons in the Transverse and Peninsular Ranges. If they want a more spectacular display, they will have to drive up to Bishop to see Quaking Aspens (plate 8B), or else go home.

At lower elevations, the dominant trees are Western Sycamore, *Platanus racemosa*, and Gooding's Black Willow, *Salix gooddingii*. Red Willow, *S. laevigata*, occurs up to about 5000 ft (1500 m), and Pacific Willow, *S. lasiandra* var. *lasiandra* occurs up to 8000 ft (2500 m) in elevation. Shrubby willows such as Hind's Willow, *S. exigua* var. *hindsiana*, occur to about 3000 ft (900 m), and Arroyo Willow, *S. lasiolepis*, occurs up to 7000 ft (2100 m) in elevation. Fremont Cottonwood, *Populus fremontii* subsp. *fremontii*, is also a common species, but it is more often found at low elevations inland than it is along the coast. Black Cottonwood, *P. trichocarpa* (figure 4.47), occurs from the foot-

hills to about 7500 ft (2300 m) in elevation. In the San Gabriel and San Bernardino Mountains, there is a variety of Black Cottonwood with narrow leaves, less than 1 in (25 mm) wide.

White Alder, *Alnus rhombifolia* (figure 4.48), occurs from about 1000 to 6500 ft (300–2000 m) in elevation. At middle elevations, where streams cut through Yellow Pine Forest, White Alder is commonly associated with Bigleaf Maple, *Acer macrophyllum* (figure 6.9), and Mountain Dogwood, *Cornus nuttallii* (figure 4.49).

At upper elevations, there is a variety of shrubby willows, including Lemmon's Willow, *S. lemmonii*; Scouler's Willow, *S. scouleriana*; and Yellow Willow, *S. lutea*. Of limited distribution in the San Gabriel and San Bernardino Mountains is Dusky Willow, *S. melanopsis*.

Quaking Aspen, *P. tremuloides* (plate [8B]), reaches its southernmost California distribution in the San Bernardino Mountains in an isolated colony just south of Heart Bar State Park, at an elevation of 7500 ft (2300 m) on Fish Creek. The only other known colony of aspens in southern California is also in the San Bernardino Mountains, southeast of Big Bear Lake. The presence of this species in these mountains is notable because it is approximately 150 mi (240 km) south of the nearest groves, in Tulare County of the southern Sierra Nevada. Perhaps even more remarkable is that at a similar distance south of the San Bernardino Mountains, even larger groves occur in the Sierra de San Pedro Mártir of Baja California. It is unfortunate that these San Bernardino Mountain populations of Quaking Aspen are now threatened by introduced beavers (*Castor canadensis*).

Of the evergreen species associated with Riparian Woodland, only California Bay Laurel, *Umbellularia californica* (figure 6.8), and Pacific Madrone, *Arbutus menziesii* (figure 6.7), occur in southern California. The odoriferous California Laurel is common in canyons as far south as the Laguna Mountains in San Diego County. Pacific Madrone has patchy distribution in canyons of southern California as far south as Mount Palomar.

ANIMALS OF CISMONTANE SOUTHERN CALIFORNIA

Animals of Riparian Woodland include a wide variety of insects, amphibians, and birds. Most of these are discussed in other parts of the book. Of particular significance to southern California is that Riparian Woodland communities have become threatened; hence, so have a number of the characteristic animal species. Channelization and flood control projects have destroyed many miles of Riparian Woodland. In addition, invasion of riparian habitats by the Brown-headed Cowbird, *Molothrus ater*, a nest parasite, has threatened a number of songbirds that require Riparian Woodland as a nesting habitat.

A specific example of a songbird that is threatened by Brown-headed Cowbirds is a small gray bird known as the Least Bell's Vireo, *Vireo bellii pusillus*, once considered a common summer visitor in southern California. Many vireos are yellowish and have a distinctive white ring around the eye. The Least Bell's Vireo has no eye ring, and its color is light gray. It is considered to be the least distinct of all the vireos. Typical of drab-colored birds that remain hidden, these vireos have a beautiful lyric song, which was formerly one of the familiar sounds of the Riparian Woodland. Typically, these birds would arrive from Mexico in the spring. They would nest in willows or wild rose bushes, remarkably close to the ground, and raise two broods of young per year. The combination of cowbirds and habitat destruction has practically eliminated this bird. Down to fewer than 300 pairs, this subspecies is now listed as endangered by the federal government. The Brown-headed Cowbird is also a documented parasite on nests of the California Gnatcatcher, a Coastal Sage Scrub species mentioned earlier that has been proposed for threatened status by the US Fish and Wildlife Service.

Other than a few creeks in the foothills of the Transverse and Peninsular Ranges, there are few pristine Riparian Woodlands of significant size left in the southern part of the state. One is found along Mono Creek near Santa Barbara, and another is along Santa Margarita Creek near Oceanside. Perhaps the most pristine is San Mateo Creek in the San Mateo Canyon Wilderness, one of the few streams in the state that flows uninterrupted from headwater to its outlet to the sea. In these areas during the summer, in order to protect the Least Bell's Vireo, federal workers trap and remove the unwanted Brown-headed Cowbirds.

Animals that live in scrub must deal with problems similar to those of animals that live in the desert. Overall, this is a hot, dry, food-poor ecosystem. As in the desert, there is an abundance of small, ectothermic animals, such as arthropods and reptiles, that have low metabolic demands and escape the harshest seasons by becoming dormant. Of the diurnal animals in Chaparral or Oak Woodland, birds are the most conspicuous. Many chaparral birds also occur in the desert, but others, such as the Western Scrub-jay, are primarily chaparral species. Most mammals are small and nocturnal. These include a number of species of rodents. The largest mammals are Mule Deer and Bighorn Sheep, and the largest predator is the Mountain Lion.

Arthropods

Insects are very common in chaparral habitats. Conspicuous are ants, grasshoppers, and butterflies, most of which are discussed in the following chapter. The most conspicuous pollinator in southern California is the European Honey Bee, *Apis mellifera*. Brought to the United States by English colonists on the east coast and Spanish missionaries in California, they have become extremely important pollinators for agricultural crops. Unfortunately, they have replaced native bees in a number of habitats. Feral bees overwinter in hollows of trees, particularly in Coast Live Oaks. In the spring, swarms of workers emerge, led by a queen, in search of a new hive, which may be another tree, a crevice in rocks, or in the walls of a house. The invasion of "Africanized Bees" has posed a problem in some areas. Africanized Honey Bees or "Killer Bees" were originally introduced to Brazil, presumably because they produce copious amounts of honey. By 1992, they expanded their ranges to reach the United States and by 1994 became established in California where they rapidly hybridized with the local Honey Bees. African Bees, which have hybridized with Honey Bees, have passed on to the offspring the habit of stinging in swarms, sometimes killing a victim. Triggered by pheromones, swarm stinging may be released by a single sting. Attempts to eliminate Africanized Bees have largely been futile, but it appears that through time this behavior has become less common. Nevertheless, the Africanized bees and their hybrids are not suited for cold climates, so their ranges remain at relatively low elevations and closer to the coast. They now range as far north as Tulare, Fresno, and San Luis Obispo Counties. They are well established in the San Joaquin Valley.

Drone Flies or Flower Flies, *Eristalis tenax*, are Honey Bee mimics. They mix in with Honey Bees and also feed on nectar. Drone Flies are so similar in appearance to Honey Bees, that most people can't tell the difference. Flies have only two wings, whereas bees have four (two pairs). Youngsters who want to terrorize their comrades, especially little girls, know how to pick up a Drone Fly by the wings, pretending they are Honey Bees. The flies are stingless. They are sometimes called "H-Bees" because the first two bands on the abdomen are connected by a vertical bar, forming the letter H. Larvae of the Drone flies live in sluggish streams or ponds. Many of the adult Flower Flies (Syrphidae) feed on aphids.

Most important among native ants is the seed-eating California Harvester Ant, *Pogonomyrmex californicus*. Based on biomass, this is probably the most common native insect in southern California. It tends to occupy sandy areas and is particularly common in grasslands or open spaces where the seeds of invasive grasses are common. This ant is the favored food of horned lizards as well. Its underground nests are easy to find by virtue of the pile of seed debris that rings the entrance to the burrow. The greatest threat to our native ants is the introduced Argentine Ant, *Iridomyrmrx humilis*. This small ant is omnivorous, feeding on almost any kind of food. It has become the most common ant in our homes and yards. Originally introduced to New Orleans in a coffee shipment, it has become one of the most common pests in the United States. In coastal areas where climate is less extreme, it has been eliminating native ant colonies, especially if there is a year-round supply of water. Elimination of the native Harvester Ant is accomplished through competition and aggressive swarming behavior.

More feared than the Argentine Ant is the Red Imported Fire Ant, *Solenopsis invicta*. An unsuspecting human who inadvertently steps on a burrow, which may be in a suburban lawn, may become victim to painful stings from a swarm of ants. These ants are moisture dependent, so lawns in public parks are good places to establish nests. Fortunately, at this time, they do not survive away from irrigated areas, so our natural landscapes should remain free of these pests.

There are a number of conspicuous butterflies in scrub habitats. The Pale Swallowtail, *Papilio eurymedon*, larva feeds on California-lilacs (*Ceanothus* spp.), and Monarchs, *Danaus plexippus*, overwinter at various coastal localities where they roost in great numbers. Popular viewing points include communal roosts in Mile Square Park in Huntington Beach. The Monarchs and their migration are discussed in detail in Chapter 7, on the Coast Ranges. A particularly attractive small butterfly is the California Dogface, *Zerene (= Colias) eurydice*, which is the California state insect. It has black-and-white forewings and yellow hindwings. It feeds on purple flowers such as the common Western Thistle, *Cirsium occidentale*. An interesting day-flying moth is the beautiful Ctenucha Moth, *Ctenucha multifaria*. Its abdomen is a brilliant metallic blue green and its head is bright red. The larvae seem to feed on grasses, but the adults can be seen in great numbers feeding on a variety of

FIGURE 8.31 Silver-backed Garden Spider, *Argiope argentata*. Note small male to upper left (photograph by Lenny Vincent).

flowers in Coastal Sage Scrub. The Quino Checkerspot, *Euphydryas editha quino*, formerly a common insect in Coastal Sage Scrub, and a victim of habitat destruction, is now listed as endangered by the federal government.

Predatory arthropods such as spiders and scorpions are also common but are far less conspicuous. In terms of biomass, spiders and scorpions are probably the most abundant predators in southern California. Orb weavers (Family: Araneidae) are the most conspicuous daytime predators in suburban gardens. They weave large orb-shaped webs in which females hang head down in the center or hub of the web. These spiders are easily recognized because they hold their legs in pairs, which may give the appearance they only have four legs. Heavy zig-zag bands of silk known as stabilimenta are often visible in the webs. It has been postulated that these visible structures are there to warn birds not to fly through the web. Another hypothesis is that the bands reflect ultraviolet light that attracts prey to the web. There are two common species of large, colorful orb weavers known as Garden Spiders. The Banded Garden Spider, *Argiope trifasciata*, has an oval abdomen that is marked on top by alternating transverse silvery-white or yellowish bands and narrow dark bands. Females are up to 1 in (25 mm) long, while the males, which are seldom seen, are about one-quarter the size. The Silver-backed Garden Spider, *A. argentata* (figure 8.31), has three pairs of lateral lobes on its abdomen. The front half of top of the abdomen is yellowish, while the back half has a triangular pattern that is dark with a few white marks. Females are slightly smaller than the Banded Garden Spider, but the small males are also seldom seen.

Cobweb Spiders (Family: Theridiidae) are very common small spiders with 32 known genera in North America. They make conspicuous webs around homes, rodent holes, rock piles, as well as tree cavities and branches. The most famous and the most feared is the Western Black Widow, *Latrodectus hesperus*, which is the only dangerously venomous spider likely to be encountered by a human in California. Its distinctive red hourglass pattern on the underside of its bulbous abdomen is its identifying characteristic. They are not likely to be encountered in natural habitats, but they are fairly common in urban settings where they may live in wood piles, old houses, or urban debris. Historically they were known to lurk in outhouses. There are five species of widow spiders in the United States. The only other one likely to be encountered in California is the Brown Widow, *L. geometricus*. It was first discovered in Los Angeles in 1999, and now it has been recorded in more than 30 cities in southern California, an amazing rate of dispersal. Much less dangerous than the Black Widow, it can still deliver a painful bite. Webs can be found easily at night in a variety of locations in suburban gardens hanging upside down with its hourglass clearly visible.

Of special interest among arthropods are the ticks and mites. Although they are more closely allied to spiders than to other arthropods, they are not predators in the typical sense of the word. Altogether, they occur in practically all habitats and rival insects in diversity. Ticks are bloodsuckers, and as such they are considered to be ectoparasites. Experienced hikers have long known the importance of checking themselves for ticks after a day in the chaparral. Recently, the spread of a tick-borne malady known as Lyme disease has made this practice even more important. Lyme disease is spread by the bite of the small Deer Tick, *Ixodes pacificus*. Symptoms of Lyme disease include a rash at the site of the bite, often target-shaped, which if left untreated is followed by an arthritis-like inflammation of joints and associated nausea and fever. The disease is caused by a bacterium, *Borrelia burdorferi*, and once the symptoms are recognized, it may be treated with antibiotics. The problem is that the symptoms may not show up for a month after the bite, by which time the victim may fail to associate the symptoms with the hike in the brush. Interestingly, the Western Fence Lizard is immune to Lyme disease, and its blood is being studied with the idea of developing a treatment for Lyme disease in humans.

Mites are very small versions of ticks. Many of them are scavengers, and a large number of them are ectoparasites on plants and animals. Close inspection of lizards in chaparral areas, for example, commonly reveals groups of small red mites clustered near the ears or in the armpits. Of recent interest is the discovery that the larvae of harvest mites or chiggers (Trombiculi-

dae) occur in Coastal Sage Scrub and adjacent grasslands. Even though some 150 species have been reported from California, they are not often encountered. In fact, it is a common belief that they do not occur here. Adult chiggers do not feed on humans, but the larvae may crawl up a person's legs and feed for a short time on skin cells. They do not burrow into the skin, as commonly believed, but when they become firmly attached to the host's skin they inject a digestive fluid that causes disintegration of skin cells. It is the digestive fluid that causes the itching and dermatitis associated by a chigger infestation. There are few reports of the chigger *Eutrombicula belkini* in southern California, from Laguna Beach, Palos Verdes, and a couple of localities in Ventura County. In one interesting instance, forensic experts were able to link a murder suspect to the scene of the crime by virtue of the chigger bites on his legs.

Amphibians

Amphibians are not common in Chaparral because they typically remain near water. One interesting exception is the Monterey Salamander, *Ensatina eschscholtzii eschscholtzii*, which inhabits deep canyons and north-facing slopes, where it lives under rocks and logs. In summer, in order to estivate, these salamanders often move into burrows or under woodrat nests. They have smooth skin and are brownish on the back. As lungless salamanders (Plethodontidae), they breathe through their skin and the capillaries in their throat.

Distribution of the genus *Ensatina* in California was mentioned in the chapters on the Sierra Nevada, the Pacific Northwest mountains, and the Coast Ranges. The range of these salamanders illustrates a rare pattern known as a *rassenkreis* (figure 8.32), a German word that means "circle of races." In California, a single species of variable coloration inhabits a range that encircles the Great Central Valley; different color variants have been described as subspecies. *Ensatina* ranges from southwestern British Columbia to extreme southern California in the

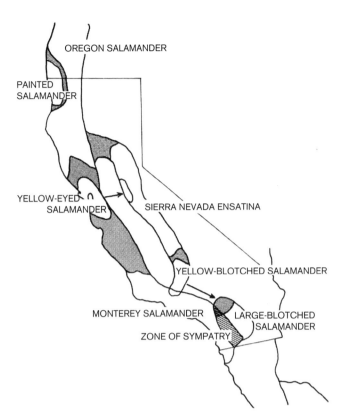

OREGON SALAMANDER

PAINTED SALAMANDER

YELLOW-EYED SALAMANDER

SIERRA NEVADA ENSATINA

YELLOW-BLOTCHED SALAMANDER

MONTEREY SALAMANDER

LARGE-BLOTCHED SALAMANDER

ZONE OF SYMPATRY

FIGURE 8.32 The *Ensatina* rassenkreis. Shaded areas indicate zones of intergradation between subspecies.

Peninsular Ranges. The ring of distribution represents a circular chain of subspecies that intergrade where ranges overlap. Coastal forms have a solid color pattern that grows lighter from north to south; these are known as Monterey Salamanders. The Sierra Nevada groups have yellow spots that become larger from north to south. These blotched forms occur at higher elevation, in association with Mixed Coniferous Forest. Similar to a pair of calipers, the two arms of distribution come together and form populations that overlap at the southern end, but they do not interbreed. In the vicinity of Mount Palomar and Julian in the Peninsular Ranges, the Monterey Salamander of lower elevation overlaps its range with the Large-blotched Salamander, *E. e. klauberi* (figure 8.33), of Mixed Coniferous Forest. The two varieties may occur under the same bit of cover, but they do not hybridize. Farther north, in the San Bernardino Mountains, a similar situation occurs with a different blotched subspecies. At the upper end of

Sawmill Canyon on Mount San Gorgonio, the Monterey Salamander hybridizes to a small degree with the Yellow-blotched Salamander, *E. e. croceater* (figure 8.33). If an extinction were to occur in northern California where the solid form grades into the spotted form, the circle of races would be broken, and we would be obligated to identify two species: a solid form and a variable blotched form. A distribution pattern of this type is important to understand because it shows that the process of speciation is dynamic and still occurring.

Also of interest in the distribution of these salamanders is a peculiar disjunction in the range of the Yellow-blotched Salamander. In southern California, these salamanders occur in the Tehachapi Mountains and on Mount Pinos to the west. They also occur in the San Bernardino Mountains to the east, but they have never been collected in the San Gabriel Mountains in between. The cause of this disjunction has not been identified.

FIGURE 8.33 Monterey Salamander, *Ensatina eschscholtzii eschscholtzii*, and Large blotched Salamander, *Ensatina eschscholtzii klauberi*, from near Julian, San Diego County.

Another lungless salamander is the Arboreal Salamander, *Aneides lugubris* (figure 7.27B). These salamanders are nearly always associated with oaks, from southern California through the Coast Ranges. In southern California, they are most often associated with Coast Live Oak, although near Crystal Lake, at 5000 ft (1500 m) in the San Gabriel Mountains, they are associated with Golden-cup Oak. They are called arboreal because they are known to climb trees and make nests in tree hollows. However, they are probably most common on the ground, where they are frequently found under the same rocks or logs as Monterey Salamanders. They are brown with small yellow spots. They are unique among salamanders by having a prehensile tail, which they curl and use to hold themselves in a tree or bush. They have very large jaw muscles, which give them an alligator-like appearance. The teeth are probably used to help enlarge cavities in wood. They are capable of inflicting a bite and drawing blood, but that rarely occurs. Another distinctive feature of these salamanders is that they have expanded toe tips, which were once thought to function as an aid in climbing. This may be true, but it is now known that the toes are loaded with capil-

laries and aid the animal in its absorption of water and oxygen. It might be stated that these salamanders breathe through their toes. They also absorb oxygen through their skin in general, and particularly through capillaries in the mouth, which they ventilate by flapping the floor of the mouth up and down.

There are two more lungless salamanders associated with oaks in southern California. These are the slender salamanders, and the evolution and distribution of this interesting group of sedentary salamanders has been discussed in detail in the chapters on the Sierra Nevada and Coast Ranges. Locally, there are two species: the Southern California Slender Salamander, *Batrachoseps major major* (= *B. pacificus*; figure 8.34) and the Black-bellied Slender Salamander, *B. nigriventris*. These salamanders are very well adapted for burrowing. They have long, wormlike bodies and very small limbs. They take refuge in any sort of crack, hollow, or burrow, and they probably do no actual digging. These are very sedentary, seldom moving far from their shelter.

The Pacific Slender Salamander is a lowland form, seldom occurring above 1000 ft (300 m) in elevation. It is the species that is frequently

FIGURE 8.34 Garden Slender Salamander, *Batrachoseps major major.*

encountered under rocks, boards, or trash cans in suburban southern California. It is the larger of the two forms, sometimes reaching 4 in (10 cm) in total length. It can be distinguished by its gray belly with small black flecks. The Pacific Slender Salamander is replaced by the Black-bellied Slender Salamander at higher elevations. In southern California, this salamander is not particularly associated with sedimentary rocks, as it is in the Coast Ranges. When the black-bellied form is found, it is in the same sort of habitat as the other lungless salamanders. This small salamander seldom exceeds 3 in (7.5 cm) in total length. Its belly is black, speckled with minute white flecks.

There are two known relict species of slender salamander in southern California. The San Gabriel Mountains Slender Salamander, *B. gabrieli*, occurs at about 5000 ft (1500 m) in the North Fork of San Gabriel Canyon, and eastward toward Waterman Canyon. It also has been found on the western end of the San Bernardino Mountains. Populations of the Black-bellied Slender Salamander occur in similar locations. Another relict species, known as the Desert Slender Salamander, *B. m. aridus,* occurs in a most unlikely location. It is known only from one locality, a palm oasis in upper

Deep Canyon, on the east side of the Santa Rosa Mountains. Presumably, during periods of heavier precipitation it was more widespread. It endures today in cracks wet by a permanent spring on a north-facing slope.

The association of these relict populations with terranes that have been translocated over long distances was discussed in relation to the Coast Ranges. A number of these associations occur in southern California as well. On Santa Cruz Island, the Channel Islands Slender Salamander, *B. pacificus*, and the Black-bellied Salamander coexist, but the two species seem to be associated with different extent terranes on each side of the Central Valley of the island. Another such region is along the San Gabriel River. Faulting in the area has brought into close proximity at least three terranes of different origin. At various points along the river are populations of distinctive native animals. For example, this is the only southern California locality where the Foothill Yellow-legged Frog, *Rana boylii* (figure 11.30A), formerly occurred in the same water with the Southern Mountain Yellow-legged Frog, *R. muscosa* (figure 11.30B). The San Gabriel Mountains Slender Salamander occurs here, and this is also the only place in the San Gabriel Mountains where the

FIGURE 8.35 Western Spadefoot Toad,
Spea (Scaphiopus) hammondii.

Yellow-blotched Salamander has been sighted. Unfortunately, this unofficial record has not been confirmed. Furthermore, along the East Fork of the San Gabriel River there is a population of the Collared Lizard, normally a desert form. The San Gabriel River is also one of the few localities in southern California that contains remnants of the native fish fauna.

Frogs and toads have no tails as adults, but their aquatic larvae, known as tadpoles, have tails. Metamorphosis of these amphibians is well known and many people have watched captive tadpoles lose their tails and grow legs. Toads are more terrestrial than frogs. They tend to have roughened, warty, skin and their hind legs are shorter that those of frogs. The webs between their hind toes are not as well developed as the webbed feet of frogs. Toads have large oval glands, known as parotoid glands, behind the eyes. These glands secrete a distasteful, sometimes poisonous mucous, as a deterrent to predation.

The North American Spadefoot Toad family (Scaphiopodidae) has only one representative in southern California, the Western Spadefoot Toad, *Spea (= Scaphiopus) hammondii* (figure 8.35). This toad has disappeared from most of the lowlands in southern California, but it still exists in coastal Orange and western Riverside County. The adults are blunt-nosed and have large golden-colored eyes with vertically ellipti-

cal pupils. They are greenish on top and whitish underneath. The hind foot has a dark-colored digging spur. This nocturnal toad spends most of the year underground. It emerges in winter when it breeds and lays eggs in temporary pools. While the adults are seldom seen, hikers sometimes see the larvae in puddles on dirt roads. They were formerly common at the mouths of creeks in the Santa Ana Mountains, but gravel mining and other activities have destroyed much of their habitat. The State of California has listed this toad as a species of special concern.

The True Toad family (Bufonidae) includes our most commonly observed amphibian, the Western Toad, *Anaxyrus (= Bufo) boreas*. This toad is fairly large, with a body length of up to 5 in. It is easily recognized by a distinct yellow stripe down its back. The parotoid glands are larger than their eyelids. These toads occur in a great variety of habitats, sometimes quite a distance from water. It is not uncommon to find these toads on wet suburban lawns at night. The breeding call is a repetitive birdlike chirp.

The Arroyo Toad, *Anaxyrus (= B.) californicus*, is an inhabitant of sandy streambeds in the foothills, in association with oaks and sycamores. During the dry season, they remain in burrows on stream terraces, but during the rainy season, they become active along streams. They are primarily nocturnal, so they are sel-

FIGURE 8.36 Common Side-blotched Lizard, *Uta Stanburiana*, in its defense posture, "doing push-ups."

dom seen. Adults can be recognized as having no middorsal stripe, and the parotoid glands are bicolored, dark in back and light in front. This species is federally listed as endangered. Its most serious threat is habitat destruction.

Two kinds of treefrogs or chorus frogs are found in southern California. The Baja California Treefrog (= Pacific Chorus Frog), *Pseudacris hypochondriaca* (= *Hyla regilla*), is the most common, occurring in a variety of freshwater habitats from the coast to the high Sierra. It is the most common frog in the western United States. These are small frogs with expanded toe tips. Acting like suction cups, these toes enable them to climb even the smoothest of surfaces. They do not need to be associated with trees. It is not uncommon to find them on the walls of the restrooms in our parks. They come in a variety of colors, but are easily recognized by their dark eye stripe. In the spring, they gather in large numbers in slow-moving pools of water. Their mating calls are easily recognized as the familiar "ribbit" sound that collectively produces the din emanating from the local canyons at night.

The California Treefrog (= Chorus Frog), *P. (= H.) cadaverina*, is most often found in the foothills along streams with large boulders. They are also found in desert Palm Oases. They have no eye stripe and they are usually grayish, matching exactly the color of the boulder on which they perch. They seldom occur in large aggregations and their call resembles that of the Pacific Treefrog, although it is a single note rather than the two-syllable "ribbit" sound.

Reptiles

Reptiles are common and conspicuous in scrub communities. Many snakes and lizards are also found in the desert. Lizards are far more common than snakes. The most common of these is the Common Side-blotched Lizard, *Uta stansburiana* (figure 8.36), but the Western Fence Lizard, *Sceloporus occidentalis*, is nearly as common. These are primarily insect eaters. Common Side-blotched Lizards are also common in desert areas and get their name from a large black blotch located just behind the armpits. These small lizards are active all year and are particularly common in rocky areas. Females are usually brownish. Males may exhibit bright blue speckles on the top surface, and the lower surface of the throat may be blue and orange.

Common Side-blotched Lizards and Western Fence Lizards are conspicuous because of their basking behavior. They appear to do push-ups as a person approaches. The underside of the chin and belly of these lizards, particularly males, is brightly colored, and this bobbing behavior is a way for the lizards to expose the color. It is a form of behavior used to defend territories and for courtship. Western Fence Lizards are also known locally as "blue-bellied lizards" because the underside is an intense iridescent blue. The color on the top surface of

FIGURE 8.37 Tiger Whiptail, *Aspidoscelis tigris*.

these lizards closely resembles that of burned Chamise stems or granite. Hence, predators, particularly mammals with limited color vision, would experience some difficulty seeing these lizards basking on these surfaces.

At upper elevations in southern California, the Western Fence Lizard is replaced by the Common Sagebrush Lizard, *S. graciosus*. This lizard ranges throughout the Great Basin and the eastern Sierra, but in southern California its range is deflected upward in the mountains along with Great Basin Sagebrush. In the Transverse and Peninsular Ranges, the Common Sagebrush Lizard is found in Upper Chaparral, Sagebrush Scrub, and Mixed Coniferous Forest. It is smaller than the Western Fence Lizard, and where the two species occur together, the Common Sagebrush Lizard can be distinguished by its rusty-colored armpits. At lower elevation, in the interior parts of the Peninsular Ranges, the Western Fence Lizard is replaced by the Granite Spiny Lizard, *S. orcutti*. This large, dark lizard is particularly common in rocky areas. In the coastal parts of the Peninsular Ranges, the Western Fence Lizard is the usual species.

Another common chaparral lizard that is also found in the desert is the Tiger Whiptail, *Aspidoscelis (= Cnemidophorus) tigris* (figure 8.37). This lizard is distinguished by its black-and-white checkered pattern and long, dark tail.

It is commonly seen foraging for insects in the litter under bushes. Those who like to capture lizards by noosing them find whiptails a frustrating challenge. They move incessantly, and even if they are snared, they usually wiggle free because their necks are thicker than their heads. On the coastal side of the Peninsular Ranges, the Tiger Whiptail is replaced by an uncommon striped form known as the Orange-throated Whiptail, *A. hyperythra* (= *C. hyperythrus*). Its distribution in southern California is the northern limit for its range, which extends southward to the tip of Baja California. Because it is an inhabitant of Coastal Sage Scrub, a community that is rapidly being replaced by pastures, agriculture, and development, the lizard is a candidate for being listed as rare or endangered. It is listed by the California Department of Fish and Wildlife as a species of special concern.

Whiptail lizards range widely across the arid parts of the southwestern United States, but they actually represent only the northernmost forms of the whiptail family (Teiidae), which extends across the tropics from temperate South America to the southern United States. In the Southwest, they are notable for a bizarre form of speciation in which all-female populations reproduce by parthenogenesis, that is, in the absence of males and without mating. It has been determined that these populations evolved in marginal habitats by hybridization between

FIGURE 8.38 Blainville's Horned Lizard, *Phrynosoma blainvillii*.

species with overlapping ranges. Genetic integrity of the hybrids is maintained because the lizards do not reproduce sexually. Furthermore, some of the populations are characterized by a rare phenomenon in animals: they have triploid (3N) chromosome numbers. Typical sexually reproducing organisms have a diploid (2N) chromosome number that is maintained indefinitely by fusion of sperm and ova that contain only the haploid (N) number of chromosomes. Nine all-female parthenogenetic populations have been identified so far.

The Blainville's Horned Lizard or Coast Horned Lizard, *Phrynosoma blainvillii* (=*Anota coronatum*) (figure 8.38), ranges throughout the state except in the humid northwest and the deserts. In the deserts, it is replaced by two other species: the Desert Horned Lizard, *P. platyrhinos*, and the Flat-tailed Horned Lizard, *P. mcallii*. These lizards are so similar in appearance that it takes experience to tell them apart. The Blainville's Horned Lizard usually has two rows of conspicuous spiny scales along the edge of its belly; the Desert Horned Lizard has only one, and the scales are smaller. The Flat-tailed Horned Lizard has a dark middorsal line. Habits of all horned lizards are similar. They frequent open spaces and sandy areas, where they eat primarily ants. They rely on protective coloration and a flattened body shape for concealment, and they usually remain motionless when approached.

In southern California, numbers of Blainville's Horned Lizards are declining rapidly. In California, the San Diego subspecies, *P. c. blainvillii*, occupies essentially the same range as the Orange-throated Whiptail and, along with some other species of Coastal Sage Scrub, is up for consideration as rare or endangered. The problem is related to habitat destruction associated with development, pastures, and agriculture. Replacement of the California Harvester Ant by the nonnative Argentine Ant is also a problem. The invasive Argentine Ants are not as palatable, and they tend to swarm and harass the horned Lizards. The situation has been complicated by the practice of fire suppression, which has allowed existing brushland to overgrow, reducing the amount of open space. The open spaces that remain, such as dirt roads, flood plains, and stream banks, may become trampled by humans and off-road vehicles. In an effort to control flooding, stream courses have been channelized and lined with concrete. Also, children love to collect and play with horned lizards. It remains to be seen whether horned lizards and humans can coexist.

Snakes are far less common than lizards. They are larger, and they are all carnivores. Many snakes of scrub communities are the same species that occur in the desert. Among those distinctly associated with scrub is the California Kingsnake, *Lampropeltis californiae*

FIGURE 8.39 California Kingsnake, *Lampropeltis californiae*. A. Banded color pattern. B. Striped color pattern.

(= *L. getula*; figure 8.39). This snake is common throughout the Southwest in many habitats, but in California it is typically a lowland, coastal species. The California Kingsnake is characterized by alternating dark and light bands or rings about the body (figure 8.39A). This pattern is found in other snakes, but the California Kingsnake is the largest of the California banded species. Color variants are found throughout its range. In southern California, for example, there is a striped phase; in this snake, a continuous stripe runs lengthwise along the top side of the body (figure 8.39B). These snakes eat small rodents, lizards, and other snakes. They are called kingsnakes because they also will eat rattlesnakes, and they are not affected by rattlesnake venom.

The California Kingsnake is replaced by the California Mountain Kingsnake, *L. zonata*, in Upper Chaparral and Mixed Coniferous Forest.

The adaptive value of a banding pattern has been questioned frequently. Several other species, such as the Long-nosed Snake (*Rhinocheilus lecontei*), the Western Shovel-nosed Snake (*Chionactis occipitalis*), and the Variable Sandsnake (*Chilomeniscus stramineus*), show similar banding. These, however, are all nocturnal burrowers, whereas the kingsnakes are usually diurnal. If a snake with a banded pattern remains motionless underbrush where filtered light makes a dark and light pattern, its pattern of alternating light and dark bands will help conceal it.

Conspicuous patterns can also serve as a warning to other organisms, a mechanism known as *aposematic coloration*. The Banded Krait, *Bungarus fasciatus*, of Southeast Asia, a venomous snake, has a black-and-white pattern that is very similar to that of the Common Kingsnake. Furthermore, studies of brightly colored snakes in Mexico and Central America confirm that mimicry occurs. There is remarkable convergence in color patterns where the ranges of nonvenomous species overlap those of venomous species such as coral snakes.

Venomous coral snakes of the southeastern United States and Mexico have banding, including red, Yellow, and black colors, that resembles the pattern of the California Mountain Kingsnake, but their ranges do not overlap. This disparity raises the question of whether mimicry could have been responsible for the bright colors of the California Mountain Kingsnake, *L. zonata*. Evidence for mimicry exists where the present range of the Arizona Coral Snake, *Micruroides euryxanthus*, overlaps the ranges of species resembling the California Mountain Kingsnake. These similarly colored species include the Long-nosed Snake and two species not found in California: the Sonoran Mountain Kingsnake, *L. pyromelana*, and the Milk Snake, *L. triangulum*. It is possible that the coloration of the California Mountain Kingsnake can be traced to a time when its range overlapped that of a coral snake. It is also possible that a common ancestor of these three species of *Lampropeltis* (the California Mountain Kingsnake, the Sonora Mountain Kingsnake, and the Milk Snake) co-occurred with a coral snake, and the events responsible for speciation of the three took place after the mimicking pattern evolved.

Another bit of evidence to support the warning function of the California Kingsnake banding pattern is that rattlesnakes instinctively avoid kingsnakes. If a kingsnake and rattlesnake are placed together, the rattlesnake retreats without rattling, which is *its* threat. Rattlesnakes have responded to the presence of kingsnake skin alone. In addition, there is evidence that rattlesnakes respond to kingsnake odor, leading to the question of whether there is odor mimicry that has escaped observation.

A color pattern that is more effective for moving through brush or grass without being seen is that of longitudinal stripes. These stripes break up solid color patterns, and they appear to remain stationary as the animal moves, particularly if it is only partly exposed. It is interesting that behavioral studies of kingsnakes show that those with the banded pattern tend to be ambush hunters, waiting motionless for prey to come near, whereas the striped forms are active foragers, frequently moving in search of prey.

The most common snake with longitudinal stripes is the California Whipsnake or Striped Racer, *Coluber (= Masticophus) lateralis*. This snake is coachwhip shaped and gets its name from its apparently rapid movement, although it probably moves no faster than about 4 mph (7 kph). It is an active forager. It moves with its head raised, and its sinewy shape and longitudinally striped elongate body give it the appearance of great speed. Distribution of this snake is closely associated with Chaparral. It is conspicuous because it is diurnal. It feeds on a variety of foods, including lizards, birds, small mammals, and insects.

Other snakes with longitudinal striping include Gartersnakes (*Thamnophis* spp.) and the Western Patch-nosed Snake, *Salvadora hexalepis*, also relatively rapid-moving snakes. Gartersnakes are found chiefly near water, but the Western Patch-nosed Snake occurs throughout desert and chaparral areas. This racerlike snake is not frequently encountered, however. It gets its name from the large, curved scale on the tip of its snout. This type of nose scale is also found on several other species of snake and is probably an adaptation for burrowing. An interesting study on Western Terrestrial Gartersnakes, *T. elegans*, which occur in a wide variety of habitats in California, showed that snakes living on the shoreline of lakes, as opposed to those in meadows with more cover, are more active and have shorter life spans. The long-lived meadow snakes actually produce

FIGURE 8.40 Gopher Snake, *Pituophis catenifer.*

more cellular antioxidants and other anti-aging enzymes characteristic of mammals.

Several snakes of southern California are divided into coastal and desert subspecies, including the Gopher Snake, the Glossy Snake, the Western Patch-nosed Snake, and the Rosy Boa. In each case, the coastal form occurs throughout the scrub communities.

The Gopher Snake, *Pituophis catenifer* (figure 8.40), is probably the most common snake in southern California. It certainly is the largest, measuring up to 8 ft (250 cm) in length. This large yellowish snake may have a diamond pattern on its back. This pattern, as well as its habit of hissing and vibrating its tail when alarmed, leads some people to mistake it for a rattlesnake. This snake is a good burrower, and it also enters the burrows of rodents such as gophers or ground squirrels. Depending on the size of the burrow, the snake may kill its prey by maneuvering past it and crushing it against the side of the burrow. Gopher Snakes also eat birds and some lizards, which they kill by constriction. It also has been reported that they eat bird's eggs, including chicken eggs.

It is not uncommon to find Gopher Snakes sunning on mountain roads, sometimes in a group of three or more. Such grouping usually occurs during spring and includes several males vying for the favors of a single female. This habit of lying on roads, plus the mistaken notion that they are rattlesnakes, has led to a decline in population numbers. These are beneficial animals that need not to be molested.

The Glossy Snake, *Arizona elegans*, resembles the Gopher Snake, but it is smaller and has smooth scales. It is more common in Coastal Sage Scrub than in Chaparral, and it tends to be nocturnal. Lizards and small mammals are its favored food.

The Rosy Boa or Three-lined Boa, *Lichanura orcutti (=L. trivirgata)*, is a slow-moving pinkish snake with a blunt nose and tail. Boas in general possess two very small hind limbs, one on each side of the vent. These vestiges indicate that snakes come from ancestors that once possessed limbs. The limbless condition is most likely an adaptation for burrowing. The Rosy Boa is becoming scarce. It usually occurs in Coastal Sage Scrub or along lower stream courses, both habitats that are disturbed by the activities of humans. Most of the time, they take shelter in rodent burrows. They are known to curl up in a ball when disturbed (figure 8.41). This extremely docile snake is favored by those who like to keep snakes as pets. Unfortunately, they do not always do well in captivity. They are usually nocturnal, feeding on small mammals, and they sometimes crawl about in the rain. They differ from most snakes by giving birth to their young alive. The Rosy Boa should not be confused with the Northern Rubber Boa, *Charina bottae*. In southern California, the Northern Rubber Boa is known only from three localities in the Mixed Coniferous Forest: Mount Pinos, Idyllwild, and near Lake Arrowhead.

The similarity between Coastal Sage Scrub and desert communities is apparent from the vegetation, but similarities between the animals also are apparent. For example, some typically desert reptiles may occur in Coastal Sage Scrub. Included here is the Desert Night Lizard,

FIGURE 8.41 Rosy Boa, *Lichanura orcutti*, in its defense posture.

Xantusia vigilis, which is associated with Chaparral Yucca on coastal slopes of the San Gabriel Mountains. Similar to its role in the desert, this lizard eats mostly termites from the dead parts of various species of *Yucca*.

Also noteworthy is the presence on the coastal side of the San Gabriel Mountains of a desert dweller, the Great Basin Collared Lizard, *Crotaphytus bicinctores* (= *C. collaris*). This lizard inhabits boulder-strewn washes in Lytle Creek and the East Fork of the San Gabriel River. These localities are adjacent to Lower Chaparral but are desertlike in their exposure. The lizards in Lytle Creek have not been seen in many years.

Other desert forms that have been recorded in Coastal Sage Scrub include the Western Banded Gecko, *Coleonyx variegatus* (figure 9.47), the Zebra-tailed Lizard, *Callisaurus draconoides* (figure 9.40B), and the Coachwhip or Red Racer, *C.* (= *M.*) *flagellum.* For unknown reasons, the Coachwhip seemed to disappear in the 1960s, coincident with the loss of Coastal Sage Scrub along the front of the Transverse Ranges. In the 1980s, however, there have been many sightings of them in the Alluvial Scrub community of Big Tujunga Wash, near the western end of the San Gabriel Mountains.

Rattlesnakes also occur from the coast to the desert. These venomous snakes possess temperature-sensitive pits on each side of their nostrils, which they use to locate endothermic prey such as small rodents or birds. Similar to the bright color of other venomous snakes, the rattle is a warning device. A rattlesnake does not want to waste its venom defending itself.

In general, rattlesnakes are not highly mobile; they are stout-bodied and incapable of rapid locomotion, relying on quick reflexes and venom to immobilize their prey. Their typical hunting behavior is to locate by odor an area with suitable prey. They detect this odor with the forked tongue. Then they remain motionless, in ambush position, near a rock or log, often with their chin resting on top of the log. Mice run along the top or side of the log and unwittingly come too near. The rattlesnake senses the mouse's body heat with its labial pits, and it strikes. Another favorite rattlesnake trick is to crawl up into a bush and remain motionless among the branches until a bird or small mammal comes near.

Rattlesnakes may group in dens for the winter, where they are able to hibernate and share each other's body heat. In the spring, they emerge from the den and migrate to hunting areas, where they will remain solitary except to mate. Rattlesnakes will migrate until they encounter an area that has an abundant supply of mice for food. They locate this site by its odor. As indicated before, odor molecules are picked up by the snake's forked tongue, but the odor is actually interpreted by two small pits,

FIGURE 8.42 Southern Pacific Ratttlesnake, *Crotalus oreganus helleri.*

known as vomeronasal organs, in the roof of its mouth. When sufficient mouse odor is detected, the rattlesnake usually takes up residence in the area for the entire summer.

The typical rattlesnake of the Chaparral is the Southern Pacific (Western) Rattlesnake, *Crotalus oreganus helleri* (figure 8.42). This snake is widely distributed in southern California, occurring in nearly all cismontane habitats. Usually it is brown, with a diamond pattern on its back, but in the Mixed Coniferous Forest it may be completely black.

In the Peninsular Ranges, and at lower elevations in general, the Red Diamond Rattlesnake, *C. ruber,* is present. This is the largest but perhaps most docile of California's chaparral rattlesnakes. This large pinkish snake is a near relative of the Western Diamond-backed Rattlesnake, *C. atrox,* another desert snake. The venom of the Red Diamond Rattlesnake is less dangerous, however.

Another pinkish rattlesnake, but without the diamond pattern, is the Speckled Rattlesnake, *C. mitchellii.* This is another species distributed in lowland desert and coastal communities in southern California. Overall, it inhabits mostly rocky areas from the southern

Sierra Nevada to Baja California. It is more common in desert areas, but it may be seen on the coastal side of the Peninsular Ranges.

A few reptiles seem to be associated particularly with Southern Oak Woodland in southern California. Unfortunately, with the encroachment of humans upon these habitats, the reptiles are far less common than they used to be, and some may have been extirpated. One such reptile is the Western Yellow-bellied Racer, *C. constrictor mormon.* This snake is associated with oaks in the Coast Ranges, and it still occurs on the Santa Rosa Plateau in the Santa Ana Mountains. It still may be present in the Mount Pinos-Gorman area and has been sighted recently on the western end of the San Gabriel Mountains near Placerita Canyon. In San Diego County, it is most often found in Riparian Woodlands. The Racer is bluish gray above and yellow on the belly. Young have brown saddles across the back and barely resemble the coloration of an adult. It feeds on salamanders and skinks.

The Western Red-tailed Skink or Gilbert's Skink, *Plestiodon gilberti,* is not often seen is southern California. The adult is brown or olive. Males in breeding condition have a red throat.

These southern races of Gilbert's Skink differ from the northern races by having pink or red tails rather than blue tails during the juvenile phase. Sometimes they coexist with the Western Skink, *E. skiltonianus*, which has a blue tail. In northern populations of Gilbert's Skink, the juveniles look very much like the Western Skink, but the two species seldom occur together. The Western Skink is more commonly seen in southern California. It commonly occurs in Coastal Sage Scrub, Southern Oak Woodland, and Mixed Coniferous Forest, where it is fed upon by the Mountain Kingsnake.

Several species of amphibians and reptiles characteristic of Southern Oak Woodland seem to have adapted to California gardens. Among these are secretive or burrowing species such as the Arboreal Salamander or slender salamanders. Of particular note is the appearance in gardens of the Ring-necked Snake, *Diadophis punctatus*, and the Northern California Legless Lizard, *Anniella pulchra*. These species prefer moist, sandy soil or leaf litter in which to burrow. Well-vegetated, watered gardens with mature shrubs and abundant litter, particularly where undisturbed native habitats occur nearby, seem to attract these species.

The Ring-necked Snake is often small, about 12–15 in (30–38 cm) long, but it may reach 30 in (75 cm). This slender snake is bluish to nearly black, with a bright yellow to orange neck ring. Its underside is orange to red. When the snake is disturbed, it coils and turns over its tail, revealing the brightly colored underside. This is an apparent warning device, as the snake is mildly venomous. Red-necked Snakes feed on earthworms, salamanders, small lizards, and insects, all of which may also appear in gardens.

The Northern California Legless Lizard looks like a small silvery snake, seldom exceeding 6 in (15 cm) in length. Even though it has no legs, close inspection will reveal that it has eyelids and ear holes, characteristics of lizards. Northern California Legless Lizards forage in the leaf litter at night for termites, small insects, and spiders. With the loss of habitat within Coastal Sage Scrub and Southern Oak Wood-

land communities, it was feared that these strange little lizards would become extirpated, and they have been listed as a Species of Special Concern by the state and federal governments. Perhaps by default, these and other delicate animals may find refuge in gardens alongside humans.

Birds

A large number of bird species are associated with scrub habitats, but most of them also occur in other communities. Many species are shared with Oak Woodland, riparian, grassland, and desert habitats. Species that are typically chaparral in distribution have certain features in common. They tend to have short wings and a long tail, which gives them maneuverability in flight. They are usually omnivorous because they must be opportunistic feeders. They eat berries and acorns when they are in season, and they eat insects, also relying on these foods for water during periods of drought.

The Western Scrub-jay, *Aphelocoma californica* (plate 19F), typifies chaparral birds in all respects but one. The Scrub-jay is bright blue and conspicuous, whereas most chaparral birds are brown. The raucous call of the Scrub-jay also attracts attention. The jay is the alarmist that calls possible danger to the attention of the entire chaparral community. Factors separating the Scrub-jay from Steller's Jay of the Mixed Coniferous Forest include water and heat. Scrub-jays are able to derive all the water they need from the food they eat, and they are extremely efficient at dissipating heat from the unfeathered portion of their feet. During the hottest part of a summer day, they remain quiet in the shade of a shrub and dissipate heat without losing water. The Island Scrub-jay, *A. insularis*, is a near relative of the Western Scrub-jay, but it is larger, bluer, and has a different call. It is found only on Santa Cruz Island. Not only is this species a California endemic, but it is a single island endemic as well.

The brown color of most chaparral birds makes them inconspicuous as they forage in

FIGURE 8.43 California Thrasher, *Toxostoma redivivum*.

the bushes, often on the ground. However, the lack of striking color poses some problems in attracting a mate. Many chaparral birds compensate for this with melodic songs, enabling them to communicate without being seen. The California Thrasher, *Toxostoma redivivum* (figure 8.43), typifies chaparral birds in this respect. It is a large brown bird that is seldom seen, but its long lyric song is a familiar sound of the Chaparral. It is easily recognized by its long, hooked beak, which it uses to thrash about in leaf litter and uncover insects.

The bird that is known as the voice of the Chaparral is the Wrentit, *Chamaea fasciata*. A person rarely spends much time in the Chaparral without hearing its distinctive call, which resembles a bouncing ping-pong ball. This smaller, long-tailed bird is seldom seen as it moves through the vegetation, gleaning insects.

Another bird with a long lyric song is the Spotted Towhee, *Pipilo maculatus*. Although its color may seem conspicuous—it has a white breast, buff sides, a black head, and black wings with white spots—the bird is actually ideally colored for an environment with filtered light. It does not form a clear silhouette because of its broken color pattern. This is an example of *disruptive coloration*, an adaptation associated with birds that spend a good deal of time on the ground beneath bushes. The Spotted Towhee and its relative the California Towhee, *Melozone crissalis* (figure 8.44), are two of the most com-

mon chaparral birds. They are ground-feeding birds with thick seed-eater beaks, but they are omnivorous. The California Towhee is a typical chaparral bird in appearance, similar in color and size to the California Thrasher. It has little to distinguish it visually other than its pale orange rump. Unlike many chaparral birds, it does not often produce a pretty song. Most of the time, it utters a single tone, "peenk," which it may repeat after a pause.

Other frequently heard birds include the California Quail, *Callipepla californica* (plate 19A), which makes a strange laughing sound ("a-ha-ha"); the Acorn Woodpecker, *Melanerpes formicivorus* (figure 7.29), which utters a "wack-a, wack-a"; the Mourning Dove, *Zenaida macroura*, with its distinctive "coo"; and the Northern Flicker, *Colaptes auratus* (figure 4.23), which utters a shrill "kee-oo." The Cactus Wren, *Camphlorhynchus brunneicapillus*, is one of the most familiar sounds of the desert, but it also occurs in Coastal Sage Scrub and Lower Chaparral. It is now a species of special concern due to its threatened habitat. Its call is a series of rapid "chuga-chuga" sounds. During spring, the call of the Great Horned Owl, *Bubo virginianus* (figure 8.45), is a familiar sound after dark. This is the familiar deep-throated "Whoo" that is associated with owls in general.

Predatory birds of Chaparral include a variety of owls and hawks. The Great Horned Owl is an important nocturnal predator. This large

FIGURE 8.44 California Towhee, *Melozone crissalis*.

FIGURE 8.45 Great Horned Owl, *Bubo virginianus*.

owl commonly hunts from a perch, locating its prey, small nocturnal mammals, primarily by hearing them. Like other owls, they have good nighttime vision, but their hearing is even better. The ring of feathers that surrounds each eye collects sound in the same way as the external ear of a mammal. Owls can hunt in total darkness, and their wing feathers are specially adapted for silent flight. Their wing bones are hollow, containing extensions of their lungs, which makes them lighter. Owls usually return to a perch to feed. After digesting their meal, they burp up a pellet containing hair and bones. When seen about the base of a tree, these owl pellets indicate that the tree is a favored owl perch. A good naturalist can identify the species of owl by the size and shape of the pellets. Pellets of a Great Horned Owl are almost as large as a chicken egg. It is not uncommon for owls to swallow their prey whole. Their pellets, therefore, commonly contain many intact bones, particularly skulls, which makes identification of prey species relatively simple. Some hawks produce pellets too, but because hawks tend to feed by tearing up the prey, their pellets contain fewer intact bones.

The most common hawks are the Red-tailed Hawks, *Buteo jamaicensis* (figure 8.46), which may be seen soaring above cliffs and hilltops. They ride on rising columns of warm air known as thermals, and they feed on a variety of prey species, including small mammals, birds, reptiles, amphibians, and insects. A mature Red-tailed Hawk has a rust-colored tail. Immature Red-tailed Hawks have brownish tails with indistinct banding. If the red tail is not visible, the hawk's best distinguishing characteristic is a dark V-shaped band across the belly. In contrast to the Red-tailed Hawk is the Red-shouldered Hawk, *B. lineatus* (figure 8.47), which usually hunts from a perch. The Red-shouldered Hawk is slightly smaller than the Red-tailed Hawk, and it has black-and-white bands on its tail.

For the most part, animals of Southern Oak Woodland are the same as those described for Foothill Woodland. Mammals and birds that eat acorns are associates, as are those that require cavities for nesting purposes. Conspicuous among these are Western Gray Squirrels, California Ground Squirrels, Acorn Woodpeckers, and Western Bluebirds. Less conspicuous are Great Horned Owls and Band-tailed Pigeons, *Patagioenas fasciata*. Many chaparral birds, during the heat of the day, take refuge in the cool canopy of the oaks.

Mammals

The small-mammal fauna of scrub communities consists mostly of nocturnal rodents. However, California Ground Squirrels, *Otospermophilus (= Spermophilus) beecheyi* (plate 18A), are diurnal. They occur from Desert Chaparral to the coast and are an important food item for predatory birds such as Red-tailed and Red-shouldered hawks.

The Eastern Fox Squirrel, *Sciurus niger*, has been introduced to California and has become established in tree-dominated parks and native Oak Woodlands where it competes with the native Gray Squirrel. In nature, Gray Squirrels

FIGURE 8.46 Red-tailed Hawk, *Buteo jamaicensis*, with a North American Deermouse.

FIGURE 8.47 Red-shouldered Hawk, *Buteo lineatus.*

feed upon fungi, nuts, acorns, and California Bay Laurel fruits. A study of the two species where they occurred together at Rancho Santa Ana Botanic Garden in Claremont found that of the 29 food items consumed by the two species, 11 were consumed by both species, but 9 were consumed only by Gray Squirrels. These included pine cones and seeds, Buckeye fruits, and fruits of California Bay Laurel. Gray Squir-

FIGURE 8.48 North American Deermouse, *Peromyscus maniculatus.*

rels also rely heavily on fungi. Acorns and walnuts were consumed by both species, and the Fox Squirrels ate a greater variety of foods, which was verified in other studies where Fox Squirrels were found to eat dry dog food and garbage. Because of its varied diet, Fox Squirrels are perceived to be a threat to the native Gray Squirrel, and in some suburban areas there have been attempts to plant trees such as California Bay Laurel that produce fruits eaten only by the Gray Squirrels.

In general, the small mammals of Coastal Sage Scrub closely resemble those of the desert. Both habitats support the North American Deermouse, *Peromyscus maniculatus* (figure 8.48), the Desert Woodrat, *Neotoma lepida* (plate 18H), and the Desert Shrew, *Notiosorex crawfordi.* There is probably a greater variety of small rodents in Coastal Sage Scrub than in the other scrub communities.

Deer Mice are distributed in nearly all communities from the desert floor to the high mountains. By allowing their body temperature to drop every day while in their burrows, they require less food, an important strategy for animals living in a food-poor habitat. This lowered metabolic rate also serves to extend their lifespan. Deer Mice live about 5 years, which is longer than other species of small rodent. Most mice live only a year or so.

Several other species of *Peromyscus* can be found in Coastal Sage Scrub. They all have hairy tails and large ears, and they differ in color, size, and tail length. Collectively, they are found in nearly every habitat in the state, and they are particularly common in scrub communities. They feed on a variety of foods, including seeds, insects, and insect larvae. They differ from meadow mice or voles (*Microtus* spp.) (figure 6.17) in that they seldom eat grass, bark, or leaves. These differences in food habits, representing examples of niche partitioning, enable these mice to coexist in a food-poor ecosystem.

A common inhabitant of Upper and Lower Chaparral communities is the California Deermouse, *P. californicus.* This mouse is more common in mature Chaparral than the North American Deermouse. The California Deermouse is one of the larger species of the genus, intermediate in size between a North American Deermouse and a wood rat. Otherwise, it resembles them both by having large ears and a hairy tail.

Woodrats (*Neotoma* spp.) are also known as packrats or trade rats. They get these names from their habit of picking up small shiny objects. If they are carrying one object and see another they like better, they will drop the first, pick up the second, and carry it back to their nest. If you happen to be living in a cabin, you may find this habit amusing or disturbing, depending on what is snatched. If a woodrat has exchanged the lid of a bean can for your $50 gold piece, you will probably be vexed. Your next step will likely be to locate the thief's nest and tear it up to see if it contains your treasure. You might find a wide assortment of things there, including other animals such as scorpions or rattlesnakes, which use woodrat nests for shelter.

Woodrats make their nests from available materials. In scrub areas, nests usually consist of a large pile of sticks. Odoriferous bits of sage or California Laurel may be used to conceal the odor of the woodrat. In desert areas, nests may be constructed entirely of pieces of cholla. Building materials may also include knives, forks, and perhaps some miner's treasure. Usually there are several false entrances, apparently to confuse predators. In the center of the mound may be several hollow areas lined with grass, string, or pieces of rags.

At higher elevations in Chaparral and in Oak Woodland, the Desert Woodrat, *N. lepida*, is replaced by the Dusky-footed Woodrat, *N. fuscipes*. From San Luis Obispo County north to Oregon, the Big-eared Woodrat, *N. Macrotis*, occupies that habitat. Wood piles made by these species may be 6 ft (2 m) high. It has been reported that smaller piles or houses constructed in the branches of shrubs, usually Scrub Oak or Coast Live Oak, may be the homes of males, whereas those on the ground are occupied by females. Large woodrat dwellings may contain several nesting rooms and food storage rooms. Urination and defecation usually takes place outside of the mound. The mounds may be many years old, forming homes for several generations of woodrats, and they may be occupied by more than one female at a time, usually an offspring of the primary resident. Many other animals share these lodges. Newts and Monterey Salamanders are known to take ref-

uge in the summer, and various snakes, including rattlesnakes may spend the winter there. Additionally various species of mice may hole up among these mounds. These woodrats subsist largely on acorns and oak leaves, much of which may be stashed in storage rooms, and much of their foraging takes place off the ground, even up in the trees. They also eat a variety of berries in season. These are larger woodrats, similar in size to the introduced Old World rats, *Rattus rattus*, which are sometimes called Roof Rats or Black Rats. Woodrats are distinguishable by their large ears and hairy tails.

In rocky areas, woodrats make their nests in small caves or hollows in rocks. The entrances to these hollows are marked by small piles of debris, including pieces of wood, urine, and feces. These piles are known as middens. If they are out of the weather, protected from rainfall, the concentrated urine apparently acts to preserve the wood and plant material. This debris may be ^{14}C dated to determine its age. It turns out that these fragments of wood may be thousands of years old, from species of plants that no longer grow in the area. As discussed in Chapter 5, this accidental bit of preservation has enabled scientists to reconstruct patterns of vegetation and climatic change dating back to the Ice Age (Pleistocene).

Woodrats are able to endure the period of summer drought by deriving water from various species of prickly pear cactus. This is true in the desert, and also in Coastal Sage Scrub. Whereas other rodents seem to experience a drop in density during the summer, woodrats maintain about the same population density all year long, probably because they are able to make use of prickly pear cactus as their sole water source. During summer months, they will actively defend patches of prickly pear.

In addition to the rodent species that are shared between the desert and Coastal Sage Scrub, there are examples of paired species, or near relatives, that occur in both places. Examples may be found among the kangaroo rats (*Dipodomys* spp.; figure 9.46A), pocket mice (*Chaetodipus* and *Perognathus* spp.; figure

9.46B), and various white-footed mice (*Peromyscus spp.*). Habits of these species are similar no matter where they occur. Kangaroo rats tend to forage for seeds in open areas between shrubs. Pocket mice forage for seeds under the canopy of the shrubs. These forms of niche partitioning will be discussed in the following chapter.

After a fire in Coastal Sage Scrub, animals associated with open areas become more common. At this time, the common small rodents include the Agile Kangaroo Rat, *D. agilis*, the Cactus Mouse, *P. eremicus*, and the Desert Woodrat. As the Coastal Sage Scrub recovers from the fire, kangaroo rats become less common, and the pocket mice and deer mice become the most common small mammals. After a fire in the chaparral communities, the Dusky-footed Woodrat and the California Deermouse disappear. At this time, Chaparral becomes occupied by an assemblage of species similar to those of Coastal Sage Scrub. There are five species of kangaroo rats that occur in Chaparral areas of the Coast Ranges, Southern Sierra Nevada, San Joaquin Valley, and Cismontane Southern California. Other species occur in the deserts and will be discussed in the next chapter.

In certain parts of Riverside County, between the Santa Ana and San Jacinto Mountains, the Agile Kangaroo Rat is replaced by Stephen's Kangaroo Rat, *D. stephensi*, an endangered species that, by virtue of its status, has prohibited some forms of development.

In certain scrub communities, bare areas occur on the periphery. This is particularly true in some Coastal Sage Scrub areas or those dominated by *Ceanothus*. Some research has demonstrated that bare areas between Coastal Sage Scrub and encroaching grassland is an expression of allelopathy perpetuated by the detritus containing odoriferous chemicals associated with the sages (*Salvia* spp.) and California Sagebrush. It has also been determined that these gaps are the result of grazing by small mammals, who pick up seeds and eat seedlings most vigorously near the existing stands of vegetation. This type of selective herbivory can also hasten the succession from *Ceanothus* to Chamise.

In Desert Chaparral, most small mammals are desert species. These include woodrats, kangaroo rats, and white-footed mice. One different species is the Piñon Deermouse, *P. truei*. This large mouse closely resembles the California Deermouse, replacing that species on the desert sides of the ranges and farther north. The Pinyon Mouse occurs in Desert Chaparral, Pinyon-Juniper Woodland, and Foothill Woodland. The large mammals of Desert Chaparral such as Mule Deer, Bighorn Sheep, Mountain Lions, Bobcats, and Coyotes are typically those also found in other chaparral communities.

In general, big fierce animals are scarce in scrub communities, just as they are in the desert. There is simply not enough biomass of animal life to support a population of large carnivores, which explains why one of the most important predatory mammals in Coastal Sage Scrub is the diminutive Desert Shrew. How do we justify a large carnivore such as a Mountain Lion, *Puma (= Felis) concolor* (figure 8.49), in a food-poor ecosystem? First, Mountain Lions prefer to eat large prey, such as Mule Deer or Bighorn Sheep, and they have large home ranges. Second, they are very good at conserving energy. They typically hunt from an ambush position, and they spend a great deal of time lying in wait. This behavior uses a minimal amount of energy, and lying in the shade helps to keep the animal cool. Ecological relationships of Mountain Lions and their favored prey Mule Deer are discussed at length in Chapter 4 on Basic Ecology.

Some serious interactions have occurred between Mountain Lions and humans. In southern California, an attack took place in Ronald W. Caspers Wilderness Park. This park, along the Ortega Highway in the Santa Ana Mountains, is surrounded on three sides by urban development, most of which originated in the mid-1980s. In 1986, a 5-year-old girl was seriously mauled by a Mountain Lion. Then in 2015, a 6-year-old boy in Cupertino was starting to be dragged off by a lion before his parents rescued him. Mountain Lion attacks on humans are not

FIGURE 8.49 Mountain Lion, *Puma concolor.*

common, but since 1750 there have been more than 65 incidents recorded in the Western Hemisphere. Very few interactions have been recorded for southern California, but most of them were in communities along the foothills, or in the Santa Monica Mountains. In 2015, lions from the Santa Monica Mountains attacked domesticated animals, including sheep, and alpacas, and a publicized attack on a koala in the Los Angeles Zoo. Apparently, civilization has encroached upon the former range of the Mountain Lions. They are unable to move farther into the national forest, because that area is already occupied by other Mountain Lions. This sort of competition, perhaps exacerbated by drought, forces the weaker Mountain Lions to take refuge in marginal habitats. There are also cases where Mountain Lions have been observed in residential areas that border open space. Usually, these appear to be young lions that may be chased from natural areas by aggression from established lions. If Mountain Lion habitat continues to be replaced at the present rapid rate, more such interactions are certain to occur.

Perhaps the most common large mammal in scrub communities is the Coyote, *Canis lat-* rans (figure 8.50). Coyotes are opportunistic feeders, eating whatever is abundant. That is one of the reasons they are so successful. They eat insects, rodents, and a wide variety of fruits. Fecal piles, called scats, may be examined to determine what animals eat. Coyote scats are among the most abundant found in scrub communities, and they nearly always contain mostly vegetable material. One authority estimated that the diet of a Coyote is 80% vegetable. The concept of the Coyote as a keystone species and its relationship to other carnivores is also discussed in Chapter 4 on Basic Ecology.

In areas where housing developments have pushed into brush-covered hillsides, interactions between humans and wild animals have become troublesome. In southern Orange County and in the foothills of the San Gabriel Mountains, Mountain Lions have been sighted in suburban areas, but most interactions are with Coyotes. American Black Bears also visit suburban communities in the foothills of the Transverse Ranges. Anecdotal accounts seem to indicate that Coyote sightings in urban areas are becoming more common. Coyotes are known to feed on pet food, domestic fruits, or

FIGURE 8.50 Coyote, *Canis latrans*.

garbage where it is accessible, but studies of scats in Thousand Oaks, Agoura Hills, and Calabasas near the Santa Monica Mountains illustrate that even in urban areas the diet of Coyotes is mostly natural. A National Park Service study in the area (1996–2004) found that urban coyotes only occasionally utilized human food sources, predominantly ornamental or nonnative fruits. Pet food or trash was only a small amount of the diet of the coyotes in this study. Interestingly, the study also found that the most important cause of coyote deaths was road kills, followed by rat poison.

Excited by the prospect of seeing wild Coyotes, some humans have taken to feeding them table scraps and dog food. Once Coyotes get accustomed to the easy calories, they may begin to demand them. When the Coyotes become too brazen or aggressive, humans stop feeding them. That is when the Coyotes get truly aggressive, particularly if they are not accustomed to feeding on their own. But, in the National Parks Service study mentioned earlier, biologists followed 110 urban coyotes using radio collars and found not one incident of coyotes attacking humans or pets. Nevertheless, a few incidents have occurred where dependent Coyotes attacked children. Four incidents of Coyotes biting humans have been documented in the Irvine area, near the Santa Ana Mountains and three people have

been bitten at a popular park and petting zoo in Montebello. In one incident, a lame Coyote was the attacker. In an undisturbed situation, the crippled Coyote may have been weeded out by natural selection, and the interaction with a human never would have occurred. In Seal Beach, there were 9 confirmed interactions with Coyotes, and 12 attacks on pets that had been confirmed since 2013. Nevertheless, one report indicated 60 attacks on pets in 2014. There is some evidence that the Coyotes have found a home on the wildlife reserve at the nearby Naval Weapons Center. At one point, the city of Seal Beach hired a pest-control company to trap and euthanize Coyotes. A backlash, however, influenced the city to stop the practice. Similarly, in 2015, in response to numerous sightings and a couple of incidents where Coyotes entered homes through open doors, Laguna Beach authorized a pest-control official to trap and euthanize brazen Coyotes. In response to numerous accounts of Coyotes killing pets, people in Newport Beach are also pressing for a program of capture and euthanization. Newport Beach also has authorized a four-color system of warnings, representing the level of threat posed by Coyotes. Similarly, Huntington Beach has recorded an increase in numbers of attacks on pets. In 2014, there were 37 attacks reported, but in 2015 there were at least 78. These interactions between humans and wild animals, whether it involves Coyotes in Irvine or American Black Bears in national parks, logically could be prevented if humans would abstain from providing the animals with easy food. Coyotes are extremely adaptable and the prospect of totally urbanized Coyotes, as unwelcome inhabitants of residential neighborhoods, is not pleasant to consider.

Other chaparral carnivores include the Common Gray Fox, *Urocyon cinereoargenteus* (figure 8.51); the Bobcat, *Lynx rufus* (= *F. rufus*; figure 8.52); and the Ringtail, *Bassariscus astutus*. These are all predators that feed on small rodents and birds. Scats and stomach contents of trapped animals have been examined to find out what these animals eat. Foxes are omnivorous, like Coyotes. However, their scats are

FIGURE 8.51 Common Gray Fox, *Urocyon cinereoargenteus.*

smaller and often contain grass. The influence of Coyotes and habitat fragmentation on the abundance of Common Gray Foxes and Domestic Cats in nature is discussed in the section on keystone species in Chapter 2.

The Bobcat is also known as the Wildcat or Lynx, not to be confused with the Canada Lynx, *L. canadensis* (= *F. lynx*), of forests in northern North America. Bobcat scats rarely contain vegetable matter, and they contain many bones. Bobcats seem to be strictly carnivorous. Contrary to popular belief, they rarely eat game birds such as quail or doves, although they do consume a variety of birds. Studies indicate that they prey primarily on small rodents and birds, although there are accounts of bobcats killing small Mule Deer. Bobcats, with their relatively short noses, do not have as good a sense of smell as some other predators, so they hunt primarily by vision.

Bobcats are probably more common than observations indicate. Estimates indicate that there is about one animal for every 1–2 mi^2 (2.6–5.2 km^2) of Chaparral. In spite of this density, they are rarely seen, as they characteristically lie quietly in wait, using their remarkable protective coloration to keep them concealed. Like Coyotes,

Bobcats are widespread, ranging from Mixed Coniferous Forest to the desert. Also like the Coyote, the Bobcat may be seen abroad in daylight. In recent years, associated with a demand for the pelts of spotted cats, Bobcats have been subjected to a significant amount of trapping, but recently the state has banned Bobcat trapping.

The Ringtail is in the Raccoon family (Procyonidae). It is not much larger than a Gray Squirrel, although its long bushy tail may give it a larger appearance. Some people call it a Ringtail Cat. Its distinctive features include enormous eyes and a bushy tail with black-and-white rings. Scats of Ringtails are seldom seen because they are usually deposited in shelters rather than along trails or roads. Ringtails and foxes eat birds by climbing into the bushes to feed, whereas Bobcats are more likely to lie in wait and then pounce. Ringtails and foxes also may take birds that are asleep or that are on the nest. Many juvenile birds meet their end this way.

Observations of hunting behavior suggest that the Ringtail holds its tail over its back when it moves through open areas. This behavior may confuse a potential predator such as a Great Horned Owl, which presumably tries to

FIGURE 8.52 Bobcat, *Lynx rufus*.

grasp the tail and gets nothing but fur. Competition between Ringtails and other predators is avoided by the Ringtail's habit of feeding very late at night. This late-night activity means that it is seldom seen by humans. Ringtails are probably more common than is believed.

Physiological studies of Ringtails indicate that they may be the most drought adapted of all mammals in North America. They can persist on water derived from food, and they conserve water by producing the most concentrated urine known among mammals. They are successful in many desert areas.

The history of the Grizzly Bear, *Ursus arctos*, in California is the story of a disaster. California is the only state in the union that has an extinct animal as its symbol. In southern California, the last grizzly was killed in Trabuco Canyon in 1908, although there also is a record of a Grizzly kill in Big Tujunga Canyon in 1916. The last Grizzly Bear in the state was killed in 1922, in Horse Corral Meadows, in the Sequoia area of the western Sierra Nevada.

The Grizzly Bear should not be confused with the American Black Bear, *U. americanus*, which is a forest bear and probably did not occur in the mountains of southern California until it was introduced in 1933. Apparently some 27 American Black Bears were brought down from Yosemite National Park and released into the San Gabriel and San Bernardino Mountains. Those bears prospered, and now there are thriving populations in the area. The population in the San Bernardino Mountains has expanded from 16 bears in 1933 to about 500 today. The population in the San Gabriels expanded from 11 to about 300. Now they are breaking into cabins and raiding trash cans here just as they do in the Sierra Nevada. They are also moving into suburban areas is the foothills. So far, they have been recorded as far south as San Diego County.

The Grizzly Bear was a chaparral and grassland bear. Similar to American Black Bears, however, Grizzly Bears were omnivorous, eating mostly small mammals, insects, grasses, berries, and various roots. The long claws and large muscle mass of the shoulders were adaptations enabling them to dig up roots, bulbs, and tubers, as well as tear up logs to get ants and termites. The ferocity of the Grizzly Bear is legendary. Bull and bear fights with the two animals roped together were usually won by the bears. Nevertheless, their extinction, encouraged by a $10 bounty, was accomplished.

The California Grizzly Bear was a subspecies of the Grizzly Bear that still inhabits portions of the northern Rockies. Of course, statements about the size of the animals often are exaggerated, but they probably averaged about 900 lb—considerably larger than American Black Bears and somewhat larger than other Grizzly Bears in the lower 48 states. Unlike American Black Bears, they probably did not hibernate, because they commonly overwintered at lower elevations than the snow-covered forest communities. Grizzly Bears primarily roamed lowland areas of the state, from foothills to the coast, feeding on what was abundant. Apparently one reason that Elephant Seals have taken to hauling out on the mainland in recent years is that in the past they would have been easy prey for Grizzly Bears. However, it was the encounters in the valleys that got them into trouble with humans. Here ranchers had encroached upon traditional Grizzly Bear hunting grounds, and cornered Grizzly Bears would attack domestic dogs and humans. To the humans of the time, it seemed it was "them or us," and the program of extermination was begun.

CALIFORNIA'S ISLANDS

From a good vantage point on a clear day, California's islands are visible from the mainland. They vary in size and distance from the mainland, but most of them are not far offshore. The California islands are continental islands as opposed to oceanic islands. Oceanic islands occur far out to sea, and most of them are volcanic in origin and were never attached to the mainland.

California's islands can be divided into three groups (figure 8.53). To the north are the Farallon Islands and Año Nuevo. These are very small islands near the central California coast that are best known for their populations of marine mammals and birds. Another group of islands is inside San Francisco Bay. Most famous is Alcatraz, previous home to a maximum security prison. The only one that has a

decent amount of native vegetation is Angel Island, the largest in the bay, and former immigration station for Japanese and Chinese immigrants. Canyons on the island are densely wooded, components of Oak Woodland and Mixed Evergreen Forest. There also are components of Northern Coastal Scrub and Chaparral. The island is now managed by the state as Angel Island State Park. In 1996, in an attempt to restore native vegetation, more than 12,000 nonnative eucalyptus trees were removed, creating the temporary appearance of a "clear-cut" on the north side of the island.

Initially, various species of *Eucalyptus* were planted all over California. In particular, Blue Gum, *E. globulus*, is the most widely planted all over the world. Introduced in 1856, it was supposed to be used for lumber, railroad ties, wind breaks, and a mechanism to reduce soil erosion, and even to promote drying of wetlands. It is probably the most common tree species on California islands, and on the mainland many communities were developed in old eucalyptus groves, some of which were planted by railroad companies. Because eucalyptus oils increase flammability and because in certain areas they have moved into native habitats, they have been classified as invasive, and in many areas such as Angel Island and San Clemente Island, there have been programs to remove them.

Off the southern California coast are the Channel Islands. They can be subdivided into two groups. The northern Channel Islands lie on an east-west line, roughly parallel to the coast south of Santa Barbara. These islands are mountaintops that project above water. In a sense, they appear as a westward extension of the Santa Monica Mountains. These are the islands that most Californians think of as the Channel Islands. From east to west, the islands are Anacapa, Santa Cruz, Santa Rosa, and San Miguel.

The southern Channel Islands lie west of Los Angeles and Orange Counties. They consist of four islands in a diamond pattern. Santa Catalina is the closest to the mainland, and San Nicolas is farthest away. Santa Barbara Island,

FIGURE 8.53 California Islands.

the smallest of the Channel Islands, lies to the north. To the south is San Clemente Island.

A small island group known as Islas Coronado lies south of San Diego. These islands actually belong to Mexico, but they are close enough that they should be mentioned here. Farther south, off Baja California, are a number of other interesting islands. These will be discussed where their animal or plant life is relevant to distributions on the California islands.

All of the Channel Islands today are managed in some form of protected status. The two outermost islands, San Clemente and San Nicolas, belong to the US Navy. The four northern Channel Islands and tiny Santa Barbara are managed as Channel Islands National Park, although the western 90% of Santa Cruz is managed by The Nature Conservancy, and the US Navy has retained the option of future use on San Miguel Island. Parts of Santa Catalina Island remain in private hands, but 86% of the island is managed by the Catalina Island Conservancy. With a permanent population of approximately 4000, Avalon is the only established community on a California Island. The Santa Catalina Island Company owns most of the land around Avalon and the Two Harbors area, as well as a horse ranch (Rancho Escondido) in the center of the island.

Examination of a fault map (figure 8.3) will reveal that many of the islands are separated from each other by major fault systems. With our present knowledge of plate tectonics, it is apparent that the islands have not always been located in their present positions, nor have they always been arranged in the pattern we see today. The islands originated much farther to the south, and they have been carried to their present latitudes by right-lateral offset along California's rifted borderland. These are extant terranes that have not yet become attached to the mainland, and because southern California no longer has a subducting margin, they probably will remain detached.

The islands are geologically complex. Most of the rock materials exposed on the surface are various kinds of sediments, illustrating that the regions formerly were underwater. The northern Channel Islands include a granitic component similar to the Transverse Ranges, which means that they were parallel to the coast during formation of the Southern California batholith. Santa Catalina and San Clemente Islands contain a fair amount of Franciscan rock, which was formed in association with the subduction trench. Layered on top of the others are many types of volcanic rock deposited from about 16 to 12 million years ago (Miocene). Paleomagnetic data in these rocks indicate that the islands have been translocated northward, and the northern Channel Islands have been rotated clockwise at least 75°. Santa Catalina also has a large component of "young" granite of about the same age as the volcanics.

The northward drift of these islands is established by a number of other geologic features (figure 8.54). For example, present-day Poway Creek lies east of Torrey Pines State Natural Reserve, in San Diego County. The alluvial outwash from this creek, when it was a much larger stream, contributed to a large underwater fan about 40 million years ago. This fan has been formed into a distinctive conglomerate rock that can be found on San Miguel Island, which is now over 100 mi (160

km) north of Poway Creek. Pebbles in the conglomerate may have a source as far south as Sonora, Mexico. This is a logical possibility, because at that time southern California west of the San Andreas fault was still attached to mainland Mexico. Rivers that began in what is now Sonora, Mexico, could have flowed directly westward into the Pacific Ocean. Poway Creek has been moved northward about 200 mi (320 km) and San Miguel an additional 100 mi (160 km). Poway gravels are also found on Santa Rosa and San Nicolas Islands, but not on Santa Cruz, Santa Catalina, or San Clemente Islands.

It is further noteworthy that the present Santa Cruz Island is a composite terrane. This means that it was formed by the fusion of two formerly separate terranes. A fault running down the center of Santa Cruz Island is marked by the central valley of the island today.

It is also significant that San Clemente and Santa Catalina formerly were a great deal larger than they are today. Landslide materials including Franciscan rocks such as Catalina schist formed distinctive deposits known today as the San Onofre Breccia. This breccia was formed underwater, but the source was a very large island, the remains of which is probably Santa Catalina. It is possible that about 25 million years ago a series of earthquakes caused the landslides that later became cemented into the breccia. This distinctive rock is found today on Santa Cruz, Island, Santa Rosa Island, Anacapa Island, Santa Catalina Island, and various points on the mainland from Point Mugu south to Oceanside (figure 8.8). The actual shoreline at the time was inland, about where the Santa Ana Mountains lie today; therefore, the eastern edge of the former island probably lay about where the mainland coastline occurs today.

It is often speculated that land bridges once connected California islands to the mainland. These land bridges presumably were formed during periods of low sea level, when a great deal of the earth's water was tied up in the polar ice caps. During maximum glaciation, shorelines were as much as 400 ft (120 m) below

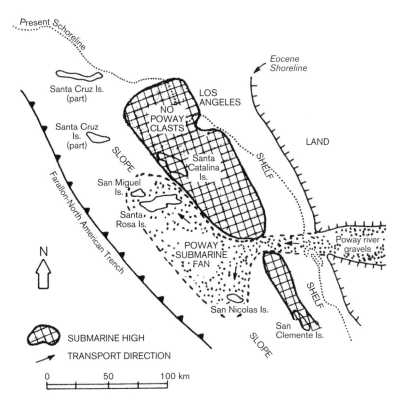

FIGURE 8.54 Reconstructed Eocene paleogeography showing the positions of islands, the Eocene shoreline, and the deposition of Poway river gravels (after Howell and others, 1974; from Rowland, S. M. 1984. Geology of Santa Catalina Island. *California Geology* 37(11):239–251).

their present position. Lowering of sea level most certainly connected Mexico's Todos Santos to the mainland. The Farallones, Año Nuevo, the islands in San Francisco Bay, and perhaps the Coronados, were attached, but it seems highly unlikely that the present Channel Islands were ever attached to the southern California mainland. If any were, they would have been attached a long time ago and a good deal farther south—most likely at a latitude presently corresponding to mainland Mexico. Maximum lowering of sea level occurred about 18,000 years ago, at which time the four northern Channel Islands were most likely a single large island, known as Santa Rosae. The channel between this island and the mainland simply was too deep for the two to be connected, however. During the lowest possible sea level,

the gap between the mainland and the island would have been at least 4.4 mi (7 km) in width.

Island Endemics

Animals and plants that inhabit islands are often relics of former widespread distributions, and because they evolve in isolation, they are often specialized. Most of the endemics mentioned here share both of these characteristics.

The implication of plate tectonics on the distribution of sedentary animals such as salamanders has been discussed frequently throughout this book. If unique salamanders were found on these islands, it would add credence to the hypothesis that distribution of specialized forms on the mainland may be related to populations on former islands that became attached to California by accretion of extant terranes that carried their plants and animals with them.

The Farallon Islands are the northernmost of California's islands, but they are composed of only 211 acres (85 ha) of land (Table 8.3). Nevertheless, 12 species of sea bird nest there, and the largest sea bird rookery south of Alaska is there. There are only 36 species of plants on the islands, 13 native and 23 introduced. The dominant plant is the Farallon Weed or Maritime Goldfields, *Lasthenia maritima*. This relative of the common California Goldfields, *L. californica* subsp. *californica*, is in the sunflower family (Asteraceae). It grows on the islands in mats over 1 ft (40 cm) thick. When it blooms, its flowers occur in great profusion, turning the landscape yellow.

There is a population of Arboreal Salamanders, *Aneides lugubris*, on the Farallon Islands. These are the salamanders frequently associated with oak trees, although there are no native oaks on the Farallones. Throughout most of its range, the salamander is characterized by small yellow spots. On the islands, the salamanders show a pattern of large spots that also typify the forms in the Gabilan Range east of Monterey. The Farallon Islands and the Gabilan Range are both west of the San Andreas fault. The Farallones are pure weathered granite, and the Gabilans also contain much granite. It has been postulated that the Gabilans and Farallones were once part of a long peninsula. It is apparent that the Gabilan Range was once an island, too. Whether or not the Farallones were ever large enough to have remained continuously above sea level is subject to some conjecture. It is possible that the salamanders were carried to the island on a log raft, or it may be that these wave-lashed outcrops are all that remains of a former large landmass. Small-spotted members of this species are also found on Santa Catalina Island, but they have not been studied thoroughly enough to determine how different they are from mainland forms.

The most sedentary of all California salamanders, the slender salamanders (*Batra-*

choseps spp.), are found on some of the southern islands. On Santa Catalina and Islas Coronado the Southern California Slender Salamander, *B. major* (figure 8.34), may be found. It is also found on Todos Santos Island near Ensenada. At the present time, it is assigned to the same subspecies that occurs on the adjacent mainland. To date, however, electrophoretic studies that could establish differentiation have not been published. An endemic species, the Channel Islands Slender Salamander, *B. pacificus*, occurs on the four northern Channel Islands. Analysis of these island forms using electrophoresis, however, shows that the island forms differ enough to indicate that they have been separated from the mainland forms by at least 4 million years. It is of special interest that Santa Cruz Island, which is composed of two terranes, has two kinds of slender salamander! The Black-bellied Slender Salamander, *B. nigriventris*, also occurs on Santa Cruz Island. It too is believed to have diverged from its mainland counterpart at least 4 million years ago.

It is tempting to search for alternative explanations for the presence of these salamanders on islands. Amphibians are unable to swim in saltwater, so any hypothesis that involves dispersal from the mainland must involve some sort of rafting. It is entirely possible that sedentary salamanders might float across on a piece of wood or debris washed out by heavy rains. This explanation might apply particularly to the presence of the Arboreal Salamanders on the islands. However, it is difficult to believe that slender salamanders, which usually occur under rocks and logs, could reach so many islands in that manner. Nevertheless, it is interesting that in 1955 a live Black-tailed Jackrabbit, *Lepus californicus*, was found floating on a kelp raft near San Clemente Island, 39 mi (62 km) from the California mainland.

Interesting evidence supporting the premise that the Channel Islands were connected to the mainland is that fossils of a pygmy elephant known as the Exiled Mammoth, *Mammuthus exilis*, have been found on Santa Cruz, Santa Rosa, and San Miguel Islands. Supporters of

TABLE 8.3
A Comparison of California's Islands

	Area, mi$_2$ (km$_2$)	Elevation, ft (m)	Distance to Mainland, mi (km)	Native Species and Subspecies of Plants[a]	Amphibians	Reptiles	Resident Birds[a]	Mammals[a]
Northern California Islands								
Farallons	0.5 (1.4)	385 (109)	20 (33)	13 (1)	1			
Ano Nuevo	0.02 (0.05)	60 (18)	0.3 (0.5)					
Northern Channel Islands								
San Miguel	14 (37)	830 (253)	26 (42)	190	1	2	15	3 (2)
Santa Rosa	84 (217)	1560 (475)	27 (44)	340 (3)	2	2	25	4 (3)
Santa Cruz	96 (249)	2470 (753)	19 (30)	540 (7)	3	6	37 (1)	10 (4)
Anacapa	1.1 (2.9)	930 (283)	13 (20)	70	1	2	14	3 (2)
Southern Channel Islands								
Santa Barbara	1.0 (2.6)	635 (194)	38 (61)	40 (1)		1	6	2 (1)
San Nicolas	22 (58)	910 (277)	61 (98)	120 (2)		2	11 (1)	4 (2)
Santa Catalina	75 (194)	2125 (648)	20 (32)	375 (3)	3	8	34	9 (5)
San Clemente	56 (145)	1965 (599)	49 (79)	233 (1)		2	24	6 (2)

[a] The number of endemic species is indicated in parentheses.

FIGURE 8.55 Two color patterns of the Island Night Lizard, *Xantusia riversiana*. A. On San Clemente Island. B. On San Clemente Island.

the land bridge hypothesis contend that it is not possible for an elephant, no matter how small, to reach an island by any means other than walking. Critics point out that Indian elephants (*Elephas maximus*) are good swimmers, and that a 4.5 mi (7 km) swim would not be out of the question. Furthermore, there are fossil remains of elephants, most of which were dwarfed, on at least 10 islands in the Mediterranean Sea.

Of special interest in the biogeography of California's islands is the presence of the Island Night Lizard, *Xantusia riversiana* (figure 8.55), on three islands. Dense populations of these lizards are found on San Clemente, San Nicolas, and Santa Barbara Islands. Electrophoretic studies show that these are clearly the most divergent of all the vertebrates on the islands. They differ enough from their mainland relatives that some authorities have placed them in a different genus (*Klauberina*).

Similar to salamanders, Island Night Lizards spend most of their time under rocks or in cracks. They are without question sedentary. Mark-and-recapture studies on San Clemente Island show that over a period of years, lizards are frequently recaptured under the same rock. Average home ranges are about 10 ft (3 m), and the greatest distance a marked lizard is known to have moved is 30 ft (10 m). The density of these lizards on San Clemente Island is truly remarkable. Studies indicate that there are over 3500 lizards per acre (1450/ha), amounting to a biomass of 55–85 lb per acre (10–16 kg/ha).

This high biomass is maintained by a lack of competitors and predators.

Mainland relatives of this lizard include the Desert Night Lizard, *X. vigilis*, and the Granite Night Lizard, *X. henshawi*. Desert Night Lizards are commonly found in association with Yucca plants, and Granite Night Lizards are found in cracks in rocks in the Peninsular Ranges. A fossil night lizard is also represented in the San Diego area in sediments determined to be at least 40 million years of age. These fossils are similar to but not the same as the island form.

Electrophoretic studies suggest that the island form has been separated from its mainland relatives for at least 10–15 million years. The two mainland forms appear to have diverged from each other about 7 million years ago. Correlating the electrophoretic data with the age of the fossils leads to speculation that the Island Night Lizards have been separated from the mainland forms for 10–40 million years. They may have been island dwellers all that time.

The night lizard family (Xantusiidae) is distributed similarly to other animals that seem to have been translocated on extant terranes. Members of this family occur in Cuba, tropical Middle America, and the arid Southwest. The Island Night Lizard may have evolved on San Clemente Island and ridden to its present locality from a former position attached to mainland Mexico. There is no evidence that San Clemente Island was ever completely submerged, and the time period of tens of millions of years is appropriate for this type of transport.

The other two islands on which the Island Night Lizard occurs are not high enough to have remained above sea level. It is therefore believed that these lizards were somehow transported to the other islands sometime in the last million years. Electrophoretic studies substantiate this idea, because there is very little difference among the lizards of the different islands. This transport would require rafting of some sort. A problem with this hypothesis is that the currents seem to move the wrong way for anything to disperse passively northward. Another hypothesis is that the lizards were carried as

FIGURE 8.56 Island Gray Fox, *Urocyon littoralis*, on San Clemente Island.

pets by Native Americans that inhabited the islands. This, of course, does not explain why the lizards are not found on all the islands.

The presence of the Island Fox, *Urocyon littoralis* (figure 8.56), on several islands also has been explained as an example of transport by Native Americans. These dwarfed foxes, also known as the Island Gray Fox, are relatives of the Common Gray Fox, *U. cinereoargenteus*, which occurs on the mainland. Island Foxes occur on the six largest islands: San Miguel, Santa Rosa, Santa Cruz, San Nicolas, Santa Catalina, and San Clemente. Each island is considered to have its own subspecies, which implies that the foxes have been isolated on the islands for quite some time. It may be that the Island Fox is more closely allied to three small species of foxes that occur in Yucatan, Mexico, and in Guatemala. If this is the case, the Island Fox may represent another example of the distribution pattern noted for the Island Night Lizard. The fact that Island Foxes have differentiated more than any other island mammals supports the idea they may have been isolated longer than we think.

If Native Americans did carry foxes around as pets, it would have been a long time ago. However, differences in the subspecies might be less significant than we think. Differences involve mainly size and color. Such differences are known to evolve rapidly, and these changes are typical of animals and plants on all sorts of ecological islands.

Studies of DNA in the chromosomes of living foxes can reveal degrees of similarity. Greater differences would imply that the populations have been separated for longer periods of time. All island foxes are significantly different from the mainland Common Gray Fox, and the degrees of differentiation among the island forms suggest that they arrived on the northern islands first. Foxes on San Clemente and San Nicolas Islands are most closely related to the foxes on the northern island of San Miguel. The foxes on Santa Catalina appear to be an amalgamation and were likely the consequence of the mixing of genetic strains from several of the other islands. There is nothing in these data to detract from the theory that the transport of the foxes occurred by humans.

The oldest fossil fox on the northern Channel Islands is from deposits dated at approximately 16,000 years. The fact that this fox was smaller than the mainland form implies that foxes arrived prior to 16,000 years ago. The earliest dates for human habitation in southern California are not clear-cut, but most authorities agree that about 10,000–12,000 years ago humans were well established here. A fire pit on Santa Rosa Island has been dated at 30,000 years, one of the oldest dates attributed to human habitation in North America. It is also difficult to determine exactly when the last period of high sea level occurred. At best guess, it can be stated that sometime during the last 500,000 years, water was high enough to cover all the islands except San Clemente, Santa Catalina, Santa Cruz, and Santa Rosa. The sea reached its lowest point some 18,000–20,000 years ago, and humans could very likely have been here then. Unfortunately, traces of human populations along the seashore would have

been obliterated by the rising sea. Whatever the time sequence, species differentiation on islands that were submerged had to occur in the last half-million years.

Historic human occupation, along with their domestic animals, took a heavy toll on the native flora and fauna of the islands. Overgrazing by horses, cattle, sheep, rabbits, and goats occurred on all of the islands at some point. Add to that, the damage associated with feral Pigs, Black Rats, and Domestic Cats on several islands. Roosevelt Elk and Mule Deer were pastured on Santa Rosa until 2011. Bison were introduced to Santa Catalina for a movie in 1924. Of the 14 that remained, they were fruitful and multiplied. At its peak, there were about 300 on the island. Overgrazing was significant. The motive for keeping them at all was that the Bison herd was a significant part of tourism on the island. It was finally determined that a herd of 150 animals was sustainable. Extra animals were exported to the Lakota Tribe in South Dakota and the remaining females were injected with a contraceptive to keep down the birth rate. Meanwhile, Santa Catalina still has a herd of Mule Deer, the numbers of which are supposed to be kept under control by hunting. Cats were a problem on several islands if for no other reason than they preyed on endangered wildlife such as Island Night Lizards and the endemic San Clemente Loggerhead Shrike, *Lanius ludovicianus anthonyi*. Feral cat populations have been brought under control on all the islands except Santa Catalina. The problem here is that residents of Avalon continue to feed the cats, and during times of drought they feed and water the Mule Deer as well. As of 2014, it was estimated that the population was 650–700 cats, with more than 70% in the vicinity of Avalon. Sterilization and/or translocation seem to have little effect on the total population.

Of the endemic endangered species on the California Islands, the story of the Island Foxes is most interesting. On Santa Catalina, introduction of canine distemper reduced the population from an estimated 1800 to about 200 in 1999. An experimental vaccine was developed and a combination of captive breeding and vaccination brought back the species from threatened extinction. Through these efforts, the population numbered 1852 in 2013, one of the most successful examples of recovery on record. Unfortunately, some animals are still being killed by vehicles on the roads, and an unusual kind of ear tumor has now appeared in some of the older animals.

The story of the foxes on the northern Channel Islands is more complicated. It involves disappearance of the Bald Eagle in association with DDT (dichlorodiphenyltrichloroethane) use prior to 1972. DDT caused thinning of the eagles' eggshells, which cracked under the weight of the incubating parents. In the absence of the Bald Eagles, *Haliaeetus leucocephalus*, the northern islands became colonized in the 1990s by Golden Eagles, *Aquila chrysaetos*. Bald Eagles primarily eat fish, and they are aggressive enough that they are able to keep Golden Eagles at bay. Golden Eagles on the other hand eat mostly terrestrial prey. As it turns out, the prey included the Island Foxes, a situation that was exacerbated with the removal of Feral Pigs. Apparently, baby pigs were important in the diet of the eagles, and with their removal from the islands the eagles increased predation on foxes. On Santa Cruz Island, the number of foxes went from 1500 to 62 by 2002. On Santa Rosa Island, the number shrank to 15. As on Santa Catalina, the program of recovery in 2002 involved vaccinations and captive breeding, but it also involved translocation of Golden Eagles to the mainland in northern California, and reestablishing native Bald Eagles. As of 2015, the number of foxes on Santa Cruz Island had grown to 1750, another profound example of a successful program of recovery. Based on their remarkable recovery, the Island Foxes have been proposed for delisting as endangered.

Reintroduction of the Bald Eagle to the Channel Islands is another example of a successful recovery program. Eagles were reintroduced to Santa Catalina in 1980. At first, biologists removed eggs from eagle nests and hatched them in an incubator, and then returned the chicks to the nests. By 2007, with DDT less of a problem, Bald Eagles were able to incubate and hatch their own chicks. In 2002, Bald Eagle chicks were reintroduced to Santa Cruz Island and they reached breeding age in 2006, at which time two chicks were hatched, and recovery was underway. Hacking of young eagles continued on the islands. As of 2015, there were 25 eagles living on Santa Catalina and about 40 on the northern Channel Islands, and five of the eight Channel Islands now have nesting Bald Eagles. Interestingly, nesting on the islands has provided a source of young eagles that have emigrated to the mainland. At least 23 sites on the mainland are now occupied by Bald Eagles, many of which appear to have come from the Channel Islands.

Most birds on the California islands are similar to their mainland relatives. Some differentiation has occurred on the subspecies level, but not to the degree typical of other vertebrates. This, of course, is complicated by the fact that birds are highly mobile. Of the birds that have become differentiated, the Island Scrub-Jay, *Aphelocoma insularis*, is the most notable. This jay differs from the California or Western Scrub-Jay, its mainland counterpart, by being bluer, larger, and having a different call. The island subspecies sounds like a Scrub-jay with laryngitis. The Island Jay is one of only two endemic birds in California, but it is also a single island endemic.

Native terrestrial vegetation of the islands consists mostly of species associated with Coastal Sage Scrub. The larger islands also contain some Chaparral and Oak Woodland. Prior to deliberate fire suppression, Coastal Grassland was a very important community on the islands. It was burned year after year by Native Americans, who depended heavily for food on seeds of wildflowers that followed the fires.

The infestation of nonnative weeds has taken its toll on the islands as well as the mainland. Of the 353 species of plants known to occur in the Channel Islands National Park, 20% are nonnative. Most of these are weeds of Mediterranean origin, but species of Australian eucalyptus have also become naturalized.

Nowhere is this more conspicuous than on Santa Catalina Island, where about 90% of the nonnative trees are eucalyptus. Ironically, those trees are now threatened by infestations of an Australian longhorn boring beetle.

The number of native species on each island is roughly proportionate to the size of the island and the distance from the mainland. The overwhelming majority of native plants on the islands are the same as those on the mainland; the number of endemics is small. Of the endemics, there are two kinds: those that occur only on islands but may appear on more than one and those that occur on only one island. Of the latter type, only San Clemente has more than 10 species. Altogether, the California islands are home to between 90 and 100 island endemics.

The largest number of island endemics is leaf succulents known as liveforevers (*Dudleya* spp.). At least 12 endemic kinds are restricted in distribution to islands. These are drought-tolerant members of the Coastal Sage Scrub community. Other groups of Coastal Sage Scrub that have diversified on islands are "Dandelions" (*Malacothrix* spp.), locoweeds (*Astragalus* spp.), and Wild Buckwheats (*Eriogonum* spp.). As examples of gigantism, a number of the Wild Buckwheats grow to large size and bear names such as *E. giganteum*, *E. arborescens*, or *E. grande*. Locoweeds, members of the Legume family (Fabaceae), are important as nitrogen fixers. Among Chaparral species, the Manzanitas (*Arctostaphylos* spp.) have diversified.

One of the most attractive island endemics is a mallow (Malvaceae) known as Malva Rosa or Island Mallow, *Malva (= Lavatera) assurgentiflora*, a drought-deciduous shrub with large, dark green, palmately lobed leaves. It may grow to tree size if it gets abundant rainfall. The flowers are strikingly beautiful, containing five violet to lavender petals with light stripes. These shrubs were originally found only on San Clemente, Santa Catalina, San Miguel, and Anacapa Islands. The Santa Catalina variety, formerly described as a different subspecies, has different-colored flowers. They have been planted on the other islands and at various localities on the

mainland. They are a very attractive shrub for landscapes along the coast. Unfortunately, due to grazing by goats and sheep, the species is becoming rare on the islands. Bush mallows also occur on some islands, but they have become so rare that the Santa Cruz Island Bush-mallow, *Malacothamnus fasciculatus* var. *nesioticus*, and the San Clemente Island Bush-mallow, *M. clementinus*, are listed as endangered. There are at least 20 endangered or threatened species on the islands.

Another distinctive shrub, also an example of gigantism, in the sunflower family (Asteraceae), is the Giant Tickseed (Giant Coreopsis), *Leptosyne (= Coreopsis) gigantea*. This shrub has succulent stems and large, finely divided leaves borne in tufts on the ends of the stems. It is another drought-deciduous species. In bloom, each shrub is covered with many large yellow sunflowers. These plants are so common that whole hillsides turn yellow. Giant Coreopsis is also found on the mainland at a few localities, but it has become a symbol of the Channel Islands, occurring on all the islands north of Santa Catalina.

A rather distinctive tree in the rose family (Rosaceae) that is found only on islands is Island Ironwood, *Lyonothamnus floribundus*. This is the only island endemic that has differentiated to the genus level. There are two subspecies on the islands. *L. floribundus* subsp. *aspleniifolius* (figure 8.57) has a fernlike appearance with pinnate leaves and occurs on Santa Rosa, Santa Cruz, and San Clemente Islands. Santa Catalina Ironwood, *L. floribundus* subsp, *floribundus*, occurs only on Santa Catalina. Its lance-shaped leaves are about 4 in (10 cm) long, and the leaf margins may be smooth or variously incised. These are both attractive trees, with red shaggy bark, shiny green foliage, and clusters of white flowers. This is another ideal plant for southern California landscapes.

A tree that many people associate with Santa Catalina Island is the Catalina Cherry, *Prunus ilicifolia* subsp. *lyonii*. Cherry Cove, a popular mooring on the east side of the island, is named in its honor. This is an example of gigantism. Catalina Cherry is related to the common

FIGURE 8.57 Island Ironweed, *Lyonothamnus floribundus asplenifolius.*

chaparral shrub Holly-leafed Cherry. Groves of Catalina Cherry also occur in canyons on San Clemente, Santa Cruz, and Santa Rosa Islands. Individual trees may be up to 50 ft (15 m) in height. The large red cherries appear in the fall and are delicious, but they have a very large poisonous seed. These cherries are an important autumn food for animals on the islands. They swallow the seed whole without releasing its load of cyanide. Native Americans ate the seeds, too, but they leached out the cyanide in the same way they leached the tannins from acorns.

It has been mentioned many times throughout this book that California is the center of diversity for pines and oaks. It is only fitting, therefore, that its islands are characterized by distinctive oaks and pines. A few specimens of Canyon Live Oak, *Quercus chrysolepis*, grow on Santa Cruz and Santa Catalina Islands. Deciduous oaks such as Blue Oak, *Q. douglasii*, and Valley Oak, *Q. lobata*, also occur in a few groves on Santa Cruz and Santa Catalina islands.

Four oaks are associated specifically with the islands. Of the scrub oaks, island scrub oak, *Q. pacifica*, is endemic to Santa Catalina, Santa Cruz, and Santa Rosa Islands where it is locally common and grows to large tree size, another form of gigantism (plate 13B). Resembling a shrub form of Interior Live Oak (*Q. wislizeni*), Santa Cruz Island oak (*Q. parvula* var. *parvula*)

is found on Santa Cruz Island, and it also occurs in Santa Barbara County on the mainland.

Island Oak, *Q. tomentella*, is a tree of north-facing slopes and canyons. It grows on Santa Cruz, Santa Rosa, Santa Catalina, and San Clemente Islands. Young twigs on this species are conspicuously hairy. Leaves are oblong, 2–3 in (50–75 mm) long, with smooth margins. The underside of the leaf is also hairy.

Another insular oak is known as *Q. xmacdonaldii*. This deciduous tree grows in similar locations as Island Oak, but it is not found on San Clemente Island. It also has hairy twigs, and the underside of the leaves is faintly hairy. It differs from Island Oak by having two to four bristle-tipped lobes on each leaf. As unlikely as it may seem, this species is believed to have arisen by hybridization between Valley Oak and Island Scrub Oak. The latter is widely distributed on these islands today, but Valley Oaks are limited and must have been far more common in the past. *Q. xmacdonaldii* may enjoy limited distribution on the mainland as well, growing in isolated localities as far south as Moro Canyon in Crystal Cove State Park, Orange County. The trees in Crystal Cove State Park are considered by some authorities to be the southernmost specimens of Valley Oak, and not Macdonald Oak.

The presence of Torrey Pine, *Pinus torreyana* subsp. *insularis*, endemic on Santa Rosa Island

FIGURE 8.58 Bishop Pines, *Pinus muricata*, on Santa Cruz Island, also showing foggy weather typical of the islands.

is particularly interesting. This large-coned, long-needled species occurs only on the mainland near Del Mar and on Santa Rosa Island. This may be the smallest distribution of any pine species in the world. The island population is characterized by larger cones. The possibility that this may be more southerly in origin is verified by circulation of eddy currents from the south within the Los Angeles Basin, and also may be related to northward translocation of Santa Rosa Island along fault systems.

Among the closed-cone pines, depending on the most recent taxonomy, two or three species have island populations. Bishop Pine, *P. muricata* (figure 8.58), a pine with two needles per fascicle, has the widest distribution, occurring on upper coastal terraces from Humboldt County to Santa Barbara County, and on Santa Cruz and Santa Rosa Islands. It is also may be found on Cedros Island off Baja California, although those trees are now considered to be a two-needle variant of Monterey Pine, *P. radiata*. The Monterey Pine, *P. radiata*, occurs naturally at three locations on the central California coast and also on Guadalupe Island (figure 7.21), a true oceanic island 157 mi (252 km) off the Baja California

coast. This volcanic island seems never to have been connected to the mainland, so all plants and animals must have arrived there by over-water dispersal. A former taxon, Santa Cruz Island Pine, *P. remorata*, which occurs on Santa Cruz and Santa Rosa Islands as well as on the mainland near Lompoc was considered to be a near relative of Bishop Pine, in which its cones grew straight out instead of reflexed against the trunk.

Distribution of closed-cone pines on islands is discussed more thoroughly in Chapter 7. It should be remembered, however, that they require fire for their cones to open, and they are dependent on fog drip for adequate moisture. They are also associated with depauperate soils such as acidified sandstones, serpentinites, and diatomites. On the islands where fire has been suppressed, many of the trees are dying of old age, and there is very little regeneration.

Principles of Island Biology

Another feature of island organisms is that they are often distinctly different in size from their mainland counterparts. The Island Fox, for example, is smaller than the mainland form. Of special interest in this respect is the fossil dwarf mammoth found on Santa Cruz, Santa Rosa, and San Miguel Islands. The rationale for dwarfism is that islands have a limited food supply. A population of small animals, requiring less food, should do better where food is scarce.

If limited food is a problem, how can it be rationalized that island animals are sometimes much larger than their mainland counterparts? On California's islands, most mammals are larger than their mainland relatives. Examples are the Big-eared Harvest Mouse (*Reithrodontomys megalotis catalinae*), Deer Mice (*Peromyscus maniculatus*), the Island Ground Squirrel (*Ostospermophilus beecheyi nesioticus*), and the Eastern Spotted Skunk (*Spilogale putorius*). Larger size is also illustrated in the Island Night Lizard. An explanation often used for gigantism on islands is that all available ecological niches may not be occupied. It is more difficult for large species to become dispersed to islands; therefore, smaller species arrive and take over the vacant niches, and in time, a small species becomes larger. The Island Night Lizard might be an example. On the mainland, night lizards are insect eaters. On islands, they eat a more varied diet, including animal and plant material. In the absence of competition from a larger omnivore, perhaps the Island Night Lizard has moved into that niche.

Gigantism on islands might also be related to a shift in life history strategy. If resources are scarce but relatively stable, the emphasis should be on survival of existing individuals rather than on the reproduction of more individuals. Animals that fit this model tend to be long-lived and large and to have poor dispersal and low reproductive rates. This is the opposite of the pioneers most likely to colonize the island in the first place. Hence, island animals seem to be larger than their mainland counterparts. One generalization for evolution on islands might be that small animals tend to become larger and large animals tend to become smaller.

Plants also show size changes on islands. The Giant Coreopsis is large on islands, and so is the Catalina Cherry. Many annual wildflowers, such as the Island Poppy, *Eschscholzia ramosa* are also larger. It also has been observed that island plants have larger leaves than their counterparts on the mainland. From comparisons of chaparral species on the mainland with those on Santa Cruz Island, it was determined that leaves were indeed larger on the island. Comparison of variables such as precipitation, average temperature, temperature extremes, and wind velocity showed no significant differences. The island had greater cloud cover and, to a degree, deeper soils. The conclusion was that plants on the island experience lower water stress, so their leaves grow larger.

Island animals and plants also seem to lose their mobility, which is coincident with the life history strategy that causes them to be larger. The dispersal mechanism that brought them to the island might become a disadvantage once a population is established. Island plants and animals do not want to be carried away from the island. Birds become flightless, and some animals become sedentary. Plants lose their mechanisms for wind dispersal. Plumes on seeds disappear, and seeds become larger.

In the absence of heavy predation, animals lose their fear, and plants lose some of their protective mechanisms, such as toxins and spines. Perhaps the larger leaves and seeds observed are also related to reduced herbivory. As long as leaves are not being preyed on by herbivores, it is more efficient to grow one set of large leaves rather than to keep replacing smaller ones.

Larger islands and islands close to the mainland tend to have more species. This concept is verified by the relative numbers of species on California's islands. The concept of a filter is also illustrated by our islands. A filter implies that in a series of islands extending in a line away from the mainland, the number of species decreases from island to island as function of distance from the mainland. With the exception of the Anacapas, which are small, the number of species diminishes from Santa Cruz to Santa Rosa to San Miguel.

Continued on p.355.

Ecological Islands

The concept of an ecological island has been discussed frequently and refers to a region with rather specific ecological parameters surrounded by a habitat of a more generalized nature. Southern California's ecologic islands, some of which have been discussed elsewhere, include areas characterized by cypresses, Knobcone Pines, and vernal pools. Included among these regions is the pavement plain with its Great Basin relicts near Baldwin Lake in the San Bernardino Mountains.

The temporary ponds that occur in areas with poor drainage are known as vernal pools or hogwallows. These pools are found also in the Great Central Valley and Sierra Nevada foothills in grassland areas, where they may be associated with Blue Oak. In southern California, they are located on Kearny and Otay Mesas in San Diego County, and on the Santa Rosa Plateau in the Santa Ana Mountains, in association with Engelmann Oak.

These pools are formed by runoff during winter rains, but the water does not percolate into the clay soil. Rather, it remains standing to form pools of various sizes, which gradually disappear by evaporation in the spring. The word *vernal* means "spring." A distinguishing feature of these pools is that circles of annual wildflowers grow around the borders and follow the water's edge inward as the pools dry up (plate 17A). Because of the high rates of evaporation, soils are usually alkaline.

The wildflowers of these pools are often unique. A list of 101 species considered typical of vernal pools was compiled. Of these, more than 70% were native ephemerals, and over half were endemic to California. Two genera of rare grasses, *Orcuttia* and *Neostapfia*, are found only in association with vernal pools, and five species of *Orcuttia* are endemic to California. A Meadowfoam, *Limnanthes alba*, is a small, delicate ephemeral that occurs in these moist areas of the Laguna Mountains. It also occurs about 800 mi (1280 km) to the north, in the Klamath Mountains. Such an incredible disjunct distribution is testimony to the relict nature of this species. The presence of summer rain in both areas may be related to such a pattern of distribution.

Many of the vernal pool species are rare or endangered. Habitat destruction due to development and grazing has caused the disappearance of many areas of vernal pools. It is to the credit of the Endangered Species Act that the remaining southern California vernal pools are protected from destruction.

A closed-cone pine associated with Upper Chaparral in southern California is the Knobcone Pine, *Pinus attenuata* (figure A). Its distribution here is extremely localized; it occurs naturally in only two areas. The larger area ranges over about 1000 acres (400 ha) at about 3500 ft (1000 m) elevation in the area around City Creek, on the south side of the San Bernardino Mountains. The other is at about the same elevation in the region of Pleasants Peak, in the Santa Ana Mountains. A third, southern stand is found near Ensenada, in Baja California. At the other localities where it occurs, it has been introduced by the US Forest Service or county road crews.

Knobcone Pine is found at isolated localities throughout the state, particularly in association with serpentine soils in the Coast Ranges, Klamath Mountains, and northwestern Sierra Nevada. In the Santa Ana Mountains, it is found on a very peculiar soil associated with hydrothermally altered serpentine. These are depauperate soils. They are fine-grained, of acidic pH (4.5–5.3), and of peculiar chemical composition. Water is not readily available to roots in these soils because the fine particles tend to retain moisture by adsorption. These trees maintain their dominance on these sites by surviving on fog drip, which has been measured at up to 4 in (102 mm) per month during spring.

These are small three-needle pines, but they are not closely related to the yellow pine group. They are more closely allied with Lodgepole Pines. The cones are crescent-shaped, and they adhere to the main branches of the tree. They remain on the tree in a closed condition for many years, in a characteristic known as *serotiny*. When the cones are heated, as by a fire, they open and shed their seeds. With precipitation and abundant light, the seeds become established, and new trees grow on the burned site. Where the trees have been introduced, there is no apparent germination of seeds after a fire. It appears that competition from rapidly growing species on sites with better soil precludes the establishment of Knobcone Pines where they are not native. They remain localized

FIGURE A. Knobcone Pine, *Pinus attenuata*. Top: Cones before a fire.
Bottom: Cones after a fire.

by virtue of their association with specialized ecological conditions, namely, poor soil, fog, and fire.

One of the most interesting ecologic islands occurs on coastal sandstone bluffs near Del Mar in San Diego County. Here is where Torrey Pines, *P. torreyana* (figure B), grow. It has been estimated that the total population here consists of less than 7000 trees. At Santa Rosa Island, about 175 mi (250 km) to the northwest, also on sandstone, is a second population of Torrey Pines, numbering about 2000 trees. This total, less than 9000 trees, makes this the rarest species of pine in the world. These trees are relicts from a former time when conditions were moister. They survive today in a region where fog drip provides the moisture and where sandstone soils restrict competition from other species.

Torrey Pine is related to Gray Pine and Coulter Pine. These are the pines with large cones, although Torrey Pines have the smallest cones of the three. These are fire-adapted pines in the sense that the cones open only partially, releasing the seeds over a period of several years. Torrey Pines resemble Gray Pines with their candelabra-like appearance, having many heavy branches in the crown. They resemble Coulter Pines by having very long, thick needles. Unlike the other two species, which have needles in clusters of three, Torrey Pines have needles in clusters of five. The needles are grayish in color, similar to those of Gray Pine. The pines on Santa Rosa Island are quite blue gray in color.

The mechanism by which a species could become distributed in such an odd manner has been the subject of much debate. It is possible that the former distribution was fairly continuous,

FIGURE B. Torrey Pine, *Pinus torreyana* subsp. torreyana, and coastal sandstone.

FIGURE C. Tecate Cypress, *Hesperocyparis forbesii*, on Silverado Sandstone in upper Coal Canyon.

and that the remaining trees are relics of that time. The distance between the Channel Islands and the mainland during the Pleistocene, when sea level was at its lowest, was about 4.5 mi (7 km). That distance would not be a significant barrier to dispersal, because cones can float that far, and birds such as California Scrub-Jays are noted for their ability to carry seeds or acorns in their throats. On the other hand, right-lateral movement of fault systems has carried land segments hundreds of miles northward, including the land on which both populations of Torrey Pines now grow. There are at least two faults lying between the populations as they occur today, and it is entirely possible that ancestors of the present Santa Rosa Island population originated south of the present population near Del Mar. If large cone size is an adaptation associated with drought conditions, it follows that the Torrey Pine would have smaller cones with a coastal origin. The fact that the island population has larger cones than those on the mainland also fits this hypothesis.

Two other conifers occur as ecologic-island populations within Lower Chaparral. These are

the local cypresses. The relic nature of cypresses and their restricted distribution is discussed in Chapter 7. A cypress resembles a juniper by its arrangement of scalelike leaves on the stem, and also by its fleshy cones. Cypress cones, which resemble small soccer balls, are larger than those of junipers, in some cases almost as large as a golf ball. They remain on a tree in closed conditions for up to 20 years, and heat from a fire causes them to open. After a fire, in a manner similar to Knobcone Pines, they drop their seeds on bare mineral soil. The seeds germinate after the first rains.

The Tecate Cypress, *Hesperocyparis (= Cupressus) forbesii* (figure C) is a variety of a Baja California species. In California, the largest stands are near the Mexican border in the Peninsular Ranges. They occur there in the Otay, Tecate, and Guatay Mountains on north-facing slopes, on alkaline clay soils derived from gabbro, a dark-colored plutonic rock rich in iron and magnesium. About 90 mi (144 km) to the north, on the northern end of the Santa Ana Mountains, another stand of Tecate Cypress occurs. Just south of where Highway 91 passes through

Santa Ana Canyon, on the northern slopes of Sierra Peak, Tecate Cypress grows in Coal, Gypsum, and Fremont Canyons. It is here that the Paleocene (60 million-year-old) Silverado Formation occurs. This formation consists of sandstone soils formed in a shallow lagoon. Coal Canyon gets its name from coal deposits in this formation. The low-grade coal was actually mined by a railroad company. The trees here are the oldest and largest of the Tecate Cypresses. Specimens over 200 years of age have been recorded. The largest, on the floor of Coal Canyon, which recently died, had a trunk circumference of 8 ft (2.8 m) and a height of 35 ft (11 m).

In order to protect the trees and several rare wildflowers, Coal Canyon has been established as a state-managed ecological reserve. Furthermore, lower Coal Canyon has been added to Chino Hills State Park. Subsequent removal of the pavement from the undercrossing of the 91 freeway completed an important corridor that now connects the Santa Ana Mountains with the Chino Hills, allowing wildlife such as Mule Deer and Mountain Lions to pass between these two areas of wildlife habitat.

Cones of Tecate Cypress from the La Brea Tar Pits, as well as fossils from the local deserts, point to a former widespread distribution. They exist today by default on clay or sandy soils where they lack competition. Associated species in the northern groves are mostly disturbed-area plants, although a depauperate mix of species from Lower and Upper Chaparral occurs in the area as well. In the northern groves, there are two associated species considered to be endangered. These are herbaceous species known as Heart-leaved Pitcher Sage, *Lepechinia cardiophylla*, and Braunton's Milkvetch, *Astragalus brauntonii*. Another interesting associated plant in this area is the Yucca-like Chaparral Nolina, *Nolina cismontana*. In the southern groves, the largest Tecate Cypress stand is on Otay Mountain. Here are six rare or endangered species, including three specialized shrubs: Mexican Flannelbush, *Fremontodendron mexicanum*; Otay Manzanita, *Arctostaphylos otayensis*; and Otay Mountain Ceanothus, *Ceanothus otayensis*.

Groves of Tecate Cypress are becoming smaller. They require fire to reproduce, but they take about 35–40 years to become mature. If fire occurs more frequently than that, or if the fire is too hot, the trees are burned before they can produce cones. Another problem can occur if a fire is followed by drought. Without the follow-up of precipitation, the seeds may die or be eaten before they can germinate. Historically, a summer storm accompanied by lightning was probably the combination of fire and precipitation that enhanced the propagation of Tecate Cypress. Interestingly, Tecate cypress is the only plant on which a rare butterfly, Thorne's Hairstreak (*Callophrys gryneus*) will lay its eggs. If Tecate Cypress were to disappear, it would probably take Thorne's Hairstreak with it.

Fire records and aerial photographs have revealed that the fire frequency on Tecate Peak in northern Baja California has increased to a 15- to 20-year interval. Similar data have been obtained for Otay Mountain in San Diego County and Sierra Peak in the Santa Ana Mountains. Apparently, these fires are associated with activities of humans. One bizarre incident in 2006 involved a prescribed burn that escaped and consumed a large portion of the Sierra Peak population. The problem here is that fires in the area previously occurred in 1988 and 2002, clearly not providing enough time for new trees to mature between burns.

The Cuyamaca Cypress, *H. (= C.) stephensonii* (figure D), occurs on gabbro soils on the southwestern side of Cuyamaca Peak in the Laguna Mountains. Once considered a localized form of the Arizona Cypress, it is now known to be more closely related to the cypress species on Guadalupe Island off Baja California. This is the rarest cypress in California. It is known only from six stands covering a total area of only 43.5 acres (17.6 ha). Illustrating how reducing the interval between fires can take a toll, fires in 1950 and 1970 extirpated one stand and reduced another by 75%. At present, the existing trees appear healthy and vigorous. It is hoped that with proper fire management on the part of federal and state agencies, the Cuyamaca Cypress will continue to thrive in this restricted area.

The Peninsular Ranges in San Diego County also contain other gabbro endemics. Certain rare plants, such as Dehesa Beargrass or Dehesa Nolina, *N. interrata*, are restricted to these soils. The state's largest stand of this yucca-like succulent is found on McGinty Mountain, near Jamul, 20 mi (32 km) east of San Diego. This region, owned by The Nature Conservancy, contains a total of seven rare and endangered plant species endemic to gabbro-derived soils.

FIGURE D. Cuyamaca Cypress, *Hesperocyparis stephensonii*, on gabbro soil in upper King Creek.

Perhaps the most spectacular ecologic island in southern California occurs at 6720 ft (2050 m) in the San Bernardino Mountains, east of Big Bear Lake. The area is known as the Baldwin Lake pebble plains. In the area, the California Native Plant Society has identified 14 rare or endangered species, 11 of which are endemic. This is the largest concentration of endemics in the state. Clay soils of the pebble plains were left behind by a Pleistocene lake. Freezing conditions in winter cause round quartzite pebbles to become pushed to the surface; hence, there is a layer of orange pebbles on the clay. This is an area where many relicts of the ice ages, most of them associated with the Great Basin, still exist. It is a habitat that superficially resembles an Alpine Fellfield. Most of the plants are low growing, and some of them are similar to cushion plants. Apparently, the area escaped glaciation, and somehow the relicts were preserved under these extremely harsh conditions.

Plants that are endemic to limestone also occur in the San Bernardino Mountains, associated with several east-west-trending ridges north of Big Bear. Four of these plants are listed in the California Native Plant Society's *Inventory of Rare and Endangered Vascular Plants of California*. The most limited in distribution is San Bernardino Bladderpod, *Physaria (= Lesquerella) kingii* subsp. *bernardina*, which occurs only in Big Bear Valley and Van Dusen Canyon, near Holcomb Valley. Cushenbury Milkvetch, *A. albens*, and Parish's Daisy, *Erigeron parishii*, are confined to lower slopes and canyons near Cushenbury Springs, just above the desert floor. Perhaps the most beautiful is Cushenbury Buckwheat, *Eriogonum ovalifolium* var. *vineum*. This plant occurs as white-leaved mats up to several feet in diameter, growing on stark white rocks. Limestone mining in the area may threaten these species in the future.

A very important concept in island biology is that once an island becomes saturated, the total number of species tends to remain the same. Even though species composition may be different at a later date, the number of extinctions should equal the numbers of colonizations. Although this is not always verifiable from comparing animals such as mammals that have limited ability to cross water barriers, those that are good dispersers fit the pattern well. Data on numbers and types of breeding land birds on the California islands, collected from 1917 to 1968, indicated that in nearly every case the total number of species remained stable, with the extinctions equal to the colonizations.

Comparison of the total number of species on an island with those in an equivalent area on the mainland usually reveals that saturation on the island occurs with fewer species. There are probably several reasons for this phenomenon. First of all, species that require large ranges are automatically selected against. Second, a defined area on the mainland includes many species that are on the fringe of their preferred habitats, and that are more common in adjacent areas. These two types of species are seldom found on islands. Perhaps of greatest importance is that the total diversity of habitats on an island tends to be reduced.

An interesting experiment in island biology is being confirmed by observations in mainland national parks. Because of a variety of disturbances on the outsides of parks, they tend to become ecologic islands that experience an overall loss of species with time. As might be expected, this "faunal collapse" is greatest in the small parks. A study of national parks of similar age in the United States and Canada shows that a number of mammal species has declined in direct proportion to the size of the park. Lassen Volcanic National Park in California, only 164 mi² (420 km²) in total area, has lost 43% of its mammal species since 1910. The provincial parks in Alberta and British Columbia, including Kootenay, Banff, Jasper, and Yoho, with coterminous boundaries, function as one large park with 8000 mi² (20,500 km²) of territory. In that large area, there has not been a single extirpation since 1903. Parks of intermediate size, such as Yosemite and Sequoia-Kings Canyon, have lost 25% and 23% of their mammal species, respectively, since their establishment in the 1890s.

9

California's Deserts

FIGURE 9.1 Colorado Desert, Yaqui Pass in Anza-Borrego Desert State Park. Cactus Scrub community in foreground.

CALIFORNIA HAS THREE main desert regions, all of which lie to the interior of major mountain ranges. Thus, California's deserts are all the product of rain-shadow effect. From north to south, the three regions are as follows: the Great Basin, east of the Cascades and the Sierra Nevada (plate 15A); the Mojave (plate 4A), to the interior of the Transverse Ranges; and the Colorado Desert (figure 9.1), lying east of the Peninsular Ranges. A fourth arid region, the southern San Joaquin Valley, lies in the rain shadow of the southern Coast Ranges. The San Joaquin Valley will be discussed in more detail in the next chapter, on the Great Central Valley, although in many respects it qualifies as one of California's deserts. Arid lands make up a sizable percentage of California's landscape.

The characteristics of a desert are as follows:

1. Low, unevenly distributed precipitation, less than 10 in (25 cm) per year

2. Temperature extremes

3. Windy (increased evaporation rates)

4. High light intensity

5. Nutrient-poor, alkaline soil

6. Low rates of primary production

CLIMATE

Traditionally, a region that receives less than 10 in (25 cm) of precipitation per year is classified as a desert. In California, many desert regions receive far less than that. Bagdad, formerly a small community in the Mojave Desert on old Highway 66, was reported to be the driest place in the United States. Its annual precipitation averaged 2.2 in (5 cm). Its truly remarkable record, however, was that from October 3, 1912, until November 8, 1914—a total of 767 d—Bagdad received absolutely no precipitation. Based on average yearly precipitation, Death Valley, at 1.5 in (4 cm) per year, officially is the driest place in the state.

Without question, California's deserts are dry, but they vary considerably in temperature extremes and the timing of precipitation (Table 9.1). For comparison, Bishop, in the Great Basin, averages 5.6 in (14 cm) of annual precipitation, much of which is snow; Barstow, in the Mojave, averages 4.0 in (10 cm), mostly winter rain; Imperial, in the Colorado Desert, averages 3.6 in (9 cm), including winter and summer rain; and Bakersfield, in the Great Central Valley, averages 6 in (15 cm) of winter rain per year.

Dryness influences all of the characteristics associated with deserts. Dry air changes temperature rapidly, because there is little water vapor to moderate the change. A range of 50°F (28°C) between nighttime lows and daytime highs is commonplace. At the community of Imperial, in the Colorado Desert, the range of temperature from the record low to the record high was a full 130°F (54°C).

Air rises when it is warmed, and it sinks when it is cooled. Uneven heating coupled with rapid temperature change causes desert air to move rapidly from place to place. Hence, deserts are windy. Wind accelerates evaporation, which adds another burden to the drought stress experienced by plants and animals.

Dry air is very clear. About 90% of available sunlight reaches the ground on a typical desert day, as compared with about 40% in a typical humid climate. This intense sunlight includes ultraviolet radiation that is capable of causing severe tissue damage to plants and animals. It also means that temperatures are very high during the long summer days, when the sun is high in the sky. The hottest temperature ever recorded was in the California desert. On July 10, 1913, at Furnace Creek, in Death Valley, the official temperature reached 134.6°F (57°C). Formerly, this temperature was believed to have been exceeded by a world record at El Azizia, Libya, in 1922. The temperature that was recorded at 136.4°F (58°C) in later years was proved to be inaccurate, and few stations in the area recorded comparable temperatures on that day. Many stations in the Death Valley area on that memorable day recorded high temperatures. In addition, for nine con-

TABLE 9.1
A Comparison of California Deserts

Desert	Features	Communities
Great Basin ("Cold Desert")	Winter precipitation, often snow	Blackbrush Scrub
		Sagebrush Scrub
	Spring growing season	Shadscale Scrub
		Alkali Sink
Mojave ("High Desert")	Winter precipitation, some snow	Joshua Tree Woodland
		Creosote Bush Scrub
	Winter, spring growing season	Shadscale Scrub
		Alkali Sink
Colorado ("Low Desert")	Winter, summer precipitation	Cactus Scrub
	Winter, summer growing season	Creosote Bush Scrub
		Saltbush Scrub
		Alkali Sink
		Desert Wash
		Palm Oasis

secutive days that same July, the high temperature was at least 125°F (52°C). The night before the record, the low temperature was 93°F (34°C).

Desert soil is also a product of the dry climate. Low precipitation coupled with a high evaporation rate causes the soil to be alkaline. Alkaline soils (aridisols) pose an osmotic stress for desert plants, which are already drought stressed. This problem is compounded by the fact that most desert soil is coarse. The soils tend to remain immature and are low in organic material. Coarse soil allows water to percolate rapidly. Therefore, after it rains, water remains available to plant roots for a comparatively short period of time.

Lack of water limits primary productivity. Photosynthesis, which requires water, is limited in desert areas; therefore, food is scarce. The entire desert ecosystem must adjust to this limitation. The total amount of biomass that can be supported in a desert ecosystem is smaller than for any other terrestrial ecosystem.

Great Basin Desert

The northernmost of California's three deserts is the Great Basin. Although this is the largest desert in North America, it enters California only along its eastern border. Two of California's natural regions are included in the Great Basin: the Modoc Plateau and what geologists call the Basin-Range Province. The Modoc Plateau of northeastern California is an undulating flatland east of the Cascades (figure 9.2). It averages 4000–5000 ft (1200–1500 m) in elevation and is drained by the Pit River. Most of the Great Basin is characterized by interior drainage in which saline lakes or dry lake beds lie on the valley floors. This is not so for the Modoc Plateau. If Goose Lake had no outlet, it would be a region of interior drainage. However, Goose Lake and the surrounding environs drain into the Pit River, which flows southwestward toward the Great Central Valley. Formerly, water from the Pit River joined the Sacramento near Redding. Now it is impounded by Shasta Dam, forming one of

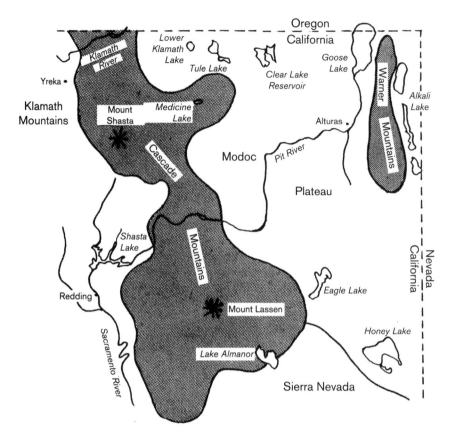

FIGURE 9.2 Great Basin Desert, Modoc Plateau, and Cascade Range.

many reservoirs that are part of the huge state water project, which ultimately carries water nearly 600 mi (960 km) to southern California. The path of the Pit River is significant to biogeographers. It has been speculated that during times of heavy precipitation associated with the Ice Ages (Pleistocene), the Pit River could have been an important avenue of dispersal for fishes and other aquatic creatures between the Sacramento River and the Great Basin.

Although there is some disagreement about age, the Modoc Plateau was formed by a series of lava flows dating back about 25 million years. Most of these eruptions took place about the time the Cascades were forming, about 5 million years ago. Actually, this plateau is no more than the southwestern corner of a large volcanic tableland that covers the northern Great Basin, including eastern Oregon, eastern Washington, southern Idaho, northern Nevada, northern Utah, and western Wyoming.

The Basin-Range Province lies south and east of the Modoc Plateau. It is composed of a series of north-south-trending valleys and mountain ranges associated with faulting and stretching of the terrain. This province occupies a large portion of the Great Basin, including Nevada, western Utah, northern Arizona, and eastern California. In California, it includes mountain ranges along the eastern border of the state. From north to south, these ranges include the Warner Mountains, Sweetwater Mountains, Inyo-White Mountains, and the Panamint Range on the western border of Death Valley. Rocks within these ranges are mostly altered ocean-floor sediments that are now exposed as a series of distinctive bands on the faulted sides of the ranges.

Not all of the Great Basin is desert. There are numerous ranges—ecologic islands—that rise above the Great Basin Desert. East of the Modoc Plateau, in the very northeastern corner of the state, is an uplifted fault block known as the Warner Mountains. The Warner Mountains, on the eastern edge of the Modoc Plateau, rise to 9722 ft (3015 m) at Eagle Peak. From a geologic perspective, this area is mostly volcanic. Among these volcanic rocks is an abundance of obsidian. The obsidian here includes the usual black variety, but much of it is a unique brown form known as mahogany obsidian. In this form, the iron is in its oxidized state. This high-quality obsidian was apparently very important to Native Americans as a material for making tools. Presumably from the Warner Mountains, it is found in chipping sites all over the Southwest, including localities in southern California.

The interesting forest of this area is discussed in Chapter 5. Subalpine trees such as stunted Whitebark Pines adorn the crests of its high peaks. With their aspen groves and pine forests, the Warner Mountains seem as if they were a part of the Rockies, out of place in California. The Washoe Pine, a high-elevation relative of Ponderosa Pine, forms a fine forest above 8000 ft (2500 m) in elevation. This is the only known location for the species in California. Indeed, with the exception of two small sites in nearby Nevada, this is its only known location in the world.

To the south, the Basin-Range Province can be appreciated best by looking eastward from the crest of the southern Sierra Nevada. Directly east of Owens Valley are the Inyo-White Mountains. Beyond are Saline Valley and Panamint Valley, then Death Valley, with the Panamint Range rising in between. On a clear day one may be able to see Charleston Peak rising to 11,919 ft (3644 m) in the Spring Mountains near Las Vegas.

Some of the highest mountains in California are in the ranges of the Great Basin. Telescope Peak in the Panamint Range is at 11,049 ft (3368 m). In the White Mountains, White Mountain Peak rises to 14,246 ft (4342 m). There is a tale formerly told in the Owens Valley that White Mountain Peak is actually higher than Mount Whitney, which would make it the tallest in the contiguous 48 states. According to the tale, that information has been suppressed by the US Forest Service because it doesn't want to deal with all the visitors it would create. According to official record, however, White Mountain is the third highest in the state, behind the Sierra Nevada's Mount Whitney, at 14,505 ft (4418 m), and Mount Williamson, at 14,375 ft (4382 m). Owens Valley is the trough between the Sierra Nevada and the White Mountains. As such, the peaks cower about 2 vertical miles on either side of the valley; hence, one popular book about the region is titled, *Deepest Valley*.

The climate of the Great Basin differs from that of other California deserts because a large proportion of its precipitation falls as snow. As the snow melts, the water percolates deep into the soil. Growth of desert vegetation in the Great Basin is delayed until spring, whereas the growth period in warm deserts occurs primarily during the winter when most of the precipitation falls.

The most common shrub in the Great Basin is Great Basin Sagebrush (plate 15A). One ecologist has calculated that all the sagebrush in the Great Basin outweighs the total mass of Sierra Redwoods, the largest living things. At lower elevations, in desert basins, other kinds of sagebrushes are found. Above the sagebrush, or often mixed with it, are pinyon pines and junipers that may form an open woodland. At high elevations, near the tops of the mountains, are found subalpine trees such as Limber and Bristlecone Pines (plate 7B). Several specimens of Bristlecone Pine from the White Mountains have been determined to be older than 4000 years. The oldest, known as Methuselah, is over 4600 years of age. Here, in one of the harshest climates of the United States, suffering drought, freezing, and gale-force winds, a forest of small trees has lived continuously since 2500 years BC. These trees are considered the oldest living things.

Throughout the Great Basin, humans have made their mark in many ways. Ghost towns, such as Bodie, attest to mineral riches that were once there for the taking. Also, large patches of green vegetation demonstrate that with imported water, or water mined from deep underlying strata, the desert is capable of supporting various kinds of agriculture. This is particularly true on the Modoc Plateau. But overgrazing, more than any other activity, has altered the area. Cattle and sheep consume, in a few years, the aboveground portions of shrubs that have taken many years to grow. The feet of cattle compact the soil, which causes increased runoff of water that formerly was allowed to percolate deep into the ground. The increased runoff is evidenced by deep, steep-sided arroyos that were unknown before the advent of humans and their domesticated beasts. Even if proper land management practices were introduced all over the Great Basin, no one knows how long it would take for the desert to recover or whether it could recover within our lifetimes. Scars on the land from long-abandoned mines are graphic examples of how slowly desert lands heal.

Directly east of the Sierra Nevada, the Great Basin is represented in California by a series of north-south-trending mountains and valleys. South of Lake Tahoe, the eastern face of the Sierra is bordered by three basins. The northernmost of these is the valley drained by the Walker River in the vicinity of Bridgeport. The east and west forks of the Walker River drain northward in to Nevada to flow into Walker Lake, another of the large lakes or closed basins of interior drainage. From Conway Summit south to Sherwin Summit is the Mono Basin-Long Valley Caldera, which is discussed in Chapter 4 in the section on volcanic rocks of the Mammoth area. South of Sherwin Summit near Tom's Place, drainage is toward the Owens River, which flows southward toward Owens Lake and the Mojave Desert.

During the Pleistocene, the Ice Age, there were periods of heavy rainfall in the deserts. The Owens River flowed southward beyond Owens Lake to terminate in China Lake, near the present town of Ridgecrest. During periods of heavy flow, water continued to flow eastward into Searles Lake, near the present town of Trona. The large tufa towers south of Trona were formed underwater. Testimony to the amount of water that formerly flowed in the Owens River is the deep gorge cut into the basaltic lava flows near the present town of Little Lake. A short distance to the east of Highway 395 is an old Indian chipping site on the edge of what must have been a spectacular waterfall. The area is known as Fossil Falls and is presently administered by the Bureau of Land Management.

The distribution of vegetation in the northern parts of the Owens Valley is typical of the Great Basin Desert in California (plate 15A). At the upper fringe of the desert in rocky soil, where the desert environment grades into the mountain environment, the dominant plant may be the Single-needle Pinyon Pine or the Sierra Juniper. At the eastern face of the Sierra Nevada, the pinyon pine belt is visible as the first band of dark-colored vegetation (figure 4.5). Below this on gravelly slopes is a community called Blackbrush Scrub, so named because on the dominant plant, Blackbrush, *Coleogyne ramosissima*, the outermost blackish stems seldom bear leaves. Along Highway 395 from Bishop to Tom's Place, there is an extensive belt of Blackbrush. Below this, and south of Bishop, is where Great Basin Sagebrush, *Artemisia tridentata*, dominates (plate 15A). The most common plant of the Great Basin Desert, Great Basin Sagebrush is typical of well-drained sandy soil. Many acres of the Great Basin are characterized by this blue-gray shrub. Where soil becomes more alkaline, in low places and south of Lone Pine, this shrub is replaced by Shadscale, *Atriplex confertifolia*, a member of the Goosefoot family (Chenopodiaceae). Saltbush is also grayish, but a darker gray. The transition from Great Basin Sagebrush to Shadscale is usually marked by a distinct color change in the vegetation. Below the Shadscale zone, soil becomes very alkaline. If the soil remains wet, a community of shrubs with succulent leaves provides the last band of vegetation before water or bare clay dominates. This band is obvious along the shores of Owens Lake.

Mojave Desert

The Mojave Desert is south of the Great Basin. In the Owens Valley, the transition becomes appar-

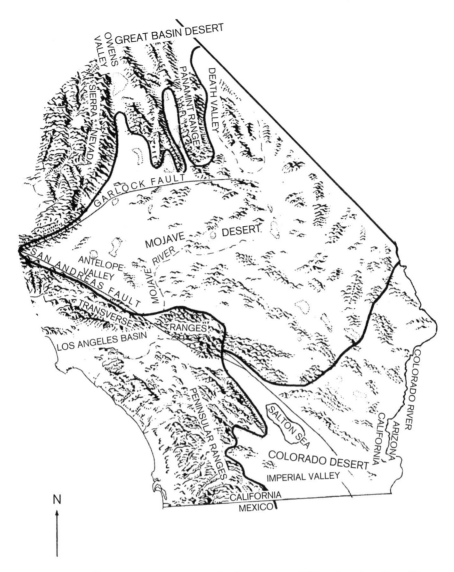

FIGURE 9.3 Southern Great Basin, Mojave, and Colorado Deserts. The northern boundary of the Mojave Desert is represented by the distribution of Joshua Trees and Creosote Bush. Note that it is north of the Garlock fault, the accepted boundary of the geologic Mojave Desert Province.

ent south of Olancha, where the Joshua Tree, *Yucca brevifolia* (plate 4A), and dark green Creosote Bush, *Larrea tridentata* (plate 15B), become the dominant vegetation, although these species also occur farther north, in the low valleys of the Great Basin. From a geological perspective, the Mojave is a large wedge-shaped basin lying between the Garlock fault on the north and the San Andreas fault on the south (figure 9.3). On the west, the two faults intersect at about a 60° angle near the town of Gorman on Interstate 5.

The Garlock fault runs northeastward along the uplifted beds of Red Rock Canyon and the southern edge of the Sierra Nevada to disappear near the southern end of Death Valley National Park. The southern boundary, the San Andreas fault, lies along the northern edge of the San Gabriel Mountains. To the northwest, it is bounded by the Tehachapi Mountains and the southern Sierra Nevada. To the east, it crosses the Colorado River to enter northwestern Arizona and southern Nevada near Las Vegas. To the east of the San Gabriels, the southern border of the Mojave is formed by the San Bernardino, Little San Bernardino, and Eagle Mountains. The geologic boundaries are so conspicuous that the Mojave is outlined boldly on relief maps and satellite photographs.

The biological boundaries of the Mojave are based primarily on temperature. Low valleys to the north of the Garlock fault are considerably warmer than the surrounding mountains, so Mojavean vegetation such as Creosote Bush extends farther north in these valleys. Thus, Death Valley with the lowest point in the United States is part of the Mojave Desert. Similarly, Panamint Valley and Eureka Valley are characterized by Creosote Bush, although they are significantly north of the Garlock fault. Nevertheless, the Mojave Desert is the "High Desert" of the weather report. Also included within the Mojave are Antelope Valley, Apple Valley, and Yucca Valley, all of which lie at about 3500 ft (1100 m) elevation. These other valleys all interconnect to form one large drainage basin. The Mojave is a large alluvial-filled basin, into which many isolated mountain ranges of diverse geologic nature are projected.

These large mountain ranges project from the gravels and alluvium of the Mojave. Ranges such as the Panamint, Clark, Providence, and New York Mountains are high enough to get considerable snowfall. The upper slopes of these ranges often bear forests of pinyon pines, the seeds of which are known as pine nuts or piñones. These seeds, which were an important staple for Native Americans, are still available in ethnic grocery stores. White Firs of Rocky Mountain affinities occur on the highest peaks.

The Mojave is a hot desert. Precipitation occurs mostly in the winter, but there are occasional summer thunderstorms. At higher elevations, snow is common. Where snow falls in the winter, and where soil is coarse and nonalkaline, vegetation is unlike that of the other desert regions. Here the dominant life-forms are the leaf-succulent yuccas, the most conspicuous of which is the Joshua Tree (plate 4A). The outline of Joshua Tree distribution is effectively an outline of the Mojave Desert. Above the Joshua Trees is the belt of pinyon pines on rocky soils and ridges. Below the Joshua Trees, similar to the Great Basin, there may be a region dominated by Blackbrush. Blackbrush is often associated with Joshua trees as well. Below the Blackbrush is the vast expanse of sandy soil occupied by Creosote Bush, the dominant shrub of America's hot deserts and the most abundant shrub in California. Creosote Bush is the dark green resinous shrub with wandlike stems that grows with orchard-like regularity on flat slopes of the desert (plate 15B). Creosote Bush fades out at the clay soils of dry lakes. Where soil becomes too alkaline, Creosote Bush is replaced by Shadscale. Cacti are more abundant in the Mojave than they are in the Great Basin, but they are mostly found on rocky slopes. Along washes, there may be thickets of trees, primarily Desert Willows, *Chilopsis linearis*, or Honey Mesquite, *Prosopis glandulosa* var. *torreyana*, both of which drop their leaves in the winter.

The Mojave is not known to everyone for its scenic beauty, yet three of the largest units administered by the National Park Service are located in this desert. Joshua Tree National Park is in the high Mojave near Twentynine Palms. It includes some of the most spectacular scenery of the high desert. Campgrounds are nestled among huge granitic boulders, and roads wind among acres of Joshua Trees. Park boundaries also include a portion of the Colorado Desert to the south. The transition between the two deserts may be viewed in a few hours by driving southward through the park.

Death Valley is the largest national park in the lower 48 states. At 282 ft (86 m) below sea level, Badwater is the lowest point in the United States. Remarkably, as the crow flies, Mount Whitney, the highest point in the lower 48 states, is only 85 mi (137 km) to the west. Telescope Peak at 11,049 ft (3368 m) in the Panamint Mountains looms to the west of Badwater, making this the steepest scarp in North America. The record high temperature of 134.6°F (57°C) was recorded at nearby Furnace Creek. Death Valley National Park is also the location for Racetrack Valley, the site of the famous sliding boulders. Its scenic badlands at Zabriskie Point and Artist's Palette are well-known landmarks.

Between the two national parks is the Mojave National Preserve, which lies roughly between Interstate 15 and Interstate 40 and is bounded on the east by the Colorado River. Included within its boundaries are some of the larger mountain ranges of the Mojave, such as the Providence and New York Mountains. The Providence Mountains with their limestone caverns, known as Mitchell Caverns, are already included in a state park. The area also includes Kelso Dunes, one of the highest fields of sand dunes in North America.

In February, 2016, three national monuments were added to the Mojave, increasing federal protection by 1.8 million acres (56.656 km²). The Mojave Trails National Monument includes 1.6 million acres (6475 km²) along Interstate 40 between Ludlow and Needles, effectively creating a corridor between Joshua Tree National Park and the Mojave National Preserve. The Sand to Snow National Monument, about 154,000 acres (623.2 km²), including Big Morongo Canyon, connects western Joshua Tree National Park with the San Bernardino National Forest. The Castle Mountains National Monument is a 20,920 acre (84.6 km²) parcel of scenic peaks north of the Mojave National Preserve.

Colorado Desert

California's "Low Desert" is the Colorado Desert. It is the California portion of a much larger region known as the Sonoran Desert. This southwestern desert region encircles the Gulf of California and includes northwestern Mexico, southwestern Arizona, southeastern California, and Baja California. Much of the Californian portion lies in the Salton Trough, an elongate valley separated from the Gulf of California by the delta of the Colorado River. This is a large, closed basin with the Salton Sea at its lowest point, 274 ft (88 m) below sea level. Before 1905, the Salton Sea did not exist. It

began to fill when an irrigation canal burst along the Mexican border. The accident created the New and Alamo Rivers, which still flow northward to the Salton Sea. After the ruptured canal was repaired, irrigation runoff from extensive agricultural operations in the Imperial and Coachella Valleys caused the Salton Sea to continue growing. Flood irrigation has converted 475,000 acres (195,225 ha) of the Imperial Valley and 75,000 acres (78,904 ha) of the Coachella Valley to farmland. At present, it is the largest inland body of water in the state; however, the Salton Sea is now getting smaller. Because the demand for Colorado River water is intense, Imperial Valley farmers have taken land out of cultivation so the water can be sold to the Los Angeles Department of Water and Power via the Colorado Aqueduct. About 300,000 acre-ft of Colorado River water are sold to the Metropolitan Water District. Reduced irrigation runoff is causing the Salton Sea to shrink and evaporation is causing the remaining water to become more saline. In an attempt to remedy the problem, the Imperial Irrigation District has been sending in water from the Colorado River, and there is a plan to send water from the Alamo River to help alleviate drying of the Sonny Bono Salton Sea National Wildlife Refuge at the south end of the lake.

The Salton Trough and the Gulf of California owe their existence to the northward displacement of lands west of the San Andreas fault. On the east side of the Salton Trough, the trace of the San Andreas fault is conspicuous along with offset stream channels and tilted and warped old lake beds. The many springs that occur along this fault are particularly visible from the air. Along this row of springs, groves of California Fan Palms stand out in stark contrast to the surrounding desert terrain.

The Salton Trough has frequently been inundated, sometimes by saltwater and sometimes by freshwater. This fact is evidenced by extensive badlands composed of highly eroded, colorful layers of clay and silt. Painted Canyon near Indio and Painted Gorge near El Centro are examples of deeply eroded lake-bed formations. A conspicuous waterline is visible at the foot of the Santa Rosa Mountains on the northwest side of the Salton Sea (figure 9.9). This line, about 200 ft (65 m) above the present Salton Sea, was formed by freshwater algae in shallow water at the edge of Lake Cahuilla. That lake dried about 500 years ago, when the Colorado River turned southward and ceased to drain into it. In Fossil Canyon, near Ocotillo, abundant marine fossils indicate that the area was once a shallow lagoon filled with saltwater.

East of the Salton Trough, the Orocopia and Chocolate Mountains rise from the alluvium. Interstate 10 as it passes between the San Bernardino and San Jacinto Mountains drops into the Salton Trough, after which it turns eastward over Chiriaco Summit passing between the Eagle Mountains to the north and the Orocopia Mountains and Chuckwalla Mountains to the south. From Chiriaco Summit eastward, the terrain of the Colorado Desert consists of long bajadas sloping toward the Colorado River. The Colorado Desert ends at the Colorado River on the east. Much of the boundary between the Mojave and Colorado Deserts is the ridge formed by the Little San Bernardino Mountains, the Pinto Mountains, and the Eagle Mountains. Joshua Tree National Park lies astride this boundary and includes a mixture of vegetation from both deserts. East of the Eagle Mountains, the boundary curves northeastward toward Needles so as to include the Turtle Mountains, the Whipple Mountains, and the Chemehuevi Mountains in the Colorado Desert.

Between the mountains, broad plains and bajadas drain into Danby, Palen, and Ford Dry Lakes. East of the Chocolate Mountains, a large flat area, the Chuckwalla Bench, drains eastward toward the Colorado River. It is designated as an Area of Critical Environmental Concern (ACEC) because of its rich flora of approximately 160 plant species, including nine cacti and rare and endangered plants and animals such as Munz's Cholla, *Cylindropuntia munzii*, Chuckwalla Cholla, *C. chuckwallensis*, and the Mohave Desert Tortoise, *Gopherus agassizi*.

The Colorado Desert experiences more summer precipitation than the northern deserts. Although yearly precipitation remains low, a significant portion of it falls in August and September, usually as thunderstorms. Data collected at the headquarters for Anza-Borrego Desert State Park indicate a yearly average of 6.9 in (16.9 cm). Most precipitation falls from December to March, with January the only month averaging more than 1 in (25 mm). On the average, thunderstorms in August and September add another inch (25 mm) to the yearly total. These thunderstorms may be severe. A total of 3–5 in (8–14 cm) of rain may fall in a few hours, washing out roads, scouring washes, and uprooting trees. These storms are often local in distribution; a few miles away there may be little or no precipitation.

Sandy flats and slopes in the Colorado Desert are very similar to those of the Mojave Desert. Creosote Bush dominates over most of the areas with saltbushes occurring where the soil becomes more alkaline. On upper rocky slopes, particularly on the eastern side of the Peninsular Ranges, there is a community of succulent and drought-deciduous plants known as Cactus Scrub (figure 9.1, plates 4B and 16A). There are many cacti in the Sonoran Desert, but the Colorado Desert differs from other subdivisions of the Sonoran by having mostly small cacti. Also in this region is a leaf-succulent plant known as Desert Agave or century plant, *Agave deserti* var. *deserti*. These agaves are noted for their tall stalks topped with large yellow flowers. Among the drought-deciduous plants, the most distinctive is Ocotillo, *Fouquieria splendens* subsp. *splendens*. Ocotillo plants are composed of many tall whiplike stems covered with long spines. In the spring, each stem terminates in a cluster of bright red, tubular flowers. The distribution of Ocotillo, effectively outlining the Sonoran Desert, extends northward nearly to Needles.

Another distinctive feature of the Sonoran Desert is the presence of many species of trees, most of them in the Legume family (Fabaceae; figure 9.4). Throughout most of the Sonoran

FIGURE 9.4 Some members of the legume family (Fabaceae) characteristic of Desert Wash (drawings by Karlin Grunau Marsh from Clarke 1977).

A. Blue Palo Verde, *Parkinsonia florida*, a drought-deciduous tree.

B. Ironwood, *Olneya tesota*, an evergreen tree.

C. Honey Mesquite, *Prosopis glandulosa* var. *torreyana*, a winter-deciduous tree.

D. Catclaw, *Senegalia greggii*, a winter-deciduous shrub.

Desert, these trees are distributed widely among Ocotillos and large cacti. In the Colorado Desert, where it is drier, these trees are primarily located along washes, but on the upper bajadas along the bases of the mountains, these trees form a woodland, known as Microphyll Woodland. Blue Palo Verde (*Parkinsonia florida*), Ironwood (*Olneya tesota*), mesquites (*Prosopis* sp.), and Smoke Trees (*Psorothamnus spinosus*) occur in these locations. Smoke Trees are nearly always leafless, and to some they resemble puffs of smoke. Their twisted, blue-gray image has become a symbol of the Colorado Desert, which is roughly outlined by the distribution of this species.

The California Fan Palm, *Washingtonia filifera* (plate 16B), is another distinctive species of the Colorado Desert, occurring in seeps and oases. From the air, groves of California Fan Palms can be seen to outline the trace of the San Andreas fault. It is along faults such as this that water seeps to the surface to form springs. Glancing eastward from Interstate 10, motorists can see lines of palm trees along the San Andreas fault near the community of Thousand Palms.

Overall, many of the species are shared with the Mojave and Arizona's Sonoran Deserts, but about 100 species of plants occur nowhere in California outside of the Colorado Desert (Mathias 1978). Many of these occur in families that are typically southern in distribution. Some are relictual, remnants of a former widespread distribution and occur in specific microhabitats (e.g., California fan palm at springs and seeps; Chuparosa, *Justicia californica*; Las Animas Colubrina or Snakebush, *Colubrina californica*; and Hall's Tetracoccus, *Tetracoccus hallii*, on rocky slopes and within washes; and Ragged Rock Flower, *Crossosoma bigelovii*, along rocky canyon walls). Some have very localized distribution, such as all-thorn (*Koeberlinia spinosa*; five populations in the Chocolate Mountains along washes and canyon walls), Emory's Crucifixion-thorn (*Castela emoryi*; 18 populations, usually associated with dry lake beds), and Littleleaf Elephant Tree (*Bursera*

microphylla; 20 populations, mostly in Anza-Borrego Desert State Park along rocky alluvial fans, washes, and slopes).

The Peninsular Ranges, along the western edge of the Colorado Desert, show distinct banding of desert communities. Above the Cactus Scrub community, there is a zone of shrubs known as Desert Chaparral, a diverse community discussed in Chapter 8. Above the scrub community is the zone of pinyon pines, marking the transition to the mountain communities. A prime example of Colorado Desert is located in Anza-Borrego Desert State Park. This is California's largest state park, and it is only a few hours by automobile from metropolitan Los Angeles or San Diego. Perhaps it is best known for spectacular displays of spring wildflowers.

Where a prevailing wind and a source of sand occur, dune fields may form. These areas are located in the Coachella Valley near Palm Desert, in Anza-Borrego Desert State Park, along Palen Lake east of Joshua Tree National Park, and in the Algodones Dunes south of Glamis. These are areas where spring wildflowers put on a spectacular visual display in those years when precipitation and temperature conditions are favorable and they are home to scores of endemic species and specialized plants. The Algodones Dunes in particular are home to the largest number of dune-endemic plants in North America (figure 9.37).

GEOLOGY OF THE DESERT

The lack of water in the desert has preserved ancient landscapes. Old rock materials that have washed away in other parts of California are still present in desert mountain ranges. In fact, geologists have deciphered the ancient history of western North America by examining rocks that are now exposed in desert areas.

Metamorphic rocks older than a billion years are found at scattered localities in the Mojave Desert. Such rock formations, including gneiss and marble, are found in the Ord and Old Woman Mountains. The Pinto Gneiss

FIGURE 9.5 Old limestones of the Ibex Hills, Death Valley National Park.

is found in Joshua Tree National Park, and at the southern end of the Marble Mountains there are old granites and schists.

From about a billion years ago to 600 million years ago, during the Proterozoic, a period of erosion stripped away many thousands of feet of rock throughout most of the Southwest. This interval of erosion produced an unconformity in which 570–700 million year old (Cambrian and Precambrian) rocks are now found lying on billion-year-old metamorphic rocks: nearly a half-billion years of rock is missing! One of the few remaining records for that period of time is approximately 10,000 ft (3000 m) of sedimentary rock exposed in the Nopah Mountains of eastern California. Here there are sandstones, shales, and limestones that fill the gap. Among the oldest life-forms are stromatolites found in the Noonday dolomite formation of the Nopah Range.

The Paleozoic Era, from about 540 to 250 million years ago, is represented by shales, sandstones, and limestones that are exposed throughout the deserts, but particularly in the Great Basin where uplift of mountains along dip-slip faults has exposed them (figure 9.5). These sedimentary rocks represent the interval of time when the western border of North America was

FIGURE 9.6 Trilobite fossil in Paleozoic shale, Saline Valley, Inyo County.

a continental shelf covered by seawater. It was the trailing edge as the continent moved eastward toward its collision with Europe. Among the oldest rocks in this sequence are shales that contain trilobites (figure 9.6), extinct marine crustacea that crept about like sowbugs on the

FIGURE 9.7 Mesozoic granite, Chuckwalla Mountains, Riverside County.

ocean floor beyond the surf zone. Trilobites are found in the Marble Mountains, the Nopah Mountains, and in Saline Valley. Also in the Marble Mountains are limestones with circular structures, known as oolites, representing the remains of blue-green algae that lived in a warm, shallow lagoon-type environment.

The Mesozoic Era, from about 250 to 65 million years ago, was a time of subduction, when the continent was moving westward over the ocean crust. The magma formed during this interval is represented by plutonic rocks throughout the west. These are the granitics of the western mountain chain. They are also the granitics of the Little San Bernardino Mountains in Joshua Tree National Park. The granitics of the Chuckwalla Mountains in the Colorado Desert (figure 9.7) are also of this age, as are those in the Granite Mountains and New York Mountains of the eastern Mojave. In the Great Basin, these granitics are found in the Alabama Hills near Lone Pine, and along lower Bishop Creek west of Bishop. In the desert climate, they form large spherical boulders that

have been weathered by oxidation to a brownish color. They are among the most popular landforms for rock climbers.

During the Cenozoic Era, from about 65 million years ago to the present, the desert area accumulated freshwater sediments and volcanic rocks, indicating that the terrain was above sea level. The oldest part of this record represents another unconformity throughout most of the west. An erosional period stripped away most of the rocks from about 65 million to about 30 million years ago. One exception to this is found on the north side of the El Paso Mountains near Red Rock Canyon. This area, uplifted along the El Paso fault, a portion of the Garlock fault system, includes a series of stream and lake deposits dated at about 58 million years of age. This area includes fossils of early crocodiles and a primitive mammal in the Golar formation. Fossils of this age are found nowhere else in California. The next oldest fossils, of primitive monkeys, rodents, and a rhinoceros, are found near San Diego in rocks about 40 million years of age. In rocks dated from about 30 million years ago to the present are found scattered fossil localities that tell us much about prehistoric California and the evolution of mammals. In northern Death Valley, in Titus Canyon, there are fossils over 30 million years old that give us a picture of the area as a wet, wooded countryside. Included here are fossils of dwarf horses and deer, as well as tapirs, rodents, rhinos, and dogs.

About 25 million years ago (Miocene), the picture of California as we see it today began to emerge. About this time, the west coast changed to become a faulted, rifted borderland. Land west of the San Andreas fault began its inexorable slide toward the north. This caused a stretching of the desert areas, a type of motion known as extension, which is accompanied by faulting and down-dropping. Faulting in the Great Basin began. The Garlock fault cut diagonally along the southern edge of the Great Basin. Mountains uplifted, and valleys sank. In the Death Valley area, for example, it has been determined that the Panamint Range moved northwestward

FIGURE 9.8 Badlands at Zabriskie Point, Death Valley National Park.

about 50 mi (80 km) from a position adjacent to the present Nopah Range. This allowed the intervening terrain to sink, forming Death Valley, at 282 ft (87 m) below sea level.

Right-lateral faulting along the San Andreas fault pulled Baja California away from what is now mainland Mexico. This rifting opened up the Gulf of California, of which the Salton Trough of the Colorado Desert is merely the northern end. It too has subsided, lying at 274 ft (88 m) below sea level. At its highest point of 105 ft (32 m) above sea level near Yuma, Arizona, the delta of the present Colorado River holds back seawater from the Gulf of California.

Northward (right-lateral) movement along the San Andreas fault may be the driving force causing left-lateral movement along the Garlock fault. The wedge-shaped basin between the San Andreas and Garlock faults, forming the main portion of the Mojave, appears to be rotating clockwise due to motion along these two faults. Studies of Paleomagnetic data in Miocene volcanic rocks indicate that about 25° of rotation has occurred in the last 15–20 million years. The Garlock fault runs northeastward from near the town of Gorman to the southern end of Death Valley, a distance of some 150 mi (240 km). Its trace is marked by an uplifted scarp that is clearly visible along the Tehachapi Mountains, southern Sierra Nevada,

Red Rock Canyon, and the El Paso Mountains. The south side of the fault has been displaced eastward about 30 mi (48 km).

The Miocene record in California's deserts is represented not only by uplift and subsidence, but also by intermittent flooding and volcanism. Tuff, ash, and other volcanics interbedded with lake bed sediments are widespread. Red Rock Canyon has the oldest beds, at about 58 million years. As indicated previously, however, there is very little record for periods before about 30 million years ago. Shales, siltstones, sandstones, and conglomerates mixed with volcanics are scattered about the southwestern deserts. Along the west side of Death Valley, there is a scenic motor route known as Artist's Drive. Here a person can see playa deposits and conglomerates hundreds of feet thick. The colorful display known as Artist's Palette is a mixture of lake beds and volcanics. In the badlands area near Zabriskie Point (figure 9.8) are thousands of feet of lake sediments spanning millions of years of deposition. These beds have been tilted vertically by uplift along the fault on the east side of Death Valley.

Eroded lake bed deposits are called badlands because they will support very little plant life (figure 9.8). Water sinks slowly, and the soil clings tenaciously to the water, so that it is difficult for plants to grow there. In the badlands east of Death Valley, there is a plant in the buckwheat family (Polygonaceae) that grows nowhere else. This rare wildflower called Golden-carpet Gilmania, *Gilmania luteola*, has a low, mat-like growth habit. Its tiny yellow flowers are only 2–3 mm in diameter. It has hard-shelled seeds that are able to withstand long periods of dormancy. It is known from only five occurrences. In those rare years when several consecutive rains succeed in wetting the clay soil, *Gilmania* puts on its display of germination, growth, and flowering.

Over the last 7 million years or so, the Salton Trough has been flooded on numerous occasions by either seawater or freshwater. In the Fish Creek and Coyote Mountains areas of the Colorado Desert, there is a series of marine deposits known as the Imperial Formation. Near the town of Ocotillo in Imperial County, these deposits have been uplifted along the Elsinore fault. These rocks, deposited about 7 million years ago, contain fossils of about 200 marine invertebrate species, as well as rare remains of marine vertebrates such as bat rays, sharks, giant barracudas, sea cows, and whales. Of the mollusks represented, 16% of the species have modern counterparts in the Caribbean, indicating that the former Gulf of California had a direct connection to the Caribbean. That is, at the time, the two continents were not connected by land. The Central American terrain that now connects North America to South America was not present.

Anza-Borrego Desert State Park contains a badlands formation known as the Palm Spring Formation. This sequence of beds, 2 mi (3 km) thick, was deposited from 1.5 to 4 million years ago in a flood plain and coastal lagoon setting. One of the most complete records of Pliocene terrestrial vertebrates in the southwestern United States is found here. The vertebrate fossils from this formation include llamas, camels, horses, donkeys, zebras, mammoths, mastodons, and sloths, as well as their predators, including lions, sabertooth cats, and bears.

The Pleistocene (1–2 million years ago) was a period of inundation in the deserts. Whereas heavy snowfall formed glaciers in the moun-tains, the deserts experienced heavy rainfall. Similar to glacial intervals, there were intervals when desert basins were large lakes. For example, in Death Valley there was Lake Manly, the waterline for which indicates it was about 600 ft (190 m) deep and nearly 100 mi (160 km) long. East of Death Valley, near the small towns of Shoshone and Tecopa, is a region of clays and silts representing old Lake Tecopa. These beds are about 400 ft (120 m) thick. They are still horizontal in orientation and have not been deeply eroded, indicating their recent deposition. Of interest here is a layer of pumice 12 ft (4 m) thick. This material, derived from volcanic ash that settled in the lake, indicates that many volcanoes must have been active in the area during the Pleistocene.

Other than the Salton Sea, which is fed today by irrigation runoff and effluent from the Mexicali area, the last major freshwater invasion of the Salton Trough ended about 500 years ago with the drying of Lake Cahuilla, which apparently was fed by the Colorado River. A waterline from that lake, 200 ft (60 m) above the water level of the present Salton Sea, is conspicuous along the base of the Santa Rosa Mountains on the west side of the Coachella Valley (figure 9.9). Fifteen rows of fish traps used by the Cahuilla Indians lie along former western shorelines. The distances between these rows suggest that they were probably constructed once a year as the lake was drying. During that time, the lake level dropped about 4 ft (1.5 m) a year. Presumably about AD 1500, the Colorado River turned southward to dump into the Gulf of California, and Lake Cahuilla lost its input of freshwater. The lake is presumed to have taken 55–60 years to dry completely.

Volcanism in desert areas is mostly of Miocene age or younger. In the Great Basin, this includes the Modoc Plateau. Miocene volcanics are also scattered throughout the Mojave and Colorado Deserts. The brilliantly colored Calico Mountains near Barstow are a mixture of volcanics and sediments of Miocene age. Perhaps some of the most interesting volcanic

FIGURE 9.9 Lake Cahuilla waterline along the base of the Santa Rosa Mountains west of the Salton Sea.

scenery is in the eastern Mojave between the Providence Mountains and New York Mountains. Here there are buff-colored rhyolitic flows alternating with ash deposits, forming spectacular steep-sided, flat-topped mesas that look like layer cakes. The hard rhyolite forms a protective cap over the softer ash deposits. A popular campsite in the Mid Hills area is known as Hole-in-the-Wall (figure 9.10), in reference to many large cavities formed as gas pockets in the tuff deposited during the time of a huge volcanic explosion, which actually carried rock fragments up to 60 ft (15 m) across, the largest ever documented. These buff-colored flows are also distinctive because they have picked up many small angular particles, forming volcanic breccias.

Younger volcanics are mostly dark-colored basalts. These rocks are abundant in Owens Valley. They began to form during the Pleistocene and have continued since then. Near Little Lake, there are about 45 different volcanic formations, including cinder cones and basaltic flows, including cliffs marked by columnar basalt. Along Highway 395 between Bishop and Mammoth is a volcanic tableland of Bishop tuff

and pumice which have been dated at 760,000 years. This is a famous deposit the source of which was an explosive eruption that resulted in the Long Valley Caldera. The Bishop tuff and extensive deposits of pumice are discussed also in Chapter 4. From Mammoth northward to the Nevada state line, most of the visible topography of the eastern Sierra Nevada is of volcanic origin (plate 6A). Rainbow Canyon is a spectacular area that lies east of Owens Valley. The highway from Owens Valley to Death Valley descends into Panamint Valley across spectacular red and yellow volcanics that border Rainbow Canyon. At one point, the road traverses the inside of several cinder cones.

There are several conspicuous volcanic cones in the Mojave extending in a line from Barstow eastward. These craters, including Amboy and Pisgah craters, probably were formed by eruptions about 2000 years ago. The basalt fields around the craters are interesting because of numerous tunnels, vents, fissures, and collapsed blisters, indicating the plastic nature of the lava as it flowed.

At the northern end of Death Valley is a peculiar volcanic formation known as Ubehebe

FIGURE 9.10 Volcanic tuff in the Mid Hills, near Hole-in-the-Wall Campground, San Bernardino County.

Crater. Ubehebe is an Indian word that means "big basket." This distinctive feature probably was formed by steam and volcanic ash that erupted about 6000 years ago and fell back upon itself, forming at least 50 layers, 80 ft (28 m) thick, of bedded sediments. All tolled there are a dozen recognizable explosion pits in the area, but Ubehebe is the most dramatic. The material is basaltic in composition, but there are no associated lava flows.

In the Colorado Desert, recent volcanics are represented by a line of pumice and obsidian domes near the southern end of the Salton Sea. This region is still mildly active. There were mud pots and steam vents in the area until about 1954, when the rise of the Salton Sea covered them. Some mud pots are still visible near the town of Niland.

Areas of recent volcanism provide the opportunity to use the earth's heat to generate electricity. The geothermal facility at Geysers

was mentioned in Chapter 7. Quite a number of geothermal facilities for desert areas are operating, and others are in the planning stages. In the Coso Range near Ridgecrest is an operating geothermal facility, and the Casa Diablo power plant near Mammoth has been operating since 1985. Several localities near Mammoth Lakes are also in the planning stages. One such proposal, near Hot Creek (figure 4.9), is controversial because the Owens River in the area is a fishery for native trout. The Casa Diablo facility and some of the controversies are discussed in Chapter 4. In Mexico, just south of the border, at Agua Prieta, there is one of the oldest functioning geothermal power plants.

One interesting feature of volcanic landscapes is that they are often associated with specific vegetation. One plant that is conspicuous in volcanic areas is a saltbush known as Desert-holly, *Atriplex hymenelytra*. This small shrub gets its common name from its small, toothed, holly-like leaves, found on no other saltbush. Its white herbage stands out starkly against dark volcanic soils.

Mining has been an important economic activity in California's deserts since the mid-1800s. Significant amounts of silver, lead, and zinc have been mined in the Inyo Mountains, Panamint Mountains, and Argus Mountains of the Great Basin. The most important silver mines in the Mojave Desert were in the Rand Mountains, south of the Garlock fault. Gold was obtained from mines in the Panamint Mountains, Rand Mountains, and Cargo Muchacho Mountains in the southeast corner of the Colorado Desert. Silver and gold also were obtained from the mine at Bodie, near Mono Lake. Small amounts of tungsten were extracted by placer mining from gravels on the southeastern side of the Rand Mountains. The largest tungsten mine in California was located west of Bishop in the Tungsten Hills and Pine Canyon. That facility is discussed in Chapter 4.

The effort to extract gold from the desert is not over. The process known as cyanide heap leaching, described in Chapter 6, may be a means of mining low-grade desert ore profitably. The oldest mine utilizing this technology in the state is the Picacho Mine, located near the Colorado River in Imperial County. Between 1980 and the mid-1990s about 9500 tons (8600 t) of ore had been mined per day, yielding about 25,000 oz (710 kg) of gold per year. When the mine was no longer profitable, in 2002 the enormous leach pads were purged of cyanide, the land reshaped, and native vegetation replanted.

The use of cyanide heap leaching for extracting gold was also being used at the Colosseum Mine on Clark Mountain near Mountain Pass and the process was shut down in 1992. A cyanide heap leach process was also used at the nearby Castle Mountain Mine. Over a million ounces of gold were produced between 1991 and 2001, and plans to reactivate the mine are ongoing. Cyanide heap leaching also was being used at mines near Randsburg and Mojave. The Briggs Mine on the west side of the Panamint Range between 1996 and 2004 produced 550,000 oz of gold. The mine is still active, although the most active heap leaching process was halted in 2004. There also has been talk of a similar operation at Bodie, a state historic park. Apparently, the plan is to remove the 400 ft (125 m) Bodie Ridge situated behind the townsite, and move the material to another location for processing. Environmentalists are opposed to this form of mining because of possible groundwater contamination and/or danger to wildlife, particularly birds and bats. Between 1980 and 1989, a total of 7609 animal deaths were reported for 97 mines in Arizona, Nevada, and California.

Among the most famous old mines was the Cerro Gordo Mine, near the crest of the Inyo Mountains. Beginning in 1860, significant amounts of silver and lead were mined, but in the early 1900s zinc became the most important product. The ore was deposited by hydrothermal solutions into Paleozoic carbonate rocks. So rich was this mine that a sizable community of miners, many of them Chinese, once occupied the area. By means of a bucket-tram system, metal was brought down the mountain to the town of Keeler, on the edge of Owens Lake. Here the ingots were loaded aboard a steamer and transported across the lake to be loaded on wagons. In later years, the ore was loaded from the steamer to a railroad. There is a story that a steamer loaded with silver once sank in Owens Lake and was never recovered. Presumably, the load was so heavy that the entire steamer and its contents sank beneath the muddy lake bottom and is still buried there, totally concealed by the sediments! Part of the mining operation at Cerro Gordo included smelting the ore. Charcoal for this process was produced in the charcoal kilns located along Cottonwood Creek on the west side of Owens Lake. Wood for the kilns was obtained from the forest on upper Cottonwood Creek.

Another famous silver mine was at Panamint city in the Panamint Mountains. The town grew to a population of 1500 in the 1870s, when the mine was most productive. A cloudburst and flash flood wiped out the town and the smelter in 1876, but by that time most of the ore had been mined. The mine was reopened for a short time in the mid-1980s when the price of silver was very high. Other famous silver mines were the Minietta and Modoc mines of the Argus Mountains. Charcoal for smelters of these silver mines was produced at beehive-shaped charcoal kilns in the Pinyon-Juniper Woodland at the northern end of the Panamint Mountains. These charcoal kilns have been restored, and the site is now a part of Death Valley National Park.

Significant amounts of iron were also mined from desert localities. The largest iron mine was in the Eagle Mountains on the northern border of the Colorado Desert. Kaiser Steel Company operated this mine until the middle 1900s. The railroad down Salt Creek Wash carried ore from this mine to the Southern Pacific Railroad on the eastern border of the Salton Sea. It is interesting that a new use for the old railroad has been proposed: to haul trash from Los Angeles to be dumped into the huge pits at the old Eagle Mountain mine. Some estimates indicate that the site could absorb Los Angeles trash for 125 years. Remains of another iron mine, the Vulcan Mine of the southern Providence Mountains, can still be seen as a "glory hole" several hundred feet deep, where it penetrates Cambrian limestones. The Vulcan Mine operated in the middle 1940s. Its close proximity to the old Union Pacific Railroad station at Kelso also makes it a potential landfill.

Most of the mining today is for sand and gravel, but minerals in salts (evaporites) are also valuable in deserts. Such mining operations occur in the salt deposits of Owens Lake, Bristol Lake near Amboy, and Searles Lake near Trona. The mine at Trona is a large operation that produces potash, a potassium mineral used in fertilizers. The town of Trona is named after a rare mineral that can be obtained from the salts of Searles Dry Lake.

Borates are other important minerals found in Searles Lake. Boron is used as an additive in jet fuel, and some borates are used in detergents and in fiberglass insulation. Boron was infused into lake water by volcanic gas emissions that occurred millions of years ago, when huge lakes filled many basins in the desert. As a result of evaporation, crystalline borate minerals may also be found among the salts of some lake beds. Other lake deposits with borate minerals are found in the Mojave Desert. Death Valley contains borate minerals known as colemanite and ulexite. Near the town of Boron is a large mine where borate minerals such as ulexite, colemanite, and kernite are profitably mined.

Plaster of Paris is gypsum, an evaporite composed of calcium sulfate. It also is an important component of various kinds of plasterboard used in construction. In the Colorado Desert, just east of Split Mountain Canyon, near Anza-Borrego Desert State Park, is a 100 ft thick (30 m) deposit of gypsum that has been mined for many years. A narrow-gauge railroad carries the gypsum south to Plaster City, where it is processed. During the Renaissance, a solid form of gypsum known as alabaster was a common material used in the carving of statuary.

Other than the saline minerals, most mining today is for talc. Talc is a magnesium mineral used in making paper, soap, and toiletry powders. It is concentrated at the contact points between certain limestones and intrusive bodies such as granite. Profitable talc mines have operated from time to time in the Cottonwood Mountains, Ibex Mountains, Argus Mountains, and around the southern end of Death Valley in the Panamint and Black Mountains.

One group of interesting elements mined in the California desert is known as rare earths. These represent a group of elements used in the glassmaking, petroleum, and metallurgy industries, particularly those involved in research and production of superconducting materials. Virtually all rare-earth production in the United States involves a mineral known as bastnaesite (= carbonatite), an igneous limestone, taken from an open pit mine near Mountain Pass, where Interstate 15 crosses the Clark Mountains between Baker and Las Vegas. Approximately 97% of United States and 33% of world rare-earth production came from this one mine in the Mojave Desert, which has varied its production in association with demand from world markets. Fears are that China could manipulate the world market.

Structure of a Desert Basin

In order to understand why desert plants grow where they do, it is necessary to become familiar with the structure of a typical desert basin. To some degree, all desert terrain is basin-like. In the Great Basin, basins are narrow and elongated. Different aspects of a slope are easily distinguishable as bands or zones of vegetation.

In most hilly terrain, water is the most important erosive force. This is true for deserts as well. The difference, however, is that most of the erosion took place thousands of years ago, during periods of heavy rainfall associated with ice ages. Desert landscapes were shaped many years ago, and they have been preserved by a lack of rainfall since that time. In this sense, deserts have fossil landscapes. Heavy rainfall occurs sporadically today, usually associated

FIGURE 9.11 Desert landforms. Alluvial fans emerging from the north side of the Avawatz Mountains merge to form an outwashed bajada. Note the parallel lines of Creosote Bush in the foreground.

with monsoonal thunderstorms. On such occasions, erosion occurs rapidly, and land is rearranged, after which it may lie for another thousand years relatively undisturbed.

The uppermost part of a desert slope is an erosion surface known as the pediment. Ped means "foot," and the term pediment refers to the foot of the mountain. There is very little soil on the pediment, as it has been carried away by wind and water. During long periods of time with no significant rainfall, the surfaces of desert rocks become darkened. This darkening, known as desert varnish or rock varnish, is due to oxidation of minerals such as iron and manganese. According to present theory, the actual varnish, less than 0.5 mm thick, involves a fine patina of clay that is deposited by wind over the rocks, and the minerals are in the clay. In exposed areas where the pH is neutral, oxidizing bacteria concentrate black-colored oxides of manganese that help to cement the clay in place. In protected areas, such as dust-filled depressions or the undersides of rocks, an alkaline pH inhibits the growth of the bacteria, and a bright orange iron oxide is formed instead. The long period of time required for the varnish to form is illustrated by locations where Native Americans have overturned or removed varnish-covered rocks to form giant intaglios on the desert floor. One such locality near Blythe, in the Colorado Desert, is believed to be at least

400 years old. Another famous site where dark rocks have been removed to form intaglios is in the Atacama Desert of Chile. Using a technique known as cation-ratio dating, petroglyphs or rock pictures in which the varnish has been chipped off in the Mojave Desert are now known to be thousands of years old.

Canyons in desert ranges cut across the pediment, and at canyon mouths are fan-shaped piles of rocks and gravel that have been carried by runoff from the mountains. These are known as alluvial fans (figure 9.11). Because these fans have remained undisturbed for thousands of years, upper surfaces on the fans have become varnished. It is not until a major storm occurs that enough water flows out of a canyon to turn over old rocks and deposit new ones. New or overturned alluvium is easy to spot because of the absence of varnish. [14]Carbon and cation-ratio dating in Death Valley have revealed that the varnish at the top or head of one alluvial fan is 50,000 years old. Varnish gradually becomes younger toward the bottom or toe of the fan. The varnish there is 14,000 years old, indicating that these were the last rocks to be deposited and darkened.

The alluvial fans spread out at their lower borders, and the edges of adjacent alluvial fans overlap (figure 9.11). This part of a slope, where alluvial fans coalesce below the pediment, is known as the bajada, which is a Spanish word

FIGURE 9.12 Skid boulders on Racetrack Playa, Death Valley National Park.

that means "lower." It refers to the lower aspect of the long slope that begins high up on the mountain bordering each valley. The bajada is composed of gravels and sand, and particle size becomes smaller toward the center of the valley. If the center of the basin is clay, it remains relatively free of vegetation and is known as a playa, which means "beach" in Spanish. A salt flat in the center of the basin is known as a salina.

The nature of a clay playa is best appreciated by a visit to Racetrack Playa, in Death Valley National Park (figure 9.12). Apparently, when the clay is wet, it is so slippery that wind is able to blow rocks around on the surface. The tracks left in the clay indicate changes in wind direction. Sometimes the tracks actually loop back upon themselves. The present theory about the tracks is that the rocks become imbedded in ice and are then blown about, as if the ice were a raft. During a period of heavy rainfall, the lake may fill to a depth of 1 or 2 ft. If freezing occurs, the top 12 in or so may turn to ice. As the lake begins to thaw, the heavy boulders are buoyed up by the ice, and wind transports these ice-boulder rafts floating on the remaining water. The fascinating part of this story is that apparently no one, in person, has ever seen the wind blow about one of these rocks, although recently remote controlled video cameras confirmed the hypothesis that rocks move while trapped in rafts of ice.

STRATEGIES OF DESERT PLANTS

In general, strategies of desert plants (Table 9.2) are based on ways to cope with drought, alkaline soil, temperature extremes, intense light, and a short growing season. In general, however, the distribution of desert vegetation (figure 9.13) reflects the water-holding capacity and alkalinity of the soil, both of which increase toward the center of a basin.

Succulents and Drought-Deciduous Plants

On the pediment and upper alluvial fans, water runs off or percolates rapidly. Therefore, it is available for a short period of time. In this region, most plants are either succulent or drought-deciduous. Succulent plants grow on the coarsest of soils. They store water, which they draw upon slowly between rainstorms. Plants such as cacti are called stem succulents because they store water in their stems. Plants such as yuccas and agaves store water in succulent leaves and are called leaf succulents. Succulents have deciduous root hairs that are shed when the soil becomes dry. When this happens, water will neither enter nor leave the roots. When it rains, the plant grows new root hairs. Growth and photosynthetic rates of succulents are slow.

TABLE 9.2
Strategies of Desert Plants

Succulent

Leaf succulent: yuccas, agaves, live-forevers

Stem succulent: cactus:

    Associated with coarse, rocky soil

    Store water; deciduous root hairs

    Store carbon at night; CAM photosynthesis

    Shallow, widespread roots

Drought dormant

Drought-deciduous: Ocotillo, Desert Brittlebush

    Associated with coarse soils

    Lose leaves during drought; dormancy maintained by roots and stems

Geophytes: Desert Lily, Coyote Melon

    Regions with predictable precipitation

    Lose aboveground parts during drought; dormancy maintained by below-ground parts (i.e., bulbs, corms, tubers)

Ephemerals (Annuals)

    Regions with unpredictable precipitation

    Entire life cycle during one season; dormancy maintained as seeds

        Winter annuals: Evening Primrose, Sand Verbena

            C_3 photosynthesis; require early winter precipitation.

        Summer annuals: Unicorn Plant, Windmills

            C_3, C_4 photosynthesis; require midsummer precipitation

    Rapid growth

    Large leaves

Drought enduring

Xerophytes

    Evergreen perennials: Creosote Bush, Great Basin Sagebrush

        Associated with medium-grade soils, lower bajadas

        Large root systems

        Small, drought-adapted leaves; lose leaves a few at a time

        Slow growth of aboveground parts; most of growth in roots

---

Drought avoiding

Associated with permanent water

Phreatophytes

    Winter-deciduous: Fremont Cottonwood, Black Willow

    Evergreen: California Fan Palm

(continued)

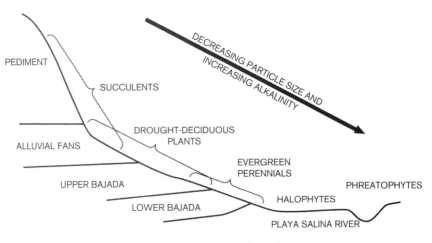

TABLE 9.2
(continued)

Drought avoiding

Large leaves

Large root systems

Desert wash plants

Winter-deciduous: Honey Mesquite, Desert Willow

Drought-deciduous: Smoke Tree, Palo Verde

Evergreen: Ironwood

Small leaves

Large root systems

Salt tolerant

Xerophytic halophytes: saltbushes

Associated with fine-grained, alkaline soil

Small leaves

Excrete salt on leaves

C_4 photosynthesis

Succulent halophytes: Pickleweed, Iodine Bush

Store water to dilute salt

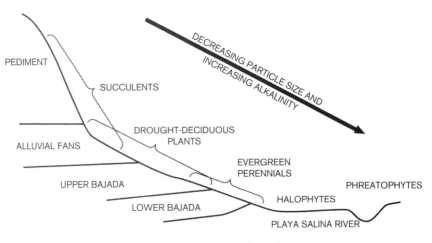

FIGURE 9.13 Generalized distribution of vegetation on a desert slope.

Drought-deciduous plants range from the pediment to the upper bajada. They drop their leaves during drought. After it rains, they rapidly grow new leaves, which are often comparatively large for desert plants. Most drought-deciduous plants drop their leaves only once a year, but Ocotillo is reported to lose and grow new leaves up to seven times a year. Drought-deciduous plants grow and photosynthesize rapidly.

Evergreen Shrubs

On the middle and lower bajada, most of the plants are evergreen, that is, they retain leaves year-round. In the Great Basin, the dominant evergreen shrub is Great Basin Sagebrush (figure 9.20A; plate 15A). In warm southern deserts, the dominant evergreen shrub is Creosote Bush. These shrubs are characterized by large root systems. They are capable of absorbing water from all levels of the soil, relying on their deepest roots during the dry season. Photosynthetic and growth rates are slow. Leaves are small and often possess a waxy or resinous coating to inhibit water loss. One of the most important characteristics of these plants is tolerance for drought. They are able to conduct photosynthesis under the highest levels of water stress known among plants. Interesting new research shows that Desert Woodrats eat the leaves of Creosote Bush, particularly during times of food scarcity. Apparently the chemicals that give Creosote Bush its resinous texture and distinctive odor are poisonous to nearly all animals except woodrats. Further research indicates that during times when air temperatures are high, the woodrats are less tolerant to the toxins, showing an unexpected potential consequence of global warming.

Halophytes

The lowermost portions of the bajada are more alkaline. Plants of these soils are called halophytes and include members of the Goosefoot family (Chenopodiaceae). Saltbushes are also evergreen shrubs, but they tolerate alkaline soils by absorbing both salt and water. The salt is then excreted upon the surface of the leaves, which is one of the reasons that most salt bushes have a silvery sheen on the leaves. As soil becomes more alkaline, the leaves on these saltbushes become more succulent. Apparently, water storage helps to dilute the salts that become concentrated in tissues. Iodine Bushes, *Allenrolfea occidentalis*, and Pickleweeds, *Arthrocnemum subterminale*, are the most succulent and the most salt tolerant, growing on the edges of salty bodies of water such as the Salton Sea.

In the center of a basin, very little vegetation occurs. Either the soil is too alkaline or there is too much clay. Clay absorbs and releases water slowly. Plants do not grow well in clay because water adheres strongly to the fine particles of soil and is unavailable to the plants. Furthermore, oxygen is unable to reach the roots, which limits cellular respiration there. Many areas of old lake beds composed of clay soils are devoid of plants. These regions, when uplifted, erode in a characteristic way, forming deeply furrowed topography. Such topography, particularly if it lacks vegetation, is known as badlands. Well-known badlands occur at Zabriskie Point (figure 9.8) in Death Valley and in Anza-Borrego Desert State Park.

Phreatophytes

Deeply rooted plants that tap permanent water are known as phreatophytes. This word means "well-plants," implying that these plants grow near a well. Where soil is unstable, or where precipitation is unpredictable, plants must be extremely deep rooted, or they are ephemeral. Sand dunes and desert washes typify unstable soils. Mesquites and Desert Willows are typical of deep-rooted plants that grow in either washes or sand dunes. Roots of mesquites may be more extensive than those of any other desert plants. They are commonly 30 ft (10 m) deep, and they have been recorded in mines as deep as 250 ft (80 m). Similar to other phreatophytes, mesquites and Desert Willows are winter-deciduous.

Phreatophytes are unable to conserve water. Many phreatophytes, such as Fremont Cottonwood, *Populus fremontii* subsp. *fremontii*, are not true desert plants. Cottonwoods actually avoid the true desert by growing only where there is abundant water. Typical of riparian plants in many parts of California, these trees are winter-deciduous, dropping their leaves in the cold season. It has been calculated that a 3-year-old Fremont Cottonwood loses through transpiration about 10.3 gal (38.9 L) per day. In comparison, 100 m² of drought-tolerant vegetation transpire only 2–3 gal of water (8–22 L) per day.

Ephemerals (Annuals)

Ephemeral or annual plants endure drought in dormant, seed form, which makes them particularly suited for survival in areas where precipitation is unpredictable. Ephemerals in deserts make up about 40% of the species, whereas in other habitats they seldom make up more than 15% of the species.

Old stands of Creosote Bush often are associated with an abundance of annuals. Little else grows with Creosote Bush in these areas. After it rains, the bare spaces between Creosote Bushes may become carpets of flowers.

Annual plants also grow in unstable soils. When wind or water moves the soil, the seeds move with it. When adequate water is available, water-soluble inhibitory chemicals are washed out of the seeds, and germination occurs. Following germination, growth is rapid; an ephemeral plant must complete its entire life cycle in one season. In California, almost all annual plants are adapted to germinate and grow during winter. Similar to alpine plants, they show a suite of characteristics designed to accelerate growth and optimize heat. Leaves grow flat on the ground in a basal rosette. Flowers and leaves track the sun. Photosynthetic rates are high. Brown-eyed Evening Primrose, *Chylismia claviformis*, a common annual plant of the Mojave and Colorado Deserts, has the highest photosynthetic rate ever measured.

In order to optimize the growing season, germination of ephemeral plants should occur early in the winter. If an inch (25 mm) or more of precipitation falls in late September or early October, inhibitors will be leached from the seeds, resulting in mass germination and a carpet of wildflowers for a spectacular spring show. In mid-October, 2015, a series of storms delivered over 3 in (75 mm) of rain to Death Valley, nearly a year's worth of precipitation in a few days. The immediate result was that roads were washed out and buildings around Scotty's Castle were flooded with mud. As a consequence, by December the desert in Death Valley was already experiencing profuse blooming, much earlier than normal. Usually the wildflower display is delayed until spring. If rain doesn't arrive until mid-December, then germination will require an inch (25 mm) of rain coupled with soil temperature greater than 50°F (10°C). Late rainfall must be coupled with warm temperature and sunny days to ensure that growth occurs rapidly, because the entire life cycle must be accomplished in a shorter time. Apparently, the seeds are also sensitive to temperature. One other opportunity for germination of winter annuals occurs if 2 in (50 mm) of precipitation occurs by early April. In the midst of a 5 year drought, heavy rains during March 1991 stimulated a spectacular bloom of some winter annuals. With abundant water, these winter annuals are able to complete an entire life cycle in an abbreviated growing season.

Precipitation during summer does not cause germination of winter annuals. Apparently, if the weather is too warm, germination will not occur; winter annuals cannot be "fooled" into germinating during the wrong season. Conversely, annuals that depend on summer precipitation cannot be "fooled" into germinating during winter. Most summer annuals grow very rapidly and complete their life cycles in only a few weeks. Summer annuals are not common in California, but they occur more frequently in the eastern parts of the Sonoran Desert, where summer rainfall is more predictable. An interesting location where both summer and winter annuals may be observed is in the new Sand to Snow National Monument on the east side of the San Bernardino Mountains near Pioneertown. One well-known weed, Barbwire Russian Thistle or Tumbleweed, *Salsola paulsenii* (= *S. tragus*), is a summer annual that has become well established in California deserts, primarily because agricultural fields are irrigated during the summer.

During a year with adequate precipitation, the number of ephemeral plants that bloom simultaneously is amazing. These annual species must compete for pollinators with each other, as well as with the perennial species that are also in bloom. Some species produce very large flowers; others produce a great number of flowers. Some species avoid competition by becoming associated particularly with certain pollinators. For example, tubular flowers that are red or yellow are frequently pollinated by hummingbirds. White flowers that bloom at night are often pollinated by moths.

The native Jimson Weed, *Datura discolor*, is a white-flowered ephemeral of the southeastern part of the Colorado Desert. It is one of some 25 species of *Datura* found throughout the world. Three other species, also known as Jimson Weeds, have been introduced to California. They are widely distributed in disturbed areas of the arid southwest, including Cismontane Southern California and the Great Central Valley. The nonnative species are perennials that die back each summer and return from root stock after spring rains. Jimson Weeds are in the nightshade family (Solanaceae). Characteristic of this family, they contain poisonous compounds—in this case, alkaloids such as atropine and scopolamine, which are sometimes called "belladonna" after a related plant, *Atropa belladonna*. The name belladonna means "beautiful lady." Apparently, European women used the sap to dilate the pupils of their eyes, a characteristic that was supposed to make them more beautiful. These chemicals are also found in some popular decongestants. These alkaloids were also used in the puberty ceremonies of several North American Indian tribes, in which

consumption of a Jimson Weed drink caused hallucinations and finally unconsciousness. The name Jimson Weed is a corruption of "Jamestown Weed," a reference to an episode that occurred in 1676 when British soldiers in Jamestown, Virginia, become intoxicated by eating a salad that contained *Datura* leaves.

Jimson Weeds have large, white, fragrant, trumpet-shaped flowers that open in the evening. They are pollinated by hawk moths or sphinx moths (*Manduca* spp.) (figure 9.38) that have remarkable proboscides (tongues) over 4 in (10 cm) in length. Apparently, the moths become addicted to the alkaloids in the nectar, which keeps them returning to the flowers for another "fix," thereby accomplishing pollination. This is another example of the interdependence of plant species and their pollinators. Furthermore, the larvae of the moths, known as tomato or tobacco hornworms, may feed on a variety of other poisonous species in the nightshade family, such as tobacco. In this way, larvae are rendered poisonous or distasteful to predators. The caterpillars of hawkmoths do not have to feed on poisonous plants. During periods of extensive germination of annual wildflowers, thousands of the larvae appear, and may denude many annuals of their leaves.

An interesting example of competition between ephemeral species occurs where the range of the Ghost Flower, *Mohavea confertiflora*, overlaps that of the Sand Blazing Star, *Mentzelia involucrata* (figure 9.14). The Ghost Flower may be considered a mimic. Its cream-colored blossom with purple dots in the center is similar in coloration to the Sand Blazing Star, but the Ghost Flower produces no nectar to reward the pollinator, a bee of the genus *Xeralictus*. The bees innately recognize the Sand Blazing Star as a rich source of nectar. Most of the time, they visit the Sand Blazing Star rather than the Ghost Flower, but when the bees are abundant and competition occurs for nectar, extra male bees visit the Ghost Flower. The Ghost Flower is thus pollinated without investing energy in the production of nectar. In this way, the Ghost Flower might be considered a pollinator parasite.

FIGURE 9.14 Examples of mimicry and pollinator parasitism.

A. The Sand Blazing Star, *Mentzelia involucrata*, is a rich source of nectar.

B. The Ghost Flower, *Mohavea confertiflora*, produces no nectar. Instead, it relies on its resemblance to the Sand Blazing Star, near which it grows, to attract pollinators.

Geophytes

The word geophyte means "earth-plant." Geophytes are plants that shed their aboveground parts during the dry season. This is a strategy that is intermediate between drought-deciduous and ephemeral plants, but it accomplishes a similar end: the plants are dormant during the dry season. Dormancy is maintained by underground organs, modified roots, or stems known variously as bulbs, tubers, corms, or rhizomes. Geophytes tend to occur in sandy, relatively stable soils where precipitation is predictable. If an inch (25 mm) of precipitation occurs early in the winter, these plants will have adequate moisture and time to produce aboveground leaves, stems, and flowers. Included among the desert geophytes are a variety of lilies, wild cucumbers, and the non-native Jimson Weeds mentioned earlier.

C_4 Photosynthesis

Three variations in the process of photosynthesis convey certain advantages to desert planes. In the type of photosynthesis that occurs in most planes, CO_2 is absorbed in such a way that the first chemical compound produced has three carbon atoms. This is known as C_3 photosynthesis. A disadvantage to this type of photosynthesis is that above 77°F (25°C) the rate of photosynthesis actually declines. One of the problems is that the enzyme controlling the process becomes inhibited by oxygen. As more photosynthesis occurs, more oxygen is produced, and the excess oxygen feeds back negatively to inhibit the reaction.

One variation of the process is known as C_4 photosynthesis. This type of photosynthesis is characteristic of many grasses, saltbushes, and summer annuals. In this variant, the first chemical compound produced has four carbon atoms instead of three, and the process is not inhibited by oxygen. It functions in low concentrations of CO_2, and it takes place at a higher temperature than the C_3 process. The ability to function at low concentrations of CO_2 means that the plant can continue to photosynthesize with its stomates nearly closed. Stomates are small pores in the leaf surface surrounded by specialized cells that allow CO_2 to enter. Water also leaves the plant through these pores. During hot weather, plants close their stomates to minimize water loss; however, this decreases the amount of CO_2 that can enter, and it traps oxygen inside the leaf. The C_4 process takes place in the outer cells of the leaves. The C_4 chemical is then passed to the cells that surround the veins of the leaves, where the standard C_3 process takes place as the carbon is donated from the C_4 chemical rather than directly from CO_2. In this way, photosynthesis operates more efficiently and at a higher temperature, to the advantage of a desert plant.

The third variation in the photosynthetic process is actually a variant of C_4 photosynthesis. It is known as crassulacean acid metabolism, or CAM photosynthesis. This process takes place in succulent plants and is named for a family of plants, Crassulaceae, that includes many leaf succulents, such as live-forevers. In CAM photosynthesis, CO_2 is fixed into a C_4 compound at night and the light-dependent C_3 process takes place by day. This photosynthetic mechanism requires more energy than either the C_3 or the C_4 process, but it enables the plant to absorb CO_2 at night and store it for use in the daytime. The advantage to this is that the plant can close its stomates during the day in order to reduce water loss. In effect, succulent plants carry on photosynthesis using stored water and stored carbon dioxide. The energy demand, however, dictates that CAM plants have slow growth rates.

BIOTIC ZONATION

Desert zonation (figures 9.15–9.17) is less a function of elevation than of soils and latitude. The influence of soils is most important locally, causing distinctive bands of vegetation. Latitude influences zonation due to its effect on temperature and day length. Northern latitudes are colder and receive more precipitation as snow; therefore, northern communities may appear at higher elevations toward the south. Growing season is postponed until spring and early summer. Very little precipitation occurs during summer, and summer photoperiods are very long during the time of greatest drought stress.

Creosote Bush Scrub

Creosote Bush, *Larrea tridentata* (= *L. divaricata*; figures 9.18A; 9.19; plate [15B]), is the dominant shrub of the Mojave and Colorado Deserts. In total, Creosote Bush Scrub covers more land than any other community in California—over 21 million acres (12,000,000 ha). Creosote Bush is most common on well-drained soils of bajadas and flats. Usually it occurs in association with White Bur-sage or Burro Bush, *Ambrosia dumosa* (figure 9.18B). Creosote Bush has dark green, resinous leaves

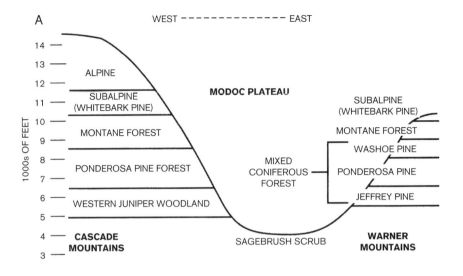

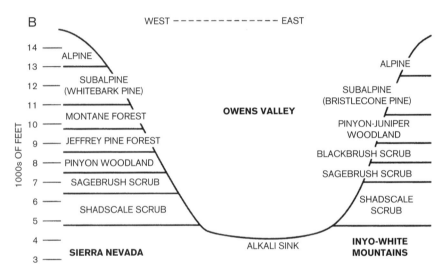

FIGURE 9.15 Biotic zonation of the Great Basin in California.

A. Modoc Plateau.

B. Northern Owens Valley.

borne on wand-like stems. Bur-sage is a short, gray, densely branched shrub with small, lobed leaves. The association of these two shrubs occurs in varying proportions, the reasons for which are not clear. It may be a function of age: Creosote Bush outlives Bur-sage. Or it may be a function of precipitation: Creosote Bush can tolerate drier soil. Another factor involves chemicals produced by the roots of Creosote Bush that inhibit growth of Bur-sage roots, a phenomenon that has been suspected but was proved only in 1991.

Creosote Bush also occurs in the Chihuahuan Desert of southern Texas and northern Mexico, and there are similar species in the deserts of South America. It appears that the dispersal of Creosote Bush has been from the south, and some evolution has occurred as it dispersed. Studies of chromosome numbers show that the Chihuahuan Desert forms have a

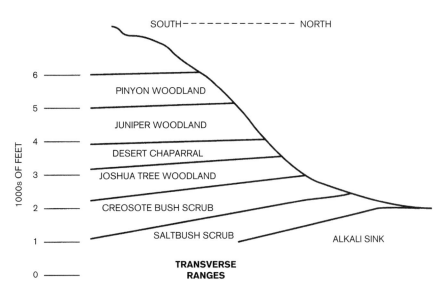

FIGURE 9.16 Biotic zonation of the Mojave Desert.

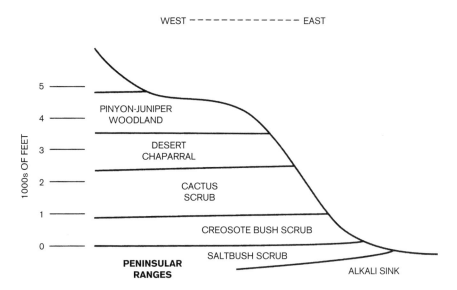

FIGURE 9.17 Biotic zonation of the Colorado Desert.

diploid (2N) number of 26. Specimens from the Sonoran Desert have 52 chromosomes (4N), and those in the Mojave have 78 (6N). This multiplication of chromosome numbers, known as polyploidy, is a common mechanism of speciation among plants. In the case of Creosote Bush, it has caused different races, but apparently they are not yet separate species.

Creosote Bush is the "champion" desert plant. It can tolerate drought stress in the form of negative water potential (pressure) more than any other shrub. Water potential is used as a measurement of water stress. It is measured in bars, with 1 bar equivalent to the pressure of the atmosphere, about 15 lb/in^2 (1.08 kg/cm^2). Typical agricultural or garden plants begin to

FIGURE 9.18 Dominant plants of Creosote Bush Scrub.

A. Creosote Bush, *Larrea tridentata*.

B. White Bur-sage, *Ambrosia dumosa*.

wilt if the negative water potential of their tis-sues exceeds 10–15 bars. Creosote Bush func-tions very well at −50 bars and has been found to function at −120 bars. This is roughly equiv-alent to a negative pressure of 1800 lb/in² (288 kg/cm²). The critical water potential, however, is about −70 bars. Below that, cellular respira-tion exceeds photosynthesis, and the plant can-not long endure.

Cell division, the addition of new cells to growing shoots, can continue to take place while Creosote Bush is under water stress. Directly after it rains, the new cells absorb water and elongate, giving the appearance of rapid growth. Lengthening of stems up to 1.25 in (3 cm) has been observed after it rains. It has been reported that the water potential must be above −24 to −28 bars for this stem elongation, as well as flowering, to take place.

The regular spacing of Creosote Bush on broad, flat terrain is remarkable; plants seem to be spaced with orchard-like regularity. It was once assumed that the spacing was due to a water-soluble chemical inhibitor produced by the leaves or roots that prevented germination of Creosote Bush seeds in the vicinity of a mature plant. In well-watered areas, plants are taller and grow closer together. In spite of evi-dence that Creosote Bush roots inhibit those of Bur-sage, no evidence has been developed to support the presence of an inhibitor that works between individual Creosote Bushes. At the present time, it is supposed that Creosote Bush is so efficient at absorbing water that germina-tion of seeds is prohibited within the scope of a mature root system. The regular interval of spacing represents the reach of root systems of adjacent plants. Because water is distributed equally over these desert flats, the shrubs pre-sumably all grow at the same rate. Thus, they are equally spaced. Because Creosote Bush is long-lived, these areas of regular spacing seem to represent even-aged stands.

Germination of Creosote Bush seeds, how-ever, is not as difficult as the paucity of Creo-sote Bush seedlings may make it seem. Appar-ently, seeds germinate abundantly following winter rains, but they die very early of water stress. This stress is accentuated if the seeds germinate within the scope of the root system of a mature plant. Perhaps more critical than that, however, is the effect of heat close to the ground. Heat accelerates water stress, and the temperature at ground level in the summer is near 160°F (71°C). Before its root system becomes established, Creosote Bush is unable to endure that kind of stress. Apparently, in nature young Creosote Bushes can only become established if there is a period of 3–5 years of cool, moist weather. Because this condition does not often occur, we do not find many young Creosote Bushes.

The age of Creosote Bushes has generated a bit of controversy. Aerial photographs show that in old stands of Creosote Bush, circles of shrubs are apparent (figure 9.19). Subsequent study shows that the bushes involved in a circle con-tain the same genetic material; that is, they are clones. There may be two to nine shrubs in these rings, all "identical twins." The circles can be explained by the habit of Creosote Bush to gener-ate new stems from an underground crown. New stems are produced from the outside of the crown, and the oldest stems are in the center. Eventually the central stems die, but the shrub continues to expand until the living stems form a ring. As this ring enlarges, it breaks up into bunches of stems, each representing a different plant. Actually, the different components of the circle originated from a single root system. The original plant had a taproot, but the different members of the clone do not have taproots.

In order to determine how rapidly these cir-cles enlarge, ¹⁴C dates were established for bits of wood found within the circle and cross-refer-enced with annual growth rings on the stems. It was found that the circles of shrubs in a clone enlarge at very slow rates. Many Creosote Bush clones about 20 ft (6 m) across were found. These were estimated to be about 3000 years of age. The oldest clone in the Mojave is located in Johnson Valley near the town of Lucerne Valley. It is 67 ft (22 m) across and has been named King Clone. It has been dated at approximately

FIGURE 9.19 A circle (clone) of Creosote Bushes.

11,700 years of age. Another clone in the Sonoran Desert near Yuma in Arizona has been dated at 18,000 years of age. If we consider that the different shrubs in the clone actually represent one living plant, then these shrubs are by far the oldest living things. They germinated during the wet years following the last Ice Age and have been living ever since. In other words, they are as old as the desert itself! To protect King Clone and its associates, The Nature Conservancy has purchased the land around it, and the area shall be maintained as a preserve.

The question of whether or not succession takes place in deserts is another controversy that can be answered in part by studying Creosote Bush Scrub at different localities. In sites that have disturbances of known age, such as areas disturbed by mining activities, there is an assemblage of plants different from mature stands. Roads at the Skidoo townsite in Death Valley have not returned fully after five decades, and tank tracks where George Patton's army practiced are still clearly visible in the eastern Colorado Desert after four decades. In areas of historic disturbance, there is a greater abundance of plants such as Desert Trumpet, *Eriogonum inflatum*; Spiny Senna, *Senna armata*; and Bur-sage. In the Colorado Desert, Teddy-bear Cholla, *Cylindropuntia bigelovii*, seems to be

an early successional species. Young Creosote Bushes are discernible because they have a single central stem. As a stand ages, the number of associated species declines, and Bur-sage and Creosote Bush remain as the dominants. In very old stands, the abundance of Bur-sage declines. Annual plants seem to be more common in old stands of nearly pure Creosote Bush, particularly where precipitation is unpredictable. Apparently, succession does occur in deserts, but the rate is so slow that it is difficult to observe within a human lifespan. Studies encompassing a 72 year period in the Sonoran Desert near Tucson show little change other than those associated with variations in precipitation.

Early successional species usually are not long-lived. They have good dispersal ability, with many possessing plumed seeds that are carried by the wind. Creosote Bush and Bur-sage are dispersed by animals. They have fruits that adhere to fur or feathers. Creosote Bush has white, hairy fruits, and Bur-sage has fruits with small spines (figure 9.18B); its name refers to the bur-like fruit. Bur-sage is in the sunflower family, but it produces inconspicuous flowers. Flowers of different sexes are borne on the plant separately, and they are wind-pollinated.

Creosote Bush produces an abundance of bright yellow flowers after the first heavy rains (figure 9.18A). Approximately 1 in (25 mm) of rain is required. Pollinators include various insects. Once the flowers are pollinated, the petals make a quarter turn so that the flowers are far less conspicuous. In this way, pollination efficiency is optimized.

Creosote Bush and Bur-sage have small leaves, typical of evergreen plants in desert areas. The high surface-volume ratio of small leaves is believed to enhance cooling by radiation without increasing water loss. Evaporative water loss is reduced by resinous or waxy coatings. Creosote Bush is renowned for the odor of its resinous coating, which is particularly pungent during and after rains, giving the entire desert a distinctive odor. As the dry season progresses, plants drop their leaves a few at a time. Bur-sage loses its leaves faster than Creosote Bush, leading some authorities to consider it drought-deciduous. Apparently, however, if all leaves are lost, the plant will not recover.

The amount of detritus from leaf fall that accumulates around the base of these shrubs is significant. A whole community of detritus feeders, including beetles and millipedes, live beneath these shrubs. In addition, when the wind blows, more organic and inorganic material is caught by the shrubs. A mound of sand may be found at the base of every shrub, and numerous seeds are brought in by the wind. Many desert rodents, such as pocket mice and kangaroo rats, make their burrows in the soil at the base of these plants and feed on the seeds that collect there. The canopy of the shrub helps to conceal foraging animals from predators. Efficient pocket mice may never have to leave the protective canopy of the Creosote Bush under which they have their burrows.

Sagebrush Scrub

Sagebrush Scrub is the dominant community of the Great Basin (plate [15A]). Some authors refer to it as Sagebrush Steppe. Great Basin Sagebrush, *Artemisia tridentata* (figure 9.20A), is the dominant shrub, although it is often joined by Bitterbrush, *Purshia tridentata* (figure

FIGURE 9.20 Dominant plants of Sagebrush Scrub. Note that both of these species have leaves with three teeth at the apex.

A. Great Basin Sagebrush, *Artemisia tridentata.*

B. Bitterbrush, *Purshia tridentata.*

9.20B), on coarse soils. In its pristine condition, this community also includes perennial bunchgrasses such as Crested Wheat Grass (*Agropyron* spp.) and fescues (*Festuca* spp.). Other species of Artemisia may also occur as subdominants. In recent years, overgrazing has altered sagebrush communities significantly. Species composition has shifted toward a greater proportion of sagebrush and less Bitterbrush, and the replacement of perennial bunchgrasses with introduced annual grasses.

Great Basin Sagebrush is also known as Big Basin Sagebrush. This plant and Bitterbrush have elongated leaves with three rounded teeth at the tip. The term tridentata of the scientific name refers to these three teeth. Great Basin Sagebrush is whitish gray, and Bitterbrush is dark green. Native browsers such as Pronghorn, Bison, and Bighorn Sheep preferred Bitterbrush, but their densities were never sufficient to deplete one of the species excessively. With the introduction of cattle and sheep, however, overgrazing has become a reality. In some areas, Great Basin Sagebrush is the only species left, and it has been pruned severely.

Overgrazing also opens the way for encroachment by introduced weedy species, such as Russian Thistle, *Salsola paulsenii* (= *S. tragus*), also called the Tumbleweed. This extremely hardy member of the Goosefoot family is a summer annual. Its life cycle in its native habitat in Asia is keyed to summer precipitation, but it survives very well in this country, particularly where irrigation of crops provides a source of summer water. Its incredible mechanism for distribution includes being tossed and tumbled by the wind; during its travel, hundreds of seeds are tossed about the landscape. In the early 1870s, Russian Thistle was introduced accidentally to central Canada by seeds that were mixed in a shipment of winter wheat. Within 25 years, it rolled and tumbled all the way to Mexico, where it thrives on summer precipitation!

In the past, fire may have been an important factor in altering species composition. Succession after fire caused a shift in species composition, favoring Bitterbrush and perennial grasses that were able to resprout from undamaged rootstock. Great Basin Sagebrush does not resprout after a fire, so it is effectively eliminated until seedlings can become reestablished, which takes about 20 years. One successional species that is common after fires or on other disturbed sites is Rubber Rabbitbrush, *Ericameria nauseosa*. Rabbitbrush is also known as Mojave Rubberbrush, a name that refers to the high rubber content of its herbage. It has been determined that if the price of rubber were to rise high enough, it would be economically feasible to harvest this species or actually cultivate it. Rabbitbrush may be nearly leafless, but when leaves are present, they are fine and resin covered. Its flowers bloom in autumn, taking advantage of pollinators during a time when very few other plants are in bloom. The bright yellow flowers of Rabbitbrush are a common sight along roads and other disturbed areas throughout the Great Basin. Rabbitbrush is another member of the sunflower family that produces wind-distributed, plumed seeds. It is a good pioneer species, as are other members of the genus *Chrysothamnus*, which are also known as Rabbitbrush.

During May 1980, Mount Saint Helens erupted in the Washington Cascades. The ash fall blanketed large portions of the Great Basin to the east. Experiments in the area showed that the blanket of ash initially caused an increase of soil moisture by inhibiting evaporation, but later increased reflection from the white surface increased temperature and transpiration rates in the vegetative canopy. Great Basin Sagebrush was able to tolerate this stress by dropping up to 75% of its leaves. Coincident water conservation allowed the development of flower buds during the dry, hot summer so that autumn flowering and seed set occurred as usual. The implication of this study is that volcanism in the Cascades favored the growth of Great Basin Sagebrush over other species, and it may have been the extensive ash fall associated with eruptions of Mount Mazama, about 6700 years ago, that favored dispersal of Great Basin Sagebrush into the rain shadow of the Cascades.

Great Basin Sagebrush is ideally suited for the climate in the Great Basin. It tolerates cold weather. When snow melts in late winter, Great Basin Sagebrush is still unable to grow new shoots and roots. The cold temperature inhibits growth until late spring, and then all of the new growth must take place in a few weeks, before soil moisture is exhausted. It is not surprising that most of the growth goes into new roots rather than aboveground parts; the plants must optimize water absorption. It is this investment of productivity into underground biomass that makes this community particularly susceptible to overgrazing. Availability of moisture dictates whether or not the plants will flower. Another member of the sunflower family, Great Basin Sagebrush has flowers that are small and inconspicuous and are wind-pollinated.

One notable characteristic of sagebrush is its odor. Odoriferous chemicals often are produced by plants to inhibit germination of competing species, and to discourage herbivory, and in the case of Great Basin Sagebrush native species of herbivores seem not to browse on it. There is one animal, however, that can defoliate vast acreage. It is the larva of the moth, *Aroga websteri*. The ultimate significance of the moth is that by increasing the amount of fuel in the form of dead foliage, it makes the community more susceptible to fires. It is possible that the moth indirectly serves to rejuvenate old stands of sagebrush by precipitating fire, which recycles mineral nutrients.

Native ephemerals are not typical of this community. Delayed germination due to snow cover and cold temperature does not favor annual species, and allelopathy from leaf litter may inhibit germination. The growing season is likely too short and cool for native ephemerals to complete their life cycle. Introduced annual grasses aided by C_4 photosynthesis and irrigation are able to hang on in spite of the climate.

The Sagebrush Scrub community is not restricted in distribution to the Great Basin. It is also found at high-elevation sites southward to Baja California. At these sites, the community is restricted to coarse, well-drained soils,

usually in flat areas locally referred to as meadows. Kennedy Meadows in the southern Sierra Nevada is one such site, at about 7000 ft (2100 m) elevation. Another large area of Sagebrush Scrub occurs in Garner Valley, at about 5000 ft (1500 m) in the San Jacinto Mountains of southern California. In these southern California localities, Great Basin Sagebrush is joined by Antelope Brush, *P. t. glandulosa*, which is the southern California equivalent of Bitterbrush Brush. Bitterbrush and Antelope Brush are in the rose family (Rosaceae). In spring, they become covered with yellow flowers with many anthers, typical of this family.

On a large flat area near Baldwin Lake in the San Bernardino Mountains is one of the largest regions of Sagebrush Scrub in southern California. Also in this region is an edaphic community known as a pavement plain, which was discussed as an ecologic island in Chapter 8. Located here is an Alpine Fellfield type of habitat at about 6720 ft (2050 m) elevation where a pocket of Great Basin species reaches its southernmost distribution. The greatest concentration of endemics in the state occurs here. These low-growing species are alpine-like but of Great Basin affinities. Surrounding the pavement plain is the Sagebrush Scrub community. Also found in the area, reminiscent of eastern Oregon and the eastern Sierra Nevada, is Sierra Juniper, *Juniperus grandis*. Another common affiliate at these mountain localities is Jeffrey Pine.

Shadscale Scrub

The Shadscale Scrub community inhabits fine-grained alkaline soils of the Mojave and Great Basin Deserts. In some areas of the Mojave, it also occurs on rocky soil with rapid drainage. The community commonly is dominated by Shadscale, *Atriplex confertifolia*, but there may be other dominant shrubs. Also in the Goosefoot family (Chenopodiaceae) are Four-wing Saltbush, *A. canescens*, Allscale Saltbush, *A. polycarpa*, and Winter Fat, *Krascheninnikovia (= Eurotia) lanata*. The saltbushes may be joined by Budsage, *Artemisia spinescens*. These shrubs resemble each other. They are usually well rounded, compact, and seldom over 2 ft (60 cm) high. The leaves are almost white and mealy surfaced. Shadscale has oval leaves, Winter Fat has linear leaves, and Bud Sagebrush has leaves with five to seven small teeth at the tip. After it loses its flowers, the sagebrush is left with naked spiny stems. It has been reported that in Utah, because of its giant salinas, there is more Shadscale than Great Basin Sagebrush. Shadscale Scrub is more tolerant of dry conditions than Sagebrush Scrub.

The least alkaline and driest areas in the community are characterized by saltbushes such as Allscale Saltbush, and Four-wing Saltbush. These are both important forage species, and they resemble each other. Where they occur together, they are difficult to tell apart. They both have narrow, elongate leaves. Four-wing Saltbush, however, has distinctive fruits with four flat, scalelike wings. Allscale Saltbush occurs in all deserts, but Four-wing Saltbush is more typical of coarser, drier soils of the Mojave and Colorado Deserts. They commonly grow on sandy mounds at the edges of playas or salinas.

Similar to Sagebrush Scrub, in this community most of the growth goes into root systems. Although these plants have remarkably shallow roots, it has been reported that the ratio of root growth to shoot growth is 7:1 in Shadscale and 4:1 in Winter Fat. As much as one-fourth of the Shadscale root system is replaced annually; therefore, more than 75% of Shadscale productivity goes into its roots. It is suspected that by dropping a portion of the root system every year, the plants conserve energy by reducing the amount of tissue that carries on cellular respiration.

Saltbushes have salt glands on their leaves. They absorb salt and water from the alkaline soil, and then they excrete the salt upon the leaves, which is one of the reasons they have a silvery appearance. Excreting salt is an energy-demanding process, and saltbushes are able to cope with this energy demand because of their high-efficiency C_4 photosynthesis, which operates at high temperatures with the stomates closed. Tearing the leaf of a saltbush and mag-nifying the cut edge with a hand lens reveals that the chlorophyll is concentrated in the cells around the veins.

Another saltbush found on these fine-grained soils is Desert-holly, *A. hymenelytra*. Mentioned previously as one of the few shrubs on soils of volcanic origin, where they stand out as white, rounded shrubs against a background of black soil. Leaves of Desert-holly are oval and toothed, resembling miniature holly leaves.

This community provides important forage for browsing animals. During the dry season, kangaroo rats are also known to eat saltbush leaves because they contain a higher amount of water than many other plants. The fastidious kangaroo rats scrape the salt off the leaves with their teeth first, because ingesting salt would complicate absorption of water.

Alkali Sink

Alkali Sink is a moist, alkaline habitat. It occurs in the bottom of basins where water remains standing for long periods of time or where the subsurface terrain is moist. The largest Alkali Sink community is on the edge of the Salton Sea in the Colorado Desert, but the community is also common at places such as Death Valley, Saline Valley, Mono Lake, and the San Joaquin Valley. These areas are covered with a layer of silt when they dry. Ephemerals are unable to become established in such a habitat, but saltbushes, with their mechanisms for coping with salt, are able to colonize the otherwise hostile habitat.

Some authors divide this community into two separate units. In areas where salinity ranges from 0.2% to 0.7% (2000–7000 ppm), it is called Saltbush Scrub, and it resembles Shadscale Scrub. Where salinity is greater, ranging from 0.5% to 2.0% (5000–20,000 ppm), it is called Alkali Sink. Actually, one community grades into the other, with the soil becoming more alkaline toward the lowest portions of a basin. The community is typical of basins in all of our deserts, especially where the soil becomes wet, at least part of the year. The

Great Central Valley also has these salt bush communities.

As soil becomes more alkaline, plants increasingly have succulent leaves. These are still members of the Goosefoot family, however, and it is interesting that they also grow along the edges of estuaries and tidal flats at the shore of the ocean. A transitional species between slightly salty and very salty soils is a conspicuous, low, blue-gray plant known as Inkweed or Bush Seepweeed, *Suaeda nigra* (= *S. torreyana*). A related plant, California Seablite, *S. californica*, is a component of the coastal Salt Marsh community. The name Inkweed comes from the black ink that can be extracted from the herbage. Cahuilla Indians in the vicinity of the Salton Sea used this ink in their artwork. This plant can tolerate salinity up to 1% (10,000 ppm). In many playa areas, this is the last species encountered before the lake bed becomes bare. Next to Highway 395, on the edge of Owens Lake, a distinct band of blue-gray vegetation marks the region occupied by Inkweed.

The "champion" desert halophyte is Iodine Bush, *Allenrolfea occidentalis*. It resembles the pickleweeds (Salicornia spp.) that also occur in coastal Salt Marshes. One species, *Arthrocnemum subterminale*, may co-occur with Iodine Bush in the Alkali Sink community. Iodine Bush and pickleweeds are nearly leafless, but their stems are composed of a series of succulent sections separated by joints. The stems appear to be composed of a series of elongate green beads. These plants have deep taproots and absorb salt and water. By storing water in their tissues, these plants dilute the concentration of salts; thus, they become more and more succulent as they grow. Small sections at the tips of the stems eventually dry out, die, and fall off; in this way, salts are shed by the plant. These plants grow on the edge of water that is over 2% salt (20,000 ppm), where no other plants can grow.

Blackbrush Scrub

At higher elevations in the Great Basin and Mojave Deserts, there is a community of drought-deciduous and succulent plants called Blackbrush Scrub. This community occurs on coarse, rocky soils of upper bajadas where a significant portion of the precipitation falls as snow. In the Great Basin, Blackbrush Scrub occurs on slopes above the Great Basin Sagebrush community. In the Mojave, it lies above Creosote Bush Scrub. In many areas, this community is composed of nearly pure stands of Blackbrush, *Coleogyne ramosissima*. The name Blackbrush refers to the appearance of the community from a distance; an entire slope may be gray black in appearance. The color is on the leafless outer stems of the shrubs.

These are drought-deciduous shrubs, although some authors would categorize them as partially deciduous. Leaves are lost from the outside first, revealing many twiggy branches with blackish bark. In the center of the bush, small green leaves are retained long into the dry season. The outer envelope of twigs forms a "lathhouse" that helps to shade leaves in the center of the plant. The twigs also form a boundary layer of trapped air that helps insulate the center of the shrub from temperature extremes.

Blackbrush is in the rose family (Rosaceae). When in full leaf, it roughly resembles Bitterbrush, with which it may co-occur. Blackbrush flowers are yellow with many stamens, but the leaves are not toothed. It differs from other members of the rose family also in that its flowers have no petals. Instead, there are four yellow sepals that resemble petals. On most plants, sepals are green.

Many members of the rose family are drought-deciduous and fruit-bearing, and they may bear names that indicate how closely related they are to domestic fruit trees. They are in the same genus (*Prunus*) as the domestic fruit trees. In various desert areas, usually on rocky slopes, Blackbrush and Bitterbrush may be joined by shrubs such as Desert Apricot, *P. fremontii* (figure 8.29); Desert Peach, *P. andersonii*, and Desert Almond, *P. fasciculata* (figure 8.30). Desert Peach and Desert Apricot are noted for their showy flowers. Desert Peach occurs from the Mojave northward into the

FIGURE 9.21 Tent caterpillars, *Malacosoma* sp., often found in species of the rose family.

Great Basin. Desert Apricot and Desert Almond were discussed as components of the Desert Chaparral community of the Peninsular Ranges, but also occur with Joshua Trees and California Junipers in the Mojave.

One way of identifying many of these plants in the rose family is that they are commonly infested with webs of tent caterpillars, *Malacosoma* spp. (figure 9.21). These caterpillars share their body heat, which is generated by twitching. By remaining together, they are able to conserve heat energy trapped within the webbed tent. Although these caterpillars seem to be associated most commonly with plants of the rose family, they may also be found on the Plateau Gooseberry, *Ribes velutinum*. Studies on the defoliation process associated with these caterpillars indicate that they seldom inhabit the same bush for more than 3 years because there is a lowering of food quality with each successive year. Furthermore, the caterpillars apparently communicate from one bush to another with airborne chemicals, presumably about the food quality. After metamorphosis, the adults live only a few hours. During that time, they mate and select a new, healthy plant on which to lay their eggs. Thus, as the food quality of one species is degraded, they move to another species. This allows recovery of the original species.

When the caterpillars are in their tents, they are conspicuous to predators. Apparently, however, they are seldom attacked. Studies of tent caterpillars in the eastern United States show that they consume cyanide compounds from the plants. When they are attacked, they regurgitate a fluid containing the toxin, and this repels the predator. This is another example of the coevolution of plants and animals that use plant toxins as a defense mechanism.

In the Mojave, Blackbrush is joined on upper coarse slopes by a drought-deciduous shrub known as Bladder-sage or Paper-bag Bush, *Scutellaria mexicana*. This shrub is leafless much of the time, and it has sinewy green stems. Most notable are its rose-colored, papery inflated pods, which remain on the shrub for long periods of time. Wind or water can carry these seedpods for long distances. White-tailed Antelope Squirrels, *Ammospermophilus leucurus* (plate 18E) are known to climb in these shrubs and extract the seeds. Bladder-sage is a member of the mint family (Lamiaceae), although it lacks the square stems and odor of most mint species.

Among the succulent plants in this community are various mormon teas (*Ephedra* spp.).

FIGURE 9.22 Three species of *Yucca* growing in the Mojave National Preserve.

A mormon tea has succulent stems and no leaves. There are separate male and female plants, each bearing cone-like reproductive structures. The seeds produced are important forage for ground squirrels. The stems possess chlorophyll, tannins, and small amounts of ephedrine, a stimulant drug related to amphetamines and one of the components in certain decongestants. Commercial ephedrine, however, is extracted from an Asian member of the group. A tea may be brewed by boiling these stems in water. Because it may appear politically incorrect to refer to a plant as "Mormon" if it contains a stimulant drug, it has been proposed to change the common name. Ironically one of the names proposed was Squaw Tea. The name Desert Tea apparently has been accepted by many persons concerned about common names that don't offend.

Another interesting succulent-stemmed shrub is Desert Rue, *Thamnosma montana*. This aromatic shrub is a member of the rue family (Rutaceae), the same family that includes citrus fruit trees. Usually leafless, its green seems to bear small glands yielding an odorous oil that is a skin irritant. When fruits are produced in spring, the plants are conspicuous. The fruits are about the size of peas and resemble miniature oranges. They are yellow green and bear many of the oil-producing glands. Native Americans claimed a number of medicinal uses for the oil.

In the Mojave Desert, the most conspicuous succulent plants of Blackbrush Scrub are three species of yuccas (figure 9.22). The Banana Yucca or Spanish Bayonet, *Yucca baccata* var. *baccata*, and the Mojave Yucca, *Y. schidigera*, are look-alikes. These are large leaf-succulent plants, with leaves often over 2 ft (60 cm) long. Mojave Yuccas are larger and can appear tree-like, with a large trunk capped by many succulent leaves. Banana Yucca gets its name from a very large fruit that was roasted and eaten by Native Americans. The two yuccas resemble each other, but where they grow together it can be seen that they are different in color. Banana Yucca is blue green, and Mojave Yucca is yellow green. Banana Yucca occurs in the eastern Mojave, and Mojave Yucca occurs on upper bajadas and rocky slopes all over the Mojave. In the Peninsular Ranges, Mojave Yucca becomes associated with pinyon pines and junipers on the desert sides of the ranges. These two species of yucca were very important to desert Native Americans. Not only did they eat the fruits, but they made a soap from the roots and used fibers from the leaves for baskets and cloth.

The third species of Yucca is the Joshua Tree, *Y. brevifolia*, which is discussed in some detail below in the section on Joshua Tree Woodland. In some localities, all three species occur together. One such location is on Cima Dome in Mojave National Preserve (figure 9.22).

FIGURE 9.23 Parry's Nolina, *Nolina parryi*, near Keys View in Joshua Tree National Park.

the Lily Family, they are now placed in the Butcher's-broom family (Ruscaceae). These large plants resemble bunchgrasses on a trunk. Some species are very yucca-like in appearance. When they are in full bloom, they may be the most beautiful of all desert plants. The flower stalks, like giant plumes covered with white flowers, are often twice as high as the rest of the plant. Parry's Nolina or Giant Nolina, *N. parryi* (= *N. wolfii*; figure 9.23), may have flower stalks 10 ft (3.5 m) high. They occur from the Dome-land Wilderness of the southern Sierra to the eastern slopes of the Laguna Mountains. They are particularly associated with the eastern Mojave Desert. Spectacular specimens may be seen in the Kingston Mountains or Joshua Tree National Park. Bigelow's Nolina, *N. bigelovii*, a smaller species, occurs primarily in the Colorado Desert at lower elevations than the other. Two of the species of *Nolina* occur in ecologic islands of the Peninsular Ranges and are discussed in Chapter 8.

When yuccas bloom, a high stalk of large white flowers projects above the leaves. White flowers are commonly pollinated at night, and the relationship between the night-flying yucca moths (*Tegeticula* spp.) and yuccas is often used as an example of mutualistic symbiosis. There is a different species of moth for each species of Yucca. Yucca moths have to pack pollen onto the stigma of yucca flowers to accomplish pollination. After that, the moth drills a small hole into the ovary of the flower, where it lays its eggs. Larvae develop inside the fruit. This is the only way that yuccas are pollinated, and this is the only place where yucca moths lay their eggs. Interestingly there is an apparent case of mimicry in the form of the Bogus Yucca Moth, *Prodoxus coloradensis*. This moth feeds on the nectar, but lays its eggs on other parts of the Joshua Tree where the larvae can feed, with no benefit to the tree.

An apparent case of convergent evolution with yuccas occurs in the beargrasses (*Nolina* spp.), although their evolutionary relationships have been subject to controversy. Formerly in

Joshua Tree Woodland

Joshua Tree Woodland is a defining community in the Mojave Desert. Distribution of the iconic trees outlines the Mojave. This community occupies the upper bajadas where snow is a common form of precipitation. There are many species in common between Blackbrush Scrub and Joshua Tree Woodland. Typical Joshua Tree habitat is well-drained loose gravel and sand, on the upper part of gentle slopes.

Joshua Trees (plate 4a; figure 9.24) form a conspicuous overstory in this community. Although they appear to dominate, in terms of actual biomass they are probably less important than some of the understory species. Joshua Trees are ecologically similar to other yucca species; for example, they are pollinated by yucca moths. They do not flower every year, but when they do, great masses of white flowers are produced. Flowering is regulated by an unknown combination of temperature and precipitation.

Germination of Joshua Tree seeds occurs in association with abundant winter precipitation,

FIGURE 9.24 Joshua Tree, *Yucca brevifolia*, Joshua Tree National Park.

but young Joshua Trees are usually gnawed off by rodents. Soil temperature is another important factor. Where soil temperature is too high the seedlings die out. It seems that the only Joshua Trees that escape are those that germinate under the protective cover of shrubs known as nurse plants, which include a variety of species. After 3–4 years, the Joshua Tree emerges from the canopy of its host and eventually replaces it.

A number of animals are known associates of Joshua Trees (figure 9.25). Yucca moths are obvious examples. Another associated insect is the Yucca Weevil, *Scyphophorus yuccae*, whose larvae kill the growing tip of a stem, causing it to branch. Flowering also causes branching. A young Joshua Tree will grow as a single trunk until either beetle larvae or flowering causes branching. After that, repeated branching is caused by the same stimuli, ultimately resulting in the largest species of flowering plant in the Mojave Desert. The largest of all Joshua Trees is located on the desert side of the San Bernardino Mountains. It is 32.3 ft (10.4 m) in height and has a crown diameter of 32.7 ft (10.5 m). Leaves of a Joshua Tree are smaller

than those of other yuccas. After the growing tip is killed, the leaves turn brown and fold backward, forming what looks like a small straw umbrella. The next year, one or two branches form below that point.

At least 25 different species of birds have been reported to use Joshua Trees for nesting. Conspicuous among these are woodpeckers such as Northern Flickers, *Colaptes auratus*, and Ladder-backed Woodpeckers, *Picoides scalaris*. There are few trees in the California desert, and Joshua Trees have the softest wood for making a nest. Woodpeckers eat termites, and termites are constant associates of Joshua Trees. Branches fall off, or strong winds blow down entire Joshua Trees, and termites live in and feed on this downed wood. A lizard known as the Desert Night Lizard, *Xantusia vigilis*, also feeds on these termites. These lizards are found under and within the fallen Joshua Tree, and they are also associated with other species of yucca. Most of this foraging occurs during daylight hours, but because the lizards rarely leave the cover of fallen boughs, it was once thought they were strictly nocturnal—hence the misnomer night lizard. Small snakes such

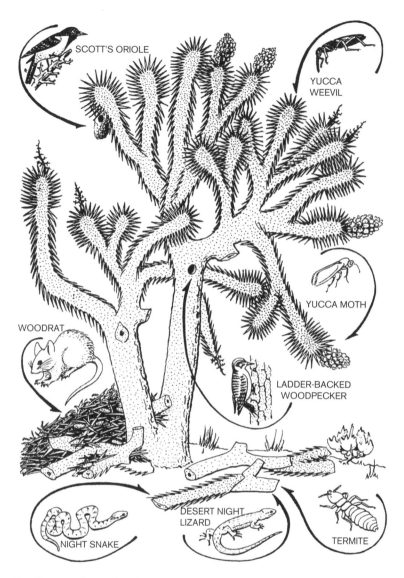

SCOTT'S ORIOLE

YUCCA WEEVIL

YUCCA MOTH

WOODRAT

LADDER-BACKED WOODPECKER

NIGHT SNAKE

DESERT NIGHT LIZARD

TERMITE

FIGURE 9.25 The small world of a Joshua Tree (from Bakker 1984).

as Desert Nightsnakes, *Hypsiglena chlorophaea* (= *H. torquata*), also live under these boughs, where they feed on the lizards and a variety of insects.

The western part of the Mojave Desert is known as the Antelope Valley. This area contains the towns of Lancaster and Palmdale, which were the fastest-growing cities in the state in the 1980s. The rapid rate of development here led to fears that Joshua Trees would be uprooted at an inordinate rate. Ordinances were passed requiring botanical reports to be prepared before desert vegetation could be destroyed. Requirements to replace Joshua Trees and other sensitive species were initiated, and a tree bank has been proposed where plants can be put into storage until they can be transplanted.

Cactus Scrub

The coarse soils of the upper bajada and rocky slopes of the Colorado Desert are inhabited by a different mix of drought-deciduous and succulent species (plate 4B, [16A]; figure 9.1). Just as Joshua Tree Woodland is truly a Mojave community, Cactus Scrub is truly a Colorado Desert community. Snow is rare in this warm desert community, and summer precipitation is significant.

The cactus family (Cactaceae) is American in origin. Succulent plants in Old World deserts resemble cacti, but these Old World plants are often in the Spurge family (Euphorbiaceae), plants characterized by milky juice. The common Christmas ornamental, Poinsettia, is an example of a euphorbia with leaves. There are native euphorbias in the New World, but there are no native cacti in the Old World.

Cacti are remnants of a tropical family. In the American tropics today, large cacti are common on dry sites within the forest. Presumably, millions of years ago, tropical forests occurred where our present deserts lie. With increasing aridity and advances of cold climate during the Ice Age, tropical forest retreated, leaving behind those species that were drought adapted. With the increasing aridity since the Pleistocene, cacti expanded their ranges from localized, warm dry refugia. The limit to northward distribution today is still cold climate. As one progresses northward or toward higher elevation, the number of species of cacti is reduced markedly.

Some authorities believe that the cactus family and the rose family are closely related. A group of primitive cacti from the Caribbean known variously as lemon vines or gooseberries (*Pereskia* spp.) show characteristics intermediate between the two families. These are long-stemmed drought-deciduous plants with leaves that arise from the stems along with spines. Superficially, they resemble Ocotillos. Rose flowers and cactus flowers are similar in that they possess many stamens. Rose flowers, however, usually have only five petals, whereas cactus flowers usually have many. The many-petaled rose flowers that adorn suburban gardens are products of horticultural tampering, and they rarely occur wild in nature.

Cactus flowers seem disproportionately large and colorful for the plant on which they grow. This illustrates an adaptation for dry climate, where precipitation is not predictable. When there is adequate rainfall, desert plants invest a major portion of their productivity on reproduction. They produce lots of flowers, and flowers are large. With all the flowers competing for pollinators, bright colors and large size are important. Very few desert flowers have odors, because odor requires moisture. The animal sense of smell requires a wet membrane for interpretation, and it apparently does not function well in dry air.

Cacti blossoms are stimulated to develop in the in the winter, in keeping with their tropical heritage. Tropical plants are short-day plants; they flower only during a short photoperiod. Physiologically, they are unable to flower during summer, when day length exceeds 14 hr. When cacti become sufficiently hydrated and day length is short, they produce buds which grow into flowers by early spring. This timing coincides with winter precipitation in our deserts. The number of flowers produced by any one cactus plant is proportional to the amount of water it receives.

Cacti show other adaptations that make them ideal for rocky soils and hot climate. The advantages to water storage and crassulacean acid metabolism (CAM photosynthesis) have already been discussed. A reduced leaf surface reduces potential water loss, but the true function of leaves that have been modified into spines is not clear. Perhaps the most useful contribution of the spines is a boundary layer of trapped air that helps to insulate the cactus from rapid temperature change. The spines also direct water toward the roots. The spines may also provide increased surface area to help dissipate heat, and they probably also produce a "lathhouse" effect, reducing the intensity of light that actually falls on the stem. Recent experiments in which spines were removed indicated that light on the stem was increased markedly. It also has been stated that spines discourage browsing by herbivores, which is probably true to some degree, although the concentration of metabolic acids associated with crassulacean acid metabolism makes cactus flesh unpalatable, if not noxious. Nevertheless, certain animals, such as woodrats, consume cactus as a water source. In times of drought, large mammals such as Bighorn Sheep and cattle will also eat cactus.

Other features that reduce light intensity on cactus plants are vertical orientation of the stem and irregularities of surface, such as lumps, ridges, and grooves. These features ensure that light strikes photosynthetic surfaces at an angle and that no surface receives direct light all of the time.

There are many species of cacti in Cactus Scrub. Common forms include barrels, hedgehogs, pincushions, chollas, prickly-pears, and beavertails. The largest of the Colorado Desert cacti is the California Barrel Cactus or Bisnaga, *Ferocactus cylindraceus* (plate 4B). It is also remarkable in terms of its heat tolerance. Experiments revealed that this species was able to tolerate a temperature of 154°F (68°C), highest recorded for any cactus in California. This form occurs in the rockiest of locations. Sometimes the only soil is found in cracks in rocks, and that is where the Barrel Cactus has its roots. The story of cutting the top off a Barrel Cactus and finding standing water within is an exaggeration. True, a considerable amount of water is stored in a Barrel Cactus, but it is contained within pulpy inner tissue. To obtain water from this tissue, a person would have to squeeze it vigorously, and the small amount produced would scarcely be palatable because of the metabolites from crassulacean acid metabolism.

Prickly-pear and beavertail are common names for a variety of cacti in the genus *Opuntia*. Prickly-pear and beavertail are flattened stem succulents. Their spines emerge in clusters from localized patches of tissue arranged symmetrically on the stem. There are two kinds of spines: long ones and short furry ones known as glochids. Glochids produce no pain when they stick to a person's skin, but there are so many of them and they are so small that they are difficult to remove. They can cause itching if they become caught in clothing.

Beavertail, *O. basilaris* var. *basilaris*, is a dominant species in the Cactus Scrub community, although it also grows at appropriate localities within Creosote Bush Scrub as far north as the Mojave Desert. This cactus is blue gray and often grows only one pad high. An obvious feature that distinguishes it from similar species of prickly-pear is that Beavertail has no long spines, only glochids. This is a favorite species for woodrats to gnaw on for a summer water source. The flowers on this species are a brilliant pinkish violet. In years with abundant rainfall, these small cacti are spectacular, with rows of large, brightly colored flowers. Some evidence indicates the number of blossoms in a given year is proportionate to the amount of precipitation.

Various species of prickly pear cactus may also occur in this community, although they are more common in the Mojave or at higher elevation. Prickly-pear cacti can tolerate more cold temperatures. These plants may be quite tall, with many stem joints. They are called prickly-pears because flower pollination is followed by production of a large red fruit that is full of sugar but covered with spines. Animals eat the fruit and then deposit the seeds at some other location in their feces. The sugar is the reward for the animal, and the bright color attracts the animal to eat the fruit. Birds and mice consume much of the fruit, but Coyotes eat the fruit whole. Coyotes are probably one of the most important dispersal agents for seeds of these cacti.

One species of this group is remarkably cold tolerant. This is the Old Man Cactus or Mojave Prickly-pear (Old Man Cactus), *O. polyacantha* var. *erinacea*. This species has very long spines that give it the appearance of having long gray hair. The abundant "hair" may provide the insulation that enables this species to survive being buried in the snow, and experiments show the spines help shade the stem. It occurs in snow-tolerant communities such as the Joshua Tree and Pinyon-Juniper woodlands as far north as the Great Basin.

Chollas (*Cylindropuntia* spp.) have spines that are very close together, representing the ultimate in the boundary layer and shading concepts. Teddy-bear Cholla, *C. bigelovii*, a dominant species in Cactus Scrub and a major indicator species for the Colorado Desert in California, appears downright furry. Its yellow thatch of spines dies at the base, giving the lower portion of the plant a blackish appearance. The joints of these cacti are fragile and easily broken. Consequently, on the ground in the area of Teddy-bear Cholla are many pieces of plants ready to "jump" on some passerby. These are the famous jumping cacti. Of course, they don't jump, but they do seem to find ways to become attached to animals, who carry them for some distance before they are able to knock them off. This is the way cholla gets dispersed. Each of the small sections is capable of growing a new set of roots, and thus the plant is able to colonize new areas. These chollas are considered pioneer species because they are particularly abundant in areas that have been disturbed. They produce small dry fruits that are usually sterile, so the species relies primarily on asexual reproduction from distribution of the stem sections that fall to the ground.

Spines on these chollas are very sharp and barbed (figure 9.26). They penetrate skin easily, but they are difficult to remove. It is a simple task for one of these fallen sections to become attached to a passing animal. The reason they seem to jump at humans is that a section of the plant is commonly picked up in the sole of one's shoe. When the person takes another step, the cactus becomes inadvertently planted in the back of the other leg. It certainly seems to have jumped there. Cholla spines also have a thin sheath around them. It is believed that this sheath is another example of a boundary layer, helping to keep the individual spines cooler and reduce water loss.

In spite of the nasty spines, some animals are able to use the cholla for their own interests. Desert Woodrats commonly gather cholla stems, which they heap into piles to form nests. The nest is in the center of the pile, and any animal attempting to dig out the woodrat is sure to experience a good deal of pain in the

FIGURE 9.26 Spines of Teddy-bear Cholla, *Cylindropuntia bigelovii*, can be a trap to species unaccustomed to them. This Clark's Nutcracker, a mountain bird, caught its feet on the barbed spines and could not escape.

process. Several species of birds also make their nests in the cholla. The Cactus Wren, for example, builds its nest of grasses within the protective arms of the cholla plant. The Cactus Wren nest is distinguishable from those of most other birds by the entrance, which is on the side of the nest.

There are also leaf-succulent species in Cactus Scrub. The most conspicuous is the Desert Agave or Century Plant, *Agave deserti*. The base of this plant is a rosette of thick, fleshy leaves. When these plants blossom, they send up a tall stalk that may grow as much as a foot (30 cm) a day. The large cream to yellow flowers are perfectly designed to be pollinated by nectar-feeding bats. In northern portions of the range of the Desert Agave, such bat species apparently no longer occur. Either they retreated southward with increasing aridity, or the plants have been carried out of their range by northward motion of the land west of the San Andreas fault. Other potential pollinators, such as moths, ants, hummingbirds, and Honey Bees, are able to extract

FIGURE 9.27 Female Carpenter Bee, *Xylocopa californica*.

the nectar without touching the anthers, that is, without accomplishing pollination. Throughout most of the Colorado Desert, therefore, the Desert Agave is left to reproduce primarily by asexual means, which it accomplishes by producing smaller plants in a circle about the base of the parent plant.

Desert Agaves and nectar-feeding bats have evolved together. With the absence of nectar-feeding bats in this area, Desert Agaves are now forced either to reproduce by clones, which could be an evolutionary dead end, or to evolve in association with a different pollinator. Evidence of this evolution can be seen in a population of Desert Agaves near Los Alamos in northern Baja California. Here large-bodied California Carpenter Bees, *Xylocopa californica* (figure 9.27), contact the anthers when they gather nectar, so some pollination takes place. Pollinations occur more frequently if anthers and stigma are shortened, so shorter reproductive parts are favored. As might be expected, anthers and stigma are significantly shorter in the flowers near Los Alamos, and as time passes, they should become shorter as the efficiency of pollination improves.

Desert Agaves are called Century Plants because they grow for many years before producing flower stalks. Contrary to popular belief, they do not wait 100 years before produc-

ing blossoms. Rather, evidence shows the interval to be about 20–25 years. During this time, water and carbohydrates are stored in the leaves. The process of flower stalk growth exhausts the entire supply of water and stored food. One scientist calculated that it takes about 40 lb (18 kg) of water to grow a flowering Agave stalk. After it flowers, the plant dies, leaving the clone of small plants at its base. Of course, if the plant has succeeded in producing seed, it has not died in vain; the expenditure of energy was worth it. In the northern part of this population, pollination rarely occurs, and the plant's death could be considered pointless. The plant has succeeded in reproducing vegetatively, however, and in time may evolve to accommodate a different pollinator, perhaps the California Carpenter Bee.

As in Joshua Trees, the rare seedlings of Desert Agave require a nurse plant. In this case, the seeds that germinate are usually under a desert bunchgrass known as Big Galleta, *Hilaria rigida*. It has been determined that Galleta Grass provides necessary shade and increased soil nitrogen for the Desert Agave seedlings.

A shrub of rocky slopes that has semisucculent leaves is Jojoba, *Simmondsia chinensis* (figure 9.28). It enjoys patchy distribution from around Twentynine Palms southward to Arizona and Mexico, pretty well outlining the

FIGURE 9.28 Jojoba, *Simmondsia chinensis.*

Sonoran Desert. It also occurs west of the crest of the Peninsular Ranges in dry Coastal Sage Scrub and Desert Chaparral. Leaves of this shrub are covered with a mealy, waxy surface and are oriented vertically, presumably to avoid the direct rays of the sun. Furthermore, the leaves track the sun. Depending on the temperature, they turn the flat surfaces away from or toward the sun's rays.

The fruits of these shrubs are borne on female plants. The fruits are about an inch (25 mm) long and contain about 50% high-quality oil, similar to Sperm Whale (*Physeter catodon*) oil. These fruits have been gathered for many years from wild plants for the purpose of extracting the oil. This enterprise was a major source of income for Native Americans in Arizona, and when tax incentives were available it was hoped by entrepreneurs and environmentalists that large-scale use of Jojoba oil would replace Sperm Whale oil. The main problem became that, typical of most desert agriculture, in order to optimize growth, the plants had to be irrigated. Also, it took a number of years before the plants produced fruit, and only female plants produced fruit. When the subsidy was discontinued, the investment was unable to produce a return sufficient to warrant the expense of leveling, road building, and irrigation. Also, if too much Jojoba oil were produced, the price would drop, and it would be economically unfeasible to continue irrigating the shrubs. Abandoned Jojoba plantations are still visible near Desert Center.

Two common drought-deciduous species are components of the Cactus Scrub community. These are Desert Brittlebush, *Encelia farinosa* (figure 9.29A), and Ocotillo, *Fouquieria splendens* (plate 4A, figure 9.29B). Desert Brittlebush is in the sunflower family (Asteraceae). When it gets adequate water, it produces a large canopy of bright yellow sunflowers at the ends of long stems. These flowers track the sun to optimize the temperature of reproductive parts. A person viewing these plants with the sun to his or her back will see every flower straight on, in a blaze of color. At the height of flowering, a broad band of bright yellow on a hillside testifies to the abundance of this plant in its preferred habitat. The bright white leaves are over an inch (25 mm) long, reasonably large for a desert plant. As the growing season progresses, smaller, hairier leaves are produced, and the larger, more water-consuming leaves are dropped. Finally, when conditions get too dry, all the leaves are dropped, and the plant goes dormant. In its summer state the plant is leafless, and the long dead flower stalks stick out all over the canopy. These stalks break easily, giving the plant its name.

Another name for this plant is Incienso. Orange crystals of resin exude from the stems, and early padres would burn this resin as an incense. Native Americans chewed the resin and smeared it warm upon their bodies to reduce pain. It was also melted and used as a varnish. This plant is a close relative of Coast Brittlebush, with which it hybridizes in low passes where their ranges overlap.

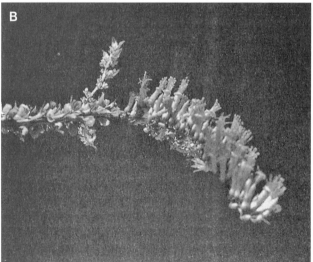

FIGURE 9.29 Drought-deciduous plants of the Cactus Scrub community.

A. Brittlebush, *Encelia farinosa*.

B. Ocotillo, *Fouquieria splendens* subsp. *splendens*.

Ocotillo is the symbol of the Sonoran Desert. In the Colorado Desert, it is distributed on rocky slopes or on gravels of the upper bajada. Bright red tubular flowers, perfect attractors of hummingbirds, occur in clusters at the end of tall, whiplike stems. When these flowers are in bloom, the plants form a band of red color on the hillsides, often just below the yellow band formed by the blooming Desert Brittlebushes.

Ocotillo is the "champion" drought-deciduous plant. It may grow and lose leaves up to seven times a year. Within 2 d after a good period of winter rain, leaf buds appear at the base of the thorns. One day later, the leaves are nearly 0.5 in (1 cm) in length. By the end of the fifth day, the oval leaves are an inch (25 mm) long and fully mature. Then the stems elongate in proportion to the amount of rain. In a good year, the stems may grow up to 3 ft (1 m). Examining the stems will indicate how much growth has been added each year, revealing a crude record of past years of precipitation. The stems also contain chlorophyll, indicating that photosynthesis may continue even when the stems are leafless. Furthermore, water is stored in the stems, meaning that Ocotillos seem to take advantage of two desert plant strategies, succulence and drought deciduousness.

On the new growth, leaves of a different shape are produced. Each leaf has a long, tapering stem (petiole) with the leaf blade folded backward at its tip. When the leaf blade is shed, the petioles become new spines. The final picture is of a plant with 25–30 long, spiny stems up to 15 ft (5 m) in length. Similar to the Desert Agave, they attract pollinators by growing their flowers high, where there is less competition.

Desert Wash

The Desert Wash or Microphyll Woodland consists mostly of deciduous, deep-rooted shrubs and trees. In the Mojave, these washes are outlined by thickets of Desert-willows, *Chilopsis linearis*. These are not true willows; they belong to a tropical family (Bignoniaceae) that includes many species of vines and trees with large, snapdragon-like flowers. Similar to cacti, these are relicts of a former time when precipitation in deserts was much greater. These trees are winter-deciduous. They are leafless during the rainy season, instead relying on water that remains deep within the sands of the wash during summer. Leaves are long and narrow when they are present. When in bloom, the tree is covered with pinkish flowers that resemble snapdragons. The flowers are similar to those of the Jacaranda, a popular ornamental tree in the same family. After blooming, the tree produces long seedpods that resemble beans; however, these are not legumes. In some areas of the Mojave, Desert Willows are joined by Honey Mesquite, *Prosopis glandulosa*, another deep-rooted, winter-deciduous species. Like Honey Mesquite, Desert Willows grow on sand dunes. For example, the trees growing partway up Kelso Dunes are Desert Willows.

In the Mojave and Colorado Deserts, Honey Mesquite, (figure 9.4) is also a component of Microphyll Woodland. This deep-rooted shrublike tree is a phreatophyte: it taps into permanent groundwater. It may also be associated with saltbushes. As discussed previously, where it grows, water is said to be no deeper than about 30 ft (10 m). These trees are often found on the edges of sand dunes, growing out of a huge mound of sand, with a base of many buried trunks (figure 9.30). There is often an abundance of dead wood beneath the mesquite canopy, and this wood is ideal for campfires, for it burns without smoke. It is also revered for its hot coals, which are ideal for broiling meat. A person wishing to gather mesquite for a campfire should keep in mind that the tree is winter-deciduous. What may appear dead may simply be leafless. The dead wood is found under the canopy, and a person gathering it is sure to become scratched and lacerated by thorns.

In the spring, the apparently dead branches come to life. Leaves are compound; each has a two-parted, wishbone-shaped stem (petiole) with many small leaflets. Mesquite is a legume

FIGURE 9.30 Honey Mesquite, *Prosopis glandulosa*, in mounds of sand near Stovepipe Wells, Death Valley National Park.

(Fabaceae), and its beans are an important staple for desert animals, including Coyotes. Native Californians depended heavily on Honey Mesquite beans for sustenance.

Desert Wash communities are most characteristic of the Colorado Desert, where they occur along the margins of arroyos, sandy washes, and sometimes occupy an entire bajada. Desert Willow also occurs in the washes, but it is joined by a number of trees in the Legume family (Fabaceae; figure 9.4). Included here are Honey Mesquite; Ironwood, *Olneya tesota*; Blue Palo Verde, *Parkinsonia florida*; and Smoke Tree, *Psorothamnus (= Dalea) spinosa*. In other parts of the Sonoran Desert, these trees are joined by large cacti such as Saguaros to form a community of drought-deciduous and succulent species that grow abundantly on the upper bajadas. Presumably this community is analogous to Cactus Scrub in the Colorado Desert, but in other parts of the Sonoran Desert, the community benefits from increased amounts of summer precipitation.

Foliage on these trees is similar in appearance. Leaves are pinnately compound, with a single petiole bounded by small leaflets arranged opposite each other. These small leaves are believed to be an adaptation to reduce dry-season maintenance costs. Reflecting slightly different microhabitats, the trees differ in deciduousness. Mesquite is winter-deciduous. Ironwood is evergreen, but it loses its leaves a few at a time during the dry season. Blue Palo Verde is drought-deciduous, and after the leaves fall, it continues to photosynthesize with chlorophyll in the bark. Smoke Trees are similar to Blue Palo Verdes but rarely have leaves.

Little-leaved Palo Verde, *P. microphylla*, is a small tree with limited distribution in California, although it is widely distributed in the Arizona portion of the Sonoran Desert. It is found where the Colorado Desert extends northward in eastern San Bernardino County, in the Whipple Mountains, west of the Colorado River. It is noteworthy that this area is also one of the few places in California where the giant Saguaro Cactus, *Carnegiea gigantea*, occurs naturally. Little-leaf Palo Verde differs from Blue Palo Verde by having a greater number of leaflets and slightly different flower, fruit, and seeds. Flowers of the Legume family are characterized by one petal, known as the banner petal, that stands upright and possesses special coloration that serves to attract insects. On Blue Palo Verde, this petal has small red dots. No such dots are found on Little-leaf Palo Verde.

FIGURE 9.31 Flower of the Blue Palo Verde, *Parkinsonia florida*, with banner petal folded down, indicating to pollinators that it has already been visited.

Furthermore, after a flower is pollinated, a change occurs that serves to signal potential pollinators that the flower has been visited. On Blue Palo Verde, the banner petal folds down and covers the throat of the flower (figure 9.31). On Little-leaf Palo Verde, the banner petal changes from yellow to white. Experiments have shown that this color change is triggered by a pH change in the flower that is activated by the act of pollination.

Seeds of the desert legumes are very important forage for animals. Native Americans also ate these seeds, which are tasty when green. Mature seeds are covered with an extremely hard coat and must be ground into a meal to be eaten. The value of this hard coat to the seed is that it prevents germination until it is cracked or broken, a process known as scarification. In nature, the mechanism that cracks the surface is rolling and tumbling of the seeds along with rocks and gravel when water rushes down washes. This ensures that seeds will not germinate unless there is enough water. Once the seeds germinate, a few leaves are produced. After that, most of the growth is in the roots, and at this time root growth is very rapid (figure 9.32). Mesquite roots can grow more than

4 in (10 cm) a day, about 10 times the rate of most plants. The green part of the plant remains small and close to the surface of the soil for several years, during which time the root system penetrates toward a potential water source. Once the roots are deep enough to provide adequate water, the upper part of the plant becomes enlarged. The deep roots of these trees serve two useful purposes: they provide for water absorption and they serve to anchor the plant in what may be a disturbed habitat. Cloudbursts can excavate the gravels of a Desert Wash, uprooting all plants but those with the deepest root systems.

Catclaw, *Senegalia greggii*, is a legume that grows on the edges of washes in the Mojave and Colorado Deserts. It is a shrub, not a tree, but otherwise it resembles the other legumes in the washes. Similar to mesquites, it is winter-deciduous. The name Catclaw refers to its dark, recurved thorns, which snag anything that brushes up against it. Some old-timers call it the "Wait-a-minute Bush." The tops of these bushes are commonly infested with Desert Mistletoe, *Phoradendron californicum*, the seeds for which are deposited there in the droppings of the Phainopepla, *Phainopepla nitens* (figure 9.33),

FIGURE 9.32 Roots of Smoke Tree seedlings.

FIGURE 9.33 Phainopepla, *Phainopepla nitens*.

a black, crested bird that perches commonly in the uppermost branches.

On the edges of Desert Washes is a community of drought-deciduous shrubs. These shrubs are able to become established on the shoulders of a wash that are disturbed by only the largest of storms. To a degree, these are successional species that are good at reseeding once a spot has been disturbed.

A common drought-deciduous shrub of the Mojave and Colorado Deserts is Cheese-bush, *Ambrosia salsola*. A successional species, it is one of the first to colonize new soil after a desert storm has reworked the soil of a wash. The resinous herbage of this shrub smells cheesy when crushed. Leaves, when present, are filamentous, resembling the stems. This is a member of the sunflower family, but it has scalelike fruit resembling a saltbush.

One of the most spectacular drought-deciduous shrubs occurring on the shoulders of a desert wash is Chuparosa, *Justicia (= Beloperone) californica* (figure 9.34). Chuparosa is the Spanish name for the hummingbird; it means "suck-rose." This shrub becomes covered with long bright red tubular flowers typical of those that are hummingbird pollinated. Birds such as Verdins (*Auriparus flaviceps*), House Finches (*Carpodacus mexicanus*), and Black-throated Sparrows (*Amphispiza bilineata*) bite off the flowers to get the nectar. Native Californians formerly ate the flowers. This is the only California member of a large family of ornamental tropical shrubs, the acanthus family (Acanthaceae).

Sand Dunes

Sand dunes require a source of sand and a prevailing wind. The source of sand could be a river or a playa. If wind blows consistently in the same direction over a playa, sand becomes deposited at its leeward end. The side of a dune toward the wind is inclined at about a 5° to 10° angle. Wind constantly carries sand from this slope to the top of the dune, where it slides down the other side. The slope out of the wind,

FIGURE 9.34 Chuparosa, *Justicia californica*, a hummingbird-pollinated flower.

known as the slip-face, is a good deal steeper, lying at a consistent 33° to 34° angle.

Major dune fields are found in each desert. These areas are ecologic islands characterized by endemic plants and animals. In California, the most spectacular dunes are the Eureka Dunes, Dumont Dunes, Kelso Dunes, and Algodones Dunes.

The Eureka Dunes are the northernmost. Eureka Valley lies between northern Owens Valley and northern Death Valley. The dune field is 3 mi (4.8 km) long and 1 mi (1.6 km) wide. These are the highest dunes in North America, reaching 680 ft (207 m) in height. There are several dune endemics, including three plants and five species of beetles. Some authorities call dune plants psammophytes, which means "sand-plants." Notable among them is the Eureka Valley Dune Grass, *Swallenia alexandrae*. Typical of dunegrasses, it keeps up with shifting substratum by spreading out with underground rhizomes. Perhaps the most spectacular endemic plant is the Eureka Dunes Evening-primrose, *Oenothera californica* subsp. *eurekensis* (figure 9.35), of the evening-primrose family (Onagraceae), so named because the

flowers open in the evening and remain open at night. Both of these plants are listed by the federal government as endangered, and in response the Bureau of Land Management has enforced a policy of no off-road vehicles in the area.

Geologists have determined that these dunes are older than 10,000 years, making them the oldest in North America. While most of the plants are indicative of the Mojave, the presence here of a large number of plant species with Great Basin affinities implies that the desert species remained in the valley during the Pleistocene. Another interesting thing about the Eureka Dunes is that no source area for the sand can be located. Old waterlines indicate that the valley once was filled with water, and there is a playa there today, but the sand particles in the dune have not been derived from the adjacent playa. The best guess is that very high winds during the ancient past carried the sand from a source to the north that today is outside the valley.

Kelso Dunes (figure 9.36), in the Mojave National Preserve, are located 35 mi (56 km) south of Soda Dry Lake, near Baker. This large

FIGURE 9.35 Endemic Eureka Dunes
Evening-primrose, *Oenothera californica*
subsp. *eurekensis*.

dune field also known as the Devil's Playground covers some 45 mi² (115 km²). Sediments carried by the Mojave River flowing toward Soda Lake are its source of sand that has been accumulating for some 25,000 years. These dunes rise 656 ft (200 m) above the surrounding desert floor, nearly as high as the Eureka Dunes. Kelso and Eureka dunes are the only two dune fields in California that "sing" when the sand is disturbed. Sometimes they are called booming dunes. This production of sound occurs when the etched surfaces of the sand grains rub on each other. The cause of the sound is not understood, but the moisture content of the sand must be less than 11% for the sound to occur.

Endemism in the Kelso Dunes is not as great as it is in the Eureka Dunes, but the field does harbor seven species of endemic insects, including the Kelso Dunes Giant Sand Treader, *Macrobaenetes kelsoensis* (a camel cricket); the Kelso Dunes Jerusalem Cricket, *Ammopelmatus kelsoensis*; and the Kelso Dunes Shieldback Katydid, *Eremopedes kelsoensis*. The Mohave Fringe-toed Lizard, *Uma scoparia*, is not endemic to the dunes, but it is rare everywhere else. A number of plants that are rare outside of dune fields are also found here.

FIGURE 9.36 Kelso Dunes, San Bernardino County.

The sensitive nature of the Eureka and Kelso dunes has prompted federal officials to close them to use by off-road vehicles. As an example of the effect of intensive use, the Dumont Dunes of the Amargosa Valley have little growing on them because they are open

FIGURE 9.37 Colorado Desert Wild Buckwheat, *Eriogonum deserticola*, with stem-like root exposed by shifting sand. Algodones Dunes, Imperial County.

for dune buggy use. On any weekend, hundreds to thousands of dune buggies and off-highway vehicles buzz around on these sands.

The Colorado Desert contains the largest continuous dune field, the Algodones Dunes. The Spanish word algodon means "cotton." These dunes get their sand from ancient Lake Cahuilla and the present-day Whitewater River. Winds blow southward, pushing these dunes farther into Baja California. The field is 40 mi (64 km) long and up to 6 mi (10 km) wide, and the dunes are as high as 300 ft (95 m). Portions of the dune field near Glamis are closed to off-road vehicles in order to protect the specialized plants and animals. This dune field contains the largest number of dune endemics in North America. There are six kinds of plants that occur nowhere else. For example, the interesting Colorado Desert Wild Buckwheat, *Eriogonum deserticola* (figure 9.37), grows here. This plant may grow up to 5 ft (160 cm) high. In this shifting sand, it continues to extend a single stem-like root in an attempt to keep the photosynthetic part of the plant above the sand. When the sand blows away, this sinuous, elongate root becomes exposed, arching up out of the sand 10–15 ft (3–5 m) away from the green part of the shrub. On some occasions, plants

such as Creosote Bush or Mormon Tea also take this form.

One of the strangest plants in the Algodones Dunes is a root parasite often associated with Colorado Desert Wild Buckwheat. This plant is called Sand Food, *Pholisma (= Ammobroma) sonorae*. Most of the plant lies beneath the sand, where it taps into roots up to 5 ft (1.5 m) down. The portion of the plant on the surface is a fleshy flower disc that resembles a partially buried bagel. Native Californians ate them raw or roasted. This nonphotosynthetic plant derives its entire sustenance from the plant it parasitizes. How seeds of this plant germinate is a mystery. The seeds are so small that they could not store enough energy for the stem of this plant to penetrate deeply enough to reach the root system of its host plant. The latest theory is that ants carry the seeds into their burrows, where they germinate in proximity to the roots they will parasitize. They may also germinate in this way in rodent burrows.

Because sand constantly shifts and dunes migrate, this is an unstable habitat, and certain ephemeral plants are typical. One of these is Sand Verbena, *Abronia villosa*. This member of the four o'clock family (Nyctaginaceae) is common along roads and in sand dunes. It is a

FIGURE 9.38 White-lined Sphinx Moth, *Hyles lineata*. This is the pollinator of many white flowers such as Jimson Weed.

trailing, viny species with brilliant rose-purple flowers borne in small, rounded clusters. This is one of the few desert species with an odor. Its sweet fragrance is characteristic of spring nights in the desert. A good place to see acres of Sand Verbena is in the dunes in Anza-Borrego Desert State Park after a good period of winter rain. At least 1 in (25 mm) of rain by late autumn is required for the best show of annual wildflowers.

The other common annual of sandy areas is Dune Evening Primrose, *O. deltoides*, which resembles the Eureka Valley Evening Primrose mentioned earlier. This prostrate, spreading annual has large white flowers with four petals. As is often the case with white flowers, these are pollinated by a moth, the White-lined Sphinx Moth, *Hyles lineata* (figure 9.38). This plant is another exception to the rule that desert flowers have no odor. Odors are more useful to plants in the desert at night, because this is the only time of day when the air is humid enough for the sense of smell to function well. Moths have a particularly good sense of smell, which helps them locate this flower at night.

Larvae of this moth are black, green, and yellow. They feed on leaves and flowers of the prim-rose, and when larvae are abundant, as occurred in 1991, most of the plants become defoliated. When this happened in the past, Native Americans would collect larvae in great numbers and eat them. When Dune Primrose plants become old, the stems have a peculiar way of rising upward and bending back to converge on themselves at the top. Plants that dry in this way form a sort of bird cage. This peculiar appearance leads to a number of its other common names. Devil's Lantern, Lion-in-a-Cage, and Basket Evening Primrose. At curio shops in the desert, these dried plants, known as bird cage plants, are available complete with an artificial bird.

Geophytes, plants that regrow from an underground storage organ, are also found in sandy areas. Desert Lily, *Hesperocallis undulata* (figure 9.39), is a large white lily common in the Mojave and Colorado Deserts. The bulbs are surprisingly deep, up to 24 in (60 cm) below the surface. Obviously, it takes a significant rainfall to reach one of these bulbs and stimulate growth. It is believed that these plants propagate primarily by seeds. A sphinx moth is one of the main pollinators. Native Californians ate the bulbs, which are reported to have an onion or garlic flavor. Sometimes

FIGURE 9.39 Desert Lily, *Hesperocallis undulata*, a geophyte.

this plant is called Ajo, which is the Spanish word for "garlic." A Desert Lily reserve is located on Highway 177 north of Desert Center.

A common member of the Gourd family (Cucurbitaceae) is a vine of sandy areas known as the Coyote Melon, *Cucurbita palmata*. This vine returns each year from a large bulbous root. Green to yellow gourds about the size of baseballs make these vines conspicuous. Native Californians ground and ate the seeds, which are quite nutritious. A similar vine, *C. foetidissima*, has a rank odor to its foliage. It is also known variously as Coyote Melon, Calabazilla, or Buffalo Gourd, and its noxious odor is the source of the Indian name Coyote Melon. According to the Indian story, Coyotes spoil anything that is good, and they urinated and defecated on the plant to give it a rank odor, hence leaving the nutritious fruit for themselves. There have been proposals to grow these plants commercially as a source of protein and cooking oil. The seeds average 31% protein and 30% oil.

Large dune fields are also water reservoirs. For this reason, it is not uncommon to find deep-rooted trees, usually associated with desert washes, growing on the dunes. Honey Mesquite (figure 9.30) and Desert Willow are found on dunes in the Mojave Desert. In the Colorado Desert, other desert wash species, such as Blue Paloverde, grow on dunes, too.

Some dune areas are characterized by specialized lizards that "swim" under the sand. These are the fringe-toed lizards (*Uma* spp.; figure 9.40A), and they occur in the Algodones Dunes, the Coachella Valley Dunes near Palm Springs, Iron Age Dunes east of Twentynine Palms, and Kelso Dunes. There are two other species in Mexico. Restricted to fine, wind-blown sand, these lizards dive headfirst in to the sand and wriggle out of sight. Adaptations to this habitat include elongated fringe scales on the toes, to facilitate running on the sand, and a number of specializations to keep sand out of body openings, such as valved nasal and ear openings, sand-trap nasal passages, a countersunk lower jaw, and overlapping eyelids.

In order to protect the endangered Coachella Fringe-toed Lizard, *U. inornata*, three tracts of land in the Coachella Valley, representing about 10% of the lizard's known habitat, are administered as preserves by a coalition of agencies including the Bureau of Land Management, the US Fish and Wildlife Service, the California Department of Fish and Wildlife, and The Nature Conservancy. The Willow Hole-Edom Hill Preserve contains 2500 acres (1000 ha), and the

FIGURE 9.40 Two species of sand-dwelling lizards.

A. Mojave Fringe-toed Lizard, *Uma scoparia.*

B. Zebra-tailed Lizard, *Callisaurus draconoides.*

Whitewater Floodplain Preserve contains 1230 acres (492 ha). The Coachella Valley Preserve, near Thousand Palms, is a 13,000 acre (5200 ha) tract of land that also includes one of the finest examples of a California Fan Palm oasis in the California desert (plate 16B). This oasis is also being used as a refuge for the endangered Desert Pupfish, *Cyprinodon macularius.*

Similar in appearance to the fringe-toed lizard is the Zebra-tailed Lizard, *Callisaurus draconoides* (figure 9.40B). Males have two dark bands on each side of the body just behind the armpits. During breeding season, the area around these bands becomes a bright blue green. Interestingly, this is one of the few species in which females take on a breeding color. They develop an orange throat patch and become orange on the sides behind the armpits. This color change in a female appears to be a signal that she already has mated and is designed not to attract a courting male. This lizard is most common in sandy and gravelly areas, but it also occurs on desert flats with a scattering of small rocks. It favors the sand of desert washes and the margins of sand dunes. Its most distinctive feature is the underside of its tail, which bears black bars in a zebra or gridiron pattern. When the lizard is at rest but alerted, it curls its tail forward to reveal the black-and-white markings. It seems to wag the tail nervously, which may function to distract a potential predator to the conspicuous markings. If the predator goes for the tail first, it can be sacrificed and later regrown. This is one of

California's fastest lizards. It has been clocked at 18 mph (29 kph), and it can run 40–50 yards at a time, stopping and veering abruptly. When running at high speed, it uses only its hind legs. It holds its forelimbs to its side and curls its tail over its back, apparently for balance, as it sprints over the sand. This is truly a remarkable sight.

Desert Pavement

The traditional understanding of desert pavement is that the wind blows away all the sand from an area and the larger pebbles and stones that remain form an armored surface preventing any more sand from being blown away. Evaporation of water from this surface leaves calcium carbonate, which can become a cement that holds the stones together, similar to icing on a cake. A new interpretation based on research at Cima Dome indicates that the stones have been there all along, and represent a trap that causes an accumulation of dust beneath the stones. Stones remain on the surface due to heave while windblown dust builds up the soil beneath the pavement.

The dark surface of a desert pavement must be one of the hottest habitats in the world and is a difficult substratum for plants. Long-lived shrubs are able to continue using water that percolates between the rocks. However, very little new establishment is possible. One interesting exception is a small annual plant in the buckwheat family (Polygonaceae) known as Devil's Spineflower, *Chorizanthe rigida* (figure 9.41). It is found in desert pavement of all California deserts, and sometimes it is the only thing growing. It must tolerate extremely high temperatures. When the plant comes up, it has a short, spiny stem with many oval leaves. After the plant dies, all that remains is a dry vertical mass of spines that may persist for years. Many people erroneously believe that this is a dead cactus. As it turns out, these dead plants retain seeds for many years releasing them when it rains. This form of seed retention or serotiny occurs in desert plants in many families, and is

FIGURE 9.41 Devil's Spineflower, *Chorizanthe rigida.*

A. As it appears with leaves, when it first grows.

B. As it appears in its dried state.

also characteristic of many fire-adapted species such as closed-cone pines.

Pavement plains are deceptively thin-skinned. A person walking upon the surface gets the illusion that it is as firm as blacktop. A vehicle such as a motorcycle, however, breaks through with little effort, and the tire track becomes an erosion channel. Wind and water carry away soil from the track. When it rains, water flows through the track and is carried away from the area, causing it to be drier than before. Plants may die as a result. It is difficult to understand what a fragile balance these desert ecosystems represent.

Riparian Woodland

Flowing streams enter the desert on the eastern or transmontane side of the Sierra Nevada, the Transverse Ranges, and the Peninsular Ranges. In addition, in various desert ranges there are springs with sufficient flow to produce permanent streams bordered by riparian vegetation. Furthermore, there are a few rivers in California deserts that maintain permanent flow over at least a portion of their courses. In the Great Basin are the East and West Walker Rivers and the Owens River. The Mojave River and the Amargosa River maintain permanent flow over a portion of their courses in the Mojave Desert, and the Whitewater River has some permanent flow in the Colorado Desert. The largest river in the California desert is the Colorado River, which flows along the eastern edge of the Mojave and Colorado Deserts, forming the border between California and Arizona.

Fremont Cottonwood, *Populus fremontii* subsp. *fremontii*, and Gooding's Black Willow, *Salix gooddingii*, are the dominant native trees in desert Riparian Woodland. Gooding's Black Willow primarily occurs below 2000 ft (900 m) in elevation and is particularly common in the Colorado River drainage. On smaller streams, Fremont Cottonwood and Gooding's Black Willow may form a gallery forest along the main flow. Along the Colorado River, however, these two species are most common in secondary channels. The establishment of seedlings depends on spring flooding, which deposits seeds in moist silt. Recruitment of seedlings is difficult along the main channel. Because the Colorado River flow is controlled by so many dams, the native Riparian Woodland along its banks is a seriously threatened community.

Narrow-leaved Willow, *S. exigua* var. *exigua*, so named for its long, narrow leaves, is a shrub that grows up to 15 ft (5 m) in height. Leaves are 2–5 in (5–12 cm) in length and about 0.75 in (8 mm) in width. It occurs below 8000 ft (2500 m) on desert streams from Imperial County north to Modoc County.

Mule Fat, *Baccharis salicifolia* subsp. *salicifolia*, is a willow-like shrub in the sunflower family (Asteraceae) that grows up to 12 ft (4 m) in height. It occurs along slow-moving streams from the Colorado Desert north to Owens Valley. Other species in this genus, such as Desert Broom or Broom Baccharis, *B. sarothroides*, and Desert Baccharis, *B. sergiloides*, are nearly leafless, carrying on photosynthesis in their broomlike stems. Desert Baccharis is a common water indicator of the Mojave and Colorado Deserts. It is a shorter plant, seldom over 6 ft (2 m) in height. Under ideal conditions, Desert Broom may reach heights of 12 ft (4 m). It occurs in the Colorado Desert east to Arizona.

Arrow-weed, *Pluchea sericea*, is another willow-like shrub in the sunflower family. Like Mule Fat, the flowers occur in clusters of small heads, with no surrounding ray flowers. Arrow-weed differs in appearance by having violet flowers and long, straight stems; these were frequently used by Native Americans to make arrows and baskets. The narrow, pointed leaves of Arrow-weed are usually less than 2 in (5 cm) in length, and the leaf blades are attached directly to the stems (i.e., without petioles). Arrow-weed and Mule Fat are probably the most common riparian shrubs in the Mojave and Colorado Deserts.

Two small ash trees are associated with moist localities in mountains of the southwestern deserts. Arizona or Velvet Ash, *Fraxinus velutina*, grows along streams as far west as the desert slopes of the Sierra Nevada. Single-leaf Ash, *F. anomala*, barely ranges into California. It occurs where streams flow through Pinyon-Juniper Woodland in the eastern Mojave Desert. The name anomala refers to the leaves of this species, which seldom occur in the typical pinnately compound arrangement. They are usually simple and oval—very un–ash-like in appearance. Oregon Ash, *F. latifolia*, has been reported to occur along the northern base of the San Bernardino Mountains, but this is most likely an erroneous reference to Arizona Ash.

Desert Oasis

There are springs and seeps in every desert region. In the Great Basin, these areas are characterized by winter-deciduous species associated with riparian habitats. Sedges, rushes, and cattails form thickets bordering the water. Riparian trees such as Fremont Cottonwood occur where their roots can reach permanent water.

In the Mojave, cottonwoods are joined by Screw Bean (Screw Bean Mesquite), *Prosopis pubescens*, named for its tightly coiled beanpod, an important forage for animals. Honey Mesquite also grows here, but it is more common away from the water.

In the Colorado Desert, all of these species are joined by the California Fan Palm, *Washingtonia filifera* (plate 16B), the only palm native to California. It is another tropical relict that has inhabited wet sites for millions of years. Its northern distribution today is probably due in part to movement northward about 200 mi (320 km) along with the Baja California peninsula, by motion along the San Andreas fault. These palm oases are located along both sides of the Salton Trough where water comes to the surface along fault systems.

Indicative of its tropical heritage, the California Fan Palm clearly has the largest leaves of California's desert plants, and this explains why it must always grow near water. One advantage to the large leaf surface is that it increases the amount of evaporative cooling that can take place. Evaporation of water cools the leaves, but it restricts the plant to areas with wet soil.

The distribution of heavy palm seeds from oasis to oasis has been the subject of considerable discussion. It cannot be explained merely from the point of view of relictism. Even during periods of heavy precipitation, palm distribution was probably spotty. The best guess today is that the seeds were carried in the digestive tracts of Coyotes. These animals are known to eat the sweet fruit, and Coyote scats filled with palm seeds are found all over the desert. However, some palm groves are found outside of the Colorado Desert. Although coyotes could have carried the seeds in a day's time from the San Andreas fault region, near Thousand Palms, to outlying localities such as Twentynine Palms, such a distribution mechanism could scarcely account for palms in the Kofa Mountains of Arizona. It would have to be a badly constipated Coyote to travel that far without a bowel movement. Perhaps birds dispersed the seeds; the American Robin has been observed to eat the palm fruit. Another explanation is that Native Californians carried the seeds and planted the palms. They not only ate the fruit, but also used fibers from the palms for cloth and mats and used the fronds for thatched roofing.

Fire is an important component of palm oases. Palms are fire tolerant, but other oasis vegetation is not. Burning off understory vegetation increases the amount of water available for the palms by decreasing transpirational water loss through the other species. Fires also open up the habitat, allowing palm seedlings to become established. Suppressing fire in these oases works to the detriment of the palm community.

Because oases often have surface water and shade is commonly available, a greater variety of animals may be observed in an oasis than in any other desert habitat. Most of these animals, however, are not restricted to an oasis. One of California's most unusual insects is a palm oasis inhabitant. This is the Giant Palm Borer, *Dinapate wrightii*. The adult is nearly 2 in (5 cm) in length, with a huge head and powerful jaws. The female lays its eggs by drilling a small hole in the wood. The large, pale yellow larvae (grubs) live inside palm trunks, boring extensive dime-sized tunnels in the wood. The actual duration of the larval stage is unknown, but one record documents emergence after 7 years. The adult emerges from its tunnel after dark, and in reverse. Apparently, it must back out of the tunnel because the abdomen is its largest part, and it would be a disadvantage for the insect to exit headfirst and become stuck by the abdomen with its legs flailing the air. Old or dead palms typically show the exit holes of the Giant Palm Borer, and there is some debate as to whether the beetle is responsible for killing the trees or whether the old trees, by virtue of their age, become probable hosts for the beetle. Nevertheless, about 45% of California Fan Palms over 35 ft (10 m) tall show the presence of beetle exit holes.

An insect that benefits from the activities of the Giant Palm Borer is the Southern California Carpenter Bee, *Xylocopa californica* (figure 9.27), the largest bee in California. The females are robust, black or metallic-blue bees. The males, of similar size, are yellow brown and rarely seen outside the burrows. They often construct their own burrows. In California Fan Palms, however, the females lay their eggs in the empty exit tunnels of the beetle. Although great numbers of these bees may be observed at a time, they are actually solitary bees, with no social life. As explained earlier, these are the bees that pollinate some Desert Agaves.

Perhaps the greatest threat to oases and riparian habitats in general is encroachment by exotic species such as tamarisks (*Tamarix* spp.). Tamarisks were imported originally from the eastern Mediterranean and east Asian regions. In the United States, they were first noted growing in the wild in 1916 following a flood on the upper Gila River in Arizona. By 1940, they dominated the riparian vegetation along that river. Now tamarisks have spread into all kinds of moist habitats throughout the Southwest.

There are two widespread forms of tamarisk. A tree also known as Salt Cedar, *T. aphylla*,

is often used for windbreaks. Occasionally, it escapes into the wild. The most common form, and the one that causes the most problems, is a shrub, Fivestamen Tamarisk, *T. chinensis* (figure 9.42). These are water-demanding species, and they are easily dispersed by the wind. Seeds are small and plumed. Once established, Chinese Tamarisks reproduce rapidly by underground shoots or rhizomes. A thicket of these plants is capable of taking over and replacing native species within a few years, after which all surface water may be dried up. It has been determined that a 3-year-old Salt Cedar transpires about 13.5 gal (49.5 L) of water per day. Oases and streamsides are the most productive desert habitats; loss of palatable vegetation and surface water can be catastrophic for animals.

In some areas, in order to restore wildlife habitat, tamarisk eradication projects involving pruning and herbicides have been undertaken. The National Park Service has expended long hours and a large amount of money removing tamarisk from riparian habitats in Death Valley. At Saratoga Springs, the home of the endemic Saratoga Spring Pupfish, *Cyprinodon nevadensis nevadensis*, all tamarisk plants have been removed. A showcase for the effect of tamarisk eradication is in Death Valley at Eagle Borax Spring, where tamarisk had dewatered the habitat. It took 10 years for the National Park Service to remove it all, but the effort restored the water and rendered the habitat once again usable by migratory water birds.

Tamarisks are not restricted to freshwater. The name Salt Cedar was coined because the trees can live in salty water, and they have cylindrical, jointed leaves resembling those of a conifer. On the surface of these leaves are salt glands, which enable the tree to excrete salt. The grayish color of the foliage is due to a coating of salt.

Limestone Endemics

Among the ecologic islands of the deserts are outcrops of limestone that harbor specialized plants and animals. These deposits are rem-

FIGURE 9.42 Saltcedar, *Tamarix ramosissima*.

nants of the time when the sea covered vast areas of the western United States. The alkaline nature of soils derived from limestone poses special problems for plants. The cluster of limestone endemics that occurs along Convict Creek on the eastern side of the Sierra Nevada is discussed in Chapter 4, and the limestone endemics in the San Bernardino Mountains are discussed in Chapter 8. The occurrence of Bristlecone Pine and several of its associates on dolomite soil in the White Mountains is discussed in Chapter 5. In the Basin-Range Province between the White Mountains and Death Valley, there are several regions of limestone endemism. Of particular interest are plants that occur on limestone cliffs in the Death Valley area. A list of the limestone endemics is impressive. Some species are widespread but only occur on limestone; others occur at only a single locality.

One of the most spectacular limestone endemics is July Gold, *Dedeckera eurekensis*. When in bloom, it becomes covered with bright yellow flowers. This shrub, in the buckwheat family (Polygonaceae), occurs in a few areas in the White Mountains and the Last Chance Mountains near Eureka Valley. It is listed as endangered by the federal government. The shrubs occur only on steep alluvial slopes below limestone outcrops, and they seem to form mounds. It turns out that all of the living shrubs are quite old, up to 400 years. There is no evidence of reproduction. It appears that the living plants date back to a time when far more precipitation occurred in the area, perhaps during the "little ice age." One interesting aspect of their growth habit is a consequence of their great age and the fact that the alluvial soil on which they grow is constantly sliding down hill: the root systems are buried in the soil several feet uphill from the aboveground parts of the plant.

The list of endemics from cliffs in the Death Valley area includes Limestone Beardtongue, *Penstemon calcareus*; Death Valley Monkeyflower, *Mimulus rupicola*; and Rock Lady, *Holmgrenanthe petrophila*. The word petrophila in the scientific name of the maurandya means "rock-loving." There are also other buckwheats, such as Jointed Buckwheat, *Eriogonum intrafractum*, and Heermann's Buckwheat, *E. heermannii*. An interesting member of the Spurge family is the Holly-leaved Tetracoccus, *Tetracoccus ilicifolius*. This is a holly-like shrub that grows in a mound up to 4 ft (1.3 m) high. It is found in the Grapevine Mountains of northeastern Death Valley. In the waterleaf family (Boraginaceae), there are two attractive herbaceous shrubs associated with limestone: Panamint Phacelia, *Phacelia perityloides,* and Death Valley Round-leaved Phacelia, *P. mustelina.*

HISTORY OF DESERT VEGETATION

The history of California's deserts is tied closely to the uplift of mountain ranges and the subsequent development of a rain shadow. The forces leading to the uplift of the Basin-Range Province, including the Sierra Nevada, began about 25 million years ago and continue today. Evidence of drying as exhibited by changes in Great Basin vegetation indicates that a significant rain shadow formed by western mountain ranges had occurred by 11 million years ago. Uplift of the Sierra Nevada, Transverse Ranges, and Peninsular Ranges accelerated about 2–5 million years ago so that during the Pleistocene they were well-developed ranges. Evidence of montane glaciers occurs as far south as Mount San Gorgonio. During the 20 or so episodes of glaciation, heavy rainfall occurred across the deserts, and large basins such as Death Valley and Panamint Valley were filled with water. Lake Manly, in Death Valley, was 600 ft (190 m) deep.

By the close of the Pleistocene, about 11,000 years ago, a drying trend began. Studies of woodrat (packrat) middens in the Southwest indicate that at that time a subalpine sort of forest, including Limber Pine and Bristlecone Pine, occurred extensively across what is now the Great Basin Desert and at higher elevations in the Mojave Desert. As mentioned previously, a woodrat midden on Clark Mountain revealed that Rocky Mountain White Fir has grown there continuously for 28,000 years. Most of the present-day Mojave and Sonoran Deserts were occupied by a Pinyon-Juniper Oak Woodland. The lower Colorado River Valley was occupied by a Mojave type of Creosote Bush Scrub that included Joshua Trees. This area may have been a desert refugium since the Pliocene, some 5 million years ago.

The presence of large grazing mammals such as horses, camels, Bison, and mammoth, as well as large browsers, including mastodons and ground sloths, implies that grasses and herbs were abundant in the Pleistocene woodlands. Perhaps there were extensive savannahs or open grasslands among the trees.

Coincident with uplift of the mountains, the ocean water off California became colder. In combination, these changes caused drying to the interior and all but eliminated summer rainfall throughout the Great Basin and Mojave areas. In the Sonoran region, where biseasonal rainfall still occurs, a woodland persisted until about 9000 years ago. The present vegetation did not appear until 4000 years ago. Some common plant communities such as Microphyll Woodland seem to have arrived about 4000–6000 years ago.

During the last 10,000 years in the Mojave and Great Basin, desert communities such as Creosote Bush Scrub and Sagebrush Scrub pushed northward to their present latitudes. Over that time, through the mechanism of polyploidy, Creosote Bush evolved into different races in each of the major desert regions. Also, ash fall from volcanic eruptions in the Cascades may have favored the invasion of Great Basin Sagebrush. While Creosote Bush Scrub and Sagebrush Scrub dispersed northward, the Subalpine and Pinyon-Juniper woodlands retreated to mountaintops. The history of the latter communities is discussed in Chapter 5.

STRATEGIES OF DESERT ANIMALS

Desert animals must adapt to three interrelated stresses: drought, temperature extremes, and sparse or unpredictable food supply—all functions of low or unpredictable precipitation.

In general, animals deal with temperature extremes by avoidance. Desert summer temperatures are the hottest of all the ecosystems. In cases of overheating, nearly all animals use some form of evaporative cooling. However, this process can be risky where water is in short supply, so keeping cool without relying on evaporative cooling is a major strategy of desert animals.

Dealing with drought is primarily a function of water conservation: keeping cool saves water. Animals accomplish this primarily through behavioral means. They avoid the heat of the day by remaining in the shade or in burrows. Activity periods are limited to dawn, dusk, or nighttime. When the climate is too severe—either too cold, too hot, or not enough food—animals go into a state of torpor; they hibernate or estivate.

One of the raw materials for photosynthesis is water. Where water is limited, photosynthesis is limited. A desert therefore is a food-poor ecosystem. Animals with low metabolic rates have an advantage; therefore, there are many cold-blooded (ectothermic) animals in the desert. Deserts are inhabited by a large number of arthropods, such as insects, spiders, and scorpions. These are animals with exoskeletons of chitin and jointed appendages. Because the skeleton is external, the legs bend at conspicuous joints, as in a suit of armor. Among the vertebrate animals, reptiles are most common. There are 53 known reptile species in the California desert: 30 lizards, 21 snakes, and 2 turtles.

Dealing with Drought

Many animal adaptations for desert existence resemble those of desert plants. Common strategies include water storage, heat avoidance, dormancy, drought tolerance, boundary layers, body orientation, and waterproof integument.

Storage of water is a cactus-like adaptation. Coupled with low metabolic rate during long periods of torpor, this is enough to get some animals through long periods of drought. For example, the Common Chuckwalla, *Sauromalus ater*, stores water in lymph spaces under the skin along the sides of its body. Some biologists refer to hibernation in reptiles as *brumation*, which is believed to be different because the state of torpor in ectotherms occurs at a lower temperature than it does in mammals and birds.

The Desert Tortoise, *Gopherus (= Xerobates) agassizii* (figure 9.43), can store about a quart (liter) of water in its body, mostly in its urinary bladder. The scientific name *Gopherus* alludes to its digging ability. The tortoise emerges from

its burrow in spring to feed on moisture-rich leaves and flowers, particularly those of ephemeral plants. It also may emerge to drink water, during rain storms. During this time, it accumulates reserves of water and fat that will carry it through a long period of dormancy including summer, fall, and winter, which may represent 95% of the time.

The Desert Tortoise is on the state list of threatened species. The gentle nature of these animals has made them desirable as pets. Some people use them for target practice. An increase in the number of Common Ravens, probably in association with open dumps, has also threatened tortoises because ravens prey on hatchlings, which still have soft shells. One researcher counted the shells of 250 young tortoises around the nest of a single pair of Common Ravens. In addition, tortoises must compete with cattle, sheep, and donkeys for food. In the face of such competition, malnutrition has cut their reproductive rate, and they are becoming rare. Some studies indicated that Desert Tortoise populations have dropped about 90% in the last 50 years.

Within its native range in the Mojave and Colorado Deserts, certain areas have been set aside as refuges for the Desert Tortoise. It is a crime to collect one or to engage in any activity destructive of its native habitat. It is hoped that such protective measures will enable the population to rebound. Captive breeding and release of juveniles may not be a good idea for tortoises, depending on where the animals were raised. It is not a good idea to release animals into the desert once they have lived for some time in a moister climate. It has been determined that contagious infections—in particular, a lung fungus—may be introduced in this way into the wild population.

As unlikely as it may seem, several species of toad are associated with deserts. The Great Plains Toad, *Anaxyrus (= Bufo) cognatus*, is able to store water in its urinary bladder equal to about 30% of its body weight. The Red-spotted Toad, *A. (= B.) punctatus*, is a common toad of desert streams. It can be recognized by a

FIGURE 9.43 Mohave Desert Tortoise, *Gopherus (= Xerobates) agassizii*. Photo of display at Twenty-nine Palms Visitor Center at Joshua Tree National Park.

number of small red spots on its back. Where its range overlaps that of other toads, it hybridizes rather freely. Several other toads have restricted distribution in the California desert. The Black Toad, *A. (= B.) exsul*, only occurs in Deep Springs Valley east of the Inyo/White Mountains. Woodhouse's Toad, *A. (= B.) woodhousii*, is distributed widely in the western United States, but it enters California only on the eastern edge of the Colorado Desert, where huge breeding choruses may be encountered after suitable rains. Its interesting call is described as sounding a bit like a snore, or the bleating of a lamb. The Sonoran Desert (Colorado River) Toad, *Incilius (= B.) alvarius*, only occurs along the Colorado River floodplain, and may be extirpated in California. It is the largest toad in California and is discussed in Chapter 11.

The "champion" desert toads are known as spadefoot toads. Couch's Spadefoot, *Scaphiopus couchi*, occurs in the southwest deserts where there are periods of summer rain. In California, they occur along the Colorado River flood plain. They come up from deep burrows after the first heavy rains of summer and congregate in tempo-

rary ponds to reproduce. Apparently, sound is the stimulus to break dormancy. The sound of thunder, in particular, brings them up. Unfortunately, they can be fooled by the loud sound of a dune buggy or motorcycle, in which case they may emerge to find no water and subsequently die.

When conditions are right, Couch's Spadefoot Toads congregate at night in large numbers at temporary ponds. Their mating calls sound remarkably like bleating lambs. Eggs are laid and fertilized in these ponds. The development of these toads is among the most rapid known for a vertebrate, for it must take place before the ponds dry up. Development from egg to larva takes about 2 d. Larvae swim about in the ponds, feeding on algae and detritus. In about 10 d, the tadpoles lose their tails and grow legs, and at this time they will be soon ready to leave the water. They will dig a burrow and become dormant as soon as it becomes too dry to remain aboveground. The total period of development from egg to dormancy may last four to six weeks.

While they are in their burrows, spadefoot toads survive on water stored in their urinary bladders. As they use up water and fat stores,

FIGURE 9.44 Southern Grasshopper Mouse, *Onychomys torridus*, and Stink Beetle. Photo of display at Twenty-nine Palms Visitor Center at Joshua Tree National Park.

they gradually lose weight. If they were to urinate, they would lose valuable water, so nitrogenous waste (urea) becomes concentrated in their bodies. These toads can tolerate a 40% weight loss, and they can concentrate urea to the point that their body fluids equal half the concentration of seawater (16,000 mg/L). By concentrating body fluids, they are able to absorb water from the soil in a plant-like fashion.

Reptiles, birds, and insects conserve water by converting their nitrogenous waste to uric acid, a substance that is not soluble in water. Therefore, they are able to excrete the waste products of protein metabolism without losing water at all. This metabolic feature is another reason that these three groups of animals are so well represented in desert and scrub communities.

Water conservation is the primary means of desert survival, but making use of every possible water source is also important. Birds and large mammals are mobile enough to move to known water sources when they need to rehydrate. Temporary pools may hold water for long periods of time after a rain. Depressions in granite may be quite large. These storage ponds are known as tinajas, which is a Spanish word for "tank." Other animals concentrate their activities near permanent water. These animals are no more desert creatures than the deep-rooted, water-loving plants known as phreatophytes.

Making use of unique sources of water is important for desert survival. Doves are unique among birds because they can drink dew; they can swallow with their heads down. Small bur-

rowing reptiles such as the Northern California Legless Lizard, *Anniella pulchra*, or the Western Blind Snake, *Leptotyphlops (= Rena) humilis*, are able to drink capillary water between sand particles. Many animals derive water from moist food. The Desert Woodrat, *Neotoma lepida* (plate 18H), eats prickly-pear cactus. Bighorn Sheep, *Ovis canadensis*, eat the rapidly growing flower stalks of agaves, a source of water and carbohydrate.

Any carnivore gains water by feeding. Because insects are very common in deserts, there are many insectivorous animals. Grasshopper mice (*Onychomys* spp.; figure 9.44) are unique rodents because they feed on insects and small mammals. There are two species in California: one in the Great Basin and one in the Mojave and Colorado Deserts. Because they are carnivores, they have large home ranges— up to 8 acres (3.2 ha).

Grasshopper mice can attack and kill mice that are twice their size. They immobilize these mice by biting through the spinal cord just behind the skull. The grasshopper mouse eats the brain first, and depending on how long it has been since the last feeding, it may or may not eat the rest of the body. Grasshopper mice even eat scorpions and stink beetles (figure 9.44). They eat scorpions by first biting off the stinger. Stink beetles (*Eleodes* spp.) are common black beetles with a large, bulbous abdomen. They are avoided by most predators because of an obnoxious odorous liquid they emit when molested. When pestered, a stink beetle puts down its head and elevates its abdomen, which is enough warning to cause any other predator to leave it alone. The grasshopper mouse grabs the beetle with its forepaws, jams its abdomen in the sand, and eats off only the head end.

One very peculiar aspect of the hunting behavior of the grasshopper mouse is that it "howls." Soon after it emerges from its burrow for an evening's hunt, it climbs to a small rise and emits a long, shrill whistle that carries for about a hundred yards (meters). Apparently, they claim territories and establish dominance with these vocalizations, in a manner similar to Wolves, Coyotes, and many birds.

Bats are also common in desert areas. Most of them obtain water from the insects they eat. They also dip into water while flying over a source of water. Their mechanism for locating flying insects with "sonar" is well known. Once a prey item is located, the bat scoops it into its mouth with the membranes between its fingers, which also serve as wings. Furthermore, unlike most mammals, bats conserve energy by allowing their body temperature to drop during the time they are in their shaded daytime roosts. In the evening, when they become active, they raise their body temperatures by shivering. Muscle contractions generate the heat necessary for them to become active, and they forage for insects at dusk, avoiding the heat of the day.

Many desert birds are also insectivorous. These include swallows, swifts, wrens, and a variety of flycatchers and gnatcatchers, which are discussed elsewhere. An interesting desert gleaner is the Verdin, *Auriparus flaviceps*. This small gray bird is noticeable by its bright yellow head and throat. It is an inhabitant of the Colorado Desert, where it often makes its nest in Smoke Trees. A Verdin nest is recognized by its spherical shape and side opening. Verdins typically glean insects in palo verde trees, although they are also known to eat nectar-laden flowers such as those that attract hummingbirds. They are able to obtain all the water they need in this way. One researcher has calculated that they eat about two-thirds of their body weight in insects each day. During summer, they spend about 50% of their active time foraging, and during

winter almost 90% of their activity is spent on foraging. This high rate of activity also generates body heat, so it is important for keeping warm in the winter. It also has been shown that the enclosed nest is a very good insulator against the cold and wind.

An interesting water source is reclaimed water vapor. Long-nosed mammals are able to reabsorb water from exhaled air. This reabsorption is enhanced by certain mammals who plug the entrance to their burrows. As a result, the humidity of the air in the burrow remains high, enabling the mammal to reclaim its water merely by breathing through its nose. Likewise, large mammals such as camels and antelopes reclaim a significant amount of water through their long, narrow nasal passages. If nothing else, the long nose that cools the air reduces evaporation.

Another means of water reclamation occurs in Greater Roadrunners, *Geococcyx californianus* (figure 9.45). Newly hatched nestlings produce urine-filled pouches that are excreted in the nest. Adult Roadrunners eat these sacs, reclaim the water, and excrete the nitrogenous waste a second time.

Roadrunners are important predators in desert and chaparral areas. The Spaniards called them Paisanos, meaning "country men." They are also known as Chaparral Cocks, in reference to their common occurrence in Chaparral. They eat insects, lizards, snakes, mice, and berries, and they are known to many as predators of whiptails, horned lizards, and rattlesnakes. The Roadrunner is the only member of the cuckoo family (Cuculidae) that occurs in the arid southwest. These crow-sized birds with long tails are best known for their habit of running rather than flying. They are able to fly, but they seem to prefer to run or glide and have been clocked at 15 mph (24 kph). An interesting habit of the Roadrunner is to bask in the sun in the morning. The bird positions itself in an opening or on a rock with its back to the sun. It fluffs its feathers and spreads its wings in order to maximize the surface for absorption. Roadrunners also have black skin

FIGURE 9.45 Greater Roadrunner, *Geococcyx californianus*. Keeping cool conserves water. This bird is exhibiting four methods of cooling itself, three of which use no water: (1) panting, (2) drooping its wings, (3) elevating itself on a rock, and (4) standing in the shade.

on their backs under the feathers, which enhances absorption of heat. This is an energy conservation device that enables them to lower their body temperature at night and warm up again without burning food calories. It has been calculated that sunning Roadrunners conserve an average of 551 cal/hr. Furthermore, they conserve heat by placing their nests in dense trees or shrubs. This placement is also vital to survival of nestlings. Two density-dependent factors also keep Roadrunner populations in line with food supply. First, the number of eggs in a nest is correlated to the immediate food supply, and second, adults eat weak or lethargic nestlings. Nothing is wasted. Another common bird that supplies itself with adequate water by being carnivorous is the Loggerhead Shrike, *Lanius ludovicianus*. This bird is conspicuous, hunting for insects, lizards, and mice from an open perch. In this position or when it flies, revealing white wing patches, it resembles a Northern Mockingbird, *Mimus polyglottos*. Close inspection, however, reveals

that the shrike has a black eye mask, and, like a raptor, it has a small hook on the tip of its beak. This bird has gained the nickname Butcher Bird by its curious habit of hanging its prey, like sides of beef, on sharp thorns or the barbs of a barbed wire fence. The fact that this bird seems to kill more animals than it could eat has puzzled biologists. It now seems clear, however, that a courting male provisions a larder and displays it conspicuously to "prove" to the females that it is a good hunter. The stored food will be used to feed the young. A highly successful male is able to mate with more than one female and supply enough food for more than one family.

The importance of metabolic water to desert animals cannot be overlooked. Waste products of cellular respiration are carbon dioxide and water. To most animals this metabolic water is insignificant, but to desert animals it contributes a significant proportion of their total water. There are several species of kangaroo rats (*Dipodomys* spp.; figure 9.46A) in California's deserts. Kangaroo rats are able to go their entire lives without drinking a drop of liquid water. They use various water conservation techniques and derive enough water solely from metabolism and food.

Kangaroo rats are more closely allied to squirrels than to rats. They are members of a family of rodents (Heteromyidae) that in desert areas have long, kangaroo-like hind limbs and fur-lined cheek pouches. They subsist nearly entirely on seeds, which they gather at night and store in their burrows. During the day, when the burrow is plugged, humidity in the sleeping chamber rises. Water vapor is reclaimed through the animal's nasal membranes, and the humidity absorbed by the seeds is reclaimed when the seeds are eaten. Inside the sleeping chamber, the temperature remains at a relatively constant 80°F (27°C) even though the temperature of the soil surface outside may rise to 180°F (82°C). The temperature in the burrow is ideal for heat and water conservation. A kangaroo rat is not required to raise its metabolic rate to heat itself, nor is it required to

FIGURE 9.46 Note the similarity between Merriam's Kangaroo Rat, *Dipodomys merriami* (A), and Little Pocket Mouse, *Perognathus longimembris* (B), two species that often coexist. The kangaroo rat hops on its hind legs and forages in the open, whereas the pocket mouse walks on all fours and forages under shrubs.

resort to evaporative cooling to dissipate heat. Kangaroo rats eat their own feces in order to reclaim water lost there. They can concentrate their urine to 30% salts and urea, reducing urinary water loss to an absolute minimum. Kangaroo rats are the "champion" desert rodents, marvels of efficiency.

Animals that store fat have another source of metabolic water. Fat helps them get through long periods of torpor by supplying calories. In addition, due to a peculiarity of fat metabolism, 1.1 g of water are produced for every gram of fat that is metabolized. In contrast, a gram of carbohydrate produces only 0.6 g of water. Camels have long been famous for storing fat in their humps.

In North America, several lizard species store fat in their tails, drawing upon it for energy and water during long months of torpor underground. For example, Gila Monsters,

FIGURE 9.47 Western Banded Gecko, *Coleonyx variegatus.*

Heloderma suspectum, primarily located in the Sonoran Desert of Arizona and Mexico, store fat in their tails. They emerge in spring to eat insects, bird eggs, and baby birds. These are the only venomous lizards in the United States, and they are brightly colored as a warning. Gila Monsters have limited distribution in California. They have been recorded in the Chocolate, Clark, Kingston, and Providence Mountains. Historically, there have been a total of 28 sightings of Gila Monsters in California.

In California, some persons confuse Gila Monsters with Common Chuckwallas, which are also large lizards with fat tails. The Common Chuckwalla's tail, however, is mostly muscle, and, as indicated earlier, they store water beneath the skin. Common Chuckwallas are primarily vegetarian. They live in rocky habitats, emerging from cracks in the morning to sun themselves. When they are sufficiently warm, they move into nearby vegetation to feed. If frightened, they dive into a crack in the rock and lodge themselves there with their muscular legs and tail, also inflating their lungs. Native Californians used to eat Common Chuckwallas. They would extract them from the cracks by tugging on a foot or the tail while poking a sharp stick into the animal's side to let out the air. Common Chuckwallas do not lose their tails as do other lizards.

The Western Banded Gecko, *Coleonyx variegatus* (figure 9.47), looks too fragile to be a successful desert lizard. Its pinkish, fine-scaled integument is nearly transparent. It has large, conspicuous eyelids over bulging eyes. Geckos are nocturnal animals that are often found on the road, where they are endangered by cars. Another threat to the animals are snake fanciers, who collect them to feed to their pets.

Geckos emerge in spring to feed on small insects. Like Gila Monsters, they store a significant amount of fat in their tails. Unlike Gila Monsters and Common Chuckwallas, they shed their tails when molested. It is more important for this lizard to escape and reproduce than it is to maintain enough fat for the next dry season. One small burrowing snake, the Spotted Leaf-nosed Snake, *Phyllorhynchus decurtatus*, seems to prey heavily on gecko tails.

In addition to conserving water and utilizing unique sources of water, some desert animals have notable tolerances for desiccation. Humans can lose up to 10% of their body weight due to dehydration, and a loss of 15–18% is usually lethal. In contrast, the Desert Bighorn Sheep can endure 20% depletion of body weight, representing a 30% loss of total body water, without showing ill effects. Old World animals such as camels and donkeys can also lose up to 30% of their body weight through dehydration. They are able to restore their body water with heavy drinking. A camel can drink 30 gal (110 L) of water in 10 min. As indicated

FIGURE 9.48 Gambel's Quail, *Callipepla gambelii*.

earlier, spadefoot toads can lose up to 40% of their body weight. Birds are the most notable in this respect because they have no urinary bladder. Doves can lose up to 40% of their body weight, and Gambel's Quail, *Callipepla gambelii* (figure 9.48), can survive a 50% weight loss due to dehydration.

Coping with Heat

Coping with heat poses a series of related problems in desert animals. Evaporation of water from a surface with a good blood supply is the most efficient mechanism for cooling an overheated animal. However, because water is in limited supply, the animal's best strategy is never to get overheated. Animals alter their daily and seasonal activity periods to avoid the heat. In general, animals regulate their body temperatures by moving in and out of the sun. Desert animals burrow or rest in the shade. To avoid the hot ground, snakes and lizards climb into the branches of vegetation. It can be cooler by 100°F (38°C) in the shade 2 ft (60 cm) above the ground.

Ectothermic animals obtain a significant amount of their body heat from the environment. In contrast, endothermic animals gain most of their heat from metabolism. Because heat in the desert is not in short supply, and

because food is scarce, ectothermic animals, with their low metabolic rates, have an advantage; they don't have to waste valuable calories keeping their bodies warm.

Heat is gained by metabolism, conduction, convection, and radiation. Metabolic heat gain is minimized in animals with low metabolic rates. Heat loss also can be enhanced by conduction, convection, and radiation. The advantage to an animal that loses heat by one of these mechanisms is that it can do so without losing valuable water.

It is often assumed that desert animals should be light colored in order to reflect light. Although it is true that dark objects absorb more heat, it is also true that they emit more heat by radiation. A phenomenon known as Allen's rule was discussed in Chapter 5, where it was pointed out that endothermic animals in cold climates tend to be white. Conversely, for ectothermic animals in hot climates, where the major source of heat is external, it is logical that diurnal animals should be light colored. One searches, then, for an explanation for the black coloration of the Common Raven, *Corvus corax* (figure 9.49). This member of the crow and jay family, Corvidae, is most common in deserts. Unlike most members of the family, Common Ravens soar hawk-like while searching for food, primarily carrion. It would seem that its dark color would enhance heat absorption by radiation. However, the Common Raven avoids overheating in a number of ways. In the first place, it is cooler high above the ground. Second, the large surface area of the bird's wings allows conduction of heat to the environment without water loss. Movement of the air over the wings transports heat away by convection. Furthermore, soaring is an effortless activity, so the bird generates minimal metabolic heat. Finally, if all these factors are not enough to keep the Common Raven cool, it simply rests in the shade, at which time its black color enhances its heat loss by radiation.

It cannot be disputed that the lifestyle of the Common Raven is successful. Ravens have expanded their ranges considerably and are now common in many habitats. The Breeding Bird Census of the US Fish and Wildlife Service indicated that the Common Raven population in the United States has been growing at about 3% per year, presumably in association with the birds' ability to thrive in open dumps. Their numbers have increased by more than 300% in the Mojave Desert. Some wildlife officials are calling for a program of poisoning and shooting to reduce their numbers in the Mojave Desert, where predation on juvenile Desert Tortoises has helped to promote a severe decline in that species.

Another factor leading to the decline of Desert Tortoises is habitat destruction. Of particular concern is the development of large-scale alternative electrical generation facilities such as solar and windmill arrays. Similarly, military bases in the deserts are expanding. One form of mitigation is to translocate tortoises from the disturbed sites to undisturbed habitat elsewhere. This is a controversial procedure that has not been proved to be successful for a number of reasons.

Antelope Ground Squirrels or White-tailed Antelope Squirrels, *Ammospermophilus leucurus* (plate [18E]), are unique among ground squirrels because they remain active all year. They are diurnal and can be seen scurrying about in the dead of summer, gathering fruit and seeds. White-tailed Antelope Squirrels look

FIGURE 9.49 Common Raven, *Corvus corax*. Note the rounded tail.

like white-tailed chipmunks. When they overheat, they can drop their body temperature up to 7°F (4°C) by pressing their bodies into cool sand in the shade of a bush. This is heat loss by conduction. If overheating is excessive, they will lose some valuable water and shift to evaporative cooling, which they accomplish by drooling, panting, and licking their forelimbs. Almost all animals resort to panting when they overheat. Evaporative cooling from capillaries in the lungs is a rapid, effective way to lose body heat.

Black-tailed Jackrabbits, *Lepus californicus* (figure 9.50), utilize radiation to cool themselves. Their long limbs and ears are ideal surfaces for dissipating heat away from their bodies. Typically, they remain in the shade on the north side of a bush facing the clear northern sky. They can continue to lose heat in this way, particularly through their ears. They enhance convective heat loss by wiggling their ears,

FIGURE 9.50 Black-tailed Jackrabbit, *Lepus californicus.*

which moves heated air away from the ear surface.

The White-tailed Jackrabbit, *L. townsendii,* of the Great Basin was discussed in Chapter 5, in the section on the alpine zone. Compared to the mountain form, the desert White-tailed Jackrabbit has longer legs and ears and a thinner body. This surface area enhances radiative heat loss, illustrating Bergmann's and Allen's rules.

Jackrabbits, also known as hares, do not live in burrows; they simply hide in the shade of a shrub. Sometimes Coyotes use this habit to their advantage. A Coyote will flush a jackrabbit from hiding, and instead of running after it, the Coyote watches where it hides next. The Coyote then flushes it again. Running repeatedly from bush to bush causes the hare to overheat, and it drops from heat exhaustion. The Coyote, having conserved its energy, is able to catch the overheated jackrabbit even though it is no match for it on a dead run. Also, the Coyote is able to utilize more efficient evaporative cooling by drooling and panting. Like all dogs, Coyotes cool themselves by evaporating saliva from the surface of their tongue.

Cottontail rabbits are true rabbits, not hares. The Desert Cottontail, *Sylvilagus auduboni,* has much shorter legs and ears than the jackrabbit. The cottontail avoids overheating by becoming nocturnal in the summer, and it lives in a burrow, rather than taking cover in a bush. Cottontails also take care of their young for longer periods of time than do jackrabbits. Juvenile cottontails are born in the burrow in a nest lined with the mother's soft belly hair, and they are nursed for several weeks before they attempt to gather their own food.

Large African mammals such as camels and elands have another trick. These animals cool themselves by perspiration when they overheat. To avoid this, they allow their body temperatures to drop as much as 12–14°F (7–9°C) during the night. During the daytime, they heat up slowly by radiation to about 108°F (42°C) before they begin to sweat. When they exhale, they reclaim water by condensation in their long nasal passages. If they overheat, water evaporates from these surfaces when they inhale. This evaporation helps to cool their blood, which is then shunted to the brain, the tissue most sensitive to overheating. This coun-

FIGURE 9.51 Desert Iguana, *Dipsosaurus dorsalis.*

try no longer has many large native desert mammals. It is unknown whether the Pronghorn or Desert Bighorn Sheep allows its body temperature to fluctuate.

Invertebrates and ectothermic vertebrates undergo a greater range of daily fluctuation in body temperature than is tolerated in endothermic animals. Similarly, ectotherms have a greater tolerance for high temperatures. The upper temperature an animal can tolerate before dying is known as its critical thermal maximum (CTM). The CTM of 122°F (50°C) for desert invertebrates that have been tested, such as scorpions, sun spiders, and stink beetles, is the highest known among animals. Not only do these animals tolerate high body temperatures, but they are also covered with a waxy integument that inhibits water loss. Among vertebrates, certain lizards of the iguana family (Iguanidae) such as the Common Chuckwalla, *Sauromalus obesus,* and the Desert Iguana, *Dipsosaurus dorsalis* (figure 9.51) can tolerate high body temperatures. In this respect, the Desert Iguana, is the "champion." It has been found to have a rectal temperature of 117°F (47°C).

Two factors lead to thermal death in animals. One is that enzymes have a narrow range of thermal tolerance. If enzymes function poorly, critical chemical reactions fail, and the animal soon dies. Lizards such as the Chuckwalla and the Desert Iguana have evolved enzymes that function over a wide range of temperatures. A second factor is that hemoglobin, the pigment in blood that carries oxygen, is also very temperature sensitive. At high temperatures, it loses its ability to transport oxygen. Such high body temperatures are also associated with increased metabolic rate. To cope with this problem, Common Chuckwallas and Desert Iguanas have an increased capacity to function in the absence of oxygen. They can function anaerobically, accumulating lactic acid in concentrations that would immobilize most animals. When it becomes cooler, they process the lactic acid by heavy breathing. The amount of oxygen required to process stored lactic acid is known as the oxygen debt.

Among birds and mammals, high body temperatures are associated with very small animals. Coincidentally, these animals are common in deserts. At 105°F (41°C), hummingbirds have one of the highest body temperatures among birds. The high metabolic rate of these small animals is necessary to produce enough heat to keep them warm in the face of their high surface-volume ratio. Living in a hot environment lowers their metabolic demand. Hummingbirds are nectar feeders, and nectar is a nutritious, high-calorie food. Fortunately for desert plants, hummingbirds do well in hot environments. The coevolution of hummingbirds and long, tubular flowers is a desert phenomenon. There is abundant nectar in desert flowers, and plants such as Ocotillo and Chuparosa are distinctly adapted to accommodate hummingbirds as pollinators. Hummingbirds minimize their metabolic demand by allowing their body temperature to drop when they are at rest, particularly at night. For every 18°F (10°C) that an animal lets its body temperature fall, there is a 50% reduction in metabolism. A hummingbird in flight increases its metabolic rate fourfold over its resting rate. It has been calculated that a hummingbird eats about 50% of its body weight in sugar each day. If it were not for its ability to drop its temperature at night, it might have to eat four times that much.

The smallest mammal in the desert is the Crawford's Desert Shrew, *Notiosorex crawfordi*. It is particularly common in the Cactus Scrub community of the Colorado Desert. These tiny carnivores must eat more than their own body weight each day. They eat mostly insects, but take mice as well. Problems and adaptations concerning metabolic rate in these animals are similar to those of hummingbirds. These ferocious little mammals are also unique in that they are venomous. This is an adaptation for eating large prey, as animals with venom are able to immobilize prey that are large enough to fight back. In the case of the shrew, practically all prey species are larger than it is.

Survival in a Food-Poor Ecosystem

The total biomass of animals in the desert is small. This is because the desert is a food-poor ecosystem. Photosynthesis produces about 890 lb per acre per year (100 g/m²/yr) of edible biomass. An acre of desert land will sustain about 89 lb (36 kg) of herbivores and less than 9 lb (3.6 kg) of carnivores. Productivity of deserts, however, may double in a wet year because of the tremendous increase in annual plants. But the animals able to respond to this "bonanza" of food must be short-lived. In long-lived animals such as Desert Tortoises, reproduction may be delayed until there is a bumper crop of annuals. In general, this kind of productivity favors small animals with short life spans, high reproductive rates, and good dispersal. Arthropods (insects, spiders, and scorpions) and reptiles are favored in deserts.

The most common herbivorous animals are insects. Because desert plants invest so much energy in reproduction, seed eaters are very common. The most common seed-eating animals in the desert are ants. Other common arthropods are leaf eaters such as grasshoppers, beetles, and bugs. Large, orange Soldier Blister Beetles (*Tegrodera* spp.; figure 9.52) eat mostly flower petals. Termites eat dead wood. Most termites in the desert are subterranean. When it rains, they build mud shells around dead sticks and other debris, encasing the wood of a dead shrub up to a foot (30 cm) or so above the ground. After they consume the wood, all that is left is the mud shell.

Some insects specialize. They are abundant only when their favored food is abundant. Among these are nectar feeders such as Honey Bees, butterflies, and moths. Juvenile forms of these insects partition resources by eating different food than the adult. The example of sphinx moth larvae was mentioned earlier.

Plants are specialized to attract certain pollinators. As has been mentioned earlier, the shape of flowers is associated with certain pollinators. For example, long tubular flowers are pollinated by moths or hummingbirds. Wide open flowers such as sunflowers are often pollinated by beetles. Colors of flowers are also associated with particular pollinators. White or pale yellow flowers attract nocturnal pollinators such as moths and nectar-feeding bats. Red or yellow flowers attract hummingbirds. Red flowers are seldom pollinated by insects because insects cannot see red. Color vision in insects is shifted to the violet end of the spectrum; they are blind to red, but they see violet

FIGURE 9.52 Soldier Blister Beetle, *Tegrodera* sp. This beetle has eaten all the petals off this Desert Brittlebush.

and mallows (*Malva* spp., *Sphaeralcea* spp.), which they defoliate during outbreaks.

Butterflies such as the Mojave Sootywing, *Hesperopsis libya*, are more typically desert in distribution. Larvae of this species feed on a variety of saltbushes (*Atriplex* spp.) and seem particularly associated with Four-wing Saltbush, *A. canescens*. This gray to blackish butterfly presumably takes advantage of its coloration to absorb heat rapidly by basking in the sun, or to lose heat equally fast by resting in the shade.

Millipedes and ground-dwelling beetles eat detritus. The most common ground beetles are in the darkling beetle family (Tenebrionidae). Distribution of this family in the United States is concentrated in the Southwest. Here there are more than 1200 species, over 400 of which occur in California. One of these, *Asbolus verrucosa*, is so common that at one locality in the northern Mojave, its biomass was calculated at nearly 2 lb per acre (275 g/ha), more than the combined biomass of mammals, birds, and reptiles in the area. Another study found a related form, *A. laevis*, in the Algodones Dunes to be so common that on one occasion 100 specimens were visible at the same time.

Darkling beetles have a number of adaptations that enable them to thrive in the desert. They are opportunistic feeders, eating all sorts of detritus. They retreat from heat by moving into cracks or under rocks, and some of them dig their own burrows. Most of them do not fly, but they use their first pair of wings, known as wing covers (elytra), as a bubble-shaped chamber to trap air. This acts as an insulation. On the ironclad beetles (*Phloeodes* spp.), the body is so hard that blows from a hammer may be needed to push an insect pin through it. Some beetles (*Asbolus* spp., *Cryptoglossa* spp.) have a distinctive way of increasing their insulative surface. They secrete a mass of microscopic waxy hairs on their backs that act as a boundary layer to trap air. This also causes them to change color. When it is hot and dry, they turn a bluish white with the waxy secretions. When it is cool or moist, they turn black, as the waxy hairs are lost. The color change also influences

and ultraviolet quite well. Vertebrates are blind to ultraviolet. It should be no surprise, therefore, that in order to attract insects some flowers incorporate ultraviolet patterns into their blossoms that are invisible to humans.

Adult butterflies take only liquid food such as nectar. When flowers become abundant, so do butterflies. Outbreaks of the Painted Lady, *Vanessa cardui*, have been recorded in the Mojave and Colorado Deserts, where motorists reported them flying northward by the thousands. Heavy rains during El Niño years stimulate such outbreaks. They fly a few feet off the ground, rising vertically to avoid obstacles. Adults continue northward toward Oregon until they die. There is no return flight. The Painted Lady is not restricted to deserts. It is said to be the most widely distributed butterfly in the world, adapting also to habitation. It is sometimes called the Thistle Butterfly because larvae feed on a variety of thistles (*Cirsium* spp.). They also feed on other disturbance-area plants, such as fiddlenecks (*Amsinckia* spp.)

their abilities to absorb or radiate heat. Many darkling beetles can tolerate a wide range of humidity without significant water loss. The most important source of water loss for the stink beetle is its defensive secretions. The defensive posture of a stink beetle, known as head-standing, in which its abdomen is raised conspicuously, is a warning to potential predators (figure 9.53). The stink beetle would rather not lose any water defending itself. This posture is so effective that it is mimicked by other beetles, even though they produce no nasty secretions.

The Desert Spider Beetle, *Cysteodemus armatus*, is a flightless beetle in the blister beetle family (Meloidae). They have a large, inflated abdomen and superficially resemble the darkling beetles. These beetles have an uncanny ability to gather a layer of pollen in pits on their elytra, which also acts as a boundary layer. These beetles often appear yellow or cream colored because of the pollen.

In their juvenile phase, periodical cicadas (Cicadidae) feed underground on fungus and juices obtained from roots. The Seventeen-year Locust of the eastern United States is actually a cicada. California species spend 2–5 years underground. When nymphs emerge, they crawl up the nearest shrub or tree, where they molt, splitting the skin down the back. When the adult emerges, its wings fill with fluids and the new skin hardens. Concealed in vegetation, the adult "sings" loudly by vibrating membranes on its underside. This noise is a territorial and mating call. After mating, eggs are laid on twigs. Newly hatched nymphs drop to the ground, burrow in, and the cycle repeats.

Internal juices of plants are nutritious, particularly during periods of vigorous growth. At this time, insects such as aphids (Apidae) become very common. They tap into the new growth with piercing mouthparts and suck the juices. Some discharge a sugary solution known as honeydew from the anus. Ants are attracted to the honeydew and may foster the spread of aphids by carrying wingless forms to new locations. Aphid life cycles are complex,

FIGURE 9.53 Stink beetle, *Eleodes* sp., in its threat position.

involving winged and wingless forms. Some reproduce parthenogenetically: eggs develop without being fertilized. Parthenogenetic forms may appear at different times, or they may alternate seasonally.

Larvae of various gossamer-winged butterflies (Lycaenidae) also produce a type of honeydew that is attractive to ants. Included in this family are small butterflies known as hairstreaks (Theclinae), coppers (Lycaeninae), and blues (Polyommatinae). A number of these are associated with deserts. The Leda Ministreak, *Ministrymon leda*, is associated with Honey Mesquite, as are Ceraunus Blue, *Hemiargus ceraunus gyas*, and Reakirt's Blue, *Echinargus isola*. The Western Pygmy-blue, *Brephidium exilis*, and the San Emigdio Blue, *Plebejus emigdionis*, are often associated with Shadscale Scrub.

The large Gray Snout-weevil, *Ophryastes desertus* (figure 9.54), also feeds on plant juices. This is a large beetle in the weevil family (Curculionidae), one of the most diverse families of organisms in the world. No other family in the animal kingdom has as many species. There are more than a thousand species in California alone.

FIGURE 9.54 Gray Snout-weevil, *Ophryastes desertus.*

Of the vertebrates that are herbivorous, most are birds and mammals, although not many can afford to be so specialized as to eat only plant material. Food is too scarce. Of the reptiles, there are a few herbivores, including the Desert Tortoise, the Desert Iguana, and the Common Chuckwalla. One problem with herbivory is that large amounts of potassium or other salts may be consumed in plant foods. Desert reptiles deal with this by expelling the excess salts through nasal salt glands and then snorting the fluid from their nostrils.

It is interesting to note that the herbivorous reptiles tend to be larger than the carnivores. Eating plant material generally produces fewer calories per bite, but much less effort is expended in gathering the food. Plant life is 10 times as abundant as animal life, and it doesn't attempt to escape. Herbivores therefore are able to move their large body masses with minimal energy expenditure, investing their calories in mass rather than locomotion. This principle also explains why grassland mammals tend to be larger than the predators. Based on these size relationships, a person might speculate that juvenile reptiles eat insects, and that their diet changes when they become larger. This is apparently not the case. It seems that whatever is the preferred diet of a reptile, it remains so throughout the life of the animal.

Of the mammals in deserts, herbivores such as rabbits and Bighorn Sheep eat leaves and twigs, but most of the herbivorous mammals are seed-eating rodents. Most herbivorous birds, such as quail and doves, are seed eaters but are not obligated to plant material, supplementing their diets with insects whenever possible.

Because seeds may be abundant in the desert, and because they are high-calorie items, they are objects of intense competition. Ants, rodents, and birds seem to compete for the same resource. How does niche partitioning avoid this problem?

Some insight into the problem of competition among seed-eating animals (granivores) can be gained by studying harvester ants. In Arizona where there is summer and winter precipitation, there are many kinds of seeds on a yearly basis. Ants are renowned for carrying loads that are disproportionately large for their body size. It has been proposed that ants, in order to be more efficient in carrying a load, might select seeds of a certain size and in this way partition the resources. In Arizona there are five to seven species of seed-harvesting ants, each of which is a different size. It appears that each species is adapted to carry seeds of a different size. In California deserts where there is only winter rain, the most common harvester ant, *Veromessor pergandei*, is polymorphic in body size. Depending on which type of seed is abundant, a different size of worker ant is apparent. This is only circumstantial evidence, but it is supported by the fact that in Arizona there are also five to seven size groups of granivorous rodents. Apparently, seed eaters in general avoid competition by gathering different sizes of seeds.

Further studies of rodents show that different species forage in different places. Kangaroo rats (*Dipodomys* spp.; figure 9.46A) forage in the open. Their bipedal form of locomotion (hopping) enables them to move rapidly and escape predators. Pocket mice (*Perognathus* spp., *Chaetodipus* spp.; figure 9.46B) superficially resemble kangaroo rats. They are in the same family (Heteromyidae), but they are smaller. In spite of having enlarged hind legs, pocket mice move around on all fours: they are quadrupedal. Pocket mice forage beneath the shrubs. When seeds are abundant, the pocket mouse need not move from under the protective canopy of the shrub, under which is the entrance to its burrow.

Seasonal variation in diet is another factor that reduces competition for seeds, particularly with birds. When seeds are abundant in spring and early summer, sparrows, doves, and quail eat primarily seeds. Later in the year, they switch to insects. Black-throated Sparrows, *Amphispiza bilineata*, in particular are noted for their fly-catching behavior during the dry season, which frees them from seed competition and provides them with a food source that is rich in water. They are also noted for eating entire flowers to obtain the nectar within.

The Phainopepla, *Phainopepla nitens* (figure 9.33), is a fly-catching species that switches to the fruit of mistletoe when it is abundant. Phainopeplas are in the family of silky flycatchers (Ptiliogonatidae), which are closely related to the waxwings. Black birds with a large, slender crest, they are conspicuous as they perch in the tops of large shrubs or trees along washes. From this perch, they search the sky for flying insects. When Desert Mistletoe comes into fruit, Phainopeplas feed heavily on the berries. The seeds pass through their digestive tracts, and while they perch they distribute the seeds to uninfested branches. It is primarily through the Phainopepla that the Desert Mistletoe, so common in Catclaw and Honey Mesquite, is disseminated.

In some parts of the Mojave Desert, there are three species of ground squirrel that seem to feed on the *same things:* the Mohave Ground Squirrel, *Xerospermophilus mohavensis*, the Round-tailed Ground Squirrel, *X. tereticaudus*, and the White-tailed Antelope Squirrel, *Ammospermophilus leucurus* (plate 18E[12E]). They scurry about in the spring, feeding on seeds, flowers, and leaves. The Antelope Ground Squirrel is active year-round, but the other two species, doubling their weight in spring, are dormant from August to March, thereby conserving a great deal of energy. The three species avoid competition by feeding together only when food is abundant. The rarest of the three species, the Mohave Ground Squirrel is now on the state list of threatened species. The Antelope Ground Squirrel hibernates in the northern part of its range. There it overlaps with only one other species, Townsend's Ground Squirrel, *Urocitellus townsendii*, another hibernator. All of the ground squirrels avoid competition with other seed gatherers to some degree by climbing into the shrubs or trees and feeding on the seeds before they fall to the ground.

Most of the carnivores in the desert are insectivorous because insects are the most common animals. The most common carnivores are probably scorpions, but spiders and centipedes are also abundant. These are all venomous predators. Scorpions inject the venom with the tail stinger, and spiders and centipedes do so by means of fangs on their jaws called chelicerae.

In California, there are no scorpions that are deadly to humans, although there are several in southern Arizona and Mexico. In most cases, a scorpion sting can be painful but is not much worse than a bee sting. Scorpions spend the day hiding under rocks or debris or in small burrows, emerging at night to forage. They do not see well, but they are effective hunters. Usually they stand and wait, sensing the approach of potential prey by feeling vibrations in the soil with their legs. The scorpion can orient to these vibrations in the same way other animals orient to sound. When the prey comes near, the scorpion grabs it with its pincers and stings it. The scorpion then crushes the immobilized prey with its pincers, tears it apart, and sucks up the fluid remains.

What wanders around the desert at night and falls prey to scorpions? Most studies indicate they eat other arthropods. A surprise that emerged from studies in the Coachella Valley is that a significant portion of food for scorpions is other scorpions. A large species, *Parturoctonus mesaensis*, makes up 95% of the scorpion population of the area, and in addition to its usual arthropod prey, it feeds heavily on three other species of scorpion. It seems to prey particularly on juvenile and male scorpions. Studies of behavior also indicate that the other species have evolved activity patterns that help them avoid predation.

Spiders usually trap their prey in a web. The prey, typically a flying insect, hits the web, causing vibrations that alert the spider. The spider then injects venom into the prey and wraps it with silk to prevent it from escaping. Regurgitated digestive enzymes liquefy the internal tissues of the insect, and the spider returns later to suck out the digested material. It does not eat the exoskeleton.

One prey item that is sometimes attacked by a spider is the millipede, a common detritus feeder. Millipedes, however, have a very interesting defense. They produce a tranquilizing chemical that puts the spider to sleep after it feeds. This does not protect the millipede that gets bitten, but the spider, if it survives its period of drug-induced sleep, will avoid millipedes as a prey species in the future. The millipede that gives up its life to a spider altruistically helps the survival of the remaining millipedes.

Few spiders in California are dangerous to humans. The only spider here that is likely to harm a human is the Western Black Widow, *Latrodectus mactans*. These spiders are most often associated with human habitation. Woodpiles and outhouses are their favorite haunts. They do not normally bite humans, and they are rarely deadly. Symptoms include pain, abdominal cramps, dizziness, nausea, and vomiting. The usual treatment is simply to relieve pain and includes tranquilizers. A typical adult human recovers fully in about 4 d.

In recent years, a legend has been building about the dangers of being bitten by the Brown Recluse or Violin Spider, *Loxosceles reclusa*. These are rare, nonnative spiders that occur only around habitation. For some reason, there is more concern about this rare, introduced species than about the four native species of *Loxosceles* that are also known as brown spiders or violin spiders. The most widely distributed is the Desert Brown Spider, *L. deserta*, which occurs in the southern San Joaquin Valley as well as the Mojave and Colorado Deserts. The other three species have localized distributions in Riverside County, Death Valley, and Palm Oases of the Colorado Desert.

There are only sporadic reports of bites from these spiders, but because the possibility exists, the symptoms of a Brown Recluse bite are worth noting. There may be a severe reaction, but it is rarely deadly. The bite may have a transient, local effect, or it may produce an ulcer that persists for several days. In some cases, severe tissue damage occurs, and skin grafting is required. Very few people get anything more than a local effect, which involves pain at the site from 2–8 hr after the bite. Treatment ranges from topical cortisone to reduce inflammation, to surgical excision of the damaged tissue.

Few predatory insects are as interesting as the praying mantis. The name refers to the manner in which they hold their front legs, as if they were praying. There are several species of these tree-dwelling predators. They grasp their prey, usually another insect, and dismember it while they eat it alive. The mating behavior of some species of praying mantis is legendary. Once the male begins to copulate, the female bites off his head. This act stops the flow of a hormone produced in the brain that inhibits copulation. Because the copulatory organ is on the posterior end of the mantis, the female is now free to eat the remainder of the male while continuing copulation. The basis of this strategy is that once the female has been impregnated, the male would merely be a competitor for food. Killing the male reduces competition for food, which enhances survival of the female and the offspring. If the female is well-fed, she will not consume the male. Although it is not common, a similar strategy may be employed by a female Western Black Widow, a behavior that is responsible for the species' common name.

Very common among carnivorous insects are red ants (*Formica* spp.). These ants are actually carrion eaters. In less than an hour, a swarm of red ants can dismember and carry away the remains of a large insect, such as a grasshopper or dragonfly. Red ants also eat the seed-gathering, black harvester ants.

Bees and wasps are related to ants. One group of wasps in which the females are flightless is known as velvet ants (*Dasymutilla* spp.). These wasps are covered with a boundary layer of long hairs. Their color ranges from bright red, to orange, to yellow, to white. They are distributed throughout the state. The most common desert forms are white and look like a bit of fluff blowing along the ground. They possess a stinger, with which they can immobilize prey, often ground-nesting bees and wasps. If molested by a human, they may produce a painful, stinging welt, but they are not dangerous.

Honey Bees are also hairy. Scientists searching for a function of the hair on animals have discovered that Honey Bees have approximately the same number of hairs (3 million) as a Gray Squirrel. While the hairs are water repellant, they are also sensory. But the primary function of the hairs may be to keep clean. Using super-slow-motion videos of pollen-covered Honey Bees and Fruit Flies, scientists observed that the bees were using their legs to bend the hairs and when the hairs were released they snapped back to their upright state effectively hurling

the pollen, catapult-like, at a great rate of speed. Also, while doing research on other insects, researchers discovered that Hairstreak butterflies (Lycaenidae) and Luna Moths (*Actias luna*), with about 10 billion hairs on their bodies, were the hairiest of all.

Among the most interesting wasps are the large, orange-winged tarantula hawks (*Pepsis* spp.). A number of wasps have evolved to be associated with certain species of prey. Tarantula hawks attack tarantulas or trap-door spiders.

Tarantulas live in a silk-lined burrow. When a prey species disturbs the web near the entrance, usually at night, the tarantula rushes out and bites it, feeding in typical spider fashion. The adult tarantula hawk feeds on nectar, usually of milkweed flowers (*Asclepias* spp.). It also walks about the desert, often at night, searching for tarantula burrows. When it finds one, it disturbs the web, causing the tarantula to rush out, and the fight begins. Once the spider is taunted into rearing back into its attack position, the wasp stings it at a precise location on the underside where it is able to pierce the spider's armor. Often the spider escapes, but if it gets stung, the wasp's venom paralyzes it without killing it. The wasp is unaffected by the spider venom. If, however, the spider bites the wasp at the same time that it is stung, its paralysis can cause the wasp to become impaled on the fangs. The wasp, unable to get free, will die.

After the spider is paralyzed, the wasp may enlarge the spider burrow by making a side chamber, it may make a new burrow, or it may use the burrow of another animal. Eventually, the wasp drags the paralyzed spider down the burrow and lays an egg upon it. The egg hatches into a white larva in a few days. Then, for about 30 d, the larva devours the paralyzed, still-living spider. When the spider is consumed, the larva pupates, and the adult wasp emerges the following year. A successful adult wasp can paralyze up to 20 spiders, ensuring that 20 new wasps will emerge the following year.

Sometimes while the wasp is underground, enlarging the chamber, another wasp will

FIGURE 9.55 Long-nosed Leopard Lizard, *Gambelia wislizenii*. Note the tail of a recently swallowed Western Fence Lizard, *Sceloporus occidentalis*, hanging from the mouth of the lizard.

locate the paralyzed spider. It will attempt to claim it, dragging the spider away from the burrow. Usually the original wasp discovers the thief, and a fight ensues. The winner will kill the other wasp and claim the spider.

These large black wasps with orange wings are conspicuous. This color is probably important for defending territories and attracting a mate. During the day, males perch on top of large shrubs, particularly along a ridge line, where they will be easily observed by flying females. When a female appears, the male courts by displaying his wings. After mating, the female presumably spends the next night on the prowl, attempting to entice a tarantula from its burrow.

Although the wasp completes its life cycle in less than a year, female tarantulas are long-lived. One female in captivity lived for 30 years. They reach maturity in about 5 years, after which they produce 50–100 babies per year. Males do not mature for 8–10 years, but they die soon after mating. The female selects a mate that is about the same size as she is. Because females continue to grow for their entire lives, females of all different sizes require males of all different sizes. Males therefore mature at various sizes and ages to provide the variety of sizes to catch the female size groups. The tarantulas commonly seen in the spring during the day, sometimes in large numbers, are usually males out searching for females.

Most insects attack other arthropods, but a few bloodsucking insects attack mammals. Mosquitoes and horseflies are well known to humans, whom they pester. A lesser-known pest is the Western Bloodsucking Conenose, *Triatoma protracta*, a large, elongated bug in the assassin bug or kissing bug family (Reduviidae). These pests usually are found in woodrat nets, but sometimes they invade cabins. They typically prey upon woodrats, but on occasion they will bite a human, producing a painful, itching bite that may form a large welt up to 3 in (7.5 cm) in diameter. Approximately 5% of those who are bitten develop an allergic reaction that requires anti-histamine treatment. In Mexico and South America, related bugs introduce a blood parasite, a protozoan known as *Trypanosoma cruzi*. The subsequent affliction is a form of sleeping sickness known as Chagas' disease. At the present time, it is unknown in California.

Among vertebrate animals in deserts, lizards, and snakes are the primary carnivores. The number of lizard and snake varieties in the desert is far too numerous for thorough discussion. Many of them also occur in Cismontane Southern California, and these were discussed along with southern California's scrub communities. Most lizards eat insects. A few carnivorous lizards, such as the Collared lizards, *Crotaphytus* spp., and the Long-nosed Leopard Lizard, *Gambelia wislizenii* (figure 9.55) eat other lizards.

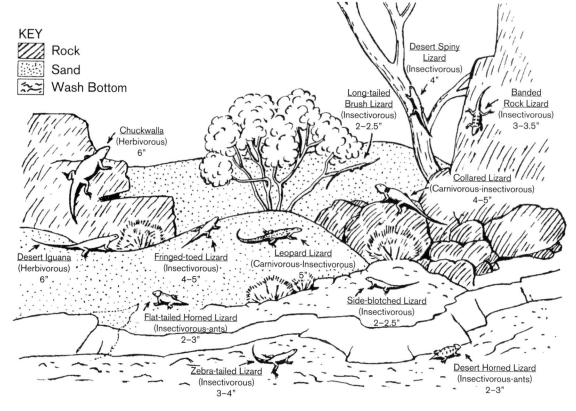

KEY

▨ Rock
░ Sand
〰 Wash Bottom

Chuckwalla
(Herbivorous)
6"

Desert Spiny
Lizard
(Insectivorous)
4"

Long-tailed
Brush Lizard
(Insectivorous)
2–2.5"

Banded
Rock Lizard
(Insectivorous)
3–3.5"

Collared Lizard
(Carnivorous-insectivorous)
4–5"

Desert Iguana
(Herbivorous)
6"

Fringed-toed Lizard
(Insectivorous)·
4–5"

Leopard Lizard
(Carnivorous-Insectivorous)
5"

Side-blotched Lizard
(Insectivorous)
2–2.5"

Flat-tailed Horned Lizard
(Insectivorous-ants)
2–3"

Zebra-tailed Lizard
(Insectivorous)
3–4"

Desert Horned Lizard
(Insectivorous-ants)
2–3"

FIGURE 9.56 Ecological segregation of lizards of the family Phrynosomatidae in the Sonoran Desert (illustration by Gene Christman from Stebbins, R.C. 1988. Reptiles: Adaptations to Desert Ecosystems. *Environment Southwest* 522:13–16; reprinted with permission).

With so many insect-eating lizards, it is not surprising that an array of mechanisms to avoid competition exists. Niche partitioning among desert lizards is based on food preference, size, and favored habitat (figure 9.56). Insectivorous sand dwellers include fringe-toed lizards (*Uma* spp.; figure 9.40A), the Zebra-tailed Lizard (*Callisaurus draconoides*; figure 9.40B), and horned lizards (*Phrynosoma* spp.; figure 8.38). Fringe-toed lizards live in sand dunes, Zebra-tailed Lizards live along washes, and horned lizards feed in sandy areas almost exclusively on ants.

Of special note are the horned lizards. There are several species in the desert, but they all have the same basic lifestyle, eating 150–200 ants per day. This diet of ants adds a load of sodium that is excreted by the nasal salt gland. Horned lizards, in turn, are preyed on by Road-

runners, but their primary enemies are off-road vehicles. It is true that a horned lizard will squirt blood from behind its eye when molested.

The Desert Spiny Lizard, *Sceloporus magister*, and the Mearn's Rock Lizard, *Petrosaurus mearnsi*, are large lizards that feed on large insects. They also occur in different habitats. The Desert Spiny Lizard primarily lives in trees, including Joshua Trees. The Mearn's Rock Lizard, as its name implies, lives in rocks.

The Long-tailed Brush Lizard, *Urosaurus graciosus*, and the Common Side-blotched Lizard, *Uta stansburiana* (figure 8.36), are small insectivores. The Long-tailed Brush Lizard lives in trees and brush. This lizard rests on branches in the shade and ambushes small insects. The Common Side-blotched Lizard also feeds on small insects, but it prefers to feed in the open.

FIGURE 9.57 Desert Night Lizard, *Xantusia vigilis*.

Banded Geckos (figure 9.47) are small, nocturnal insectivores. Peninsular Leaf-toed Geckos, *Phyllodactylus xanti*, are also small, nocturnal insectivores, but they live and feed almost entirely under rock flakes. The Desert Night Lizard, *Xantusia vigilis* (figure 9.57), is a small insectivore that feeds primarily on termites under the fallen boughs and trunks of yuccas.

All snakes are carnivores, eating all forms of insects and vertebrates. Because of this, snakes are scarce. They make up only a portion of the less than 9 lb per acre (1 g/m²) carnivore biomass. Because they are not common, and because they are easy to maintain, they have become desirable pets to some people. Snake collecting at night along desert roads has become a hobby and a vocation for some people. Unfortunately, this collecting has seriously threatened many species of snake, and it is now illegal to collect or sell wild amphibians and reptiles. Offspring of only three kinds of captive snakes may be sold: Rosy Boas, California Kingsnakes, or Gopher Snakes. Collecting native reptiles or amphibians requires possession of a valid California Fishing Permit. A bag limit of two animals per day is legal. Some species may not legally be collected because of their protected status, or because they are rare in certain regions. Rattlesnakes require no permit, but the limit is also two in possession. Captive snakes may not be released into the wild.

Of great interest to many people are the venomous snakes. In California, all truly venomous snakes are in the rattlesnake family (Viperidae). About six different species of rattlesnake are present in the deserts. A couple of others occur in marginal habitats and are found occasionally in what might be called a desert habitat. Three species might be found in coastal southern California.

Rattlesnakes are pit vipers. They have a temperature-sensitive pit beside each nostril, and this gives them "stereo" heat perception useful in locating endothermic (warm-blooded) prey and predators. Typical of venomous animals in general, they can immobilize prey species that are large enough to do them harm. In typical snake fashion, jaws disarticulate to allow the prey to be swallowed, even though it may be larger than the snake's head. Teeth curve backward, and each half of the lower jaw moves independently. The snake, when it swallows, moves one half of its lower jaw at a time. In this way it can "crawl" around its dinner. The rattle is a threat; nothing more. Its function is to warn some foolish animal that is too big to eat. The snake has no intention of wasting its venom to protect itself except as a last resort. Any human threatened by a rattlesnake simply should walk slowly away. Most bites occur by accident, when the snake is stepped on, or when it is being handled. Legends have arisen about the speed of a rattlesnake when it strikes. Experiments at the University of Louisiana have demonstrated they literally are quicker than the blink of an eye, which takes 202 milliseconds. Average rattlesnakes strike within 50–90 milliseconds and accelerate at 193 m/s (432 mph) creating 28 gs of force in the process.

One of the most interesting rattlesnakes is the Sidewinder, *Crotalus cerastes* (figure 9.58). Sidewinding is a form of locomotion that many kinds of snakes use to move over loose sand. Sidewinders use this type of locomotion on all surfaces. They are usually found in sandy areas, where, similar to other rattlesnakes, they lie and wait for small animals, usually rodents. Sidewinders have a small horn over each eye that acts like a gable on a roof. It keeps sand out of the eyes of the snake as it lies in wait, buried

FIGURE 9.58 Sidewinder, *Crotalus cerastes.*

except for its nostrils and eyeballs. A parallel development has occurred in an African pit viper. This species looks and acts like an American Sidewinder, but it has no rattle.

The Mojave Rattlesnake, *C. scutulatus*, is sometimes known as a Mojave Green even though it is rarely greenish. This rattlesnake is notable for its potent venom. In most rattlesnakes, the venom has a hemolytic action; it digests tissue. Mojave Rattlesnakes have a venom that has hemolytic and neurotoxic properties, making this snake a bit more dangerous than other species. It is important to realize, however, that rattlesnake bites are rarely deadly, unless they are close to the head. Most bites are on the leg or on the hand from handling the snake. Symptoms include pain and swelling, and sometimes weakness and nausea. The latter two symptoms may be the result of anxiety. Under some circumstances, there is significant tissue loss. Not all bites inject venom, and all persons do not react the same way. No first aid should be administered unless the victim is more than 3 hr from a hospital—a rare circumstance in California. Once in the hospital, the victim can be watched. If a serious reaction begins to develop, an antivenom (antivenin) will be administered. If a victim is more than 3 hr from a hospital, the best treatment is to splint the limb and wrap it to inhibit lymph circulation beneath the skin. By no means should

a tourniquet be used or should the wound be cut and blood sucked out. Chilling or cryotherapy also has been proved to have little value.

The Western Diamondback, *C. atrox*, our largest western rattlesnake, can measure 90 in (229 cm) in length. Found primarily in the Colorado Desert, it can be identified by its distinctive pattern of black-and-white rings on the tail (figure 9.59). The Speckled Rattlesnake, *C. mitchellii*, is primarily a rock dweller in the Mojave and Colorado Deserts, but it is also found on the coastal side of the Peninsular Ranges. Its color pattern is highly variable, including dark gray to shades of pink or orange, with dark bars across its back. Northern populations have been described as the Panamint Rattlesnake, *C. stephensi*, which appears to be a dark-colored Speckled Rattlesnake.

A few rear-fanged snakes occur in California. Among these are California Lyresnakes, *Trimorphodon biscutatus*, and Desert Nightsnakes, *Hypsiglena chlorophaea* (figure 9.60). Strictly speaking, these are venomous snakes, but they are not common and are not likely to bite a human. Even if a person is bitten, the location of the fangs at the rear of the mouth precludes the introduction of much venom. There are no California records of a human experiencing difficulty from the bite of a rear-fanged snake. They are found under debris such as fallen Joshua Tree boughs or under rock flakes on cliffs.

There are no large carnivores in the desert, with the occasional exception of a Mountain Lion, simply because there is not enough food to support a population of large animals that eat nothing but other animals. With the desert capable of supporting less than 9 lb per acre, the range of a 100 lb carnivore would have to be spread over 11 acres, and there would be no food left for any other carnivorous species, even an insect. There are carnivorous birds such as Golden Eagles, hawks, and owls. Their home ranges by necessity are quite large, which is not usually a problem for a bird. Common Poorwills, *Phalaenoptilus nuttallii* (figure 9.61), nighthawks, *Chordeiles* spp., swifts (Apodidae), and bats (Chiroptera) are insectivorous. They

FIGURE 9.59 Western Diamond-backed Rattlesnake, *Crotalus atrox*. Note black and white rings on the tail.

avoid competition by foraging at different elevations, and at different times of day.

It was mentioned earlier that Roadrunners allow their body temperatures to drop during the night, thus conserving metabolic energy. A more extreme example of this phenomenon is exhibited by the Poorwill, which hibernates during the winter, conserving a great deal of energy during the time that flying insects are least abundant. The Poorwill was the first bird discovered to be a hibernator. When emerging from hibernation a Poorwill might lie in warm sand and spread is wings in order to maximize absorbed solar radiation (figure 9.61).

It is now known that swifts are also hibernators. White-throated Swifts are widespread in the state, even nesting on high-rise buildings in urban areas. In deserts, they are conspicuous flying about cliffs. They often attract attention by their high-pitched squeaks as males chase females. Apparently they mate in mid-air, falling downward, while contacting each other, and pulling out of the fall before they reach the ground. During winter, they may gather by the hundreds in communal hibernacula such as large cracks in rocky cliffs.

The larger mammals with reputations for being carnivores are actually omnivorous. These include foxes, Badgers, and Coyotes. The Kit Fox, *Vulpes macrotis*, is small, about the size of a house cat. Its size is an adaptation to a food-

FIGURE 9.60 Desert Nightsnake, *Hypsiglena chlorophaea*.

poor environment. It takes a smaller biomass of food to support a population of Kit Foxes than it does to support, for example, a population of Common Gray Foxes, *Urocyon cinereoargenteus*. Kit Foxes eat insects, scorpions, mice, and fruits. They are nocturnal and apparently very curious, because they are common visitors to campsites. Often they approach closely enough to be clearly visible from the light of a campfire. Their large ears indicate that hearing is important for their foraging, but they probably are just as important for radiating heat.

The Coyote, *Canis latrans* (figure 8.50), has a worse reputation among ranchers than any

FIGURE 9.61 Common Poorwill, *Phalaenoptilus nuttallii*, basking in the sun after emerging from hibernation.

other animal. They have been shot, trapped, and poisoned, but somehow they endure. The problem is clear. The Coyote has a reputation for preying on lambs and calves, not to mention house cats and small dogs. Vast acreage of desert is leased for grazing. Of 171 million acres administered by the Bureau of Land Management (BLM, hereafter), 150 million are authorized for grazing. Coyotes are opportunistic, feeding on whatever is abundant. They will take meat when they can get it. If an opportunity arises, they will take a lamb or calf, particularly if there is a family of pups to be fed. Coyotes are normally solitary; the only time they hunt in packs is during winter when food is scarce. Grouping up is more common in the Great Basin than in other deserts.

As of 1982, the US Fish and Wildlife Service had an annual budget of $8 million for animal control. The bulk of that money was spent on Coyote control, and most of that was spent in the western deserts to protect cattle and sheep. Several authors have calculated that, from an economic point of view, it would be cheaper to pay a rancher directly for lost livestock than to continue attempting to reduce Coyote populations. Coyotes are marvelously density-dependent. A female is capable of producing two to eight pups, depending on the size of the Coyote population. When population density is reduced, Coyotes respond by having larger families.

Coyotes are a natural control on other pests, such as rats, mice, and rabbits. Jackrabbits are also remarkable examples of density-dependent animals. If predators are removed, the jackrabbit population explodes, and the rabbits invade fields and eat crops. Even where they don't eat crops, jackrabbits can have a significant effect on desert vegetation. Another calculation showed that killing five Coyotes increases the jackrabbit population to the degree that they eat the same amount of forage as a steer. In the state of Arizona, 1864 Coyotes were killed in 1969, the last year that these data were published. Competition from increased jackrabbits potentially cost ranchers grazing land that would support an additional 373 cattle. In other words, killing Coyotes actually reduced the cattle herd by increasing competition from jackrabbits. It actually cost the ranchers more money than if they had allowed the Coyotes to remain.

Until the early 1900s, Gray Wolves, *C. lupus*, also roamed the Great Basin, feeding on Pronghorn and Bison. Conversion of the Great Basin to pasture nearly eliminated the large native herbivores, and the predator extermination program finished off the Gray Wolf. Gray Wolves today are on the list of endangered species, occurring only in a few protected habitats along the US-Canadian border and in national parks such as Glacier and Yellowstone. Gray Wolf reintroductions have been successful in those parks. The wolf family that has become established in northern California is discussed in Chapter 6. The former Gray Wolf population of the Great Central Valley will be discussed in the next chapter.

Overgrazing by cattle and sheep is a reality. Only 20% of BLM land is not overgrazed. Environmentalists argue that grazing fees are so low that it represents a subsidy to the cattle industry. Ranchers meanwhile contend that they are better stewards of the land and that they sustain rural economies. Scientists contend that livestock browse selectively, shifting the composition of natural vegetation toward unpalatable species; livestock enhance the distribution of weeds. Compaction of soil by livestock causes increased erosion. Old photographs taken at the turn of the century show that the southwest desert had a different appearance before the introduction of livestock. Steep-sided, dry arroyos are caused by increased runoff associated with overgrazing. There were few arroyos prior to the introduction of livestock.

An unfortunate consequence of overgrazing is the depletion of habitat for native sagebrush inhabitants such as the Greater Sage-Grouse, *Centrocercus urophasianus*, which is now recognized as a threatened species. It's range in California includes the Modoc Plateau and the Mono Basin. The problem is that the desert's low productivity will sustain very few livestock. On a sustainable basis, it takes about 10 acres (24.7 ha) to support a single steer or five sheep, and this disregards the existence of all other herbivorous animals, including native insects, lizards, rodents, birds, and mammals. In actual practice, for deserts to support cattle with minimum impact, there should be no more than one animal for every 70–100 acres (30–40 ha). Deserts simply are not appropriate for livestock grazing. The number of cattle grown on desert ranges is but a fraction of total beef production in this country. Most cattle in deserts are brood stock for production of calves. In California, a mere 0.3% of beef production occurs in the desert. If cattle grazing here were stopped entirely or greatly reduced, it would have minimal impact on beef production, but it would be a step toward recovery of our desert ecosystems. Abandoning desert grazing would also save the millions of dollars per year now spent trying to control Coyotes.

The damaging effect of overgrazing and other disturbances on desert soil has been well documented, but only recently has it been shown that the effect on cold deserts such as the Great Basin is even more dramatic than was previously thought. Disturbance of the soil surface can destroy a microbiotic crust that includes stabilizing filaments of cyanobacteria (blue-green algae)

and nitrogen-fixing lichens. This filamentous mat in a sponge-like way absorbs water and prevents erosion. It takes about 10 years for these mats to form, without which wind and water carry away the soil. Another problem with overgrazing is associated with feral animals, those that were domesticated but have gone wild. These include Feral Horses or Mustangs, *Equus caballus*, in the Basin-Range Province, particularly in the Modoc National Forest, and Feral Ass (Burro or Donkey), *E. asinus*, in the Mojave. In 1985, the number of Feral Horses and Burros in the western states was estimated at more than 60,000. Compared to the 4.1 million domestic livestock animals, that number seems barely significant.

Until 1971, government agencies hunted feral animals to keep them under control. Sometimes meat was sold for pet food; other times, it was exported to Europe, where it is consumed by humans as a delicacy. In 1971, Congress passed the Wild Horse and Burro Act, which made it a federal crime to kill or even molest these animals.

The number of Feral Horses is increasing slowly. In California, the problem evidently has not reached crisis proportions, and it may not. The total number of Feral Horses in California has been estimated at about 2100, and research shows that the total number of animals has remained stable for many years. There may be as many as 750 Feral Horses in the Coso and Argus Ranges on land managed by the Naval Weapons Center China Lake. The largest single herd, containing about 150 animals, is located in the Montgomery Pass area, east of Mono Lake.

The main problem is that ranchers view Feral Horses as direct competition for cattle. Ranchers want Feral Horses removed, but the herds roam across state lines and occur on property managed by different government agencies. The BLM established a program to round up Feral Horses and have them adopted, and from time to time the US Navy contributed Feral Horses to this pool from their herds. Only about one-third of the animals are adopted. As of 2015, it was estimated that there were about 47,329 horses in the wild, which was estimated to be about twice the carrying capacity of western lands. In the late 1980s, the federal government was spending about $17 million a year and in 2015 holding costs were about half of the BLM's $72 million budget. The BLM decided to waive the $125 adoption fee if a rancher would graze the Feral Horses on his property. After a year, if the Feral Horses were not adopted, the rancher could keep them. What aggravated some people is that most of the Feral Horses were not adopted, and after the waiting period was over, some ranchers illegally sold them. Some of them became rodeo stock, but others were sold for meat. In 1988, the BLM suspended the fee-waiver arrangement, but there are still problems with horses being sold for food. It is illegal to kill horses on US soil, but it is not illegal in Mexico. Apparently a rancher in Colorado had exported 1794 horses to Mexico, which were subsequently slaughtered and the meat sent to European markets.

The Burro problem is another matter. As of 1986, their numbers were continuing to increase dramatically. With no natural enemies, Burros have overgrazed and compacted vast stretches of Mojave terrain to the detriment of native plants and animals. Burros eat about 10 lb (4 kg) of vegetation per day. One recent estimate for Saline Valley put the Burro population at one per acre (0.4 ha), which, similar to cattle, would put them over carrying capacity, even if there were no other herbivores in the area. In the Death Valley region, experimental exclosures were built to keep Burros out of certain areas. Three times as many plants were found inside the exclosures as on the outside, where the Burros grazed. Furthermore, the ratio of dead shrubs outside to dead shrubs inside was 27:1. As of 2015, about 10,800 Burros still roamed wild on BLM land. Currently the number of Wild Horses and Burros exceeds carrying capacity by about 35,000 animals. At that time, the number of unadopted or unsold horses and burros being held by the BLM was about 47,000.

One of the animals most seriously affected by Burros has been the Desert Bighorn Sheep, *Ovis canadensis nelsoni*. Although this is the most common of the three types of Bighorn Sheep in California, there is cause for alarm when one considers the rate at which the Burro population is increasing. Competition for food is an obvious problem, but of greater concern is that aggressive packs of Burros take over water holes in the summer. They foul the water with waste products, and they will not let the sheep drink. Sheep populations in some areas have dwindled rapidly. The Peninsular Bighorn Sheep, formerly a separate subspecies from the Desert Bighorn is now the same subspecies, but considered a separate population. The Peninsular population now numbers about 800 animals. It has been threatened by habitat destruction and the introduction of pneumonia, a respiratory disease introduced by domestic sheep. One of its important refuges is Anza-Borrego Desert State Park. The name "Borrego" is Spanish for the Bighorn Sheep.

10

The Great Central Valley

FIGURE 10.1 Rolling hills near Taft, western San Joaquin Valley. These introduced Mediterranean grasses have largely replaced native bunchgrasses.

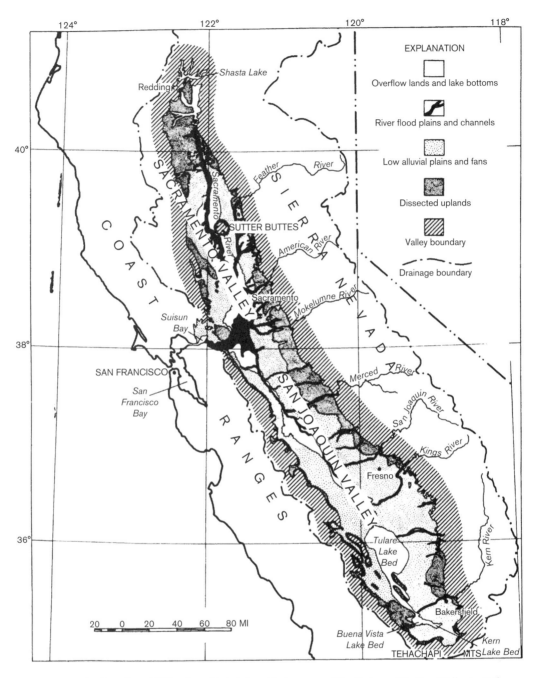

FIGURE 10.2 The Great Central Valley (from Hackel, O. 1966. Summary of the Geology of the Great Valley. In Bailey, E. H., ed., *Geology of Northern California*. Bulletin 190. California Division of Mines and Geology).

THE GREAT CENTRAL VALLEY, also referred to as the Central Valley or the Great Valley, is an elongate depression that lies between the Coast Ranges and the Sierra Nevada. It is about 430 mi (690 km) long and about 75 mi (120 km) wide. At its extreme northern and southern ends, the elevation is about 400 ft (120 m). At its center, east of San Francisco Bay, it is slightly below sea level. The valley floor is composed of thousands of feet of

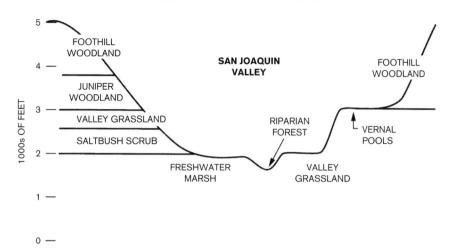

WEST – – – – – – – – – – – – EAST

FIGURE 10.3 Biotic zonation of the San Joaquin Valley.

sediments deposited by runoff from the surrounding mountains. Below these sediments lie important petroleum and natural gas deposits. In an area near Bakersfield, oil wells dot the landscape for many miles.

The Great Central Valley is actually two large valleys lying end to end, each drained by a major river. North of San Francisco Bay, the Sacramento Valley is drained by the Sacramento River. To the south, the San Joaquin Valley is drained by the San Joaquin River. The confluence of these two rivers occurs east of San Francisco Bay. This area, the Sacramento-San Joaquin Delta, was formerly a massive wetland. It is now one of California's important agricultural areas. Groundwater from the delta is pumped into the California Aqueduct and pushed southward (uphill) all the way to southern California. So much water is pumped by this system that the State Water Project is the largest user of electricity in the state. Irrigation water is also provided by the Central Valley Project, which captures streams of the western Sierra Nevada. By virtue of all this water, the arid San Joaquin Valley has been altered from its native grassland to one of the nation's richest agricultural areas, and the San Joaquin River today is largely agricultural runoff for

most of the year. It is difficult to establish precise boundaries for the Great Central Valley (figure 10.2) because at its edges the biological communities, such as Valley Grassland, grade into Foothill Woodland or Oak Woodland in an irregular manner. (Foothill Woodland is discussed in Chapters 4 and 7.) Elevation considerations are not precise either, because the floor of the valley grades into the foothills over a series of terraces (figure 10.3), and a number of low-lying adjacent valleys of the Coast Ranges and Sierra Nevada foothills are dominated by biological communities that are typically associated with the Great Central Valley. In general, the borders of the Great Central Valley are formed by the zone where alluvial soils grade into bedrock features and the landscape is dominated by Foothill Woodland.

This region of the state, more than any other, has been altered by activities of humans. In its pristine state, it contained three primary communities of plants and animals: Valley Grassland, Freshwater Marsh, and Riparian Woodland (figure 10.4). In 1987, a report entitled *Sliding toward Extinction* was prepared at the request of the California Senate Committee on Natural Resources and Wildlife and commissioned by the California Nature Conservancy.

Freshwater Marshes in California

Riparian Woodland in the Central Valley of California

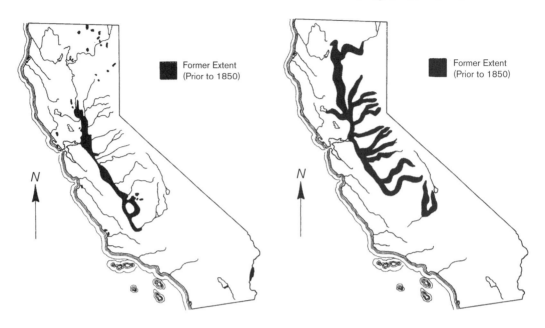

Central Valley Grassland

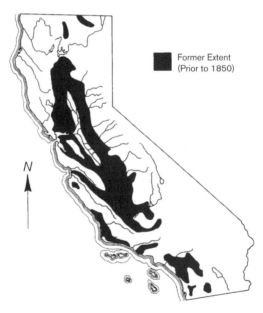

FIGURE 10.4 Former distribution of native communities in the Great Central Valley (from Seligmann, P. 1981. Changing California. *Nature Conservancy News* 31:18–23; reprinted with permission).

In this report, the state of natural communities in the Great Central Valley is clearly depicted. Valley Grassland, formerly a mixture of herbaceous annual wildflowers and grasses, was once the most common community. This was California's prairie, but it was virtually barren in the dry season. There were more than 22 million acres (9,000,000 ha) present in the early 1800s, and it is now 99% gone. In low spots, there were numerous Freshwater Marshes fed by winter precipitation and runoff from the mountains. About 94%, or 3,750,000 acres (1,540,000 ha), of this community have been lost. Cutting through the valley were large rivers fringed with a gallery forest of Riparian Wood land. About 89%, or 819,000 acres (334,000 ha), of it are lost.

The Great Central Valley is one of California's arid regions. As such, it shares many characteristics with the deserts. Desert communities such as Saltbush Scrub and Alkali Sink are distributed extensively in low-lying areas throughout the southern San Joaquin Valley.

CLIMATE

The climate of the Great Central Valley promotes widespread grasslands. Precipitation occurs during winter months; it is reduced because of the rain shadow of the Coast Ranges. To the north, in the Sacramento River Valley, average annual precipitation is moderate. To the south, it is reduced to desert proportions. At four weather stations, from north to south, average annual precipitation has been measured as follows: Redding, 38 in (95 cm); Sacramento, 18 in (46 cm); Fresno, 10 in (24 cm); and Bakersfield, 6 in (15 cm).

Grasslands of the world go by a variety of names such as prairie or plains, but the term "steppe" is often applied as well. These regions have many characteristics in common, including plant-animal interactions and conversion to agriculture and pastures. Precipitation in California is Mediterranean, with the rainy season in the winter, but that is not true throughout the world

where summer precipitation is often the case, as it is to some degree in the central United States. Nevertheless, grasslands occur where precipitation would be classified as arid to semiarid.

Cold-air drainage from the surrounding mountains becomes trapped, forming a persistent inversion layer in the Great Central Valley. During winter, this is manifested in a dense, ground-hugging fog known as tule fog. In association with these dense fogs, massive, chain-reaction traffic accidents occur nearly every winter along Interstate 5 or Highway 99. Sometimes the sun fails to penetrate for weeks on end. During summer, the inversion layer "burns off," but photochemical smog becomes a problem. Summer days are hot and hazy.

GEOLOGY

The Great Central Valley is a huge basin filled with sediments. Sands and gravels over 30,000 ft (9500 m) deep lie upon Sierran basement rocks that extend downward at an angle from the western slope of the Sierra Nevada. The deepest region of sediments is off-center, near the eastern edge of the Coast Ranges; presumably, there is a deep fault system in that area. A thrust fault zone occurs along the western edge of the valley. The underlying basement rocks simply are a continuation of the landform that makes up the Sierra Nevada.

The deepest parts of the gravels and sands are marine sediments that have accumulated since late Jurassic time, about 145 million years ago. Apparently, the sea retreated from the valley at about the same time that the southern Coast Ranges were uplifted, and during the long history of accumulation of marine sediments, the basement rock continued to subside. During most of the Pleistocene, the area was occupied by shallow brackish and freshwater lakes, particularly in the southern San Joaquin Valley.

During the last 5 million years, sediments accumulated as alluvial deposits washed out of the mountains. These deposits are only a few thousand feet (1000 m) deep over most of the valley floor. Depths up to 10,000 ft (3300 m) occur in some places, indicating that the valley floor was not always as monotonous as it is today.

The only topographic relief that occurs in the valley today is near Marysville, where Sutter Buttes project above the valley floor to an elevation of 2100 ft (640 m). These are plug domes of andesite that are about the right age to be the southernmost of the old Cascade volcanoes. Their origin, however, is controversial.

Trapped beneath the sediments of the San Joaquin Valley is one of California's richest oil fields. Oil has been the keystone of Kern County's economy since the 1920s. In 2008, Kern County had 42,000 oil wells that produced 68% of California's oil production. Add to that another 2000 wells in Fresno County and California ranks as the fourth highest oil producer in the United States; over 70% of California's production comes from the San Joaquin Valley. In 2012, California produced 197 million barrels of crude oil, which represents a continuing trend of decline in production. San Joaquin Valley crude oil is very thick, making it a high-cost, low-price product, and much of it is trapped in accordion-like folds in Monterey shale, which makes extraction difficult and subject to fracking. Its production therefore is highly subject to the vagaries of world oil prices. When oil prices drop, it is not economically feasible to pump oil in the San Joaquin Valley. The resulting layoff of thousands of employees sends the Kern County economy into a slump, with unemployment rates sometimes rising to 20%. When world oil prices go up, the economy revives. Steam must be pumped into the ground to thin the oil and make it easier to pump. The plus side to this is that the steam is used to generate electricity before it is pumped into the ground, an example of cogeneration.

BIOTIC ZONATION: NATIVE VEGETATION

Valley Grassland

Valley Grassland once covered all well-drained areas in the Great Central Valley as well as larger valleys in the Coast Ranges and cismontane southern California. As indicated earlier, only about 1% of the grassland today could be considered pristine. Even in its disturbed form, Valley Grassland is still one of California's most common communities, covering over 22 million acres (8,980,000 ha) of land (figure 10.1). Pristine grassland, however, was dominated by herbaceous wildflowers. While perennial bunchgrasses did occur, they were more common along the coast, in the foothills of the Sierra, and the desert. Bunchgrasses are perennial grasses. They die back each year but return from roots with winter precipitation. The annual forbs or wildflowers return from seeds each year. The bunchgrasses include Needle Grasses, *Stipa* (*Nassella*) spp., Three-awn Grasses (*Aristida* spp.), Blue Grasses (*Poa* spp.), and Wild-Rye (*Elymus* spp.). By far, the dominant species was Purple Needle Grass, *S. (N.) pulchra*. Pristine grassland also contained some native annual grasses. Early descriptions of the valley vegetation describe abundant wildflowers in the winter but a dry barren landscape in the summer. The annual wildflowers include many that have been described for other areas of the state, including Lupines, California Poppies (plate 11C), Goldfields, and Purple Owl's-clover.

Geophytes include species that return from underground bulbs. Among the lilies, the mariposa lilies (*Calochortus* spp.) are the most beautiful. It is not certain how they got the name mariposa, which means "butterfly" in Spanish. The allusion could be a reference to the shape and color of the flower petals, or it could refer to some of the common pollinators. These plants usually grow about a foot (30 cm) tall, and they are usually capped by a single three-petaled blossom about 2 in (5 cm) across. Depending on the species, petal color varies from white or violet to yellow or orange, and the base of each petal usually has a blotch of vivid color—purple, red, or orange.

Members of the amaryllis family have flowers in starburst clusters known as umbels. Sepals and petals are the same color, so it appears that each flower has six petals. The Brodiaea family, Themidaceae (= Amaryllidaceae) includes numerous species, varying in size from Blue Dicks, *Dichelostemma capitatum*, to White Brodiaea, *Triteleia hyacinthina*. Blue Dicks usually has a few bright blue flowers in a dense cluster less than an inch (2.5 cm) across. The plant is usually no more than a foot tall (30 cm). White Brodiaea, in contrast, may be 6 ft (2 m) in height. The surrounding vegetation dictates the height to which it will grow. In the foothills, this species may grow among shrubs, in which case it grows up through them in order to flower above. A cluster of 4–10 white flowers, each over an inch (2.5 cm) in diameter, caps the elongated stem.

Perhaps the most spectacular of the wildflowers are lupines (*Lupinus* spp.); California Poppies, *Eschscholzia californica* (plate 11C); and Purple Owl's-clover, *Castilleja exserta* (= *Orthocarpus purpurascens*; plate 20H). Descriptions of fields of wildflowers by early writers such as John Muir are common in the literature. Perhaps the best is that of John Steinbeck. In his book *East of Eden*, he describes flowers in the Salinas Valley:

FIGURE 10.5 Goldfields, *Lasthenia californica*, and Purple Owl's-clover, *Castilleja exserta*.

On the wide level acres of the valley the topsoil lay deep and fertile. It required only a rich winter of rain to make it break forth in grass and flowers. The spring flowers in a wet year were unbelievable. The whole valley floor, and the foothills too, would be carpeted with lupins and poppies. Once a woman told me that colored flowers would seem more bright if you added a few white flowers to give the colors definition. Every petal of blue lupin is edged with white, so that a field of lupins is more blue than you can imagine. And mixed with these were splashes of California poppies. These too are of a burning color—not orange, not gold, but if pure gold were liquid and could raise a cream, that golden cream might be like the color of the poppies. When their season was over, the yellow mustard came up and grew to a great height. When my grandfather came into the valley the mustard was so tall that a man on horseback showed only his head above the yellow flowers. On the uplands the grass would be strewn with buttercups, with hen-and-chickens, with black-centered yellow violets. And a little later in the season there would be red and yellow stands of Indian paintbrush. These were the flowers of the open places exposed to the sun.

Other carpets of wildflowers may include various members of the sunflower family (Asteraceae). Among these are small species such as Goldfields, *Lasthenia californica* (= *L. chrysostoma*; figure 10.5), that occur in untold millions. Often mixed with Goldfields are larger members of the sunflower family, such as Tidy-tips, *Layia platyglossa*, whose name refers to the fact that each yellow petal is tipped with white. Tarweeds and thistles bloom long after the grassland becomes dry. By blooming during summer, they capitalize on a lack of competition for pollinators. Common Madia (Tarweed), *Madia elegans*, stands about 2 ft (60 cm) high. It has yellow blossoms about 2 in (5 cm) across, and each petal has three small lobes at the tip. These plants have glandular hairs that produce a tarlike substance. This resinous material helps to retard water loss during the long, hot summer. Also to help retard water loss, flower heads close up about midday.

FIGURE 10.6 Artichoke Thistle, *Cynara cardunculus.*

They open after dark and remain open during the next morning.

Thistles (*Cirsium* spp.) are prickly plants that usually have purple flowers. These are sunflowers without the usual circle of ray flowers around the edge. The disc flowers in the center provide the color. In some overgrazed areas, the only thing that seems to remain is Artichoke or Cardoon, *Cynara cardunculus* subsp. *cardunculus* (figure 10.6). This plant was introduced in the 1920s from Europe, where it was used for food. It first appeared near Benicia in the inner Coast Ranges and has spread into overgrazed areas of the western Great Central Valley. It is also widespread in similar areas in southern California. If harvested in the bud stage, this plant provides an edible artichoke. When it goes into bloom, its purple flowers put on a spectacular display. Cattle and sheep avoid eating it, so once it becomes established, it spreads rapidly to infest entire pastures. Its beauty is the only redeeming feature of overgrazing.

The Carrizo Plain lies in a broad, flat valley west of Taft. It was established as a national monument in 2001. The Temblor Range, a low range of hills uplifted along the San Andreas fault, lies between the Carrizo Plain and the Great Central Valley. Some people therefore would contend that the Carrizo Plain is merely one of the valleys of the southern Coast Ranges. It is truly a Valley Grassland ecosystem, however. It is home for some of the Great Central Valley's endangered species, such as the San Joaquin Kit Fox, *Vulpes macrotis mutica*, and the Blunt-nosed Leopard Lizard, *Gambelia sila*. Moreover, although it has suffered the same sorts of disturbances as the Great Central Valley, it also has areas that contain native Valley Grassland in a relatively undisturbed condition. The California Nature Conservancy has purchased 82,000 acres (32,800 ha) of this valley with the ambitious intention of restoring it to the largest single tract of Valley Grassland in the state. Large grazing animals such as the Pronghorn, *Antilocapra americana*, and Tule Elk, *Cervus elaphus nannodes*, also have been reintroduced.

Corral Hollow is another valley of the inner Coast Ranges that contains vestiges of native plants and animals characteristic of the Great Central Valley. Situated about 8 mi (13 km) southwest of Tracy, in the southwestern portion of San Joaquin County and the eastern edge of Alameda County, the area has been used primarily for grazing sheep and cattle. Some of it has been converted to the Carnegie State Vehicular Recreation Area, a motorcycle playground. A portion of Corral Hollow was purchased by the federal government as part of a nuclear testing facility known as the Lawrence Livermore National Laboratory. In 1974, part of the test facility was determined to be surplus property. Particularly because of its interesting population of amphibians and reptiles, the California Department of Fish and Game expressed interest in obtaining the parcel. The title was transferred to the department in 1976, and it is now operated as an ecological reserve.

The Corral Hollow Ecological Reserve is a 99 acre (24 ha) site composed of approximately 30% Riparian Woodland and 70% grassland. It is the home of a wide variety of amphibians and reptiles. The California Red-legged Frog (*Rana draytonii*), the Foothill Yellow-legged Frog (*R. boylii*), and the Northern Western Pond Turtle (*Actinemys marmorata*) have a refuge here in the riparian habitat. Seven species of reptile reach their northernmost distribution in this area. These species are the Common Side-blotched Lizard (*Uta stansburiana*), Gilbert's Skink (*Plestiodon gilberti*), the San Joaquin Coachwhip (*Coluber flagellum ruddocki*), the Glossy Snake (*Arizona elegans*), the Long-nosed Snake (*Rhinocheilus lecontei*), the Western Black-headed Snake (*Tantilla planiceps*), and the Desert Nightsnake (*Hypsiglena chlorophaea*).

Furthermore, Corral Hollow is one of the places featuring serpentine soils and its unique vegetation. It is also a locale for some plant species, such as the Desert Olive (*Forestiera pubescens*) and Iodine Bush (*Allenrolfea occidentalis*), that usually are associated with desert habitats. Honey Mesquite, *Prosopis glandulosa*, also has been recorded there, but much of it has been uprooted by ranchers. There may still be some of these plants located on private property.

Of particular interest in Corral Hollow is the presence of a species in the Borage family (Boraginaceae) that is on the state and federal endangered lists: the Large Flowered Fiddleneck, *Amsinckia grandiflora*. Its entire known distribution consists of about 250 plants located on a half-acre (625 m²) site within the confines of the Lawrence Livermore National Laboratory. This is a robust ephemeral species associated with fine-grained soils. It stands about 2 ft (60 cm) tall and bears bright, showy, orange-red flowers about 0.75 in (14–18 mm) in length. The name fiddleneck refers to a scroll-like arrangement of flowers that resembles the carving at the end of a violin neck. It is a characteristic of the Waterleaf family (Boraginaceae), as exemplified by *Phacelia* as well.

Vernal Pools

Vernal pools or "hogwallows" occur where the hardpan is close to the surface. There are three kinds of pools in the Great Central Valley: valley pools, pools of volcanic areas, and terrace pools. Valley pools, most common in low places of the San Joaquin Valley, occur in basins or valleys in saline or alkaline soils. Vegetation of these areas includes typical salt-tolerant plants such as Salt Grass, *Distichlis spicata*.

One of the most interesting plants that is widespread in vernal pools is a spore-bearing plant known as Quillwort, *Isoetes howellii*. Inclusion of this aquatic plant in the family Isoetaceae aligns it with so-called "primitive" plants such as club mosses and horsetails. The name Quillwort refers to its rather yucca-like appearance, a mass of long quills. What makes it especially interesting is that it is an aquatic plant that uses crassulacean acid metabolism, a type of photosynthesis usually associated with drought-adapted succulent plants. Apparently, the mechanism is important in conserving CO_2. During a typical day, the dense vegetation of a vernal pool takes up all available dissolved CO_2. The Quillwort compensates for this depletion by storing CO_2 as a C_4 acid when it is available, continuing photosynthesis for the rest of the day by drawing upon its stored product.

Pools of volcanic areas are found throughout the state. They are typified by those on the Vina Plains Preserve, in Tehama County, and Mesa De Colorado, in Riverside County. Floristically, they are similar to the terrace pools.

Terrace pools occur on some of the oldest soils in the state: ancient flood terraces on higher ground. Water accumulates during the winter because it is unable to percolate into the ground. During spring, the rim of a pool becomes encircled by wildflowers, *Downingia* sp. (plate [17A]). As the water evaporates, the ring of wildflowers moves inward. Toward the center of the area, the soil becomes increasingly alkaline, favoring specialized species of plants. Endemism is high in these ecologic islands. All five species of Orcutt grasses (*Orcuttia* spp.) are restricted to vernal pools.

Associated with vernal pools on these terrace soils are peculiar mounds known as "mima mounds," up to 6 ft (2 m) in height. The terrain has a rolling, mounded appearance with vernal pools in the low spots. There are many theories about the origin of these mounds, including wind deflation of old stream channels and the piling of soil around old shrub fields. Soil scientists list several places in the world typified by mima topography, but there is no satisfactory explanation for the mounds because of the diversity of climates and geologic processes in the localities where they occur.

Vernal pool vegetation attains its highest development on terrace soils of the east side of the Great Central Valley. Many of these areas have been destroyed by activities of humans. The report *Sliding toward Extinction* indicated that 66%, or nearly 2.8 million acres (1,131,000 ha), of Central Valley vernal pools has been lost. The best pools today are on the higher terraces. Similar pools occur in other parts of the Great Central Valley, the Coast Ranges, and the mesas of the Peninsular Ranges. Vegetation of the vernal pools is discussed in more detail in Chapter 8.

Freshwater Marsh

Many wetland habitats were part of the pristine conditions in the Great Central Valley. Enormous marshes occurred where runoff from the mountains accumulated. Tulare County gets its name from the tules that were abundant in the area. Tule fog gets its name from the fact that thick fog often hung over these marshes.

Wetland habitats characterized by a mixture of water and emergent vegetation are known variously as swamps, marshes, and bogs. These terms mean different things to different people. Swamps have the greatest proportion of open water and are characterized by various species of trees that tolerate flooding of their roots. True swamps do not occur in California, but spring flooding of riparian vegetation creates a similar habitat. Bogs are the wetlands characterized by the least amount of water. These are habitats composed of soggy or supersaturated substrates. They are said to be dystrophic, because the high acidity of the water prohibits water-soluble nitrates from forming. The vegetation therefore suffers from nutrient deficiencies. What water is present is often coffee brown in color due to accumulated decay products. In California, sphagnum bogs of moist northwest forests are a good example of this habitat. Vegetation of these bogs is discussed in Chapter 7.

Freshwater marshes typically occur in flatlands where water accumulates in shallow depressions. True freshwater marshes are no longer common in California. Formerly, they were widely distributed in basins where input of stream water was roughly balanced by evaporation. Marsh habitat commonly bordered large, shallow lakes.

In California's virgin waterscape, there were three large lakes in the San Joaquin Valley. Tulare Lake, Buena Vista Lake, and Kern Lake, with their bordering marshes, sloughs, and connecting channels, formed the largest wetland habitat in the state (figure 10.1). The water source for these lakes was runoff from the Sierra Nevada, particularly by means of the Kern, Tule, Kaweah, and Kings Rivers. According to estimates, there were over 2100 mi (3360 km) of shoreline marsh habitat before draining and damming "dewatered" them.

Tulare Lake represented the largest freshwater lake west of the Mississippi River. It covered over 700 mi^2 (1800 km^2) and was fed by the four major rivers just mentioned. The lake was named for its extensive bed of tules. Early

FIGURE 10.7 Freshwater marsh dominated by cattails, *Typha* spp.

accounts indicated that tules formed a belt 100 yards (meters) wide around the edge of the lake and extended southward for 15 mi (24 km).

Other extensive freshwater marshes occurred in desert regions such as the Owens Valley, and along the Mojave and lower Colorado Rivers. In southern California, marshland was extensive throughout the Los Angeles basin in association with the Los Angeles, San Gabriel, and Santa Ana Rivers. The region of Los Angeles known as La Cienega was formerly a freshwater marsh. (La Cienega is Spanish for "the marsh.")

Regardless of the surrounding terrain, freshwater marshes throughout the state were remarkably similar. Where marshes still remain, studies of habitat and native plants give us an idea of what the extensive marsh systems were like. One such area is the Creighton Ranch Preserve, a 3200 acre (1280 ha) relict of Tulare Lake. Another is the San Joaquin Marsh Reserve in Orange County near the University of California, Irvine. This 202 acre (80 ha) region of wetlands is part of the University of California Natural Reserve System. In addition, various marshland reserves are operated by the Department of Fish and Wildlife as waterfowl management areas. Gray Lodge is a beautiful marsh reserve north of Sutter Buttes. Its 8400 acres (3400 ha) form some of the most intensively used and developed wetlands in the Pacific Flyway. Ash Creek Wildlife Area covers more than 14,100 acres (5755 ha) on the Modoc Plateau in northeastern California. At its heart is Big Swamp, 8000 acres (3265 ha) of natural wetlands that are closed to entry so that wildlife will remain undisturbed.

Marsh vegetation is distributed in distinct bands. Ecological requirements for each of the plant types are precise enough that they occur in single-species clumps or zones that appear as bands on the edge of the water mass. Farthest out in the water are floating plants called macrophytes, a word that means "large plants." Floating macrophytes give way to rooted macrophytes in the shallow water, and some of these remain submerged. A distinct progression of rooted plants then occurs from the shallow water to drier land.

Freshwater marshes are dominated by reed-like plants that grow in water-saturated soil. Various species occur throughout different parts of the state, but they essentially fall in four groups. Most common are rushes (*Juncus* spp.) and bulrushes (*Scirpus* spp.). Locally common are sedges (*Cyperus* spp.) and cattails (*Typha* spp.; figure 10.7). It is not always easy to tell these plants apart because they all have a similar, grasslike appearance. Cattails are the easiest to identify. They are the tallest, growing up to 10 ft (3 m) in height. They have long bladelike leaves and dense flower spikes that look like long brown sausages. They cannot tolerate deep water; therefore, they occur at the outer fringes of the marsh. Sedges and rushes are shorter, up to 3 ft tall (1 m). Leaves, if present, are usually located at the base of the plant. Above the leaves, a solid stem extends. Sedges have triangular stems, whereas most rushes have round stems. (Sedges have edges, and rushes are round!) Bulrushes are the forms commonly called tules, and they are related to sedges. As such, many of them have triangular stems, although the stems vary from round to triangular. They grow up to 6 ft (2 m) in height, and their flowers are borne atop leafless stems in umbels.

On higher ground in the marshes are various water-loving trees such as willows. Gooding's Black Willow, *Salix gooddingii*, is the most common willow in marsh habitats of the San Joaquin Valley, although Red Willow, *S. laevigata*, and Pacific Willow, *S. lasiandra* var. *lasiandra*, also occur on the edges of marshes. Some of these trees may branch close to the ground, resembling large shrubs. They have elongate, lance-shaped leaves. Pacific Willow has bright red branchlets. Yellow branchlets of Black Willows snap off easily and may float to a location where the water is shallow enough for them to root. Plumed seeds of willows are easily carried by the wind, which also aids dispersal.

On higher ground is a group of shrubs that superficially resemble willows and that typically occur where the water is more alkaline. One common form known as Mule Fat, *Baccharis salicifolia* subsp. *salicifolia*, is distributed throughout the Great Central Valley and southward along the coast. To the east, it occurs in desert areas from the Owens Valley southward. These plants are willowlike in appearance, and sometimes they are called Seep-willows, but they are in the sunflower family (Asteraceae). Flowers are small and white. The plants do not possess brightly colored ray flowers, only disc flowers. As in willows, their plumed seeds are wind disseminated.

In the Great Central Valley, a plant known as Button Bush or California Button Willow, *Cephalanthus occidentalis*, is a large shrub associated with these marshes. Its lance-shaped

FIGURE 10.8 Yerba Mansa, *Anemopsis californica*.

leaves have smooth margins and are opposite or whorled on the stem. The flowers occur in a spherical head up to an inch (25 mm) in diameter. This is another example of a plant left over from a more tropical period. It belongs to the madder family (Rubiaceae), which is widespread in the tropics today. The family includes many plants of economic importance, such as coffee and cinchona, the plant from which quinine is derived. It also includes ornamental plants such as gardenias.

The only California species in the Lizard's-tail family (Saururaceae) is found where water may be somewhat alkaline. This is Yerba Mansa, *Anemopsis californica* (figure 10.8). This plant has strap-like leaves and often grows about a foot (30 cm) high in pure stands on soggy ground. It is spectacular in blossom, producing a vertical cluster of white flowers. All of the plants seem to bloom at the same time, producing spectacular fields of white flowers. These plants occur in localized masses throughout the state from inland valleys to coastal estuaries, wherever appropriate brackish habitat occurs. In Spanish, Yerba Mansa means "gentle or mild herb," and the plant has been used for diseases of skin and blood.

FIGURE 10.9 Water hyacinth, *Eichhornia crassipes*, an introduced floating plant that can become a pest.

In areas of open water, there are low-growing aquatic plants that emerge only a few inches above the water. Important among these is Water Cress, *Nasturtium officinale* (= *Rorippa nasturtium-aquaticum*), an edible member of the mustard family (Brassicaceae). This tasty herb, with small circular leaves, grows where water is cool and flowing. In some areas, it is cultured for salads. Water Cress decomposes rapidly. In autumn when there is a major die-off, 80% of the nitrogen is returned to the ecosystem within three weeks. Pacific Marsh Purslane, *Ludwigia palustris*, is an herb similar in appearance and may grow with Water Cress. The purslane differs by having its leaves arranged in opposite pairs along the stem.

Also in open water may be dense mats of floating plants such as Water-fern, *Azolla filiculoides*, or duckweeds (*Lemna* spp.). Water-fern is a small, mosslike fern with many branched stems covered with minute, overlapping, bilobed leaves. Water-fern contains a symbiotic cyanobacterium (blue-green alga), *Anabaena azollae*, a nitrogen fixer that helps enrich the whole marsh ecosystem. For this reason, in some areas Water-fern is harvested and dried for cattle feed. It also may be used as a fertilizer for fields of rice, a cultivated marsh plant.

Duckweeds occur as minute, floating, leaf-like structures that totally cover the water, giving it the appearance of a closely mowed dichondra lawn. The suspended roots of duckweeds dangle in the water and are often covered with nitrogen-fixing bacteria. Like Water-fern, duckweeds help to enrich the system. Decomposition of these mat-forming plants may add considerably to the organic matter that accumulates on the bottom as detritus. If the water is warm, the bottom muds may become anaerobic, forming a thick, black, gelatinous ooze that may bubble hydrogen sulfide (H_2S) gas.

Large numbers of nonnative floating plants such as Water hyacinth, *Eichhornia crassipes*, sometimes accumulate in these open water areas (figure 10.9). In tropical areas, these plants become so abundant that they interfere with navigation. They may represent the most troublesome aquatic weed in the world. In California, they are locally abundant in the Great Central Valley, the Los Angeles Basin, and a few localities in San Diego County. They have long oval leaves about 4 in (10 cm) long. The stems (petioles) of the leaves are inflated, which keeps them afloat. During spring and summer, many blue flowers are borne in a spikelike cluster. When these plants are blooming, it is a beautiful sight, for they cover large areas of water. At such a time, it is difficult to perceive the species as a noxious introduced pest that crowds out native plants. These plants are renowned for their rapid growth and rapid rate of absorption of dissolved nutrients. Some authorities have suggested that Water hyacinth represents a possible solution to the rapid accumulation of organic matter in sewer outfalls. After the plants absorb the nutrients, they could be dried and used as a source of organic biomass for bacterial decomposition and generation of natural gas (methane).

Another floating plant that may be introduced is the Yellow Pond-lily, *Nuphar polysepala*. These plants occur naturally in ponds and slow-moving water in the northern two-thirds of the state, but because of their attractive large flowers and large floating leaves, they have been introduced into a great variety of quiet waters. Again, the problem is that they crowd out native vegetation and contribute greatly to the amount of organic matter that accumulates on the bottom of the marsh.

Marsh ecosystems are known to be among the most productive in the world. An abundance of photosynthesis results in a large biomass of vegetable material per year, most of which feeds back into the system as detritus. Activity of microorganisms recycles material back into water-soluble chemicals that reenter the water mass. Rooted aquatic plants absorb the nutrients from the sediment and directly from the water. The comparative roles of direct absorption and root absorption have not been thoroughly studied, but some evidence indicates that the roots may be more important for their anchoring quality than for absorption of nutrients.

Animals of the Freshwater Marsh

High productivity means that a marsh ecosystem is capable of supporting a large biomass of consumers. Most animals, however, are unable to feed on the vegetative parts of marsh plants, so a large proportion of the edible biomass winds up as detritus that feeds back into the system through decomposition. The higher plants in the marsh, however, produce an abundant seed crop that is an important food for birds. In addition, shelter provided by the dense foliage is important for bird nests, as well as homes for small invertebrates. There is an abundance of aquatic invertebrates in marshes; these are discussed in detail in the following chapter.

The most conspicuous animals in marshes are the birds. Migratory species such as swans, geese, ducks, and shorebirds visit wetlands of the Great Central Valley in great numbers. However, they have suffered a decline because of the activities of humans; old reports indicate that water birds covered plains and waters in countless numbers from October to April. Many of the birds of Freshwater Marshes also occur in Salt Marshes. The discussion of marsh birds occurs in the section on estuaries in Chapter 11.

Many mammals are attracted to water, particularly in hot, dry places such as the Great Central Valley. Most of these mammals, however, are not truly marsh species. One exception is the Common Muskrat, *Ondatra zibethicus*. Common Muskrats are ratlike rodents, about half the size of an average house cat. They have rich brown fur and a long, scaly, sideways-flattened tail, which they wiggle from side to side to help them swim. They eat tules, rushes, cattails, and some aquatic animals. In some places, they have been introduced to help keep the water clear of aquatic weeds. They make their homes of aquatic plants, often with the entrance underwater. These conical piles of mud and vegetation may project 2–3 ft (1 m) above the water. They may live in holes with submerged entrances dug in the banks of ditches and ponds. This practice can cause considerable damage if the holes are dug in the levees of ponds or irrigation ditches.

Common Muskrats are important fur-bearing animals. The fur is of average quality and has been marketed as "Hudson Seal." Common Muskrat hats have enjoyed some popularity. In California, the original range of the Common Muskrat was along the Colorado River and on the Modoc Plateau. It has been introduced in many parts of the state, primarily for its fur. Now it is abundant in the Great Central Valley. Common Muskrats are also edible, and in some southern states the meat is sold as "marsh hare."

The American Beaver, *Castor canadensis*, occurs in streams and small lakes throughout the northern two-thirds of the state. They are not specifically marsh animals, but their occurrence in the Sacramento-San Joaquin Delta is of significance. In that area, they typically live in burrows in the levees, and some of the levee breaks are due, at least in part, to the activity of American Beavers. Their activities as dam builders are discussed in Chapter 11.

Riparian Woodland

The most extensive Riparian Woodlands in the state formerly occurred in the Great Central Valley. Often known as gallery forests, they bordered the large rivers that cut across the valley—particularly the San Joaquin and Sacramento Rivers and their tributaries. Riparian Woodlands reached their greatest development on the natural stream terraces or levees that rose from 3 to 20 ft (1–6 m) above the streambed. In some areas, the woodlands were 10 mi (16 km) wide. Historical accounts suggest that in the middle 1800s there were about a million acres (400,000 ha) of Riparian Woodlands in the Great Central Valley. Today's estimate of 102,000 acres (41,600 ha) may be generous, and much of this Riparian Woodland has been seriously degraded. Only about 1%, or 10,000 acres (4000 ha), could be considered pristine.

Vegetation of the Riparian Woodlands consists of water-loving shrubs and trees. Because distribution of water may be patchy, these species are usually equipped with good mechanisms for dispersal, such as plumed seeds.

Among the shrubs are Mule Fat and several species of willows (*Salix*). Most spectacular in the pristine woodlands were the trees. These were mostly winter-deciduous species and included Western Sycamore, *Platanus racemosa*; Box Elder, *Acer negundo*; Fremont Cottonwood, *Populus fremontii* subsp. *fremontii*; and the three species of willow mentioned in association with marshes. Perhaps the most impressive tree in the Riparian Woodland is Valley Oak, *Quercus lobata* (figure 10.10). There are records of Valley Oaks 27 ft (9 m) in circumference. These trees, along with Western Sycamores, grow more commonly on the higher terraces, or in deep valley soils. The largest remaining stand of Valley Oak is in The Nature Conservancy's Cosumnes River Preserve, between Stockton and Sacramento. This 1000 acre (405 ha) preserve includes not only Riparian Woodland but endangered marshland as well.

The fact that these trees drop their leaves in the winter is testimony that the environment along stream courses is similar to that of the eastern United States. There is abundant water, and winter temperatures are cold—a very "un-California-like" climate. These trees are relicts from a time when California's climate was very different. Winter temperatures along these streams are a product of cold-air drainage, short day length, and frequent fog. The average low temperature during winter ranges from 32°F to 38°F (0–4°C), and may dip as low as 15°F (−9°C). Some of California's rarest, most endangered plant species are restricted to Riparian Woodlands of the Great Central Valley. One example is the Woolly Rose-mallow, *Hibiscus lasiocarpos* var. *occidentalis* (= *H. californicus*). This is an ephemeral plant, which is a rare strategy for a riparian community. Flowers of this species are white with a crimson center. It is restricted to undisturbed slough banks from the Sacramento Delta north to Butte County.

Animals of Riparian Woodland

Riparian Woodlands are among the most productive habitats in the state, capable of

FIGURE 10.10 A. Deciduous Valley Oaks, *Quercus lobata*, in Malibu Creek State Park. B. Valley Oak leaves.

supporting a large biomass of animals. Most of the animals are visitors from neighboring habitats seeking water, food, and shelter, but the total list of species is impressive. One researcher listed 39 species of mammals, 6 species of frogs, 7 species of lizards, 6 species of snakes, and a turtle for the Sacramento Valley Riparian Woodland. Sixty-nine species of breeding birds have been recorded for the Riparian Woodland of the Sacramento Valley. Of that number, 21 species have their primary affinity with Riparian Woodlands. The number of insect species must be enormous. Seventeen species of butterflies alone have been described as native to the Sacramento Valley Riparian Woodlands. Most of these species are described elsewhere in this book, but a few that are characteristic of lowland watercourses will be mentioned here.

INSECTS Of the enormous number of insects associated with Riparian Woodlands, two are worth mentioning because they are listed as threatened by the federal government. They are the Delta Green Ground Beetle, *Elaphrus viridis*, and the Valley Elderberry Longhorn Beetle, *Desmocerus californicus dimorphus*. The former is now known only from two regions of vernal pools in Solano County. These bright green 0.25 in (6 mm) ground dwellers (only one has ever been observed to fly) scurry along the edges of vernal pools in search of females or soft-bodied prey, including small gnats, beetle larvae, and springtails. The Valley Elderberry

Longhorn Beetle is known only from the Riparian Woodlands of Sacramento, Yolo, and Merced Counties. It requires the Blue Elderberry, *Sambucus nigra* subsp. *caerulea* (= *S. mexicana*) as a host plant.

A number of butterflies are characteristic of these lowland Riparian Woodlands, and many of these are also conspicuous around cities. Groups of butterflies that often are associated with specific groups of plants are called *guilds*. The largest and most common is the Western Tiger Swallowtail, *Papilio rutulus*. The larvae are bright green with big "false eyes" behind the head that may serve to intimidate predators. Caterpillars feed on many riparian species, including willows, cottonwoods, alders, and Western Sycamores.

Lorquin's Admiral, *Limenitis lorquini*, is an attractive, dark butterfly with broad white bands on its wings and orange wing tips. It resembles its relative, the California Sister, *Adelpha californica*, which occurs along foothill and mountain streams in the vicinity of oaks. Larvae of Lorquin's Admiral feed on leaves and other riparian plants, and they overwinter in a case made from a leaf.

Another common butterfly is the Mourning Cloak, *Nymphalis antiopa*. This butterfly is black to brownish purple with a cream-colored border to the wings. Adults hibernate under pieces of bark or other debris during winter. The black, hairy larvae feed on a variety of plants, including various species of imported elms (*Ulmus* spp.). Along streams, they feed on willows. The Satyr Comma, *Polygonia satyrus*, is one of the orange butterflies with brown spots. Similar to other anglewings, the edges of the wings appear torn and ragged. When the Satyr Comma is at rest on a tree trunk with its wings closed, it is nearly impossible to see because its color blends perfectly with the bark. The larvae of these butterflies feed on Hoary Nettle, *Urtica dioica* subsp. *holosericea*. The larvae also create a tent or shelter of the nettle leaves by partially cutting the midrib from beneath, allowing the leaf to droop downward. By using silk threads, a larva draws the margins of the leaves together from beneath. These tents superficially resemble those of the Red Admiral or Alderman, *Vanessa atalanta*, whose larvae feed on the same plant. Larvae of the Red Admiral, however, draw the margins of the leaves upward so that the lower surface is exposed. The Red Admiral has red banding on its wings, and the tips of the wings are black with white spots.

BIRDS The Yellow-billed Cuckoo, *Coccyzus americanus*, has been described as a barometer for conditions in Riparian Woodlands. Its distribution is closely restricted to cottonwood-willow forest, and its decline has paralleled the decline of riparian vegetation in general. Some authorities relate the decline of the Yellow-billed Cuckoo to heavy use of DDT (dichlorodiphenyltrichloroethane) and other pesticides, which began in the late 1940s. This bird fed on insects and, in some areas, on caterpillars of the Nevada Buckmoth, *Hemileuca nevadensis*. These caterpillars are black with yellowish stripes down the back and sides. They have yellowish or brownish branched spines that can produce a stinging rash. The black-and-white adults fly in the fall. These caterpillars were so abundant that they defoliated the Black Willows every year. In the late 1950s and 1960s, Nevada Buckmoths became rare, and for some years the willows were not defoliated. The return of the Nevada Buckmoth in 1969 again caused defoliation, but the Yellow-billed Cuckoo apparently has not returned to most of its former range.

This bird is grayish brown above and white below. Its wing tips are reddish, and the underside of its tail is banded with black and white. In the early 1900s, it nested in Riparian Woodlands of coastal streams from San Diego County north to Sonoma County and throughout the Great Central Valley. Its distribution in the Great Central Valley today is probably restricted to a few pairs along the Sacramento River between Colusa and Red Bluff. Nine to 10 pairs may occur in the Riparian Woodland along the South Fork of the Kern River above

Lake Isabella, and there may be a few pairs nesting along the Colorado River. It is listed as threatened by the state government.

The Yellow-breasted Chat, *Icteria virens*, has also declined in numbers. The largest of California's wood warblers (Parulidae), the Yellow-breasted Chat is about 6 in (15 cm) in length. It is olive green on the back and has a bright yellow breast. A black stripe bordered with white runs from the bill through its eye, and the eye is ringed with white. It resembles the smaller Common Yellowthroat, *Geothlypis trichas*, which has no eye ring and a smaller bill. The bill of the Yellow-breasted Chat seems rather large and its tail rather long for a warbler.

The Yellow-breasted Chat feeds on insects or, occasionally, fruits and seeds. This is a summer visitor to Riparian Woodlands, where it prefers tangles of willows and other thick streamside vegetation. Even when it was a more common visitor, it was an inconspicuous bird because it kept to the brush. It is probably best known for its variety of noises, which, like those of the Northern Mockingbird, *Mimus polyglottos*, are uttered at all times of the day and night. Also like the Northern Mockingbird, the Yellow-breasted Chat is a good mimic. To its benefit, this is one bird that is able to cope to some degree with the nest parasitism of the Brown-headed Cowbird, *Molothrus ater*. When the Brown-headed Cowbird lays an egg in the nest of the Yellow-breasted Chat, the mother has been observed to destroy the egg of the invader along with its own, but the chat is then free to lay more eggs.

Other birds that have declined in abundance along with Riparian Woodlands are tree-nesting ducks such as the Wood Duck, *Aix sponsa*; the Common Merganser, *Mergus merganser*; the Red-breasted Merganser, *M. serrator*; and the Hooded Merganser, *Lophodytes cucullatus*.

The male Wood Duck (plate 19C) is probably the most beautiful bird in the state. In the words of W. L. Dawson, an early California ornithologist, the male Wood Duck is "of almost indescribable elegance." It is the only duck in the state that has a long, slicked-back crest on its head. Like "racing stripes," white lines from the bill and eye run backward along the crest. The eye is orange, and the bill is orange, white, and black. The head is iridescent, appearing green, purple, blue, or black, depending on the light.

Wood Ducks formerly ranged through the Riparian Woodlands of the northern part of the state. They are no longer common in the Great Central Valley, although they still range to some degree along streams into the foothills where appropriate habitat is present. On the verge of extinction in the early 1900s, they have recovered to their present rare state primarily through the efforts of captive breeders. They seem to prefer quiet, undisturbed woodlands.

Wood Ducks are agile fliers, dodging effortlessly through the trees. They perch on branches with ease. They feed heavily on acorns in season; otherwise, they forage in water for aquatic plants and insects. They nest in the hollows of trees, particularly oaks, willows, and Western Sycamores. Often they nest in deserted woodpecker holes, such as those of the Pileated Woodpecker, *Dryocopus pileatus*. Preferred nest sites overlook the water so that the ducklings, when they hatch, are able to tumble from the nest to the water. If the nesting tree is away from the water, the young birds tumble to the carpet of leaves beneath the tree. If the distance is too great, the mother carries the ducklings to the ground and leads them to the nearest water. Another threat to this species has been the spread of European Starlings, *Sturnus vulgaris*, which invade the nest cavities and replace the native birds.

Mergansers are fish-eating ducks. They have narrow, serrated bills with a hook on the end and were formerly common along lakes and rivers with wooded shorelines. The Hooded Merganser was never abundant, but the Common and Red-breasted mergansers have declined in abundance along with Riparian Woodlands. The Red-breasted Merganser is more likely to be observed year-round in brackish water or sloughs along the coast.

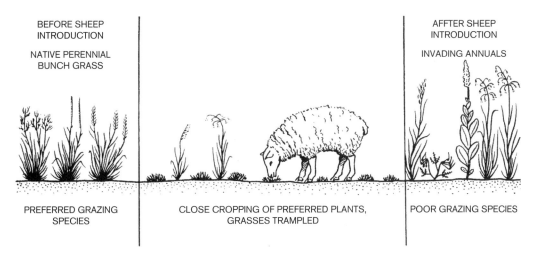

BEFORE SHEEP
INTRODUCTION

NATIVE PERENNIAL
BUNCH GRASS

AFFTER SHEEP
INTRODUCTION

INVADING ANNUALS

PREFERRED GRAZING
SPECIES

CLOSE CROPPING OF PREFERRED PLANTS,
GRASSES TRAMPLED

POOR GRAZING SPECIES

FIGURE 10.11 History of alien grass invasion in native grassland (from Bakker 1984).

Disappearance of Native Vegetation

The demise of native vegetation in the Great Central Valley goes back to the period of Spanish missions in the 1700s. Originally, there were some 23 million acres (9 million ha) of California prairie. Its disappearance is not a consequence of grazing per se, because Valley Grassland evolved for millions of years in association with large mammalian grazers. Rather, the problem is overgrazing. Whereas native large mammals migrated seasonally, and numbers were kept in check by natural ecological constraints, domestic mammals may occur year-round or in numbers that far exceed carrying capacity. The Spanish introduced small, long-horned cattle and small, long-legged sheep known as Churro. As the missions spread northward, the numbers of livestock increased phenomenally. At the height of the mission period, there were about 400,000 head of cattle and 300,000 sheep, but mostly along the coast. In addition, there were large numbers of horses and mules, many of which had gone wild.

Old records document a drought between 1828 and 1830, when no rain fell in coastal California for 24 months. Such a drought would be a hardship for the perennial bunchgrasses and annual forbs, and would favor establishment of introduced annual grasses (figure 10.11). This kind of drought would also promote overgrazing. There would be no water for plant regeneration, and introduced livestock would eat every available scrap before starving. It has been estimated that 40,000 horses and cattle perished during that drought.

The late 1840s saw the decline of Mexican influence in California and the onset of the Gold Rush. Vast numbers of livestock were introduced to feed the forty-niners. By 1862, there were about 3 million cattle and 9 million sheep. During that year, unbelievably heavy precipitation flooded the Great Central Valley, forming a lake about the size of Lake Michigan. The prairie was underwater, and livestock drowned. Another 2 year drought, and overgrazing reduced the Great Central Valley to a dust bowl. By the time normal precipitation returned to the valley, during the winter of 1864–1865, the numbers of livestock had declined markedly.

Cattlemen became discouraged by the vagaries of climate; not so the sheepherders. Whereas the number of cattle declined to a mere half-million by 1870, the sheep population was increasing markedly. Sheep required less water, and they could be herded easily, to take advantage of available forage. At this time, it was common practice to herd sheep into meadows of the Sierra Nevada during summer. In fact, the great John Muir was originally such a shepherd. It has been estimated that by 1875 there were 5.5 million sheep. By that time, damage to rangeland was so complete that the Great Central Valley never recovered, and total livestock never again reached those numbers.

Where did the introduced grasses come from (figure 10.12). They were of Mediterranean origin, introduced as seeds in the coats and digestive tracts of introduced livestock. They were also introduced as livestock food brought from the Mediterranean along with the animals. The history of the replacement of native vegetation by introduced grasses is recorded in the adobe bricks that the Spanish and other early settlers used to build dwellings. Dry grass was used as a binding to help hold together the clay in the bricks. Botanists, by identifying species of grasses in bricks of known age, have been able to document the replacement of species.

Herbaceous species other than grasses were also introduced. Although the history of introductions is not always clear, many of these, such as sow thistles (*Sonchus* spp.), may have been introduced as food plants. These ephemerals have large, lyre-shaped leaves that were eaten raw or boiled. Other introduced food plants were various members of the geranium family (Geraniaceae), such as Redstem Filaree or Storksbill, *Erodium cicutarium*. It appeared in California early enough that it became a favored food of the Native Californians, who ate it raw or boiled.

Other food plants were introduced by the missionaries. For example, some of the members of the mustard family (Brassicaceae) that are so common today owe their introductions to missionaries. The mustard genus Brassica includes a number of species with small, yellow, four-petaled flowers. These mustards cover many acres in the spring, and the leaves may be eaten raw or boiled. Actually, all members of the genus Brassica in California, including a number of important cultivated food plants, are introduced. For example, broccoli, cabbage, cauliflower, brussels sprouts, kale, and kohlrabi are all variants of the same species, *Brassica oleracea*. Turnips are cultivated varieties of the common Field Mustard, *B. rapa*.

Radish, *Raphanus sativus*, is another introduced member of the mustard family. The flowers have four whitish petals with pink tips. Its greens are eaten in the same way as the others, or the young seeds may be eaten in salads. The root of the Radish is too tough to be eaten like the domestic radish, although it has a similar flavor. The domestic radish is a cultivated variety of the same species.

Horehound, *Marrubium vulgare*, is an introduced member of the mint family (Lamiaceae) that has been used for medicinal purposes. It is boiled to produce a bitter fluid that is sweetened and formed into candy used to soothe sore throats and coughs.

The period of the late 1800s in the Great Central Valley was marked by a new form of land use. Large mechanized wheat farms extended from Fresno to Chico. For the most part, these wheat farms depended on natural precipitation, a technique known as dry farming. On the west side of the Great Central Valley, farmers continued to raise wheat by this method at least until the late 1920s. The beginnings of irrigated agriculture go back to the 1850s, but its widespread use came later. Around the turn of the century, wheat prices began to drop. Improved irrigation systems and refrigerated rail cars ushered in the era that continues today.

Most of California's multibillion dollar agricultural industry is concentrated in the Great Central Valley, although the Imperial Valley is a major source of garden vegetables. Encouraged by cheap available land, inexpensive water, and government price supports, California farmers have prospered. In a single year, the income from California's agriculture is greater than the total value of all the gold mined since 1848. As of the mid-1980s, dairy products topped the list, with beef, grapes, rice, and cotton close behind. About half was attributed to dairy and beef production. The cheapest way to feed grazing animals is to allow them to eat what grows naturally. Nevertheless, hay and alfalfa production,

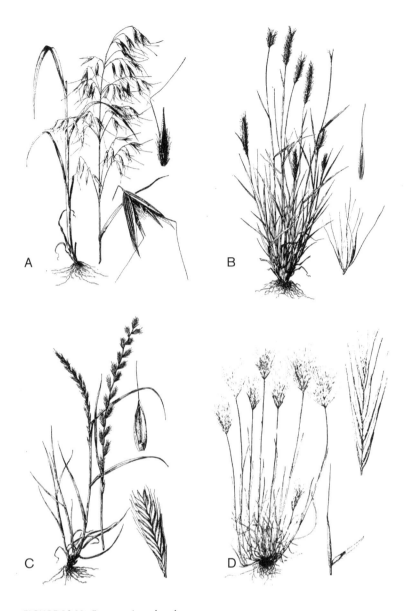

FIGURE 10.12 Common introduced grasses.

A. Wild Oats, *Avena fatua*.

B. Meadow Barley, *Hordeum brachyantherum*.

C. Rye Grass, *Festuca perennis*.

D. Red Brome, *Bromus madritensis* subsp. *rubens*.

ADAPTED FROM: Schoenherr et al. 1999.

for feeding the animals, also ranks high in terms of state agricultural income. By the 1990s, marijuana, almonds, and hay had moved up. In 2014, the top 10 money producers in order of value were milk, almonds, cattle, grapes, strawberries, lettuce, walnuts, tomatoes, pistachios, and hay. In fact, California produces more than 90% of US agricultural exports of various nut products, wine, and tomatoes.

Another aspect of California's agriculture is that most of it can be classified as "cash crops." These are crops that are not required by most

FIGURE 10.13 Overgrazed pastureland near Panoche, San Benito County. Note the steep-sided arroyo indicative of soil compaction.

people for everyday existence; they are part of the luxuries of life. They are things we do without when money becomes scarce, so demand for these crops fluctuates with the economy. Farmers who grow these crops depend on inexpensive land, various subsidies, and cheap water, delivered by either the state's California Aqueduct or the Federal Central Valley Project. Many farmers cannot survive periods of low demand or when water is rationed. When demand is high, they reap enormous profits. Periods of drought are sometimes difficult for these farmers. Almonds and grapes are among the most water demanding. During the extensive drought of the 2000s, some farmers were cutting down dying almond orchards. A complicating factor was that the chipped wood, which could be sold to biomass energy plants, also was accumulating because the cost of electrical energy from biomass was more expensive than that produced by heavily subsidized large-scale solar plants. The biomass-burning electrical power plants were going off line in the face of competition from solar facilities, so farmers couldn't even sell their wood.

So, climate change in the form of drought has a profound influence on the kinds of crops raised in the Central Valley. Equally important, however, is that the winter minimum temperatures have increased markedly. Farmers refer to the length of time that the temperature drops below 45°F (7.2°C) as *chill hours*. The reduction in number of chill hours per year mostly affects fruit production on vine and tree products, with walnuts being the most sensitive. The continuing trend of fewer winter chill hours could have a profound influence on future agriculture in the Great Central Valley.

Consequences of poor agricultural practices can be seen all over the valley. Overgrazing has compacted the soil, encouraging gully (arroyo) formation (figure 10.13). Winds blow soil away from land that has been ground into dust by too many hooves. On lands that are irrigated for crop production, signs of overuse include soil erosion and salt buildup. A 1985 report from the American Farmland Trust seated that about one-third of California's 31 million acres (12.4 million ha) of crop and rangeland suffer from excessive soil erosion. Another report stated that by the year 2000, over one-third of the state's 35 million acres (12.5 million ha) of farmland would be destroyed by salt. Another 44,000 acres (17,600 ha) per year are gobbled up by urbanization.

The valley's difficulties are compounded by water problems. Relying on groundwater pumping during times of drought has caused widespread subsidence. The massive groundwater overdraft of 1.5 million acre-feet per year is causing water tables to drop, promoting draining of wetlands, drying of springs, and subsidence. Marshes and Riparian Woodlands have suffered from this dewatering. Subsidence of the land in some parts of the San Joaquin Valley is as much as 30 ft (10 m). Lowering of the water table means that wells must be dug deeper, which increases the energy cost to pump the water.

Riparian Woodlands also have suffered from dewatering due to dam building and channelization. Furthermore, trees have been removed for firewood and wood chips, and to increase available land for agriculture or pasture. Today the Sacramento-San Joaquin Delta is made up of diked "islands" covered with agriculture. These islands are frequently lower than the surrounding watercourses, which make up about 1000 mi (1600 km) of braided channels.

Mass transport of water through aqueducts adds to the problems. Irrigation of big tracts of land, using water subsidized by public funds, has encouraged large-scale agribusiness oriented toward immediate profit, rather than long-term land use. California has more than 81,000 farms that use 80% of the water. Only 1% of American farmland is found in the Central Valley, but it grows about 25% of the nation's food. In 1981, two-thirds of the state-irrigated land in the southern San Joaquin Valley was managed by only eight companies, three of which were oil companies. This type of agriculture involves large-scale energy input for pumping and farm machinery. Heavy use of chemical fertilizers, herbicides, and pesticides brings high productivity for a short time. Human labor is minimized, and small family farms with owners who depend on long-term productivity are forced out of business.

Leach water containing a variety of chemicals pollutes groundwater. Banned pesticides such as DDT and dieldrin have accumulated at high levels in the sediments of the San Joaquin River and its tributaries. This problem is even more severe on the west side of the valley where marginal lands are used for agriculture. In this area, a few hundred large landowners obtain most of their water at subsidized rates from the federally operated Central Valley Project. In this area, a shallow hardpan causes accumulation of groundwater just beneath the surface. This accumulation concentrates chemicals and salts in the root zone of plants. To help solve the problem, the Bureau of Reclamation built a large drain to carry away water from some 180,000 acres (72,000 ha) of farmland where rising groundwater was a problem. The canal known as the San Luis Drain was originally proposed to carry water 188 mi (300 km) to San Francisco Bay. After the state spent $40 million and built only 82 mi (130 km) of canal, the project was halted. Money became scarce, and environmentalists objected to the prospect of the wastewater being discharged into San Francisco Bay. The canal ended at a marsh known as Kesterson National Wildlife Refuge, and between 1978 and 1981, water quality in the refuge worsened. Birds that hatched there showed congenital defects. In 1983, 246 birds were found dead, and 106 were deformed. The culprit was determined to be high levels of selenium. The selenium was leached out of the soil in the agricultural area and had accumulated in the water of the refuge. In 1986, the drains were closed. It appeared that the farmland was doomed to accumulate salty groundwater. The government apparently chose to reopen the drains, renew the contract to deliver water to the district at a fraction of the true cost and transfer wastewater to the Tulare Lake Basin, but the problem of Kesterson was not resolved. The reservoir was dried up in 1986 and later covered with $6 million worth of fill dirt. The problem now is that pools of rainwater in the low areas draw up toxic selenium through the dirt. Furthermore, evaporation ponds in the Tulare Lake Basin have now become the cause of dead and deformed birds.

These problems are not restricted to the Great Central Valley. California's Imperial and

Coachella Valleys are part of an agricultural empire that includes adjacent Mexico and southwestern Arizona where about 90% of America's fall and winter produce is farmed. Problems with overuse of pesticides in the area led to highly publicized massive epidemics of immune pests. Runoff from these fields enters either the Salton Sea or the lower Colorado River. The selenium accumulation has caused health advisories to be posted regarding consumption of fish from the Salton Sea. In addition, there has been a documented crash in numbers of breeding birds such as egrets, herons, and cormorants. The number of documented nests dropped from over 1700 in 1987 to 19 in 1990. Similarly, the number of overwintering American White Pelicans, *Pelecanus erythrorhynchos*, dropped from 40,000 in 1987 to 650 in 1991. It is not certain what caused the decline, but increasing salinity, pesticide residues, heavy metals, selenium, and boron, all products of agricultural runoff, have been implicated.

The disappearance or collapse of Honeybee colonies on a national scale has become a concern for agriculture. In particular, the widespread use of neonicotinoid pesticides, which are nicotine-imitating chemicals, has been implicated. In response, California has prohibited their use on almond crops, and restricted their use on citrus and other fruit trees during their blooming periods. California, however, still allows use of the pesticide on wine grapes, raisin grapes, tomatoes, and cotton, as well as widespread use in the urban pest control industry. Environmentalists are also concerned about the effect on native bee and butterfly pollinators.

When activities of humans cause lands to become desertlike, it is called desertification. Desertification in the Great Central Valley is continuing. All the symptoms are there: salty soil, erosion, gully formation, subsidence, and replacement of native vegetation by weeds. The causes are overgrazing, poor agricultural practices, groundwater overdraft, and poor water management in general. A comprehensive plan, involving environmentally sound principles, will have to be enacted eventually, or agri-culture in California will suffer the same fate as that of the Fertile Crescent in the Tigris and Euphrates Valleys. In the sixth century BC, that was the richest agricultural region in the world. Now it is a salty wasteland.

ANIMALS OF THE GREAT CENTRAL VALLEY

Animals of Freshwater Marsh and Riparian Woodland have already been discussed. The discussion here will be mostly on the animals of the native and altered grassland. Productivity in Valley Grassland is not high. This is a drought-adapted, and therefore food-poor, ecosystem. The animal biomass supportable by such a grassland is between that of the desert and that of the scrub communities. Consequently, insects such as ants and grasshoppers are the most common animals. Among vertebrates, reptiles are most common. Many species that occur in deserts and scrub communities also occur in grasslands: these have been discussed in the chapters on deserts and Cis-Montane Southern California.

Reptiles

Some of the reptiles in the valley are unique. Among these is the Blunt-nosed Leopard Lizard, *Gambelia sila*. A similar species, the Long-nosed Leopard Lizard (*G. wislizenii*), occurs in southwestern deserts. The Blunt-nosed Leopard Lizard is an associate of Alkali Sink, dry wash, and bunchgrass habitats in the southern part of the Great Central Valley and adjacent foothills. It is now federally listed as endangered, another victim of habitat destruction. It is formally protected in the Pixley National Wildlife Refuge. This is a large-headed, carnivorous lizard that feeds on insects, lizards, and small mammals. Its spotted pattern is well suited to its habit of waiting for prey in the shade of a bush.

Birds

Among birds of the Great Central Valley, those associated with wetlands were once most

FIGURE 10.14 Turkey Vulture, *Cathartes aura*.

FIGURE 10.15 Golden Eagle, *Aquila chrysaetos*.

numerous. Ducks, geese, and shorebirds abounded. They will be discussed in Chapter 11. These and other game birds, such as grouse and quail, were hunted nearly to extirpation. Introduced species, such as Chukar, *Alectoris chukar*, and Ring-necked Pheasants, *Phasianus colchicus*, compounded the problem by creating competition for the little remaining habitat. Grouse and quail retreated primarily to the mountains.

The story of the California Condor is outlined in Chapter 7. Its relative the Turkey Vulture, *Cathartes aura* (figure 10.14), is still a familiar sight in the Great Central Valley. This carrion eater is not as large as the California Condor, but it has a wingspread of 6 ft (2 m). No bird other than the Golden Eagle, *Aquila chrysaetos* (figure 10.15), reaches this size in the Great Central Valley today. The Turkey Vulture can be easily distinguished from most hawks because it holds its wings in a V shape, noticeably above the horizontal. Golden Eagles also hold their wings in a V when they soar.

The Turkey Vulture, like the California Condor, has a red, naked head. It is believed that the value of this featherless skin is to dissipate heat; a blackbird is apt to absorb heat by radiation.

Another possible role of a featherless head is related to the carrion-feeding habit. The back of the head is one of the places where a vulture has trouble cleaning its feathers. Perhaps by not having feathers on its head, it avoids accumulating bacteria and decayed meat from the carcasses upon which it feeds.

Adaptations for feeding on carrion are apparent in Turkey Vultures. Similar to hawks or owls, they have a hooked beak, which they use to tear meat from a carcass. Unlike hawks or owls, their feet are weak, not equipped for grasping prey. Their feet are more like those of a chicken. One interesting adaptation is that a Turkey Vulture has a sense of smell, whereas most birds do not. The Turkey Vulture is particularly sensitive to the odor of rotting flesh. What it detects are chemical compounds known as mercaptans. These are sulfur compounds with pungent odors. This ability of Turkey Vultures was discovered when it was noted that soaring Turkey Vultures could be used to

FIGURE 10.16 White-tailed Kite, *Elanus leucurus* (photo by Charles Z. Leavell).

FIGURE 10.17 American Kestrel, *Falco sparverius*.

locate leaks in gas lines. Natural gas in nature is odorless, but the gas company adds a mercaptan to natural gas so that leaks in homes will not go undetected.

Predatory birds (raptors) include the same species that were discussed in Chapter 8. Red-tailed Hawks and Red-shouldered Hawks are common. Particularly notable among the raptors are those that have adapted to and benefited by activities of humans. Of particular interest is the White-tailed Kite, *Elanus leucurus* (figure 10.16). Kites get their name from their habit of hovering with flapping wings, giving the impression of a kite on a string. This is an adaptation for hunting in an open habitat. They search the ground for rodents and insects, swooping down from the hovering position. These beautiful birds had largely disappeared, but they have made a remarkable comeback. An estimated 50 pairs remained in the state in the early 1950s. Now they are a common sight, hunting for prey on the edges of fields and freeways. Because they have pointed white wings with black tips, they look a bit like a gull hovering next to the road.

Another conspicuous raptor that has benefited from human activities is the American Kestrel, *Falco sparverius* (figure 10.17). This is a bird of agricultural lands and parks. California's smallest falcon, it also hunts while hovering, although an alternative habit is to hunt from a perch. These birds are quite commonly seen perched on power pole, power lines, and fence posts. These elevated positions give them a good view of the edges of fields and roadways, where mice and insects abound. Sometimes they will also take small birds. The American Kestrel can be told from other raptors by its small size and pointed wings. When it first lands on a perch, it pumps its tail up and down as if trying to catch its balance. These are pretty birds, with blue-gray wings and a reddish tail. Their face has two black bars on a white background, which is particularly conspicuous on males.

Few predators are more conspicuous in these disturbed areas than White-tailed Kites and American Kestrels. It is interesting to note, however, that Burrowing Owls, *Athene cunicularia* (figure 4.24), also may be found along roads, particularly where they have been able to move into abandoned ground squirrel burrows. In fact, Burrowing Owls have mostly disap-

FIGURE 10.18 American Crow, *Corvus brachyrhynchos*.

peared from former coastal localities, and their primary habitat is now agricultural areas in the Great Central and Imperial Valleys, where they are threatened by rodent poisons and cultivation of habitat. They have been listed by the California Department of Fish and Wildlife as a Species of Special Concern, and in 2013 four breeding pairs were put under full protection at the Naval Weapons Station Seal Beach. The Common Barn Owl, *Tyto alba*, is another example of a raptor that has adapted to human habitats. It is nocturnal, hunting more by sound than by vision. Its daytime roost is usually in a dark place in city and farm buildings. These pale-colored owls are seldom seen, but they are perhaps the most common owl. On occasion, they are found dead on the road after being hit by a car or truck at night while swooping along a roadway.

Throughout the book, many communities have been said to contain a member of the crow and jay family (Corvidae) that is noisy and conspicuous. In agricultural areas and urban woodlands, it is the American Crow, *Corvus brachyrhynchos* (figure 10.18), that fills this role. These birds are smaller than a Common Raven but nevertheless are up to 18 in (45 cm) in total length. In flight, it can be told from a Common Raven by its smaller size and squared, fan-shaped tail; a Common Raven has a more

pointed or wedge-shaped tail. Common Ravens are often solitary, whereas American Crows usually occur in flocks. These are noisy, raucous birds that are famous for raiding farmers' crops. Their "cah, cah" is a familiar sound to many people. Many authorities consider crows and ravens to be the most intelligent birds in the world. Many experiments have supported this hypothesis, particularly their ability to construct tools. The American Crow's habit of carrying walnuts high into the sky and cracking them by dropping them on pavement has given them the reputation of being clever. Some biologists argue whether this behavior constitutes tool use. One researcher at the University of California at Davis suggested that American Crows purposely drop walnuts on the road in front of cars and, in so doing, have invented the world's first gasoline-powered nutcracker.

Most farmers consider American Crows to be pests. They have been shot and poisoned, but still they abound. Like the Coyote, they have thrived in spite of humans, adapting well to urban and rural woodlands. According to the Federal Breeding Bird Survey, California's population of American Crows is growing at the healthy rate of 4% per year. At that rate, the population will double every 17 years.

Another member of this family that is conspicuous in grasslands is the Yellow-billed

Magpie, *Pica nuttalli* (plate 19H). As indicated in the chapter on the Coast Ranges, this is one of two birds only found in California. The other is the Santa Cruz Island Jay. These are beautiful birds with very long tails. They are nearly as large as an American Crow. They have black-and-white markings, and the black of their wings and tail has iridescent green highlights. The bill is bright yellow. These birds are common in much of the Great Central Valley and in valleys in the Coast Ranges and foothills of the Sierra Nevada. They often feed in flocks, but not as conspicuously as American Crows. They frequently nest in groups, in groves of trees. At the northern end of the Great Central Valley, a similar bird, the Black-billed Magpie, *P. hudsonia*, is an occasional winter visitor. This is a bird of northern prairies and the Great Basin. It occurs on the Modoc Plateau and south to the Owens Valley along the eastern side of the Sierra Nevada. Sometimes it wanders into the Great Central Valley in the winter. Except for its black bill, it is nearly identical in appearance to the Yellow-billed Magpie (plate 19H).

Mammals

In a wide-open habitat such as a grassland, there are basically two strategies for native herbivorous mammals. Either they are burrowers or they are large grazing species that take refuge in herds. Of the burrowers, the Great Central Valley includes its assemblage of rabbits, hares, ground squirrels, gophers, mice, and kangaroo rats. Characteristics and strategies of these mammals have been discussed in other parts of the book. The Great Central Valley also includes endemic species. However, because of the vast amount of habitat destruction, some of them, such as two of the kangaroo rats, have become endangered species. Others, such as the California Ground Squirrel, are probably more common because of their ability to colonize disturbed sites, such as abandoned fields and road cuts, and because of the increased abundance of seeds associated with introduced annual weeds.

Of greater interest has been the fate of California's large herds of grazing animals. This story is not unique to California; the fate of animals in the Great Plains has been similar. Early accounts of Bison herds in the midwest are well known. Similar accounts for the Great Central Valley paint a picture of large herds of Mule Deer, Pronghorn, and Tule Elk darkening the landscape.

The accounts of deer are of interest. These animals have now vanished from the valley, and there is no way to know exactly which subspecies was involved. Presumably, it is the Mule Deer (figure 4.40), which now migrates back and forth between foothill and forest. The influence of humans on early deer herds is discussed in Chapter 4.

The Pronghorn, *Antilocapra americana*, formerly was considered to be the only living member of the family Antilocapridae. The Pronghorn is not a true antelope, such as those that live in Africa, but it is now classified in the family that includes African antelopes (Bovidae) as well as domestic cattle, sheep, and goats. Animals that live in wide open habitats usually occur in herds, they see well, and they can run fast. Similar to African antelopes, Pronghorns are able to run very rapidly. They are among the fastest animals in the world. Speeds of 65 mph (104 kph) have been recorded, but their ability to maintain high speed for a sustained time is phenomenal. They have been observed to travel 35 mph (53 kph) for 4 mi, and 55 mph (88 kph) for 0.5 mi.

Pronghorns stand about 3 ft (1 m) high at the shoulder. They are tawny with white patches on the face, chest, and rump. The bright color of the rump enables the herd to keep together when running. Both males and females have horns, which are shed in midwinter, leaving a bony core. New skin soon covers the bone, and new horns are replaced from the tip downward. The animal gets its name from a single branch, or prong, that grows forward on each horn.

The former range of the Pronghorn included all open terrain in California, including parts of the deserts. The western portion of the

FIGURE 10.19 Tule Elk, *Cervus elaphus nannodes* (photo by Michael G. Webb).

Mojave is still called Antelope Valley, in reference to former habitation by the Pronghorn. Pronghorns were rare in the Great Central Valley by 1875. People did not want herds of Pronghorns in their wheat fields. The only place in the state where they still occur naturally is in sagebrush country on the Modoc Plateau. Efforts have been made to reintroduce them to appropriate habitat within their former range. They have been reintroduced in Mono County, the Antelope Valley, and to the Carrizo Plain in San Luis Obispo County.

The Tule Elk or Tule Wapiti, *Cervus elaphus nannodes* (figure 10.19), disappeared with the Pronghorn. The biology of this species, endemic to the Central Valley, is similar to that of the Roosevelt Elk (Wapiti) of the northern coastal forests, but the Tule Elk is much smaller. Some authorities claim that the dwarf size is merely a function of poor food. Other authorities claim that the two will not hybridize, and therefore they are separate species.

Unlike the case with the Pronghorn, as humans destroyed the prairie, Tule Elk retreated to the vast Freshwater Marshes that covered hundreds of thousands of acres. In the tules, higher than a man's head, they made a labyrinth of trails. Then the wetlands began to be drained and the land cleared. The Tule Elk

retreated even farther. By 1885, there was only one band of animals left, near Lake Buena Vista in the southern corner of the San Joaquin Valley. In 1904, it was estimated that only 145 animals remained, although at its smallest, the population may have numbered fewer than 10, with a single breeding male. A rancher in the area, Henry Miller, took pity on the small elk and established a refuge on his ranch. A number of transfers were attempted, but the only band that prospered was in Yosemite National Park. In 1933, the 26 Tule Elk remaining in Yosemite were transferred to the Owens Valley. The next year, 28 from the refuge were added to the herd. There they prospered, returning from the brink of extinction. This is one of the success stories for principles of wildlife management, although the prospect of a genetic bottleneck arising from too few progenitors is presently under study.

The number of Tule Elk in the Owens Valley had risen to 300 by 1961, at which time the California Department of Fish and Game decided that the range was overstocked. Part of the problem was that the animals had a propensity for knocking down fences and raiding fields of alfalfa. The Department of Fish and Wildlife recommended a hunt, to kill 150 (half) of them. Screams of protest from animal lovers succeeded in aborting the hunt, but an over-stocked range did not help the Tule Elk. In 1971, a state law was passed that forbade hunting of Tule Elk until their number reached 2000. To alleviate the problem of an overstocked range, animals were tranquilized and transported to other parts of the state and placed in other refuges. There are estimated to be about 600 in the Owens Valley, and the wild population in the state exceeds 4000. They may be observed in the Carrizo Plain National Monument, Point Reyes National Seashore, Coyote Ridge in Santa Clara Valley, and in Owens Valley.

What predators kept these herds from over-populating before humans began to hunt them? Early accounts indicated that Mountain Lions, Grizzly Bears, and Gray Wolves, *Canis lupus*, were also inhabitants of the Great Central Valley. Grizzly Bears probably were not too successful in capturing Tule Elk or Pronghorn. Mountain Lions certainly were important, but they probably were more common in the Riparian Woodlands and preyed on deer.

It is likely that Gray Wolves were an important component of the ecosystem. Wolves hunt in packs. Usually, they take lame or stray animals, and they commonly feed on squirrels and mice. Nevertheless, a pack of Gray Wolves could follow a herd of Tule Elk or Pronghorn and crop those that tire first. One of the interesting accounts of the early animal life of the Great Central Valley is that of the legendary James C. "Grizzly" Adams. He and other early settlers have left us with numerous accounts of the Gray Wolves in the Great Central Valley. Authentic Gray Wolves were trapped in the Providence Mountains in 1922 and in Lassen County in 1924. After that, they were presumed to be extinct. In 1962, however, a wolf was killed near Woodlake on the edge of Sequoia National Park. Identification of this animal was

FIGURE 10.20 San Joaquin Kit Fox, *Vulpes macrotis mutica.*

verified by qualified mammalogists, but it is now believed that the animal was an escaped pet of Asiatic origin. The skin and skull are in the Museum of Vertebrate Zoology at the University of California in Berkeley. The reappearance of a family of Gray Wolves in northern California is described in Chapter 6.

Coyotes and foxes also occur in the Great Central Valley, and with the demise of the Gray Wolves and Grizzly Bears, in accordance with mesopredator release, their numbers, particularly Coyotes, likely have increased. Notable among these canids is the San Joaquin Kit Fox, *Vulpes macrotis mutica* (figure 10.20). Similar to the Desert Kit Fox, these small foxes live mostly on small rodents. In the Great Central Valley, they are an important natural control on nonnative rodents such as Black Rats, *Rattus rattus*, that invade grain crops. Habitat destruction, trapping for fur, and deliberate poisoning have driven these beautiful animals to the brink of extinction. Of further concern is new evidence that Coyotes may be preying on them. If this proves to be the case, it will be a sticky problem for management policy. The San Joaquin Valley Kit Fox is now listed as endangered by the federal government.

11

Inland Waters and Estuaries

FIGURE 11.1 Tarns in glacially scoured depressions. View northeastward from Bishop Pass, Inyo National Forest.

IN GENERAL, CALIFORNIANS have not been pleased with the natural distribution of water. Rather than moving population centers to locations where water is readily available, governmental agencies have chosen to move the water. California has the largest aqueduct system in the world. In the process, formerly large lakes such as Owens Lake is now dry, and the largest lake in the state, the Salton Sea, was formerly a large dry salt flat that is now filled with agricultural runoff. Nevertheless, there are thousands of natural lakes and streams in California, particularly in the Sierra Nevada.

The study of the physical, chemical, and biological properties of inland waters is known as limnology. In general, inland waters include freshwater lakes and rivers, but some lakes are very saline, particularly in deserts. Because of its varied climate and topography, California has a great diversity of aquatic habitats distributed over 2674 mi² (6845 km²) of inland waters.

Various classifications for inland waters have been established. Some are based on geologic processes; others are based on factors such as size, appearance, location, biota, or chemical constituents of the water. Various governmental agencies, such as the state Department of Fish and Wildlife, have established classifications of aquatic systems based on mixtures of characteristics. One such classification lists 23 kinds of lakes or impoundments and 27 kinds of streams in California.

LAKES

Lakes owe their origin to geologic processes. A lake is a low spot in the terrain that captures and holds water. Lakes are often classified into different types based on the geologic nature of the basin. Limnologists have recognized at least 11 categories, each of which has been further subdivided. Nearly every type of lake that has been described can be found somewhere in California (figure 11.2).

Glacial Lakes

The carving and scouring action of a glacier as it passes over bedrock leaves behind an irregular terrain. Low spots that become filled with water are glacial lakes called tarns. The Sierra Nevada abounds in such lakes (figure 11.3). Basins formed in canyons, with a high, amphitheater-like back wall, are called cirques, and tarns in these basins are called cirque lakes. The southernmost cirque lake is Dollar Lake on Mount San Gorgonio. A series of these lakes, one above the other, is known as a paternoster, a word that refers to a rosary or string of beads. Paternosters of cirque lakes occur in nearly every canyon of the Sierra Nevada, particularly on the steep east side of the range. Perhaps the most spectacular of these is on the east side of the western divide near Sawtooth Pass. Southwest of Black Rock Pass, three lakes, carved into granite, lie below in a staircase (plate 17B).

Most tarns are small mountain lakes that lie in depressions produced by glacial scouring. Large, glacially carved valleys commonly have tarns scattered about on their bottoms. One of the largest lakes in the Sierra Nevada, Thousand Island Lake, at the head of the San Joaquin River, is a huge but shallow tarn (plate 17C). The view down on a Sierran basin from a pass usually includes several tarns (figure 11.1).

Other lakes of glacial origin owe their existence to the moraine left behind when the glacier retreated. Where active glaciers still occur, a moraine lake filled with azure water can be found at the lower end of the glacier. The lake at the toe of the Palisade Glacier is a good example. The color of the water is due to tiny particles of ground rock suspended in the water. The Big Pine Lakes below Palisade Glacier are examples of cirque lakes in a paternoster, filled with such water.

A moraine may occur at the outlet of a tarn, forming a dam that enables the lake to hold more water. Other moraines may form nearly the entire lake basin. These large moraine lakes are often found at the mouths of canyons on the east side of the Sierra. Convict Lake is one of

FIGURE 11.2 Lakes and reservoirs of California (from Hill, M. 1984. *California Landscape: Origin and Evolution.* Berkeley: University of California Press).

the best examples of a moraine lake. Also, circular lakes known as kettles sometimes occupy a hollow in a large moraine field. Such lakes are common in Minnesota and Wisconsin, where huge continental glaciers formerly occurred, but they are present in mountain regions as well. A kettle forms in a depression left behind by the melting of a small fragment of glacier. In

the southern Sierra Nevada, there are two well-known lake basins in which most of the lakes are kettles. These basins, in the Golden Trout Wilderness, are known as the Cottonwood and Rocky Basins. They lie at about 11,000 ft (3300 m), southwest of the town of Lone Pine. Tulainyo Lake, lying at 12,800 ft (3925 m) north of Mount Whitney, is also a kettle lake. This is

FIGURE 11.3 Sallie Keys Lake, Sierra National Forest. A typical Sierra Nevada tarn.

the highest lake in the Sierra Nevada. To the east of it is a smaller, unnamed kettle lake at the base of Mount Russell.

Tectonic Lakes

Tectonic lakes lie in basins formed by warping, folding, fracturing, and faulting of the earth's crust. Fault-block mountains of the Great Basin are oriented in such a way the mountains alternate with valleys that may contain impounded water. This is known as horst-graben topography, and this is why geologists call the Great Basin the Basin and Range Province. A lake formed in such a basin is known as a graben lake. The name *graben* is German for "grave," implying that the lake lies in a low place. These grabens are formed by a combination of uplift and subsidence, usually along a dip-slip fault. One side of the fault rises, forming a mountain range, and the other side sinks. The consequence of this geologic activity is a valley that may contain a lake.

All of the large lakes that formerly lay in valleys of the Great Basin were graben lakes. These particular graben lakes were also examples of pluvial lakes, because they existed primarily during the periods of heavy rainfall that occurred during the Pleistocene. The lake in Saline Valley today is such a graben lake, but it is quite small compared to the one that occurred there during the Pleistocene. On the Modoc Plateau today, Goose Lake is an example of a large graben lake.

One of the most spectacular graben lakes in the world is Lake Tahoe (plate 3B), which lies in a valley between the Sierra Nevada and the Carson Range. The valley floor lies on a block that sank between two fault systems, forming a steep-walled, flat-bottomed lake basin. Subsequent damming by a lava flow at one end deepened the lake considerably. At a depth of 1643 ft (501 m), Lake Tahoe is the largest, deepest lake in California, and, next to Crater Lake in Oregon, the second deepest in North America. The largest mountain lake in North America, Lake Tahoe covers 122,000 acres (49,780 ha) and has a 72 mi (115 km) shoreline.

When an earthquake occurs, movement of the land may create—literally overnight—a new lake basin. These earthquake lakes often lie directly over a fault. The source of water could be springs that upwell along the fault zone. These lakes are often called sag-ponds, because

FIGURE 11.4 Mono Lake, a closed-basin, hypersaline lake.

the depression was formed by subsidence. In Inyo County, west of Highway 395 south of the town of Lone Pine, there is a sag-pond known as Diaz Lake. This lake appeared following the famous 1872 earthquake on the Lone Pine fault.

Another sag-pond, known as Lost Lake, lies on the San Andreas fault at the mouth of Lone Pine Canyon near the eastern end of the San Gabriel Mountains. Jackson Lake is another sag-pond on the same fault, about 17 mi (11 km) to the west. It is in a popular resort area on Highway 138 in the San Gabriel Mountains. Perhaps the largest sag-pond in California is Lake Elsinore occurring on the Whittier-Elsinore fault east of the Santa Ana Mountains. Today it is kept full with imported water from the Colorado River.

Sometimes crustal warping is enough to create a depression that will hold water. Thurston Lake on the Modoc Plateau is such a lake. Water in this lake is also impounded by a volcanic mudflow.

Landslide Lakes

Many lakes throughout the world owe their existence to impoundment of stream valleys by rockslides or mudflows. Sometimes the event trigger-

ing the rockslide is an earthquake. These lakes are often short-lived because the river, once it is impounded, overflows the dam, and subsequent erosion carries away material, restoring the area to a river. If the dam is substantial enough, a landslide lake may last for many years. On the Kern River at the southern end of the Sierra Nevada are two landslide lakes, known as Kern and Little Kern Lakes. The event that precipitated the two landslides was probably an earthquake on the Kern fault, along which the river flows.

Perhaps the most famous landslide lake is Mirror Lake, in Yosemite National Park. This lake, beneath the north face of Half Dome, is an impoundment on Tenaya Creek and has been filling with sediment for many years. Formerly, the National Park Service maintained it as a lake by dredging out the sediment. Now, however, it is National Park Service policy not to tamper with nature. Mirror Lake soon will be a meadow, and photographs of Half Dome reflected in the lake will become antiques.

Volcanic Lakes

Volcanoes often occur along faults, so it is not surprising that a combination of volcanic and tectonic activity might cooperate to create a lake basin. A lava dam at the north end of Lake Tahoe is partly responsible for its great depth. The Truckee River flows over the dam at the outlet of the lake.

Clear Lake is a large lake that lies in a valley of the Coast Ranges north of San Francisco. This is the largest lake in California that lies entirely within state boundaries. Two arms of the lake lie in grabens, but water is impounded by a lava dam at its southern end. It is a relatively shallow lake for its size, only about 30 ft (9 m) deep. Clear Lake is famous in California for its fishing and tourism. It is also famous worldwide as one of the first localities where DDT (dichlorodiphenyltrichloroethane) worked its way up the food chain and reached sufficient concentration to kill aquatic birds. Much of the concept known as biological magnification was formulated from data gathered there.

In the Cascade and Modoc Plateau regions, there are numerous lakes formed by volcanic processes. In Lassen Volcanic National Park, Snag Lake was formed by a lava dam, Hat Lake was formed by a volcanic mudflow, and Manzanita Lake was formed by a landslide of volcanic rocks associated with the 1914–1920 eruptions.

Another kind of volcanic lake is a caldera, which results when the summit of a volcano collapses. This occurs when the reservoir of molten magma empties from beneath. The deepest lake in the United States, Crater Lake, Oregon, lies in a caldera. It is nearly 2000 ft (610 m) deep. Perhaps the best example of a caldera lake in California is Medicine Lake, on the Modoc Plateau, just south of Lava Beds National Monument. The original caldera was about 6 mi (10 km) long, 4 mi (6 km) wide, and 500 ft (150 m) deep. Later eruptions built a ring of eight small volcanoes, nearly obscuring the basin. Medicine Lake lies in the remainder of the basin. The Long Valley Caldera, which contains the city of Mammoth Lakes on its southwest edge, contained a large lake during pluvial times (Pleistocene). At the present time, Lake Crowley, a reservoir on the Los Angeles Aqueduct system, lies in the bottom of the Long Valley Caldera.

A volcanic crater that has not collapsed may also hold water. Technically, this is not a caldera. Such a lake is found in Crater Butte, in Lassen Volcanic National Park.

Mono Lake (figure 11.4), on the east side of the Sierra, is the remainder of pluvial Lake Russell, which during the Pleistocene was about 900 ft (300 m) deep. This is also known as a closed-basin lake, which means that it lies in a region of interior drainage. All the water from the surrounding mountains has been draining into it for years, and there is no outlet. Evaporation causes water to be lost, but the dissolved minerals accumulate. Mono Lake therefore is hypersaline. It is the oldest lake in California and has been accumulating salts for thousands of years. It is three times saltier and 80 times more alkaline than the ocean. The tufa towers along its shoreline are the product of precipitation of limestone caused by freshwater bubbling up through the brine. The fact that the towers were formed underwater and now stand high and dry is testimony to a former lake level. Similar towers, now high and dry, can be found south of Trona near Searles Dry Lake. Mono Lake today is about 14 mi (23 km) long and 10 mi (1.6 km) wide. In this century, its greatest depth was 169 ft (51.5 m), and it has lowered considerably due to evaporation. After 1941, when diversions of water to the Los Angeles Aqueduct began, the water level was dropping. Between 1941 and 1988, its level dropped 40 ft (13 m) because its primary source of freshwater was taken away. Under the public trust doctrine established by the US Supreme Court, the California Supreme Court in 1983 was able to classify Mono Lake as a navigable body of water and therefore challenge water rights held by the Los Angeles Department of Water and Power (DWP). After numerous law suits, among them when lawyers for California Trout sued over the impact of water diversions on fish in Rush Creek, by 1990 they had obtained a permanent injunction that established minimum stream flow, particularly below Grant Lake, the main reservoir and diversion point for Rush Creek water. Finally in 1994, the California Water

Resources Control Board ordered that the elevation of Mono Lake be restored in order to fluctuate around an elevation of 6392 ft (1948 m). According to the ruling, the DWP must cease diversions if the water level is projected to drop to 6377 ft. At its low in 1982, elevation of the lake surface was at 6372 ft (1942 m). The DWP has had to restrict its diversions, which has not always worked in favor of agriculture in Owens Valley, because the department increased ground water pumping to make up for the loss. Additional lawsuits have been filed, most of which forced the DWP to mitigate for its water diversions, which succeeded in restoring stream flow to the lower Owens River and flooding portions of dry Owens Lake to keep down the dust, which was deemed a source of air pollution. North of Mammoth Mountain is a chain of volcanic domes known as the Mono-Inyo Craters that were formed by eruptions along a fissure. Mono Craters at the northern end include the two islands, Paoha and Negit, in the center of Mono Lake. At the southern end, the Inyo Craters include Obsidian Dome, Wilson Butte (plate 6A), and two explosion pits that contain small volcanic lakes. Of interest here is that one of the lakes is blue in color, indicating that it is essentially nutrient poor or oligotrophic, while the other is green or in a state of eutrophication, as explained later in this chapter. The apparent explanation for the difference is that the green lake has more vegetation on its rim, which is a source of detritus that contributes to the nutrient load.

Fluviatile Lakes

Fluviatile lakes lie in depressions caused by flowing water. When rivers flow over flat gradients, they meander back and forth across a floodplain. The meanders constantly change direction, and old meanders that have become separated from the present flow may still contain water. Because of their shape—rounded on one side—they are known as oxbow lakes. Such lakes are present in the Great Central Valley in association with large rivers. Three good examples are Pear Slough, near the San Joaquin River; Murphy Lake, near the Feather River; and Horseshoe Lake, near the Sacramento River.

Lakes can also be formed by flowing water when sediment produced by a tributary dams up the main river. These are known as fluviatile-dam lakes. The best example is Tulare Lake, which was formed when sediment deposited by the Kings River dammed a portion of the San Joaquin Valley. The Salton Sea basin was also formed in this way by damming from the Colorado River Delta. The huge Lake Cahuilla, which formerly occupied the basin, lay in a graben dammed by the Colorado River Delta.

Shoreline Lakes

Shoreline lakes are those impounded by barriers of sand. In California, these occur within a mile or so of the ocean. The sand accumulates by the combined action of wind and water. If a sandspit gets large enough, it can block the entrance to an estuary, allowing it to become a freshwater lake free of tidal flushing. Sometimes a large dune field forms a barrier that traps water. In the large Guadalupe-Nipomo Dunes field, south of Pismo Beach, there are several such bodies of freshwater. Oso Flaco Lake is a popular example (figure 11.5).

Solution Basins

Some rocks are water-soluble. Pits and depressions may occur in granite where water-soluble minerals have become dissolved. In deserts, these depressions catch rainwater and provide an important water source for animals. These depressions are known as *tinajas*, which is a Spanish word for "tanks." Where limestone is the bedrock, water can create large cavities. In the Sierra Nevada, Boyden, Crystal, and California Caves are limestone caverns. Black Chasm Cavern in the Sierra foothills east of Pine Grove has been established as a National Natural Landmark by the National Park Service. Mitchell Caverns, administered as a California State Park in

FIGURE 11.5 Oso Flaco Lake, in San Luis Obispo County, is a shoreline lake impounded by the Guadalupe-Nipomo Dunes.

the Providence Mountains, is a limestone cavern in the Mojave Desert. If the roof of a cavern collapses, a subterranean lake may be exposed. Devils Hole, in Death Valley National Park, is a small solution basin that was formed in this way.

Reservoirs

Reservoirs are man-made bodies of water formed by damming rivers. True reservoirs are primarily for storing water, which is released at a later time for irrigation, drinking, or industrial use. Some reservoirs, however, are constructed primarily to generate hydroelectric power. This is one of the cheapest ways to generate electricity. For this purpose, electric companies own and operate several large dams on the western side of the Sierra Nevada. These include Lake Thomas A. Edison, Huntington Lake, Florence Lake, Shaver Lake, and several others. After the water leaves Lake Thomas A. Edison on Mono Creek, it passes through eight different power plants before flowing out to Millerton Reservoir on the floor of the Great Central Valley. On the east side of the Sierra, several watersheds have dams. The dams on Bishop Creek were built by Southern California Edison. Waugh Lake, Gem Lake, and Agnew Lake on

Rush Creek are power reservoirs operated by the Los Angeles Department of Water and Power.

California does not have a shortage of reservoirs, we have a shortage of water. There are more than 1400 dams in the state. There are 36 reservoirs that hold at least 200,000 acre/feet and 11 can hold about 1 million acre/feet. One acre/foot is equivalent to water use for two average households for a year. One acre/foot is equivalent to 325.851 gallons. Many reservoirs serve multiple purposes. They may be used simultaneously for drinking, irrigation, recreation, and to generate electricity. Nearly every river in California has been dammed for flood control. These reservoirs, of necessity, must not be allowed to hold large amounts of water, or they would be worthless during a flood. Often downflow from these dams the streams are lined with concrete to prevent erosion and facilitate flow. In recent years, there has been pressure from environmentalists to restore riparian habitat in the channelized watercourses. Restoration of parklands along the Los Angeles River is a notable example. Other rivers such as the San Joaquin and Napa Rivers are also undergoing renovation.

Many dams and impoundments in California are associated with major aqueducts (figure 11.6 and color plate). There are 18 reservoirs on

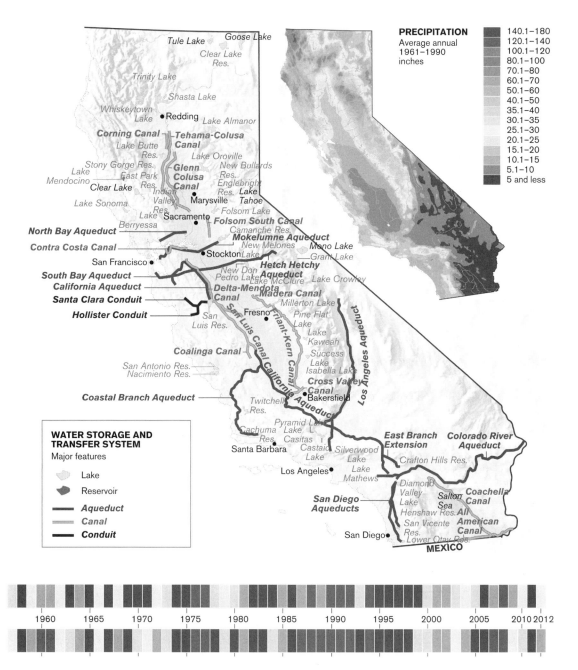

FIGURE 11.6 California has the most intricate aqueduct system in the world. The aqueducts shown here include state projects such as the California Aqueduct and Federal Projects such as the Coachella Canal and the canals of the Central Valley. The Colorado River Aqueduct is part of the Metropolitan Water District, the Los Angeles Aqueduct belongs to the Los Angeles Department of Water and Power, and the Hetch Hetchy Aqueduct was built for the City of San Francisco. See color version in insert.

the California Aqueduct alone. Lakes Shasta and Oroville are large reservoirs on the California Aqueduct system. The dam on Lake Oroville is the highest in the United States. During heavy rains of January, 2017, water had to be released from the reservoir and the spillways began to erode, threatening the integrity of the dam itself. Towns downstream on the Feather River were evacuated as a precaution. On the east side of the Sierra Nevada, reservoirs such as Grant Lake and Lake Crowley are part of the Los Angeles Aqueduct system. Another series, including Lakes Havasu and Mathews, is part of the Colorado Aqueduct administered by the Metropolitan Water District of Southern California. In addition, many artificial impoundments in southern California were created merely to provide recreation or for their visual quality. Some of these are associated with exclusive residential communities.

The earliest claim for Colorado River water goes to the Imperial Valley Water District. Water is diverted at Imperial Dam into the All-American Canal north of Yuma, Arizona. According to the legal principle, "First in time, first in right," the Imperial Valley gets nearly three times the amount of water from the Colorado River than does the Metropolitan Water District. Seven states take water from the Colorado River. The Imperial district gets more water than all but Colorado, and it receives about 70% of California's allocation of Colorado River water. Historically, this water was used to irrigate cropland that produced most of California's winter vegetables, about 75 different crops, including a large portion of California's alfalfa. Because San Diego is "water starved," in 2003 the Imperial board approved selling 500,000 acre-feet of their 3.1 million acre-foot allotment to San Diego. The water would be removed upstream at Lake Havasu and transported by means of the Colorado Aqueduct to Los Angeles, and then south through the San Diego Aqueduct. In response, the Imperial Valley would take out of cultivation 50,000 acres (20,230 ha) of farmland. By fallowing so much land, it has reduced the irrigation runoff that historically was the source of water

for the Salton Sea. As a result, water level the Salton Sea is lowering and salinity is increasing. Wildlife biologists fear that the result of increased salinity will interfere with the large number of breeding birds that inhabit the area.

Perhaps the most controversial dam project in California was the construction of O'Shaughnessy Dam in Hetch Hetchy Valley on the Tuolumne River. The spectacular scenery of the valley frequently was compared to Yosemite Valley, 20 mi to the south. Since the dam was inside Yosemite National Park, it required an Act of Congress, which was signed by Woodrow Wilson in 1913. John Muir and the Sierra Club fought the project for 7 years to no avail. John Muir, in one of his famous quotations stated, "Dam Hetch Hetchy! As well dam for water tanks the people's cathedrals and churches, for no holier temple has ever been consecrated by the heart of man." The dam was completed in 1929 and enlarged in 1934. Water from the reservoir is the primary water source for the San Francisco Bay area and includes a series of hydroelectric plants. Still today there is pressure among environmentalists to have the dam removed and the valley restored.

Not only dams block spawning runs for anadromous fishes such as salmon, but also as time passes nearly all dams become filled with sediment, which reduces their ability to control floods or store water. Hetch Hetchy is not the only dam proposed for removal. Several dams on the upper Klamath River have been proposed for removal as will be explained later in this chapter. At this point, the largest dam removal project underway in California is the San Clemente Dam on the Carmel River in the Big Sur area of the Coast Range. The 20-story concrete dam was built in 1921, but it has become so filled with sediment that it is no longer useful. Removal was begun in 2013, a project which will open up a 25 mi (40 km) downstream stretch of river that will once again allow Steelhead to return to historical spawning grounds. Interestingly, the project involves rerouting the river and restoring the sediment-filled basin to a meadow.

Other animals create dams, too. Of greatest significance is the activity of the American Bea-

TABLE 11.1

A Comparison of Oligotrophic and Eutrophic Lakes

Characteristic	Oligotrophic	Eutrophic
Age	Young	Old
Nutrient state	Poor	Rich
Clarity	Clear	Cloudy
Color	Blue	Green to brown
Depth	Deep	Shallow
Temperature	Cold	Warm
Dissolved oxygen	High, well distributed	Low, only near surface
Total dissolved solids	Low	High
Sediment	Sparse, coarse	Deep, muddy
Locality	Mountains	Valleys
Fish	Trout	Catfish

ver, *Castor canadensis*. Native distribution of this animal in California includes the rivers of the Great Central Valley and the Modoc Plateau. In large rivers, they usually do not build dams, but in smaller rivers their activities can seriously alter natural stream flow. Where they have been introduced into the Sierra Nevada, they fall aspens and willows, producing dams that may flood meadows or cause sedimentation that covers gravel used by spawning trout. In the late 1800s, Beaver pelts were very valuable, and the quest for Beavers was a major factor that helped to open up the west. Beavers were almost extirpated in California during that time. In recent times, natural reproduction and numerous introductions have enabled the Beaver to establish itself in many parts of the state, usually to the detriment of the native flora and fauna. California's southernmost groves of Quaking Aspen, *Populus tremuloides*, in the San Bernardino Mountains, may be in jeopardy because of the introduced Beavers.

Lake Succession

As lakes become older, they go through a process leading to their own demise. By their very nature, lake basins accumulate material and become filled with sediment. A lake ultimately becomes a meadow. Virtually all nutrients come from outside a lake, and as a lake becomes older, water becomes enriched with these minerals and nutrients. This process is known as eutrophication; the term *trophic* means "nourish."

The nutrient state of a lake can also be used as a basis for its classification (Table 11.1). A eutrophic lake is one that is nutrient rich. Usually, it is a fairly old lake. Young lakes are nutrient poor and are called *oligotrophic* lakes, which means "few nutrients." As a lake ages, it goes through a series of stages, including oligotrophic, mesotrophic, eutrophic, and finally senescent.

Most mountain lakes are typical oligotrophic lakes. They contain clear, blue water and may have very little sediment on the bottom. While there is a great deal of variation in shape and size, typical oligotrophic lakes are deep, clear, cold, and nutrient poor. The water contains low concentrations of dissolved minerals and a high concentration of dissolved oxygen. Fish such as trout, with small heads and small gill surfaces, would be content here.

Eutrophic lakes usually occur in lowland areas. Having had long periods of time to accumulate materials, they contain water that is murky. They are green to brownish green in

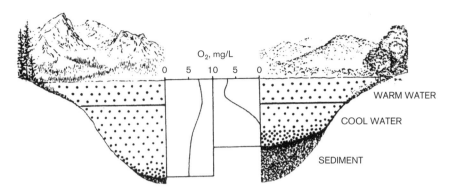

FIGURE 11.7 Diagrammatic comparison of an oligotrophic lake (left) and a eutrophic lake (right) during summer stratification. Horizontal lines indicate the regions of greatest temperature and density differences associated with a thermocline. Dots indicate distribution of an equal amount of plankton and particulate matter. Graphs indicate differences in dissolved oxygen (after Cole 1988; reprinted with permission).

color. The sediment of such a lake is rich in organic material, forming a muddy ooze. The abundance of particles in the water helps to absorb and hold heat, so the water tends to be warm and oxygen poor. Trout would not be happy here because they cannot tolerate concentrations of dissolved oxygen lower than 5 mg/L. Eutrophic lakes contain large-headed fishes with abundant gill surface, including catfish, carp, bass, and panfish such as bluegill or crappie, none of which is native to California. The Sacramento Perch, which will be discussed later, is one of the few fishes of this type that is native.

In California, most eutrophic lakes become stratified during the summer (figure 11.7). This means that there is a distinct layering of water of different temperature and density. The upper portion of the lake is warm, but the deeper parts are much cooler. If a person treads water in such a lake, it is not uncommon for his or her feet to be in cold water while the rest of the body is warm. The region where the water changes temperature over a relatively short range of depth is known as the thermocline. Above the thermocline, water is warm. Below the thermocline, it may be as much as 45°F (25°C) cooler. Often the temperature drops over 0.5°F/ft (1°C/m). This temperature change is analogous to the inversion layer that occurs in the air of a basin. In a lake, the cold water below the thermocline has greater density, and it is not

disturbed easily. Above the thermocline, the water is less dense and may be stirred by the wind. This stirring mixes oxygen with the water, but only above the thermocline.

A person would think that the deeper water, by virtue of its colder temperature, would hold more dissolved oxygen. What happens is that bacteria living in the rich sediment deplete the oxygen. Because the oxygen cannot be stirred downward below the thermocline, the deep water may actually become anaerobic. Fishes therefore are forced to occupy the warmer water near the surface.

In the warmer water, fishes are placed in double jeopardy. Warm water increases their metabolic race, so they require more oxygen when less is available. Actually, the problem is more complex. Fishes must ventilate their gills by pumping water over the gill membranes. They do this with a bellows-like action of the bony operculum that covers the gills. Muscle activity to operate the operculum requires fully one-fourth of the fish's metabolism. In other words, because water is heavy, a fish uses one-fourth of its food and oxygen to pump water over its gills. When a fish lives in warm water, its metabolic rate is high, but there is less oxygen dissolved in the water; therefore, the work required to ventilate its gills is disproportionately large, requiring an even greater amount of oxygen. This represents triple jeopardy.

Water Quality

Various characteristics of lake water may be measured as indicators of eutrophication. Aquatic biologists want to know the nutrient state of a lake so they may estimate what kinds of organisms live there and what sort of biomass and growth rates should be expected. Indicators of water quality are also used to determine whether or not the water is appropriate for introduction of certain fish species. In areas of habitation, these types of measurements are used to determine if the water is suitable for human consumption. Humans prefer to drink oligotrophic water.

A measure of the quantity of organic material in water is the biological oxygen demand (BOD). This is a measure of the oxygen consumed by microorganisms in a certain period of time. The organic material is food for microorganisms; therefore, if lots of organic material is in the water, lots of oxygen is used. Such a test involves keeping a sample of the water in a sealed container in the dark. Dissolved oxygen in the water is measured before and after storage. This measure of water quality is usually expressed as the decrease in oxygen, in milligrams per liter (mg/L), over a 5 d period.

Another indicator of eutrophication is the amount of photosynthesis that occurs in phytoplankton over some period of time. Known as primary production, this measurement is accomplished by introducing a measured amount of radioactive CO_2 to a sample of water. The amount of radioactive carbon (^{14}C) absorbed by the plankton is measured with a Geiger counter, and it usually is expressed in milligrams of carbon per unit time in proportion to surface area (e.g., mg $C/m^2/d$). For comparison, in 1974, data for Clear Lake, a large eutrophic lake in the Coast Ranges, indicated that phytoplankton absorbed 433 mg $C/m^2/d$. In Lake Tahoe, a large oligotrophic lake, during the same season and year phytoplankton absorbed only 174 mg $C/m^2/d$. On a yearly basis, oligotrophic lakes fix from 0.7 to 25 g C/m^2, whereas eutrophic lakes fix from 75 to 250

g C/m^2. Heavily polluted or senescent lakes might fix from 350 to 700 g $C/m^2/yr$.

Measuring BOD and primary productivity requires specialized equipment and skilled technologists. The best results are obtained from analyzing water samples in a laboratory. Persons working in the field are always looking for simple techniques that lead to similar conclusion; by comparing data from various indicators of water quality, limnologists have decided that a good estimate of eutrophication may be obtained by measuring the amount of dissolved material in the water. This bit of information may be obtained in several ways. The most accurate measure is total dissolved solids (TDS), which is the same as salinity. Obtaining this information accurately involves weighing a sample of water, then evaporating the water, burning organic material out of the residue, and weighing the ash. The amount of ash per liter of water is then calculated and expressed in milligrams per liter (mg/L) or equivalently, in parts per million (ppm).

In a study of 112 lakes at various elevations in the Rocky Mountains, TDS values varied from an average of 14.9 mg/L for alpine lakes to 202.9 mg/L for lakes in the prairie. These are reasonably accurate indicators of eutrophication, and they indicate that elevation plays a role. Lakes at lower elevation have greater nutrient input and longer seasons of productivity.

In the Sierra Nevada, similar values were calculated for a series of lakes in the Convict Lake Basin. Bunny Lake, at an elevation of 11,000 ft (3354 m), had only 5 mg/L TDS. Lake Crowley, at 6800 ft (2073 m), had 200 mg/L. In Bunny Lake, the small quantity of nutrients dictates that growth rates are slow. The oldest Brook Trout on record, 24 years old, was captured there. It was 8 in (20 cm) long, and it weighed only 0.25 lb (115 g). In contrast, fishes in Crowley Lake grow from 1 oz (20.5 g) to 1 lb (456 g) in a single year. At this lake, 2 lb (1 kg) fishes 15 in (38 cm) long may be no older than 1.5 years.

Because TDS is a measure of the amount of dissolved minerals in water, hydrologists have developed an even simpler way to measure it.

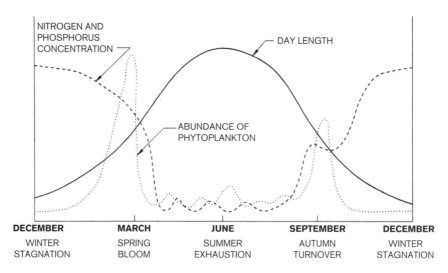

FIGURE 11.8 Seasonal cycles of temperate waters.

They determine how well the water conducts a current of electricity. This measure of water quality is called conductance. The more minerals in the water, the better it conducts electricity. The attractiveness of this method is that any person with a few minutes of training can carry a battery-powered instrument to a body of water and measure its conductance. The water need not be returned to a lab for analysis, so the investigator gets an instantaneous reading. For all practical purposes, conductance will reveal what an investigator needs to know about the eutrophic state of a body of water.

Seasons in a Lake

In high-mountain lakes, there are only two seasons: winter and summer. During winter, the duration of ice cover is the critical factor. Because wind is the source of oxygen for lake water, a lake that is covered by ice for too long cannot sustain a fish population. The larger the lake, the greater the oxygen reservoir; therefore, large, deep lakes are able to maintain fish populations under ice for a longer period of time. A shallow lake will not sustain a fish population during an average winter.

In typical eutrophic lakes, there are four seasons (figure 11.8). During summer, concentrations of phytoplankton are low because dissolved nutrients such as nitrates (NO_3) and phosphates (PO_4) have been exhausted. Plankton that were abundant in spring have died and settled to the bottom. This season may be referred to as summer exhaustion. At this time, bacteria in the sediment decompose the dead organisms, depleting the oxygen in the deeper part of the lake. If oxygen concentrations in the deep become too low, decomposition is carried on by anaerobic bacteria that may produce toxic by-products, such as hydrogen sulfide (H_2S). Decomposed organisms also return soluble nitrates and phosphates to the lake water at the bottom.

As autumn approaches, day length becomes shorter, and the sun moves to a lower position in the sky. As a result, surface water receives less warming each day. Air temperatures also begin to drop. Water at the surface of the lake cools first. When the temperature at the surface is cooler than that of the deep water, surface water begins to sink, producing a stirring effect known as the autumn turnover. When this happens, nutrient-rich water from the bottom rises to the top. If the deep water has been anaerobic, this turnover may be accompanied by a fish kill. The deep water rising to the surface may be so oxygen poor and loaded with hydrogen sulfide that it is insufficiently diluted, a combination that may be lethal to fish.

The upwelling of nutrients often causes an autumn plankton bloom, but it is short-lived because day length is too short and water temperature too cool to maintain phytoplankton for very long. Then the period of winter stagnation begins. Eutrophic water is clearest during this time. This stage occurs even if there is no ice cover.

If there is a cover of ice during winter, a second turnover during spring may occur when the ice melts. Ice, at 32°F (0°C), is less dense than water that is slightly warmer, which is why ice floats. When the ice melts, the cold water produced sinks to the bottom, initiating a spring turnover. Water at the bottom of a lake is the densest and usually is about 39°F (4°C).

During spring, day length increases. The sun rises in the sky and shines more directly on the lake, and air temperature increases. As the surface water becomes warmer, phytoplankton become more numerous, absorbing abundant nitrates and phosphates. This is the spring bloom. Lake water at this time may turn from clear to cloudy green in a few days. Algae that are capable of nitrogen fixation appear first. These are cyanobacteria or cyanophyta (blue-green algae). They are followed by green algae (chlorophyta) and small animals (zooplankton). Stimulated by abundant food and warm water, large aquatic animals such as frogs and fish begin to breed. Juveniles hatch and thrive on the abundant food. When the nitrates and phosphates become exhausted by the phytoplankton, there is a population crash, followed by clearing of the water. This is the beginning of a new phase of summer exhaustion, another critical time for frogs and fishes. When food becomes scarce, only the hardiest will survive; often these are the oldest and largest. All of the dead organisms, from plankton to vertebrates, will sink to the bottom, contributing to the organic ooze in which bacteria are busy decomposing and recycling.

Cultural Eutrophication

If nutrients such as nitrates and phosphates are added to a lake artificially, eutrophication may occur at an accelerated rate. This type of artificial eutrophication can be serious enough that an oligotrophic lake can become green and murky in a short period of time. If the process goes far enough, deep water becomes anaerobic, and a fish kill could result. This has occurred in the Great Lakes, where sewer discharge carries nitrogen from human urine and phosphorus from detergents.

Accelerated eutrophication in Lake Tahoe is of serious concern. A combination of development, gambling casinos, ski resorts, and great beauty has attracted large numbers of visitors to the area. The lake's remarkable clarity is one of the reasons for its striking beauty. In the 1950s, objects on the bottom could be viewed to depths of 130 ft (40 m). In the 1960s, clarity began to be lost, and attached algae began to grow on rocks around the lake's margin. Between 1960 and 1980, primary production in phytoplankton more than doubled, from about 40 g C/m²/yr to over 80 g C/m²/yr. In an attempt to halt the eutrophication process, the Environmental Protection Agency sponsored construction of an advanced sewage processing plant. This involved tertiary treatment of the water to remove nitrates and phosphates as well as organic material. In addition, all processed sewer water was pumped out of the basin. The Lake Tahoe tertiary treatment plant, however, has been taken off line in favor of advanced secondary treatment that leaves nitrogen and phosphorus in the water, and that water is used for irrigation. Nevertheless, eutrophication in Lake Tahoe continued. The source of nutrients was determined to be the forest ecosystem in the basin. There was so much construction that soils were being disturbed at an unprecedented rate. Runoff from the forest carried nutrients that normally would cycle through the forest ecosystem. The outcome was eutrophication of the lake water in spite of the advanced sewer system. A moratorium on development in the 1980s stirred a storm of protest. A compromise has been negotiated among private-property owners, the US Forest Service, and the states of California and Nevada. Meanwhile in 1978 another tertiary treatment facility was developed north of Lake Tahoe, near Truckee, but that processed water is injected into the ground. In recent years, there have been a number of projects, some proposed, and some already in operation, that are designed to use advanced water processing techniques to reclaim wastewater or desalinate seawater. One of the earliest desalination plants was built by Southern California Edison on Santa Catalina Island. In exchange for allowing more development, the electric company agreed to increase the water supply through seawater desalination. During years of adequate precipitation, however, the plant is shut down. In order to make it financially feasible, San Diego County agreed to buy water from a new desalination plant in Carlsbad, which ironically added to the supply they get from the Metropolitan Water District of Southern California, and in so doing gave them a surplus during a time of drought and water rationing.

Cultural eutrophication may also affect streams and lakes in regions of clear-cut logging. Again, it is runoff from disturbed land that creates the problem. This is particularly serious in regions of heavy precipitation, such as the Klamath Mountains.

STREAMS

Flowing water may be characterized in several different ways. Terms such as *river, creek,* or *brook* are based on relative size. However, it is foolish to quibble because the terms mean different things depending on the region of the country. A river in California might be called a creek in the eastern part of the United States.

Types of Flowing Water

Sometimes rivers are described according to permanence of flow. A permanent stream flows year-round. An intermittent stream dries up during a portion of the year. An interrupted stream is one that flows aboveground in portions of its run and underground in others. This vocabulary can reveal a great deal about a stream. A fisherman who knows that a certain stream is interrupted will not be dismayed to find a dry stream channel. A walk upstream will, sooner or later, lead to a region where there is permanent water. A wise hiker will read the symbols on a topographic map in order to know, before he or she arrives, whether a stream is permanent, intermittent, or interrupted. Most desert streams are intermittent. Most streams in southern California are interrupted.

Sometimes the appearance of the flow is used as a basis for comparing streams. Water that flows smoothly is said to be laminar, a situation that occurs in large rivers where the water flows slowly. Most rivers are characterized by various degrees of turbulence, reflecting a combination of speed and irregularities of the stream bottom.

Another means of comparison is based on the rate of fall, a combination of speed and turbulence based on stream gradient. Water moves fastest when it flows straight down without friction on the streambed. This is a waterfall. All cliffs over which a stream falls, however, are not vertical. Water falling over a surface that is not vertical is very turbulent as it moves over and around irregularities in the streambed. At some arbitrary point, it is called a cataract rather than a waterfall. Turbulent flow over a nearly horizontal surface is known as a rapids.

The Sierra Nevada abounds in waterfalls, cataracts, and rapids. For example, some of the most spectacular waterfalls in the world are found in Yosemite National Park. Two of the 10 highest waterfalls in the world are in Yosemite Valley. At 2425 ft (739 m), Yosemite Falls is currently recognized as the fifth highest in the world, although some people quibble because it falls in two steps divided by a cataract. Angel Fall in Venezuela, at 3212 ft (979 m), is the highest. The eighth highest waterfall is Sentinel Fall in Yosemite Valley. It drops 2000 ft (609 m) in several steps. Ribbon Fall in Yosemite Valley drops 1612 ft (491 m) in a single drop. It ranks among the top 20 in total height, but among the top 10 in highest single fall.

Falling water directs a great deal of energy at the stream bottom. In a waterfall, nothing remains that is not attached. In a cataract, only large boulders remain. In a rapids, large and small boulders remain in the flow. As the stream gradient lessens, smaller particles remain in the bed, grading from gravel to sand to mud. In a mud-bottom pool, water barely flows.

The speed of falling water depends on the stream gradient. If it moves too fast, it carries along not only sediment, but organisms as well. The optimal flow for a typical stream is only about 1–3 ft (1 m) per second.

To a biologist, velocity of stream flow in itself is not a particularly useful item of information. It is more important to think in terms of volume of water flow per unit of time, a measure known as discharge. It is typically expressed in cubic feet per second (cfs). A rate of 1 cfs is equivalent to 28.3 L/s. The amount of discharge varies with the width, depth, velocity, and sediment load of a stream. Niagara Falls, on the St. Lawrence River between New York and Canada, is an impressive sight. Its discharge has been estimated at about 220,000 cfs. The Mississippi River at New Orleans looks big but is not as impressive as Niagara Falls. However, discharge of the Mississippi at New Orleans has been estimated at 1,740,000 cfs. The difference is that it is deeper.

When large volumes of water are involved, discharge is often expressed in acre-feet per year. One acre-foot covers an acre (0.4 ha) to a depth of 1 ft (30 cm). The Mississippi River discharges 450 million acre-feet per year. In the west, the largest river is the Columbia. Its discharge has been measured at 141 million acre-feet per year. In California, typical discharges per year of the major rivers are as follows: Sacramento River, 17.1 million acre-feet; Klamath River, 12.5 million acre-feet; Colorado River, 10.5 million acre-feet; and the San Joaquin River, 3.3 million acre-feet. Although the Klamath River is dammed in its upper reaches, most of the flow in California is undammed. Is it any wonder why agricultural interests view the Klamath as a wasted river? On the other hand, salmon and salmon fishermen are delighted that the lower Klamath is a wild river.

A number of years ago, limnologists attempted to devise a system of classification that would remove most of the subjective judgment from the vocabulary. This led to the concept of stream order, a classification of flowing water on the basis of number of tributaries, drainage area, total length, and age of water. First-order streams are the beginnings of a river system. These are the youngest segments, where the water begins to flow; they have no tributaries. Second-order streams are formed by the junction of two first-order streams. A third-order stream is fed by four first-order streams, fourth-order stream is fed by eight first-order streams, and so on. Tenth-order rivers are rare. The Mississippi is probably the only one in North America.

Obviously, correct classification of a stream requires careful examination of a detailed topographic map, which would be complicated by intermittent streams and would be only as reliable as its level of detail. Unfortunately, the system of stream orders is so complex that only limnologists can communicate with each other, and even they cannot agree on whether the Mississippi is a 10th-, 11th-, or 12th-order river. Think of what would happen to literature if a romantic scene were described as taking place on the banks of a third-order stream.

California's rivers (figure 11.9) are divided into seven major drainage systems. The Sacramento–San Joaquin system has the largest watershed and the greatest discharge, but the Klamath system carries a greater volume of water in proportion to its watershed. Conversely, in the southwest deserts there are numerous streams with large watersheds but low discharge. The system of stream orders would be meaningless for comparing these three areas of the state.

Formation of Stream Channels

At upper elevations, stream channels and valleys have a U-shaped bottom, the result of glacial sculpturing. In the Sierra Nevada, U-shaped valleys and stream channels occur all

FIGURE 11.9 California rivers (from Hill 1984).

across the high country. As the waters flow toward lower elevations, they pass through V-shaped river valleys. A river, as it flows, cuts a notch into its bed. As the notch becomes deeper, weathering, gravity, and side streams work to eliminate the steep walls of the notch. The result is the typical V-shaped valley. Prehistorically, the U-shaped valleys of the high country began as V-shaped valleys but were reshaped later by passage of a glacier through the notch. The scenery of Yosemite Valley was formed in this way (plate 3A). The large waterfalls that pour into the valley from all sides flow from the mouths of hanging valleys. The glaciers in the hanging valleys were smaller and did not cut as deeply into the bedrock as the glacier flowing

FIGURE 11.10 Nevada Fall on the Merced River, Yosemite National Park.

down the main channel. Hence, the water from the tributaries today plunges over a cliff before joining the main channel. Several miles below the entrance to Yosemite Valley, the Merced River flows through a V-shaped gorge. This gorge is below the point that was reached by the Pleistocene glaciers.

The physics of water as it cuts a stream channel have been studied thoroughly. A stream may be thought of as having the "goal" of flowing in a smooth arc from its headwater to its mouth. At its upper end, where less water is involved, the curve is steep. At its lower end, the curve flattens out. If an ideal curve is achieved, the stream uses all its energy to carry sediment over a smooth bed. If the gradient of the hill lessens, the energy in the water decreases, and the stream responds by dropping some of its load of sediment. This causes the stream channel to become filled, and the stream gradient is increased. If the stream flows over a high spot, such as a waterfall, it

picks up energy and cuts away at the headwall. In this way, a waterfall moves up valley, leaving behind a steep gorge. This is exactly the way the gorge of the Merced River below Vernal and Nevada Falls (figure 11.10) was formed. These spectacular waterfalls are still cutting away upstream.

Limnologists construct stream profiles. A stream at base level flowing over its ideal curve is said to be at grade. Where a waterfall flows over a cliff, it is above grade and picks up energy. Where the water flows into a lake, it is below grade; it loses energy. By building a dam, engineers create a situation where the water is placed artificially above grade. The energy gained is used to generate electricity as the water falls. Of course, the reservoir above the dam is below grade, so it is doomed to become filled with sediment.

At the stream's lower end, where the grade is flattened, Coriolis effect begins to have a profound influence on the flow of water. In the Northern Hemisphere, flowing water turns to the right. This causes streams to flow to the right across a floodplain, such as the Great Central Valley. Eventually, however, the flow will reach a point where if it were to continue flowing to the right, it would be forced to flow uphill. That is not possible, so the stream turns to the left, flowing downhill until Coriolis effect once again draws it to the right. The result is a stream channel that meanders back and forth across a floodplain.

Fluctuations in flow associated with normal variations in precipitation cause a river to flood and subside. The result of flooding is deposition of new topsoil, which is one of the reasons that the Great Central Valley has been such a rich agricultural region. Overflow and deposition form natural levees, creating the terraces upon which natural vegetation establishes itself. Overflowing also supplies water to floodplains, which can help recharge the water table. As a river erodes its channel on the outside of a bend, it deposits sediment on the inside. The meander in this way moves down the valley, depositing new soil behind it.

Controlling rivers with dams and concrete channels disrupts the natural order of things. Without rejuvenation, soil is lost, not gained. Fertilizer must be applied artificially. All the symptoms of desertification discussed in the previous chapter may be related to poor agricultural practices based on short-term financial gain. Accumulation of salt, erosion, groundwater overdraft, and subsidence are problems involving agriculture and water management that must be faced in the near future.

ESTUARIES

An estuary is a place where freshwater mixes with seawater. This occurs where a river flows into a bay. In California, the origin of such a bay is usually associated with a barrier beach bar, a sandspit formed by the longshore current. The type of bay formed by a drowned river mouth is less common in California because, for the most part, the coastline here is rising rather than subsiding. San Francisco Bay is a notable example of a drowned river mouth. Tomales and Bolinas Bays, along the San Andreas fault near Point Reyes, are drowned rift valleys.

Estuaries are among the most productive ecosystems on earth. Salt marsh alone contributes about 3000 gC/m²/yr (16,700 lb/acre/yr), but including the contribution of phytoplankton the amount is doubled. There is not a great diversity of habitats, however. As a result, there is not as large a number of species as there is in the rocky intertidal zone, but the number of individuals of each species is enormous. Diversity there is manifestation of *edge effect*. There is a boundary between the land and water, and there is also a boundary between freshwater and seawater that moves back and forth with the tides. These boundary regions are characterized by species from each of the adjacent habitats.

Birds are extremely common in estuaries, particularly during the winter. They feed and breed at higher latitudes during the summer during the long days of high primary production. In the winter, they migrate southward to habitats such as estuaries and tropical forests

that have abundant food. Coexistence of all those species is enhanced by behavioral and anatomical forms of niche partitioning.

Freshwater is lighter than seawater, so it flows over the top of seawater in the estuary. Because of Coriolis force, freshwater also tends to flow closer to the coast on the north side (the right side) of the estuary. A wedge of water where salinity changes rapidly from freshwater to seawater (0.0–3.5% TDS), moves back and forth with the tides. Free-swimming organisms will remain in the region that suits them best, but organisms in the mud may be subjected to dramatic changes in salinity as this wedge of water passes over them. It appears that a critical region for organisms occurs where the salinity varies from 0.05% to 0.08%. Upstream from that region, freshwater species increase in number, and below that region there is an increase in numbers of marine species. Surprisingly few species occur at the region of transition.

The consequence of this variation is that estuaries are characterized by four primary communities: Freshwater Marsh, Salt Marsh, Mud Flats, and Open Water. The Freshwater Marsh was discussed in Chapter 10. Attention here will be directed to the other three habitats.

Salt Marsh

As in the Freshwater Marsh, there are bands of vegetation based on the amount of submergence tolerated by the plants (figure 11.11). Among the rooted macrophytes, the plant that occurs farthest into seawater is Eelgrass, *Zostera marina*. Eelgrass grows from the low-tide level to depths of about 20 ft (6 m). This is the same species that grows with surfgrasses in the rocky intertidal zone. However, it tends to be more filamentous in appearance in estuaries. Eelgrass traps sediment and is a substrate for attachment of algae and bacteria. Grazers such as snails feed on the algae and bacteria as well as the Eelgrass itself. Herbivorous fishes also feed on Eelgrass. The clumps of Eelgrass are an important nursery for fish and invertebrate larvae. Physiologically, Eelgrass is well suited for

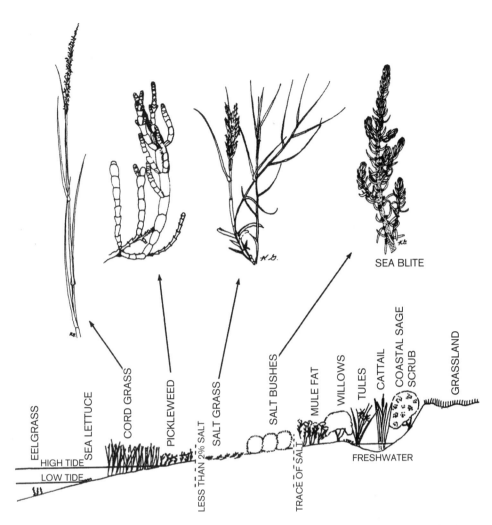

FIGURE 11.11 Salt Marsh zones (drawings by Karlin Grunau Marsh from *The Mountains and the Wetlands*, Orange County Land Environment Series No. 2, 1975).

alternating inundation with freshwater and seawater. The plants excrete salt as the tide goes out and absorb salt as the tide comes in.

The emergent plant that grows farthest into seawater is California Cord Grass, *Spartina foliosa* (figure 11.12). This is a plant up to 3 ft (1 m) in height that may be half submerged. It spreads through the salty mud by means of buried runners (rhizomes). California Cord Grass grows like rice in a paddy, and it is able to tolerate its harsh habitat by means of several important adaptations. California Cord Grass stems are hollow and air filled, a feature that enables oxygen to reach submerged roots. Also,

similar to many salt-tolerant plants (halophytes), California Cord Grass is equipped with salt glands to excrete excess salts. It is also interesting to note that California Cord Grass is one of the very few cool-climate plants that have C_4 photosynthesis. This type of high-efficiency photosynthesis occurs in desert halophytes such as saltbushes that must photosynthesize and excrete salts in a hot, dry climate. In order for this plant to cope with drought stress, it is probably important that it be able to photosynthesize with stomates partly closed, and this is enhanced by high-energy C_4 photosynthesis. California Cord Grass is also equipped with

FIGURE 11.12 Upper Newport Bay, the largest extent of natural Salt Marsh in southern California.

nitrogen-fixing bacteria in its roots; therefore, its foliage is an important source of protein for the marsh ecosystem.

California Cord Grass is the greatest contributor to the high rate of primary production in an estuary ecosystem, although it does not make its contribution in the expected way. Very few organisms are equipped to eat the salty foliage. California Cord Grass adds to the ecosystem through detritus. Root systems trap sediment and detritus. Microorganisms consume the detritus and contribute to the dissolved organic matter that enters the water. Worms and snails feed directly on the detritus, and nutrients enter the estuary food web. Fishes and birds feed on the snails and worms, and nitrogenous wastes of the animals return to the system to complete the cycle. When the tide goes out, soluble nutrients are carried out to sea by lighter surface water. It is this outwelling that contributes nutrients to sandy beaches and the rocky intertidal region. However, the heavier saltwater carries sinking nutrients back upstream on incoming tides and deposits it in the muds of the Salt Marsh. In this way the circulation of freshwater and saltwater creates a nutrient trap.

Inland from the California Cord Grass is a belt of succulent plants. Of these, the pickleweeds and glassworts (*Salicornia* spp.) are the most common. Roots of pickleweeds are covered by seawater only during the highest tides, a habitat similar to the Alkali Sink community discussed in the chapter on deserts. A pickleweed absorbs water and salt. The salty water is stored in its tissues, which helps to maintain a favorable flow of water into the plant even though it is rooted in salty soil. The plant is leafless. Its stem resembles a series of small green pickles. The section at the tip of the stem becomes filled with salt and drops off, recycling the salt back into the ecosystem.

Associated with the pickleweeds may be other halophytes such as Salt Grass, *Distichlis spicata*, and salt-tolerant shrubs such as California Seablite, *Suaeda californica*. Similar to plants in the desert, these shrubs excrete salt upon their leaves through salt glands. Because salt marshes are among our most endangered habitats, they have their share of endangered species. The Salt Marsh Bird's beak, *Chloropyron* (= *Cordylanthus*) *maritimum* subsp. *maritimum*, is a paintbrush-like species that is federally listed as endangered.

Diversity in the Salt Marsh is maintained by environmental perturbations such as freshwater flooding and tidal flushing. Plant productivity is high, and growth of new plants is rapid. Detritus enriches the muds. Dense root systems trap the detritus and mud, which allows more nutrients to accumulate. A natural catastrophe such as a flood may scour an area, and the cycle of regeneration repeats itself. Some people call this a "build-and-tear" ecosystem.

Animals are not numerous in Salt Marshes. The kinds of animals vary from aquatic to terrestrial, but no type is particularly conspicuous. In the water, there is a variety of immature and adult aquatic invertebrates, including arthropods, worms, and snails. When the tide comes in, fishes such as the California Killifish, *Fundulus parvipinnis*, move among submerged stems and feed on detritus and invertebrates. The Bay Pipefish, *Syngnathus leptorhynchus*, hides in the Eelgrass, feeding on small crustaceans. Also when the tide is in, burrowing detritus feeders and filter feeders, many of which are worms, go to work. Among these are polychaete worms similar to those that live on sandy beaches. Filter-feeding tube worms build their tubes of cemented sand grains, fitted together as precisely as the rocks in a stone wall. This is remarkable considering that the tubes are composed of very fine particles and are only one grain thick. One worm in captivity was observed to require two months to build only 0.75 in (19 mm) of tube. Each tube is the life work of its resident.

There are few terrestrial animals in a Salt Marsh. The vegetation is not very nutritious, so few organisms eat it. Apparently, grasshoppers are able to tolerate a diet of Salt Marsh vegetation, but most of the animals that live in a Salt Marsh feed upon seeds or the aquatic animals.

The largest Salt Marsh in the state occurs in San Francisco Bay. In this marsh, there is an endemic mammal, the Salt-marsh Harvest Mouse, *Reithrodontomys raviventris*. This mouse builds its nest in the vegetation in a birdlike manner. It is nocturnal, coming out at night to feed on seeds. Even though this small mouse has a high metabolic rate, it is able to conserve energy by huddling in its insulated nest. Salt-marsh Harvest Mice are truly unique in their ability to survive on seawater. Because Salt Marshes are endangered habitats, distribution in the southern part of San Francisco Bay is now limited to narrow strips, and it should be no surprise that the Salt-marsh Harvest Mouse is listed by the federal government as an endangered species. Due to the filling and diking of various portions of San Francisco Bay, the range of this unique mouse has declined to a small population on a few isolated islands in the remaining marshland.

There are very few resident bird species in Salt Marshes. Of the five species and subspecies that nest here, four are listed as threatened or endangered by the California Department of Fish and Wildlife. Three of these are in the rail family (Rallidae). These are narrow, slab-sided birds that move through marsh vegetation without disturbing it. The expression "skinny as a rail" refers to the appearance of a rail in front view. The only rail that is not threatened is the Sora, *Porzana carolina*, which is equally at home in a freshwater marsh.

Ridgway's Rail (= California Clapper Rail), *Rallus obsoletus* (= *R. longirostris*), is the largest of these rails. They are more often heard than seen, and their distinctive call is the basis for their former name. It is a series of 10 or more sharp "kek, kek, kek" noises that sound a bit like clapping. They resemble chickens with a short, upturned tail and light rump. They are active at dawn and dusk, feeding on insects and other invertebrates. At this time, they may venture onto Mud Flats to feed.

Three species of these rails in California are listed as endangered. The two that inhabit coastal Salt Marshes are the California Clapper Rail, *R. obsoletus*, which occurs from San Francisco north, and the Light-footed Rail, *R. o. levipes*, which occurs from Santa Barbara south. The third variety, the Yuma Rail, *R. o. yumanensis*, frequents Freshwater Marshes in the southwest deserts. All three of these forms are victims of habitat destruction, including damming, diking, draining, pollution, and predation by introduced Black Rats, *Rattus rattus*. In spite of protection, it appears that the Light-footed Rail is in serious trouble. It was listed as endangered in 1970, and in 1974 the California Department of Fish and Wildlife estimated that there were from 500 to 750 birds left. By 1986, only 143 pairs of the birds were left, and five marshes within its range contained none.

The other threatened rail that occurs in Salt Marshes is a sparrow-sized bird known as the Black Rail, *Laterallus jamaicensis*. It is small and secretive; therefore, it is seldom seen by humans. Black rails also range into Freshwater Marshes at several inland localities. They have been observed in the San Joaquin and Imperial Valleys.

Belding's Savannah Sparrow, *Passerculus sandwichensis beldingi*, is also listed as endangered. This is a dark-colored, heavily striped sparrow that occurs most commonly in the pickleweed habitat from southern California to Baja California. It feeds primarily on seeds but will take insects when they are available. This bird is unique among the songbirds of California because it can drink seawater.

The fifth resident bird of the Salt Marshes is the Marsh Wren, *Cistothorus palustris*, a fairly common bird in marshes throughout the United States. It is a small, long-billed bird with a distinctive white eye stripe. It is secretive, but during the breeding season it will sing day and night, uttering a series of loud, rapid notes and rattles. It moves about incessantly, feeding on insects it gleans from the vegetation.

Of migratory birds that spend time in marshes, none is more at home than the American Bittern, *Botaurus lentiginosus*. This member of the heron family (Ardeidae) has a brown and yellow streaked breast. During daylight hours, bitterns usually remain motionless with their beaks pointed straight up. The streaking on the breast provides perfect protective coloration. They also have been reported to sway back and forth when the wind blows and remain still when the wind quits. Bitterns are most active at dusk or during moonlight. They feed on snails, frogs, fish, and even small mammals if they can get them. The sound of a courting American Bittern on a spring moonlit night has been described as a "thunder pump." Males make an outrageous booming sound accompanied by an inflated breast and peculiar posturing. American Bitterns seldom breed in southern California, but this mating ritual may be observed farther north.

Another bird that spends most of its time in the marshes is the Northern Harrier, *Circus cyaneus*. This raptor was formerly known as the Marsh Hawk, in reference to its preference for this habitat. These hawks are most easily recognized by their light-colored rump and habit of flying low over the marsh. They seldom soar in the typical hawk-like manner. Northern Harriers are not restricted to Salt Marshes. They occur in a variety of open habitats, including agricultural fields, and they have even moved into desert areas where agriculture is important. Their numbers have increased since the banning of DDT.

Most birds seen from time to time in the marshes are merely using the marsh for a temporary refuge or resting place. For example, during high tide it is very common for birds that feed on the Mud Flats to wait in the marshes for the tide to go out.

Mud Flats

When the tide is in, Mud Flats are covered with water. At this time, a variety of creatures become active, feeding on the surface of the mud or emerging from their tubes to feed on small planktonic organisms. When the tide is out, many kinds of birds venture onto the mud to feed.

Various kinds of algae grow on the surface of the mud, but this is not the primary food for the Mud Flat community. The contribution of these algae is greatest during spring and summer, when long day length enables more photosynthesis to occur. The primary food is detritus. Of secondary importance is the food provided by plankton or that which is in the water as dissolved organic matter.

A variety of animals inhabit tubes and burrows in the mud. Most famous of these is the Innkeeper or Fat Innkeeper, *Urechis caupo*, a specialized worm. The name Innkeeper refers to cohabitation of the burrow by other animals, such as worms, crabs, and fishes, which may eat the Innkeeper's "table scraps," another form of commensalism. One of the best-known cohabitants is the Arrow Goby, *Clevelandia ios.* This 2 in (5 cm) fish uses the burrow primarily as a refuge. It forages on the surface of the mud and retreats to the burrow when danger is near. As many as five gobies may be found at a single time in an Innkeeper's burrow.

Also common in the mud is a variety of shrimps. Among these is the Bay Ghost Shrimp, *Neotrypaea californiensis.* When the tide is in, these shrimps propel through their burrows a current of water, from which they extract plankton and detritus. The digging of these animals is important to the community because it helps to overturn and oxygenate the mud.

The most conspicuous animals of the mud flats are the birds. Most are migratory shore birds that spend the spring and summer months feeding and reproducing in northern latitudes, where the long photoperiod provides abundant food. During winter months, these birds migrate southward by the thousands to highly productive habitats such as estuaries or tropical forests. There may seem to be too many birds to be fed in one habitat, particularly where there is not an abundance of conspicuous prey species. However, careful inspection will show that these birds are equipped with bills of different shapes and lengths, and with legs of different lengths, a means of niche partitioning. Different birds wade in different depths of water, and the different shape and length of each bill enables different birds to probe the mud for different food items.

One strategy is to wade in shallow water and feed on fish. This is the strategy practiced by members of the heron family (Ardeidae). The largest of these birds is the Great Blue Heron, *Ardea Herodias,* which stands about 4 ft (120 cm) tall and has a 6 ft (2 m) wingspan. They often stand motionless and spear fish quickly with their long bill. When the tide is in, they retreat to the marsh. In the early 1900s,

Great Blue Herons were hunted for their long blue-gray plumes, and their numbers diminished radically. Since the hunting ceased, they recovered remarkably, and now they are one of the most common wetland birds.

The Great Blue Heron is sometimes called a crane, but true cranes are rare along the coast, although Sandhill Cranes, *Antigone (= Grus) canadensis,* may be observed at some inland aquatic habitats. Herons fly with their necks pulled back, and cranes fly with their necks outstretched. The expression "to crane one's neck" refers to the way a crane flies. Juvenile Great Blue Herons disperse widely in summer, occurring from the coast to the desert, wherever there are fish to eat. They are even observed in habitats away from wetlands. It is not uncommon to see a Great Blue Heron in a pasture, presumably hunting mice. This is the bird for which many places were named. For example, Crane Flat in Yosemite National Park was named for the occasional presence there of the Great Blue Heron. John Muir mentioned seeing the "blue crane" there.

Egrets are also in the heron family. There are three kinds in California, all of which are white. The largest is the Great or Common Egret, *A. alba,* a bird nearly as large as the Great Blue Heron. It can be distinguished from other egrets by its large size and yellow bill. The Snowy Egret, *Egretta thula,* is smaller, but it is nearly 3 ft (1 m) tall. The Snowy Egret has a black bill and black legs. In the adult, the feet are bright yellow. Apparently, the Snowy Egret uses its yellow toes to lure curious fishes. Snowy Egrets can be seen wading in the shallows, pushing their feet ahead of them, stirring the sediments and attracting small fishes to come take a look. The smallest egret is the Cattle Egret, *Bubulcus ibis,* an Old World species that has become naturalized in North America. Breeding males are quite attractive, with buffy-yellow plumes on their head and back. They are most commonly seen in pastures or agricultural regions, where they feed on insects. They are also common on the Channel Islands. Cattle Egrets have a yellow bill, and their legs are yellowish to pinkish. Herons and egrets also have been common at the Salton Sea; however, in recent years the US Fish and Wildlife Service has documented a precipitous decline in numbers. Increased salinity as well as pesticides and heavy metals are implicated.

A strategy practiced by various shorebirds is walking or wading at the water's edge to feed on organisms that live in the mud. Depending on the length and shape of their bills, they feed on different organisms. Among the most diverse of these birds are those in the sandpiper family (Scolopacidae). Curlews and Whimbrels are mottled, brown birds with very long, downcurved bills. The Long-billed Curlew, *Numenius americanus*, is the largest of the shorebirds, nearly 2 ft (60 cm) in length. Its downcurved bill may be 8 in (20 cm) in length. The Long-billed Curlew walks along the shore probing deeply into the mud. One of its favorite foods is ghost shrimp. The Whimbrel, *N. phaeopus*, is slightly smaller and can be distinguished by its shorter bill and dark head stripes. It feeds in a similar way, but on different food.

The Marbled Godwit, *Limosa fedoa*, is similar to the Whimbrel in size and appearance, except that its bill turns up at the end. Also, the bill is dark at the tip and pinkish at the base. In mixed flocks with the godwits and curlews occur various other sandpipers in assorted sizes. The smallest is the Least Sandpiper, *Calidris minutilla*. Its total length is only about 6 in (15 cm). Its distinctive feature is that it has yellow legs. The Western Sandpiper, *C. mauri*, is very similar in appearance, but it has black legs and a longer bill. Sandpipers are most conspicuous in flocks feeding just above the water's edge. Sanderlings, *C. alba*, are common where there is surf. They run back and forth in front of the swash and backwash. Characteristic of many sandpipers, when they fly, they call to each other with loud peeps. Some people call all sandpipers "peeps." Red Knots, *C. canutus*, are the largest of the peeps. Their claim to fame is the longest yearly migration of any shore bird, traveling 9300 mi (15,000 km) from their breeding grounds in the Arctic to Tierra Del Fuego. In California, they are most commonly seen in the San Francisco and San Diego areas on their way north.

Also included in the sandpiper group are Dunlins, *C. alpina*; Surfbirds, *Aphriza virgata*; snipes *Gallinago* spp.; yellowlegs, *Tringa* spp.; Wilson's Phalaropes, *Phalaropus tricolor*; tattlers *Heteroscelus* spp.; turnstones, *Arenaria* spp.; dowitchers, *Limnodromus* spp.; and Willets, *T. semipalmata*. There is no point in describing all of these, but a few additional comments are in order. Turnstones feed as their name suggests: they flip over stones to feed on organisms that take refuge there. They are also quite common on rocky shores where there is surf. Dowitchers usually feed while wading. They are known as "sewing machine birds" because their heads move up and down rapidly as they walk along probing into the mud. Phalaropes are unique in that typical roles of the sexes are reversed. Females are more brightly colored and males tend the nest. Females may court several males, and in so doing increase the number of hatchlings. They winter along the coast but breed at interior lakes in northern California, including Mono Lake and Lake Crowley. Willets are among the most common shorebirds. They are medium sized, about 15 in (38 cm) in length, and are gray with a dark bill. The most distinctive feature about their color is the white wing patches, known as flash marks, that show when they fly. White patches in the wings occur in other shorebirds too, but they are most distinctive in the Willets. If a predator were to attempt to grab a single Willet from a flock, the mass of confusion caused by the sudden appearance of all the white flash marks is enough to enable the entire flock to escape.

Similar to the members of the sandpiper family are birds in the plover family (Charadriidae), the most distinctive of which are the Semipalmated Plover, *Charadrius semipalmatus*, and the Killdeer, *C. vociferus*. Compared to many other shorebirds, these are short legged and short necked. Both of these birds have distinctive dark neckbands; the Semipalmated Plover has one, and the Killdeer has two. The loud call of the Killdeer sounds like "kill-dee," which is the basis for its name. It is a very common bird in all kinds of wetlands, from the beach to mountain meadows. A Killdeer makes its nest on the ground. The eggs are protectively colored, but certain predators, such as foxes and Coyotes, attempt to prey on eggs, babies, or nesting adults. The adult is famous for a broken-wing act, designed to fool the predator into thinking the bird is hurt. The limping bird is actually leading the predator away from its nest. The Snowy Plover, *C. nivosus*, is a federally threatened species that nests on the ground in coastal sandy areas. It is smaller than the other plovers and its neck ring is incomplete. In southern California, they are threatened by development, beach goers, and Coyotes that raid their nests. Fences have been erected to protect the nesting grounds in the vicinity of the University of California at Santa Barbara and in the Seal Beach National Wildlife and Bolsa Chica areas. These sites are also protected sites for nesting Least Terns.

Birds of the avocet family (Recurvirostridae) are characterized by a narrow bill and long legs. The American Avocet, *Recurvirostra Americana* (figure 11.13) is easily recognized by its distinctive black, narrow, upturned bill. In winter plumage, it is a gray bird with distinctive black-and-white wings. Breeding adults in spring and summer have a rusty-colored head and neck. Adults are about 18 in (45 cm) in length. These birds are very much at home on the Mud Flats, where they are among the most common birds. They walk all over the mud or in shallow water, feeding by sweeping their bills from side to side.

Related to the American Avocet is the Black-necked Stilt, *Himantopus mexicanus*. This bird is distinctly bicolored—black above, including its neck, and white beneath. It has bright red legs. These birds are common in a variety of wetland habitats, particularly marshes. They usually feed while wading.

During low tide, out on the mud, American Avocets are joined by a few other common birds. Among these, no bird is more conspicuous than the American Coot or Mud Hen, *Fulica americana*. This bird looks like a black duck, but it is a short-legged member of the rail family. It differs from a duck by having a relatively short, white bill that extends to its forehead. Also, instead of a web connecting its toes, a coot has flattened, lobed toes that aid in swimming and walking on soft mud. The bird's feet are conspicuously large, and as it walks about on the mud it makes a sucking sound. American Coots make their nests of California Cord Grass or other reeds that float when the tide is in. They are common in all quiet-water habitats. Many people view them as pests because they may be the only water bird in the area, and they are aggressive. This is particularly true in the artificial lakes and ponds on golf courses, or in "planned" communities that include a pond or lake. Coot feces are a source of eutrophication, causing these ponds and lakes to turn green or emit foul odors. Various schemes to exterminate coots from these areas have been attempted, to no avail.

There are many kinds of ducks (Anatidae) in a typical estuary, and it is unnecessary to describe them all. There are two basic kinds of ducks: dabblers, also known as bay ducks, and diving ducks. Dabblers consume algae, detritus, and snails. Even though some authorities would call them birds of open water, they are more correctly birds of mud flats. They feed in shallow water by tipping tail-up in order to reach the bottom (figure 11.13). Some of them also walk out on the mud when the tide is out. Dabblers differ from diving ducks in their ability to become airborne. Light-bodied dabblers can spring directly into the air, but the heavier diving ducks have to run along the surface of the water for some distance in order to reach sufficient speed to fly.

The Northern Shoveler, *Anas clypeata*, is a duck that may be seen walking about on the mud or in shallow water. These may be distinguished from other ducks by their large, spoon-shaped bills. No other duck has a bill that is larger than its head. Females are a mottled brown, and males have a green head and buff-colored sides. They are distinctive, but no more so than when they wander across the mud with their necks extended, sloshing their bills back and forth in the mud.

FIGURE 11.13 Black-necked Stilt, *Himantopus mexicanus*, and American Avocet, *Recurvirostra americana*. These two birds, in the same family, have differently shaped bills, indicating that they are adapted for different feeding modes.

FIGURE 11.14 Northern Pintails, *Anas acuta*, dabblers in their upturned positions feeding on the bottom.

Dabblers (*Anas* spp.) include Northern Shovelers, Gadwalls (*A. strepera*), Northern Pintails (*A. acuta*; figure 11.14), and Mallards (*A. platyrhynchos*), as well as several types of teals and wigeons. The Northern Pintail is the most common duck in the state, but the Mallard is probably the most familiar. The long, pointed tail is the distinguishing feature of the Northern Pintail. The Mallard resembles the Northern Shoveler in that the male has a green head. However, the Mallard's bill is much smaller and yellow. Male Mallards have a buff-colored breast, whereas on the Northern Shoveler, the buff color is on the side. Mallards also have orange feet. Mixed in with flocks of Mallards may be domestic ducks that have gone wild—in other words, feral ducks. These white ducks are European in origin, but they will

FIGURE 11.15 Ruddy Duck, *Oxyura jamaicensis*, male in breeding plumage; a diving duck.

hybridize with Mallards. The result is that along with typical Mallards there may be hybrid ducks with a variety of color patterns, from white to blotched.

Open Water

The Open Water habitat is characterized by a typical aquatic food chain. Nutrients are absorbed by phytoplankton, which are fed upon by zooplankton. Plankton are fed upon by filter-feeding or top-feeding fishes such as Topsmelt, Striped Mullet, and perhaps California Killifish. These fishes in turn may be fed upon by marine mammals, predatory fishes, or predatory birds. The birds are the most conspicuous.

Birds may hunt from a floating position, or they may dive into the water from the air. Of those that float, most numerous are the ducks. Diving ducks, also known as Pochards, spend most of their time on the water. They dive to feed on fishes, crustaceans, and various kinds of aquatic larvae. These ducks are good swimmers, but they are awkward on land. Many species have their legs placed well back on their bodies so that locomotion on land is nearly impossible. Among common ducks that dive for food are Canvasbacks (*Aythya valisineria*), Redheads (*A. americana*), Ring-necked Ducks (*A. collaris*), Buffleheads (*Bucephala albeola*),

and Ruddy Ducks (*Oxyura jamaicensis*; figure 11.15), as well as several kinds of scoters (*Melanitta* spp.), scaups (*Aythya* spp.), and mergansers (*Mergus* spp.).

Scaups, Buffleheads, and Ruddy Ducks often form mixed flocks floating on the water. They are similar in profile—small, chunky ducks with short necks. Scaups have iridescent feathers on the head that, depending on the light, range from purple to green. Females have a white patch around the bill, and males are white on their sides. Buffleheads have a large white patch on each side of the head. Male Ruddy Ducks also have white cheeks, but they are not apt to be confused with any other type of duck. Ruddy Ducks float with their stiff tail feathers pointing upward. In good light during the breeding season, it can be seen that males are brownish red and have a bright blue bill.

Scoters are thick-billed ducks with very large nostrils. They seem more at home fishing and diving outside the surf line than they do in bays, but some species are fairly common in bays or inland waters. California has three kinds of scoters. The Surf Scoter, *M. perspicillata*, has the widest distribution, but the Black Scoter, *M. americana*, may be locally common during winter. Males are black with an orange bill. The Surf Scoter may be distinguished by white patches on its forehead and the back of its neck. The white-winged Scoter, *M. fusca*, is black with a white line on its wing and a white patch below its eye.

Mergansers are fish-eating ducks. Unlike typical ducks, they have long, thin, serrated bills, ideal for grasping prey. There are three kinds of mergansers in California, but the one most common in estuaries is the Red-breasted Merganser, *M. serrator*. This is another green-headed duck, but its shaggy double crest, white neck ring, and thin bill should distinguish it from any other type of duck. As in many other ducks, females are brown.

Geese are also in the duck family, but they are larger than ducks and have very thick bills. Geese are winter visitors in wetlands all over the state. Migrating flocks of geese are recognizable by their familiar V formation. The most common species is the Canada Goose, *Branta canadensis*. It can be recognized by its large size, up to 45 in (114 cm) in length, and its distinctive coloration. Both males and females are pale breasted and have a black neck and head with a broad white ring across the chin. Large flocks of Canada Geese commonly feed in grasslands and cultivated fields along with domestic livestock.

Geese that are occasionally observed in estuaries include the Greater White-fronted Goose, *Anser albifrons*, and the Snow Goose, *Chen caerulescens*. The Snow Goose is a large white bird with pink feet. In California, they tend to occur inland rather than along the coast. They winter in the wildlife refuges of the Great Central Valley in northern California or in the Salton Sea area.

In flight, it is possible to confuse Snow Geese with American White Pelicans, *Pelecanus erythrorhynchos*. Both are white with black wing tips. Pelicans, however, are much larger, and their massive bills are usually visible from a distance. With wings spanning more than 8 ft (2.8 m), they are the second largest birds in North America. American White Pelicans also overwinter at the Clear Lake National Wildlife Refuge or the Salton Sea, but they are much more common to the east. Like the herons and

egrets, numbers of American White Pelicans at the Salton Sea are declining. A similar decline was documented for the Clear Lake National Wildlife Refuge. On the other hand, the number of locations, from the deserts to the coast, where overwintering American White Pelicans have been documented has increased markedly. They are classified as a species of special concern by the California Department of Fish and Wildlife.

California's typical coastal pelican is the Brown Pelican, *P. occidentalis*. Juveniles are uniformly brown, and adults have a white head that becomes yellow during the breeding season. A pelican's large size and enormous bill are easy features to recognize and should keep it from being confused with other birds. Brown Pelicans dive into the water from the air, capturing fish in their pouch. The size of the pouch enables them to feed on fairly large fishes such as Striped Mullet. The crash dive of a Brown Pelican is a thing to behold. It looks so awkward that it is difficult to believe they don't get hurt. A pelican seems to stop short as it hits the surface because the bill becomes distended by the water and acts like a parachute. In contrast to the awkward-looking dive, the graceful flight of pelicans as they glide in formation, inches above a wave, is one of the most beautiful sights a beach walker can experience.

Brown Pelicans nest on the islands off southern California and Baja California. They were severely threatened by DDT in the 1960s, and subsequently the species was listed as endangered. In 1971, at the height of their decline, 552 pairs of birds nested on Anacapa Island. Only one bird hatched that year. Not only did DDT interfere with egg production, but also the eggs that were laid had thin shells. A large bird such as a pelican crushes a thin-shelled egg when it sits on it.

After the banning of DDT, Brown Pelicans from Mexico began to move into California, and by the late 1980s the population had recovered remarkably. In 1985, about 6500 newborn pelicans survived on Anacapa Island, and over 4000 fledged in 1986 and 1987. The remarka-

FIGURE 11.16 Double-crested Cormorant, *Phalacrocorax auritus*, drying its wings.

ble recovery of this species has led some authorities to recommend that its listing be downgraded. It was federally listed as endangered from 1970 to 2009. On the other hand, it has been pointed out that about 85% of California's Brown Pelicans nest on the Islas Coronado, about 6 mi (10 km) from the coast near Tijuana, Mexico. In that area, beyond the jurisdiction of the US Fish and Wildlife Service, pelican and gull eggs are gathered for human consumption. Without protection from "eggers" in Baja California, the population in the United States probably would not be as safe as it appears.

Cormorants are similar to pelicans in that they have webbed feet with four toes and a membranous pouch under their beak. The cormorant's beak, however, is much shorter. There are three kinds of cormorants along the coast, but only two are common. These are large black birds with a visible hook on the tip of the beak. When they swim or perch, they are easily recognized by their beaks, which are pointed upward at an angle. The membranous pouch of the Double-crested Cormorant, *Phalacrocorax auritus*, is orange, whereas that of Brandt's Cormorant, *P. penicillatus*, is blue. Cormorants hunt while swimming. They dive and chase fish. Because they spend so much time underwater, they often stand on a perch with their wings outspread (figure 11.16). Not only does this help to warm them, but it dries their wing feathers as well. Unlike most aquatic birds, cormorants have no oil gland with which to waterproof their feathers.

Grebes (Podicipedidae) also hunt while swimming. There are five different species of grebes in California's estuaries, four of which are small. In winter plumage, they are all rather similar in appearance, with short, rounded bodies and short tails. Even though they seem duck-like, the bill and feet are nothing like those of a duck. Grebes have a slender pointed beak, which they use to catch small fish, crustaceans, and various kinds of larvae. Their feet are not webbed but have flattened, lobed toes similar to those of an American Coot. Their feet are much smaller than those of a coot, however, and unlike a coot they are very awkward on land. The state's largest grebe is the Western Grebe, *Aechmophorus occidentalis*, which is about 2 ft (60 cm) long. This is a black bird with a long white neck and throat. It has a thin yellow bill and red eyes. These are common winter visitors in estuaries. When they dive, they remain underwater for several minutes. It is difficult to believe that they can hold their breath so long or swim so far underwater.

Loons resemble large grebes, but they are in a different family (Gaviidae). Loons are almost 3 ft (90 cm) long. They ride low in the water and have short necks. Unlike most birds, they have dense bones, and this added weight enables them to dive easily. They have been caught in fishermen's nets 200 ft (60 m) below the surface. In California, loons are usually seen in winter plumage. At this time, they resemble the Western Grebe, but they are not as dark, and the neck is considerably shorter. There are three kinds of loons that may occur in California's waters, but the Common Loon, *Gavia immer*, is most likely to be seen. Loons breed in the north. The yodeling call of the male loon is usually uttered at night and is the sound that many people identify with the north woods. Loons are usually silent throughout the winter, but their unusual cry may be heard in the spring before they migrate northward.

Mergansers, cormorants, grebes, and loons hunt in a similar manner, but the species have different bill sizes, can swim to different depths, and remain under for different lengths of time. In this way, they partition their food supplies and reduce competition.

The most conspicuous birds of the California coast are the gulls and their allies (Laridae). For the most part these are fish eaters, but their scavenging habit is well known. In many coastal areas, large flocks migrate back and forth every day from the estuary to the dump. During winter, all along the California coast, flocks of mixed species may be observed. Gull species may be distinguished from each other by the color of their bills and legs, as well as differences in plumage. In the spring, most gull species migrate northward to reproduce, but the Western Gull, *Larus occidentalis*, remains to nest along the California coast, particularly on the Channel Islands, where there are no predators. It can be recognized by its dark gray wings and pinkish feet. A variant of this species that often occurs during winter at the Salton Sea is known as the Yellow-footed Gull, *L. livens*. As its name implies, it has yellow legs. It breeds in the Gulf of California.

During winter, the two most common gulls of estuaries are the Ring-billed Gull, *L. delawarensis*, and the California Gull, *L. californicus*. Ring-billed Gulls have a black ring around their bill, light gray wings with black wing tips, and yellow feet. California Gulls look like Ring-billed Gulls except that they lack the black ring and have a red spot on the lower part of the beak. For breeding, Ring-billed and California Gulls migrate to various lakes of the Great Basin in the United States and Canada. The

California Gull has a breeding population at Mono Lake, where it has been threatened in recent years by lowering of the water level. Diversions of water to Los Angeles have caused the lake to become so low that the island on which the gulls were breeding became connected to the mainland. Coyotes and other predators walked onto the island and ate the eggs and baby gulls. A series of court battles between environmentalists and the Los Angeles Department of Water and Power has established that enough water must be allowed to enter Mono Lake so that the island remains an island.

The California Gull has been immortalized in Salt Lake City. A statue there commemorates an incident in which a flock of California Gulls arrived in time to consume a swarm of Mormon Crickets, *Anabrus simplex*, that threatened the Mormon settlers' first crop.

Terns are in the same family as gulls, and many of them look like forked-tail gulls. Most terns can be distinguished from gulls because they fly with their beaks aimed downward; gulls fly with their beaks straight out. Terns feed on fish that they spot from the air. Unlike the behavior of pelicans, their dive into the water is very graceful. Often, terns hover over the water before dropping on their prey, and in this respect they resemble kites and kingfishers. Three kinds of terns might be seen along the coast. The most common is probably Forster's Tern, *Sterna forsteri*. In winter plumage, it is pale gray with a black eye stripe. In breeding plumage, the top of the head is black, and the bill and feet are red.

The only bird in the area with which Forster's Tern could be confused is the Least Tern. The Least Tern, however, is only about 9 in (22 cm) long. It is another endangered species. Its problem is that it nests in small colonies in an endangered habitat, that is, open sand, at the highest point before dune vegetation begins. As indicated in the discussion of the Snowy Plover, fences have been placed around nesting areas to protect birds and nests from predators and beach goers.

FIGURE 11.17 Osprey, *Pandion haliaetus,* a fish-eating hawk.

Perhaps the most spectacular bird of Open Water is the Osprey, *Pandion haliaetus* (figure 11.17), a fish-eating hawk. From beneath, it appears to be all white, but it is dark brown on the top. It has a white head with a conspicuous dark eye stripe. Ospreys are fun to watch as they fly over the water, plunging feet-first to grasp fish in their talons. When they fly back to a perch to feed, they turn the fish in their feet so that it aims headfirst into the wind. They construct large nests made of sticks, usually in trees or on top of poles.

Formerly these birds were conspicuous all along the coast and around lakes in northern California. Like other raptors, they experienced a serious decline in association with the widespread use of DDT. They are making an important recovery. Ospreys now construct nests on tufa towers in Mono Lake. This is a strange place for an Osprey nest, for Mono Lake is fishless; the birds would have to feed elsewhere. They are also observed feeding at various lakes on the eastern side of the Sierra Nevada. Another interesting incident occurred in Newport Harbor, where an Osprey constructed a nest in the mast of a sail boat. The owner objected, so the Department of Fish and Wildlife erected a tall post nearby and moved the Osprey nest there. The birds abandoned it.

They knew where they wanted to live! Attempts to establish Ospreys on Santa Catalina Island are presently underway.

Estuaries are among our most threatened habitats. Originally there were about 300,000 acres (120,000 ha) of coastal marshes. Due to filling and development, about 80% of it is gone. San Francisco Bay contains about 90% of California's remaining Salt Marsh habitat. The remaining 10% is made up of about 25 Salt Marshes scattered between Oregon and Mexico. Even though San Francisco Bay is the state's largest Salt Marsh, 60% of it also has been lost. By the early 1960s, filling had reduced the bay from 780 to 550 mi² (2000–1400 km²). Of what remains, the Suisun Marsh, at 87,000 acres (34,000 ha) is the largest Salt Marsh in the United States. It was preserved through the Suisun Marsh Preservation Act. It and the Bay Conservation and Development Act of 1969 are models of what grassroots efforts can bring about.

North of San Francisco, the largest Salt Marsh was formerly in Humboldt Bay. Draining and filling nearly destroyed it, but a grassroots movement among Arcata's citizens produced a remarkable turnaround. Today there are 94 acres (40 ha) of wetlands, fed largely by sewer effluent. By trapping and utilizing various organic materials and nutrients, marsh plants such as cattails act as a "natural" sewage treatment facility, and wildlife have returned.

In southern California, wetlands have fared more poorly. South of Morro Bay, 90% of this habitat has been destroyed. Of what remains, the largest area is in Upper Newport Bay (figure 11.12). Of the Salt Marsh in Newport Bay, 740 acres (296 ha) is administered by the California Department of Fish and Wildlife as an ecological preserve, and Orange County manages another 134 acres (54 ha) as the Upper Newport Bay Nature Preserve.

Tidal flushing is restricted in many southern California estuaries. A full Salt Marsh including significant California Cord Grass colonies requires tidal flushing. The large estuaries with significant tidal flow include Tijuana Estuary, Newport Bay, Anaheim Bay, Mugu Lagoon, Carpinteria Salt Marsh, and Goleta Slough. Estuaries with varying degrees of tidal flow such as San Mateo, Santa Margarita, San Luis Rey, Batiquitos, San Marcos, San Diego, and San Dieguito Creeks occur in San Diego County. Bolsa Chica, currently the largest estuary under restoration, had a new opening to the sea restored in August 2006, ushering in a new era of tidal flushing.

Loss of wetland habitats has taken its toll on wildlife. For example, there has been a 50% decline in California's waterfowl population just since the 1970s. Some birds, such as the Fulvous Whistling-duck, *Dendrocygna bicolor*, and the Hooded Merganser, *Lophodytes cucullatus*, were never common, but now they are rare. This loss is exacerbated by the fact that some 40,000 birds die each year from diseases. Some of these diseases, such as waterfowl botulism and avian cholera, have been recognized in California for over 100 years. On the other hand, duck virus enteritis was introduced from Europe with domestic waterfowl and has been known to occur in the United States only since 1967. Crowding of birds on dwindling habitat, particularly where they must share resources with domesticated waterfowl, could lead to devastating outbreaks of this deadly viral infection.

In addition to these infectious diseases, there are deaths associated with pollutants such as selenium, lead, or boron. Lead poisoning takes the greatest toll and results from shotgun pellets ingested by the bird while sifting marsh mud for food or grit. Some authorities contend that 80,000–150,000 birds die of lead poisoning in California every year, but Department of Fish and Wildlife personnel claim that the correct number is closer to 6000 deaths per year. Whatever is the correct number, the combination of habitat loss and disease has had a severe impact on waterfowl in the state.

Reconstruction and reclamation attempts have restored viable wetland habitats in Elkhorn Slough near Monterey and Bolsa Chica near Huntington Beach. At Elkhorn Slough, over 3000 acres (1225 ha) of marshland and adjacent habitats consisting of federal, state, and private holdings have been preserved and are under protection. The Bolsa Chica reserve represents the largest marshland reconstruction project ever undertaken. With the completion in 2006 of a connection to the open sea that allows direct tidal flushing for the first time since 1899, it is anticipated that a full-scale salt marsh, complete with California Cord Grass, will repair itself with time.

Even though sanctuaries and ecological preserves have been established to protect the state's remaining estuaries, they are still threatened by water diversions and pollution. Water diversions are a problem principally for San Francisco Bay. At the present time, diversion of freshwater from the Sacramento-San Joaquin delta has robbed San Francisco Bay of at least half of its freshwater input, increasing salinity and allowing intrusion of saltwater into surrounding wetlands and farmland. Various proposals to divert a large portion of the Sacramento River around the delta are perceived as major threats to water quality in the bay. Residual concentrations of DDT and heavy metals are still detected in samples of wildlife in California. Sewer outfalls are also a problem in some areas where the high concentration of organic waste causes a septic zone with a paucity of species.

ANIMALS OF INLAND WATERS

In lakes and rivers, there are three sources of nutrients: photosynthesis, material carried by downflow in streams, and material from the land. An analysis of the relative importance of these three sources has indicated that most nutrients enter the system from the outside. Leaf litter and other debris washed off the land or dropped from trees represent the greatest amount of input in streams and lakes. This is especially important where alders, with nitrogen-fixing bacteria in their roots, are part of the riparian community. About one-third of the biomass is tied up in this detrital form. Most of that material goes through a decay cycle before it becomes available for typical animal consumption.

Because most of the biomass in an aquatic system is involved in the decomposition process, the greatest amount of organic material in the system lies in the sediment or becomes dissolved in the water. The dissolved organic matter (DOM) is distributed nearly equally through the water and sediment and represents about half of the available food. If half the organic material in the system is dissolved and one-third is tied up in detritus, only one-sixth of the organic biomass is left to become incorporated into living organisms. Furthermore, the bulk of the living organisms are either microscopic algae or bacteria; therefore, it should be no surprise that animals are not conspicuous in many aquatic ecosystems.

In stream ecosystems, the amount of organic material washed in is roughly equivalent to the amount washed out. The contribution of photosynthesis is small because only attached microorganisms and mosses are not carried away. As material is carried downstream in a river, however, the system becomes more productive. The contribution of photosynthesis increases because there are more nutrients, and the stream is wide enough for light to shine directly on the water.

The amount of fine particulate organic matter (FPOM) and dissolved organic matter increases in downstream areas.

Aquatic insects that metamorphose into flying adults cause a net loss of organic matter for aquatic ecosystems. These organisms do most of their feeding in the water, and when they emerge, they carry their organic compounds into the terrestrial system. In one study of a desert stream, 97% of the emergent insect biomass was transferred to neighboring terrestrial ecosystems.

In lakes, attached and floating microorganisms play an important role. The contribution of photosynthesis increases in direct proportion to the nutritional state, or degree of eutrophication. In marshes, the photosynthetic contribution provided by macrophytes (rooted plants) is the greatest of any aquatic ecosystem. Nevertheless, the role of decomposition in the system also increases proportionately. In a senescent system, there is so much bacterial activity that little or no oxygen is left for large animals. The consequence of this trend is that large animals such as fishes and frogs are more conspicuous in lakes at the lower stages of eutrophication. In marshes, nearly all the biomass is tied up in producers and decomposers (microconsumers).

Invertebrates

With so much organic matter represented by water-soluble molecules and microorganisms, a specialized group of small invertebrates has evolved that is able to harvest this resource. Of particular interest are the crustaceans and aquatic insects, many of which spend at least part of their life cycles in freshwater. These animals belong to the phylum Arthropoda. Two-thirds of all the animal species in the world are arthropods.

Based on feeding behavior, there are four ecological types of arthropods (figure 11.18). Shredders feed on coarse particulate matter such as leaf litter. Grazers scrape algae off rocks. Collectors are organisms that gather

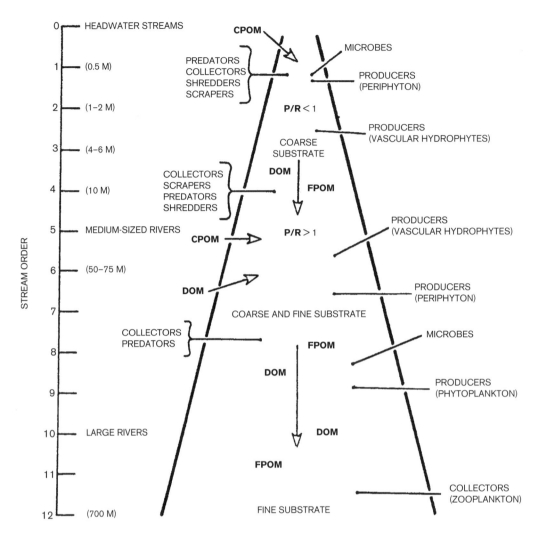

FIGURE 11.18 Sierra Nevada stream profile showing an expanding continuum from headwaters to a large river up to 700 m wide (from Bowler 1988).

ABBREVIATIONS: P/R, ratio of primary production to respiration; CPOM, coarse particulate organic matter; FPOM, fine particulate organic matter; DOM, dissolved organic matter.

plankton and floating organic matter. Predators feed on the other three types.

In upper reaches of streams, shredders and collectors are most common. Downstream, grazers become more common. In midsized streams, collectors and grazers form the bulk of the invertebrates. Downstream in large, slow-moving rivers, collectors make up about 90% of the invertebrates; the remainder are predators.

In lakes and ponds, phytoplankton such as diatoms and green algae are an important food resource. Grazers on this resource include small shrimplike creatures known collectively as crustaceans (figure 11.19). Among the small crustaceans in the animal plankton (zooplankton) are three basic types. A fourth group of crustaceans is composed of large bottom dwellers such as crayfish.

Among the three groups of crustaceans in the zooplankton, one is composed of the branchiopods, a word that means "gill-foot." Included among branchiopods are fairy shrimp

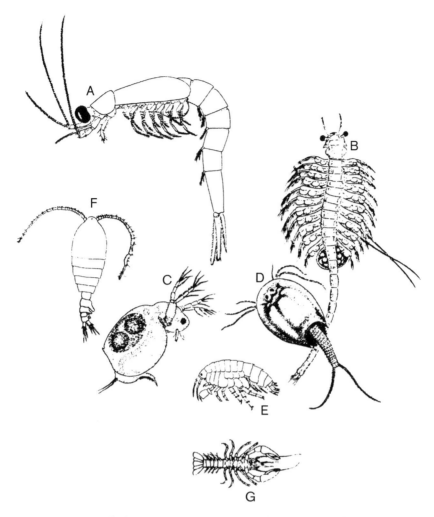

FIGURE 11.19 Some freshwater crustaceans. A. Opossom shrimp. B. Brine shrimp.
C. Water flea. D. Tadpole shrimp. E. Amphipod. F. Copepod. G. Crayfish.
SOURCE: From Cole 1988 (reprinted with permission).

and brine shrimp. These are very important animals in saline lakes such as those in the desert. These small, shrimplike creatures swim on their backs, feeding on small suspended particles. In Mono Lake, an endemic brine shrimp, *Artemia monica*, is an important component of the food base that supports a large breeding colony of California Gulls, *Larus californicus*, as well as numerous migratory birds. Diversions of water to the Los Angeles Aqueduct have caused significant lowering of the water level in Mono Lake. It appears that the coincident increase in salinity associated with evaporation may negatively affect the brine shrimp population and thus reduce the food supply for the birds. Furthermore, after the lake level dropped, Negit Island became connected to the mainland. It was on this island that the gulls laid their eggs. Between 1982 and 1985, the combination of increased salinity, reduced populations of brine shrimp, and exposure of the eggs and chicks to terrestrial predators contributed to a reproductive failure in the California Gulls. Heavy precipitation from 1983 through 1986 restored water to the lake, and the gulls reproduced again. Drought in the late

1980s and early 1990s lowered it again. El Niño conditions in 1997, 2015, and 2017 helped the lake recover. The controversy over Mono Lake continues.

Brine shrimp are also quite common in temporary ponds. Their fertilized eggs are capable of withstanding drying and high temperatures. They lie in the dried muds until the next rainstorm forms a new pond. These are analogous to annual plants, and brine shrimp accurately could be thought of as "annual" animals.

The largest members of the branchiopods are tadpole shrimps (*Triops*). Like brine shrimp, they are annual animals, found in a number of playas in the southwestern deserts. They may be 2 in (5 cm) or more in length. They have a large carapace that covers the front half of their body, so that they superficially resemble a tadpole swimming in the water. These animals have biting mouthparts, and they can chew on detritus or prey on other animals, including brine shrimp.

Other small branchiopods in the zooplankton are Cladocera. This order includes water fleas (*Daphnia*), which are often sold in tropical fish stores for live fish food. Water fleas also have a carapace, but it covers the back part of the body except for the head. Reproduction in water fleas is interesting. During the spring bloom, they reproduce rapidly by parthenogenesis, that is, from unfertilized eggs. These offspring are all females. Later in the year, some males are produced. Fertilized eggs carried by females that overwinter are very resistant to cold and even drying. When optimal conditions return the following year, the population reappears seemingly from nowhere.

The second group of crustaceans in the zooplankton are the copepods, a name that means "oar-foot." There are over 10,000 known species. Most of them occur in the sea where they are extremely common, but there are three common types in freshwater. Copepods are shrimplike in appearance, without a carapace. Their bodies are fatter at the head end, tapering to a thin tail composed of a pair of appendages.

They are quite small, seldom over a few millimeters long. One group of copepods, known as anchor worms (Lernaeopods), is parasitic on fishes. They could be a problem if they appeared in a small population of fishes, such as some of the rare or endangered species. Cladocera and copepods are the most common members of the zooplankton community and represent an important link in the food chain between microscopic and visible life forms.

The third group of planktonic crustaceans is known as the ostracods, a word that means "shelled." These tiny creatures have unsegmented bodies enclosed in a pair of hinged valves that form a carapace. This gives them the appearance of tiny, transparent clams. Their antennae stick out of the carapace, and they use them for swimming. Actually, these are not abundant in the plankton community. Most commonly, they live on the bottom, feeding on detritus. Nevertheless, they also form a link in the food chain that supports visible animals.

The largest and most diversified group of crustaceans is known as the malacostracans, which means "soft-shelled." There are four different groups of malacostracans in freshwater, three of which include small but visible shrimplike creatures. These groups are known as mysids, isopods, and amphipods. The mysids include a strong-swimming lake dweller, one of the opossum shrimps, *Mysis relicta*. This species occurs normally as a component of deep northeastern oligotrophic lakes such as the Great Lakes. It is a relict of marine origin that became isolated in freshwater during or soon after the Pleistocene. It feeds on zooplankton and is itself fed upon by fishes. Adults are about 1 in (25 mm) long. Because it is an easily transported organism, it has been introduced far outside of its normal range as a forage species for game fish. The assumption was that introducing opossum shrimp would allow a greater biomass of predatory fishes such as lake trout to be supported. Obviously, introducing another link in the food chain means that fewer large fishes, not more, can be supported on the same food base. The increase in size and abundance of game fish is a temporary phenomenon and lasts only until the food base is exhausted.

The introduction of opossum shrimp into Lake Tahoe in 1969 had a profound effect on the natural zooplankton community. By the early 1970s, opossum shrimp had increased dramatically, and cladocerans had practically disappeared. In the absence of cladocerans, copepods also increased in numbers.

Studies of zooplankton in Lake Tahoe indicated that they migrate on a daily basis. They migrate away from the shoreline, and also up and down. At night, they move to the surface to feed on phytoplankton. When it is light, they move to deeper water to avoid predation. Weak swimmers, such as some of the copepods, are unable to expend the energy that it takes to penetrate the dense water of the thermocline in the top 60 ft (20 m) of the lake.

The opossum shrimp migrate to the bottom, to a depth of nearly 1000 ft (300 m) every day, but during summer they remain below the thermocline, avoiding the upper, warmer waters. This behavior allows some of the copepods to escape predation above the thermocline. Other copepods, particularly in their larval stages, remain in very deep water, below the greatest concentration of opossum shrimp. The result of these behaviors is a layering of crustacean species promoted in particular by predation of the introduced form. This layering of animal life can deplete oxygen in regions of high density. It can also localize phytoplankton where there is less predation. Uptake of nitrogenous wastes produced by the zooplankton can also increase phytoplankton in certain layers. This is just another example of unexpected consequences, and demonstrates the problem of introduced nonnative species.

Isopods are small, flattened, creeping crustaceans. Among the terrestrial members of this group are the familiar sowbugs or pill bugs. In aquatic systems, isopods such as *Asellus* are most commonly found in streams, where they feed on detritus. They are particularly abundant where leaves have accumulated, and they can be found by separating wet leaves or looking under small stones on the stream bottom.

Amphipods are important components of stream bottom and lakeshore habitats. They are thin, laterally flattened forms known as scuds, side swimmers, or freshwater shrimp. They feed on fallen leaves and detritus caught in algal mats or clumps of aquatic moss. They are more characteristic of lowland habitats, but they are locally common in mountain or high-elevation lakes. The amphipod *Gammarus* was found to represent about half of the bottom fauna of Mildred Lake, a medium-sized but shallow lake at 9800 ft (3200 m) elevation in the Convict Lake basin. By comparison, in Bright Dot Lake, at 10,500 ft (3400 m) in the same basin, cladocerans were found to be the most common crustaceans at a depth of 50–60 ft (12–20 m).

Isopods and amphipods are important to the aquatic ecosystem because they are shredders. Shredding increases the surface area for action of microorganisms, which greatly accelerates the recycling of organic material.

The fourth group of malacostracans is the group known as decapods. The name means "10-legged" and refers to the organisms' five pairs of walking legs. These are mostly predatory bottom dwellers. In California, this group is represented by freshwater shrimp and crayfish.

Only a few species of freshwater shrimp have been recorded in North America, but three of these are endemic to California. The Pasadena Freshwater Shrimp, *Syncaris pasadenae*, has been extinct since the 1930s. It formerly occurred in low-gradient streams near Los Angeles. The California Freshwater Shrimp, *S. pacifica*, now known only from a few streams in Marin, Napa, and Sonoma Counties, is on California's list of endangered species. These nearly transparent shrimp grow up to 2 in (5 cm) in length. They live in shallow, low-gradient streams with good riparian cover and undercut banks. The Hilton Shrimp, *Palaemonetes hiltoni*, is a rare brackish water species of southern California. Unfortunately, the habitats of these three species are seriously degraded.

TABLE 11.2
Aquatic and Semiaquatic Insects

Order	Common Name	Active Aquatic Stages
Collembola	Springtails	Immature, adult
Ephemeroptera	Mayflies	Nymph
Odonata	Dragonflies, damselflies	Nymph
Orthoptera	Grasshoppers, crickets, katydids	None (semiaquatic)
Plecoptera	Stoneflies	Nymph
Hemiptera	True bugs	Nymph, adult
Neuroptera	Spongillaflies, dobsonflies	Larva
Trichoptera	Caddisflies	Larva, pupa
Lepidoptera	Moths	Larva
Coleoptera	Beetles	Larva, adult
Hymenoptera	Bees, wasps	Larva (parasitic)
Diptera	True flies	Larva

SOURCE: After Cole 1988.

Crayfish are lobsterlike animals up to 8 in (20 cm) in length. Common in almost all warm-water habitats, they ingest crustaceans, worms, insect larvae, and small fishes. They also feed readily on carrion. In this respect, they are the scavengers of the system. Crayfish are preyed on by large fish, frogs, and humans. Crayfish may be caught with a fishing rod and a piece of bacon or liver for bait. No hook is required, because crayfish will cling so tenaciously to the bait that they can be dragged from the water. The tails and claws of crayfish provide a tasty treat. In some places, crayfish are sold under the name *langostino*, which means "little lobster." Another name for them is crawdads.

The center of crayfish evolution in the United States is in the southeast, where there are more than 40 species. In California, there are only 4 native species, all in the genus *Pacifastacus*. Due to habitat destruction, an endemic species, the Shasta Crayfish (*P. fortis*), is on California's list of threatened species. A very similar species, *P. nigrescens*, is endemic to streams of the San Francisco area. The most common crayfish in the state is an introduced species known as the Swamp Crayfish, *Procambarus clarki*. This is one of the species of crayfish commonly dissected in biology laboratories.

Insects are the most common arthropods. On a worldwide scale, about three-fourths of all animal species fall in this group. It is primarily a terrestrial group, however. They have not succeeded in invading the marine environment to a significant degree, and only 3% of them are aquatic at some point in their life cycles. Nevertheless, there are about 30,000 species of aquatic insects, distributed among 13 orders (Table 11.2).

The evolutionary story of insects is interesting. Insects evolved first on land and then invaded water. It appears that they first invaded water in the cool headwaters of streams. From there, they spread to warmer, lower reaches of streams and then to ponds and lakes. This last step required the evolution of tolerance for higher temperatures. Temperature is very important to the distribution of insects. Most evidence indicates that insects are sorted into niches within their habitats primarily on the

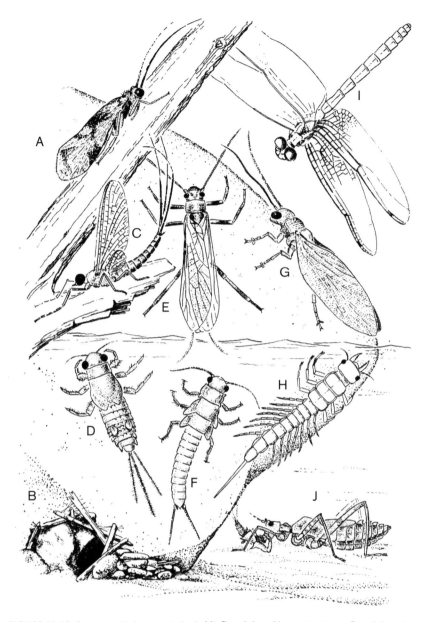

FIGURE 11.20 Some aquatic insects. A, B. Caddisfly (adult and larva). C, D. Mayfly (adult and nymph). E, F. Stonefly (adult and nymph). G, H. Dobsonfly (adult and larva). I, J. Dragonfly (adult and nymph).

SOURCE: From Cole 1988 (reprinted with permission).

basis of preferred temperature rather than competition for food or other resources.

Five groups (orders) of insects have at least one aquatic stage in nearly all their species (figure 11.20). In these groups, the adults are terrestrial flying forms. They reproduce by laying their eggs in the water, where the immature forms, known as larvae or nymphs (naiads), develop. These five groups are caddisflies (Trichoptera), mayflies (Ephemeroptera), stoneflies (Plecoptera), dobsonflies (Neuroptera), and dragonflies (Odonata).

Adult caddisflies, mayflies, and stoneflies are similar in appearance, with large, membranous wings. Usually the adult stage is of short duration. The larvae or nymphs spend their lives underwater, coming out of hiding to feed on detritus or algae at night. They commonly live in streams, so they are equipped with hooks and are often flattened or streamlined to withstand the current. The emergence of adults is known as "the hatch." At this time, trout fishermen attempt to catch fish by offering dry flies that mimic whatever is hatching.

Some larvae of caddisflies are known as case builders. They pick up bits of debris from the bottom and build protective shells around themselves. Different species in different habitats use different materials. Some caddisfly larvae cut up bits of leaves, others use pine needles, and some use small stones. One species in the southwest deserts builds a small coiled shell of stones so that it comes to resemble a small snail. In mountain lakes, the forms that use pine needles are conspicuous. A person peering into the shallow water of a high Sierra lake may be surprised to see what appears to be a 1–2 in (25–50 mm) bundle of pine needles lurching slowly along the bottom. Other caddisfly larvae are known as net spinners. These forms live in streams and build sticky nets to trap materials that drift with the current.

Dobsonflies are closely related to antlions and lacewings (Neuroptera). Adults are about 2 in (5 cm) long, and they superficially resemble dragonflies and damselflies, but dobsonfly wings are much longer than their bodies. The larvae, known as hellgrammites, can be found under stones at the edges of streams and ponds. They look like underwater centipedes and have large, chewing mouth parts, which they use to chew detritus or insect larvae. Hellgrammites live 3 years in the water before emerging as adults.

Dragonflies and damselflies are related. They are included in the same order, Odonata, but they differ in appearance and habits. Adult dragonflies always keep their wings extended. These are large-bodied, brightly colored insects with very large eyes. They have two pairs of wings with conspicuous veins and crossveins.

They often fly over water, where they prey on flying insects such as mosquitoes. They mate in midair. Sometimes they may be seen dipping their tails into the water as they lay their eggs. Dragonfly nymphs or naiads are large, predatory, grub-like creatures. They may spend up to 5 years in the water, gradually metamorphosing to an adult. When they emerge from the water, they molt for a final time into the winged stage.

Damselflies in flight look like slender-bodied dragonflies. Unlike dragonflies, when they land they rest with their wings folded back. Many are found in the same sorts of habitats as dragonflies, and they may be conspicuous, flitting over mountain lakes. They mate in midair, after which they lay their eggs in the stems of aquatic plants. They do this by crawling headfirst down the stem until they are underwater, and they pierce the stem and lay the eggs in the tissues of the plant. After the eggs hatch, the nymphs spend their immature period crawling about the lake bottom eating detritus.

True flies (Diptera), true bugs (Hemiptera), and beetles (Coleoptera) also have their aquatic members. These large orders of insects are very common in terrestrial habitats, but because they are such large groups, they are important components of aquatic communities as well. Aquatic larvae of true flies represent a very large portion of the aquatic insect biomass, and they occur in a wide variety of microhabitats. Flies with aquatic larvae include mosquitoes, midges, gnats, and blackflies (figure 11.21). As adults, they are all small, delicate insects with a single pair of wings. The name *Diptera* means "two wings." Most insects have four wings (two pairs).

Mosquitoes (Culicidae) are among the best-known insects. Male and female adults feed on nectar and other plant juices. In most species, females have two stomachs. The second stomach is for blood that they obtain from a vertebrate animal. They require at least one blood meal to stimulate ovarian development and reproduction. Because they may fly from person to person, sucking blood, they carry several important diseases, the best known of which is malaria.

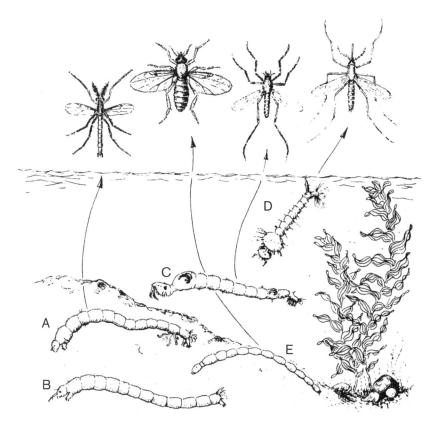

FIGURE 11.21 Some aquatic flies. A, B. Nonbiting midges. C. Phantom midge. D. Mosquito. E. Biting midge (from Cole 1988; reprinted with permission).

Although malaria is not a major problem in California today, it was epidemic in the mining camps of the 1850s, when it was also known as "miner's fever." Until about 1880, there were frequent sharp, local epidemics. After that time, the disease began a slow decline, interrupted by local outbreaks associated with the introduction of new irrigation systems or mining operations. For example, in the early 1900s, an area encompassing 13 counties in the Great Central Valley had a malarial death rate twice that of Mississippi, which was then considered one of most malarial states in the nation. In the year 1916, those counties, composing an area half the size of Mississippi, had a malarial death rate of 14.2 per 100,000 persons. Around 1920, the disease began a comparatively rapid decline, in spite of the continued abundance of one of the known vectors, Freeborn's Mosquito, *Anopheles freeborni*. Three other species of

Anopheles are known to occur in California, but only one other, *A. punctipinnis*, was considered an important carrier of malaria. It may be that this species was formerly more common than Freeborn's Mosquito, and as it declined in abundance in association with alterations of wetland habitats, so did the incidence of malaria. In recent years, using genetic engineering, scientists have inserted genes into *Anopheles* mosquitoes that prevent them from passing on the malaria parasite, which could be a major breakthrough for controlling the disease on a worldwide scale.

Different species of mosquitoes carry different diseases. The only mosquito-borne disease that is significant in California today is western equine encephalitis, sometimes incorrectly referred to as sleeping sickness. This disease is carried primarily by *Culex tarsalis*. Another species of mosquito, *Aedes triseriatus*, is known to carry a similar disease, called California encephalitis, which was first described from Kern County. Some mosquito abatement districts in California keep chickens in pens in strategic locations. They periodically sample the blood of these chickens to learn if mosquitoes in the area are carrying encephalitis. In 2013, the Yellow Fever Mosquito, *A. aegypti*, was discovered in the San Joaquin Valley. This is a significant mosquito throughout the world because it transmits yellow fever, as well as dengue fever, two potentially lethal diseases. So far, none of the trapped mosquitoes has been carrying disease, but the potential has mosquito abatement authorities concerned.

Mosquito abatement districts also engage in spraying programs or introduce small predatory fish species to permanent standing water to control the numbers of mosquito larvae. These districts also promote the concept of source reduction, which is essentially an attempt to reduce the amount and duration of standing water in an area. Some wetlands are drained and converted to fertile land for agriculture. In this way, the standing water for mosquitoes is removed, but so is the habitat for many other water-dependent organisms.

Because mosquitoes are widely considered to be such pests, much time and money has been spent to determine how to control them. On a worldwide scale, beginning in the 1940s, DDT was used principally for the purpose of controlling mosquitoes that carried malaria. Widespread use of the chemical for that purpose was quite effective until the mosquitoes evolved immunity. The only mosquitoes reproducing were those not killed by DDT. The impact of the indiscriminate use of DDT is now a well-known story.

To keep mosquitoes from biting humans, a number of attempts have been made to develop a mosquito repellant. The most successful has been a chemical known as toluamide or DEET, which stands for one of the forms of toluamide. Mosquitoes locate potential blood sources by a combination of heat, odor, and carbon dioxide. Toluamide works by blocking the receptors for carbon dioxide on the proboscis of the mosquito. The mosquito may find you by the heat you produce, but it won't recognize you as a potential blood meal if it can't detect the odor of the CO_2 emitted by your skin!

Mosquito larvae are known as wigglers. These are small wormlike creatures that hang just below the surface of the water and breathe with an air tube that projects toward the surface. When the water is disturbed, the larvae wiggle and squirm. Presumably this wiggling helps them move to a new location and distracts predators that would attack from beneath. Many aquatic predators feed on these larvae. Where mosquitoes occur naturally, there are usually predators specialized for eating mosquito larvae. In natural bodies of water, therefore, it should be unnecessary for mosquito abatement districts to introduce mosquito-eating animals, in particular the nonnative Mosquitofish, *Gambusia affinis*, which is known to eliminate native fishes by a variety of means.

A family closely related to mosquitoes includes species of nonbiting flies known as midges or gnats. The phantom midges, *Chaoborus* spp., are of interest because their transparent larvae are predatory members of the zooplankton communities of some large lakes, such as Clear Lake in the northern Coast Ranges. In fact, these larvae are the only insect members of the zooplankton. Because the larvae are nearly invisible and the adult seems to appear from nowhere, they are known as phantom midges.

Clear Lake is heavily eutrophic. In fact, it is not very clear, suffering from periodic algal blooms, particularly blue-green algae (Cyanobacteria), which contribute to anaerobic sediments and fish-kills. In addition, tourists and summer residents have long complained about the enormous numbers of Clear Lake Gnats, *C. astictopus*. At the height of the emergence season, up to 3.5 billion flies may leave the lake on a single evening. In an attempt to control these nonbiting midges, deemed a nuisance because of their swarming around lights and stoves, in the early 1960s the people of Clear

Lake authorized the application of DDT to the surface of the lake. The result was an ecological disaster. Concentrations of DDT increased as the chemical passed through the food chain. Fish-eating birds, such as grebes, began to die. It was determined that the cause was DDT poisoning, and this was one of the first documented cases of biological magnification. Such evidence ultimately led to the banning of DDT

Another attempt to control the Clear Lake Gnat also was a biological disaster. This episode involved the introduction of nonnative species of fish. Bluegill (*Lepomis macrochirus*) were introduced in the 1940s, after which the native minnow, the Clear Lake Splittail (*Pogonichthys ciscoides*), suffered a precipitous decline. Finally, in 1967, the Mississippi Silverside (*Menidia audens*) was introduced. The silverside replaced the Bluegill *and* the native splittail. The Clear Lake Splittail officially has been pronounced extinct. Overall, the history of introductions and other manipulations of the Clear Lake ecosystem have not served well for native species. Over the years, at least 39 species of fishes have been recorded from Clear Lake. As of 2008, there still were four native species hanging on, including the endemic Clear Lake Tule Perch, *Hysterocarpus traskii lagunae*, the endemic Clear Lake Hitch, *Lavinia exilicauda chi*, the Sacramento Blackfish, *Orthodon microlepidotus*, and the Prickly Sculpin, *Cottus asper.*

Larvae of the other aquatic flies are most often found in the sediment of lakes and rivers. Of particular importance are the nonbiting midges (Chironomidae) that produce great swarms of adults that swirl about and follow humans as they walk along a lakeshore. The adults, which look like very small mosquitoes, are short-lived and have reduced mouthparts; they do not feed. The swarms are mating swarms, composed entirely of females. Solitary males select a female from the swarm, and mating takes place in a nearby shrub.

Larvae of these nonbiting midges, called bloodworms, form a highly significant portion of the biomass in the sediment of many lakes.

Bloodworms get their name from their red color. This is one of the few insects to possess hemoglobin, the red respiratory pigment found in the blood cells of vertebrate animals. Bloodworms have no blood cells, but the pigment is suspended in their "blood." These larvae may be quite abundant in the sediment under nearly anaerobic conditions. The affinity of the pigment for oxygen gives them the ability to capture what oxygen molecules are available. Chironomid larvae (bloodworms) are extremely important as forage for fishes. In some situations, gut contents of fishes are composed of nothing but these larvae.

Closely related to the nonbiting midges are the biting midges (Ceratopogonidae), also known as punkies or "no-see-ums." These are the tiny bloodsuckers that sneak up on you during the heat of the day. One species is known in desert areas and along the north coast as the Bodega Gnat, *Leptoconops kerteszi*. Their orange-colored larvae seem to prefer sandy soil on the edge of a tidal marsh or saline lake. Another species is the Valley Black Gnat, *L. torrens*. Larvae of these midges prefer finely compacted clay soils in the western part of the San Joaquin and Santa Clara Valleys.

Black flies (Simuliidae) are small, compact flies that vary in color, frequently black, gray, or yellowish tan. Females are bloodsuckers. In California, they seem mostly to annoy wild or domestic birds and mammals, and when they emerge in spring or summer, they may bite humans. Black flies seem not to be affected by the usual repellents. A black fly bite is recognizable by a small red hole in the center of the swelling. The bites itch a great deal, and some sensitive individuals develop large welts. In some parts of the world, these animals carry tiny parasitic worms that cause river blindness (onchocerciasis), but this is not a problem in California.

Black fly larvae occur in flowing water, attached to stones, vegetation, or other objects, often in the swiftest parts of streams. They are most conspicuous in clear mountain streams where water flows in a sheet over a smooth rock surface. Here they are visible as rows of small, black, wormlike objects, resembling moss in ridges. They attach by means of a sucker near the tail end, with their heads oriented downstream. Larvae possess a pair of large mouth brushes, which are spread out to strain food particles. A person who takes the time to sit near a mountain stream and watch these larvae will see that they are able to move about on the rock without being carried away. They do this with a looping motion involving the sucker on one end and a leg on the other. They also produce small silken threads that enable them to cling spiderlike to rocks in the current.

To the uninformed, bugs (Hemiptera) and beetles (Coleoptera) superficially resemble each other. They are easy enough to distinguish, however. An informed observer may retort, "That's not a bug, it's a beetle." Coleoptera means "sheath-wing," and on beetles the horny front wings, known as elytra, meet in a straight line down the back. In contrast, on bugs the front wings are thick and leathery at the base, and the tips are membranous and overlap. Hemiptera means "half-wing." In addition, true bugs possess a conspicuous triangular scale on the back, known as a scutellum, located at the attachment of the front wings.

Adults and immature stages of beetles and bugs are associated with freshwater (figure 11.22). Most species are terrestrial, but these groups are so large that the aquatic species form a significant part of the animals in aquatic ecosystems. The adults are air breathers. Either they carry air down with them or they return to the surface frequently to breathe.

Adult beetles and bugs form a significant portion of the biomass in most shallow water, where they play important roles at intermediate stages of food chains. Nymphs (naiads) and larvae are important components of the bottom fauna.

Water bugs occupy a wide variety of habitats. They are found in saltwater pools, hot springs, mountain lakes, and large rivers. In general, however, they are characteristic of shallow, quiet waters. In particular, they inhabit ponds, backwaters, and shoreline habitats of

lakes. Most water bugs overwinter as adults. They lay eggs in the spring and go through their developmental phases during summer. Their eggs are sticky and when laid become glued to various objects in the water. They do not lay floating eggs, as do many other aquatic insects.

Giant water bugs (Belostomatidae) are large, flattened predators. The largest aquatic insect a person is likely to encounter is a giant water bug known as *Lethocerus americanus*, sometimes called the Electric Light Bug, because it will fly to lights at night. In California, it may be over 2 in (45–60 mm) in length. Giant water bugs capture and suck dry a variety of prey, including insects, tadpoles, and small fishes. They are also known as fish killers and toe biters.

Giant water bugs may be encountered in any shallow warm-water pool or stream, from the desert to the coast. Usually they lie concealed in shallow water among mats of leaf litter or other debris. In two genera, *Abedus* and *Belostoma*, the eggs when laid are deposited on the back of the male. What better way of protecting them? There seems to be no predator that would dare approach one of these fierce insects. There must be something that eats giant water bugs, however, because they have defense mechanisms other than biting. When they are disturbed, they become rigid and feign death. If removed from the water, they also may squirt an obnoxious liquid from the anus.

In spite of their reputation, in some parts of the world giant water bugs are eaten by humans. In Asia, one species, *L. indicus*, is eaten roasted or boiled and is said to have a strong flavor like that of Gorgonzola cheese. In Thailand, many sauces and curries are flavored with this insect. In Mexico, water bugs of the genus *Belostoma* are either roasted or used in stews, sometimes mixed with other insects.

Most numerous of the aquatic bugs are water boatmen (Corixidae). They get their name from their oar-like hind legs, which they use to propel themselves through the water. While underwater, they carry a bubble of air on their underside and use it as a source of oxygen.

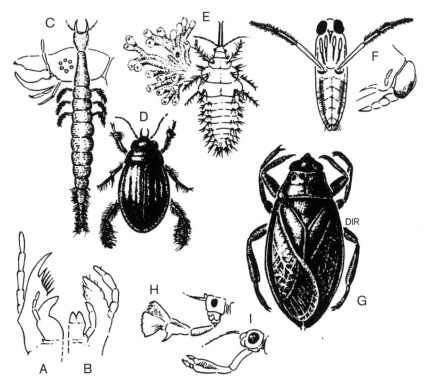

FIGURE 11.22 Predaceous aquatic insects and some comparisons (From Cole 1988; reprinted with permission).

A. Mouthparts of predatory stonefly naiad, *Isoperla*.

B. Mouthparts for shredder-scraper stonefly naiad, *Pteronarcys*.

C. Larva of predatory diving beetle, *Dytiscus*, and side view of a larva's head.

D. Adult diving beetle, *Dytiscus*.

E. Spongilla fly larva, *Climacia*, a piercing predator, and its freshwater sponge prey.

F. Underside of piercing predatory backswimmer, *Buenoa*, and side view of its head, showing the sharp rostrum.

G. Adult giant water bug, *Belostoma*.

H, I. Side view of head of dragonfly naiads, *Macromia* and *Aeshna*, showing grasping mouthparts.

This bubble gives the underside of the body a silver appearance.

Water boatmen play an important role in aquatic communities because they are shredders of leaf litter and debris. Other food for these bugs consists of small midge and mosquito larvae. Furthermore, they are fed upon by many large fishes, and they are the preferred diet item of fishes in some habitats.

Water boatmen are distributed from below sea level in Death Valley to alpine lakes in the Sierra Nevada. Many species are restricted to certain habitats and therefore may be used as ecologic indicators. For example, one group (*Trichocorixa*) occurs in briny habitats, such as the salt pools on the south shore of San Francisco Bay. They also occur in the alkaline closed-basin lakes of the southwest deserts. They are the dominant insect in the Salton Sea, and they also are found at Badwater, in Death Valley. In the Salton Sea, they are an important forage item for game fish. Another group

(*Corisella*) is an indicator of septic conditions. These are found in Little Borax Lake of the northern Coast Ranges and in sewage-oxidizing ponds.

Occurring in many habitats with water boatmen are bugs known as backswimmers (Notonectidae), which get their name from the fact that they swim on their backs (upside down). They are longer and narrower than water boatmen, and their backs (undersides) are shaped like a boat bottom. These fast-swimming predators feed on anything they can overpower, including crustaceans, larvae of aquatic insects, and small fishes. A person choosing to handle a backswimmer should expect a very sharp bite.

Another widespread family is known as creeping water bugs (Naucoridae). They can be recognized by their flattened shape and their mottled, yellow to green color pattern. They are predatory and will bite humans. This group of bugs is not well known because they are often concealed among stones or vegetation. Typically they inhabit flowing water at low elevations. One species, *Ambrysus mormon*, has been collected along the shoreline of Clear Lake, Eagle Lake in Lassen County, and Lake Tahoe. This is the dominant insect species in the saline waters of Pyramid Lake, Nevada, and it is common in most large streams of the Great Central Valley and northern Coast Ranges. It is replaced to the south and in the desert by other, similar species.

Other water bugs are the sticklike water scorpions (Nepidae), which lie in wait for prey among tangled debris. On the surface, long-legged water striders (Gerridae) or riffle bugs (Veliidae) may be found. Bugs in these two families move about on the water surface, feeding on small prey items. They are kept from sinking by the presence of many tiny water-repellent hairs on their feet. The effect of increased surface area of the feet of these bugs can be best appreciated by watching the shadow of a water strider on the bottom of a shallow pool. On the shadow, it appears that each foot terminates in a large circular disc that looks like a catcher's mitt.

Throughout the state, along the edges of streams and ponds or on lakeshores and beaches, inconspicuous bugs such as shore bugs (Saldidae) or toad bugs (Gelastocoridae) may occur. Shore bugs may be overlooked because they run or fly quickly at the slightest disturbance. Some are secretive, hiding in vegetation or shady spots. Shore bugs are scavengers and predators, eating a variety of living and dead organisms on the damp soil. Toad bugs are broad, flattened bugs with a warty appearance. Their mottled coloration enables them to blend in with the sand along the edges of streams and ponds. These bugs seem to hop like a toad when approached. Their short, grasping front legs are used to catch prey such as other insects.

There are about 10 families of water beetles. The most common are the diving beetles (Dytiscidae). Adults of this family spend most of their time underwater, obtaining air by surfacing rump first or backing into bubbles formed by aquatic plants. They carry the air under their elytra, and they swim with oar-like hind legs. Both larvae and adults are predacious. Eggs are laid on or in underwater plants. When the larvae, sometimes called water tigers, mature, they leave the water to form a pupa in damp soil nearby. Diving beetles are most common in ponds and slow-moving streams, where they inhabit weedy shallows. There are a few species in fast-moving streams, saline pools, and hot springs.

A group of beetles very similar to the diving beetles often occurs in the same water, particularly in marshes. These are the water scavenger beetles (Hydrophilidae). Water scavenger beetles differ slightly in appearance, being flat rather than rounded on the underside. Also, their antennae are club-like instead of thread-like. More importantly, they are herbivorous rather than predatory. They swim more slowly than diving beetles, using their legs alternately as if they were walking underwater. A person watching these two kinds of beetles in a pool would notice that diving beetles come to the surface for air rump first, but the scavenger beetles break the water with their antennae. Life histories of the two kinds of beetles are similar. Larvae of both groups are predatory.

TABLE 11.3
Fishes in Major Drainage Systems of California

	Klamath	Sacramento-San Joaquin	Lahontan	Death Valley	Colorado River	Southern California	All California
Endemic species	6	17	5	8	6	2	25
Primary freshwater species	10	20	7	3	4	4	34
Secondary freshwater species	7	8	1	4	1	0	17
Anadromous species	10	11	0	0	0	1	15
Catadromous species	0	0	0	0	1	0	1
True marine species	1	3	0	0	0	2	3
Introduced species	17	36	14	25	25	30	49

NOTE: These numbers may change with time as extirpations occur, new species are discovered, or taxonomic changes occur.

A unique group of beetles is the water pennies (Psephenidae). In this group, only the larvae are aquatic. They are small, oval, flattened creatures that cling to rocks in well-aerated flowing water, often near waterfalls. They occur in waters of varying chemical composition, but seldom above 6000 ft (1800 m) in elevation. In headsprings where CO_2 concentration is quite high, they may seem to be the only living organisms. The adults are small black beetles, sometimes found on the rocks near a stream.

Among the invertebrates that inhabit freshwaters throughout the state, snails are perhaps the least studied. There are many species, and there is a high degree of endemism, which makes studies of snail distribution important to understanding the historic distribution of freshwater. There are too many species to discuss here, but essentially there are two types of aquatic snails: those with gills and those with a saclike lung.

Snails with a lung are called pulmonate snails. Most of them are terrestrial, but some are aquatic. All snails with gills are aquatic, and they can be distinguished from pulmonate snails by the presence of an operculum, which is like a trapdoor that may be pulled over the opening when they are out of water. In this way, they are able to retreat inside their shell and tolerate temporary drying of habitat. Nearly all snails are herbivores or detritus eaters, harvesting their food with a rasp-like tongue (radula) that is used to shred vegetation. This is such a ubiquitous group that it is too bad scientists don't spend as much time learning their natural histories as they do classifying them. The Trinity Bristle Snail, *Monadenia setosa*, is the only mollusk on the state list of threatened species.

Fishes

Freshwater fishes are the most common aquatic vertebrates (Table 11.3). They range from coastal estuaries to alpine streams to desert springs. There are three kinds of fishes in freshwater. Primary freshwater fishes are those that cannot tolerate saltwater. Secondary freshwater fishes are those that usually live in freshwater but can tolerate saltwater. Marine fishes are those that live in seawater, although some may be found in coastal streams a considerable distance inland.

Primary Freshwater Fishes

Of the native primary freshwater fishes, nearly all fall in three families (figure 11.23). They are either minnows (Cyprinidae), suckers (Catostomidae), or trout (Salmonidae). Minnows and suckers make up the bulk of the native fishes. This may come as a surprise to a person accustomed to catching bass, Bluegill, crappie, or catfish, but these are all species introduced to California as game fish. Approximately 50 introduced species inhabit freshwater in California, often to the detriment of native species. Distribution of native primary freshwater fishes is of great interest to biogeographers because studies of modern-day distribution can lead to broad generalizations about ancient waterways and help to explain the relative positions of ancient landmasses.

The minnow family is the largest family of native freshwater fishes. They usually live in flowing water and, as such, are streamlined with elongated bodies and large fins. They are found all over the state, but there is a good deal of endemism. Of the 15 native species, only 2 are found outside of a single drainage.

The most widespread species is the Speckled Dace, *Rhinichthys osculus* (figure 11.23). These are slender minnows with a speckled pattern on the back. They are small, seldom reaching over 4 in (10 cm) in length. These minnows are usually riffle fish. That is, they occur in streams where the bottom is rocky and the water is turbulent. Their mouths are directed downward, which is an adaption for feeding on the bottom. They eat various kinds of small invertebrates such as snails and insect larvae. In spite of their obvious association with cool flowing water, they also occur in a variety of other situations, such as Lake Tahoe, where they live along the shoreline, and desert streams, such as the Owens River. In Death Valley, they occur in the outflows of desert springs.

The question arises: How does a freshwater fish become distributed throughout so many freshwater habitats? It is certainly true that during periods of heavy rainfall, various freshwater habitats, including those of the deserts, were interconnected. However, it is difficult to explain how a freshwater fish can cross a mountain range. The answer seems to be what is known as headwater stream capture. Streams that are above grade eat backward into their beds. At the summit, between two drainages, headwaters erode toward each other. If one stream cuts into the rock faster than the other, it is possible for the stream on one side to intercept the stream from the other side of the ridge. In so doing, it also may intercept its fishes. Distribution of Speckled Dace all over the west seems best explained by headwater stream capture. These fish are most common in the upper, fast-moving reaches of streams, and they occur on both sides of major drainage divides, such as the Rocky Mountains. In California, they are on both sides of major ranges such as the Sierra Nevada and the Transverse Ranges.

Another widespread minnow is the Tui Chub, *Gila bicolor* (figure 11.23). Native distribution of this species includes all drainages except the Colorado River and coastal southern California. This is a species of slow-moving water. It may occur in association with the Speckled Dace, but chubs prefer sandy or mudbottom pools. They are most common in relatively lowland habitats—weedy shallows of lakes or sluggish rivers. They are also found in Lake Tahoe and some other large lakes, such as Eagle Lake in northeastern California. They are larger than dace, reaching 14 in (40 cm) in length. Their heavier body is typical of the form associated with flowing but slow-moving water. These are opportunistic feeders, consuming everything from aquatic insects to detritus.

Tui Chubs have undergone considerable differentiation throughout their range, indicating a highly adaptable, plastic species. Many subgroups have been identified, and two coexist in Lake Tahoe, theoretically by feeding on different foods.

Of particular interest are the Blue Chub, *G. coerulea*, of the Klamath drainage, and the Arroyo Chub, *G. orcutti*, a southern California species. These are relatives of the Tui Chub that are

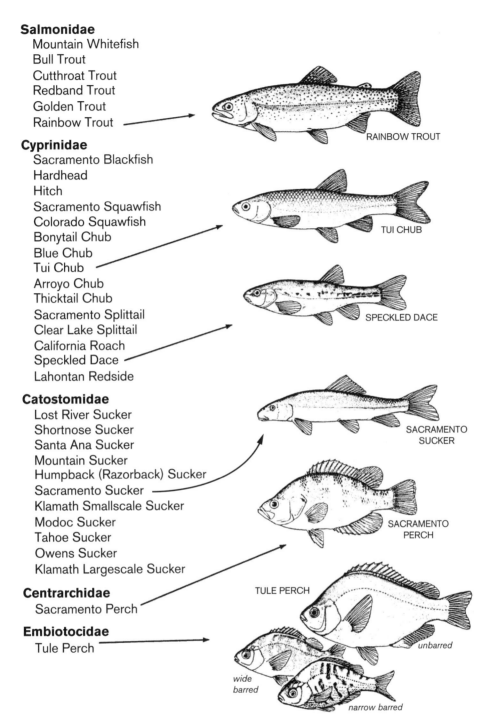

Salmonidae
  Mountain Whitefish
  Bull Trout
  Cutthroat Trout
  Redband Trout
  Golden Trout
  Rainbow Trout

RAINBOW TROUT

Cyprinidae
  Sacramento Blackfish
  Hardhead
  Hitch
  Sacramento Squawfish
  Colorado Squawfish
  Bonytail Chub
  Blue Chub
  Tui Chub
  Arroyo Chub
  Thicktail Chub
  Sacramento Splittail
  Clear Lake Splittail
  California Roach
  Speckled Dace
  Lahontan Redside

TUI CHUB

SPECKLED DACE

Catostomidae
  Lost River Sucker
  Shortnose Sucker
  Santa Ana Sucker
  Mountain Sucker
  Humpback (Razorback) Sucker
  Sacramento Sucker
  Klamath Smallscale Sucker
  Modoc Sucker
  Tahoe Sucker
  Owens Sucker
  Klamath Largescale Sucker

SACRAMENTO SUCKER

SACRAMENTO PERCH

Centrarchidae
  Sacramento Perch

TULE PERCH

Embiotocidae
  Tule Perch

unbarred

wide barred

narrow barred

FIGURE 11.23 Primary freshwater fishes (illustrations by Doris Alcorn from McGinnis, S.M. 1984. *Freshwater Fishes of California*. Berkeley: University of California Press).

different enough to be identified as distinct genus, which implies that they have been geographically separated for thousands of years. Interesting theories involving plate tectonics have been advanced to account for these differences. Consider that the Klamath Mountains were once in line with the northern Sierra Nevada, much farther east of their present location. There is also evidence that the Klamath region has experienced about 70° of clockwise rotation, with over one-third of that occurring during the last 20 million years. It is highly probable that waters that now flow directly to the Pacific were once part of the ancient Sacramento River system. Both Tui Chubs and Blue Chubs occur in the Klamath system today. Could the Blue Chub represent early separation of fish populations associated with translocation and rotation of the Klamath Mountains? Perhaps the Tui Chub became introduced more recently through mechanisms of headwater stream capture.

The populations of the Arroyo Chub in coastal southern California are of interest because closely allied chubs occur today on the coastal side of Sonora, Mexico. The Arroyo Chub most likely rode the landmass to its present location from a position that was to the south, when the Baja California peninsula was attached to what is now mainland Mexico.

Dispersal of chubs to Mexico is part of a long and complicated history. It appears that during the Miocene, less than 25 million years ago, chubs began their odyssey at the upper reaches of the Colorado Plateau, in a region that is now the eastern part of the Great Basin in Utah. They crossed the Rockies and, through headwater capture, entered the upper reaches of the Rio Grande. They swam down the Rio Grande to Mexico and then southwestward up the Rio Conchas to the mountains of western Mexico. Headwater stream capture then carried them to the rivers of Mexico that drained westward toward the Pacific. Thereafter, northward displacement along the San Andreas fault brought them to their present location, on the coastal side of the Transverse and Peninsular Ranges.

The final leg of this journey from Mexico may have involved about 90° of clockwise rotation of the San Bernardino Mountains. Arroyo Chubs also occur in the upper reaches of the Mojave River, on the desert side of the San Bernardino Mountains. If such rotation occurred, it could mean that the present-day Mojave River formerly flowed westward on the coastal side of the mountains, and was turned to its present drainage away from the coast by rotation of the mountain range. Evidence for rotation of the Transverse Ranges lies in magnetic lines of force trapped in volcanic rocks during the Miocene. The reason there is no evidence of rotation in the San Bernardino Mountains is that there are no suitable outcrops of Miocene volcanic rocks, but if rotation occurred in the other Transverse Ranges, it is likely to have taken place in the San Bernardino Mountains as well. Some authorities contend that the chubs probably were introduced. Other authorities speculate that they could have gotten there by headwater capture from the Santa Ana River.

Whatever is responsible for coastal species on the desert side of the San Bernardino Mountains, other aquatic species have similar distribution. A small fish, the Threespine Stickleback, *Gasterosteus aculeatus*, is present in coastal streams and also on the desert side in the Holcomb Creek and Baldwin Lake areas. Likewise, the Northern Western Pond Turtle, *Actinemys marmorata*, has its only desert populations in the Mojave River.

Two subspecies of the Tui Chub are also found in desert rivers. The Owens Tui Chub, *G. b. snyderi*, formerly occurred throughout the Owens River. Threatened with extinction by dewatering of its habitat, it is now listed by the federal government as endangered. The Mohave Tui Chub, *G. b. mohavensis*, biochemically one of the most distinctive subspecies, is also listed as endangered. It formerly occupied the lower reaches of the Mojave River. It was almost lost through hybridization with the Arroyo Chub, after which its total range was a spring system and pond next to Soda Lake that was operated as a "health resort" known as

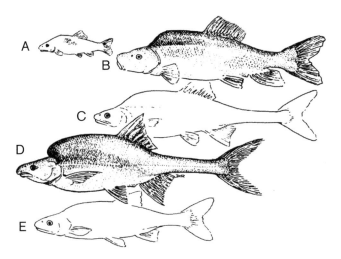

FIGURE 11.24 Fishes of fast-flowing waters of the Colorado River drainage. A. Mountain Sucker, *Catostomus platyrhynchus*. B. Razorback Sucker, *Xyrauchen texanus*. C. Bonytail Chub, *Gila elegans*. D. Humpback Chub, *Gila cypha*. E. Colorado River Roundtail Chub, *Gila robusta*.

SOURCE: From Cole 1988 (reprinted with permission).

Zzyzx. The facility is now operated as a center for desert research in collaboration with the Bureau of Land Management, National Park Service, and a consortium of California state universities.

Coexisting in Lake Tahoe with Tui Chubs and Speckled Dace is the most beautiful of the minnow species in California, the Lahontan Redside, *Richardsonius egregius*. It is native to Great Basin streams in northeastern California. Both sexes of this species bear bright golden and red lines on the sides of their bodies. In Lake Tahoe, they occur in large schools, and in early summer they can be seen in shallow water, where they spawn. These are small minnows, rarely over 3 in (8 cm) in total length. Presumably, the three minnow species avoid competition through niche partitioning, but they all spawn in the shallows at the same time of year, so occasional hybrids are produced.

Chubs of the Colorado River system illustrate another example of the plasticity of genetic systems (figure 11.24). A series of closely related forms are adapted for varying degrees of turbulence and speed in stream water. These forms have been separated into species and subspecies on the basis of their anatomy. At this time, three species are recognized for the mainstream Colorado River.

Roundtail Chubs, *G. robusta*, which occur in slow-flowing water of the Colorado River, look very much like Tui Chubs. They are streamlined (i.e., trout-shaped). In areas of rapidly flowing water, chubs develop a suite of characteristics that apparently enables them to cope with the swift flow. The Bonytail Chub, *G. elegans*, shows these characteristics to a moderate degree. These fish have an elongated, narrow posterior body and greatly enlarged fins. In addition, an enlarged hump is present on the head. Water flowing over this hump may serve to thrust the fish against the bottom so that it is able to maintain position in the swift flow. These adaptations reach their greatest extreme in the Humpback Chub, *G. cypha*, which inhabits the swift-flowing, turbulent waters of Grand Canyon. Unfortunately, the native distribution of these chubs has been influenced by severe alterations to the Colorado River. Damming and channelization have all but eliminated the turbulent flow and the spawning grounds for the native fishes. The Bonytail Chub is now probably extinct in California, although it still occurs in the upper Colorado River in Utah. It is on the federal list of endangered species.

Parallel development of a large hump behind the head has occurred in the Razorback Sucker, *Xyrauchen texanus* (figure 11.24). These large suckers are also inhabitants of turbulent water in the mainstream Colorado River and are listed as endangered by the state of California and the federal government. In spite of the scientific name for this species, which implies they are from Texas, the species is restricted in distribution to the Colorado River system.

The most spectacular of the fishes in the minnow family also formerly occurred in the lower Colorado River. This is the Colorado Pikeminnow (= Squawfish), *Ptychocheilus lucius*. Most people would never believe that this is a minnow. Fish larger than 5 ft (160 cm) in length formerly were caught by fishermen, and weights up to 100 lb (45 kg) were recorded. Like the Bonytail Chub and the Razorback Sucker, the Colorado Squawfish was unable to adapt to the series of impoundments on the lower Colorado River. It still occurs in the upper Colorado River, but the extraordinarily large sizes have not been recorded there. The Colorado Pikeminnow is also on the federal list of endangered species. A similar fish, the Sacramento Pikeminnow, *P. grandis*, also occupies the Sacramento-San Joaquin river system. The Sacramento Pikeminnow is also large, but specimens over 3 ft (1 m) in length are not common. With the disappearance of the Colorado Pikeminnow, this is now California's largest native minnow species.

Pikeminnows are big-river fish. They occupied the position of apex carnivore in the pristine river systems. In a manner similar to salmon, they would migrate upstream to lay their eggs in gravels of smaller streams. Unlike salmon, they did not die after spawning, nor did they swim all the way to the ocean to spend their adult lives.

The demise of California's native fishes has accompanied changes in big-river systems that result from the activities of humans. Damming of rivers blocked spawning runs and changed characteristics of flow. Rivers formerly experienced seasonal fluctuations of temperature, turbulence, and silt load; now they are monotonously clear and cold. The water in flowing portions of the rivers comes from the bottoms of the reservoirs, and fluctuations in flow occur on a daily basis rather than seasonally. Finally, the large number of imported game fish has caused near elimination of the native species,

many of which are classified as rare, threatened, or endangered. Reestablishment of Colorado Pikeminnow, Razorback Sucker, and other native fishes is a major responsibility of government agencies such as the California Department of Fish and Wildlife and the US Fish and Wildlife Service.

The greatest assemblage of minnows occurs today in and about the Sacramento-San Joaquin river system. Before it became a disturbed system, there were at least nine species of minnows nicely separated through niche partitioning. Introduction of many nonnative species, including five kinds of minnows, created serious problems for the native fishes. Introduced species, coupled with channelization, draining, damming, and pollution, led to the demise of the native assemblage.

In the system's pristine condition, Sacramento Pikeminnows were top carnivores, swimming up and down the large rivers. Hitch, *Lavinia exilicauda*, are plankton feeders that inhabit slow-moving, deep-water streams. Their ecologic niche has been largely taken over by the introduced Threadfin Shad, *Dorosoma petenense*, a member of the herring family (Clupeidae). The Hardhead, *Mylopharodon conocephalus*, is a large bottom dweller that feeds on invertebrates and aquatic plants. Its ecologic niche is threatened by introduced Inland Silversides, *Menidia beryllina*; Common Carp, *Cyprinus carpio*; and Goldfish, *Carassius auratus*.

In smaller streams, the Speckled Dace inhabited many habitats that have been taken over by introduced Red Shiners, *Cyprinella lutrensis*; Golden Shiners, *Notemigonus crysoleucas*; and Fathead Minnows, *Pimephales promelas*. The Tui Chub inhabited similar habitat in the Pit River portion of the drainage. The Speckled Dace still dominates in its favored habitat: riffles and rocky areas in smaller streams.

In backwaters, the Thicktail Chub, *G. crassicauda*, wandered through tule beds, feeding on detritus and invertebrates such as mosquito larvae. The Sacramento Splittail, *Pogonichthys macrolepidotus*, lived in deeper water of sloughs and backwaters. The Sacramento Blackfish, *Orthodon microlepidotus*, was a filter feeder in fertile warm backwaters. Destruction of wetlands through draining and channelization has pretty much destroyed the habitat of these backwater species. The Thicktail Chub is extinct, and the Splittail is only locally common.

In the intermittent tributaries that may become reduced to interrupted pools, the California Roach, *L. symmetricus*, feeds on algae. Along with the Speckled Dace, this is one of the most widespread species in the state. It can tolerate water temperature up to 95°F (35°C) and dissolved oxygen down to 1.0 ppm. It was formerly common in many habitats, but has been extirpated from them by introduced predatory fishes.

Suckers (Catostomidae) are the second most common fishes in California. They are closely related to minnows and show similar patterns of distribution. These fishes are adapted to a variety of habitats, but their mouths are specialized for feeding on the bottom. Those that occur in streams have hard edges on their jaws for scraping algae, small invertebrates, and detritus off of rocks. The largest, as discussed previously, is the Razorback Sucker of the Colorado River. In the Sacramento-San Joaquin river system, there are four subspecies of the Sacramento Sucker, *Catostomus occidentalis* (figure 11.23). Now largely absent from lowland habitats, this species has thrived in reservoirs and streams with regulated flow. This is a slow-growing species that eventually reaches a size up to 18 in (0.5 m).

There are 11 native species of suckers in California, most with localized distributions. The Tahoe Sucker, *C. tahoensis*, is the counterpart of the Sacramento Sucker in the lakes and streams of the Tahoe Basin. The Mountain Sucker, *C. platyrhynchus*, occurs in the Lahontan drainage east of Lake Tahoe; the Owens Sucker, *C. fimeiventris*, has adapted to reservoirs on the Owens River and is common in Lake Crowley; and the Santa Ana Sucker, *C. santaanae*, occurs in four streams of the Los Angeles Basin: the Santa Clara River, Big Tujunga Creek, the upper San Gabriel River, and the upper Santa Ana River. In these locations, they coexist with other native fish such as the Arroyo chub and the Speckled Dace. Based on threats from habitat destruction and introduced species, the Santa Ana Sucker has been federally listed as a threatened species. The Modoc Sucker, *C. microps*, is an endangered species that occurs in Johnson Creek, Modoc County.

There are four species of sucker in the Klamath drainage. They are the Lost River Sucker, *Deltistes luxatus*; the Shortnose Sucker, *Chasmistes brevirostris*; the Klamath Largescale Sucker, *C. snyderi*; and the Klamath Smallscale Sucker, *C. rimiculus*. The Klamath Smallscale Sucker is the smallest of the four species, and it is the only one that is still common. The other three are large, deep-water species, and their habitats have been threatened by draining of large lakes. The Lost River Sucker and the Shortnose Sucker are listed as endangered species.

Suckers are considered ugly by many people, and there is a mistaken idea that if suckers are present, something is wrong with the water. Nothing could be farther from the truth. In fact, they are fairly good to eat. It has been said that many Indian groups preferred suckers to trout as a food source. Because of habitat destruction, seven native species are endangered or threatened.

The most romantic of California's native freshwater fishes are in the trout family (Salmonidae). There are five native species of trout and one whitefish. These large, colorful fishes are the epitome of game fish in California. Many meals feature trout, many holidays are organized around trout fishing, and the largest fraction of the budget for the California Department of Fish and Wildlife is devoted to raising and planting trout in California waters.

Similar to minnows and suckers, trout are streamlined in shape, implying that they evolved in flowing water. Trout are basically stream fishes. Their habit of maintaining position in quiet water, downstream from a large boulder, is an adaptation for feeding. When a food item is carried downstream, the trout darts from the quiet water to catch it. The impression that high-country lakes have always been the home of trout is not true. Even though trout are widely distributed in streams and lakes today, the only high-country lake that originally contained members of the trout family is Lake Tahoe.

Lake Tahoe originally was the home for two members of the trout family, although there is some debate as to how common they actually were in the lake itself. Rivers and streams on the east side of the Sierra Nevada in the Lake Tahoe region are part of what is known as the Lahontan drainage. Lake Lahontan was an enormous body of water that occupied, during pluvial times, most of the northwestern part of Nevada and northeastern California. That drainage today includes Pyramid Lake in Nevada. In California, remnants of the drainage include Lake Tahoe, as well as Eagle Lake and Honey Lake, large saline lakes in northeastern California. Streams and rivers on the eastern side of the Sierra in that region are part of the Lahontan system. The major river in the system is the Truckee, which carries water from Lake Tahoe to Pyramid Lake.

The Mountain Whitefish, *Prosopium williamsoni*, is a small-mouthed, large-scaled member of the trout family. It has widespread distribution throughout the Great Basin, but in California it is limited to the Lahontan system. It occurs generally in lakes and deep pools from about 4500 to 7000 ft (1300–2100 m) in elevation. In Lake Tahoe, they are found mostly in deep water, except in the spring, when they move into shallow water to spawn. They feed on the organisms that inhabit the bottom. They are tasty fish, and they achieve reasonable size—about 18 in (50 cm) in length. For some reason, they are not relished as a game fish by California fishermen. In other parts of the United States, whitefish are considered a prized game fish and a delicacy.

In similar water with the Mountain Whitefish, the Lahontan Cutthroat Trout, *Oncorhynchus (= Salmo) clarkii henshawi*, formerly occurred. This trout was so common that it once supported a commercial fishery. For example, in 1900 over 9600 lb (4360 kg) of this species were harvested commercially from the Truckee River. This is a typical trout, but it is renowned for its large size and the ease with which it is caught. Salmon-sized cutthroats over 3 ft (1 m) in length, weighing up to 30 lb (13.6 kg), formerly were caught in Pyramid Lake. A Lahontan Cutthroat is recognizable by its olive to yellow color and many black spots. In particular, these trout can be distinguished from Rainbow Trout because the spotting occurs on the belly. Cutthroats get their name from a red slash that occurs on each side, just under the lower jaw. A highly colorful subspecies with very little spotting is known as the Paiute Cutthroat, *O. (= S.) c. seleniris*. It occurs in Silver King Creek, in the Carson River drainage just south of Lake Tahoe. Both of these subspecies are listed as threatened by the federal government. A third subspecies of cutthroat, the Coast Cutthroat Trout, *O. (= S.) c. clarkii*, occurred in small coastal streams. Like steelhead, they migrated back and forth to the sea.

Other than California Golden Trout, *O. (= S.) mykiss aguabonita*, of the Kern Plateau, Mountain Whitefish and two subspecies of cutthroat were the only high-country trout before humans started introducing them. With the exception of Lake Tahoe, as mentioned previously, lakes of the high Sierra were fishless. Most of the lake basins were the product of Pleistocene glaciation. After the glaciers retreated, trout were unable to make it to these lakes because the terrain was too steep. Waterfalls were barriers to the dispersal of trout.

The Coastal Rainbow Trout, *O. m. irideus* (= *S. gairdneri*; figure 11.23), originally was a fish of foothill and coastal streams from Los Angeles County northward. It was first discovered and described in 1855 from specimens obtained from the Redwood Creek drainage of eastern San Francisco Bay. Rainbows that move into seawater to spend their adult lives are called Steelhead Trout and are exactly the same subspecies as Rainbows that do not migrate to the sea. Like salmon, in order to reproduce, Steelhead return to the stream where they originally hatched. Their southernmost distribution in the early 2000s was recorded in San Mateo Creek in northern San Diego County. The population persisted for several years until low water conditions probably led to its extirpation.

Rainbows occur in a wide variety of colors and patterns. They have adapted well to rearing in hatcheries, and have been distributed far beyond their native range. Most of the hatchery trout raised in California are Rainbows, but they have been hybridized with other forms to the degree that pure Rainbow Trout may be difficult to find.

The Redband Trout, *O. (= S.) m. "gairdneri"*, is of Great Basin affinities. It is considered to be an inland version of the Rainbow Trout. These trout are found in isolated pockets throughout the western Great Basin. In California, true Redband Trout are found in Sheephaven Creek, Siskiyou County. Apparently, former distribution included the Pit River that drains the Modoc Plateau. They were also found in Goose Lake at the head of the Pit River and in the upper Klamath drainage. This pattern of distribution is important, as it shows the relationship of Great Basin species to those of the Sacramento River. The implication is they evolved in the Great Basin and made their way to the Sacramento-San Joaquin system by traveling down the Pit River (or its predecessor), which drains the Modoc Plateau.

Redband Trout resemble Rainbows, but they tend to be pale yellow with a brick-red stripe on the side. In contrast, Rainbows tend to have a pink stripe on a silvery background. Redband and Rainbow Trout are so closely related that they interbreed freely. It appears that, at some point, Rainbows that were able to disperse through seawater moved up the Sacramento-San Joaquin system and swamped, through hybridization, the genetic identity of most trout populations.

The California Golden Trout is presumed to be a highly derived Redband Trout that evolved in isolation on the Kern Plateau of the southern Sierra Nevada. These fish are native to high-elevation streams such as Golden Trout Creek, the South Fork of the Kern River, and the Little Kern River, which cut through the large meadows of the Kern Plateau. The latter population has been described as a separate subspecies, *O. m. whitei*, and has been listed as threatened by the federal government. Because of its beauty, the California Golden Trout has been named the state freshwater fish. Breeding males have a bright golden color on the lower sides, and the belly and cheeks are a bright red, as is the lateral band. The assumption is that California Golden Trout reached the southern Sierra by means of the San Joaquin River. Ancestors in the San Joaquin were similar to Redband Trout. At times of high water on the floor of the Great Central Valley, the Kern, draining into Tule Lake, could overflow into the San Joaquin, and the trout could make their way eventually all the way up the Kern to the high country. If glaciation pushed them out of high-country basins, they remained in southern refugia such as Golden Trout Creek and the South Fork of the Kern. They were not able, however, to invade the high-country lakes. Distribution of trout to the thousands of high lakes in the Sierra Nevada is entirely the result of human efforts. The Rainbow Trout of the mainstream Kern River has also been described as a separate subspecies, the Kern River Rainbow Trout, *O. (= S.) m. gilberti*.

The story of California Golden Trout and its introduction to high-country lakes is particularly interesting. In 1876, there was a sawmill on Cottonwood Creek, an east-side stream of the southern Sierra Nevada near Lone Pine. At the same time, across the Owens Valley in the Inyo Mountains, the Cerro Gordo Mine was operating. This was a large, profitable enterprise that required charcoal to extract gold and silver from the ore. Logs cut in Cottonwood Creek were converted to charcoal in a set of kilns located on the shore of Owens Lake. Hundreds of people were employed in these operations. To provide fresh fish for his employees, the owner of the Cottonwood sawmill introduced California Golden Trout to Cottonwood

Creek. The initial stocking was two trout carried in a coffeepot over the pass from Mulkey Creek, a tributary of the South Fork of the Kern River. The trout prospered, supplying food for the employees.

In 1891, in order to increase the sawmill's capacity, Cottonwood Creek was diverted to a new channel. As the old channel dried up, California Golden Trout trapped in the pools were harvested. Some were kept alive and introduced to the Cottonwood Lakes, a fishless basin. So began the practice of stocking high-country lakes with California Golden Trout.

Since 1918, California Golden Trout have been introduced by the Department of Fish and Wildlife into lakes and streams all over the Sierra, usually at elevations above 9000 ft (2700 m). California Golden Trout also have been introduced to other western states. California Golden Trout require clear, cold water. Generally, they will not tolerate water temperatures above 70°F (22°C).

The Department of Fish and Wildlife harvests eggs and sperm from the California Golden Trout at Cottonwood lakes. These were fertilized artificially, and the hatchlings formerly were raised in the Mount Whitney Hatchery near Independence. In the 1990s, genetic analysis showed that the Cottonwood Lakes fishes had been corrupted through hybridization with Rainbow Trout, so the hatchery procedure was terminated. In 2008, a flood damaged a large portion of the Mount Whitney Hatchery and it was taken out of service. The historic building is manned today by volunteers for interested visitors. The Kern River Fish Hatchery today is dedicated to propagation of the Kern River Rainbow Trout and the California Golden Trout. The populations of California Golden Trout in the upper Kern River are pure, but they are still threatened by introductions of other trout species, such as Brown Trout, *S. trutta*. Barriers to upstream migration of nonnative trout have been constructed at various points on the South Fork of the Kern.

For many years, the California Department of Fish and Wildlife has been busily introduc-ing new fish species. Brown Trout were first introduced to North America in 1883. In 1894, a strain from Scotland known as the Loch Leven Trout was introduced to California. The more common German Brown Trout was introduced in 1895 and is now common in waters all over the state. Brown Trout grow to large size. They are bottom feeders, and, to a large degree, they eat other fishes, a habit that has led to the reduction of other fish species that occur in the same water. Introductions of Brown Trout have been implicated in the extinction of the native Bull Trout and reduction in numbers of California Golden Trout and the rare Modoc Sucker.

In 1872, the first Brook Trout, *Salvelinus fontinalis*, were imported from New Hampshire and Wisconsin. Native to the eastern United States, Brook Trout belong to a group of trout known as chars. Brook Trout is one of the few species of trout that does well in the high-country lakes. Brown Trout are known to hybridize with Brook Trout, and the sterile hybrid is known as the Tiger Trout, in reference to the distinctive banding on their sides. Although the two species are not related closely enough to produce fertile offspring, the hybrids do possess what is known as hybrid vigor. Where Tiger Trout have been introduced, they are known for their voracious feeding.

Chars (*Salvelinus* spp.) can be distinguished from other trout by their large mouth and pointed nose. The state's native char is known as Bull Trout, *S. confluentis*. The Bull Trout is on the state list of endangered species, although it probably has been extirpated in California. It formerly occurred in the upper Sacramento drainage in the McCloud River. There are two dams on the drainage today, and nonnative Brown Trout occur there as well. In other parts of its range, the Bull Trout was eliminated through hybridization with introduced Brook Trout. Hybrids are sterile. Some authorities contend that a contributing factor to the demise of the Bull Trout was the loss of salmon in the Sacramento River. Apparently, salmon fry were an important food item. Its distribution today is in large bodies of water in the upper Great

Basin, particularly in Idaho and Montana, where they grow to 3 ft (1 m) in length and may weigh 30 lb (14 kg). The Bull Trout is another example of a Great Basin species with its California distribution in the Sacramento River drainage.

There are also tales that sea-run Dolly Varden Trout, *S. mama*, occasionally enter the Klamath River. The popular name of this species apparently refers to Dolly Varden, a character from a Charles Dickens tale, *Barnaby Rudge*, whose clothing was brightly colored and was a popular style in the 1870s. In reference to its bright color and spotting, the name *Dolly Varden* or *Calico Trout* was used for the fish in those days.

Neither Brook Trout nor California Golden Trout grow very large in typical high-country lakes, but not all species of trout can survive and reproduce in the high country. Nevertheless, anglers yearning for large trout sometimes conduct unauthorized stockings of fishes such as Rainbows, Cutthroats, or Brown Trout, which are known to achieve large size.

Indiscriminate or unauthorized stockings can spell disaster for native fishes. California Golden Trout in particular are threatened by the introduction of other fishes. Alien trout may become cannibalistic and eat the young of the native species. If Rainbows or Cutthroats are introduced, they hybridize with the California Golden Trout, and the genetic integrity of the species is lost. Brook Trout normally do not hybridize, because they spawn in the fall whereas all of the native trout spawn in the spring. Brook Trout, however, are known to overproduce offspring, which causes serious competition for available resources, leading to stunted growth. Brook Trout are especially susceptible to stunting of their own populations, not to mention the effect that their overproduction has on the native species.

If introduced species of fishes were not enough of a problem for native California Golden Trout, Beavers have been introduced to many trout streams. Damming activity of the Beavers has destroyed spawning habitat by allowing the accumulation of sediment over the clean gravels where eggs are laid.

In response to these problems, the Department of Fish and Wildlife began a program of remediation. Introduced species were removed by poisoning, and California Golden Trout were reintroduced. Barriers to upstream dispersal were constructed to prevent natural encroachment by species from downstream warm-water habitats. Special angling regulations were adopted to prevent overharvesting. Beavers were trapped and removed and their dams destroyed. Thousands of dollars were spent to protect the California Golden Trout. After all, it is the official freshwater fish of a state already stung by extinction of the California Grizzly Bear, adopted in 1953 as the state mammal. How would it look if the state were unable to protect its unique animals from extinction?

There are only two species of native primary freshwater fishes that are not in the minnow, sucker, or trout family. They are found in the Sacramento-San Joaquin drainage, where, through niche partitioning, they have coexisted with all the minnows and suckers. This is the largest drainage in the state, and it has the largest assemblage of species. The Sacramento Perch, *Archoplites interruptus* (figure 11.23), formerly inhabited sloughs, sluggish water, and lakes of the Central Valley floor. This is a relict species, the only member of the sunfish family (Centrarchidae) that occurs west of the Rocky Mountains. It is a deep-bodied, "slab-sided" fish with a large head and gills. It is adapted to moving through tules and aquatic vegetation in warm-water habitats. Large specimens may grow to 2 ft (60 cm) in length. Introduction of other sunfishes, such as Bluegill (*Lepomis macrochirus*) and crappies (*Pomoxis* spp.), has practically eliminated the Sacramento Perch from its native habitat. Evolving in a habitat with little competition, these perch are no match for other aggressive fishes. Ironically, introduction of this species into the Great Basin, far outside of its native range, has preserved the species. It is a common fish in Lake Crowley. Unlike most members of the sunfish family, Sacramento

Perch can tolerate alkaline waters. This feature, plus their large size, makes them a desirable game fish in waters that are hostile to most sunfish. To preserve the species in California, the Department of Fish and Wildlife has tried to encourage their propagation in California farm ponds.

The other native primary freshwater fish is the Tule Perch, *Hysterocarpus traski* (figure 11.23), the only freshwater member of the surfperch family (Embiotocidae). They are small—up to 6 in (15 cm) in total length—and they are slab-sided. They occur naturally in slow-moving waters in association with aquatic plants. Like the Sacramento Perch, they feed primarily on aquatic invertebrates. Formerly they inhabited the San Joaquin Valley marshes and many tributaries of San Francisco Bay. They are still locally common in the Sacramento Valley and some large lakes such as Clear Lake, but their overall distribution is greatly reduced. They are victims of habitat degradation, particularly drying of marshes, increased turbidity, pollution, and reduced cover.

Secondary Freshwater Fishes

Secondary freshwater fishes are able to tolerate saltwater, but they spend most of their lives in freshwater. Because of their ability to tolerate a wide range of salinity, they are said to be euryhaline. Two groups of native fishes make up the bulk of these species: the sculpins and Killifish families (figure 11.25).

The largest group of secondary freshwater fishes is the Sculpins (Cottidae). There are eight species of *Cottus* in California streams. This group has its origins in the north. It entered California by dispersal down the coast or from the northern Great Basin via the Pit River system. California has 10 species of sculpin associated primarily with freshwater. This is a small part of what is primarily a marine family. These are bottom fish that seldom exceed 4 in (10 cm) in length. Their brown, mottled coloration conceals them as they lie motionless on the bottom. They have large, flattened heads and large, fanlike pecto-

ral fins. These adaptations help them maintain position on the bottom, even in fast-flowing water.

The most abundant sculpin in California is the Prickly Sculpin, *C. asper* (figure 11.25). It is a fish of northern affinities, ranging from Alaska to the Ventura River in southern California. They occupy a wide range of bottom habitats, from saltwater to freshwater, and occur in all sorts of water quality and flow. Most typically, they inhabit low-elevation, moderate-sized streams with abundant sand-bottom to mud-bottom pools. They feed on aquatic invertebrates that inhabit the bottom, particularly immature aquatic insects.

The reproductive behavior of Prickly Sculpins is interesting. They migrate downstream to breed, sometimes into saltwater. Fishes that migrate to saltwater to reproduce are said to be catadromous. Males construct a nest, usually under a large flat rock or other object on the bottom. These days, it may be in a beer can. Eggs are attached to the ceiling of the nest. The male defends the eggs and, with his pectoral fins, keeps a current of water flowing through the nest. After the fry hatch, for a few weeks they are members of the plankton community. They settle to the bottom and move slowly upstream to where their adult life will be spent.

Other species of sculpins have more restricted distribution, but it is not uncommon for similar species to occur together, making identification nearly impossible. Some species occur in a single drainage, and a few, such as the Paiute Sculpin (*C. beldingi*), have limited distribution in California. In the Pit River, three species may overlap.

The Paiute Sculpin occurs in Lake Tahoe and its tributaries and is locally common in streams on the east side of the Sierra. In streams, they occupy riffles, where they lie in wait until prey species come close enough to be grabbed. In Lake Tahoe, they feed mostly at night. During daylight hours, they remain motionless in shallow water, relying on their coloration to conceal them.

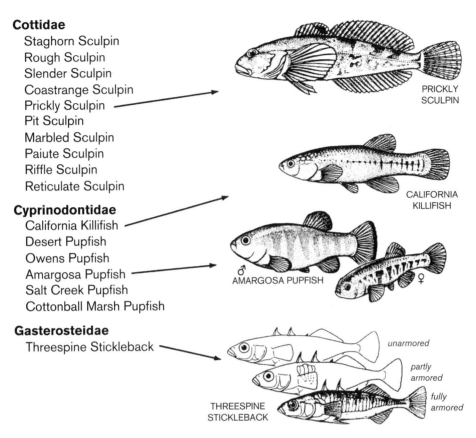

Cottidae
- Staghorn Sculpin
- Rough Sculpin
- Slender Sculpin
- Coastrange Sculpin
- Prickly Sculpin
- Pit Sculpin
- Marbled Sculpin
- Paiute Sculpin
- Riffle Sculpin
- Reticulate Sculpin

Cyprinodontidae
- California Killifish
- Desert Pupfish
- Owens Pupfish
- Amargosa Pupfish
- Salt Creek Pupfish
- Cottonball Marsh Pupfish

Gasterosteidae
- Threespine Stickleback

PRICKLY SCULPIN

CALIFORNIA KILLIFISH

AMARGOSA PUPFISH

♂ ♀

THREESPINE STICKLEBACK

unarmored

partly armored

fully armored

FIGURE 11.25 Secondary freshwater fishes (illustrations by Doris Alcorn from McGinnis, S.M. 1984. *Freshwater Fishes of California*. Berkeley: University of California Press).

Paiute Sculpins are the most abundant fish in the Lahontan system. In Sagehen Creek, densities of six adults per square meter have been recorded. Because this is also a trout stream, relationships between sculpins and trout have been studied. It has been determined that the diets of Paiute Sculpins and trout overlap by only 20%. Trout eat mostly drift organisms, and sculpins eat mostly what is on the bottom. In Lake Tahoe, there is even less overlap (figure 11.26). Moreover, a large part of the diet of trout is sculpins, so the ecological role of sculpins is as a forage fish. It is the link between the bottom-dwelling organisms and the top carnivores.

The Pupfish family (Cyprinodontidae) reached California from the south. They are found worldwide in tropical freshwater and marine habitats. The Killifish Family (Funduli-

dae) is closely related to pupfishes. Because of their bright coloration and interesting behavior, many members of this family are popular aquarium fishes. The California Killifish, *Fundulus parvipinnis* (figure 11.25), is a marine species that lives in brackish lagoons and estuaries as far north as the mouth of the Salinas River, near Monterey. In a few localities, they have moved into freshwater. In San Juan Creek, in southern Orange County, an entirely freshwater population formerly existed near the town of San Juan Capistrano. This population has been extirpated by habitat destruction. San Juan Creek is now a concrete-lined ditch along its lower course. A freshwater species still survives near San Ignacio, Baja California.

Of greater interest, perhaps, are the Pupfish (*Cyprinodon* spp.; figure 11.25) that inhabit

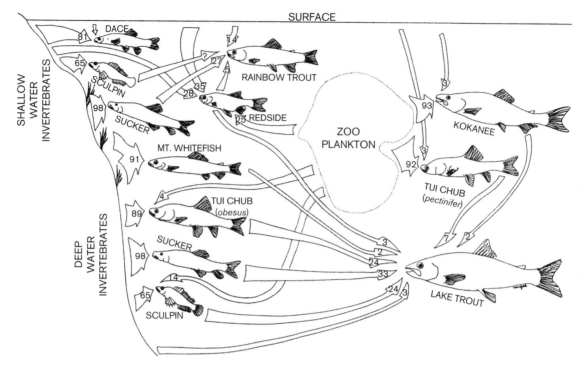

FIGURE 11.26 Feeding relationships of Lake Tahoe fishes (modified from Miller 1951). Numbers represent percentage of the diet by volume (Kokanee data are from Cordone et al. 1971).

SOURCE: Moyle, P. B. 1976. *Inland Fishes of California.* Berkeley: University of California Press.

warm, often saline bodies of water in the desert. In California, there are five native species. Including the Death Valley system as a whole, there are seven species and six subspecies of Pupfish and Killifish in California and nearby Nevada. Each of these forms is characterized by a rather restricted distribution. The Devils Hole Pupfish, *C. diabolis*, is a species 1 in (25 mm) long that occupies a collapsed limestone cavern with about 180 ft² (20 m²) of surface water. This is the smallest known range of any vertebrate animal in the world. While this area is technically in Nevada, it is in a detached portion of Death Valley National Park that lies a few miles east of the California-Nevada border. Devils Hole is actually part of the largest oasis system in the Mojave Desert. This area is known as the Ash Meadows National Wildlife Refuge, and it contains 20 endemic species. This is the largest number of endemic plants, fish, and invertebrates for an area this size in the continental United States. It is the home of

four endemic Killifishes that are federally listed as endangered. As of the fall of 2013, the Devils Hole population had become reduced to 35 individuals. At that time, scientists moved fertilized eggs to the laboratory and alternative ponds. Unfortunately, the refuge ponds at cooler temperature, more dissolved oxygen, and more food, produced fish that were more robust and actually had morphological changes, so they looked different from the Devils Hole fish. A large artificial tank with a 100,000 gal (380,000 L) of water was constructed in Ash Meadows where three generations of Devils Hole Pupfish are being protected. Meanwhile, the population in Devils Hole has been slowly recovering. As of fall 2015, the population had grown to 131 fish, making this the rarest fish in the world, and maintaining its status as a federally and state listed as endangered species.

The species with the largest natural distribution in the Southwest is the Desert Pupfish, *C. macularius.* It was formerly common in sloughs and backwaters associated with the lower Colorado River and the Gila River in Arizona. This is the largest of the North American Pupfish. Although they may reach 3 in (75 mm) in length, they are seldom more than half that size. Typical of their preference for quiet water and aquatic vegetation, they are deep-bodied and slab-sided.

Tolerance for environmental extremes is a notable feature of the Desert Pupfish. This is important because desert habitats experience wide variations in temperature, salinity, and dissolved oxygen. The critical thermal maximum of 112°F (44.6°C) for this species is the highest ever recorded for a species of fish. This ability to tolerate hot water also enables them to live in hot springs. In such a habitat, the Pupfish may feed on blue-green algae that live in water hotter than its critical thermal maximum. It does this by hovering in water as hot as it can tolerate and then darting into the hotter water for a quick bite of food. The thermal minimum of the Desert Pupfish is 39.9°F (4.4°C). The Owens Pupfish, *C. radiosus*, lives under ice at Fish Slough, north of Bishop.

Also recorded for the Desert Pupfish is the lowest tolerated minimum for dissolved oxygen, at 0.13 mg/L, and there is some evidence from Salt Creek that the fish become anaerobic as overnight recordings of dissolved oxygen have reach 0.0 mg/L. The species' range of tolerance for salinity is also high. Adult Desert Pupfish tolerate water from distilled to 70 g/L (twice the concentration of seawater). The Cottonball Marsh Pupfish, *C. salinus milleri*, a resident of Death Valley, holds the record for the highest salinity tolerated by a fish. Living fish were observed in saline pools on the floor of Death Valley at five times the concentration of seawater.

Introduction of nonnative species and habitat destruction have threatened the fishes of the desert. Of the 18 kinds of California fishes listed as threatened or endangered, 10 are considered to be desert species. Four kinds of Pupfish are on the list. The Cottonball Marsh Pupfish and the Tecopa Pupfish, *C. nevadensis*

calidae, are on the state list of threatened and endangered species. The Tecopa Pupfish is now extinct, however. The Owens Pupfish and the Desert Pupfish are listed by the federal and state governments as endangered.

The range of the Desert Pupfish has been diminished to one general locality in the United States, the area around the Salton Sea. Natural populations occur in two streams that are tributaries of the Salton Sea. The species is found in San Felipe Creek on the west side and Salt Creek on the east side of the Salton Sea (figure 11.27). Formerly, they were the most common fishes in shoreline pools of the Salton Sea, but they have been seriously threatened in that habitat by interactions with introduced species of fishes. Apparently as the Salton Sea increases in salinity, the Pupfish are recovering in shoreline habitats, and they often show up in irrigation drains at various points around the sea. In order to protect this endangered species, a series of refugia have been established on state land around the Salton Basin. A population closely related to the Desert Pupfish, the Quitobaquito Pupfish, *C. eremus*, is native to Quitobaquito Spring, in Organ Pipe Cactus National Monument in Arizona, and in Sonoyta Creek across the border in Sonora, Mexico.

Of course, the present Salton Sea is not a natural body of water, owing its continued existence to irrigation runoff. In prehistoric times, the Salton Trough was occupied by Lake Cahuilla, which dried about the year 1500, taking with it what was essentially an assemblage of fishes native to the Colorado River. The cause of its drying was a change in the direction of flow of the mouth of the Colorado River. It had been dumping into the Salton Trough, but as a result of southward meandering its flow was diverted into the Gulf of California. Estimates indicate that Lake Cahuilla took about 55–60 years to dry completely. Between 1840 and 1905, at least seven times the Salton Sea basin was reflooded by overflow from the Colorado River. The present body of water began with flooding of the Colorado River and rupturing of earthen irrigation ditches along the Mexican

FIGURE 11.27 Desert Pupfish, *Cyprinodon macularius*, habitat in upper Salt Creek, Riverside County.

border. The Desert Pupfish was reintroduced to its former range at that time.

An interesting series of old waterlines lie on the west side of the Salton Sea. The highest waterline (figure 9.9) is marked by a visible crust formed by precipitation of travertine ($CaCO_3$) in association with algae that lived along the shoreline. Lower waterlines are marked by rows of fish traps that were constructed by Cahuilla Indians. These traps consist of rows of rocks in a V shape. Fish that entered the structures were trapped by closing of the apex with a large rock. The rows of traps are nearly 5 ft (160 cm) apart in elevation, which corresponds to the known evaporation rate of water in the area today. Fifteen rows of traps mark the fall of water over a 15 year period. The absence of traps beyond that suggests that the lake became too saline for most of the large fish species preferred by the Native Californians. As the lake evaporated, it became unsuitable for the fishes that were more at home in freshwater of the lower Colorado River.

Runoff of irrigation water from fields in the Coachella and Imperial Valleys continues to add water to the Salton Sea, but that input has been reduced by transfers of Colorado River water to the Metropolitan Water District of Southern California, resulting in lowering of the water level and increasing salinity. The introduction of game fish from the Gulf of California was successful, and a good sport fishery was established, although it became threatened by increasing salinity and toxins such as selenium. Tests conducted by the US Fish and Wildlife Service revealed concentrations of toxic substances comparable to the ill-fated Kesterson National Wildlife Refuge, mentioned in Chapter 10. With lowering of the lake level and increased salinity, the target of fishermen at the Salton Sea today is the primary fish species remaining, a salt-tolerant hybrid of two African nonnative species. Whereas predation probably eliminated pupfish from the lower Colorado River, in the Salton Sea it was the introduction of small nonnative fishes that led to the decline. At least 29 fish species have been recorded from drains and canals in the Salton Sea area. Only one of these was native—the Desert Pupfish.

Subtle behavioral interactions with two species seem to have caused the problem. Sailfin Mollies, *Poecilia latipinna*, were introduced about 1964. These are aquarium fish, and no one is certain how they were introduced. Male Sailfin Mollies seem to court any fish that is about the right size, regardless of species or sex. By 1979, these represented 98% of the fishes in shoreline pools. By sheer numbers, they interfered with mating and courtship of the pupfish, which apparently escaped by moving into shallow water in the canals.

However, in 1973 the Coachella Valley Water District introduced a "weed-eating" fish to the canals in the hope that it would control the aquatic weeds. This fish, the Redbelly Talapia or Zill's Cichlid, *Tilapia zilli*, is an aggressive African cichlid, but adults cannot live in water less than a few inches deep. Pupfish can live and breed in water about an inch (25 cm) deep, but juvenile cichlids in this shallow water can interfere with reproduction in the following way. A dominant male pupfish has a bright blue color and patrols a territory that it keeps clear of other males and other fishes. Females are lured into the area, where they lay eggs in the sand or in algal mats. The male responds by releasing sperm over the eggs and then vigorously guarding the spawn. However, when he is busy chasing intruders, subdominant males, which look like females, enter the spawning ground in an attempt to fertilize an egg or two. These are the "sneak" maters. In an attempt to maximize their contribution, they also eat some of the dominant male's spawn. Likewise, when the dominant male discovers another male's spawn, he eats it. The result is that if too much interference occurs, males eat each other's spawn, and reproduction is curtailed. Apparently, swarms of juvenile cichlids promote this kind of interference, and pupfish reproduction fails to keep up with the natural death rate.

The two habitats where Desert Pupfish are hanging on, San Felipe and Salt Creeks, are small, interrupted streams. They are not ideal for pupfish, because there is not always quiet water and there are extreme variations in flow. As buffers against extinction, the species has several refugia in the Colorado Desert (plate 16B).

How did all these Pupfish of isolated drainages reach their habitats in the southwestern deserts? They all occur in tectonically active areas and may have experienced some degree of translocation along with migrating terranes. However, it is simpler to provide aquatic connections between the various localities. Similar to killifish, it seems that the pupfish's centers of origin are from the south. In some instances, they are estuary dwellers, similar to the California Killifish. It appears that the Desert Pupfish or a very similar species inhabited the Bouse Embayment, a Mio-Pliocene estuary that covered what today is the lower Colorado River and the Salton Trough. Fossil pupfish of about that age are known from as far north as Nevada. Similarly, Fossil Killifish have been found as far north as Lake Lahontan, and fossil sticklebacks also occur throughout the Southwest.

Retreat of the marine water from the Bouse Embayment could have left the pupfish isolated in streams and springs. A basic hypothesis is that during the Pleistocene, heavy precipitation in desert areas caused the formation of a series of interconnected lakes and streams with the general direction of flow toward Death Valley (figure 11.28). Connections to the Colorado River have not been established, but dispersal throughout the system might have occurred at that time. Some recent evidence about lake levels during Pleistocene cast doubt about the connections, and a Miocene connection from Owens Valley to Death Valley has been proposed. Evidence for this connection lies in basalt flows from west to east that might have followed ancient stream channels. Mitochondrial DNA evidence indicates divergence of various species through isolation occurred about 3–2 million years ago, which is much later in time than a proposed connection to Owens Valley. The implication is that speciation among the pupfishes continued to occur during the last 20,000 years, accompanied by reduction of precipitation and stranding in isolated desert spring systems. It appears that all of the Death Valley pupfishes evolved from the Desert Pupfish or a similar ancestor from the south. Of particular interest in this hypothesis is that the Owens Pupfish is the most closely

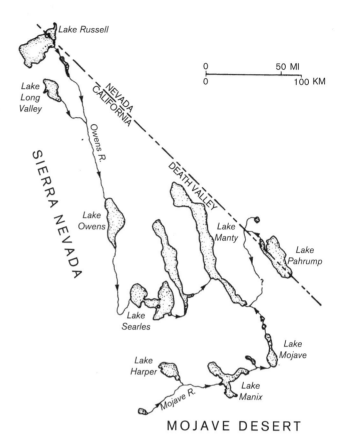

FIGURE 11.28 Proposed Mono-Death Valley drainage system during the Ice Age, Pleistocene (from Hill 1984).

MOJAVE DESERT

related to the modern Desert Pupfish, yet its range, in the upper Owens Valley, is the farthest away.

An alternative hypothesis is that, similar to the chubs, pupfish were carried from latitudes of present-day Mexico in springs and streams west of the San Andreas fault. During pluvial times, they dispersed through the interconnecting system of desert lakes. The absence of pupfish in Baja California and the Peninsular Ranges, however, speaks against this hypothesis.

The stickleback family (Gasterosteidae) is the third group of secondary freshwater fishes in California. There is one species in California today, the Threespine Stickleback (*Gasterosteus aculeatus*; figure 11.25), although the fossil record indicates that in Miocene times, about 20 million years ago, there might have been several species. The family is abundant in Europe, northern Asia, and North America. It

appears that in California its origin is from the south, in a pattern similar to that of the Killifish family.

Sticklebacks are distributed in coastal streams and estuaries up and down the coast of California, and they are common throughout the Great Central Valley. Seldom over 3 in (75 cm) in length, they have three sharp dorsal spines and a series of armored plates on each side. These features may help protect them from predation.

Sticklebacks have been divided into three subspecies on the basis of habitat. Anadromous populations spend most of their time in saltwater and migrate into freshwater to spawn. They have been identified by some authorities as a subspecies. Those fish that spend their entire lives in freshwater have been named another subspecies. The anadromous form has plates of armor completely along each side, whereas the freshwater forms are armored only in the head regions. Finally, in the Los Angeles basin, there were populations that were unarmored. These have been identified as the third subspecies. They occurred as far south as San Juan Creek, where they were associates of the freshwater Killifish. The Unarmored Threespine Stickleback, *G. a. williamsoni*, is now listed as endangered by the federal government. It occurs primarily in protected habitat in the Santa Clara and Cuyama Rivers. The population in the Santa Clara River is in the way of a proposed planned community. Part of the proposal involves potential translocation of the fish.

Interesting research on sticklebacks showed that after the 1964 earthquake in Alaska, populations became isolated in freshwater ponds after which fish rapidly evolved genetically and anatomically. This evidence documents rapid evolution of this fish species. Overall changes occurred within 50 years, but in some populations changes occurred within decades.

Recent studies have indicated that the degree of armoring is a function of predation pressure, and this has clouded the naming of subspecies. When the armored varieties are raised in habitats with no predators, they produce little or no armor. The mechanism for this variation has not been worked out, but it bears testimony to the genetic plasticity of this group.

Of special interest is the presence of sticklebacks in the upper reaches of the Mojave River drainage in Holcomb Valley, on the desert side of the San Bernardino Mountains. This is similar to the pattern of distribution of Chubs and Pond Turtles. Some authorities contend that those on the desert side were introduced. However, those in Holcomb Valley are darkly pigmented, indicating that they represent a special variety that may have evolved in isolation. A population of the unarmored variety was also discovered in a small creek near Big Bear Lake. The fact that this is an endangered subspecies disturbed plans for a major housing development in the area.

Breeding in sticklebacks has been studied heavily. Males develop a bright red belly in the breeding season. They defend territories among beds of aquatic plants. Fighting behavior in the males is stereotyped so that they do not kill each other. It has been determined that the signal to begin fighting is the sight of the red belly. The fight is stopped when the loser stands on his head to conceal the color, analogous to the submission posture assumed by deer and elk. Experimenters were able to elicit fighting behavior in male sticklebacks by placing in the water practically any object that was red on the bottom. Similarly, any object that resembles a female with a swollen belly will stimulate courtship.

Overall, breeding behavior is similar to that of sculpins. Males build nests, which they defend vigorously. After eggs are laid in the nest, the male causes water to circulate over them by fanning with his pectoral fins. After the eggs hatch, the male guards the school of fry. If one of the juveniles wanders from the school, the male picks it up in his mouth and spits it back into the school.

In spite of the armor, these are important forage fish. They are fed upon by trout and large chubs as well as predatory birds.

Marine Fishes

Saltwater fishes that appear in freshwater are of three types. There are the anadromous species, which breed in freshwater; catadromous forms, which breed in saltwater; and the true marine species, which may have a single member that has become adapted to freshwater (figure 11.29).

The most economically important anadromous fishes are salmon, which are in the same genus as trout. There are five species in California waters, although two of them make up the bulk of the catch. Chinook and Coho Salmon are the most common. Sea-run Rainbow Trout, known as Steelhead, are another important anadromous fish.

Chinook Salmon, *Oncorhynchus tshawytscha* (figure 11.29), are also known as King Salmon. These are the largest of the Pacific salmon, reaching 31 in (80 cm) in length. The California record for weight of these salmon is 85 lb

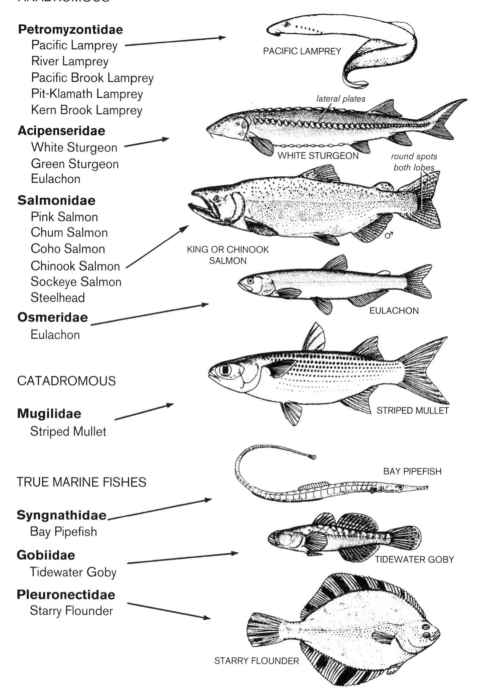

ANADROMOUS

Petromyzontidae
    Pacific Lamprey
    River Lamprey
    Pacific Brook Lamprey
    Pit-Klamath Lamprey
    Kern Brook Lamprey

Acipenseridae
    White Sturgeon
    Green Sturgeon
    Eulachon

Salmonidae
    Pink Salmon
    Chum Salmon
    Coho Salmon
    Chinook Salmon
    Sockeye Salmon
    Steelhead

Osmeridae
    Eulachon

CATADROMOUS

Mugilidae
    Striped Mullet

TRUE MARINE FISHES

Syngnathidae
    Bay Pipefish

Gobiidae
    Tidewater Goby

Pleuronectidae
    Starry Flounder

PACIFIC LAMPREY

lateral plates

WHITE STURGEON

round spots both lobes

KING OR CHINOOK SALMON

EULACHON

STRIPED MULLET

BAY PIPEFISH

TIDEWATER GOBY

STARRY FLOUNDER

FIGURE 11.29 Marine fishes found in freshwater (illustrations by Doris Alcorn from McGinnis, S.M. 1984. *Freshwater Fishes of California*. Berkeley: University of California Press; Bay Pipefish from Hinton, S. 1987. *Seashore Life of Southern California*. Berkeley: University of California Press).

(38.6 kg). These are the least abundant of the Pacific salmon, but in California they are the most common. In good years, they make up 90% of the commercial salmon catch. In addition, up to 130,000 are caught each year by sport fishermen.

In years past, Chinook Salmon spawned as far south as the Ventura River. Today they are concentrated from the Sacramento River northward. Historically, the Sacramento-San Joaquin system supported a substantial run of Chinook Salmon, but most of the historic salmon habitat in that river system is gone. Formerly, most spawning took place in the autumn when the first rains increased stream flow and lowered the water temperature to 77°F (25°C) or less. Interestingly, there are also spring runs and a unique winter run of Chinook Salmon in the Sacramento River. For maximum egg survival, water temperature must be less than 57°F (19°C). Eggs laid in the autumn hatch in spring and embryos remain in stream gravel for a few weeks. When hatchlings emerge from the gravel, they migrate downstream and enter the sea at about two months of age. This rapid exit from freshwater is an adaptation to California's Mediterranean climate. It is associated with the tendency for California streams to experience very low flow or drying in the summer. This phenomenon seems to explain why Chinook Salmon, the least abundant of the Pacific salmon, are the most common in California. Among salmon, they stay in freshwater for the shortest period of time. These autumn runs are a thing of the past in most rivers because dams now regulate the flow, in some cases blocking the salmon runs completely.

The interesting spring and winter runs for salmon in the Sacramento River is likely due to cold water released from Shasta Dam in order to provide consistent supplies for irrigation. This water comes from the bottom of Shasta Reservoir and is therefore cold. This cold water creates havoc for many of the native species of freshwater fishes that require warm water to spawn, but it has stimulated increased spring and winter spawning runs in Chinook Salmon.

Although there always have been some fish that would spawn in the upper, perennially cool portions of the drainages, there are now more fish that run upstream in midwinter and spawn in spring. Eggs laid in the spring hatch in late summer. This has created three major seasons for salmon hatching, which has the potential to greatly increase productivity throughout the state. Two other Chinook populations occur in California; fall-run salmon in coastal streams from Cape Blanco in Oregon south to San Francisco, and fall- and spring-run Chinooks in the Trinity and Klamath Rivers. The Sacramento winter-run population is classified as endangered, whereas the Sacramento spring-run, and the coastal fall-run populations are classified as threatened.

Adult Chinooks spend 5–7 years at sea, after which they return to their home stream to spawn. They die soon after. Elaborate experiments on migrating salmon have determined that they recognize their home stream by its odor. Furthermore, they migrate up and down the coast using visual cues such as polarized light and, possibly, familiar submarine landmarks.

Coho Salmon, *O. kisutch*, are also known as Silver Salmon. These may reach a length of 2 ft (60 cm), and the California record for weight is 22 lb (10 kg). Unlike Chinook Salmon, Coho Salmon spend at least a year in freshwater before they move out to sea. They spawn in cooler, smaller streams than the Chinook Salmon do, and they are absent from intermittent streams. At sea, they spend most of their time near the coast within 100 mi (160 km) or so of their home stream. Range of the species in the east Pacific is from Alaska to central California. In the west Pacific, they range from northern Japan to Russia. Two populations occur in California. The northern population ranges from Cape Blanco, Oregon, to the Mattole River in northern California. The central California population extends from Punta Gorda, near the Mattole River, to the San Lorenzo River in Santa Cruz County.

There are two kinds of male spawners. Sneak-mating males, known as jacks, grow and mature rapidly. These males are smaller than the large, hook-jawed males that return from the sea after 3–5 years. These large males actually grow and mature more slowly than the jacks, but by defending territories they contribute more offspring to future generations. The sneaks, on the other hand, by reproducing early without going to sea, are more likely to live long enough to reproduce. Evidence also shows there are four kinds of juveniles depending on where they mature and how fast they grow. Presumably this gives the species a form of genetic variation which increases survival potential of juveniles under variable environmental conditions.

Coho are the easiest salmon to raise in hatcheries. Furthermore, they can spend their entire lives in freshwater. Larger and larger numbers of hatchery-raised salmon are being released in California rivers. The result of this program is that, in spite of habitat degradation, particularly because of logging, the total number of Coho Salmon seems to be increasing. Coho Salmon "Silvers" are also being planted in many large lakes and reservoirs, where they feed on introduced schooling fishes such as Threadfin Shad. These salmon now occur in catchable numbers in lakes such as Lake Berryessa and in reservoirs such as Lake Mead on the Colorado River.

The decline in numbers of salmon has been a significant factor in the economic collapse of the Pacific Northwest. The cause of the decline is controversial. Loggers and dam builders blame fishermen, and vice versa. Commercial fishermen blame Native Californians, and vice versa. All are to blame to some degree. The combination of overfishing and habitat destruction has reduced the salmon harvest to a fraction of its former numbers. Warm water associated with the El Niño condition also has an effect, particularly in the Klamath drainage, and warm water released from low reservoirs during drought of the 1980s practically wiped out the Sacramento River winter runs. Coho Salmon of the central California coast have been classified as endangered, and the northern population is classified as threatened.

The Klamath River and its tributaries remain the most productive salmon fishery in California. Prior to 1900, migrating fish could travel all the way to its headwaters. Today seven dams on the Klamath system are near its headwaters. These are part of a hydroelectric project that supplies significant electricity to the area. The uppermost dams also divert water for irrigation in northern California. Four of the dams are in northern California near the Oregon border. In 2001, which was a dry year, the federal government reduced irrigation deliveries to farmers in northern California. In response, the formers threatened to open irrigation gates by force. In 2002 to pacify farmers, the government increased the release of water for irrigation, the consequence of which was reduced flow in the lower Klamath River, resulting in a major die-off of salmon and steelhead. One estimate was that 30,000 dead fish were found floating in the lower Klamath. In 2010, environmentalists, commercial fishermen, private anglers, and Native Californians tribes hammered out an agreement to remove the four California dams, which would open up about 300 mi of river for salmon migration. The final project has been held up in congress because part of the agreement involved transferring an unrelated 200,000 acres (81,000 ha) of national forests to northern counties in California and Oregon, which could open up the land to commercial interests such as logging and agriculture. Frustrated by inaction of congress, in April 2016 a pact was signed by the secretary of the interior and the governors of California and Oregon, which enables the project to go forward without requiring congress to sign off on the removal. Iron Gate Dam, the lowermost of the dams, regulates the main flow. Nevertheless, from there to the coast the Klamath flows as a wild river, and the fall run of salmon is still significant. The Trinity River, a major branch of the Klamath, and the Smith River system, to the north of the Klamath, are also classified as wild and scenic rivers.

The American salmon fishermen and the federal government have worked out fishing quotas in an attempt to prevent overfishing. The quota system, however, is complicated by Russian and Asian fishing boats, which are less likely to respect our attempts to limit harvest. Furthermore, Canadian interests must be respected. Last but not least is the issue of native fishing rights. Native Californians are allowed to fish without regard to seasons, limits, or techniques, so long as their catch is for subsistence only. In effect, this tends to promote illegal sales of salmon by Native Californians, who view it as their inalienable right to catch salmon whenever, wherever, and however. Obviously, for the benefit of all concerned, there must be limits on the total harvest. Details of those limits must be worked out cooperatively and adhered to, or else all concerned parties, especially the fish, will lose.

Of special interest among anadromous species are the lampreys (Petromyzontidae), eel-like fishes with no paired fins. They are remnants of a very primitive group of jawless fishes, the Agnatha. There are six species in California, and their distribution is similar to that of salmon. Two of the species are predatory. These are the Pacific Lamprey, *Entosphenus tridentatus* (=*Lampetra tridentata*) (figure 11.29), and the River Lamprey, *L. ayersii*. The Pacific Lamprey is the largest, reaching 2 ft (60 cm) in length. The River Lamprey is half the size.

The life history of these lampreys is interesting because the larval stage is long and the adult stage is short. Larvae are sedentary filter feeders. They live for 5–7 years in mud-bottom streams with their heads projected into flowing water. They metamorphose when they are but a few inches in length. The two predatory species migrate to the sea, where they prey on soft-bellied fishes such as salmon or herring. As adults, they have a suction cup for a mouth, which they use to attach to the belly of a fish. With their rasp-like tongue, they scrape a hole in the fish and suck out body fluids. Some authorities prefer to call these fish parasites because they do not eat their prey in its entirety, and sometimes the prey species is not killed. As adults, they live in the sea for 1 or 2 years, after which they return to freshwater to spawn. Similar to salmon, they die after spawning.

Recent surveys by ichthyologists show that the Pacific Lamprey seems to be retreating from its former range in southern California. Today its basic range is no farther south than Big Sur, yet as recently as January, 2017, it was found in the Santa Ana River.

The four species of brook lamprey remain in the freshwater their entire lives. Most have restricted distribution. The Western Brook Lamprey, *L. richardsoni* (= *pacifica*), enjoys the widest distribution, occurring in streams along the north coast and the Sacramento-San Joaquin Delta. They also have been reported in the Los Angeles basin. These are not predatory, because adults do not feed, living only long enough to reproduce. Larvae live about 4 years as filter feeders. In order to reproduce, they metamorphose at about 8 in (20 cm) in length. Similar to the strategy of many aquatic insects, the adult stage is for reproduction only.

The sturgeon family (Acipenseridae) includes two anadromous species in California. The Green Sturgeon, *Acipenser medirostris*, is not common in freshwater, although some specimens are found in the Klamath, Sacramento, and San Joaquin Rivers. The White Sturgeon, *A. transmontanus* (figure 11.29), spawns in these large rivers. This is America's largest freshwater fish, reaching lengths in excess of 11 ft (4 m). They grow slowly and live up to 100 years. They become sexually mature at about 13 years of age. They are cherished for their tasty roe, which is American caviar.

Because of their long life and late maturity, these fish do not respond well to environmental disturbance. They were common "nuisance" fish in the late 1800s, but an acquired taste for caviar and sturgeon flesh nearly extirpated them by 1909, when commercial fishing for them was halted. They have recovered slowly, and it is now legal to fish for them as a game fish.

Sturgeon are cigar shaped with large diamond-shaped scales. Their mouths are underneath a broad, flat, shark-like nose. They have barbels with taste buds near the mouth, which they use to search in the mud for food. They have a mouth that resembles a flexible suction hose, which they use to slurp up invertebrates and carrion in the mud. Fishermen in the Sacramento-San Joaquin Delta fish for them using small grass shrimp as bait. Sometimes, because of its value as a delicacy, all they keep is the egg mass, which they call "black gold." Also because of the value of the caviar, a program has been initiated to raise sturgeons in fish hatcheries.

The smelt family (Osmeridae) includes a number of small schooling species that are important forage fishes in bays and lagoons. The Eulachon, or Candlefish, *Thaleichthys pacificus* (figure 11.29), is the largest of our smelts, reaching a total length of 12 in (30 cm). This is an anadromous form that in spring spawns in the Klamath River and a few small streams in Humboldt and Del Norte Counties. In spite of their oily flesh, many people enjoy eating them. It is legal to fish for them with a dip net. Native Americans not only ate them, but also dried and burned them as if they were candles. They are important forage species for salmon and sturgeon. Other smelt species are also sometimes found in freshwater. For example, the Delta Smelt, *Hypomesus transpacificus*, occurs in freshwater of the lower Sacramento and San Joaquin Rivers. They were once one of the most common fishes in the upper San Francisco Estuary, but their populations since the 1980s have fluctuated greatly. Their distribution in the bay is related to the amount of freshwater flowing into the estuary. They prefer a zone of mixing between freshwater and seawater. This makes them sensitive to the amount of water that is diverted for the massive pumps that feed California's aqueduct systems. During low flow, the fish move up river where they may be sucked up by the pumps and there is reduced feeding area to support the population. Interestingly, during wet years and periods of high flow, adult smelt and their plankton food supply is washed out of the system, also lowering the breeding population. In addition, invasive plankton-feeding organisms such as nonnative clams, mussels, and copepods have reduced the food supply for the smelt. Finally, predation by an introduced game fish, the Striped Bass, *Morone saxa-*

tilis, became an issue. As a consequence of these problems, the Delta Smelt has been federally and state listed as a threatened species, which has required that, during times of drought, freshwater diversions be reduced to protect the fish populations. This has become a very controversial issue between agricultural interests, which depend on pumping of freshwater, and environmentalists who want to protect the ecological integrity of San Francisco Bay.

The Longfin Smelt, *Spirinchus thaleichthys*, primarily a northern California species, was once one of the most common smelt in San Francisco and Humboldt Bays. In light of its decline, it has been declared by the state as a species of special concern. In San Francisco Bay, they are concentrated in the southern part of the estuary. They spawn in freshwater, particularly in the San Joaquin drainage. Causes of its decline are unknown, but probably related to reduced flows of freshwater. Another species, the Wakasagi, *H. nipponensis*, was introduced to a few reservoirs as a forage fish and subsequently found its way into most northern coastal rivers.

The mullets (Mugilidae) are estuary fishes. In the lower Colorado River, they are catadromous, returning to seawater to reproduce. California's only species, the Striped Mullet, *Mugil cephalus* (figure 11.29), is conspicuous in estuaries and bays. For unknown reasons, they are often seen leaping from the water. They reach about 2 ft (60 cm) in length, and when a fish this size leaps clear of the water and lands on its side with a resounding smack, it is conspicuous. They feed on detritus and algae, so they are seldom caught on a hook, but they are tasty eating. They invaded the Salton Sea soon after it began to fill, and by 1915 they were so numerous that a successful commercial fishery developed for them. They were netted in great numbers. Commercial fishing ceased in the early 1950s. They are now extinct in the Salton Sea, presumably because it is too saline for the eggs.

Of the true marine fishes that occur in freshwater, there are three families with a single member that may have incidental distribution in freshwater. Bay Pipefish, *Syngnathus leptorhynchus* (figure 11.29), are small, green, elongated fish that live in eelgrass. They are in the seahorse family (Syngnathidae), and, similar to seahorses, eggs are incubated in brood pouches of males. They have been found living in beds of filamentous green algae in the Navarro River in Mendocino County.

The flounder family (Pleuronectidae) is a group of bottom-dwelling flatfish. They are notable for having both eyes on the same side of their head, so they may lie on their sides, flat on the bottom. The Starry Flounder, *Platichthys stellatus* (figure 11.29), is relatively common in freshwater in the Sacramento-San Joaquin Delta and various coastal streams. They have even made it into San Luis Reservoir by way of the California Aqueduct. When found in freshwater, they are not always in good condition.

The goby family (Gobiidae) is a group of bottom-dwelling fishes that superficially resemble the sculpins. Unlike the sculpins, these fishes have a peculiar suction cup formed by fusion of the pelvic fins on the front half of their underside. This suction cup helps to hold them on rocks in the intertidal zone, where there may be waves to contend with. In freshwater, the suction cup helps hold them on the bottom in flowing water. The Tidewater Goby, *Eucyclogobius newberryi* (figure 11.29), is the only native species that enters freshwater. They are endemic to coastal streams in California having a wide range of distribution from San Diego to Arcata, but they are not common except in the Morro Bay area. They have disappeared from many coastal localities including San Francisco Bay, but most of their disappearance has been south of Point Conception. The state declared them protected in 1987, and they were listed as endangered by the federal government in 1994. A sister species, the Arrow Goby, *Clevelandia ios*, prefers totally marine habitats in estuaries. Interestingly, the Tidewater Goby prefers lagoons in which there may be seasonal absence of tidal flushing. Dispersal of both species requires tidal flushing. The introduced Yellowfin Goby, *Acanthogobius flavimanus*, is becoming increasingly common in freshwater in California. It is native to the coast of Japan, Korea, and China. Since its first discovery in 1963, in the Sacramento-San Joaquin Delta, it has become the most common bottom fish in San Francisco Bay. It probably arrived from Asia, attached by its sucker to the bottom of a boat.

The Longjaw Mudsucker, *Gillichthys mirabilis*, lives primarily in mud bottom, shallow parts of estuaries, usually in saltwater or brackish water, although they can live up to a week in freshwater. Their natural distribution is in the northern end of the Gulf of California and from Bahía de Magdalena in Baja California to Tomales Bay. They were introduced to the Salton Sea in 1950 by the California Department of Fish and Wildlife (formerly California Department of Fish and Game), presumably as part of a program to develop an ecosystem that would support a sport fishery in that body of water. They are able to tolerate salinity as high as 8.2% and water temperature as high as 95°F (35°C), and have become common in shoreline pools and canals around the Salton Sea. They are commonly used as bait for sport fish in the Colorado River. This is an air-breathing fish that is able to gulp air and absorb oxygen in a vascularized chamber at the back of their throat. They can also slither across mud flats in an attempt to find water after being stranded.

Amphibians

There are many unique, interesting species of amphibians in California, all limited to some degree by the distribution of water. Because salamanders are largely terrestrial, they have been discussed in association with the specific regions where they are common. California frogs, on the other hand, are largely aquatic. Their larvae (tadpoles) develop in water, and the adults leave water only temporarily.

There are only six native species in the "true" frog family (Ranidae) that enjoy widespread distribution in California. The state's largest native frogs are the red-legged frogs. The Northern Red-legged Frog, *Rana aurora*, occurs in the Coast Ranges from Del Norte to Mendocino

County. The California Red-legged Frog, *R. draytonii*, has populations in the Cascades, Sierra Nevada, and the Coast Ranges south of Mendocino County. There are only two populations known to occur south of Santa Barbara and six in the Sierra Nevada. Due to habitat loss and competition from Bull Frogs, it has been listed as threatened by the federal government and is now the official California state amphibian.

Body length of these frogs may reach 5 in (13 cm). These are primarily pond frogs. Formerly they were common in lowland, slow-moving waters and marshes throughout the state except for the Great Central Valley and deserts. Breeding behavior of red-legged frogs is typical of Ranids. For protection from predators, they mate at night. In late winter, when day length begins to increase, males gather in ponds and begin to "sing." These breeding choruses are designed to attract females. If a female is ready to breed, she will call back. Frogs hear well, possessing large eardrums on the sides of their heads. A male is able to locate the position of the female by sound alone. Upon hearing her call, he will swim toward her. Upon reaching her, he climbs on her back and grasps her by the armpits. This is called amplexus. During the breeding season, males have an enlarged nuptial bump on the first finger of each hand. This bump is apparently important for tactile stimulation of the female's armpits, causing her to lay eggs. From a position on her back, the male is able to shed sperm over the eggs as they are laid. Floating masses of fertilized eggs develop into tadpoles.

Communication between male and female is largely acoustic. A male attempting to locate the female that responded to his call will grasp anything in the water that is the proper size. If he grasps a female that is not ready to shed eggs or if he grasps another male, the unwilling partner will utter a warning croak. If no warning croak is uttered, the male assumes he has grasped a willing female. Unfortunately, if he grasps a rock, log, or beer can of the right size and texture, he has no choice but to hang on and wait for something to happen. Fortunately, if nothing has happened by dawn, the clear light of day will reveal the mistake, sending the frustrated lover into hiding until the next night. In frog mating systems, the rule of thumb is, "If an object is silent, too large to eat, but not large enough to eat you, hang on for dear life." California's largest frog is the introduced Bullfrog, *R. catesbeiana*. This is the frog whose booming call is sometimes described as "jug-o-rum." This species is native east of the Rockies, but because frog legs are a delicacy, it has been introduced all over the west. These frogs may reach 8 in (20 cm) in body length. They are highly predatory, feeding on crayfish, frogs, and fishes. Sometimes they even eat small snakes! Unfortunately for the native Red-legged Frog, the two species occur in the same habitat. What habitat destruction has not done, the Bullfrog has through resource utilization and predation. The Red-legged Frog is now becoming scarce.

At higher elevations, the Red-legged Frog is replaced by the Foothill Yellow-legged Frog, *R. boylii* (figure 11.30A). The body length of these frogs is seldom over 3 in (8 cm). The underside of the legs and the belly are yellow. This is a frog of rivers and streams. Its distribution lies in the northern Sierra and the Coast Ranges, but is now apparently absent from former localities in southern California. It occurs in riffles, where it rests with head and body out of the water, counting on its mottled yellow coloration to conceal it. If alarmed, it dives into the water and remains motionless at the bottom of the pool. For inexplicable reasons, this frog species has disappeared from its known habitat in the San Gabriel Mountains and southern Sierra Nevada. One possible cause of its decline in the western Sierra Nevada is blow-over of pesticides and herbicides from agriculture in the Great Central Valley. Organophosphate residues have been found in the air, water, and snow in the Lake Tahoe area and Sequoia National Park. Half of the Sierran treefrogs and their tadpoles in Yosemite National Park had pesticide residues in their bodies.

In the Sierra Nevada above 6000 ft (1800 m), the Foothill Yellow-legged Frog is replaced by the Sierra Nevada Yellow-legged Frog, *R. sier-*

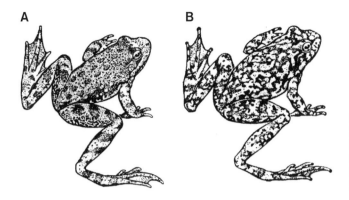

FIGURE 11.30 A. Foothill Yellow-legged Frog, *Rana boylii*. B. Southern Mountain Yellow-legged Frog, *Rana muscosa* (from Stebbins, R.C. 1951. *Amphibians of Western North America*. Berkeley: University of California Press).

rae (figure 11.30B). In southern California, from the southern Sierra Nevada, to the San Gabriel Mountains to Mount Palomar, a related species, the Southern Mountain Yellow-legged Frog, *R. muscosa*, occurs in the habitat of the Foothill Yellow-legged Frog. At two localities in the San Gabriel Mountains, Bear Creek and the North Fork of the San Gabriel River, the two species formerly coexisted. Unfortunately for the Foothill Yellow-legged Frog, much of its habitat in the lower reaches of the San Gabriel drainage has been destroyed by dams and off-road vehicles. The Foothill Yellow-legged Frog has not been seen in the San Gabriel Mountains since the early 1960s.

The Sierra Nevada Yellow-legged Frog, the common frog of the high Sierra, has habits similar to those of the Foothill Yellow-legged Frog. It lives in a harsh habitat up to 12,000 ft (4000 m) in elevation, and it breeds when lakes and streams are ice free. Tadpoles are then faced with a serious problem: They must grow and metamorphose in an abbreviated season. Because the growing season is so short in the high country, tadpoles may overwinter for 2 years, undergoing metamorphosis during their third season. Postponing metamorphosis provides another advantage for these frogs, because the larvae are better able than the adults to tolerate conditions of low oxygenation under the ice. In an attempt to optimize temperature conditions that will accelerate growth, larvae spend daylight hours basking in shallow water, usually on granite slabs, where the water tem-perature may warm considerably. They feed on algae that accumulate on these rocks as well. At night, they move to deeper water, where heat is retained. While basking in shallow water, the tadpoles are protected from predation by fishes, but when they move to deeper water, they are fed on by trout. In shallow water, where they are relatively free from predation by trout, Yellow-legged Frogs may become very common. Trout fishermen are aware that a lake with a large frog population is seldom good for fishing, and it has been documented that frog populations decline after trout are introduced. The Mountain and Sierra Nevada Yellow-legged Frogs are classified as endangered by the federal government. In order to protect them from predation, much to the dismay of anglers, certain Sierra Nevada populations have had trout removed completely by public agencies associated with wildlife man-agement. In 2016, a controversial federal ruling which established critical habitat for the frog, that included some reservoirs reachable by roads, raised concern among resort owners that fish will be removed from the reservoirs, and in so doing damage their businesses.

Adult yellow-legged frogs produce an odor-ous mucous that helps protect them from pre-dation, but they still are fed upon by fishes, Gartersnakes, and birds. Brewer's Blackbird, *Euphagus cyanocephalus*, and Clark's Nut-cracker, *Nucifraga Columbiana*, have been observed to gather at the water's edge and feed upon the frogs as they emerge from the water during metamorphosis. Tadpoles may also be fed upon by California Gulls, *Larus californicus*, and Great Blue Herons, *Ardea herodias*.

The Oregon Spotted Frog, *R. pretiosa*, occurs on the Modoc Plateau in cold streams and pools. This is a frog of the northern Great Basin. Its range barely extends into California. A relative of the Oregon Spotted Frog is the Cascades Frog, *R. cascadae*. This mountain frog occurs in habitats similar to those of Yellow-legged Frogs, ranging from Mount Lassen northward in the Cascade Mountains. When disturbed, unlike Yellow-legged Frogs they swim rather than remaining motionless on the bottom.

The most common frog in the United States is the Northern Leopard Frog, *Lithobates (= R.) pipiens*. The natural range of this frog does not reach California, but it has been introduced and is now common in some localities. There are actually four species of Leopard Frogs the may occur in California. The Southern Leopard Frog, *L. (= R.) sphenocephala*, is likely the species that persists in the Santa Ana River, and the Rio Grande Leopard Frog, *L. (= R.) berlandieri*, was introduced into the Lower Colorado River and apparently has moved into the irrigation systems of the Imperial Valley. Because the Northern Leopard Frog is the familiar laboratory frog, it has been introduced all over the state. Feeling pity for them, students, faculty, and researchers release them in nature when they have outlived their usefulness in the lab. This is a very bad practice. It introduces not only an aggressive competitor, but many unwanted diseases as well. The Lowland Leopard Frog, *L. (= R.) yavapaiensis*, was found in California only where it occurred in isolated, sporadic localities in the desert. A small population in San Felipe Creek on the southwestern side of the Salton Sea represents an endangered species, although it may be extinct.

Another introduced frog that has created havoc for native species is the African Clawed Frog, *Xenopus laevis*. This is a member of a family of tongueless frogs (Pipidae) that is not even native to North America. The species was originally imported for the purpose of conducting human pregnancy tests. When injected with urine from a pregnant woman, a female African Clawed Frog responds to hormones in the urine and ovulates. It is now known that a similar reaction occurs in many native species as well. Because it is completely aquatic, the African Clawed Frog also became popular as an aquarium species. It is easy to maintain, and the species in captivity is known to live 5 years.

The sources of African Clawed Frogs that have been released into the wild are not known, but once they become established they multiply rapidly and become major predators. As adults, they seem to have no natural enemies. Their skin secretions are distasteful, if not poisonous, to all known potential predators. In reservoirs, large predatory fishes keep them under control by eating the tadpoles, but in streams or small ponds the frogs take over. One of the most serious situations occurred in the Santa Clara River of Los Angeles County, critical habitat for the endangered Unarmored Threespine Stickleback. African Clawed Frogs appeared in 1978 in a tributary of the Santa Clara River. A decade later, they had moved into the area occupied by sticklebacks and threatened to wipe out the population. The California Department of Fish and Wildlife has begun a program of trapping to remove the frog. It is hoped that, at the very least, they can keep the population small enough to prevent obliteration of a fish that is listed as endangered by the state and federal governments.

Many people confuse frogs with toads. "True" toads are in the family Bufonidae. Frogs are mostly aquatic, but toads as adults are mostly terrestrial. Toads have rough skin that is nearly waterproof, and most can also be distinguished by a thin yellow stripe down their backs along the midline. Their hind legs are shorter than those of frogs, and when they are on land they commonly walk rather than hop. They hop mainly when they are excited. In addition, the web between their toes is not nearly as well developed as it is in frogs.

The Western Toad, *Anaxyrus (= Bufo) boreas*, occurs statewide in all habitats except the southern deserts. It is the common toad found in residential gardens. South of Kaiser

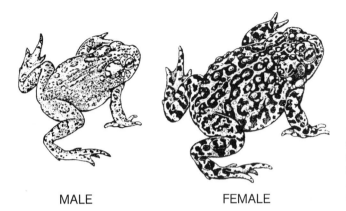

MALE   FEMALE

FIGURE 11.31 Yosemite Toad, *Anaxyrus canorus*. Note the different size and appearance of the two sexes (from Stebbins, R.C. 1951. *Amphibians of Western North America*. Berkeley: University of California Press).

Pass in Fresno County, it occurs all the way up into the alpine zone. Similar to frogs, they breed in water. During the dry season, in order to estivate, they may bury themselves in loose soil or enter rodent burrows.

In desert areas, there are several species of toads with localized distribution. These have behavioral habits very much like those of the spadefoot toads, *Spea* (= *Scaphiopus*) spp., that were discussed in Chapter 9. The largest "desert" toad is known as the Sonoran Desert (Colorado River) Toad, *Incilius (= B.) alvarius*. It occurs on floodplains along the lower Colorado River and may be up to 6 in (15 cm) in body length. It is renowned for its toxic secretions. Native animals know instinctively to avoid these creatures, but occasionally a domestic dog will bite one. The result is copious frothing of the mouth and temporary paralysis. It could seriously harm a small dog. The Colorado River Toad is becoming scarce, and may be extirpated in California. It is apparently a victim of the dams that no longer allow periodic flooding.

A truly remarkable toad of the high Sierra Nevada is the Yosemite Toad, *A. (= B.) canorus* (figure 11.31). These toads inhabit high-country meadows from about 6500 to 11,500 ft (2000–3500 m). They breed in temporary ponds fed by snowmelt and in the warm shallows of permanent lakes. One of the places where they have been studied is Dana Meadows, near Tioga Pass. This is the same area where populations of Belding's Ground Squirrels were studied.

Unlike other frogs and toads, Yosemite Toads are diurnal. This change in behavior is necessitated by the cold alpine nights. Yosemite Toads are the only toads with distinct sexual dimorphism. The males and females differ in size and color. Adult females are about 3 in (75 mm) in body length, and they are gray or brown with distinct dark blotches on the back. Males are slightly smaller, and they are olive green with only faintly visible blotches. The different colors may be associated with protective coloration, or they may be related to reproduction.

Perhaps the different coloration of the two sexes is related to visual mating cues, a process associated with daytime mating. Females spend daylight hours in meadows or foraging at the meadow's edge under willows, where the blotched pattern may help conceal them. Males spend much of their time floating in shallow water or sitting at the water's edge attempting to entice females to mate. Similar to other true toads, males utter a musical trill. Females attracted by the call move to ponds, and mating proceeds in a manner similar to that described for frogs. Unlike other frogs and toads, there are silent males, too. The silent males are "sneak" maters. They sit or float near the calling males, and when receptive females approach the calling males, the silent males attempt to intercept them.

Yosemite Toads hibernate in rodent burrows or in cracks under rocks and logs, emerging when the snow melts from their winter refuges. They are most vulnerable when they cross snowfields. At this time, they are quite conspicuous to predators such as California Gulls and Clark's Nutcrackers. The noxious secretions that discourage most predators seem not to bother these birds. Nutcrackers have been observed to wipe the toads repeatedly on the ground before swallowing them. Presumably, this removes some of the secretion.

Walking on snow seems as if it would impose a hardship for an ectothermic creature such as a toad. Yosemite Toads cope with the problem in a way that might be expected: They walk on their toes with their bellies well elevated. Once off the snow, they bask in the sun to restore body heat.

After mating, eggs develop quickly in warm, shallow water and hatch in about 10 d. After they hatch, tadpoles move and feed in the ponds. Similar to the tadpoles of the Sierra Nevada Yellow-legged Frog, they gather in the shallows during daylight hours and move to deeper water at night. It takes seven to eight weeks for metamorphosis to occur. After that, there may only be a few weeks of vigorous feeding before the adults must enter hibernation. Adults eat a variety of arthropods, including things such as millipedes, spiders, and bees that other predators find distasteful.

During their larval stages, toads are preyed upon by birds and fishes. In temporary ponds where fishes are absent, major predators are diving beetles and dragonfly larvae. The major cause of death, however, is dehydration. Often the pond dries out before they have had enough time to metamorphose.

In order for toads to avoid death by desiccation, early breeding is a necessity. On the other hand, if toads emerge and breed too early, an unexpected spring snowstorm can cause high mortality. Once the males break hibernation, they do not have the energy stores to reenter the dormant state. One feature of a toad population that serves as insurance against mass mortality is that all of them do not emerge from hibernation at the same time. During a year when spring arrives early and the weather remains favorable, toads that emerge early have an advantage. A late spring storm favors those that emerge later. Variation in the biological clock that regulates hibernation is a necessity in dealing with such vagaries of climate. Similar to the Yellow-legged Frogs, Yosemite Toad populations are on the decline. In fact, amphibian populations in general seem to have declined since the 1960s. Although these declines are not easy to explain, they will bear watching by the scientific community. Like canaries in a mine, loss of amphibians may be a signal that serious widespread environmental degradation is occurring. Threatened by cattle grazing and predation, in 2014 the Yosemite Toad was officially classified as threatened by the US Fish and Wildlife Service.

Chorus Frogs or Treefrogs (Hylidae) are also common in aquatic habitats. These are small frogs, about 2 in (5 cm) in total length. They are easily recognized by their long legs and adhesive toe pads. These frogs are agile jumpers. A person driving along a canyon road on a rainy night might be surprised to see these little frogs crossing the road in giant leaps that may carry them over 3 ft (1 m) into the air.

The most common treefrogs have been split into four species. The Northern Pacific Treefrog bears the original name *Pseudacris (= Hyla) regilla*, and is found only in the Pacific Northwest. The species formerly known as the Pacific Treefrog or Chorus Frog, *P. (= H.) regilla*, is now known as the Baja California Treefrog, *P. hypochondriaca*, and the Sierra Nevada form is now the Sierran Treefrog, *P. sierra* (figure 11.32) They all have a black eye stripe and occur in a variety of colors, but most often they are greenish. In a few minutes, they are able to change color to match their surroundings. These are the most commonly heard frogs in California. Their call may be described as a loud two-part sound, "kreck-ek." It is the call of these frogs that has been joked about. Many people think the call sounds like "rib-bit."

Treefrogs occur in every freshwater habitat from the coast to over 11,000 ft (3350 m) in the mountains. In spite of its name, this is chiefly a

FIGURE 11.32 Sierran Treefrog, *Pseudacris sierra.*

ground dweller. Its usual habitat is in low vegetation along the borders of slow-moving water such as marshes, ponds, and roadside ditches. In alpine meadows, they may be found in habitats similar to the Yosemite Toad. In the Sierra Nevada, they are even known to inhabit talus on the edges of alpine lakes. Treefrogs breed in the spring, although breeding choruses in southern California have been heard as early as December. Mating behavior is similar to that of other frogs. These treefrogs usually call from a floating position near the shore, in shallow water.

The other treefrog in southern California is known as the California Treefrog or Chorus Frog, *P. (= H.) cadaverina.* A southern California species, it occurs in canyons of the Transverse and Peninsular Ranges from the foothills to about 3000 ft (900 m) elevation. It is less common on the desert sides of the mountains, but it does occur in palm oases in the Colorado Desert. Adults of this species are usually light gray and exhibit marvelous protective coloration. They tend to rest in depressions on large rocks, a situation that gives these boulders a warty appearance. Close inspection is required to discern that the lumps on the rocks are actually frogs.

At some localities in southern California, the two treefrogs occur together. Simultaneous breeding choruses have been observed in the western San Gabriel Mountains and the Santa Susana Mountains. At these localities, the Pacific Treefrog calls from a floating position in the water, and the California Treefrog calls from the water's edge. The Baja California Treefrog utters a short, duck-like quack. It seldom produces a two-part sound similar to that of the Pacific Treefrog.

Formerly, the Baja California Treefrog was considered to be the same species as the Canyon Treefrog of Arizona and New Mexico. An interesting investigation into speciation revealed that the California populations were, in fact, a different species. The two species are nearly identical in appearance, and even experts have difficulty telling them apart. The two populations, however, do not overlap. Most of the Colorado Desert lies between ranges of the two species. In nature, it would not be possible for a Baja California Treefrog to meet a Canyon Treefrog; therefore, there is no opportunity for them to breed with each other. If they could interbreed and produce fertile offspring, we would be obligated to call them one species even though their ranges do not overlap. On the other hand, if they are identical in appearance and their ranges do not overlap, what is there to make us believe that they are not the same species with a divided range? In listening to the calls of the two species, experts noted that the Canyon Treefrog makes a 1 to 3 s whirring noise on one note, rather than the short burst of sound uttered by the California species. As might be expected, recordings of the calls played for members of the other species fail to attract attention. If their mating calls are different, they can't attract a mate, and they won't reproduce. But suppose they did mate. Could they produce fertile offspring? Artificial insemination studies in the laboratory solved the question. Eggs fertilized by sperm from the opposite species failed to develop. The two frogs of identical appearance (*cryptic species*) are therefore two separate species.

One other amphibian should be mentioned. This is the Coastal Tailed Frog, *Ascaphus truei*, sometimes incorrectly known as the Bell Toad. Their range and habitat, in rapidly flowing coastal streams from Washington south to Mendocino County, are similar to those of the Olympic Salamander. They are very primitive frogs. What appears to be a tail is actually a copulatory organ that the male uses to transfer sperm to the cloaca of the female. These are unique frogs because fertilization is internal. After birth, larvae cling to rocks in fast-moving water with a suction cup sort of mouth. By this means, they are even able to maintain position in waterfalls. This is a marvelous mechanism for larvae to develop in an environment free of predators.

Reptiles

Few reptiles are characteristically riparian in distribution. In California, there are two general kinds: Gartersnakes and turtles. Gartersnakes, (*Thamnophis* spp.) (= *Nerodia* spp.), get their name from their resemblance to a colorful garter. Most of them are of moderate size, although large ones reach about 4 ft (130 cm) in total length. They usually have a yellow stripe down the back and two lateral stripes. Their heads are slightly wider than the body, which leads some people to confuse them with rattlesnakes. Because they swim well, some uninformed persons call them water moccasins. Water moccasins do not occur west of Texas.

Gartersnakes usually occur around or in water. They feed on all sorts of animals, including fishes, salamanders, frogs, toads, small mammals, and birds. Sometimes they remain motionless underwater, waiting for a fish or tadpole to come close enough to be nabbed. Very few animals prey on Gartersnakes. When they are molested, they curl around the molester and defecate. Mixed with the excrement is material secreted from musk glands. The combination produces an odorous excretion designed to convince the predator that this item is too disgusting to eat. It is likely that the bright colors are part of a warning system to advertise that fact.

Classification of the different Gartersnakes in California has been a complex problem. There are five common species, all of which overlap their distribution in various parts of the state. In the Klamath Mountains, the Northwestern Gartersnake, *T. ordinoides*, reaches its southernmost distribution west of the Cascades in California. These snakes have two basic color patterns, striped or solid colored. Similar to the different patterns in California Kingsnakes, the striped forms are highly motile and the solid-colored forms tend to be sedentary. Gartersnakes are absent from the desert regions except for one species. The Checkered Gartersnake, *T. marcianus*, has moved into the Colorado Desert from the east by following irrigation ditches and canals.

It would be time-consuming to describe each of the four common species. One that lacks the central stripe down its back is the Sierra Gartersnake, *T. couchii*. This snake ranges throughout the state except in the deserts. It is the most common one in southern California. Above about 8000 ft (2400 m), particularly in the Sierra Nevada, the only Garter-

FIGURE 11.33 Northern Western Pond Turtle, *Actinemys marmorata*, to right and nonnative Red-eared Slider, *Trachemys scripta elegans*, sharing a basking perch.

snake is the Western Terrestrial Gartersnake, *T. elegans*. In spite of its name, it occurs in and around water. The most colorful of the group is the Common Gartersnakes, *T. sirtalis*. These snakes usually have red on the sides and may have blue-green bellies. They are most common in the Coast Ranges, although they range as far south as San Diego County. The Two-striped Gartersnakes, *T. hammondii*, also has a coastal distribution, but it has no central stripe, and many have no stripes at all.

There is only one native species of turtle in California. The Northern Western Pond Turtle, *Actinemys (= Clemmy) marmorata* (figure 11.33), occurs in streams, marshes, and ponds of the Coast Ranges, Transverse Ranges, Peninsular Ranges, and the Great Central Valley. In the desert area, they only occur in the Mojave River. These are sedentary animals, spending lots of time basking. They feed on aquatic plants, insects, and carrion. Distribution of this turtle in southern California, in association with destruction of riparian habitats, is becoming seriously diminished. It is listed as a species of special concern by the state and federal governments.

Other aquatic turtles in California have been introduced. In some cases, the introductions have occurred because people have released unwanted pets. It is sometimes possible to find a Snapping Turtle, *Chelydra serpentina*, a large-headed turtle with a sharp ridge along its back whose powerful jaws are capable of inflicting a serious wound. Apparently, it has not yet established a breeding population in California. The Common Slider or Red-eared Slider, *Trachemys scripta elegans*, resembles the native pond turtle, but the slider has a red line on each side of its face. Breeding populations of this turtle have become established at a variety of localities in California, and competition with the native Northern Western Pond Turtle may be the cause of the latter's decline. The Painted Turtle, *Chrysemys picta*, sometimes available in pet stores, also resembles the Northern Western Pond Turtle. Although this species is sometimes found in the wild in California, breeding populations apparently have not become established.

The other turtle is the Spiny Softshell, *Apalone spinifera*, a species introduced into the lower Colorado River. This is a pancake-shaped turtle up to a foot (30 cm) in length with a long protrusible neck. One of these turtles may be seen floating along with the current, with only its head above water. They have a long, narrow, flat-tipped nose with two large nostrils at the tip. In quiet backwaters, they will bury themselves in the mud. From time to time, in order to breathe, they extend their long necks to break the surface. They feed on all sorts of slow-moving aquatic animals.

EPILOGUE

WHAT DOES THE FUTURE HOLD?

California is a marvelous place, with a greater range of landforms, a greater variety of habitats, and more kinds of plants and animals than any area of equivalent size in North America. Similarly, there are more endemic plants and animals here than anywhere in the United States. When Europeans first arrived in California, it was a scene of unparalleled, almost unimaginable, natural richness. Unfortunately, this richness coupled with the region's pleasant Mediterranean climate soon attracted an unprecedented number of people, and humanity has encroached upon the landscape to such a degree that California is the state with the greatest number of endangered species.

Nevertheless, California still has the greatest amount of open space and roadless terrain in the lower 48 states, and an increasingly devoted group of citizens attempting to protect it. These people have won great victories, but many Californians still fail to appreciate the intricacy and beauty of the natural order of things, and that human activities are capable of spoiling it all.

The most serious threat to California's natural ecosystems is continued habitat fragmentation, urbanization, and development, a consequence of continued population growth. California's Mediterranean climate along with abundant economic opportunities will continue to attract new residents and businesses. As long as growth, as opposed to sustainability, drives our economy, we will be faced with increased pressure on the environment. Our only protection from these factors that threaten natural ecosystems is our system of habitat preserves, parks, and protected areas. We are still fortunate that

over half of California is public land, and at the present time nearly 47%, representing 49 million acres (200,000 km²), is in a state of total protection; 14.36% is designated wilderness. Our most serious problem with these public lands is the island effect produced as urbanization encroaches. Islands support fewer species and less diversity. Visitation in our national and state parks remains high, but it is important that our populace is educated to the importance of wild lands, and appreciation of these lands is a major function of visitation. There is still a major movement among certain pressure groups to transfer portions of federal and state property to local city or county control in order to promote economic activity such as electrical energy generation, grazing, mining, or logging. Citizens who believe in the importance of undisturbed landscapes will have to be diligent in resisting these special interest groups.

Aside from direct habitat destruction and degradation, the most serious threats to ecosystems is climate change, including drought and increased temperatures. Plants may be directly killed by drought, or in a weakened condition they may be invaded by pathogens that ultimately cause them to die. Our forest trees can be weakened by drought or ozone damage, both of which allow bark beetle infestations and widespread stands of dead or dying trees. In many parts of the state, we are already seeing large plants such as conifers and Joshua Trees dying at the lower reaches of distribution, with an increase of germination at the upper portions of their range. This change in distribution is accompanied by gradual northward dispersal to cooler, moister climate. If the amount of summer fog is diminished, will this force a northward retreat of our Coastal Redwood forests? Selective dying of sensitive or less tolerant plants also causes changes in species composition of natural communities. Loss of diversity clearly upsets stability.

Climate change also influences animal distribution and species composition in a manner similar to plants. Animals, however, are generally mobile and can move as the climate changes. Animals are important to ecosystems for all the reasons described in this book. Among other things, animals are involved in dispersal and pollination of plants. Climate has been changing for millennia, and changes in plant and animal distribution over time are well documented. The difference today is that the rate of change is unprecedented. Questions arise as to whether changes in animal and plant distribution can keep up with the changes in climate. If they can't keep up, we may witness significant extinction events.

Climate change and global warming include a whole suite of environmental changes:

- Increased sea level and coastal flooding
- Increased ocean temperature and acidity
- Change in precipitation regimes (wet areas get wetter, dry areas get drier)
- Increased drought and fires
- Change in seasons (early spring, longer summer)
- Reduced mountain snowpack and stream flow
- Extinction of sensitive species
- Northward and elevational shifting of biotic zones

Combustion of fossil fuels provides the power for transportation, generation of electricity, and heat for industry and residential use. Greenhouse gases such as methane (CH_4), carbon dioxide (CO_2), carbon monoxide (CO), and oxides of nitrogen (NOx) are all associated with combustion of fossil fuels. Any attempt to reduce production of those gases will involve a number of lifestyle changes for California citizens, but if we are going to heed the warnings of scientists they are changes we Californians must face.

Associated with human activity and climate change are problems such as air pollution and water pollution. Activated by solar radiation, pollutants such as hydrocarbons (CnH_2n) and oxides of nitrogen are converted to ozone (O_3), a component of polluted air that causes chlorotic decline, particularly in conifers, which is exacerbated by drought and encourages invasion by bark beetles. Contributing to the problem is that oxides of nitrogen combine with water in the air, causing acid rain and fog. Other nonnative pests are attacking trees in our Oak Woodlands and Riparian communities. Hopefully some of the trees will possess a natural resistance to invasive

pests and repopulate the communities through natural selection.

The invasion of natural communities by invasive, mostly nonnative, species is a serious problem in some parts of the state. Our native grasslands have been taken over by a suite of nonnative, mostly Mediterranean, weeds that thrive in our climate. Invasive annual plants such as Black Mustard, Star Thistle, and Artichoke Thistle, as well as nonnative grasses have completely replaced native species in many locations. Various grasses and African Mustard have invaded many of our desert habitats, filling in bare spaces with flammable annual vegetation. Already weakened by drought, the dead thatch between plants increases the probability of a serious fire spreading through communities such as Creosote Bush Scrub that normally are not susceptible to fire. Unlike Chaparral, Coastal Scrub, and Mixed Coniferous Forest in which many species are fire adapted, most desert communities take hundreds of years to recover from a fire. Water-demanding shrubs such as Tamarisk have invaded our desert wetlands, causing springs to dry up in many locations. Changes in precipitation regimes could also have a serious effect on winter annual wildflowers in our deserts, particularly the Colorado Desert where winter precipitation is decreasing but summer precipitation is increasing.

Scrub communities such as Chaparral are also plagued by air pollution and invasive species. Ozone and oxides of nitrogen influence Chaparral plants through dry nitrogen fallout, acid rain, and fog. Invasive shrubs such as Fennel, Wild Tobacco, Pampas Grass, French Broom, and Spanish Broom are replacing native plants in many scrub habitats.

Also associated with climate change is increased fire frequency and intensity. According to meteorologists, by 2015 the fire season in the United States was 78 d longer than in the 1970s. While these trends are not always associated with heat and drought, the fact remains that communities that are normally fire adapted suffer when fires occur with greater frequency. These fires are exacerbated by wind, and periods of wind, formerly associated with autumn Santa Ana conditions, now occur during the summer as well. In Chaparral communities, fire frequency and severity can be related to encroachment of human habitation, not only by building more structures, but also by construction of roads and increased automobile traffic. In forested communities, policies of fire suppression on public lands have caused an increase of small trees and shrubs that carry fire into the crowns of mature trees, which normally would have remained close to the ground. Periodic low-intensity ground fires clear out such vegetation without burning the mature trees.

An increase in sea level is often associated with problems in low-lying coastal communities, but what will be the effect on estuaries? Salt Marsh plants are restricted by certain tolerances to salinity. As seawater encroaches, will Salt Marsh and Freshwater Marsh plants be able to move farther inland faster than rising sea level?

Distribution and availability of water may become one of the most vexing problems in the future. Exacerbated by drought and lack of snowpack in the mountains, California's multimillion-dollar agricultural industry is sure to suffer. Already agriculture uses around 80% of California's water, yet it accounts for less than 3% of California's economy. At the same time, this represents about a quarter of the nation's food in the form of vegetables, nuts, and beef. California has a shortage of water, not reservoirs or dams. Increasing the number of dams will not help if the amount of precipitation is reduced. Recent droughts already have caused serious lowering of water levels in reservoirs. Pumping of groundwater is already an issue in the form of drying wells and ground subsidence. During the drought in 2014 and 2015, so much water was diverted from the delta for human use that the Chinook Salmon spawning runs failed in the Sacramento River.

Can we depend on technology to rescue us from prolonged drought? Unfortunately, many decisions are based on political forces that steer the direction. Long-distance transport of water is an energy-intensive operation. Our aqueduct system already is the largest user of electricity in the state, and unless we can shift a very large proportion of our electricity production away from fossil fuels, we will be faced with increased air pollution. Diverting our rivers into the aqueduct system may temporarily satisfy our agricultural needs and water for urban society, but it does not help fisheries and water quality downstream. Some areas have constructed water desalination systems, another energy-demanding process. Desalination of seawater also has an added environmental cost in disposal of brine produced in the process. On the other hand, a similar process can produce a potable product from urban wastewater without so many disposal problems.

California is already attempting to move away from fossil fuel for electrical energy production, but it doesn't seem to want to abandon the distribution of electricity through the major electrical grid operated by large electric companies. Similarly, electric cars presently rely on fossil fuels to generate the electricity, which does nothing to reduce global warming. Huge wind farms are not an attractive component of the landscape and windmills take a toll on birdlife and bats. Large solar arrays on desert land unfortunately disturb huge swaths of desert habitat, removing native plants and displacing animal inhabitants. While rooftop solar installation on buildings is an option, it is not favored by the large electric companies who still have to maintain the distribution grid. Installation of solar arrays on already disturbed private land could be part of the solution, and somehow encouraging commuters to abandon automobiles, and switch to mass transit also should be pursued.

So, California will remain one of the most attractive places on earth. We are blessed with an abundance of public land that should remain in a state of preservation for a long time. Our system of National, State, and County Parks are treasures and among the most popular in the United States. Our biggest problem is that our increasing human population is putting a strain on our environment and natural resources. While slowing population growth is the ultimate remedy, we hope that an educated populace will see the importance of protecting the natural landscape that makes California such a special place.

Humans are merely one of the living organisms in the system, but unlike other organisms, they are capable of thinking and analyzing. Those who choose not to think about natural treasures have already altered the face of California forever. This book is dedicated to an attitude that is essential if future generations are to enjoy and appreciate California's natural beauty and diversity. It is also dedicated to the thoughtful people who are willing to work together to preserve the natural order of things and restore the land from damage that thoughtless development has brought.

NOTEWORTHY PUBLICATIONS

GENERAL REFERENCES

Airola, D. A., and T. C. Messick. 1987. *Sliding toward Extinction: The State of California's Natural Heritage*. Prepared at the request of the California Senate Committee on Natural Resources and Wildlife. Commissioned by the California Nature Conservancy. Prepared by Jones and Stokes Associates, Sacramento.

Anderson, B. R., 1975. *Weather in the West*. Palo Alto, CA: American West.

Bailey, R. G. 1983. Delineation of Ecosystem Regions. *Environmental Management* 7: 365–373.

Bailey, R. G. 1995. *Description of the Ecoregions of the United States*. 2nd ed. Miscellaneous Publication 1391. Washington, DC: USDA Forest Service.

Bakker, E., 1984. *An Island Called California*. Berkeley: University of California Press. First published 1971 by University of California Press.

Barbour, M. G., J. M. Evans, T. Keeler-Wolf, and J. Sawyer. 2016. *California's Botanical Landscapes: A Pictorial View of the State's Vegetation*. Sacramento, CA: California Native Plant Society.

Barbour, M. G., B. Pavlik, F. Drysdale, and S. Lindstrom. 1993. *California's Changing Landscapes*. Sacramento: California Native Plant Society.

Betancourt, J. L., T. R. Van Devender, and P. S. Martin. 1990. *Packrat Middens: The Last 40,000 Years of Biotic Change*. Tucson: University of Arizona Press.

Brewer, W. H. 1997. *Up and Down California*. 4th ed. Berkeley: University of California Press.

Buckman, H. O., and N. C. Brady. 2015. *The Nature and Properties of Soils*. New York: Macmillan.

Colinvaux, P. 1978. *Why Big Fierce Animals Are Rare*. Princeton, NJ: Princeton University Press.

Cox, C. B., and P. D. Moore. 2010. *Biogeography: An Ecological and Evolutionary Approach*. 8th ed. New York: John Wiley and Sons.

Cunningham, L. 2010. *A State of Change: Forgotten Landscapes of California*. Berkeley, CA: Heyday.

Dasmann, R. F. 1965. *The Destruction of California*. New York: Macmillan.

Department of Fish and Game. 2003. *Atlas of the Biodiversity of California*. Sacramento, CA: Resources Agency, California Department of Fish and Game.

Miller, G. T. 2014. *Living in the Environment*. Belmont, CA: Wadsworth.

Mooney, H., and E. Zavelta, eds. 2016. *Ecosystems of California*. Berkeley: University of California Press.

Ricklefs, R. E., and G. L. Miller. 2000. *Ecology*. New York: W. H. Freeman and Co.

Smith, R. L., and T. L. Smith. 2007. *Ecology and Field Biology*. San Francisco, CA: Benjamin Cummins.

Vessel, M. F., and H. H. Wong. 1987. *Natural History of Vacant Lots*. California Natural History Guides No. 50. Berkeley: University of California Press.

Walker, R. A., and S. K. Lodha. 2013. *The Atlas of California: Mapping the Challenges of a New Era*. Berkeley: University of California Press.

Whitaker, R. H. 1970. *Communities and Ecosystems*. New York: Macmillan.

GEOLOGY

Alt, D. D., and D. W. Hyndman. 1975. *Roadside Geology of Northern California*. Missoula, MT: Mountain Press.

Brooks, R. R. 1987. *Serpentine and Its Vegetation: A Multidisciplinary Approach*. Portland, OR: Dioscorides Press.

Cooper, J. D., R. H. Miller, and J. Patterson. 1986. *A Trip through Time: Principles of Historical Geology*. Columbus, OH: Charles E. Merrill.

Ernst, W. G. 1981. *The Geotectonic Development of California*. Englewood Cliffs, NJ: Prentice-Hall.

Guyton, B. 1998. *Glaciers of California*. Berkeley: University of California Press.

Harden, D. H. 1998. *California Geology*. Upper Saddle River, NJ: Prentice Hall.

Hill, M. 1984. *California Landscape: Origin and Evolution*. California Natural History Guides No. 48. Berkeley: University of California Press.

Kruckeberg, A. R. 1984. *California Serpentines: Flora, Vegetation, Geology, Soils, and management Problems*. Berkeley: University of California Press.

Meldahl, K. H. 2015. *Surf, Sand, and Stone: How Waves, Earthquakes, and Other Forces Shape the Southern California Coast*. Berkeley: University of California Press.

Norris, R. M., and R. W. Webb. 1990. *Geology of California*. New York: John Wiley & Sons.

Oakeshott, G. B. 1978. *California's Changing Landscapes*. New York: McGraw-Hill.

Pipkin, B. W., and D. D. Trent. 2014. *Geology and the Environment*. Pacific Grove, CA: Brooks/Cole.

PLANTS

Abrams, L., and R. S. Ferriso. 1923–1960. *Illustrated Flora of the Pacific States*. 4 vols. Stanford, CA: Stanford University Press.

Axelrod, D. I. 1976. *History of the Coniferous Forests, California and Nevada*. University of California Publications in Botany No. 70. Berkeley: University of California Press.

Baldwin, B. G., D. H. Goldman, D. J. Keil, R. Patterson, T. J. Rosatti, and D. H. Wilken, eds. 2012. *The Jepson Manual: Vascular Plants of California*. 2nd ed. Berkeley: University of California Press.

Barbour, M. G., J. H. Burk, and W. D. Pitts. 1987. *Terrestrial Plant Ecology*. Menlo Park, CA: Benjamin Cummings.

Barbour, M. G., J. M. Evans, T. Keeler-Wolf, and J. O. Sawyer. 2016. *California's Botanical Landscapes: A Pictorial View of the State's Vegetation*. Sacramento: California Native Plant Society.

Barbour, M. G., T. Keeler-Wolf, and A. A. Schoenherr. 2007. *Terrestrial Vegetation of California*, 3rd ed. Berkeley: University of California Press.

Benson, L. 1982. *The Cacti of the United States and Canada*. Stanford, CA: Stanford University Press.

Brown, D. E. 1982. Biotic Communities of the American Southwest-United States and Mexico. Special Issue, *Desert Plants* 4:52–57.

Chabot, B. F., and H. A. Mooney. 1985. *Physiological Ecology of North American Plant Communities*. New York: Chapman & Hall.

Cheatham, N. H., and J. R. Haller. 1975. *An Annotated List of California Habitat Types*. Berkeley: University of California Press. Unpublished report.

Clarke, C. B. 1977. *Edible and Useful Plants of California*. California Natural History Guides No. 41. Berkeley: University of California Press.

Crampton, B. 1974. *Grasses in California*. California Natural History Guides No. 33. Berkeley: University of California Press

Dallman, P. R. 1998. *Plant Life in the World's Mediterranean Climates*. California Native Plant Society. Berkeley: University of California Press.

De Nevers, G., D. S. Edeliman, and A. Merenlender. 2013. *The California Naturalist Handbook*. Berkeley: University of California Press.

Elias, T. S. 1987. *Conservation and Management of Rare and Endangered Plants*. Sacramento: California Native Plant Society.

Faber, P. M., ed. 2005. *California's Wild Gardens*. Berkeley: University of California Press.

Griffin, J. R., and W. B. Critchfield. 1976. *The Distribution of Forest Trees in California*. USDA, Forest Service Research Paper PSW-82.

Grillos, S. J. 1966. *Ferns and Fern Allies of California*. California Natural History Guides No. 16. Berkeley: University of California Press.

Holland, V. L., and D. J. Keil. 1995. *California Vegetation*. Dubuque, IA: Kendall Hunt Publishing Co.

Ingram, S. 2008. *Cacti, Agaves, and Yuccas of California and Nevada*. Los Olivos, CA: Cachuma Press.

Johnston, V. R. 1994. *California Forests and Woodlands: A Natural History*. Berkeley: University of California Press.

Keeley, J. E., ed. 1993. *Interface between Ecology and Land Development in California*. Los Angeles: Southern California Academy of Sciences.

Keeley, J. E., M. Baer-Keeley, and C. J. Fotheringham, eds. 2000. *2nd Interface Between Ecology and Land Development in California*. Open-File Report 00-62. Sacramento, CA: US Geological Survey.

Keeley, J. E., W. J. Bond, R. A. Bradstock, J. G. Pausas, and P. W. Rundel, eds. 2012. *Fire in Mediterranean Ecosystems*. Cambridge: Cambridge University Press.

Lanner, R. M. 1999. *Conifers of California*. Los Olivos, CA: Cachuma Press.

McMinn, H. E. 1964. *An Illustrated Manual of California Shrubs*. Berkeley: University of California Press.

McMinn, H. E., E. Maino, and H. W. Shepherd. 1963. *Pacific Coast Trees*. Berkeley: University of California Press.

Minnich, R. A. 2008. *California's Fading Wildflowers: Lost Legacy and Biological Invasions*. Berkeley: University of California Press.

Munz, P. A. 1961. *California Spring Wildflowers*. Berkeley: University of California Press. First published 1959 by University of California Press.

Munz, P. A., and D. D. Keck. 1968. *A California Flora and Supplement*. Berkeley: University of California Press.

Noble, P. S., ed. 2002. *Cacti: Biology and Uses*. Berkeley: University of California Press.

Orrnduff, R., P. M. Faber, and T. Keeler-Wolf. 2003. *Introduction to California Plant Life*. California Natural History Guides No. 35. Berkeley: University of California Press.

Orr, R. T., and D. B. Orr. 1979. *Mushrooms of Western North America*. California Natural History Guides No. 42. Berkeley: University of California Press.

Pavlik, B. M., P. C. Muick, S. Johnson, and M. Popper. 1991. *Oaks of California*. Los Olivos, CA: Cachuma Press.

Perry, B. 1992. *Landscape Plants for Western Regions: An Illustrated Guide to Plants for Water Conservation*. Claremont, CA: Land Design Publishing.

Plumb, T. R. 1980. *Proceedings of the Symposium on the Ecology, Management, and Utilization of California Oaks*. General Technical Report PSW-44. Albany, CA: Pacific Southwest Forest and Range Experiment Station, USDA Forest Service.

Raven, P. H., and D. I. Axelrod. 1978. *Origins and Relationships of the California Flora*. University of California Publications in Botany No. 72. Berkeley: University of California Press.

Richardson, D. M., ed. 1998. *Ecology and Biogeography of Pinus*. Cambridge: Cambridge University Press.

Sawyer, J. O. T. Keeler-Wolf, and J. M. Evans. 2009. *A Manual of California Vegetation*. Sacramento: California Native Plant Society.

Smith, Jr., J. P., and R. York. 1988. *Inventory of Rare and Endangered Vascular Plants of California*. Special Publication No. 1. Sacramento: California Native Plant Society.

Spellenberg, R. 1979. *The Audubon Society Field Guide of North America Wildflowers Western Region*. New York: Alfred A. Knopf.

Standiford, R. B. 1991. *Proceedings of the Symposium on Oak Woodlands and Hardwood Rangeland Management*. General Technical Report PSW-126. Albany, CA: Pacific Southwest Forest and Range Experiment Station, USDA Forest Service.

Stuart, J. D., and J. O. Sawyer. 2001. *Trees and Shrubs of California*. California natural History Guide No. 62. Berkeley: University of California Press.

Warner, R. E., and K. M. Hendrix, eds. 1984. *California Riparian Systems: Ecology, Conservation, and Productive Management*. Berkeley: University of California Press.

ANIMALS

Adams, R. J. 2014. *Field Guide to the Spiders of California and the Pacific Coast States*. Berkeley: University of California Press.

Banks, C. R., R. W. McDiarmid, and A. L. Gardner. 1987. *Checklist of Vertebrates of the United States, the U. S. Territories, and Canada*. Resource Publication 166. Washington, DC: US Department of the Interior, US Fish and Wildlife Service.

Behler, J. L. 1979. *The Audubon Society Field Guide to North American Reptiles and Amphibians*. New York: Alfred A. Knopf.

Borrer, D. J., and R. E. White. 1970. *Field Guide to the Insects of America North of Mexico*. Boston, MA: Houghton Mifflin.

Dasmann, R. F. 1964. *Wildlife Biology*. New York: John Wiley & Sons.

Dunn, J. L., and J. Alderfer, eds. 2006. *Field Guide to the Birds of North America*. 5th ed. Washington, DC: National Geographic Society.

Eder, T. 2005. *Mammals of California*. Edmonton, AB: Lone Pine Publishing International.

Flores, D. 2016. *Coyote America: A Natural and supernatural History*. NY: Basic Books.

Garth, J. S., and J. W. Tilden. 1986. *California Butterflies*. California Natural History Guides No. 51. Berkeley: University of California Press.

Hall, E. R. 1981. *Mammals of North America*. New York: John Wiley & Sons.

Hogue, C. L. 1993. *Insects of the Los Angeles Basin*. Los Angeles: Natural History Museum of Los Angeles County.

Ingles, L. G. 1965. *Mammals of the Pacific States*. Stanford, CA: Stanford University Press.

Jameson, E. W., Jr., and H. J. Peeters. 1988. *California Mammals*. California Natural History Guides No. 52. Berkeley: University of California Press.

Jennings, M. R. 1987. *Annotated Check List of the Amphibians and Reptiles of California*. Special Publication No. 3. Van Nuys, CA: Southwestern Herpetologists Society.

Klauber, L. M. 1982. *Rattlesnakes*. Abridged ed. Berkeley: University of California Press.

Merritt, R. W., and K. W. Cummings. 1978. *An Introduction to the Aquatic Insects of North America*. Dubuque, IA: Kendall/Hunt.

Peterson, R. T. 1990. *A Field Guide to Western Birds*. Boston, MA: Houghton Mifflin.

Powell, J. A., and C. L. Hogue. 1979. *California Insects*. California Natural History Guides No. 44. Berkeley: University of California Press.

Robins, C. R., R. M. Bailey, C. E. Bond, J. R. Brooker, E. A. Lachner, et al. 1980. *A List of Common and Scientific Names of Fishes from the United States and Canada*. Special Publication No. 12. Washington, DC: American Fisheries Society.

Sibley, D. A. 2003. *Field Guide to Birds of Western North America*. New York: Alfred A. Knopf.

Small, A. 1975. *The Birds of California*. New York: Collier Books.

Small, A. 1994. *California Birds: Their Status and Distribution*. Vista, CA: Ibis Publishing Company.

Smith, A. C. 1961. *Western Butterflies*. Menlo Park, CA: Lane.

Stebbins, R. C. 1954. *Amphibians and Reptiles of Western North America*. New York: McGraw-Hill.

Stebbins, R. C. 2003. *Field Guide to Western Reptiles and Amphibians*. Boston, MA: Houghton Mifflin.

Stebbins, R. C., and S. M. McGinnis. 2012. *Field Guide to the Amphibians and Reptiles of California*. Berkeley: University of California Press.

Steinhart, P. 1990. *California's Wild Heritage: Threatened and Endangered Animals in the Golden State*. San Francisco, CA: Sierra Club Books.

Tilden, J. W., and A. C. Smith. 1986. *Field Guide to Western Butterflies*. Boston, MA: Houghton Mifflin.

Udvardy, M. D. F. 1977. *The Audubon Society Field Guide to North American Birds (Western Region)*. New York: Alfred A. Knopf.

Whitaker, J. O. 1980. *The Audubon Society Field Guide to North American Mammals*. New York: Alfred A. Knopf.

SIERRA NEVADA

Arno, S. 1973. *Discovering Sierra Trees*. Yosemite, CA: Yosemite Natural History Association, Sequoia Natural History Association.

Aune, P. S. 1992. *Proceedings of the Symposium on Giant Sequoias: Their Place in the Ecosystem and Society*. Albany, CA: Pacific Southwest Research Station.

Barbour, M. G., N. H. Berg, T. G. F. Kittel, and M. E. Kunz. 1991. Snowpack and the Distribution of a Major Vegetation Ecotone in the Sierra Nevada of California. *Journal of Biogeography* 18:141–149.

Barbour, M. G., and R. A. Woodward. 1985. The Shasta Red Fir Forest of California. *Canadian Journal of Forestry* 15:570–576.

Beedy, E. C., and S. L. Granholm. 1985. *Discovering Sierra Birds*. Yosemite, CA: Yosemite Natural History Association, Sequoia Natural History Association.

Botti, S. J. 2001. *An Illustrated Flora of Yosemite National Park*. Yosemite Village, CA: Yosemite Conservancy.

Browning, P. 1986. *Place Names of the Sierra Nevada*. Berkeley, CA: Wilderness Press.

Gaines, D. 1988. *Birds of Yosemite and the East Slope*. Lee Vining, CA: Artemisia Press.

Graf, M. 1999. *Plants of the Tahoe Basin*. Sacramento: California Native Plant Society.

Grater, R. K. 1978. *Discovering Sierra Mammals*. Yosemite, CA: Yosemite Natural History Association, Sequoia Natural History Association.

Gruell. G. E. *Fire Sierra Nevada Forests*. Missoula, MT: Mountain Press Publishing Company.

Grinnel, J., and T. I. Storer. 1924. *Animal Life in the Yosemite*. Berkeley: University of California Press.

Guyton, B. 1998. *Glaciers of California*. Berkeley: University of California Press.

Haller, J. R., and N. J. Vivrette. 2011. Ponderosa Pine Revisited. *Aliso* 29:53–57.

Harvey, H. T., H. S. Shellhammer, and R. E. Stecker. 1980. *Giant Sequoia Ecology*. Washington, DC: US Department of the Interior, National Park Service.

Hill, M. 1975. *Geology of the Sierra Nevada*. California Natural History Guides No. 37. Berkeley: University of California Press.

Horn, E. L. 1976. *Wildflowers 3: The Sierra Nevada*. Beaverton, OR: Touchstone Press.

Huber, N. K. 1987. *Geologic Story of Yosemite National Park*. Geological Survey Bulletin 1595. Washington, DC: US Department of the Interior.

Huning, J. B. 1978. *Hot Dry Wet Cold and Windy: A Weather Primer for the National Parks of the Sierra Nevada*. Yosemite, CA: Yosemite Natural History Association, Sequoia Natural History Association.

Jenkins, O. P. 1948. *Geologic Guidebook along Highway 49: Sierran Gold Belt, the Mother Lode Country*. Bulletin 141. Sacramento: Division of Mines, California Department of Natural Resources.

Johnston, V. R. 1970. 1998. *Sierra Nevada: The Naturalist's Companion*. Rev. ed. Berkeley: University of California Press.

Laws, J. M. 2007. *Laws Guide to the Sierra Nevada*. San Francisco: California Academy of Sciences.

Longstreth, C. 2014. Managing Burned Landscapes in the Sierra Nevada: Back to the Future (Slowly). *Femontia* 42:7–13.

Moore, J. G. 2000. *Exploring the Highest Sierra*. Stanford, CA: Stanford University Press.

Morgenson, D. C. 1975. *Yosemite Wildflower Trails*. Yosemite, CA: Yosemite Natural History Association.

Muir, J. 1961. *The Mountains of California*. Garden City, NY: Doubleday.

Munz, P. A. 1963. *California Mountain Wildflowers*. Berkeley: University of California Press.

Niehaus, T. F. 1974. *Sierra Wildflowers*. California Natural History Guides No. 32. Berkeley: University of California Press.

Parker, A. J. 1994. Latitudinal Gradients of Coniferous Tree Species, Vegetation, and Climate in the Sierran-Cascade Axis of Northern California. *Vegetatio* 115:145–155.

Peterson, P. V., and P. V. Peterson, Jr. 1975. *Native Trees of the Sierra Nevada*. California Natural History Guides No. 36. Berkeley: University of California Press.

Rose, C. and S. Ingram. 2015. *Rock Creek Wildflowers*. Sacramento, CA: California Native Plant Society.

Schaffer, J. P. 1997. *The Geomorphic Evolution of the Yosemite Valley and Sierra Nevada Landscapes: Solving the Riddles in the Rocks*. Berkeley, CA: Wilderness Press.

Smith, G. S. 1976. *Mammoth Lakes Sierra*. Palo Alto, CA: Genny Smith Books.

Stebbins, C. A., and R. C. Stebbins. 1974. *Birds of Yosemite National Park*. Yosemite, CA: Yosemite Natural History Association.

Stocking, S. K., and J. A. Rockwell. 1969. *Wildflowers of Sequoia and Kings Canyon National Parks*. Yosemite, CA: Sequoia Natural History Association.

Storer, T. I., R. Usinger, and D. Lukas. 2004. *Sierra Nevada Natural History*. Berkeley: University of California Press. First published 1963 by University of California Press.

Sumner, L., and J. S. Dixon. 1953. *Birds and Mammals of the Sierra Nevada*. Berkeley: University of California Press.

Thomas, J. H., and D. R. Parnell. 1974. *Native Shrubs of the Sierra Nevada*. California Natural History Guides No. 34. Berkeley: University of California Press.

Verner, J., and A. S. Boss. 1980. *California Wildlife and Their Habitats: Western Sierra Nevada*. General Technical Report PSW-37. Albany, CA: Pacific Southwest Forest and Range Experimentation Station, USDA Forest Service.

Weeden, N. F. 1986. *A Sierra Nevada Flora*. Berkeley: Wilderness Press.

Weise, K. 2013. *Sierra Nevada Wildflowers. A Falcon Guide*. Guilford, CT: Globe Pequot.

Wenk, E. 2015. *Wildflowers of the High Sierra and John Muir Trail*. Birmingham, AL: Wilderness Press.

Whitney, S. 1979. *A Sierra Club Naturalist's Guide to the Sierra Nevada*. San Francisco, CA: Sierra Club Books.

MOUNTAINTOPS

Arno, S. F., and R. P. Hammerly. 1984. *Timberline, Mountain and Arctic Forest Frontiers*. Seattle, WA: The Mountaineers.

Chabot, B. F., and W. D. Billings. 1972. Origins and Ecology of the Sierran Alpine Flora and Vegetation. *Ecological Monographs* 42:163–199.

Hall, C. A., Jr. 1991. *Natural History of the White-Inyo Range, Eastern California*. Berkeley: University of California Press.

Hall, C. A., Jr., and V. Doyle-Jones, eds. 1988. *Plant Biology of Eastern California. Natural History of the White-Inyo Range Symposium*. Vol. 2. Oakland, CA: Regents of the University of California.

Hall, C. A., Jr., V. Doyle-Jones, and B. Widawski, eds. 1991. *Natural History of Eastern California and High Altitude Research: White Mountain Research Station Symposium*. Vol. 3. Oakland, CA: Regents of the University of California.

Hall, C. A., Jr., and D. J. Young. 1985. *Natural History of the White-Inyo Range, Eastern California and Western Nevada and High Altitude Physiology. University of California White Mountain Research Station Symposium*. Vol. 1, *August 23–25*. Oakland, CA: Regents of the University of California.

Lanner, R. M. 1981. *The Pinon Pine*. Reno: University of Nevada Press.

Lanner, R. M. 1996. *Made for Each Other: A Symbiosis of Birds and Pines*. Oxford: Oxford University Press.

Lanner, R. M. 2007. *The Bristlecone Book: A Natural History of the World's Oldest Trees*. Missoula, MT: Mountain Press.

Lloyd, R. M., and R. S. Mitchell. 1973. *A Flora of the White Mountains, California and Nevada*. Berkeley: University of California Press.

Maloney, P. E. 2011. Population Ecology and Demography of an Endemic Subalpine Conifer

(*Pinus balfouriana*) with a Disjunct Distribution in California. *Madroño* 58:234–248.

Marchand, P. J. 1991. *Life in the Cold: An Introduction to Winter Ecology*. Hanover, NH: University Press of New England.

Rundel, P. W. 2011. The Diversity and Biogeography of the Alpine Flora of the Sierra Nevada, California. *Madroño* 58:153–184.

Zwinger, A. H., and B. E. Willard. 1972. *Land above the Trees: A Guide to American Alpine Tundra*. New York: Harper & Row.

PACIFIC NORTHWEST

Ferlatte, W. J. 1974. *A Flora of the Trinity Alps of Northern California*. Berkeley: University of California Press.

Gillett, G. W., J. T. Howell, and H. Leschke. 1961. *A Flora of Lassen Volcanic Park, California*. University of San Francisco Press. Reprinted from the Wasmann Journal of Biology.

Grescoe, A. 1997. *Giants: The Colossal Trees of Pacific North America*. Boulder, CO: Roberts Reinhart Publishers.

Sawyer, J. O. 2006. *Northwest California: A Natural History*. Berkeley: University of California Press.

Sawyer, J. O. 2007. Why Are the Klamath Mountains and Adjacent Northern Coast Floristically Diverse? *Fremontia* 35:3–11.

Van Pelt, R. 2001. *Forest Giants of the Pacific Coast*. Seattle: University of Washington Press.

Wallace, D. R. 1983. *The Klamath Knot: Explorations of Myth and Evolution*. San Francisco, CA: Sierra Club Books.

Whitney, S. 1989. *A Sierra Club Naturalist's Guide to the Pacific Northwest*. San Francisco, CA: Sierra Club Books.

COAST RANGES

Baldocchi, D. D., and K. Xu. 2007. What Limits Evaporation from Mediterranean Oak Woodlands: The Supply of Moisture in the Soil, Physiological Control by Plants or the Demand by the Atmosphere? *Advances in Water Resources* 30:2113–2122.

Becking, R. W. 1982. *Pocket Flora of the Redwood Forest*. Covelo, CA: Island Press.

Borchert, M. I., N. D. Cunha, P. C. Krosse, and M. L. Lawrence. 1993. *Blue Oak Plant Communities of Southern San Luis Obispo and Northern Santa Barbara Counties, California*. General Technical Report PSW-139. Albany, CA Pacific Southwest Research Station, USDA Forest Service.

Bowen, O. E. 1959. *Rocks and Minerals of the San Francisco Bay Region*. California Natural History

Guides No. 5. Berkeley: University of California Press.

Cooney-Lazaneo, M. B., K. Lyons, and H. King. 1981. *Plants of Big Basin Redwoods State Park and the Coastal Mountains of Northern California*. Missoula, MT: Mountain Press.

Cooper, W. S. 1967. Coastal Sand Dunes of California. *Geological Society of America Memoir* 104:1–147.

Cylinder, P. D. 1995. The Monterey Ecological Staircase and Subtypes of Monterey Pine Forest. *Fremontia* 23:7–13.

Evans, J. G. 1988. *Natural History of the Point Reyes Peninsula*. Point Reyes, CA: Point Reyes National Seashore Association.

Evans, J. M., A. Klein, J. Taylor, D. Hickson, and T. Keeler-Wolf. 2006. *Vegetation Classification and Descriptions of the Clear Creek Management Area, Joaquin Ridge, Monocline Ridge, and Environs in San Benito and Western Fresno Counties*. California Report to USDI, Bureau of Land Management, Hollister District, California. Sacramento: California Native Plant Society, California Department of Fish and Wildlife.

Evarts, J., and M. Popper, eds. 2001. *Coast Redwood: A Natural and Cultural History*. Los Olivos, CA: Cachuma Press.

Ferris, R. S. 1968. *Native Shrubs of the San Francisco Bay Region*. California Natural History Guides No. 24. Berkeley: University of California Press.

Fox, W. W. 1976. Pygmy Forest: An Ecological Staircase. *California Geology* 29:4–7.

Gilliam, H. 1962. *Weather of San Francisco Bay Region*. California Natural History Guides No. 6. Berkeley: University of California Press.

Griggs, G., K. Patsch, and L. Savoy, eds. 2005. *Living with the Changing California Coast*. Berkeley: University of California Press.

Hensson, P., and D. J. Usner. 1993. *The Natural History of Big Sur*. Berkeley: University of California Press.

Hoover, R. F. 1970. *The Vascular Plants of San Luis Obispo County, California*. Berkeley: University of California Press.

Howard, A. D. 1979. *Geologic History of Middle California*. California Natural History Guides No. 43. Berkeley: University of California Press.

Howell, J. T. 1970. *Marin Flora*. Berkeley: University of California Press.

Jenkins, O. P. 1941. *Geologic Guidebook of the San Francisco Bay Counties*. Bulletin 141. Sacramento: California Division of Mines and Geology.

Lentz, J. E. 2013. *A Naturalist's Guide to the Santa Barbara Region*. Berkeley, CA: Heyday Books.

Marianchild, K. 2014. *Secrets of the Oak Woodlands: Plants and Animals among California's Oaks.* Berkeley, CA: Heyday Books.

Metcalf, W. 1959. *Native Trees of the San Francisco Bay Region.* California Natural History Guides No. 4. Berkeley: University of California Press.

Peattie, B. 1946. *The Pacific Coast Ranges.* New York: Vanguard Press.

Pickart, A. J., and J. O. Sawyer. 1998. *Ecology and Restoration of Northern California Coastal Dunes.* Sacramento: California Native Plant Society.

Plumb, T. R., and A. P. Gomez. 1983. *Five Southern California Oaks: Identification and Postfire Management.* General Technical Report PSW-071. Berkeley, CA: Pacific Southwest Forest and Range Experimental Station, USDA Forest Service.

Raiche, R. 2009. The Cedars: Sonoma County's Hidden Treasure. *Fremontia* 37(2):3–15.

Schoenherr, A. A. 2001. California Redwood: What is the State Tree? Fremontia 29: 34–35.

Sharsmith, H. K. 1965. *Spring Wildflowers of the San Francisco Bay Region.* California Natural History Guides No. 11. Berkeley: University of California Press.

Sharsmith, H. K. 1982. *Flora of the Mount Hamilton Range of California.* Sacramento: California Native Plant society.

Sholars, R. E. 1982. *The Pygmy Forest and Associated Plant Communities of Coastal Mendocino County, California.* Mendocino, CA: Black Bear Press.

Smith, A. C. 1959. *Introduction to the Natural History of the San Francisco Bay Region.* California Natural History Guides No. 1. Berkeley: University of California Press.

Smith, C. F. 1976. *A Flora of the Santa Barbara Region.* Santa Barbara, CA: Santa Barbara Museum of Natural History.

Thomas, J. H. 1961. *Flora of the Santa Cruz Mountains of California.* Stanford, CA: Stanford University Press.

VanDyke, E., K. D. Holl, and J. R. Griffin. 2001. Maritime Chaparral Community Transition in the Absence of Fire. *Madroño* 48:221–229.

CISMONTANE SOUTHERN CALIFORNIA

Allen, R. L., and F. M. Roberts, Jr. 2013. *Wildflowers of Orange County and the Santa Ana Mountains.* Laguna Beach, CA: Laguna Wilderness Press.

Bailey, H. P. 1966. *Weather of Southern California.* California Natural History Guides No. 17. Berkeley: University of California Press.

Beauchamp, R. M. 1986. *A Flora of San Diego County, California.* National City, CA: Sweetwater River Press.

Belzer, T. J. 1984. *Roadside Plants of Southern California.* Missoula, MT: Mountain Press.

Booth, E. S. 1968. *Mammals of Southern California.* California Natural History Guides No. 21. Berkeley: University of California Press.

Bowler, P., and S. Brown, eds. 1987. *California Oak Heritage Conservation Conference.* Irvine, CA: Sea and Sage Audubon Society.

Carlquist, S. 1965. *Island Life.* Garden City, N.Y.: Natural History Press.

Clarke, H. 1989. *An Introduction to Southern California Birds.* Missoula, MT: Mountain Press.

Clarke, O. F., D. Svehla, G. Ballmer, and A. Montalvo. 2007. *Flora of the Santa Ana River and Environs: With References to World Botany.* Berkeley, CA: Heyday Books.

Dale, N. 1986. *Flowering Plants: The Santa Monica Mountains, Coastal and Chaparral Regions of Southern California.* Santa Barbara, CA: Capra Press.

De Gouvenain, R. C., and A. M. Ansary. 2006. Association between Fire Return Interval and Population Dynamics in Four California Populations of Tecate Cypress (*Cupressus forbesii*). *The Southwestern Naturalist* 51:447–454.

De Lisle, H. D., Southwestern Herpetologists Society, and Society of the Study of Amphibians and Reptiles. 1986. *The Distribution and Present Status of the Herpetofauna of the Santa Monica Mountains.* Special Publication No. 2. Van Nuys, CA: Southwestern Herpetologists Society.

Dunne, G., and J. Cooper. 2001. *Geologic Excursions in Southwestern California.* Fullerton, CA: The Pacific Section, Society for Sedimentary Geology.

Garrett, K., and J. Dunn. 1981. *Birds of Southern California.* Los Angeles, CA: Los Angeles Audubon Society.

Hall, C. A., Jr. 2007. *Introduction to the Geology of Southern California and Its Native Plants.* Berkeley: University of California Press.

Halsey, R. W. 2005. *Fire, Chaparral, and Survival in Southern California.* San Diego, CA: Sunbelt Publications.

Jaeger, E., and A. C. Smith. 1966. *Introduction to the Natural History of Southern California.* California Natural History Guides No. 13. Berkeley: University of California Press.

Junak, S. 2008. *A Flora of San Nicolas Island California.* Santa Barbara, CA: Santa Barbara Botanic Garden.

Junak, S., T. Ayers, R. Scott, D. Wilken, and D. Young. 1995. *A Flora of Santa Cruz Island.* Santa Barbara, CA: Santa Barbara Botanic Garden in collaboration with the California Native Plant Society.

Keeley, J. E., W. J. Bond, R. A. Bradstock, J. G. Pausas, and P. W. Rundel. 2012. *Fire in Mediterranean Ecosystems: Ecology, Evolution,* and Management. NY: Cambridge University Press.

Lathrop, E. W., and R. F. Thorne. 1985. *A Flora of the Santa Rosa Plateau.* Special Publication No. 1. Claremont: Southern California Botanists.

Latting, J., ed. 1976. *Plant Communities of Southern California.* Special Publication No. 1. Sacramento: California Native Plant Society.

Lemm, J. M. 2006. *Amphibians and Reptiles of the San Diego Region.* Berkeley: University of California Press.

Lenz, L. W., and J. Dourley. 1981. *California Native Trees & Shrubs for Garden & Environmental Use in Southern California.* Claremont, CA: Rancho Santa Ana Botanic Garden.

Lozinsky, R. P. 2012. *Our Backyard Geology in Orange County, California.* Boston, MA: McGraw-Hill Learning Solutions.

Moran, R. 1996. *The Flora of Guadalupe Island, Mexico.* Memoirs of the California Academy of Sciences Number 19. San Francisco: California Academy of Sciences.

Munz, P. A. 1974. *A Flora of Southern California.* Berkeley: University of California Press.

Perry, B. 1981. *Trees and Shrubs for Dry California Landscapes.* San Dimas, CA: Land Design.

Peterson, P. V. 1966. *Native Trees of Southern California.* California Natural History Guides No. 14. Berkeley: University of California Press.

Quinn, R. D., and S. C. Keeley. 2006. *Introduction to California Chaparral.* Berkeley: University of California Press.

Raven, P. H. 1966. *Native Shrubs of Southern California.* California Natural History Guides No. 15. Berkeley: University of California Press.

Raven, P. H., H. J. Thompson, and B. A. Prigge. 1986. *Flora of the Santa Monica Mountains, California.* Special Publication No. 2. Claremont: Southern California Botanists.

Rundel, P. W., and R. Gustafson. 2005. *Introduction to the Plant Life of Southern California.* Berkeley: University of California Press.

Schoenherr, A. A. 1976. *The Herpetofauna of the San Gabriel Mountains* Special Publication No. 1. Van Nuys, CA: Southwestern Herpetologists Society.

Schoenherr, A. A., ed. 1990. *Endangered Plant Communities of Southern California.* Special Publication No. 3. Claremont: Southern California Botanists.

Schoenherr, A. A. 1996. *Coyote. Zooscape on the Rio Grande.* Albuquerque, NM: New Mexico Zoological Society.

Schoenherr, A. A. 2011. *Wild and Beautiful: A Natural History of Open Spaces in Orange County.* Laguna Beach, CA: Laguna Wilderness Press.

Schoenherr, A. A., D. Clarke, and E. Brown, eds. 2005. *Docent Guide to Orange County Wilderness.* Laguna Greenbelt, The Nature Conservancy, Orange County Wild, County of Orange Harbors, Beaches, and Parks.

Schoenherr, A. A., C. R. Feldmeth, and M. J. Emerson. 1999. *Natural History of the Islands of California.* Berkeley: University of California Press.

Sharp, R. P., and A. F. Glazner. 1993. *Geology Underfoot in Southern California.* Missoula, MT: Mountain Press Publishing Company.

Syphard, A. D., D. R. Volker, T. J. Hawbaker, and S. I. Stewart. 2009. Conservation Threats due to Human-Caused Increases in Fire Frequency in Mediterranean Climate Ecosystems. *Conservation Biology* 23:758–769.

Weigand, P., ed. 1998. *Contributions to the Geology of the Northern Channel Islands, Southern California.* Bakersfield, CA: Pacific Section American Association of Petroleum Geologists.

Westman, W. E. 1981. Diversity Relations and Succession in Californian Coastal Sage Scrub. *Ecology* 62:170–184.

CALIFORNIA'S DESERTS

Baldwin, B. G., S. Boyd, B. J. Ertter, R. W. Patterson, T. J. Rosatti, and D. H. Wilken. 2002. *The Jepson Desert Manual: Vascular Plants of Southeastern California.* Berkeley: University of California Press.

Benson, L. 1982. *The Cacti of the United States and Canada.* Stanford, CA: Stanford University Press.

Benson, L., and R. A. Darrow. 1981. *Trees and Shrubs of the Southwest Deserts.* Tucson: University of Arizona Press.

Brown, G. W., Jr. 1974. *Desert Biology.* 2 vols. New York: Academic Press.

Cloudsley-Thompson, J. L., and M. J. Chadwick. 1964. *Life in Deserts.* Philadelphia, PA: Dufour Editions.

Cornett, J. W. 1987. *Wildlife of the North American Deserts.* Palm Springs, CA: Nature Trails Press.

Cowles, R. B., and E. S. Bakker. 1977. *Desert journal.* Berkeley: University of California Press.

Dawson, E. Y. 1966. *Cacti of California.* California Natural History Guides No. 18. Berkeley: University of California Press.

DeDecker, M. 1984. *Flora of the Northern Mojave Desert, California.* Special Publication No. 7. Sacramento: California Native Plant Society.

Fulton, R. 1984. Floral Morphology and Pollination in *Agave deserti* Englm. *Crossosoma* 10:1–11.

Goerrissen, J., and J. M. Andre, eds. 2005. *Sweeney Granite Mountains Desert Research Center 1978-2003: A Quarter Century of Research and Teaching.* University of California natural Reserve Program.

Grayson, D. K. 2011. *The Great Basin: A Natural History.* Berkeley: University of California Press.

Hunt, C. B. 1975. *Death Valley: Geology, Ecology, and Archaeology.* Berkeley: University of California Press.

Ingram, S. 2008. *Cacti, Agaves, and Yuccas of California and Nevada.* Los Olivos, CA: Cachuma Press.

Jaeger, E. 1933. *The California Deserts.* Stanford, CA: Stanford University Press.

Jaeger, E. 1950. *Our Desert Neighbors.* Stanford, CA: Stanford University Press.

Jaeger, E. 1967. *Desert Wildflowers.* Stanford, CA: Stanford University Press.

Jefferson, G. T., and L. Lindsay. 2006. *Fossil Treasures of the Anza Borrego Desert.* San Diego, CA: Sunbelt Publications.

Kirk, R. 1973. *Desert: The American Southwest.* Boston, MA: Houghton Mifflin.

Larson, P. 1977. *A Sierra Club Naturalist's Guide to the Deserts of the Southwest.* San Francisco, CA: Sierra Club Books.

Lengner, K. E. 2013. *A Prehistory and History of the Death Valley Region;s Native Americans and the Environments in Which They Lived.* Death Valley National Park, CA: Deep Enough Press.

Lengner, K. E. 2013. *A Trip Through Death Valley's Geologic Past: The Magnificent Rocks of Death Valley.* Death Valley National Park, CA: Deep Enough Press.

Louw, G. N., and M. K. Seeley. 1982. *Ecology of Desert Organisms.* New York: Longman.

Mabry, T. J., H. Hunziker, and D. R. Difeo, Jr. 1977. *Creosote Bush.* New York: John Wiley & Sons.

MacKay, P. 2003. *Mojave Desert Wildflowers. A Falcon Guide.* Guilford, CT: Globe Pequot Press.

Mathias, M. E. 1978. The California Desert. *Fremontia* 6:3–6.

Miller, A. H., and R. C. Stebbins. 1964. *The Lives of Desert Animals in Joshua Tree National Monument.* Berkeley: University of California Press.

Monson, G., and L. Sumner. 1980. *The Desert Bighorn.* Tucson: University of Arizona Press.

Morhardt, E., and S. Morhardt. 2004. *California Desert Flowers.* Berkeley: University of California Press.

Munz, P. A. 2004. *Introduction to California Desert Wildflowers.* Berkeley: University of California Press.

Noble, P. S., ed. 2002. *Cacti: Biology and Uses.* Berkeley: University of California Press.

Pavlik, B. 2008. *The California Deserts: An Ecological Rediscovery.* Berkeley: University of California Press.

Phillips, S. J., and P. W. Comus, eds. 2000. *A Natural History of the Sonoran Desert.* Tucson: Arizona Sonoran Desert Museum Press.

Pianka, E. R. 1986. *Ecology and Natural History of Desert Lizards.* Princeton, NJ: Princeton University Press.

Reimold, R. J., and W. H. Queen. 1974. *Ecology of Halophytes.* New York: Academic Press.

Remeika, P., and L. Lindsay. 1992, *Geology of Anza Borrego: Edge of Creation.* San Diego, CA: Sunbelt Publications.

Robichaux, R. H., ed. *Ecology of Sonoran Desert Plants and Plant Communities.* Tucson: University of Arizona Press.

Rundel, P. W., and A. C. Gibson. 1996. *Ecological Communities and Processes in a Mojave Desert Ecosystem: Rock Valley Nevada.* Cambridge: Cambridge University Press.

Schmidt-Nielsen, K. 1964. *Desert Animals: Physiological Problems of Heat and Water.* London: Oxford University Press.

Sharp, R. P., and A. F. Glazner. 1997. *Geology Underfoot in Death Valley and Owens Valley.* Missoula, MT: Mountain Press Publishing Company.

Simpson, B. B. 1977. *Mesquite.* New York: John Wiley & Sons.

Smith, G., ed. 1995. *Deepest Valley: A Guide to Owens Valley Its Roadsides and Mountain Trails.* Rev. ed. Reno: University of Nevada Press.

Sowell, J. 2001. *Desert Ecology: An Introduction to Life in the Arid Southwest.* Salt Lake City: University of Utah Press.

Spellenberg, R. 2003. *Sonoran Desert Wildflowers. A Falcon Guide.* Guildford, CT: Globe Pequot Press.

Stein, B. A., and S. F. Warrick. 1979. *Granite Mountains Resource Survey.* Publication No. 1. University of California–Santa Cruz Environmental Field Program.

Stewart, J. M. 1993. *Colorado Desert Wildflowers.* Albuquerque, NM: Jon Stewart Photography.

Stewart, J. M. 1998. *Mojave Desert Wildflowers.* Albuquerque, NM: Jon Stewart Photography.

Stone, R. D., and V. A. Sumida. 1983. *The Kingston Range of California: A Resource Survey.* Publication No. 10. University of California-Santa Cruz Environmental Field Program.

Trent, D. D., and R. W. Hazlett. 2002. *Joshua Tree National Park Geology.* Twentynine Palms, CA: Joshua Tree National Park Association.

Zabriskie, J. G. 1979. *Plants of Deep Canyon and the Central Coachella Valley, California*. University of California-Riverside, Philip L. Boyd Deep Canyon Desert Research Center.

GREAT CENTRAL VALLEY

Barbour, M. G., A. Solomeshch, C. Witham, R. Holland, R. MacDonald, S. Cilliers, J. A. Molina, J. Buck, and J. Hillman. 2003. Vernal Pool Vegetation of California: Variation within Pools. *Madroño* 50:129–146.

Barry, W. J. 1972. *The Central Valley Prairie*. Sacramento, CA: California Department of Parks and Recreation.

Garone, P. 2011. *The Fall and Rise of the Wetlands of California's Great Central Valley*. Berkeley: University of California Press.

Germano, D. J., G. B. Rathbun. L:R. Saslaw, B. L. Cypher, E. A. Cypher, and L. M. Vredenburgh. 2011. The San Joaquin Desert of California: Ecologically Misunderstood and Overlooked. *Natural Areas Journal* 31:138–147.

Jain, S. 1976. *Vernal Pools: Their Ecology and Conservation*. Publication No. 9. University of California-Davis, Institute of Ecology.

Mason, H. L. 1957. *A Flora of the Marshes of California*. Berkeley: University of California Press.

Minnich, R. A. 2008. *California's Fading Wildflowers: Lost Legacy and Biological Invasions*. Berkeley: University of California Press.

Sands, A. 1977. *Riparian Forests in California*. Special Publication No. 15. Davis: Davis Institute of Ecology, University of California.

Twisselman, E. 1967. *A Flora of Kern County*. San Francisco, CA: University of San Francisco.

Wester, L. 1981. Composition of the Native Grasslands in the San Joaquin Valley, California. *Madroño* 28:231–241.

INLAND WATERS

Behnke, R. J. 1992. *Native Trout of Western North America*. Monograph 6. Bethesda, MD American Fisheries Society.

Bowler, P. A. 1988. Thoughts about Subtle Impact Potentials of Headwater Sited Small Hydroelectric Developments or River Processes. Pp. 13–18. In Bradley, E., ed. *Proceedings of the State of Sierra Symposium, 1985–1986*. San Francisco, CA: Pacific Publishing Company.

Cogswell, H. L. 1977. *Water Birds of California*. California Natural History Guides No. 40. Berkeley: University of California Press.

Cole, G. A. 1994. *Textbook of Limnology*. Long Grove, IL: Waveland Press. First published 1983 by C. V. Mosby.

Dahl, T. T. 1990. *Wetlands Losses in the United States, 1780s to the 1980s*. Washington, DC: US Fish and Wildlife Service.

Ferren, W. R. 1985. *Carpinteria Salt Marsh*. Publication No. 4. Santa Barbara, CA: The Herbarium, Department of Biological Sciences, University of California.

Goldman, C. R., and A. J. Horne. 1983. *Limnology*. New York: McGraw-Hill.

Jacobs, D., E. S. Stein, and T. Longcore. (2010) 2011. *Classification of California Estuaries Based on Natural Closure Patterns: Templates for Restoration and Management*. Technical Report 619. Revised. Costa Mesa, CA: Southern California Water Research Project.

Macan, T. T. 1974. *Freshwater Ecology*. New York: Halstead Press.

McGinnis, S. M. 2006. *Field Guide to Freshwater Fishes of California*. Berkeley: University of California Press.

Minckley, W. K., and J. E. Deacon, eds. 1991. *Battle against Extinction: Native Fish Management in the American West*. Tucson: University of Arizona Press.

Moyle, P. B. 2002. *Inland Fishes of California*. Berkeley: University of California Press.

Naiman, R. J., and D. L. Soltz, eds. 1981. *Fishes in North American Deserts*. New York: Wiley-Interscience.

Norment, C. 2014. *Relicts of a Beautiful Sea: Survival, Extinction, and Conservation in a Desert World*. Chapel Hill, N.C.: University of North Carolina Press.

Oglelsby, L. D. 2005. *The Salton Sea. Memoirs of the Southern California Academy of Sciences*. Vol. 10. Los Angeles: Southern California Academy of Sciences.

Palmer, T. 1982. *Stanislaus: The Struggle for a River*. Berkeley: University of California Press.

Pennack, R. W. 1953. *Fresh-Water Invertebrates of the United States*. New York: Ronald Press.

Reisner, M. 1986. *Cadillac Desert: The American West and Its Disappearing Water*. New York: Viking Press.

Rinne, J. N., and W. L. Minckley. 1991. *Native Fishes of Arid Lands: A Dwindling Resource of the Desert Southwest*. Technical Report RM-206. Fort Collins, CO: Rocky Mountain Forest and Range Experiment Station, US Forest Service.

Schoenherr, A. A. 1988. A Review of the Life History and Status of the Desert Pupfish,

Cyprinodon macularius. Bulletin of the Southern California Academy of Sciences 87:104–134.

Schoenherr, A.A. 2001. When Wetlands Are Not Enough. *California Wild, Magazine of the California Academy of Sciences* 54:32–53.

Soltz, D.L., and R.J. Naiman. 1978. *The Natural History of Native Fishes in the Death Valley System.* Los Angeles, CA: Natural History Museum of Los Angeles County.

Usinger, R.L. 1956. *Aquatic Insects of California.* Berkeley: University of California Press.

Whitton, B.A. 1975. *River Ecology.* Berkeley: University of California Press.

Winkler, D.W. 1977. *An Ecological Study of Mono Lake, California.* Publication No. 12. Davis: Davis Institute of Ecology, University of California.

Zedler, Z.B. 1982. *The Ecology of Southern California Coastal Salt Marshes: A Community Profile.* FWS/OBS-81/54. Washington, DC: Biological Services Program, US Fish and Wildlife Service.

INDEX

Note: Plants and animals are indexed under common and scientific names. If you search for a common name and do not find an entry, look for a group name (for example, beetles or firs) or under "American," "California," "Pacific," "common," "northern," "southern," or "western." For example, there is no entry for "crow," but there is one for "American Crow." There is no entry for "kingsnake," but under "snakes," there are several kingsnake subentries. Page number followed by (f) and (t) indicates figure and table, respectively. Color plates are referred as *pl./pls.* followed by their respective numbers in italics.